THE INTERSTELLAR MEDIUM IN GALAXIES

ASTROPHYSICS AND SPACE SCIENCE LIBRARY

A SERIES OF BOOKS ON THE RECENT DEVELOPMENTS
OF SPACE SCIENCE AND OF GENERAL GEOPHYSICS AND ASTROPHYSICS
PUBLISHED IN CONNECTION WITH THE JOURNAL
SPACE SCIENCE REVIEWS

PROCEEDINGS
VOLUME 161

THE
INTERSTELLAR MEDIUM
IN GALAXIES

INVITED TALKS PRESENTED AT THE SECOND WYOMING CONFERENCE,
HELD AT GRAND TETON NATIONAL PARK,
WYOMING, U.S.A., JULY 3–7, 1989

Edited by

HARLEY A. THRONSON, Jr.

*Wyoming Infrared Observatory, Laramie, U.S.A.
and Royal Observatory, Edinburgh, U.K.*

and

J. MICHAEL SHULL

*Joint Institute for Laboratory Astrophysics,
University of Colorado, U.S.A.*

KLUWER ACADEMIC PUBLISHERS
DORDRECHT / BOSTON / LONDON

Library of Congress Cataloging in Publication Data

Teton Conference (2nd : 1989 : Grand Teton National Park, Wyo.)
   The interstellar medium in galaxies : invited talks presented at
the Second Teton Conference, held at Grand Teton National Park,
Wyoming, USA, July 3-7, 1989 / edited by Harley A. Thronson, Jr. and
J. Michael Shull.
     p.   cm. -- (Astrophysics and space science library ; 161)
   ISBN-13:978-0-7923-0760-0   e-ISBN-13:978-94-009-0595-5
   DOI: 10.1007/978-94-009-0595-5

   1. Galaxies--Congresses. 2. Interstellar matter--Congresses.
3. Astrophysics--Congresses.   I. Thronson, Harley A. II. Shull, J.
Michael. III. Title. IV. Series: Astrophysics and space science
library ; v. 161.
QB856.T48 1989
523.1'125--dc20
                                                              90-34937
ISBN-13:978-0-7923-0760-0   CIP

Published by Kluwer Academic Publishers,
P.O. Box 17, 3300 AA Dordrecht, The Netherlands.

Kluwer Academic Publishers incorporates
the publishing programmes of
D. Reidel, Martinus Nijhoff, Dr W. Junk and MTP Press.

Sold and distributed in the U.S.A. and Canada
by Kluwer Academic Publishers,
101 Philip Drive, Norwell, MA 02061, U.S.A.

In all other countries, sold and distributed
by Kluwer Academic Publishers Group,
P.O. Box 322, 3300 AH Dordrecht, The Netherlands.

Printed on acid-free paper

TABLE OF CONTENTS

PREFACE

The interstellar medium (ISM) in galaxies is the stage upon which the evolution of these grand systems is played out. Their gas and dust are enriched by material lost by dying stars, and new generations of stars and planets are born from giant clouds swept along in the great dance of galactic motion. In recent years, the material between stars has become increasingly subjected to detailed scrutiny by astronomers. The International Ultraviolet Explorer extended our ability to obtain spectra of the material toward hot stars in the Milky Way, and the Hubble Space Telescope promises to carry this capability even further. With the development of millimeter- and centimeter-wave telescope arrays and sensitive single dishes, we are now able to survey a large variety of galaxy types. Important to our deepening understanding of the ISM was the discovery of strong x-ray emission from large numbers of nearly-normal galaxies. The *Infrared Astronomical Satellite* results showed how important it is to include the emission from the solid state component of the ISM in our understanding of large-scale energy balance.

At the close of our 1986 Tetons summer school on the astrophysical processes within the ISM, a number of us began to discuss which topics might be most timely for a meeting three years later. Since the emphasis during the 1986 meeting was on the Milky Way, it was natural that one early contender for a topic was the interstellar medium within other galaxies. With the results from instruments spanning almost the entire electromagnetic spectrum, and with new understanding of the extragalactic processes that govern the appearance of galaxies, we believed the time was appropriate to summarize our knowledge of the extragalactic ISM. We made an effort not to emphasize one particular wavelength range and to invite speakers who could give a balanced review of current beliefs and problems within a field. The 1986 speakers were almost entirely American. However, much of the most exciting new work presented in 1989 was from our foreign colleagues.

There were about 230 registered participants for our 1989 conference at Jackson Lake Lodge. This may have been the largest American gathering of professional astronomers, other than general meetings such as those of the American Astronomical Society or the International Astronomical Union. We take particular pleasure in the youth of the participants in Wyoming: the astronomers who attended our conferences will be teaching and carrying on research for years to come.

Together with long hours at a word processor, money is the backbone of a conference. We were fortunate to be assisted by several sources throughout the year-long preparation for our 1989 meeting. The University of Wyoming supported all our advertisements and mailings, supplied our secretarial and logistical staff, and assured us of sufficient resources to guarantee travel support for some key speakers. We thank Dr. Ralph DeVries, the Vice-President for Research, Dr. Walter Eggars, the Dean of Arts and Sciences, and Dr. Glen

Rebka, Chairman of the Department of Physics and Astronomy. We solemnly listened to lectures from each of these gentlemen every time that they were asked for more money, but they always found more for our meeting. In the Department of Physics and Astronomy, we thank Marce Mitchum, Evelyn Haskell, and Kristy Anderson, the people who got things done and did not have to ask how.

It is a pleasure to thank both the National Science Foundation and the National Aeronautics and Space Administration for their support. We were able to help around 35 younger astronomers to attend the conference, due almost entirely to an NSF grant. We got the credit for passing out someone else's money, so we must thank Dr. James Wright at NSF, which we do with pleasure. As with the proceedings from our 1986 conference, *Interstellar Processes*, NASA supported the publication of a less-expensive, soft-cover version of this current book that we hope will be particularly accessible to younger scientists. Drs. Larry Caroff and Fred Gillett at NASA Headquarters made this possible.

We also thank Gert Kiers at Kluwer Publishing, who has guided us through a pair of proceedings now. It has also been a pleasure to work with the staff of Jackson Lake Lodge, including Dorris Stalker and Wally Young. We also thank our organizing committees, which put in a lot of work, despite insufficient recognition.

Above all, we thank all the participants of the meeting for making its organization as pleasant a task as long hours on the telephone can be. As editors, we are pleased with the quality of the work produced by the individual authors of the sections in this book. Such quality reflects the dedication to astronomy that makes working with our colleagues so challenging and rewarding.

We also appreciate the hospitality of the Royal Observatory Edinburgh, where this book was edited.

Harley A. Thronson, Jr. January, 1990
Royal Observatory Edinburgh
    and
University of Wyoming

J. Michael Shull
University of Colorado, Boulder

Figure 1: Grand Teton National Park, Wyoming

Be glad of life!
Because it gives you the chance to love and work,
To play and to look up at the stars.

Henry Van Dyke

The Interstellar Medium in Galaxies - An Overview

G.R. Knapp
Princeton University Observatory
Princeton, NJ08544, USA

ABSTRACT. An attempt is made to summarize recent observational developments on the subject of the interstellar medium in galaxies, with emphasis placed on global properties. After a summary of the properties and distribution of the ISM in the solar neighborhood and in the plane of the Galaxy, some results from the most important observational probes (HI, CO and infrared) are described. A recent development is the observation of the ISM in galaxies of all morphological types, early to late; these developments are summarized and the properties of galaxies of different types compared to each other. Finally, the origin of radio galaxies, the effect of environment and the prospect for direct observations of the evolution of the ISM in galaxies are discussed.

1. Introduction

Pace the organizers of this conference, the title of this contribution is changed by the omission of a single word; many of the advances in the understanding of our own and of other galaxies have come about through the comparison of observations of different galaxies including our own, and this short review will follow a similar line of attack. I am charged with the task of discussing: why is the study of the interstellar medium so important? (or, to quote directly from the organizing committee, why should we care?) And this really means: why should those of our colleagues who are not at this conference care? The answer, in a nutshell, is that at one time *all* galaxies, or at least the parts we can observe, consisted entirely of interstellar gas; understanding the interstellar medium is thus literally fundamental to understanding galaxy formation, galaxy evolution and cosmology. Further, the physics of the interstellar medium is classical physics; the physical processes which produce the radiation we observe are well understood and observations of the interstellar medium are a very powerful probe of the physical conditions in a galaxy.

The fundamental nature of interstellar medium studies has been elegantly described by Herbig (1977) and Tinsley (1980). In his summary remarks for I.A.U. Symposium 75, "Star Formation", Herbig noted:" Let me say at first - - how struck I am by the delicate symbiosis that exists between the stars and the interstellar medium, how each is nourished by the other, and how the Galaxy as we know it is entirely a consequence of this balance and interplay. It is interesting to speculate how, if one were able to tinker with just one parameter of this beautiful machinery, the whole ecosystem - - might find a new balance

3

H. A. Thronson, Jr. and J. M. Shull (eds.), The Interstellar Medium in Galaxies, 3–37.
© 1990 *Kluwer Academic Publishers.*

point and become unrecognizable to us." This eloquent statement of the role of small- and large-scale evolutionary processes should be kept in mind throughout this conference.

One of the most interesting developments in recent years has been the wealth of information gathered on the properties of the interstellar medium in two types of galaxy which are markedly different from the classical spiral type which has heretofore dominated interstellar medium studies. The ISM in very early-type (elliptical) and very late-type (dwarf irregular) galaxies has finally yielded to observational pressure and the material is available for a first characterization of the global properties of the ISM in these galaxies, its distribution among various components and its relation to stellar evolution processes. Recent results on such galaxies are much in evidence at this conference, and will allow us to think about the nature of the "balance point" reached under conditions which are quite different from those in spirals.

A second very important observational development has been the work on the effect of environment on galaxy properties and ISM properties. In the Virgo cluster, galaxies are observed in the act of having their ISM stripped from them. Collisions of gas-rich galaxies are apparently responsible for the spectacularly luminous starburst galaxies whose extensive study has been made possible by, above all, the success of IRAS. Some dwarf systems are apparently currently at an early generation of star formation. The enormous increases in the sensitivity and linearity of optical detectors has allowed the study of galaxies at a range of epochs, with the very surprising result that major star formation in at least some ellipticals is relatively recent. In all of these observations one can see the effects of "tinkering" with the machinery which makes a galaxy run.

Herbig's statement can be put in diagrammatic form, such as that shown in Figure 1, adapted from Figure 1 of Tinsley's (1980) wonderful review article on galactic evolution. The main processes which affect the interstellar matter content of a galaxy are: (1) the gas removal mechanisms; star formation, galactic winds and stripping: (2) the gas accumulation mechanisms; gaseous inflow, mass loss from evolving stars and mergers with other galaxies. A currently open question is the contribution to the present-day gas content of a galaxy from galaxy formation studies, i.e. the effect of the initial collapse of the primordial gas cloud from which the galaxy was made. The conventional wisdom is that the lack of angular momentum in proto- elliptical galaxies caused all of the presently-observed stars to have formed in essentially a free-fall time, leaving little or no residual gas, while angular momentum caused the formation of the disk in spiral galaxies. The properties of the present-day ISM in dwarf galaxies bear directly on these questions.

The formation of stars from the ISM is a process which converts mass to energy, causes the galaxies to shine and allows us to observe them. I have added one more box to Figure 1, the swallowing of gas by a central galactic engine (black hole). While this is probably not a significant sink for the ISM (although interesting limits can be set by the present-day mass distributions in the center of galaxies) it provides a far more efficient conversion of mass to energy that the processes of star formation and stellar evolution, allowing the observation of galaxies at very large distances and very early times.

The topics with which this conference is concerned are those given double outlines in Figure 1. In particular, we are concentrating on "mass and composition of the ISM". The interstellar gas comes in several flavors. The bulk of the mass is apparently in the form of HI and H_2; the other components of the ISM, at least in the Galaxy, contain insignificant fractions of the mass but are of primary importance for the energy transport and cycling of the ISM. Most of the heavy atoms are in particulate form as interstellar dust,

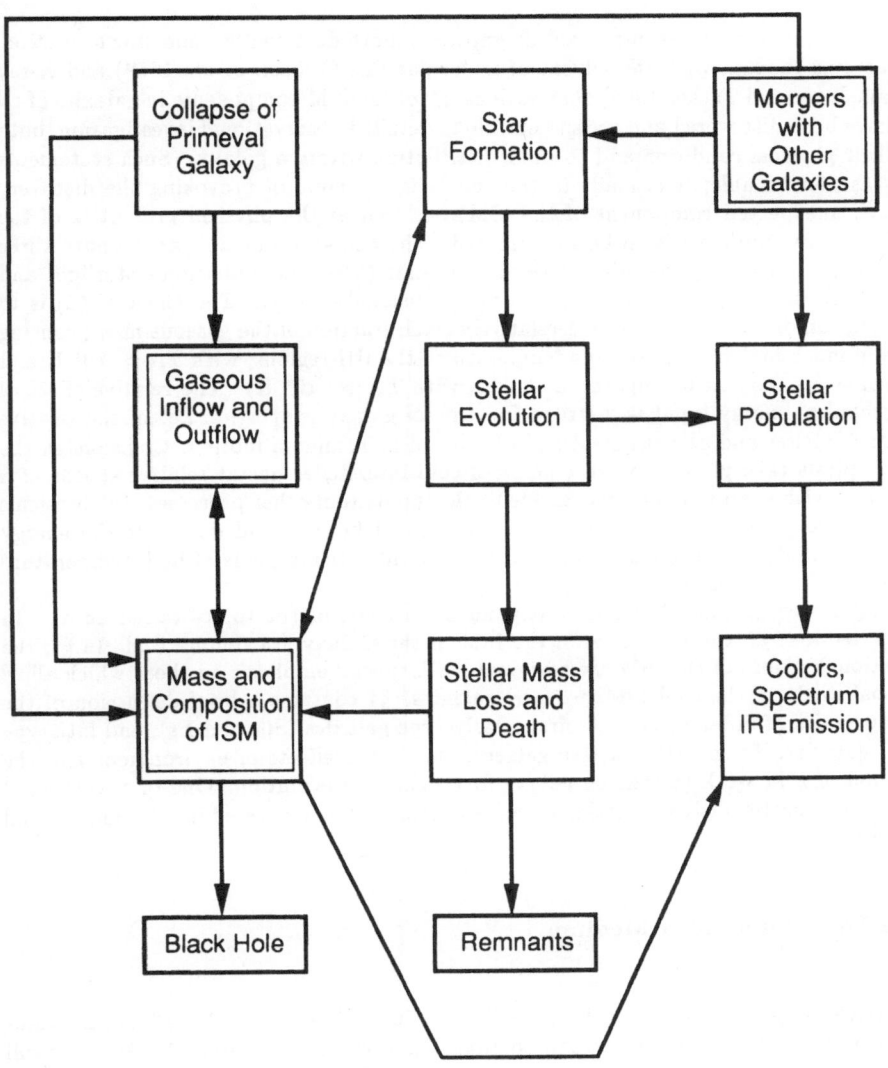

Figure 1. Flow chart showing the principle processes affecting the observed properties of a galaxy, from Tinsley (1980). The primary processes affecting the properties of the interstellar gas are shown by double outlines.

whose IR emission is a most important diagnostic of both dust content and star formation. It appears as though, with the advent of molecular line (Wilson *et al.* 1969) and X-ray (Forman, Jones and Tucker 1985) observations, all of the ISM components in galaxies of all types have been discovered and are susceptible to detailed observational investigation, both as to their physical condition and to their distribution within a galaxy. (Such statements as this ask for trouble; it is made in the superstitious hope of provoking the discovery of a new, unexpected component of the ISM). As well as the physical properties of the ISM the energy input needs to be investigated. The energy input appears to have three main contributions : (1) the microwave background: (2) stellar radiation - starlight and supernovae/cosmic rays: and (3) mechanical (collisional) energy. The effect of (2) is to dump large amounts of energy into a relatively small fraction of the gaseous mass, causing regions of much higher than average temperature (the HII regions, with $T_e = 10^4$ K and the coronal gas, heated by supernova shocks, with $T_e > 10^5$ K). The relative effect of the mechanical energy input is a strong function of galaxy properties, i.e. of the relative amounts of stellar and gaseous mass and of ordered and random motion. Collisions in the disks of spirals take place between clumps of cold interstellar gas at relative speeds of a few km s^{-1} with energy input comparable to the input from other processes. In ellipticals collisions take place at relative speeds of hundreds of km s^{-1} and dominate the energy input to the ISM, with the result that most of the interstellar gas is of high temperature and emits at X-ray energies.

This review is thus intended to expand upon some of the topics raised above. In the next section, §2, the properties of the ISM in the Galaxy are summarized. In §3, the observational probes of the ISM are reviewed, with special emphasis on those which allow the global study of the cold ISM in many galaxies. §4 contains a brief discussion of the ISM in galaxies of different types; spirals, early-type galaxies (S0s and Es), and late-type (dwarf) galaxies. §5 mentions active galactic nuclei, the effects of environment and the connection of ISM work to that on galaxy formation and evolution. One or two topics I consider to be particularly interesting or controversial will be discussed in a bit more detail throughout the text.

2. The Local Interstellar Medium

The physical properties of the local interstellar medium, observationally defined, are summarized in Table 1; this table is adapted from the reviews by Draine (1989) and Shull (1989). The local interstellar medium within a few hundred parsecs of the Sun is the only part of the Universe where we can use all available techniques to observe all of these components in coexistence. The distribution of those components containing a large fraction of the mass (the HI and H_2) or of the energy (the HII regions and supernova remnants) can be traced across the Galaxy and are therefore those which can readily be observed in other galaxies.

Table 1. The Local Interstellar Medium

Component	f_v	$< n_H >$ cm^{-3}	T_e K	f_M	Probes
Hot ICM	0.5	0.005	5×10^5	0.001	[OVI]; Xrays
Warm ICM	0.5	0.3	8000	0.1?	HI, Hα
Warm HI					
HI Clouds	0.05	5-20	10-100	0.4	HI,etc
H$_2$	0.005	300	5-30	0.5	CO
HII	0.001	$10 - 10^4$	10^4	0.02	Hα, etc.
SNR					radio, X-ray
Dust	1.0		5 - 50	0.01	IR, extinction

The hot intercloud medium is observed by UV line absorption in lines of [OVI] and other highly-ionized species (Jenkins and Meloy 1974; Jenkins 1978, Spitzer 1989), and via soft X-ray emission (McCammon et al. 1983). This component is in pressure equilibrium with the warm intercloud HI and the cool HI clouds. The [OVI] absorption is locally ubiquitous (Jenkins 1978) so that the volume filling factor is large, though the total mass fraction is low. This gas is explained as interstellar gas heated to high temperatures by supernova shocks (McKee and Ostriker 1977). It is not observable in other spiral galaxies; the sensitivity and spectral resolution of the available X-ray observations are insufficient to detect it and to separate its emission from that of stellar X-ray sources (Fabbiano 1989). Its presence in other galaxies may be indirectly inferred from the properties of the interstellar HI (Rupen 1989; E. Brinks, this conference; R. Braun and E. Brinks, this conference).

The warm ionized gas reveals its presence by local diffuse Hα emission (Reynolds 1988; Cox and Reynolds 1987), via pulsar rotation and dispersion measures (Lyne and Smith 1989) and via interstellar scattering of emission from extragalactic radio sources (Readhead and Duffet-Smith 1975). While the local volume density of this gas is small, its z-extent is large:

$$n_e = 0.04 \operatorname{sech}^2(z/900 \text{ pc}) \text{ cm}^{-3} \tag{1}$$

and this component contributes a surface density perpendicular to the disk of 2×10^{20} cm^{-2}, a significant fraction of the local interstellar mass.

The neutral atomic gas appears to consist of two components, the HI clouds with temperatures of < 100 K and the warm intercloud medium with spin temperatures > 1000 K and probably up to 8000 K. These two components are in rough pressure balance with each other and with the hot intercloud gas and are described by the McKee-Ostriker (1977) three-phase model of the ISM. The HI appears to be in two layers (evidence for a third HI component?); the bulk of the HI has $z_o = 260$ pc (using the definition of Equation (1)), with about 15 % in a layer with higher velocity dispersion and with $z_o = 500$ pc (Lockman 1984). The total HI surface density is 6 $\times$ 10^{20} cm^{-2}

The relationship between the warm ionized medium, whose total mass is about a third of that of the HI, and the HI intercloud medium is unclear, since their z-distributions are different. This could be caused by the vertical distribution of the halo hot stars, so that an increasing fraction of the HI is ionized with z.

Molecular Gas is traced by emission in the rotational transitions of many molecules, of which by far the most widely used is CO. The molecular gas, the *HII regions* and the *supernova remnants* are not for the most part in pressure equilibrium with the HI gas or with the hot intercloud medium. The interstellar medium, Galaxy wide, is distributed in a thin disk (ignoring the large molecular gas and star forming complexes in the galactic center region). If the galactic distance scale is taken to correspond to a Sun-center distance of 8.5 kpc, the total interstellar gas content of the Galaxy is about 5×10^9 $M_\odot$ of which about half is in HI and half in H_2. However, the distributions of these two components are dissimilar. The total gas surface density decreases approximately exponentially with radius with an e-folding length of about 3.5 kpc (Gordon and Burton 1976). Most of the H_2 lies between radii of about 3 and 7 kpc, while the HI distribution is quite flat with radius out to about 15 kpc, where its surface density begins to drop; much of the HI lies outside the solar circle. Likewise, the HI layer is several times thicker than that of the molecular gas - the full width at half maximum is 430 pc for HI at R_o (Knapp 1987; van Steenberg and Shull 1988) and 110 pc for H_2 at R_o, (Cohen and Thaddeus 1977; Stark 1979; Dame and Thaddeus 1985; Knapp 1987). Thus the higher the density of the gas, the more likely it is to be molecular.

The *dust* is in particles of typically a few tenths of microns in size and contains most of the heavy elements. The dust appears to be ubiquitous in the ISM and to be well mixed on all scales, with the exception of the hot phase (cf. the discussion of the local ISM by Lockman 1989). The dust is heated by energetic stellar photons and radiates in the far IR. The detailed properties of dust in the Galaxy and in other galaxies are discussed by Draine (these proceedings) and Rowan-Robinson (this meeting).

A most important and fundamental question is how and why the interstellar gas cycles among these various components, and what are the time scales for it to do so. Indeed, many of the conclusions about the state of the interstellar gas rest, explicitly or implicitly, on assumptions about these processes. A review of the current wisdom on this subject would take a book; my impression from the literature is that there is little consensus. At various places in this review I will discuss one small aspect of this question, and attempt to summarize the often contradictory evidence. The first such discussion I want to make concerns the question of the origin of interstellar dust. This subject is discussed in detail by Draine (1989). Interstellar dust particles are apparently well mixed with the gas on many scales with an abundance of about 100:1 by mass. While formation mechanisms for dust are not well understood theoretically, the existence of at least one important source of dust is well established observationally. Highly evolved luminous red giant stars on the AGB are observed, via their molecular line emission, IR excesses and IR molecular band features, to be losing gas and dust with a ratio similar to that seen in the ISM (Rowan-Robinson and Harris 1983a,b; Knapp 1985). Further, observations of thermal emission from SiO (Morris *et al.* 1979) show that 99% of the silicon is depleted onto grains. UV observations of the interstellar Si abundance (Bohlin *et al.* 1983) suggest that 5 - 25 % of the cosmic Si is in the gas phase, and this is consistent with the pattern of mass loss by evolving stars, where about 70% of the stellar mass return to the ISM appears to be from red giants with cool envelopes and 30% is from OB and WR stars, where the mass is returned in gaseous rather than partly solid form (Knapp, Rauch and Wilcots 1989; Bieging 1989). Statistical studies of mass loss from AGB stars suggest that these objects deposit $\approx$ 0.5 $M_\odot yr^{-1}$ Galaxy wide into the ISM, so that the time scale for dust replenishment is $\sim 10^{10}$ years.

In regions of the ISM which suffer by fast supernova shocks, the dust is completely destroyed, and indeed cosmic abundances of several elements normally heavily depleted onto grains are observed towards the Vela SNR (Jenkins, Wallerstein and Silk 1984). Models of the impact of SN shocks suggest that such shocks pass through a given region of the ISM every 10^8 years or so (Seab and Shull 1985; McKee et al. 1987), so that the primary location for dust formation is elsewhere than AGB stars. The obvious site is molecular clouds; there is no direct evidence for dust formation in these objects, although there is for grain growth (e.g. Carrasco, Strom and Strom 1973). In any case, if dust is destroyed on these short time scales and is made in molecular clouds, then, depending on how the ISM cycles among the molecular clouds, HI and the hot diffuse gas, the typical lifetimes of the molecular clouds would be a few times 10^6 to 10^7 years (Draine 1989). At cycling rates like these, the molecular and atomic gas clouds would never reach chemical or physical equilibrium. The assumption of equilibrium is implicit in the use of CO observations to estimate the molecular cloud mass, as will be discussed further below.

The rapid destruction of dust and its rapid formation in molecular clouds seems unlikely to me for several reasons. First is the appeal to Occam's razor; approximately the right amounts of dust are observed to form in AGB stars. Second is the great constancy of the gas to dust ratio in several different regions; along the lines of sight to nearby OB stars (Bohlin et al. 1975); along the lines of sight to distant bright halo objects such as globular clusters (e.g. Knapp and Kerr 1974) and in dense molecular clouds, where the dust is observed by its far-IR emission (Keene 1981). Third is the likelihood that dust in molecular clouds is well protected against supernova shocks, as shown by the presence of molecular gas in the vicinity of supernova remnants (Herbst and Assousa 1977, DeNoyer and Frerking 1981). It seems more likely that completely destructive supernova shocks are less pervasive than thought (either because they vent into the halo if the interstellar pressure is low or are contained if the pressure is high - see the later discussion below and Cox, this volume), or that dust can survive most interstellar shocks.

It is obvious that a lot more work on this problem is necessary. Much more work needs to be done at high spectral resolution of the interstellar depletions of species thought to be important constituents of dust (cf. the discussion by Joseph 1988). An evaluation needs to be made of how representative are the sight lines towards early-type stars, which must have a significant effect on the ISM in their immediate vicinity, and of how typical is the region of the ISM in the immediate solar vicinity - there is strong evidence that the ISM within a few tens of parsecs of the Sun is of anomalously low density and high temperature (Cox and Reynolds 1988; Lockman 1989). The contributions of various sources of mass return (AGB, OB/WR stars) need to be measured more carefully. Finally, an evaluation of how one might measure or set limits on the dust formation rate in molecular clouds should be made. Observations of the gas to dust ratio, and of the pattern of heavy element depletion, should provide much understanding of how the interstellar medium cycles among its various components.

3. Observing the Interstellar Medium

3.1 HI

Atomic hydrogen can be observed locally using either the Lyα line (in absorption) or the 21 cm line (in both emission and absorption). When made along the same lines of sight, the two types of observation give extremely good agreement (Lockman 1989). For galaxies in the nearby universe, the probe is the 21 cm line because it is seen in emission and maps can be made; at large redshifts, only Lyα emission or absorption is available observationally.

The physics of the 21 cm line are discussed by Purcell and Field (1956) and reviewed by Kulkarni and Heiles (1988). It is by far the most powerful and best- understood probe available for the study of the interstellar medium. From the point of view of the 21-cm hyperfine transition, HI is a two-level atom; further, the line has a characteristic temperature hν/k = 0.07 K which is much less than interstellar temperatures. The combination of these two properties means that, for an optically thin source, the observed intensity is directly proportional to the number of HI atoms regardless of temperature. HI clouds are usually quite optically thin, despite the great abundance of HI, because the line is highly forbidden. Simple models (e.g. Bajaja et al. 1984) suggest that self-absorption effects even in edge-on galaxies are unlikely to exceed 10 - 20%. Because the line is so highly forbidden, the transition is collisionally (or via Lyα) excited at very low densities (10^{-4} to 10^{-5} cm^{-3}) so that the emission can be traced to large radial distances in galaxies. The line is at a wavelength where the atmospheric extinction is negligible. In summary, the total HI content of a galaxy can be measured to about 20% accuracy (ignoring of course the distance uncertainty, a can of worms which is fortunately not the subject of this conference), and is:

$$M_{HI} = 2.36 \times 10^5 D^2 \sum S\Delta V \quad M_{\odot} \tag{2}$$

where the distance D is in megaparsecs and the integrated line flux $\Sigma S \Delta V$ is in Jy $\times$ km s^{-1}. HI observations of several thousand galaxies have been made; an up-to-date catalog is available (Huchtmeier and Richter 1989). Observations of the 21 cm line have yet further advantages. At this wavelength there are a large number of strong radio continuum sources (supernova remnants, pulsars, HII regions, galactic nuclear radio sources, quasars) which allow detailed studies of both column density and spin temperature via absorption observations. HI gas is cold, and the lines are narrow. This, together with two useful properties of radio lines, that they are far from being limited by photon noise and are measured by heterodyne techniques, allowing very accurate frequency measurement, means that almost arbitrarily high frequency resolution and accuracy are achieved for velocity and velocity dispersion measurements (the accuracy is usually determined by intrinsic source properties). This is of course greatly exploited in studies of galaxy dynamics and large scale structure (Tully and Fisher 1977; Giovanelli and Haynes 1985).

Total flux measurements of galaxies, normalized by the optical luminosity, show that M_{HI}/L_B increases systematically from about 0.05 $M_{\odot}/L_{\odot}$ for early-type spirals (Sas) to about 1 $M_{\odot}/L_{\odot}$ for Magellanic irregulars (Roberts 1975; Shostak 1978), with a probable value of 0.12 for the Galaxy. Synthesis maps of spiral galaxies detect HI to large radii (see for example the classic work of Bosma 1981) and allow the measurement of their rotation curves. Almost thirty years after the first detection of the HI 21 cm line, exciting discoveries continue to be made about the structure and physical properties of the ISM in the disks of spirals, as discussed by Brinks in this volume. An additional most encouraging development is the detection of HI from early-type (E and S0) galaxies. These will be discussed in §4, where the global properties of the ISM in 'non-traditional' (i.e. non-spiral) galaxies are discussed.

Some outstanding problems raised by HI observations of galaxies remain to be examined. These include: (1) the nature of the high-velocity clouds and the infall of gas into the Galaxy. Observations of stars of different ages suggest that the galactic metallicity has changed little over the past 5×10^9 years, yet stellar evolution processes such as supernova activity continue to dump enriched material into the galactic disk at a steady rate. A global galactic accretion rate of about 1 $M_\odot$ yr^{-1} of primordial gas is required to accomplish this (Tinsley 1980). Where is it? Are we in fact observing this in at least some of the high velocity clouds? This is an old problem, probably complicated by several origins for galactic HI which is moving very differently from allowed galactic rotation, but no satisfactory (or at least generally accepted) solution has been found to date. (2) What is the structure of the diffuse ISM? The ratio of mass in clouds and the intercloud HI? Is this the same all over the Galaxy? What is the temperature structure of the HI? Why does the peak HI brightness temperature seen in the Galaxy (about 135 K) have such a constant value across the Galaxy? What is the mass spectrum of the HI clouds (cf. Dickey *et al.* 1983)? Detailed answers to these questions are elusive. The ubiquity of the galactic HI is a hindrance here; elucidating the structures is difficult and observations of the low-level wide velocity dispersion HI are bedevilled by stray radiation scattered into the beam from other parts of the Galaxy (Lockman *et al.* 1986; Stark *et al.* 1989b). This information is needed for understanding the more fundamental problem of the cycling of gas through the different components of the ISM. (3) How do HI diameters of galaxies evolve with time? HI disks are sometimes observed to several Holmberg radii but are often much less extended than this (Bosma 1982). In particular, the currently-observed HI extents of spiral galaxies are insufficient to account for the observed frequency of damped Lyα absorption systems in quasar spectra (Wolfe, this conference). Were HI disks much larger in the past (cf. Gunn 1981) or is much of the HI in the Universe in a different type of galaxy such as the common, but faint, dwarf galaxies (e.g. Tyson 1988)?

3.2 H$_2$

As all the world knows, H$_2$ is not directly observable except under limited conditions, which, while of great interest to local abundance and excitation problems (UV absorption and IR emission lines) give no information about the total amount of H$_2$ and its distribution in the ISM of a galaxy (except to reassure us that H$_2$ does indeed exist in the Universe). Failing a direct measurement, rotational lines of molecular trace species, most commonly CO, are observed.

CO line emission has been very extensively used as a tracer of H$_2$ in the Galaxy and in several hundred others. What might at first appear to be a disadvantage, that its emission is heavily saturated, is probably something of an advantage because it measures the surface density and hence, perhaps, the number, of an ensemble of molecular clouds (Morris and Lo 1978; Dickman *et al.* 1986), and may be relatively insensitive to abundance variations. If CO is to be used as a tracer of molecular hydrogen content for a galaxy, then, the emission from this optically thick line must be calibrated in the time honored astronomical tradition. This has been done in a variety of ways by using observations of the CO emission in the plane of the Galaxy.

CO emission is ubiquitous in the galactic plane (Gordon and Burton 1976). Detailed surveys of the galactic plane in the ^{12}CO and ^{13}CO J = 1 - 0 lines (Clemens et al. *1986*, Dame et al. 1986; Stark et al. 1988) led to the discovery of the Giant Molecular Clouds (GMCs) (Solomon et al. 1972) with M $\sim$ 10^6 M$_\odot$, and to the mapping of the dense molecular gas distribution in the galactic plane. On the assumption of a constant ratio of molecular gas mass to CO line intensity, various empirical methods have been applied to the data to extract the constant of proportionality. Those methods which have been applied on a Galaxy-wide scale include the use of γ − ray emission to trace the hydrogen nuclei (Bloemen et al. 1986; Bhat et al. 1986) and comparison of the cloud masses estimated from the virial theorem with the CO line luminosity (Solomon et al. 1987; Young and Scoville 1982; Dame et al. 1986). These methods suggest that there is indeed a large scale proportionality between these quantities and that the CO luminosity of a galaxy can be used to estimate its total gas content if it is assumed that the galaxy is similar to our own. The proportionality used is

$$N(H_2) = X \times 10^{20} I_{CO} \quad cm^{-2} \qquad (3)$$

where I_{CO} is the ^{12}CO$(1-0)$ line flux in K $\times$ km s^{-1}. Values of X between 1 and 6 have been estimated by various authors, but cluster around X = 3. Scoville (1989) has summarized the statistical results on the mass function of molecular clouds in the plane of the Galaxy; he finds

$$N(M) \sim M^{-1.6} \qquad (4a)$$

and

$$n(H_2) \sim 290 \left(\frac{D}{20pc} \right)^{-0.75} \quad cm^{-3} \qquad (4b)$$

and, since the mass function is fairly steep, much of the molecular gas is in GMCs of mass $\geq 10^5$ M$_\odot$.

I find the result in Equation (4b) to be a curious one, for it is difficult to imagine how a locally-determined quantity like particle density can be affected by cloud diameter. It *is* approximately the behavior that one might expect if molecular clouds grow by collision (Cowie 1980; Jog and Ostriker 1988; Elmegreen 1989; Gammie and Ostriker 1989), since part of the pre-collision kinetic energy is retained as internal energy in the composite cloud, causing it to bloat. Since the cloud filling factor and cloud-cloud velocity dispersion are small, the cloud collision time is about 10^8 years, so that if the cloud mass spectrum is determined by cloud-cloud interactions, we require molecular clouds to be long lived. In addition, Scoville (1989) points out that the large molecular clouds are self-gravitating (indeed this assumption provides one important method for determining the $N(H_2)/I_{CO}$ conversion factor). None of these statements and assumptions are well established, and all have been the subject of much controversy.

Critiques of the use of formulae like equation (3) have been made on empirical and theoretical grounds by Williams (1985), Rickard and Blitz (1985), Polk et al. (1988), Israel (1988) and Maloney and Black (1988). Its application to CO emission in the Galaxy and others like it seems reasonable, since the various determinations give quite similar answers and the virial theorem seems to hold for clouds over a wide range of masses. This is not the place to rehash this issue, a subject which in any case would take another book, but perhaps it is worth mentioning a few recent developments. It has now become possibly to spatially resolve molecular clouds in other galaxies. On the assumption of virial equilibrium, values of $I_{CO}/M(H_2)$ much lower than those in the Galaxy are found for M33 (Wilson and Scoville, this conference), the Large Magellanic Cloud (Cohen et al. 1987; Nyman 1989) and NGC55 (Dettmar and Heithausen 1989). To be sure, the discrepancy for NGC55 could be significantly reduced by the adoption of a different Hubble parameter, and all of these are dwarf galaxies of metallicity considerably lower than that of the Galaxy however, these results should provide a strong caution against using a single conversion ratio between CO intensity and molecular gas content for galaxies of all types.

A second strong caveat comes from observations of the galactic center region, where the molecular clouds are quite different from those in the galactic plane; in particular, tidal stress is partly responsible for the line widths and sets lower limits on the densities of clouds which can survive (e.g. Bally et al. 1988). Also, the massive Sgr B molecular cloud complex has properties different from those of the galactic plane clouds. The global CO line flux from a galaxy contains contributions from the central regions as well as the disk, so could well have a different ratio of molecular cloud mass to CO intensity from that found for disk clouds alone.

It is easy to imagine that observational and analysis effects such as distance uncertainties, the difficulty of defining molecular cloud edges, and so on could lead to relationships like those in Equation (3). A new characterization of the galactic molecular cloud population is underway using the Bell Laboratories ^{13}CO galactic plane survey (Stark et al. 1989c). This survey is much more closely sampled than previous galactic plane surveys and uses a relatively optically thin line, so should make the identification of clouds more straightforward.

A recent observational development which is of considerable interest for studies of the local ISM as well as for galactic structure and molecular gas in galaxies is the extension of the cloud mass spectrum to lower masses, from 100 $M_\odot$ to $<$ 1 $M_\odot$ (Blitz, Magnani and Mundy 1984; Jannuzzi et al. 1988; de Vries, Heithausen and Thaddeus 1987; Keto and Myers 1986; Knapp and Bowers 1988; Guhathakurta and Tyson 1989). These clouds are detected weakly in the CO line and in the IRAS cirrus, and typically have visual extinctions in the range 0.1^m to 2^m. These are translucent clouds (van Dishoeck and Black 1986) whose chemistry is strongly driven by photodissociation of molecules by the UV radiation field. A CO map of one of these clouds is shown in Figure 2.

Comparison of the masses determined by the CO intensity or the IR emission shows that these cloudlets, unlike the GMCs, are very much out of virial equilibrium and are either transient objects, expanding into the ISM or are in pressure equilibrium with it. If this latter were the case, the interstellar pressure would have a value of p/k = 2 - 3×10^4 cm^{-3} K. The CI absorption line observations by Jenkins, Jura and Loewenstein (1983) give pressure values $p/k = nT$ in the range 10^3 to 10^4 cm^{-3} K, lower than the value requires to pressure bind the small clouds. However, a total interstellar pressure (thermal, turbulent, cosmic ray, magnetic) considerably higher than the values derived from the CI observations is required to hold the HI disk in hydrostatic balance to the scale height observed in the galactic plane. At R_o, with z_o = 260 pc, and HI pressure of 1 - 2 $\times 10^4$ cm^{-3}.K is required, even with a local midplane mass density as low as 0.06 $M_\odot$ pc^{-3} (Knapp 1987); this is just about the value required to bind the small molecular clouds, and is the value given by the mean HI density $(0.5 - 1$ cm$^{-3})$ and the total line of sight velocity dispersion (10 to 13 km s^{-1}) (cf. Maloney 1988; Blitz 1989).

If the small molecular clouds are really pressure bound, several interesting possibilities arise. They may be fairly long lived, and provide the material which is swept up to form

14

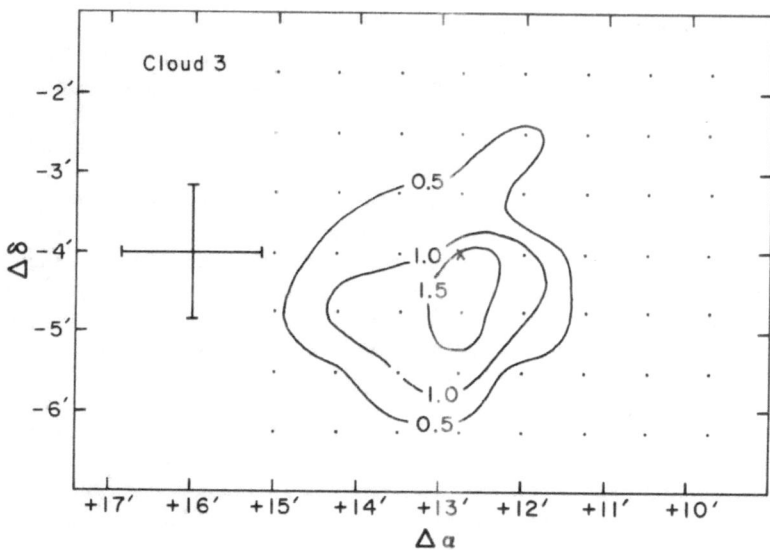

Figure 2. CO(1-0) observation of a small (diameter < 0.5 pc, M $\sim$ 0.2 M$_\odot$) molecular cloud in Orion, made with the Bell Labs 7 meter telescope. The cross shows the half-power beamwidth; the dots the positions observed; and the coordinate offsets are with respect to the position of α Ori $\alpha = 05^h 52^m 28^s, \delta = +7°13'58''(1950)$.

the larger clouds - i.e. pressure fluctuations in the ISM become dense enough to self-shield, molecules form, and these small fragments coagulate into larger clouds. There is the additional possibility that these small clouds could be the site of single, efficient, low mass star formation.

Most of the small molecular clouds are likely to be embedded in larger HI envelopes, as is observed in several cases (Weiland et al. 1986). Indeed, much of the galactic HI could well be associated with molecular clouds; molecular emission does tend to occur at the position and velocity of maximum HI emission. On a small as well as large scale, then, the more dense the gas is the more likely it is to be molecular. Photodissociation of molecular gas in the outer layers of small clouds could account for much of the galactic HI (Shaya and Federman 1987).

Since they are but weak CO emitters, the small clouds can only be observed in the solar neighborhood. An extrapolation of the local volume density to the entire Galaxy suggests that there may be as much as 0.5 to 1 $\times 10^9$ M$_\odot$ of molecular gas in these clouds, which is a non-negligible fraction of the mass of the ISM.

These clouds also provide information about the lifetime of large molecular clouds. The presence of significant amounts of molecular gas between the spiral arms in the galaxy (Solomon et al. 1987) and the cloud-cloud agglomeration timescale have been used to argue for long-lived molecular clouds (Scoville 1989). The smaller scale height for large clouds (Stark 1979) also argues for dynamical relaxation and therefore large lifetimes. On the other hand, Blitz (1989) points out that it is likely that large molecular clouds as well as small ones have considerable HI envelopes (cf. also the observations of Blitz and Thaddeus 1980; Read 1980, 1981; Goss et al. 1983; and Wannier, Lichten and Morris 1983). By contrast with the above claim for the Galaxy, Stark (1985) finds that the arm-interarm density contrast in the CO line is > 25:1 in M31. In the Rosette molecular cloud the gas is clumpy and the low density interclump gas is expanding away from the region (Blitz and Stark 1986). The high CI abundance in molecular clouds found by Phillips and Huggins (1981) argues for a short time scale for the exposure of molecular gas to the interstellar radiation field. All of these observations suggest that molecular clouds are short-lived and embedded in HI envelopes; in this case, they are hardly in equilibrium and all bets about a simple mass measurement are off. The real situation is probably a lot more complicated. Perhaps molecular clouds have a greater probability of surviving as their masses grow since the surface area to mass ratio decreases, and they lurk at z = 0 like the huge, hungry old pike at the bottom of the pond in every fishing story.

The HI - H_2 connection is worth much more observational effort; this is the vital interface in the ISM because these two components comprise most of the mass of the ISM. This is not easy to do because the ubiquity of the HI makes the delineation of individual structures ambiguous. But what is really needed is a four-component model of the ISM, comprising the hot medium, the HI, the H_2 and the HII regions. The first of these is important for tracing the input of mechanical energy from the dominant source, supernova blast waves, whose passage converts all other components instantly to hot diffuse gas. The last is important because it traces the dominant radiative input to the ISM, the high energy photons from hot young stars which control the HI/H_2 interface.

3.3 Radio Continuum and IR Emission

Radio continuum emission from galaxies comes in two types; powerful emission due to nuclear activity (jets, core sources etc.) which is primarily associated with high luminosity early type galaxies, and weaker emission from the disks of spirals (Dressel and Condon 1978). Both sources of emission have non-thermal spectra (the thermal radiation which is seen from galactic HII regions and, with care and interferometry, from HII regions in nearby spirals contributes little to the global flux). The emission from spiral disks is attributed to energetic particles interacting with the galactic magnetic field and injected by supernovae (e.g. Condon et al. 1982) - the radio spectral index is typically -0.5 to -1.

Infrared emission has two main sources - direct starlight at short wavelengths and radiation from interstellar dust, warmed by starlight, at long wavelengths. About 50 % of the radiation from a typical spiral galaxy is emitted at wavelengths longer than 5μ (de Jong et al. 1984). Work on long wavelength emission was changed fundamentally and forever by the success of IRAS (Neugebauer et al. 1984) and its huge data base and catalog. The first IRAS observations of spirals published by de Jong et al (1984) showed that S_{60}/S_{100}, where S_n is the flux density in Jy at wavelength n in microns has a value of about 1/3, showing that the source of the emission is cool and is interstellar dust. de Jong et al. proposed a two component model for the emission as arising from warm dust associated with star forming molecular clouds and cool dust associated with HI (the analog of the galactic cirrus, Low et al. 1984). Boulanger and Perault (1988) showed that some of the blue light escapes molecular clouds, leading Devereux and Eales (this conference) to propose that the dust in both the low and high density regions is heated by blue light from young stars.

Reliable dust temperatures are not available from the flux density ratio at the two longest IRAS bands, 60μ and 100μ, because of significant contribution to the 60μ flux density from non-thermal emission from small grains (this process dominates the dust emission at the two shorter IRAS wavelengths, 12μ and 25μ) (Draine and Anderson 1985) without a detailed model of the dust emission. Preliminary analysis of the galactic plane emission observed by the Berkeley- Nagoya microwave background experiment (Matsumoto et al. 1988) shows that the galactic dust temperature is 19 K. At this temperature, and with a long wavelength dust emissivity law of $Q_\nu \sim \nu^1$, the emitted flux density peaks at about 200μ. Observations at wavelengths longer than the IRAS wavelengths are thus vital for the characterization of the dust emission from a galaxy. The pioneering observations of M51 (Smith 1982), NGC6946 (Smith et al. 1984), of four Virgo cluster spirals by Stark et al. (1989a) and of field spirals by Eales et al. (1989) provide the first direct measurements of the dust temperatures and masses for normal galaxies and confirm the model of Draine and Anderson (1985).

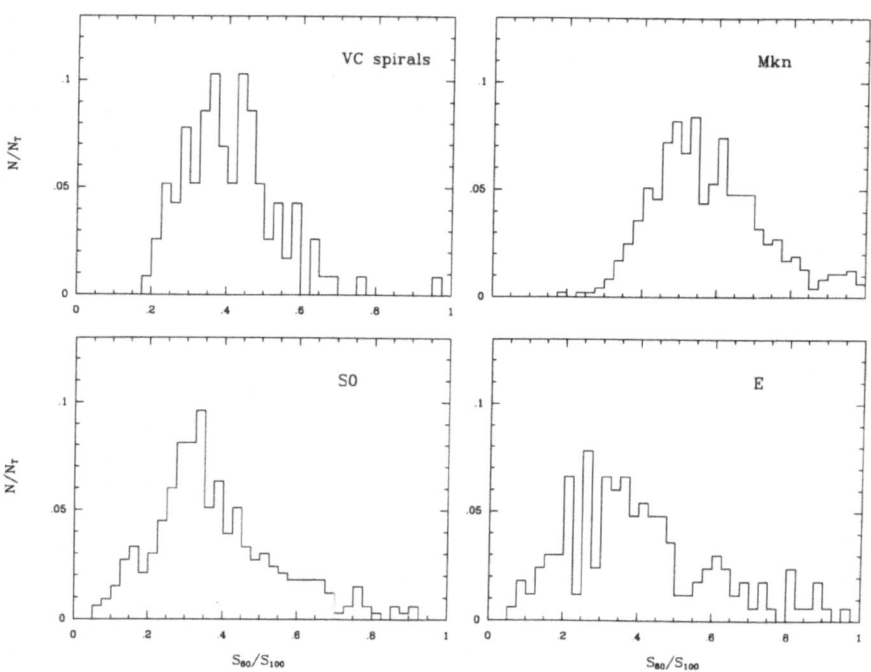

Figure 3. Distributions of the IR spectral index S_{60}/S_{100} for (a) normal Virgo Cluster spirals (b) Markarian galaxies (c) S0 galaxies (d) elliptical galaxies.

Despite this, the ratio of the 60μ and 100μ flux densities can be a diagnostic of the UV radiation field and thus current star formation in a galaxy. Figure 3 shows the IR

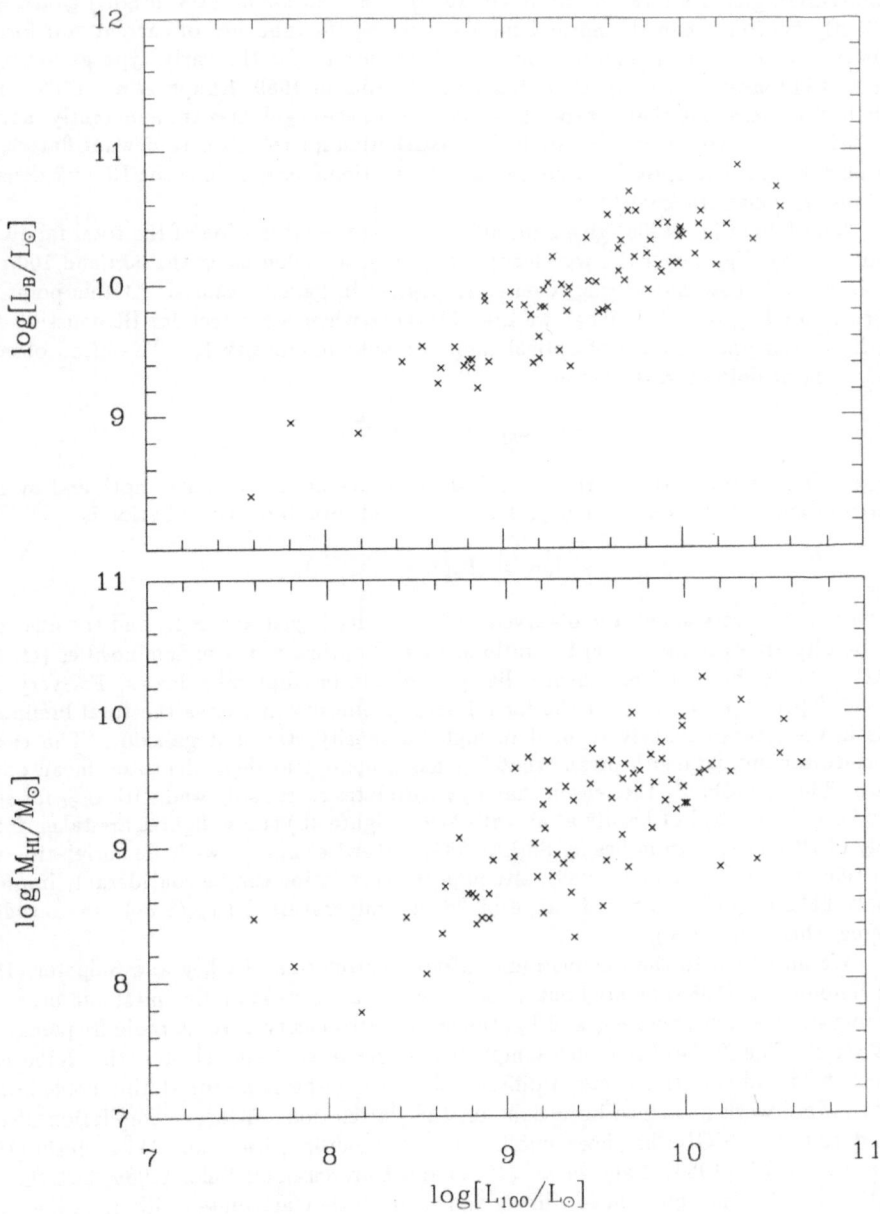

Figure 4. (a) Blue luminosity versus 100μ luminosity for Sc galaxies (b) HI mass versus 100μ luminosity for Sc galaxies (Sarlin 1988).

spectral index distributions for samples of Virgo cluster spirals by Helou *et al.* (1988) and of Markarian galaxies (Kunth and Sevre 1985). The Markarian (UV bright) galaxies are distinctly "hotter" than the normal spirals, showing the influence of current star forming activity. Also shown in Figure 3 are the distributions for the early type galaxies, S0s and Es (Thronson and Bally 1987; Bally and Thronson 1989; Knapp *et al.* 1989). Both distributions resemble that for spirals, showing that these galaxies are apparently 'normal' in UV flux per amount of ISM (while the distribution for the Es is somewhat flatter, this can be attributed to spread introduced by observational error, since the IR flux densities for these galaxies are small).

Lonsdale *et al.* (1985) give a formula allowing the estimation of the total far-IR flux from a galaxy, F_{FIR}, over the wavelength range 40μ to 1000μ using the 60μ and 100μ flux densities, and these have been extensively applied in galaxy studies. At this point it is worth reminding ourselves what we are observing when we detect far-IR emission from a galaxy. Imagine a source of optical radiation with luminosity L_o. The mean observed luminosity at optical wavelengths is

$$L_{opt} = L_o\, e^{-<\tau>} \qquad (5)$$

where $< \tau >$ is the dust absorption optical depth averaged over wavelength and over the source - dust distribution geometry. The re-emitted total infrared radiation is

$$L_{IR} = L_o(1 - e^{-<\tau>}) \qquad (6)$$

When $< \tau >$ is very small, the observed IR luminosity is just $< \tau > L_o$ and the dust mass can be directly estimated from the ratio of IR and optical fluxes or luminosities (cf. Jura 1982). This is the situation which is likely to obtain in elliptical galaxies. For very large optical depth, $L_{IR} = L_o$ and the far infrared luminosity measures the total luminosity. This is the situation likely to hold in high luminosity starburst galaxies. The case of intermediate optical depth means that L_{IR} has a non-linear dependence on mean optical depth. This is probably the reason that L_{IR} correlates reasonably well with L_{opt} for spiral galaxies (Figure 4a) but hardly at all with M_{HI} (Figure 4b) (these figures are taken from a study of IR emission from Scs by Sarlin 1988). Likewise, L_{IR} shows little correlation with L_{CO} (e.g. Solomon and Sage 1988), although the correlation can be considerably improved if only galaxies in a restricted range of "dust temperatures" (S_{60}/S_{100}) are considered (Young, this conference).

We now turn to the famous radio-infrared correlation. Dickey and Salpeter (1984) and Helou *et al.* (1985) pointed out a very tight linear correlation for spiral and other star forming galaxies between L_{FIR} and L_ν, the luminosity density at some radio frequency (1.4 or 5 GHz). The Dickey/ Salpeter sample is galaxies in the same cluster, the Helou *et al.* sample field and cluster objects at different distances - the same correlation holds in both cases. More work has since been done on this correlation. A large compilation of total flux densities at 5 GHz has been published by Wunderlich, Klein and Wielebinski (1987) while Cox *et al.* (1988), Eales *et al.* (1989) and Devereux and Eales (1989) find that the relationship does not quite have a linear slope, which they attribute to illumination effects such as those simplified by Equations 5 and 6 above.

A slightly different version of this correlation than that usually published is shown in Figure 5, where the radio luminosity density at 5 GHz is plotted against the IR luminosity density in each of the four IRAS bands. The data points in Figure 5 are taken from

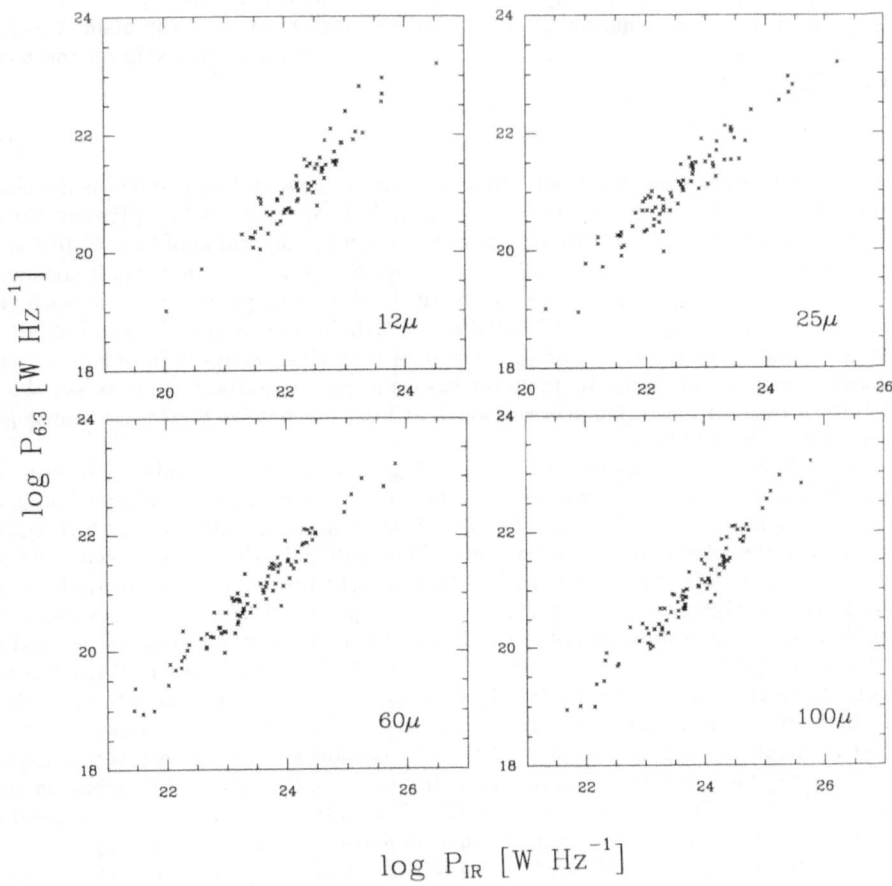

Figure 5. Radio luminosity density (at 5 GHz, 6.3 cm) versus IR luminosity density for the four IRAS bands for spiral and starburst galaxies.

Wunderlich, Klein and Wielebinski (1987), and data for one dwarf elliptical and three high-luminosity starburst galaxies are added. Figure 5 shows that this correlation holds not just at the long IRAS wavelengths but at all. It is clearly one of those 'the bigger it is, the bigger it is' correlations, i.e. both axes are measuring the same thing. But what is 'it'? Almost certainly 'it' is massive star formation, manifesting itself both in the galaxy luminosity and in the injection of energetic particles into the ISM, as shown by the following simple argument (cf. also Cox *et al.* 1988). The total luminosity of a region which is steadily forming stars at rate ψ is

$$L_\star(L_\odot) = \int_{m_1}^{m_2} \phi(m)L(m)\tau(m)\psi dm/\tau_1 \tag{8}$$

where $\phi(m)$ is the IMF, m_1 and m_2 are the masses of the lowest and highest mass star formed, $\tau(m)$ is the main sequence lifetime for a star of mass m, τ_1 is the main sequence lifetime of a star of mass m_1 and L(m) is the mass-luminosity relationship on the main sequence. The supernova rate r is

$$r = \int_{m_s}^{m_2} \psi \phi(m)/\tau_2 \qquad (9)$$

where m_s is the lowest mass star which will undergo supernova explosion and τ_2 is the main sequence lifetime of the most massive star formed. Taking $m_s = 8\ M_\odot$ (Tinsley 1975), $m_1 = 2\ M_\odot$, the Miller-Scalo (1979) IMF and the energy per supernova of $L_{SN} = 10^{51}$ ergs released over a timescale τ_2 and as radio energy with $S_\nu \sim \nu^{-1}$, so that νS_ν is constant, equations (8) and (9) give $L_{SN}/L_* = 1.5 \times 10^{-6}$. This is in good agreement with the observed ratio of $\nu L_\nu/L_{IR} = 2 \times 10^{-6}$ found at 1.4 GHz by Devereux and Eales (1989). A much more careful treatment of this problem than that given above is in order, because the observed radio to infrared ratio probably has some quite important things to say about the details of the star formation process, which at least for massive stars seem amazingly constant from galaxy to galaxy.

Figure 6 reproduces one of the panels of Figure 5 with several galaxies of interest marked. It is clear from the above discussion that either the radio or the infrared emission measure the massive star formation rate. NGC2366 is an apparently normal SBb spiral galaxy yet has the lowest star formation rate of any galaxy in the sample. NGC4194 on the other hand is classified as a nearby Magellanic dwarf irregular yet is currently forming stars at a rate 10^3 higher than is NGC2366. Mk231, Arp220 and NGC6240 are examples of the highly luminous starburst galaxies - the data in Figure 6 suggest a total star formation rate of 150 $M_\odot$ yr^{-1} for Arp 220 (if stars down to 0.1 $M_\odot$ are included). If the CO-to-molecular hydrogen conversion ratio found for the Galaxy is used, the CO data of Sanders *et al.* (1988) give a gas exhaustion timescale for Arp 220 of only 7×10^7 years.

As a complete contrast, consider NGC855 at the low end of the observed emission. This is a nearby bluish dwarf elliptical (Burstein *et al.* 1987) with weak IR emission and several $\times 10^7\ M_\odot$ of HI (Walsh *et al.* 1989b). The galaxy also has a weak, extended non- thermal radio source. The IR and radio luminosities of the galaxy suggest a star formation rate of about 0.02 $M_\odot$ yr^{-1}. CO is not detectable from this galaxy, so that the gas exhaustion time scale is 2 - 4 $\times 10^9$ years, similar to the value for the Galaxy - this is because NGC855, like the Galaxy, has much of its gas where the star formation isn't. Figure 7 shows a Thuan-Gunn g-band CCD image of the galaxy (Wallington *et al.* 1989). This high dynamic range image shows normal outer contours but several knots at the center whose position coincides with that of the radio continuum emission.

This discussion, and Figure 6, show that star formation is essentially a small scale process. It apparently happens when an aggregate of ISM, maybe as low as 10^5 to $10^6\ M_\odot$ in mass, decides to form stars. NGC855 has two or three such regions; Arp 220 has 20,000 to 30,000, i.e. the entire interstellar medium is forming stars, and each aggregate is doing pretty much the same thing. Why? This is attributed to density wave spiral arms in spiral galaxies, though a growing body of opinion (e.g. Elmegreen and Elmegreen 1987) holds that spiral arms are irrelevant. But why isn't NGC2366 forming stars more rapidly? How do the starburst galaxies work? My own opinion is that while it is easy to agree with the statement "stars form in spiral arms" this is in no sense an answer to the question "why do stars form?". It looks as though the answer is rather to be found in the detailed physics of individual molecular clouds.

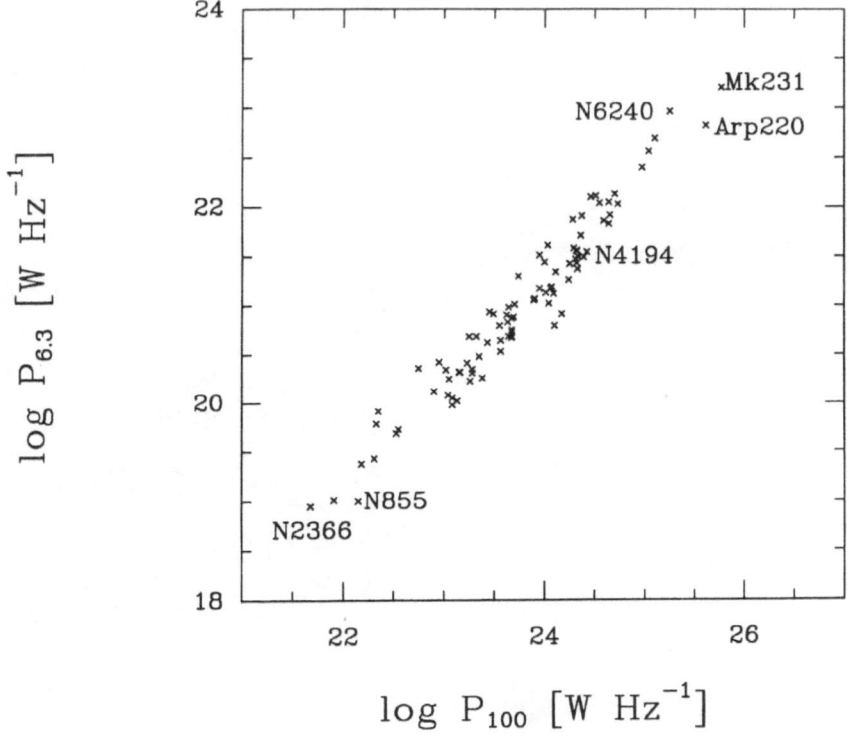

Figure 6. Radio luminosity density (at 5 GHz, 6.3 cm) versus 100μ luminosity density for spiral and starburst galaxies. The data points corresponding to NGC2366 (SBb spiral), NGC855 (dwarf elliptical), NGC4194 (Magellanic irregular) and Arp220, Mk231 and NGC6240 (starburst galaxies) are marked.

4. The Interstellar Medium in Galaxies

4.1 SPIRAL GALAXIES

The vast majority of the work done on the ISM and star formation has been done on spirals, and these galaxies are extensively reviewed at this conference, so this section will be short. Total HI flux densities are available for several thousand spirals (Huchtmeier and Richter 1989) and CO flux densities for several hundred spirals (Young and Scoville 1982; Young et al. 1986; Kenney and Young 1988; Verter 1988; Stark et al. 1986; Stark et al. 1987; Young et al., this conference). These observations show that for spirals of reasonably high luminosity like our own, the CO/HI flux ratio suggests that $M(HI) \sim M(H_2)$.

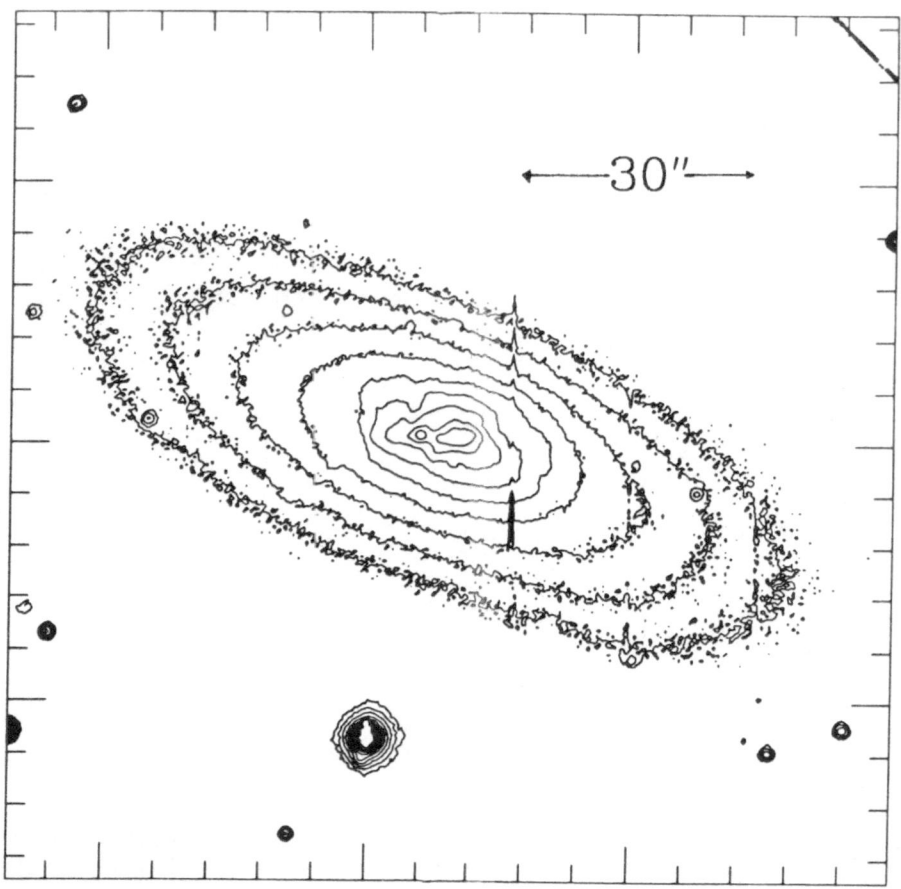

Figure 7. Thuan-Gunn g-band image of the dwarf elliptical NGC855 at $\alpha = 02^h11^m10^s, \delta = +27°38'36''$ (1950).

The CO distributions are more concentrated to the inner regions of spiral disks and the molecular gas is apparently much more intimately connected with star formation than is the HI (Scoville 1989). Verter (1988) has found that the CO flux has a maximum value for galaxies of intermediate morphological type (Sb).

   HI maps of spirals have been made for many years at high spatial resolution primarily using the Westerbork and VLA synthesis instruments. CO maps of comparable resolution are now becoming available and many of them are to be presented at this symposium. The first spiral galaxy mapped in detail in the CO line is M31; the CO emission is closely associated with the spiral arms, and the CO velocities show density wave streaming of about 30 km s^{-1} across the arms (Ryden and Stark 1986). The CO emission appears to come from giant molecular clouds quite similar to those in the Galaxy.

M31 presents a real mystery, however. Its HI surface density distribution is quite flat with radius (Roberts 1975) as is that for the Galaxy. But the CO distribution is quite unlike that in the Galaxy (or in M51); the *total* gas density does not increase towards smaller radii. In the inner regions of the Galaxy (R = 4 to 6 kpc) the H$_2$ surface density is 5 to 10 times the HI surface density, while in M31 the H$_2$ surface density is only about 1/5 that of the HI and closely follows the HI spiral arms. Indeed, in M31 the total gas surface density peaks at a large radius, about 10 kpc; this ring is clearly seen in the IRAS maps of the galaxy (Habing *et al.* 1984). M31 and the Galaxy, which apparently have similar masses and morphological types and dominate the mass of the Local Group, apparently have quite different ISM contents and distributions. The solution to this local mystery could well be of considerable importance for galaxy evolution.

4.2 EARLY TYPE GALAXIES

These galaxies fall into two distinct types morphologically and dynamically. The S0s have both a hot stellar component (the bulge) and a cold stellar component (a disk), but with no spiral structure; the ellipticals are diskless systems. For a long time these galaxies were thought to be essentially devoid of interstellar matter and star formation, a conclusion based on the failure of early searches to find anything and on evolutionary considerations; the colors of these galaxies indicate old stellar populations, and galaxy formation scenarios suggest that they (especially Es) formed all of their stars during the initial free fall.

It has also long been known that some of these galaxies contain trace amounts of gas; optical emission lines and faint dust lanes are seen in a minority of them. The observational situation has improved dramatically in recent years. All of the ISM components seen in spirals have been detected in several, and sometimes many, of these systems, using the observational probes used to investigate spirals, due to the enormous sensitivity increases at radio, infrared, X-ray and millimeter wavelengths.

A major difficulty in this little subspeciality has been the lack of a reliable optical catalog. Since the conventional wisdom has it that these galaxies contain no ISM, the discovery of substantial amounts of cold gas in these galaxies often provokes the reaction "that cannot be a true E/S0". Often, indeed, further investigation shows that the galaxy is a misclassified spiral, though not always, and there is now a substantial number of early-type galaxies with well-established detections of cold gas. This problem also plagues studies of spiral galaxies, though it is not so forcibly brought to the attention of the practitioners. For example, there is the aforementioned problem with M31, and there are several bright apparently bona-fide spirals in which little or no HI is detected (e.g. Shostak 1978). Optical observing sensitivity and linearity have also improved enormously over the past few years, and perhaps it is time to produce a new state of the art optical data set for nearby galaxies.

The really interesting thing about early-type galaxies is not that they contain cold gas just like everybody else; rather, it is that the hot component of the ISM, detected by its X-ray emission dominates the ISM just as the dynamically hot stars dominate the structure of the galaxy. In these galaxies, the ISM has reached a quite different equilibrium from that in spirals.

4.2.1 *S0 Galaxies.* Hot gas is detected in these galaxies by X-ray emission (Forman, Jones and Tucker 1985; Canizares, Fabbiano and Trinchieri 1987), warm ionized gas via optical line emission (Pogge and Eskridge 1987; Wrobel and Heeschen 1988); cold diffuse gas by HI (summarized by Wardle and Knapp 1986); dust via optical absorption and IR emission (Dressel 1988; Bally and Thronson 1989; Knapp et al. 1989); molecular gas via CO emission (Wiklind and Henkel 1989; Sage and Wrobel 1989; Thronson et al. 1989); and disk emission at radio wavelengths (Wrobel and Heeschen 1988; Walsh et al. 1989a).

Among the interesting characteristics of S0 galaxies found in this work are: (a) the L_{CO}/L_B and $M(HI)/L_B$ ratios have a very much wider range than found for the typical spiral class; at the low end, these values are lower than for any spiral and at the high end as high as the most gas-rich spirals. A lot of this could be due to difficulties with the morphological typing, but it suggests that for these galaxies, and for the ellipticals, the ISM is no longer in smooth equilibrium with the stars. (b) the ratio of CO and HI fluxes is significantly higher for S0s than for spirals. (c) The X-ray luminosities indicate masses of hot gas in the range 10^8 to 10^9 $M_\odot$, about the amount expected to have been accumulated from stellar mass loss. (d) the distribution of the optical emission line intensity shows that star formation is in extended regions of size 1 to several kpc in the inner regions of the galaxies. (e) for those galaxies for which it can be measured, the radio to infrared luminosity ratios are similar to those seen in spirals, showing a common origin in the star formation process.

4.2.2 *Elliptical Galaxies* Again, all components of the ISM have been observed in these galaxies. The hot gas is seen via its X-ray emission (Forman, Jones and Tucker 1985; Canizares, Fabbiano and Trinchieri 1987) in amounts consistent with mass loss from evolved stars. Optical line and radio continuum emission is seen and shows two sorts of behavior. In low luminosity ellipticals, the optical emission lines are HII-region like and the radio continuum weak, extended and similar to that produced by star forming activity. In high luminosity ellipticals, the emission line spectra resemble those seen from AGNs and the radio continuum is much more powerful and due to a non-thermal nuclear point source (Sadler 1987; Walsh et al. 1989a; Sadler et al. 1989). Thus low-level star formation is apparently present in the inner regions of low-luminosity ellipticals. If it is present in high-luminosity ellipticals, it is quite swamped by the powerful emission from the active galactic nuclei in these systems.

Cold material has also been detected in a fair number of elliptical galaxies. About 40% have dust absorption (Sadler and Gerhardt 1985; Bertola 1987). A similar fraction of bright ellipticals is detected at long IRAS wavelengths (Knapp et al. 1989). HI emission has been detected from about 15% of nearby ellipticals (summarized by Knapp, Turner and Cunniffe 1985) and CO emission has been seen in several Es (Wiklind and Rydbeck 1986; Sanders and Mirabel 1985; Phillips et al. 1987).

The origin of the cold gas in Es is a long-standing problem. The kinematics and distribution of the cold gas are usually quite different from those of the stars, suggesting an external origin (a merger event) for the cold gas. Such events, which presumably happen to galaxies of all types, are more obvious for Es because of the paucity of the ambient ISM and because the accretion of cold gas can give a new lease of life to the central dormant engine in these dense systems, causing them to become active galaxies and radio sources (Gunn 1979).

My feeling is that the evidence is moving against this view. As ISM is detected in more and more Es (albeit in very small amounts), it seems increasingly likely that it is intrinsic rather than due to a recent event. Many gas-rich early type systems, like NGC855 discussed above, are isolated. Further, detailed probing of the structure and kinematics of Es shows that they are not quite such simple systems as they appear- they have counter-rotating cores (Franx and Illingworth 1988), shells (Malin and Carter 1983) boxy isophotes and small internal disks (Lauer 1985; Bender and Möllenhoff 1987). A comparison of these various features with indicators of cold ISM shows no correspondence (Thronson, Bally and Hacking 1989; Walsh and Knapp 1989), suggesting that the features are not due to an identifiable accretion event.

The masses of cold gas in these systems are quite small compared to those of the hot gas (1% - 10% or so) and they are indeed in a very different state from spirals. In detail, however, the hot and cold gas show no correspondence. Figure 8 shows the infrared versus X-ray dependence at two IRAS wavelengths, 12μ and 100μ, for E and S0 galaxies. The 100μ emission is likely to be measuring the cool ISM; the 12μ emission measures the mass injection rate from the evolving stars in the galaxy (Jura et al. 1987). There is no correlation for the hot and cold ISM (Figure 8b); however, Es, though not S0s, have their X-ray emission weakly correlated with the 12μ emission. These results suggest that the X-ray emission may be correlated with the current injection rate of gas into the galaxy, and that the X-ray emission may be due to hot outflowing gas.

4.3 GAS RICH DWARFS

At the opposite end of the Hubble sequence are the gas rich dwarf irregular galaxies, whose prototypes are the Magellanic Clouds. These objects are of interest because they are numerically the most common systems in the Universe and because they may contain a significant fraction of the remaining neutral hydrogen. Galaxies like the Magellanic Clouds have $M(HI)/L_B \sim 1\ M_\odot/L_\odot$ and lower metallicity than the Galaxy (by about a factor of 10) so have undergone fewer generations of star formation. This raises the possibility of using these galaxies to attempt to understand whether there is a threshold density of gas for star formation, what triggers it, and how metallicity evolves with time.

Large scale surveys of the HI content of dwarf galaxies have been available for many years (Fisher and Tully 1976). Hydrogen masses as low as a few $\times 10^7\ M_\odot$ have been found for these systems, and $M(HI)/L_B$ increases as L_B decreases. This naturally raises the suggestion that there are intergalactic HI clouds devoid of stars, i.e. containing primordial material. Searches for such clouds have so far proved negative (Lo and Sargent 1977; Sargent and Lo 1985). Several very low mass HI rich dwarfs have been found, but all are associated with stars. Since then, VLA maps of nearby galaxy groups almost always turn up one or two HI rich dwarfs, but all found so far are associated with stars (e.g. Guhathakurta et al., this conference).

Many dwarf galaxies, such as NGC1569 (Young, Gallagher and Hunter 1984) apparently form stars in bursts at intervals of one or two $\times 10^9$ years. The reasons for this are not obvious but may well be related to the low total mass and gas mass; in these systems star formation may not be able to reach equiilibrium with the ISM and may oscillate (H. Thronson, 1989, useful suggestion).

CO emission is detected weakly or not at all in irregular galaxies despite the observation of extensive star formation (Thronson and Telesco 1986), leading to the suggestion that there is little molecular gas and that stars form directly from HI clouds in these systems. An alternative suggestion, however, is that the CO intensity per mass of molecular

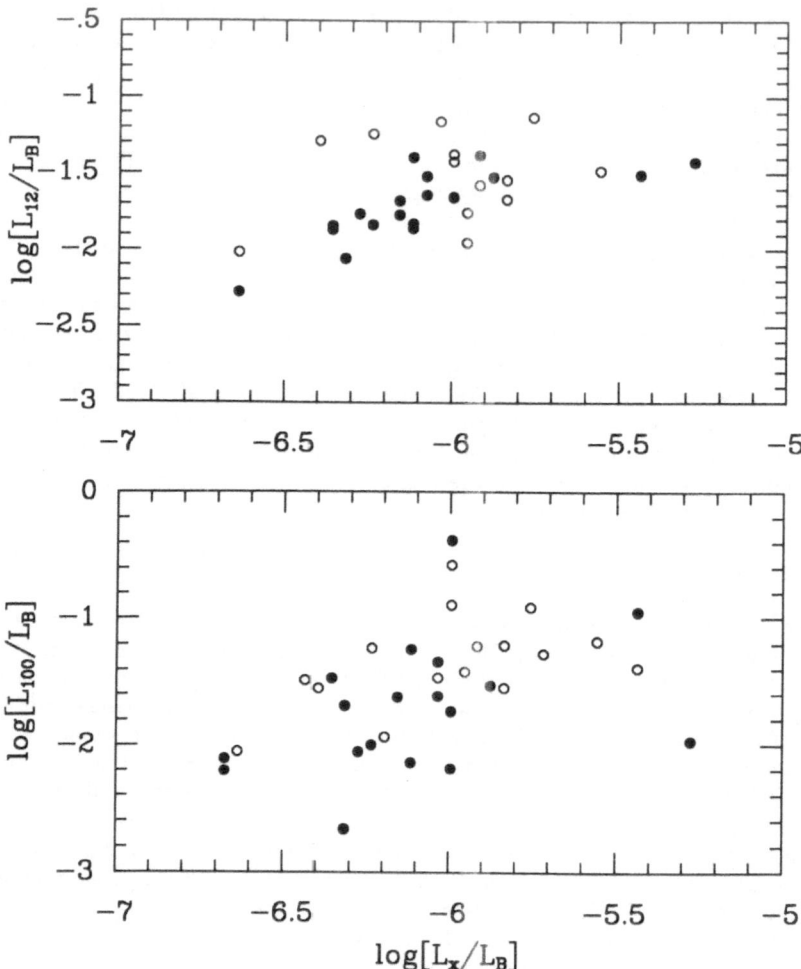

Figure 8. Ratio of infrared to blue luminosity versus X-ray to blue luminosity for S0 galaxies (open symbols) and elliptical galaxies (filled symbols): (a) upper panel, at 12μ; (b) lower panel, at 100μ.

hydrogen is much lower in these systems due to their lower metallicity (Thronson *et al.* 1987; Israel 1988). Weak CO emission has now been detected in several dwarf systems (Tacconi and Young 1985, 1987; Young, Gallagher and Hunter 1984; Kenney and Young 1988; Young *et al.* 1988; Cohen *et al.* 1988; Nyman 1989; Dettmar and Heithausen 1989). Mapping of individual clouds in two very nearby dwarf galaxies, the Large Magellanic

Cloud and NGC55 allows the calculation of molecular cloud masses from the virial theorem (Cohen *et al.* 1988; Nyman 1989; Dettmar and Heithausen 1989), and it is indeed found that $I_{CO}/M(H_2)$ has a value 1/5 to 1/20 of that in the Galaxy. A curiousity of the work on NGC55, though, is that the IR luminosity per mass of the molecular clouds is apparently similar to that in the Galaxy; in lower metallicity systems, one would expect less dust and perhaps less IR emission. The infrared/non-thermal radio relationship should be examined rather carefully for these galaxies.

In any case, the LMC map by Cohen *et al.* (1988) shows that $M(H_2) \sim M(HI)$, that the molecular clouds are in huge irregular complexes rather than spiral arms and that star formation indicators (HII regions, SNR) are found at the peaks of the molecular, not the atomic emission. In dwarf galaxies as in spirals, at least massive star formation takes place in the molecular clouds.

5. Radio Galaxies, Clusters and Galaxy Formation

The powerful radio frequency emission from radio galaxies and quasars has been recently discussed by Perley and Bridle (1984); the accepted model for the production of the enormous radio luminosities of these systems is the conversion of mass to energy by a central black hole at high efficiency (10%) (Lynden-Bell 1969) - these mechanisms are discussed in detail by Begelman, Blandford and Rees (1984) and Rees (1984). The galactic hosts of these radio sources are high luminosity early-type systems - the likelihood of a galaxy being a radio source increases rapidly with optical luminosity, and the fuel is likely to be gas. Indeed, the cosmological distribution of these radio sources (Rees 1984) shows that they were much more numerous in the past ; presumably, the fuel supply is exhausted for most nearby elliptical galaxies.

Observational support for this picture has come from a variety of work. Mirabel (1989) and Mirabel *et al.* (1989) found large amounts of HI and CO in several nearby quasars and radio galaxies. A search for HI absorption towards the nuclei of a large sample of low-z radio sources by van Gorkom *et al.* (1989) found that about 25% of these galaxies have one or more HI absorption features near the systemic galaxy velocity; there is strong statistical evidence that the HI absorption components arise from infalling gas and that the gas in these galaxies is in disks. Among the radio sources in which absorption was detected is 3C236, whose enormously extended radio jet structure makes it the largest coherent object in the Universe. The inferred infall rates are 10^{-2} to 10^{-1} $M_\odot$ yr^{-1}, more than enough to fuel the observed radio luminosity.

It has long been recognized that radio galaxies and quasars also have peculiar emission in other spectral regions (UV excesses, X-ray emission, IR excesses etc.). Golombek, Miley and Neugebauer (1988) analyzed the IRAS data for a large number of radio galaxies and conclude that the IR luminosity correlates with indicators of nuclear activity (radio emission, optical line emission etc.). These relationships are shown in a slightly different form in Figure 9, which shows the 60μ luminosity density versus the 5 GHz luminosity density for the low redshift radio galaxies from the sample of van Gorkom *et al.* (1989). Included on this plot are data for some examples of radio-loud quasars (Neugebauer *et al.* 1986) and the data for spiral and starburst galaxies from Figures 5 and 6. There are a couple of interesting points to note about the data in Figure 9. First, the detection rate by IRAS is much higher (60% versus 45%) for the radio galaxies than for "normal" ellipticals,

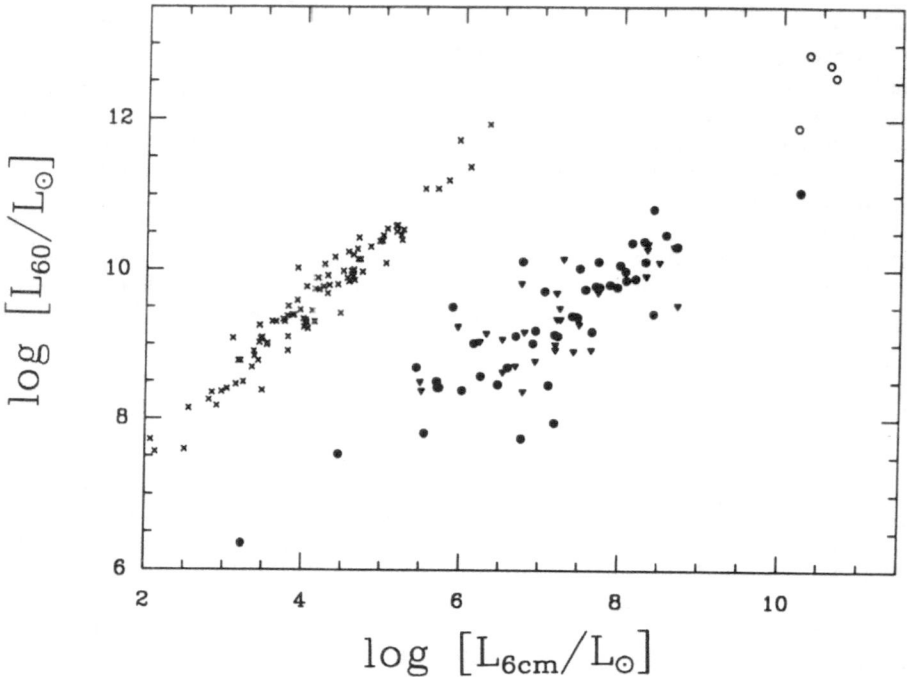

Figure 9. Luminosity density at 60μ versus luminosity density at 5 GHz, 6 cm, for starburst and spiral galaxies (x), radio galaxies (•) and quasars (o). Upper limits are indicated by triangles.

despite the fact that the optical fluxes of the radio galaxy sample are several magnitudes fainter than those of the elliptical galaxy sample. Second, the radio galaxies and quasars have about the same range of IR luminosities as do the spirals, but have a ratio of radio to IR luminosity three orders of magnitude higher on average - the two samples occupy quite distinct regions of Figure 9.

What is the source of the IR luminosity from the radio galaxies? A first suspicion is that it is the high-frequency tail of the non- thermal radio emission. However, the fact that the IR luminosities cover a similar range for the radio galaxies as for the spirals suggests that it might rather be thermal. The spectrum of a nearby radio galaxy (3C120) from radio to IR is shown in Figure 10 (Knapp, Bies and van Gorkom 1989) - cf. also the spectra of several nearby radio galaxies presented by Wrobel, Miley and Neugebauer 1986). The IR emission is clearly in excess over the extrapolated radio flux and has the spectral signature (not unambiguously, though) of emission from cool dust. Since the IR luminosities of the radio galaxies are similar to those of spirals, it is not unreasonable to suppose that they contain similar amounts of ISM.

Optical imaging of radio galaxies (Heckman *et al.* 1986) shows that many of them

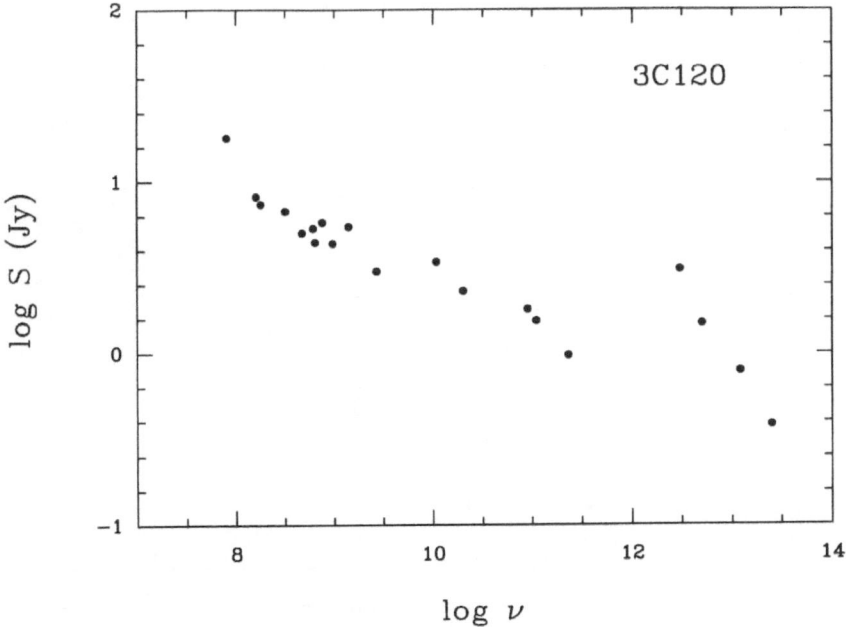

Figure 10. Continuum spectrum of the radio galaxy 3C120. The four points at the highest frequencies give the IRAS data.

have peculiar and disturbed morphologies; this is also the case for the high luminosity starburst systems (Sanders *et al.* 1988). These data then suggest that the merger of two spirals produces a starburst while the merger of a spiral and an elliptical produces a radio galaxy. It has been suggested on the basis of the enormous luminosities and of the space density of the starburst systems that they are proto quasars (Soifer *et al.* 1987; Sanders *et al.* 1988). However, while it is true that the starburst galaxies are powerful radio sources, Figure 9 shows that this can all be accounted for by star formation and that there is no sign that these objects contain radio-active nuclei (at least not yet). Thus apparently not all mergers lead to the formation of a massive black hole such as those apparently present in the centers of luminous ellipticals, and it is possible that such black holes are formed only during the initial galaxy formation process. I cannot think of an observational test of this suggestion, but it would be an interesting thing to find out.

This raises the subject of the function of the environment of a galaxy in determining its properties. The striking difference in galaxy populations in and out of dense clusters has led to two scenarios for galaxy evolution. Dressler (1984) has suggested that the density of the cluster environment affects the initial formation of a galaxy, with high density galaxies (Es) formed in high-density environments. Gunn and Gott (1972) and Melnick and Sargent (1977) have suggested that early-type galaxies are formed in clusters when their ISM is

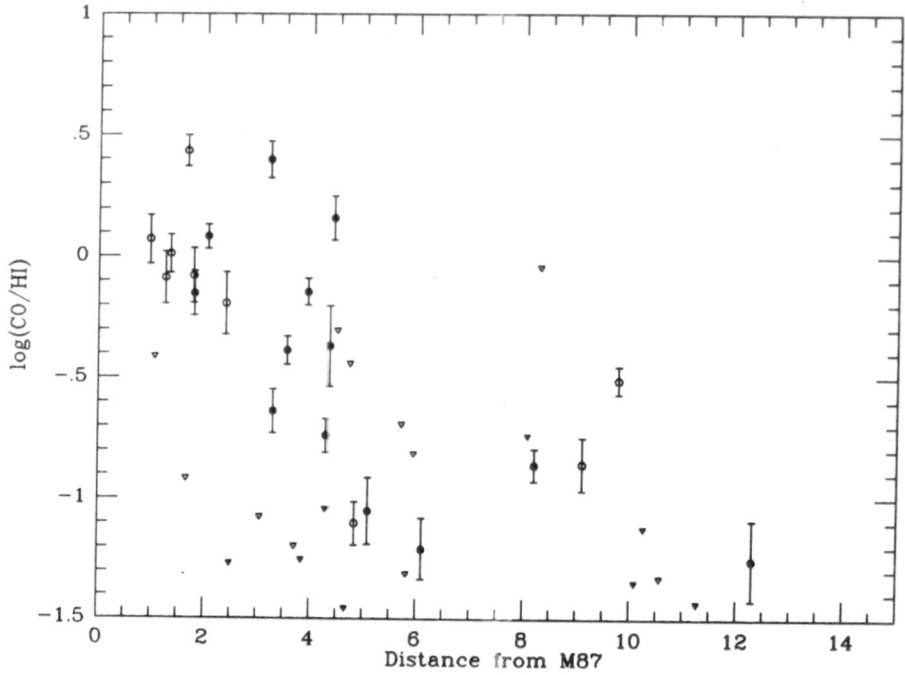

Figure 11. Ratio of CO to HI fluxes for spiral galaxies in the Virgo Cluster versus angular distance from M87 in degrees (Stark *et al.* 1986).

stripped by passage through the dense intergalactic gas in the cluster.

The gas content of galaxies of similar morphological type in and out of clusters bears directly on this question and on that of the evolution of galaxies in clusters (see below). Measuring the HI deficiency of galaxies in and out of clusters has proven difficult because of the intrinsically wide range of $M(HI)/L_B$ for spirals of a given morphological type but has now been well established for several clusters (Haynes, Giovanelli and Chincarini 1984). The effect of the cluster environment on HI content has been dramatically established for the Virgo Cluster, however, and these results are of great interest for considerations of the effect of the environment on galaxy evolution. van Gorkom and Kotanyi (1985) present VLA maps of the HI distribution of spiral galaxies in the Virgo Cluster which clearly show that the HI diameters (normalized to the optical diameters) decrease with decreasing projected distance from M87, the central galaxy of the Virgo Cluster (cf. also Warmels 1986). Many of the spiral galaxies in the Virgo Cluster have also been observed in the CO(1-0) line, and the ratio of CO to HI flux has a value 10 to 20 times higher for spirals within 4° of M87 than it does for 'normal' 'field' spirals, as shown in Figure 11, from Stark *et al.* 1986) (see also Kenney and Young 1986, 1989). Comparison of CO and HI fluxes with luminosity shows that almost all of this effect can be accounted for by the removal of HI rather than the conversion of HI to H_2. This result can easily be explained by ram pressure sweeping (Gunn and Gott 1972); the mass per unit area for molecular clouds is far higher than for HI clouds, and the molecular clouds tend to be found in the inner regions of the galactic disks where the mass surface density is higher. At the typical speeds with which galaxies move through the inner regions of a cluster and for typical intergalactic densities in clusters, the HI is expected to be swept out while the molecular clouds are not. This means that spiral galaxies may be able to retain a significant fraction of their ISM during passage through a cluster.

Galaxies are generally thought to form at early epochs. The conventional picture for the formation of the brightest galaxies, the ellipticals, has them forming all of their stars in a free-fall time $(G\rho)^{-1/2}$ to avoid the formation of a disk - this implies a star formation rate of several 100 $M_\odot$ yr^{-1}. Much of what we know about early galaxy formation comes from work on high redshift radio galaxies, whose powerful radio emission alerts us to their presence. For example, the colors of the radio galaxy 0902+34 ($z = 3.4$) suggest that, while there is a large current rate of star formation (100 $M_\odot$ yr^{-1}), most of the mass (90%) has already formed into stars some 1 - 2×10^9 years previously (Lilly 1988). This puts the first epoch of star formation well out of the observational range of current or conceivable HI and CO observations.

However, there is growing evidence that significant star formation in early type systems also takes place at much more recent epochs - this is the Butcher-Oemler (1984) effect. The enormous increases in the sensitivity and linearity made available for optical observations by CCD devices now make possible photometry and spectroscopy of galaxies to $z = 1$, and extensive studies of the populations of galaxies in clusters are underway (Dressler, Gunn and Schneider 1985; Lavery and Henry 1986; Gunn and Dressler 1988). These observations show significant galaxy evolution at recent epochs; the fraction of galaxies in clusters with colors like those of present-day E/S0 systems drops from 95% at $z = 0$ to 70% at $z = 0.5$ (Gunn and Dressler 1988).

Galaxies in clusters at $z = 0.5$ show three types of spectra; normal E type spectra, indicating little star formation over the past 5×10^9 years; E + A star spectra, indicating a large burst of star formation some 10^9 years ago, and active galaxies with emission lines and blue colors, possible indicating current star formation. Gunn and Dressler (1988) attribute these bursts to the greatly increased pressure on the galactic molecular clouds in the cluster environment (cf. the discussion on the Virgo Cluster above). The residual small amounts of cold ISM seen in present-day E and S0 galaxies also support the idea that these systems have undergone continued star formation to the present epoch.

Is there any hope of observing cold ISM in galaxies at $z = 0.5$? The present observational sensitivity is not sufficient, but such observations are within the reach of present technology. The lowest practical system temperature available at 1 GHz on the telescope is about 20 K. With such sensitivity, Arecibo could reach $z = 0.2$ in about one hour of observing (for a galaxy containing 5×10^9 $M_\odot$ of HI), while the VLA could reach this redshift in about 10 hours. A galaxy with $M(H_2) = 10^{10}$ $M_\odot$ could be detected in the CO(2-1) line in about an hour of observing with a 100m millimeter telescope. The CI and CII fine structure lines are likely to be the strongest lines in the submillimeter spectra of galaxies and may be detectable with present sensitivities in highly redshifted galaxies and quasars. The use of the Universal time machine to study the evolution of the ISM in galaxies is thus something which is just about within reach and to which we can look forward in the next ten or twenty years.

6. Acknowledgements

I would like to say many thanks to Harley Thronson and to the organizing committee for the invitation to this conference, and to H. Thronson, C. Joseph, C. Gammie, A. Stark and J. Gunn for much very useful and interesting discussion. This research is supported by the N.S.F. via grant AST86-02698 and by NASA under the IRAS Extended Mission project via grant R010-87.

References

Bajaja, E., van der Berg, G., Faber, S.M., Gallagher, J.S., Knapp, G.R. and Shane, W.W. 1984, *Astron. Astrophys.* **141**, 309.

Bally, J., Stark, A.A., Wilson, R.W. and Henkel, C. 1988, *Ap.J.* **324**, 223.

Bally, J., and Thronson, H.A. 1989, *A.J.* **97**, 69.

Begelman, M.C., Blandford, R.D., and Rees, M.J. 1984, *Rev. Mod. Phys.* **56**, 255.

Bender, R., and Möllenhof, C. 1987, *Astron. Astrophys.* **177**, 71.

Bertola, F. 1987, in I.A.U. 127, "Structure and Dynamics of Elliptical Galaxies", ed. T. de Zeeuw, D. Reidel Co., p135.

Bhat, C.L., Mayer, C.J., and Wolfendale, A.W. 1986, *Phil. Trans. Roy. Soc. London A* **319**, 249.

Bieging, J.H. 1989, in "The Evolution of the Interstellar Medium", ed. L. Blitz, (A.S.P. Conference Proceedings) (in press).

Blitz, L., in "Evolution of the Interstellar Medium" ed. L. Blitz, A.S.P. Conference Proceedings (in press).

Blitz, L., Magnani, L. and Mundy, L. 1984, *Ap.J. (Letters)* **282**, L9.

Blitz, L., and Stark, A.A. 1986, *Ap.J.(Letters)* **300**, L89.

Blitz, L., and Thaddeus, P. 1980, *Ap.J.* **241**, 676.

Bloemen, J.G.B.M. *et al.* 1986, *Astron. Astrophys.* **154**, 25.

Bohlin, R.C., Hill, J.K., Jenkins, E.B., Savage, B.D., Snow, T.P., Spitzer, L., and York, D.G. 1983, *Ap.J. (Suppl.)* **51**, 277.

Bosma, A. 1981, *A.J.* **86**, 1825.

Boulanger, F., and Perrault, M. 1988, *Ap.J.* **330**, 964.

Bridle, A.H., and Perley, R.A. 1984, *A.R.A.A.* **22**, 319.

Burstein, D., Davies, R.L., Dressler, A., Faber, S.M., Stone, R.P.S., Lynden-Bell, D., Terlevich, R.J. and Wegner, G, 1987, *Ap.J. (Suppl.)* **64**, 601.

Butcher, H., and Oemler, A.G. 1984, *Ap.J.* **285**, 426.

Carrasco, L., Strom, S.E., and Strom, K. M. 1973, *Ap.J.* **182**, 95.

Clemens, D.P., Sanders, D.B., Scoville, N.Z., and Solomon, P.M. 1986, *Ap.J. (Suppl.)* **60**, 297.

Cohen, R.S., and Thaddeus, P. 1977, *Ap.J. (Letters)* **207**, L189.

Cohen, R.S., Dame, T., Garay, G., Montani, J., Rubio, M. and Thaddeus, P. 1988, *Ap.J. (Letters)* **331**, L95.

Condon, J.J., Condon, M.A., Gisler, G. and Puschell, J.J. 1982, *Ap.J.* **252**, 102.

Cowie, L.L. 1980, *Ap.J.* **236**, 868.

Cox, M.J., Eales, S.A., Alexander, P. and Fitt, A.J. 1988, *M.N.R.A.S.* **235**, 1227.

Cox, D.P., and Reynolds, R.J. 1987, *Ann. Rev. Astron. Astrophys.* **25**, 303.

Dame, T.R., and Thaddeus, P. 1985, *Ap.J.* **297**, 751.

Dame, T.M., *et al.* 1987, *Ap.J.* **322**, 706.

DeNoyer, L.K., and Frerking, M.A. 1981, *Ap.J. (Letters)* **246**, L37.

Dettmar, R.-J., and Heithausen, A. 1989, preprint.

Devereux, N., and Eales, S.A. 1989, *Ap.J.* **340**, 708.

Dickey, J.M., Kulkarni, S.R., van Gorkom, J.H., and Heiles, C.E. 1983, *Ap.J. (Suppl.)* **53**, 591.

Dickey, J.M., and Salpeter, E.E. 1984, *Ap.J.* **284**, 461.

Dickman, R.L., Snell, R.L., and Schloerb, F.P. 1986, *Ap.J.* **309**, 326.

van Dishoeck. E.F., and Black, J.H. 1986, *Ap.J. (Suppl).* **62**, 109.

Draine, B.T. 1989, in "The Evolution of the Interstellar Medium" ed. L. Blitz, (A.S.P. Conference Proceedings) (in press).

Draine, B.T. and Anderson, N. 1985, *Ap.J.* **292**, 494.

Dressel, L.L. 1988, *Ap.J. (Letters)* **329**, L69.

Dressel, L.L., and Condon, J.J. 1978, *Ap.J. (Suppl.)* **36**, 53.

Dressler, A. 1984, *A.R.A.A.* **22**, 185.

Dressler, A., Gunn, J.E. and Schneider, D.P. 1985, *Ap.J.* **324**, 70.

Eales, S.A., Wynn-Williams, C.G. and Duncan, W.D. 1989, *Ap.J.* **339**, 859.

Elmegreen, B.G. 1989, *Ap.J.* (in press).

Elmegreen, D.M. and Elmegreen, B.G. 1987, *Ap.J.* **314**, 3.

Fabbiano, G. 1989, *Ann. Rev. Astron. Astrophys.* (in press).

Fisher, J.R., and Tully, R.B. 1976, *Astron. Astrophys.* **53**, 397.

Forman, W., Jones, C. and Tucker, W. 1985, *Ap.J.* **293**, 102.

Franx, M., and Illingworth, G.D. 1988, *Ap.J.* **327**, L55.

Gammie, C.F., and Ostriker, J.P. 1989, in preparation.

Giovanelli, R., and Haynes, M.P. 1985, *A.J.* **90**, 2445.

Golombek, D., Miley, G.K., and Neugebauer, G. 1988, *A.J.* **95**, 26.

Gordon, M.A., and Burton, W.B. 1976, *Ap.J.* **208**, 346.

van Gorkom, J.H., Knapp, G.R., Ekers, R.D., Ekers, D.D., Laing, R.A., and Polk, K.S. 1989, *A.J.* **97**, 708.

van Gorkom, J.H., and Kotanyi, C.G. 1985, in "Proceedings of the ESO Workshop on the Virgo Cluster of Galaxies", ed. O. Richter and B. Binggeli, ESO Proceedings #20, p61.

Goss, W.M., Retallack, D.S., Felli, M., and Shaver, P.A. 1983, *Astron. Astrophys.* **117**, 115.

Guhathakurta, P., and Tyson, J.A. 1989, *Ap.J.* (in press).

Gunn, J.E. 1979, in "Active Galactic Nuclei", ed. C. Hazard and S. Mitton, Cambridge University Press, p213.

Gunn, J.E. 1981, in "Astrophysical Cosmology", ed. H.A. Brück, G.V. Coyne and M.S. Longair, *Pont. Acad. Scient. Scripta Varia* **48**, p233.

Gunn, J.E., and Dressler, A. 1987, in "Towards Understanding Galaxies at High Redshift", ed. R. Kron and A. Renzini, Kluwer Academic, p227.

Gunn, J.E., and Gott, J.R. 1982, *Ap.J.* **176**, 1.

Habing, H.J. *et al.* 1984, *Ap.J. (Letters)* **278**, L59.

Haynes, M.P. Giovanelli, R., and Chincarini, G.L. 1984, *A.R.A.A.* **22**, 445.

Heckman, T.M., Smith, E.P., Baum, S.A., van Breughel, W.J.M., Miley, G.K., Illingworth, G.D., Bothun, G.D., and Balick, B. 1986, *Ap.J.* **311**, 526.

Helou, G., Khan, I.R., Malek, L., and Boehmer, L. 1988, *Ap.J. (Suppl.)* **68**, 151.

Helou, G., Soifer, B.T., and Rowan-Robinson, M. 1985, *Ap.J. (Letters)* **298**, L7.

Herbig, G.H. 1977, in I.A.U. Symposium No. 75 "Star Formation" ed. T. de Jong and A. Maeder, (D. Reidel Pub. Co.), p283.

Herbst, W., and Assousa, G.E. 1977, *Ap.J.* **217**, 473.

Huchtmeier, W.K., and Richter, O.-G. 1989, "A General Catalog of HI Observations of Galaxies", Springer-Verlag.

Israel, F.P. 1988, in "Millimetre and Submillimetre Astronomy" ed. R. Wolstencroft and W.B. Burton, Kluwer Academic, p281.

Jannuzzi, B.T., Black, J.H., Lada, C.J. and van Dishoeck, E.F. 1988, *Ap.J.* **332**, 995.

Jenkins, E.B. 1978, *Ap.J.* **219**, 845.

Jenkins, E.B., Jura, M. and Loewenstein, M. 1983, *Ap.J.* **270**, 88.

Jenkins, E.B., and Meloy, D.A. 1974, *Ap.J. (Letters)* **193**, L121.

Jenkins, E.B., Wallerstein, G. and Silk, J. 1984, *Ap.J.* **278**, 649.

Jog, C., and Ostriker, J.P. 1988, *Ap.J.* **328**, 404.

de Jong, T., *et al.* 1984, *Ap.J. (Letters)* **278**, L67.

Joseph, C.L. 1988, *Ap.J.* **335**, 157.

Jura, M. 1982, *Ap.J.* **254**, 70.

Jura, M., Kim, D.-W., Knapp, G.R., and Guhathakurta, P. 1987, *Ap.J. (Letters)* **312**, L11.

Keene, J.B. 1981, *Ap.J.* **245**, 115.

Kenney, J.D., and Young, J.S. 1986, *Ap.J. (Letters)* **301**, L13.

Kenney, J.D., and Young, J.S. 1988, *Ap.J. (Suppl.* **66**, 261.

Kenney, J.D., and Young, J.S. 1989, (in preparation).

Keto, E.R., and Myers, P.C. 1986, *Ap.J.* **304**, 466.

Knapp, G.R. 1985, *Ap.J.* **293**, 273.

Knapp, G.R. 1987, *P.A.S.P.* **99**, 1134.

Knapp, G.R., Bies, W.E. and van Gorkom, J.H. 1989, in preparation.

Knapp, G.R., and Bowers, P.F. 1988, *Ap.J.* **331**, 974.

Knapp, G.R., Guhathakurta, P., Kim, D.-W. and Jura, M. 1989, *Ap.J. (Suppl.)* **70**, 257.

Knapp, G.R., and Kerr, F.J. 1974, *Astron. Astrophys.* **35**, 361.

Knapp, G.R., Rauch, K.P. and Wilcots, E.P. 1989, in "The Evolution of the Interstellar Medium" ed. L. Blitz (A.S.P. Conference Proceedings), in press.

Knapp, G.R., Turner, E.L., and Cunniffe, P.E. 1985, *A.J.* **90**, 454.

Kulkarni, S. and Heiles, C. 1988, in "Galactic and Extragalactic Radio Astronomy" ed. G.L. Verschuur, K. I. Kellerman and E. Bouton, Springer-Verlag.

Kunth, D., and Sèvre, F. 1985, in "Star Forming Dwarf Galaxies" ed. D. Kunth, T.X. Thuan and J. Tranh Than Van, Éditions Frontières, p331.

Lauer, T.R. 1985, *M.N.R.A.S.* **216**, 429.

Lavery, R., and Henry, P.C. 1986, *Ap.J. (Letters)*, **304**, L5.

Lilly, S. 1988, *Ap.J.* **333**, 161.

Lo, K.-Y., and Sargent, W.L.W. 1977, *Ap.J.* **227**, 756.

Lockman, F.J. 1984, *Ap.J.* **283**, 90.

Lockman, F.J. 1989, in "The Evolution of the Interstellar Medium", ed. L. Blitz, (A.S.P. Conference Proceedings), in press.

Lockman, F.J., Jahoda, K., and McCammon, D. 1986, *Ap.J.* **302**, 432.

Lonsdale, C.J., Helou, G., Good, J., and Rice, W. 1985, "Catalogued Galaxies and Quasars Observed in the IRAS Survey", JPL D-1932.

Low, F.J., *et al.* 1984, *Ap.J. (Letters)* **278**, L19.

Lynden-Bell, D. 1969, *Nature* **233**, 690.

Lyne, A.G. and Smith, F.G. 1989, *M.N.R.A.S.* **237**, 533.

Malin, D.F., and Carter, D. 1983, *Ap.J.* **274**, 534.

Maloney, P. 1988, *Ap.J.* **334**, 761.

Maloney, P., and Black, J.H. 1988, *Ap.J.* **325**, 389.

Matsumoto, T., Hayakawa, S., Matsuo, H., Murakami, H., Sato, S., Lange, A.E., and Richards, P.L. 1988, *Ap.J.* **329**, 567.

McCammon, D., Burrows, D.N., Sanders, W.T., and Kraushaar, W.L. 1983, *Ap.J.* **269**, 107.

McKee, C.F., Hollenbach, D.J., Seab, C.G. and Tielens, A.G.G. 1987, *Ap.J.* **318**, 674.

McKee, C.F., and Ostriker, J.P. 1977, *Ap.J.* **218**, 148.

Melnick, J., and Sargent, W.L.W. 1977, *Ap.J.* **215**, 401.

Miller, G.E., and Scalo, J.M. 1979, *Ap.J. (Suppl.)* **41**, 513.

Mirabel, I.F. 1989, *Ap.J. (Letters)* **340**, L13.

Mirabel, I.F., Sanders, D.B. and Kazès, I. 1989, *Ap.J. (Letters)* **340**, L9.

Morris, M., and Lo, K.-Y. 1978, *Ap.J.* **223**, 803.

Morris, M., Redman, R., Reid, M.J. and Dickinson, D.F. 1979, *Ap.J.* **229**, 257.

Neugebauer, G., *et al.* 1984, *Ap.J. (Letters)* **278**, L1.

Neugebauer, G., Miley, G.K., Soifer, B.T. and Clegg, P.E. 1986, *Ap.J.* **308**, 815.

Nyman, L.-Å. 1989, in "Submillimetre and Millimetre Astronomy" ed. A. Webster, (Kluwer Academic), in press.

Phillips, T.G., and Huggins, P.J. 1981, *Ap.J.* **251**, 533.

Phillips, T.G., Ellison, B.N., Keene, J.B., Leighton, R.B., Howard, R.J., Masson, C.R., Sanders, D.B., Veidt, B. and Young, K. 1987, *Ap.J. (Letters)* **322**, L73.

Pogge, R.W., and Eskridge, P.B. 1987, *A.J.* **92**, 291.

Polk, K.S., Knapp, G.R., Stark, A.A. and Wilson, R.W. 1988, *Ap.J.* **332**, 432.

Purcell, E.M., and Field, G.B. 1956, *Ap.J.* **124**, 542.

Read, P.L. 1980, *M.N.R.A.S.* **192**, 11.

Read, P.L. 1981, *M.N.R.A.S.* **194**, 863.

Readhead, A.C.S., and Duffet-Smith, P.J. 1979, *Astron. Astrophys.* **42**, 151.

Rees, M.J. 1984, *A.R.A.A.* **22**, 471.

Reynolds, R.J. 1988, *Ap.J.* **333**, 341.

Rickard, L.J., and Blitz, L. 1985, *Ap.J. (Letters)* **292**, L57.

Roberts, M.S. 1975, in "Galaxies and the Universe", ed. A. Sandage, M. Sandage and J. Kristian, University of Chicago Press, p309.

Rowan-Robinson, M., and Harris, S.E. 1983a, *M.N.R.A.S.* **202**, 767.

Rowan-Robinson, M., and Harris, S.E. 1983b, *M.N.R.A.S.* **202**, 797.

36

Rupen, M.P. 1989, Ph.D. Thesis, Princeton University.

Ryden, B.S., and Stark, A.A. 1986, *Ap.J.* **305**, 823.

Sadler, E.M. 1987, in I.A.U. Symposium 127 "Structure and Dynamics of Elliptical Galaxies", ed. T. de Zeeuw, D. Reidel Co., p125.

Sadler, E.M., and Gerhardt, O. 1985, *M.N.R.A.S.* **214**, 177.

Sadler, E.M., Jenkins, C.R., and Kotanyi, C.G. 1989, *M.N.R.A.S.* (in press).

Sage, L.J., and Wrobel, J.M. 1989, *Ap.J.* (in press).

Sanders, D.B., and Mirabel, I.F. 1985, *Ap.J. (Letters)* **298**, L31.

Sanders, D.B., Soifer, B.T., Elias, J.H., Madore, B.F., Matthews, K., Neugebauer, G. and Scoville, N.Z. 1988, *Ap.J.* **325**, 74.

Sargent, W.L.W., and Lo, K.-Y. 1985, in "Star Forming Dwarf Galaxies", ed. D. Kunth, T.X. Thuan and J. Tran Thanh Van, Éditions Frontières, p253.

Sarlin, S. 1988, unpublished.

Scoville, N.Z. 1989, in "Evolution of the Interstellar Medium" ed. L. Blitz, A.S.P. Conference Proceedings, in press.

Seab, C.G., and Shull, J.M. 1985, in "Circumstellar and Interstellar Dust", NASA Conference Proceedings No. 2403.

Shaya, E.J., and Federman, S.R. 1987, *Ap.J.* **319**, 76.

Shostak, G.S. 1978, *Astron. Astrophys.* **68**, 321.

Shull, J.M. 1989, in "The Evolution of the Interstellar Medium", ed. L. Blitz, *A.S.P. Conference Proceedings* (in press).

Smith, J. 1982, *Ap.J.* **261**, 463.

Smith, J., Harper, D.A. and Loewenstein, R. 1984, in Proceedings of the 10th Anniversary Symposium: Kuiper Airborne Observatory, ed. H. Thronson and E. Erickson, NASA-Ames Publications, p277.

Soifer, B.T., Houck, J.R. and Neugebauer, G. 1987, *A.R.A.A.* **25**, 187.

Solomon, P.M., Rivolo, A.R., Barrett, J. and Yahil, A. 1987, *Ap.J.* **319**, 730.

Solomon, P.M., and Sage, L.J. 1988, *Ap.J.* **334**, 613.

Solomon, P.M., Scoville, N.Z., Jefferts, K.B., Penzias, A.A. and Wilson, R.W. 1972, *Ap.J.* **178**, 125.

Spitzer, L. 1989, *Roy. Dan. Acad. Arts and Letters (Symposium in honor of Bengt Strömgren)* (in press).

Stark, A.A. 1979, Ph.D. Thesis, Princeton University.

Stark, A.A. 1983, in 'Kinematics, Dynamics and Structure of the Milky Way', ed. W.L.H. Shuter (D. Reidel Co.), p127.

Stark, A.A. 1985, in "I.A.U. Symposium 106: The Milky Way Galaxy" ed. H. van Woerden, W.B. Burton and R.J. Allen (D. Reidel Co.) p445.

Stark, A.A., Bally, J., Knapp, G.R., and Wilson, R.W. in "Molecular Clouds in the Milky Way and External Galaxies" ed. R.L. Dickman, R.L. Snell and J. Young, Springer-Verlag Lecture Notes in Physics **315**, 303.

Stark, A.A., Davidson, J.A., Harper, D.A., Pernic, R., Loewenstein, R., Platt, S., Engargiola, G., and Casey, S. 1989a, *Ap.J.* **337**, 650.

Stark, A.A., Elmegreen, B. G., and Chance, D. 1987, *Ap.J.* **322**, 64.

Stark, A.A., Gammie, C.F., Bally, J., Linke, R.A., and Heiles, C.E. 1989b, in preparation.

Stark, A.A., Gammie, C.F., Bally, J. and Wilson, R.W. 1989c, in preparation.

Stark, A.A., Knapp, G.R., Bally, J., Wilson, R.W., Penzias, A.A., and Rowe, H.E.

1986, *Ap.J.* **310**, 660.

van Steenberg, M.E., and Shull, J.M. 1988, *Ap.J. (Suppl.)* **67**, 225.

Tacconi, L., and Young, J.S. 1985, *Ap.J.* **290**, 602.

Tacconi, L., and Young, J.S. 1987, *Ap.J.* **322**, 681.

Thronson, H.A., and Bally, J. 1987, *Ap.J. (Letters)* **319**, L63.

Thronson, H.A., Bally, J. and Hacking, P. 1989, *A.J.* **97**, 363.

Thronson, H.A., Tacconi, L., Kenney, J.D., Greenhouse, M.A., Margulis, M., Tacconi-Garman, L., and Young, J.S. 1989, *Ap.J.* (in press).

Thronson, H.A., and Telesco, C.M. 1986, *Ap.J.* **311**, 98.

Thronson, H.A., Hunter, D.A., Telesco, C.M., Harper, D.A., and Decher, R. 1987, *Ap.J.* **317**, 180.

Tinsley, B.M. 1975, *P.A.S.P.* **87**, 837.

Tinsley, B.M. 1980, *Fundamentals of Cosmic Physics* **5**, 287.

Tully, R.B., and Fisher, J.R. 1977, *Astron. Astrophys.* **54**, 661.

Tyson, N.D. 1988, *Ap.J. (Letters)* **329**, L57.

Verter, F. 1988, *Ap.J. (Suppl.)* **68**, 129.

de Vries, H.W., Heithausen, A., and Thaddeus, P. 1987, *Ap.J.* **319**, 723.

Wallington, S., *et al.* 1989, in preparation.

Walsh, D.E.P., and Knapp, G.R. 1989, in preparation.

Walsh, D.E.P., Knapp, G.R., Wrobel, J.M., and Kim, D.-W. 1989a, *Ap.J.* **337**, 209.

Walsh, D.E.P., van Gorkom, J.H., Bies, W., Katz, N., Knapp, G.R. and Wallington, S. 1989b, submitted to *Ap.J.*.

Wannier, P.G., Lichten, S.M. and Morris, M. 1983, *Ap.J.* **168**, 727.

Wardle, M.J., and Knapp, G.R. 1986, *A.J.* **91**, 23.

Warmels, R.H. 1986, Ph.D. Thesis, University of Groningen.

Weiland, J.L., Blitz, L., Dwek, E., Hauser, M., Magnani, L. and Rickard, L.J. 1986, *Ap.J. (Letters)* **306**, L101.

Wiklind, T., and Rydbeck, G. 1986, *Astron. Astrophys.* **164**, L122.

Wiklind, T., and Henkel, C.R. 1989, *Astron. Astrophys.* (in press).

Williams, D.A. 1985, *Q.J.R.A.S.* **26**, 463.

Wilson, R.W., Jefferts, K.B. and Penzias, A.A. 1969, *Ap.J. (Letters)* **161**, L43.

Wrobel, J.M., and Heeschen, D.S. 1988, *Ap.J.* **335**, 677.

Wrobel, J.M., Neugebauer, G., and Miley, G.K. 1986, *Ap.J. (Letters)* **309**, L11.

Wunderlich, E., Klein, U., and Wielebinski, R. 1987, *Astron. Astrophys. Suppl.* **69**, 487.

Young, J.S., Claussen, M.J., Kleinmann, S.G., Rubin, V.C., and Scoville, N. 1988, *Ap.J.(Letters)* **331**, L81.

Young, J.S., Gallagher, J.S. and Hunter, D.A. 1984, *Ap.J.* **276**, 476.

Young, J.S., Schloerb, F.P., Kenney, J.D., and Lord, S.D. 1986, *Ap.J.* **304**, 443.

Young, J.S., and Scoville, N.Z. 1982, *Ap.J. (Letters)* **260**, L11.

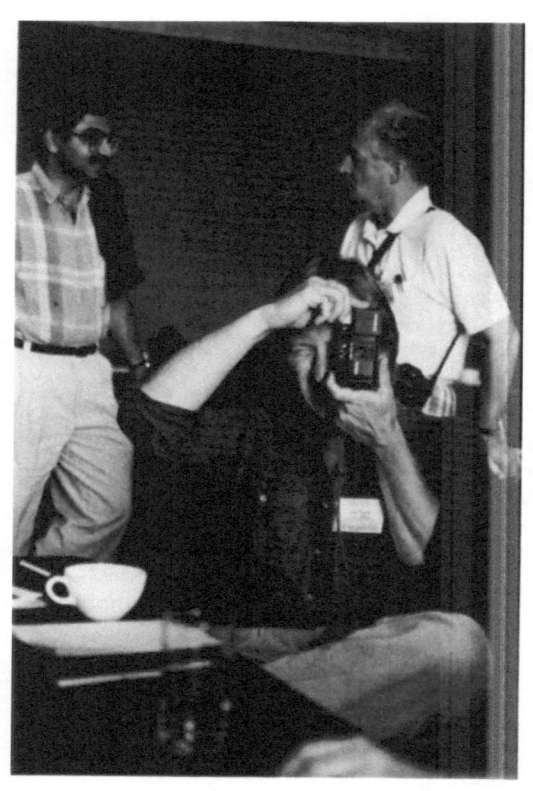

The Cool Phase of the Interstellar Medium: Atomic Gas

Elias Brinks
Royal Greenwich Observatory and
National Radio Astronomy Observatory

ABSTRACT. The purpose of this review is to discuss high resolution HI observations of a sample of nearby galaxies and to explain what they can teach us about the composition, structure and energy balance of the Interstellar Medium or ISM. The first part presents a summary of what is known about the ISM of the Galaxy and is followed by an overview of the current theoretical models. It continues by showing that it has become feasible to extend studies of the local ISM to nearby spirals, thus avoiding the well known problems associated with Galactic studies such as distance ambiguities and line of sight confusion. The second part is devoted to a discussion of the individual galaxies, addressing both the global properties of the ISM as well as its detailed structure, focussing on the properties of HI shells and supershells.

1. Introduction

The study of the interstellar medium (ISM) in galaxies has reached a very exciting stage as it has become feasible with present day instruments to observe nearby galaxies at similar resolutions as has been hitherto reserved to studies of the Galaxy. By looking at nearby systems it is now possible to complement our very detailed but necessarily restricted view of the local ISM. The restrictions are mostly caused by our unfavourable position. As we are located within the disk we suffer from line of sight confusion and are hindered in determining accurate distances within the Galaxy. In addition, at optical and shorter wavelengths the line of sight in the disk becomes prohibitively small.

In this review I will mainly deal with the ISM as it is traced by its major constituent, neutral atomic hydrogen or HI. Recently two outstanding reviews have been written by Kulkarni and Heiles (1987, 1988) who describe in detail the diffuse ISM in our Galaxy and the rôle which is played by HI. Another fine review concentrating on the global HI properties of galaxies was written by Giovanelli and Haynes (1988). I will assume that the reader is familiar with these reviews. What I hope to achieve is to describe the structure of the ISM in several of the nearby galaxies which were described by Giovanelli and Haynes, but in as much detail as possible and in the terms used by Kulkarni and Heiles.

In order to limit the scope of this paper I will refrain from describing the kinematics

H. A. Thronson, Jr. and J. M. Shull (eds.), The Interstellar Medium in Galaxies, 39–65.
© 1990 *Kluwer Academic Publishers.*

and large scale dynamics of galaxies nor will I discuss at any length spiral structure or spiral density waves. Obviously, as this review is concerned with the atomic gas it will be dealing with the extreme Population I or, in other words, the disk component of spiral galaxies. As a result I will ignore their bulges. For the same reason I will not cover E and S0 galaxies which are really in a class of their own. High sensitivity surveys have recently revealed the presence of detectable amounts of HI in several early type systems. There are many indications that a lot of this gas is due to infall, i.e., it is not necessarily remnant material from which the galaxy was formed (but see also the contribution by G. Knapp in this volume). For a statistical analysis the reader is referred to Knapp et al. (1985) and Wardle and Knapp (1986).

There are, of course, other ways to probe the atomic gas component such as via emission lines in the infrared which can be used to study the higher density neutral gas which occurs near photodissociation regions (see for example Wolfire et al. 1989). A more direct way to determine hydrogen column densities is by observing the Lyman-α absorption line in the UV. This technique has been successfully applied to look for neutral hydrogen in the Galaxy (Shull and Van Steenberg 1985), the LMC (Koornneef 1982) and the SMC (Bouchet et al. 1985). The advantage of this technique is that HI column densities can be determined which are independent of saturation effects or spin temperature. The disadvantage is that it only samples HI along lines of sight towards individual objects. Also, one will need powerful UV satellites in order to push this method to the nearest galaxies.

The structure of this review is as follows. It starts with a brief description of the basic radiation transfer equations which are needed to interpret HI observations. After that the main properties of some of the instruments which have opened up this new field of research are discussed. This is followed by a description of the structure of the ISM as we now know it, mainly from observations of the Galaxy. Finally a description is given of the new HI observations of about a dozen nearby galaxies, explaining what they have taught us thus far and indicating which questions one might hope to get answered in the foreseeable future.

2. Basic Radiative Transfer

There are numerous textbooks which derive the basic radiative transfer equations for the 21-cm line of neutral hydrogen, mostly following Wild (1952). For the sake of completeness and to avoid ambiguities due to a different choice of notation I will summarize the most significant formulae involved, following the notation used by Kulkarni and Heiles (1988).

The starting point is the equation of radiative transfer:

$$\frac{dI(\nu)}{ds} = j(\nu) - \kappa(\nu)I(\nu) \tag{1}$$

where $I(\nu)$ is the specific intensity at frequency ν, $j(\nu)$ the emissivity, $\kappa(\nu)$ the absorption coefficient and s the distance along the line of sight. Let us introduce the usual definitions for the optical depth $d\tau(\nu) = \kappa(\nu)ds$ and for the source function $S(\nu) = j(\nu)/\kappa(\nu)$. In thermodynamic equilibrium Kirchoff's law applies so that, using the Rayleigh-Jeans approximation for the Planck law, the source function at radio wavelengths can be written as:

$$S(\nu) = \frac{2\nu^2 k}{c^2} T_s \tag{2}$$

Here c is the speed of light in vacuum, k Boltzmann's constant and T_s the spin temperature. In an equivalent way one can define a brightness temperature T_B which corresponds to the temperature which a black body would have at frequency ν if it were to emit the specific intensity $I(\nu)$. Now we can rewrite the equation of transfer as:

$$\frac{dT_B}{d\tau(\nu)} = T_s - T_B(\nu) \tag{3}$$

This equation can be solved in the case of a homogeneous isolated HI cloud embedded in a background radiation field with a brightness temperature of T_{bg}. Employing e^τ as an integrating factor one finds the following important relation:

$$T_B(\nu) = T_{bg}(\nu)e^{-\tau(\nu)} + T_s(1 - e^{-\tau(\nu)}) \tag{4}$$

Let us assume that one can ignore the background radiation ($T_{bg} \ll T_s$). In the case of an optically thick cloud ($\tau(\nu) \gg 1$) the formula for the brightness temperature simply reduces to $T_B = T_s$, or the measured brightness temperature is equal to the spin temperature, or characteristic temperature, of the cloud. If the cloud is optically thin ($\tau(\nu) \ll 1$) we find that $T_B = \tau(\nu)T_s$. It is important to realise that $\tau(\nu)$ is directly related to the absorption coefficient per H atom which is dependent on atomic constants only. Then

$$\tau(v) = \frac{N_H(v)}{1.823 \times 10^{18}T_s} \tag{5}$$

In this relation N_H is defined as the column density of HI along the line of sight in atoms cm^{-2} and the dependence on frequency has been changed for dependence on velocity, expressed in km s^{-1}. In general the profile of the HI line is determined by its Doppler width which in turn is dominated by the macroscopic motion of the gas. Combining all this, we find in the optically thin case the following useful relation:

$$N_H = 1.823 \times 10^{18} \int T_B dv \tag{6}$$

When reporting HI column densities and HI masses based on this formula, radio astronomers implicitly assume that the emission comes from an optically thin medium. However, HI absorption studies of the Galaxy have indicated that an appreciable fraction of the gas is optically thick so that column densities derived under the assumption of $\tau \ll 1$ represent a lower limit to the true HI content.

Now let us look at what happens when $T_{bg} > T_s$. In general an HI spectrum is measured with respect to the background emission,

$$\Delta T_B(v) = T_B(v) - T_{bg} = (T_s - T_{bg})[1 - e^{-\tau(v)}] \tag{7}$$

If there is a strong background source present a spectrum along the line of sight towards this source will show an absorption feature. By measuring a spectrum along a nearby line of sight which presumably has an identical *emission* spectrum, and assuming that the brightness temperature, T_{src}, of the source has been determined seperately, one can determine the optical depth profile

$$\tau(v) = -\ln\left(1 + \frac{\Delta T_{B,on} - \Delta T_{B,off}}{T_{src}}\right) \tag{8}$$

Once the optical depth profile is known the spin temperature can be derived directly via $T_s(v) = \Delta T_{B,off}/(1 - e^{-\tau(v)})$. As explained in Kulkarni and Heiles the spin temperature is in general a column-density weighted harmonic-mean temperature. Therefore, and to enable a meaningful comparison between different regions in the Galaxy, a better parameter is the integrated optical depth which was introduced by Crovisier (1981) and which is defined as:

$$\Delta V_{abs} = \int (1 - e^{-\tau(v)})dv \qquad (9)$$

Correspondingly one can define the mean spin temperature as:

$$< T_s >= \int T_B(v)dv/\Delta V_{abs} \qquad (10)$$

In the case that the spin temperature would drop as low as the 2.7 K cosmic background temperature, no HI emission could be observed above the background and clouds such as the giant HI ring in Leo detected by Schneider et al. (1983), but with even lower densities, could go unobserved. However, it seems that this situation is unlikely as pointed out by Schneider et al. (1989) for the M96 group and by Corbelli and Salpeter (1989), who report on an extensive HI absorption line survey of High Velocity Clouds (HVCs) and a study of HI absorption towards bright extragalactic sources whose lines of sight pass close to galaxies. They find no evidence for HI column densities of about 5×10^{18} atoms cm^{-2} or larger which are hidden from detection in emission because their spin temperature approaches the microwave background temperature (the subthermal HI effect). Secondly, they infer from their data a lower limit for T_s of several hundred Kelvin for clouds with column densities of the order of 10^{20} atoms cm^{-2} with some evidence for cold HI being present in directions with larger column densities (cf. the review by J. Kenney in this volume for more information on this subject).

3. High Resolution Imaging Facilities

The two telescopes which have been instrumental in obtaining high resolution radio maps of nearby galaxies are the Westerbork Synthesis Radio Telescope (WSRT) operated by the Netherlands Foundation for Research in Astronomy (NFRA) and the Very Large Array (VLA) of the National Radio Astronomy Observatory (NRAO)*. No doubt, the compact array of the Australia Telescope (AT) will soon begin to supplement our knowledge by providing high resolution observations of galaxies which are located in the southern hemisphere.

In order to study the structure of the ISM it is necessary to work at linear resolutions of the order of 100–200 pc. Even for galaxies as nearby as M31, M33 or M81 this corresponds to angular resolutions of 10–20 arcsec, clearly only within the realm of synthesis telescopes. Table 1 lists some of the characteristic properties of the WSRT, VLA and AT and gives representative figures for the angular and velocity resolution and sensitivity which can be reached during a typical observing session. These figures are strictly for guidance only and are likely to change rapidly as more sensitive receivers and increasingly

---

* The NRAO is operated by Associated Universities, Inc., under cooperative agreement with the National Science Foundation.

Table 1: Characteristics of the WSRT, VLA and AT-Compact Array at 1420 MHz.

| Instrument | WSRT | VLA | | | | AT |
Configuration		A	B	C	D	Compact Array
Number of elements	14	27				6
Antenna diameter (meter)	25	25				22
Primary beam (arcmin)	37.6	30				35
System temperature (K)	55	60				18
Sensitivity* (mJy beam^{-1})	2.0	0.7				1.1
(K)	7.2	220	28	2.7	0.22	16
Longest Baseline (km)	2.8	36.4	11.4	3.4	1.03	6.0
Shortest Baseline (km)	0.036	0.68	0.21	0.073	0.033	0.030
Synthesized beam (arcsec)	$13 \times 13 / \sin \delta$	1.4	3.9	12.5	44	$6.5 \times 6.5 / \sin \delta$
Largest structure (arcmin)	14	0.6	2	7	15	12

*The sensitivity in mJy beam^{-1} corresponds to the one sigma *rms* noise reached after a 12^h period. The frequency resolution of respectively 19.5, 12.2, and 15.6 kHz converts to a velocity resolution of 2.5–4.1 km s^{-1}. The sensitivity in terms of surface brightness, expressed in Kelvin, assumes that the emission is distributed uniformly and fills the beam.

complicated correlators are brought on-line. General information about synthesis imaging can be found in the volume on *Synthesis Imaging in Radio Astronomy* (Perley *et al.* 1989).

All instruments are extremely sophisticated and there usually is a wide choice in baseline coverage, defining the lowest and highest spatial frequency for which one is sensitive, and velocity resolution and coverage. In general the VLA is the preferred instrument for sources which lie within the declination band $-30° < \delta < 30°$ as the East-West oriented WSRT and AT rapidly loose resolution in the declination direction when approaching the equator.

4. The Structure of the Interstellar Medium of the Galaxy

4.1 THE COMPOSITION OF THE ISM

The composition of the ISM is dominated by hydrogen which contributes 90% of all interstellar matter in either its molecular, neutral or ionised form. Based on Spitzer (1978) one can produce the following breakdown for the composition of the ISM:

- H_2, HI, HII, e$^-$
- He
- other atoms and molecules
- dust
- cosmic ray particles and magnetic fields

Averaged over the Galaxy, about equal amounts of molecular and neutral hydrogen are found; each component comprises $\sim 4 \times 10^9$ M$_\odot$. The ratio of molecular to atomic gas varies as a function of position, the molecular gas being concentrated in giant molecular cloud complexes which are most abundant in the molecular ring. There is a steep decrease of molecular material as a function of radius and atomic hydrogen dominates outside the solar circle. Hydrogen in its ionised form has until recently been solely associated with HII regions, contributing about 1% in mass. However, as argued by Kulkarni and Heiles (1987) there now is ample evidence that diffuse ionised hydrogen is ubiquitous in the solar neighbourhood and that there might be about 10^9 M$_\odot$ of it present in the Galaxy.

Helium is usually considered to be uniformly mixed with hydrogen. It makes up $\sim 8\%$ of the ISM by number, but contributes $\sim 40\%$ by mass. This point is often overlooked, especially by radio astronomers when they quote the mass of neutral gas in a galaxy based on observations of the 21-cm line of HI. Other elements, such as C, N, O, Ne and Fe make up only minor fractions of at most 10^{-3} to 10^{-4} by number. All other atoms and molecules can be considered as trace elements. This, of course, does not in any sense reduce their tremendous value for probing the physical conditions of the ISM. Dust particles and grains are thought to contribute 1–2% of the mass of a typical interstellar environment. Lastly, the ISM is permeated by a magnetic field of a few μG which constrains the motion of cosmic ray particles, mainly protons.

4.2 THE STRUCTURE AND PHASES OF THE ISM

The various components which make up the ISM in the Galaxy can find themselves in any of the following five phases (Mihalas and Binney 1981, Kulkarni and Heiles 1988):

- **Molecular Medium (MM)**. Typical values for the temperature, volume density and volume filling factor are $T \simeq 20$ K, $n > 10^3$ cm^{-3}, $f < 1\%$. The MM is characterised by cold dense molecular clouds which are mostly gravitationally bound. Although, on average, this phase contains as much mass as the atomic hydrogen, it occupies only a very small fraction of the ISM.
- **Cold Neutral Medium** (CNM; $T \simeq 100$ K, $n \simeq 20$ cm^{-3}, $f \simeq 2 - 4\%$). The CNM is distributed in rather dense filaments or sheets, occupying a minor fraction of the ISM. The CNM is most readily traced by HI measured in absorption.
- **Warm Neutral Medium** (WNM; $T \geq 6000$ K, $n \simeq 0.3$ cm^{-3}, $f \geq 30\%$). This phase provides the bulk of the HI seen in emission line surveys.
- **Warm Ionised Medium** (WIM; $T \simeq 8000$ K, $n \simeq 0.3$ cm^{-3}, $f \geq 15\%$). Until recently this phase was mainly associated with HII regions, but observations by Reynolds (1984) have convincingly shown that a considerable fraction of the ISM is filled with ionised gas.
- **Hot Ionised Medium** (HIM; $T \simeq 10^6$ K, $n \simeq 10^{-3}$ cm^{-3}, $f \leq 50\%$). The hot gas which is produced by supernova explosions has a long cooling time and consequently a large fraction of the ISM is filled with this "coronal" gas.

As mentioned above, the molecular material is mostly confined to molecular clouds which are held together by gravitation. The cold, warm and hot phases are in global pressure equilibrium. The filling factor for each of the phases is highly uncertain, as is the topology of the ISM (see also section 4.3). The values listed above represent what are currently thought to be the most realistic estimates. It is in trying to determine these values that observations of nearby galaxies will be most useful.

In general, the ISM has structures on all scale lengths, from smaller than 1 pc to larger than 1000 pc. HI observations of the Galaxy have painted a picture of a violent ISM. Heiles (1979, 1984) has shown that there exist dozens of HI shells and supershells, some of them showing clear signs of expansion. The energetics involved suggest that the HI shells are the result of the cumulative effect of supernovae going off within their parent OB association. The shells are the swept up matter whereas the inner portions of these shells are probably filled with HIM. This picture of the ISM is confirmed by HI observations of M31 (Brinks and Bajaja 1986) and other nearby galaxies which will be described in more detail below. The features which Heiles calls supershells are too energetic to be explained in this way (see for a thorough discussion the review article by Tenorio-Tagle and Bodenheimer 1988) and it might be necessary to invoke a different mechanism, such as the infall of material from outside the galaxy.

4.3 MODELS DESCRIBING THE ISM

Several models have been proposed over the years to explain the ever more detailed observations of the ISM. The first attempt to explain the different absorption and emission characteristics of the ISM was due to Clark (1965) whose description was formalised by Field, Goldsmith and Habing (1969) in their famous two-phase model for the ISM. In modern terms, they describe the ISM as consisting of a cold and a warm neutral phase in static equilibrium. Cox and Smith (1974) soon realised that the ISM is far from static and that supernovae dominate the energy input. They noticed that the cooling time of the HIM created by SN explosions is comparatively long and that the supernova blown bubbles can link up to form a tunneling network.

In order to better understand Cox and Smith's model and subsequent descriptions of the ISM it is perhaps useful to review briefly the evolution of a region of massive star formation or OB association. The most massive stars, i.e., $M > 25$ $M_\odot$, firstly ionise their immediate surroundings and blow an initial cavity through the action of their stellar winds. Each star quickly evolves and explodes as a supernova after some 5×10^6 year, depositing about 10^{51} erg in the ISM, partly in the form of kinetic energy, enlarging the wind blown cavity. Less massive stars with $M > 8$ $M_\odot$ evolve more slowly and will go off as supernovae after about 5×10^7 year. Although the most massive stars have a large influence on their surroundings via photoionisation and stellar winds, averaged over a typical galaxy, the larger number of less massive stars makes that the effects of supernova explosions dominate (assuming a normal Initial Mass Function or IMF, of course). An instructive calculation of the energy balance of the local ISM has been given by Abbott (1982). A more complete account is given by Tenorio-Tagle and Bodenheimer (1988) who review the multitude of models which describe the evolution of an ensemble of O and B stars and the effects which they have on the surrounding ISM.

McKee and Ostriker (MO, 1977) incorporated the idea put forward by Cox and Smith into their famous three phase model, in which the cool, warm, and hot phases of the ISM are in global pressure equilibrium. The MO model proved a major step forward and has been very successful in explaining several key features of the ISM, e.g., the occurrence of highly ionised lines such as the OVI line which was observed in absorption by the *Copernicus* satellite, the interstellar pressure, and the observed cloud dispersion. However, partly as a result of increased observational efforts over the last decade, it has come under serious criticism (cf. Shull 1987). Some of the main shortcomings of the MO model are:

- the assumption of a uniform distribution of supernova explosions. This is a gross simplification as in reality supernovae, especially those of Type II, tend to cluster. Also, the adopted supernova rate by MO is quite high.

- the topology for the WIM and WNM is radically different from the predicted onion skin model in which a cool cloud is surrounded by a warm neutral envelope which in turn is surrounded by a warm ionised layer, the whole being immersed in the HIM. HI observations rather suggest the HI to be in sheets or filaments. Moreover, the inferred filling factor of the WNM is much higher than that predicted.
- the predicted filling factor for the HIM is too high in comparison with HI observations of the Galaxy and other nearby spirals (see e.g., Heiles (1987) for a discussion on the Galaxy and Brinks and Bajaja (1986) who reach a similar conclusion for M31).

These findings have prompted considerable efforts to try to improve upon the MO model and to gain a better understanding of the ISM. Ikeuchi *et al.* (1984) propose a set of models based on MO which incorporate time evolutionary effects. They show that depending on the supernova rate and ambient density the ISM can either be in a three phase or a two phase state, or even cycle between two states. Heiles (1987) explicitly takes into account the difference in distribution between Type I and Type II supernovae and uses a more reasonable value for the supernova rate. In a more recent effort Heiles (1989) reviews the observational evidence related to the number of supernovae per OB association and takes into account the effect which the thickness of the HI layer has on the confinement of expanding HI shells. He includes in his calculations the full thickness of the disk, including the low density part which decreases exponentially with increasing distance from the plane, known as the Lockman disk (Lockman 1984).

A possible way to avoid creating models which have too high a filling factor for the HIM is to vent hot material into the halo via a galactic fountain (Bregman 1980, Cox 1981, Corbelli and Salpeter 1988). A particularly interesting model for the ISM in this respect is the one proposed by Norman and Ikeuchi (1989) in which they combine the virtues of the MO model with those of the galactic fountain models. Norman and Ikeuchi show that in their chimney model they can allow for a two phase and a three phase regime à la MO by simply changing the SN rate. In their model most of the hot phase resides in the halo of the Galaxy.

It will be clear that the main problem, in fact, does not lie with a lack of models and ideas, but with the fundamental limitations posed by our unfavourable position within the disk of the Galaxy. In order to test the validity of the improved descriptions for the ISM it will be necessary to look at nearby galaxies in as much detail as possible.

5. The ISM in nearby galaxies

Before continuing it is perhaps best to first define what is meant with the term nearby. Present day instrumentation allows angular resolutions of typically 10 arcsec or slightly better. To study the ISM in the line of HI one would like to have at least 200–300 pc linear resolution. This translates to a distance limit of about 4–6 Mpc. A quick survey of for example the nearby galaxy catalogue published by Kraan-Korteweg and Tammann (1979, see also Huchtmeier and Richter 1988), excluding from their list all dwarf irregulars and gas poor systems, and selecting only those systems which are suitably oriented shows that there are some two dozen nearby galaxy candidates. Below I will discuss the currently available results for several of these. In this discussion I will concentrate on the morphology of the HI distribution and I will make some excursions to other areas of interest such as the spin temperature, volume density and velocity dispersion of the gas.

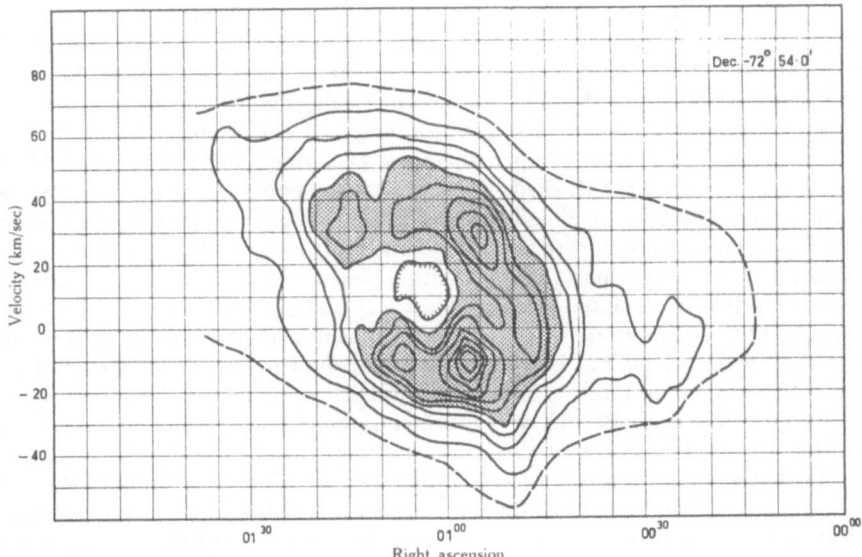

Figure 1: Position velocity diagram through one of the three SMC HI shells which were discovered by Hindman (1967). Contour levels are at 5, 10, 15... × 1.75 K brightness temperature. The shell has a diameter of 1.8 degree, corresponding to 1600 pc, and an expansion velocity of 23 km s^{-1}.

5.1 THE MAGELLANIC CLOUDS

The Magellanic Clouds are the odd ones out, and for two reasons. Firstly they are dwarf companion systems to the Galaxy with which they are in obvious interaction and as such their ISM might not be representative of a typical galaxy. Secondly, they are close enough, at a distance of about 50 kpc, that they can be meaningfully studied by a single dish telescope such as the Parkes 64-m dish. I include them in this discussion because they form a valuable step between the very detailed view which is obtained in Galactic studies and the necessarily more crude pictures which can be obtained of the nearest galaxies. Their importance will only increase once AT maps of parts of the LMC and SMC become available.

The first maps which had enough sensitivity and resolution to reveal structure in the ISM were obtained by McGee and Milton (1966), who mapped the LMC, and by Hindman (1967) who studied the SMC. Both studies were carried out with the Parkes telescope which has a resolution of 14.6 arcmin at 21-cm. They used a velocity resolution of 7.8 km s^{-1}. The angular resolution corresponds to linear sizes of about 230 pc at the distance of the LMC. The characteristics of the HI gas in the clouds are generally similar to those of the Galaxy. The measured one-dimensional velocity dispersion is about 11 km s^{-1} and the average volume density of HI lies around 1 atoms cm^{-3}. A plausible value for the spin temperature is about 200 K with some areas in the SMC, despite the large beam and consequently large area over which the ISM is averaged, showing brightness temperatures as high as 150 K.

Not only were the Magellanic Clouds the first extragalactic systems in which neutral hydrogen was detected (Kerr *et al.* 1954), they were the first galaxies to show HI shells. Hindman, in his 1967 paper, points out three distinct HI shells in the SMC for which he measures the position, central velocity, expansion velocity, diameter and amount of HI mass associated with the shell. Figure 1 shows an example of one of his shells. He tentatively explains their origin as due to a super-supernova explosion. To my knowledge Hindman was the first to recognise the HI shells which were found only much later in surveys of the Galaxy by Heiles (1979) as such and the first to provide their most likely interpretation. However, regarding the SMC there has arisen some controversy and it should be mentioned that several groups prefer to interpret the multiple peaked velocities in the SMC as evidence that it is being tidally disrupted (see e.g., Martin *et al.* (1989) and references therein).

In the LMC, mainly due to the efforts of Meaburn (1980) who imaged giant and supergiant shells in the light of Hα, the existence of the HI shells and their relation with other constituents of the ISM has been well established. Meaburn lists nine supergiant shells which together cover 12–15% of the surface of the LMC. Their diameters range from 600 to 1400 pc. The largest complex, LMC4, coincides with the Shapley Constellation III. He estimates that energies of order $10-200 \times 10^{51}$ erg are required to produce these structures. There is no doubt that the HI holes are regions in which gas has really been removed. The neutral gas cannot have been entirely ionised as the predicted Hα flux would exceed the observed emission by a large factor. Also, the HI cannot have turned into H$_2$. In that case one would expect a detectable ^{12}CO signal, even taking into account that the metallicity of the LMC is only about one fifth of the Solar neigbourhood (see Cohen *et al.* (1988) for the first complete ^{12}CO survey of the LMC). In addition one would expect that the presence of molecular material would be associated with a high value for the optical extinction which is in contradiction with what is actually found (Dopita *et al.* 1985).

The parameters describing the HI shells in the Magellanic Clouds are very similar to those of the Galaxy. The advantage of looking at the LMC is that one can study the relation between these shells and for example optical data and that a direct picture of their spatial distribution can be obtained. Such a comparison shows that the Hα supergiant shells lie along the inside of the HI shells. In turn numerous giant Hα shells, powered by young OB associations, can be found lining the inside of the supergiant shells. The OB associations have to be younger than about 5×10^6 year. The stars at the very centre of a supergiant shell tend to be older, about 2×10^7 year. Meaburn (1980) suggests that this is all evidence of star formation propagating from the centre of the shell outwards, giant Hα shells chiseling out a supergiant shell. This process continues until the HI shell breaks out of the disk and the excess pressure is lost.

Some interesting follow-up work has been done by Dopita *et al.* (1985) and Meaburn *et al.* (1987), both groups reporting on new HI data covering either a small region of the LMC or crosscuts through a series of shells. Dopita *et al.* studied in detail Shapley Constellation III which coincides with LMC4. They support the idea that self propagating star formation has determined its structure. They estimate that between 32 and 100×10^{51} erg are involved. The reason why LMC4 is such a near perfect shell is attributed to the fact that Magellanic Irregulars tend to show hardly any differential rotation. Furthermore, they claim, on the basis that 15% of the surface area of the LMC is covered by HI shells and using some assumptions for the time needed to cycle the gas through the various phases of the ISM, that the LMC is in a steady state of star formation. Both Clouds have recently been reobserved using once more the Parkes telescope but with vastly improved sensitivity and velocity resolution. Rohlfs *et al.* (1984) remeasured the inner 6° × 6° of the LMC and Mathewson *et al.* (1988) reobserved three 40 arcmin diameter areas in the SMC. But because the angular resolution is the same as in the previous surveys not much more can be learned regarding the structure of the ISM. It will be up to the AT to provide the necessary increase in angular resolution.

5.2 MESSIER 31

The WSRT survey of M31 (Brinks and Shane 1984) has thrown new light on the nearest normal spiral which closely resembles the Galaxy in shape and size. The HI observations were done at a resolution of 24 × 36 arcsec which corresponds to a linear resolution of 80 × 120 pc. The effective velocity resolution was 8.2 km s^{-1}. Although at about ten times the distance of the Magellanic Clouds, the availability of synthesis telescopes made it possible to study M31 in at least as much detail. In order to determine beam averaged values for the parameters describing the physical state of the gas, such as the spin temperature, velocity dispersion, gas volume density, etc., Brinks and Burton (1984) developed a model for the HI distribution of M31 in which they took the basic radiative transfer into account. The model calculations implied beam averaged values for M31 which are broadly consistent with values derived for the Galaxy from early single dish surveys. They found the following values for the:

- spin temperature, $< T_s > = 125$ K
- one dimensional velocity dispersion, $\sigma = 8$ km s^{-1}
- one sigma scale height, h = 120 pc
- gas volume density, $n_H = 0.6$ atoms cm^{-3}

It should be emphasized that these values are beam averages, representing the mean value over an elliptical area of 80 × 120 pc. Moreover, due to the high inclination of M31 of 77° the quoted values are averages along the line of sight as well which traverses an effective one sigma scale height of ∼ 500 pc.

A new survey of the northern half of M31 by Braun (1989) with the VLA at a resolution of $\Delta\alpha \times \Delta\delta \times \Delta V = 10" \times 10" \times 5.1$ km s^{-1} will eventually allow more specific values to be determined. An interesting new result which has already been extracted from this impressive data set and presented at this meeting (Braun and Walterbos 1989) is the finding that there exist clumps of HI with brightness temperatures up to 180 K. Braun and Walterbos argue that the peak brightness temperatures in M31 represent the upper envelope of optically thick kinetic temperatures. This is based on the experience in the Galaxy where peak brightness temperatures of 135 K, which are found in the direction of spiral arm tangents, are interpreted as representative for the harmonic mean of the spin temperature, $< T_s >$. The value for M31 is significantly higher than that for the Galaxy. On the other hand, M33 shows peak brightness temperatures of only 95 K (Deul and van der Hulst 1987). It should be kept in mind, however, that this value is most likely a lower limit to the intrinsic peak brightness temperature due to the relatively large size of the beam which was used to map M33. The value for the HI spin temperature depends on many processes, such as the heating due to photoelectric emission from dust and cooling by heavy elements. Important parameters in that respect are the metallicity, grain depletion, ionization rate, and gas pressure. Braun and Walterbos argue that the complex interplay between these factors can be parameterized in terms of the current and past star formation

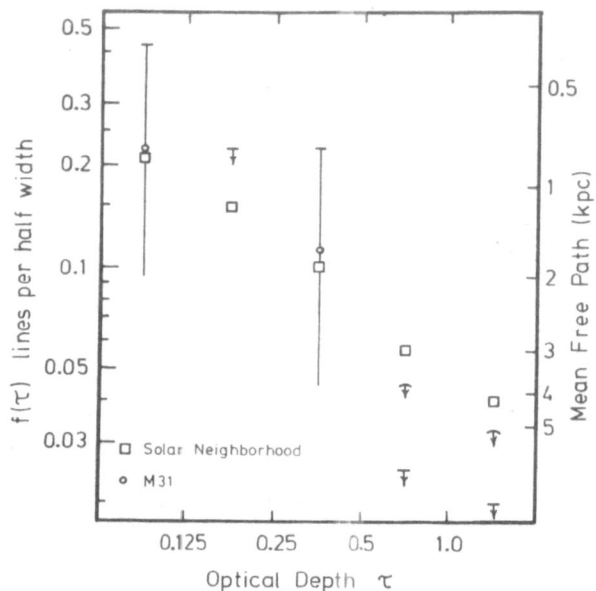

Figure 2: The differential number density of absorption lines in M31 compared to that measured in the Galaxy. The left hand scale indicates the probability of detecting a line per half width of the disk as a function of the optical depth of that line. The general agreement between M31 and the Galaxy suggests an identical abundance of absorbing clouds (Dickey and Brinks 1988).

rates. If this is correct, the differences in peak brightness temperatures in the Galaxy, M31, and M33 reflect the relative star formation rate in these galaxies. Braun (1989) also points out that many HI spectra show deviations from a Gaussian profile such as steeper wings and a flatter top which indicates that HI has an appreciable optical depth and that at places an amount of HI equivalent to 50% of the detected amount may be hidden. In fact, it is only with the present high spatial *and* velocity resolution that this shows up.

A different approach to probe the physical parameters of the ISM is by HI absorption measurements. The snag here is that, compared to similar such studies in the Galaxy, there are very few extragalactic background sources of a sufficient strength which happen to shine through the HI disk of nearby galaxies. Nevertheless, Dickey and Brinks (1988), using the VLA, made a successful attempt to detect HI absorption towards M31. Because of the very high resolution it was possible to eliminate from the absorption measurements all traces of 21-cm emission. The results are summarised in Figure 2 which compares the number density of absorption lines per unit path length as a function of optical depth in M31 with similar data for the Galaxy. The observations indicate that within the errors the absorbing clouds in M31 are as abundant as in the Galaxy. The mean spin temperature of the gas, $< T_s > \approx 300$ K, is similar or higher than the local value. The fraction of HI which is in the cool (CNM) phase is about 25% or less, much higher than predicted by the MO model, and implies that in M31 the mixture of cool and warm phases is similar to that in the solar neighbourhood. Moreover, as the star formation rate in M31 is much lower than in the

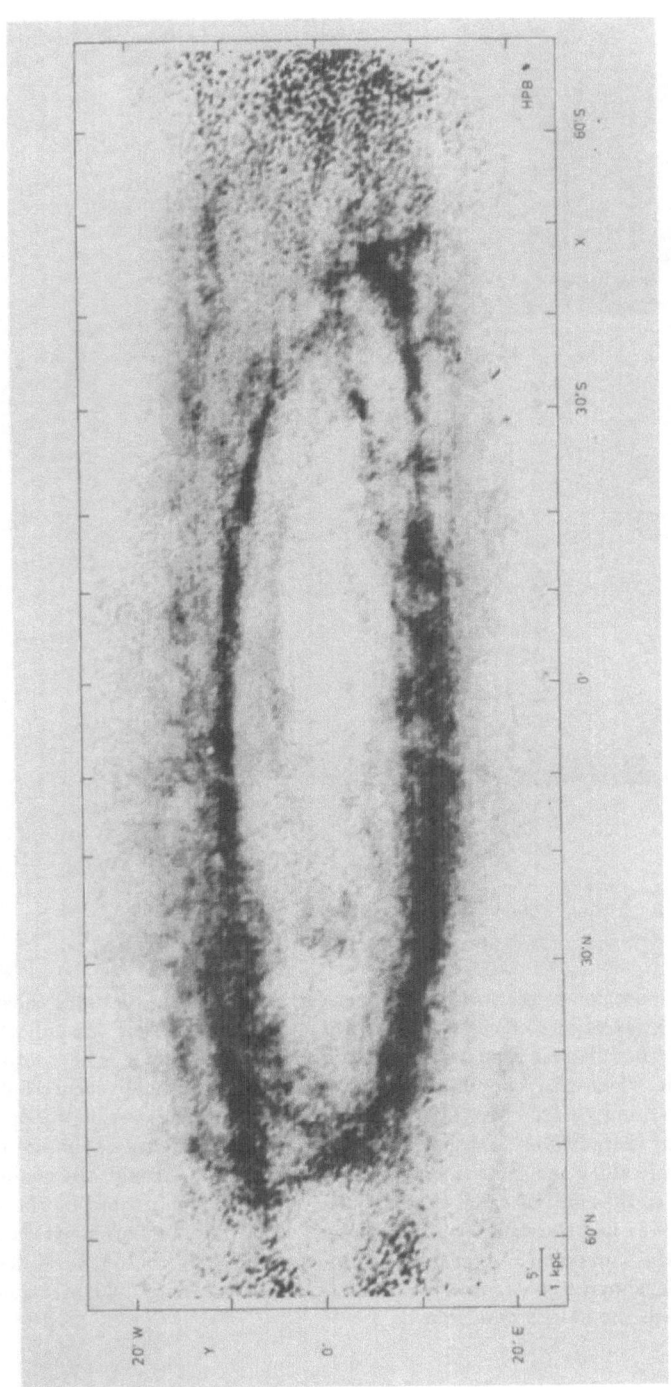

Figure 3:

Integrated surface brightness map of M31 at a resolution of 24×36 arcsec from Brinks and Shane (1984). Dark shading corresponds to HI emission. Column densities in the pronounced ring at 10 kpc correspond to about 5×10^{21} atoms cm^{-2}. The rectangular X,Y-coordinate system is aligned along the principal axes and the linear scale and size of the beam are indicated. The white spot near 7.5N and 12.1W is caused by absorption against a bright far-extragalactic radio continuum source. Due to the limits of the survey the noise increases rapidly towards the northern and southern ends of the major axis.

52

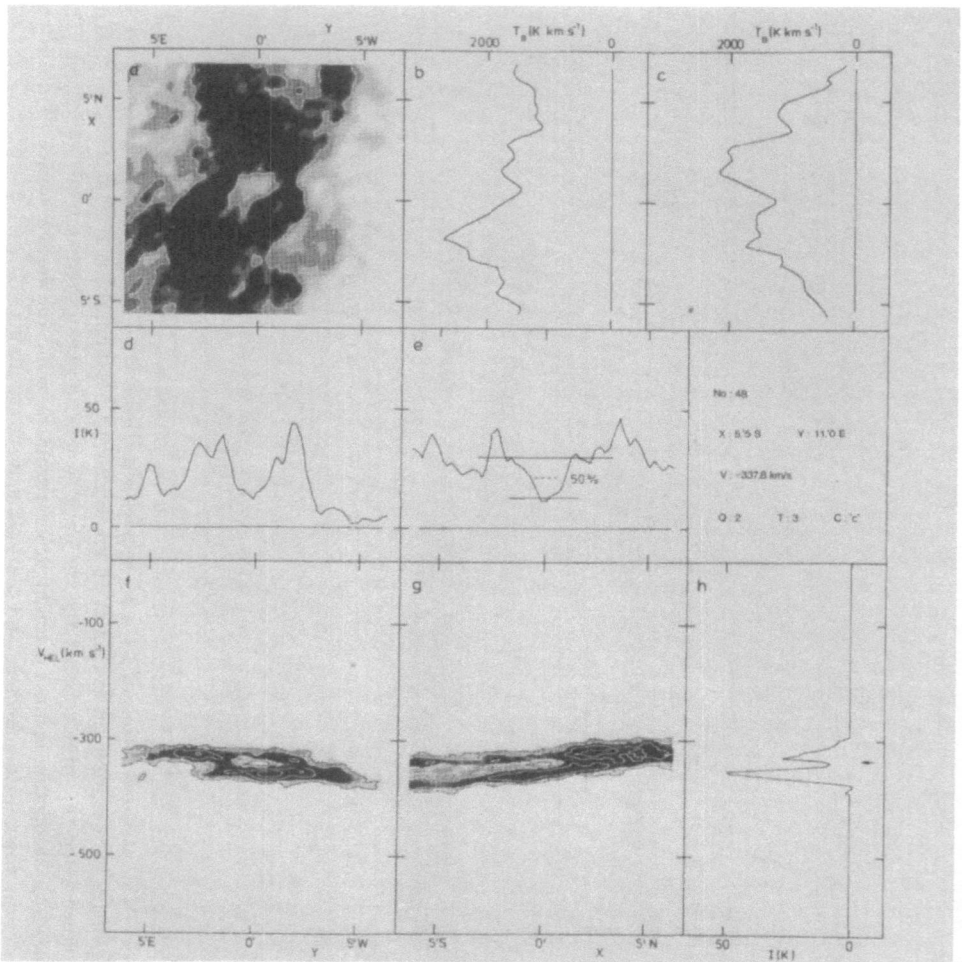

Figure 4: Collection of pictures taken from Brinks and Bajaja (1986) showing the detailed characteristics of one of the holes in M31. The panels show the following: **a)** expanded view of the channel map at which the hole is best seen. The contour level is drawn at 12.0 K which corresponds to the level at which the hole reaches 50% of its minimum depth; **b)** and **c)** are cross-cuts plotted on the same linear scale, made parallel to the X- and Y-axes respectively, through the total HI surface brightness map of M31; **d)** and **e)** are cuts made along the same lines and on the same scale through the channel map shown in panel a); to the right of panel e) some information is listed regarding the position of the hole within M31 and the velocity of the channel map; **f)** and **g)** are position-velocity maps made along the same cuts and on the same linear scale. The contour levels are at 10, 20, 40, and 60 K; **h)** shows a spectrum through the centre of the hole. The arrow indicates the velocity at which the hole is best seen.

Galaxy, these observations prove that the mixture of phases is not strongly dependent on the precise values for the interstellar pressure or supernova rate and that the WNM is a robust phase (Cowie 1987).

The M31 HI integrated surface brightness map which is shown in Figure 3, as well as the individual channel maps show a stunning amount of detail in the form of loops, arcs and filaments. Brinks and Bajaja (1986) have compiled a list of 141 roughly elliptical features which show a relative absence of neutral gas and which they call HI holes. Figure 4 shows a blowup of one such hole. The properties of the holes in M31 can be summarized as follows. Their sizes range from 100 pc, the lower limit imposed by the resolution limit of the survey to 1000 pc. Some holes show a clear shell structure although for most holes the shell, if present, is washed out due to the limited resolution. The majority show deviations in the local velocity field which can be ascribed to expansion velocities of the order of 10–30 km s^{-1}. The amount of HI which appears to be absent as compared to neighbouring areas is 10^4 to 10^7 M$_\odot$. Kinematic ages for the M31 holes range from 2.5 to 30×10^6 year. The estimated energy requirements are 10^{49} to 10^{53} erg. Based on the observed number of holes and some assumptions, Brinks and Bajaja estimate a Type II supernova rate for M31 of 10^{-14} pc^{-3} yr^{-1}, an order of magnitude lower than the corresponding local value for the Galaxy. Also, if the holes trace the distribution of the HIM, the hot gas in M31 does not form a tunnelling network.

None of the M31 holes would be classified as an HI supershell. This could either reflect the much lower star forming activity or be related to the fact that no high velocity clouds have been detected thus far near M31. As is the case in the LMC it is clear that the HI holes are regions in which the gas has been removed. Although Hα emission is associated with many of the holes, the observed emission measure falls far short of what would be expected if all the gas were ionised. Following the same argument, the HI cannot have been all transformed into H$_2$. The most likely explanation is that evolving OB associations through the combined effects of photoionisation, stellar winds and supernova explosions shape the ISM. A comparison between the location of the HI holes and OB associations shows indeed that there is a correlation. Contrary to the LMC, the disk of M31 is in differential rotation. Paloš et al. (1989) studied the effects which shear has on the evolution of the shells and presented a parametrisation for the evolution of their shape under the influence of shear. They predict that matter will pile up at the tips of the hole which can lead to molecular cloud formation and eventually star formation by the time an HI hole closes onto itself, thus allowing for propagating star formation. Their calculations show that the efficiency with which this occurs depends, among other things, on the radial distance from the centre of the galaxy.

5.3 MESSIER 33

The next nearest galaxy to be discussed is M33. This system was studied with the WSRT by Deul and van der Hulst (1987) and Deul (1988). They used a resolution of $\Delta\alpha \times \Delta\delta \times \Delta V = 12" \times 24" \times 8.2$ km s^{-1}, corresponding to spatial details of 40×80 pc. Figure 5 shows the integrated surface brightness map. Although M33 is smaller than M31 it is presently much more active in forming stars. Deul and den Hartog (1989) made an analysis similar to the one discussed above for M31 and came up with the following results. M33 contains some 148 HI holes larger than 40 pc, the resolution limit. Their diameters range from 40 pc to 1000 pc, the larger holes lying preferentially at larger galactocentric distances. The swept

54

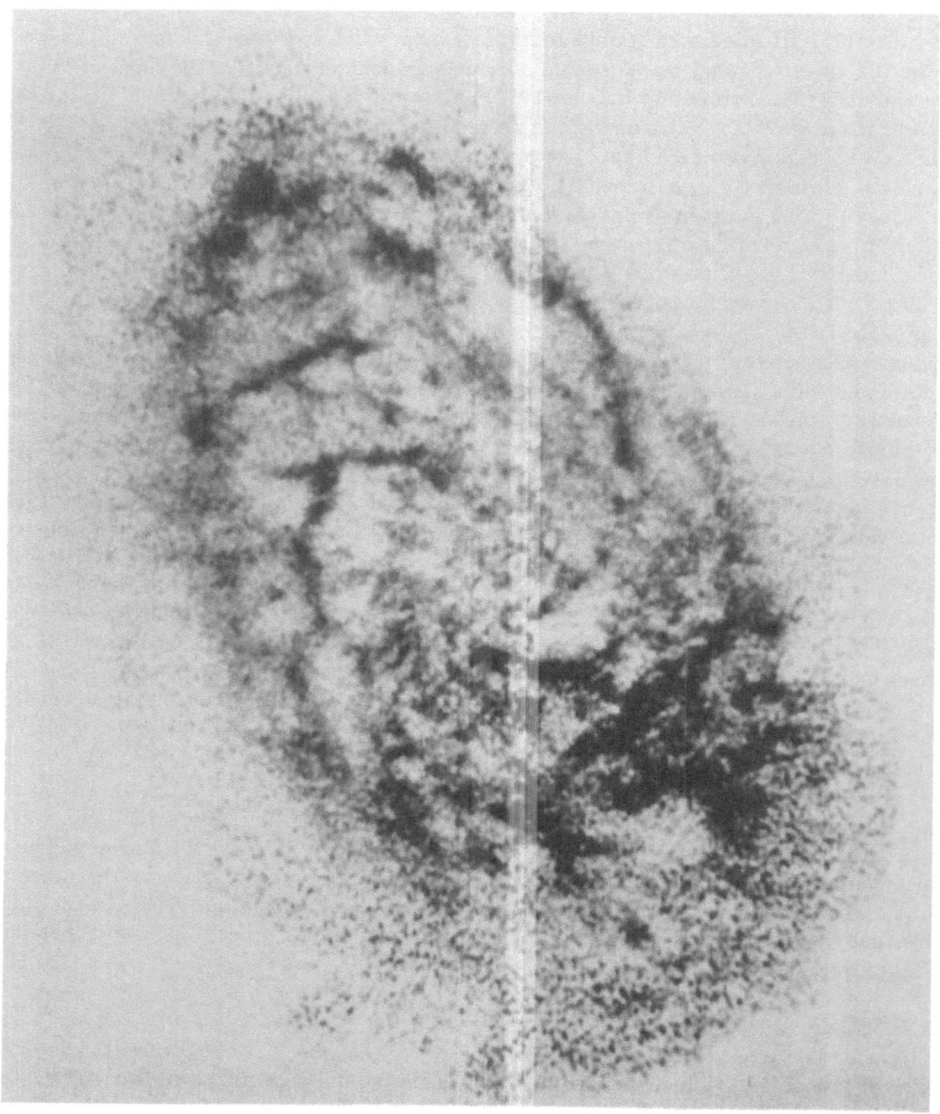

Figure 5: Radio photograph of the HI column density distribution in M33 at full resolution. The column densities range from 1.5×10^{20} to 5.5×10^{21} atoms cm^{-2}. Note the enormous amount of detail and the many loops, arcs and filaments. (Photo courtesy of E. Deul).

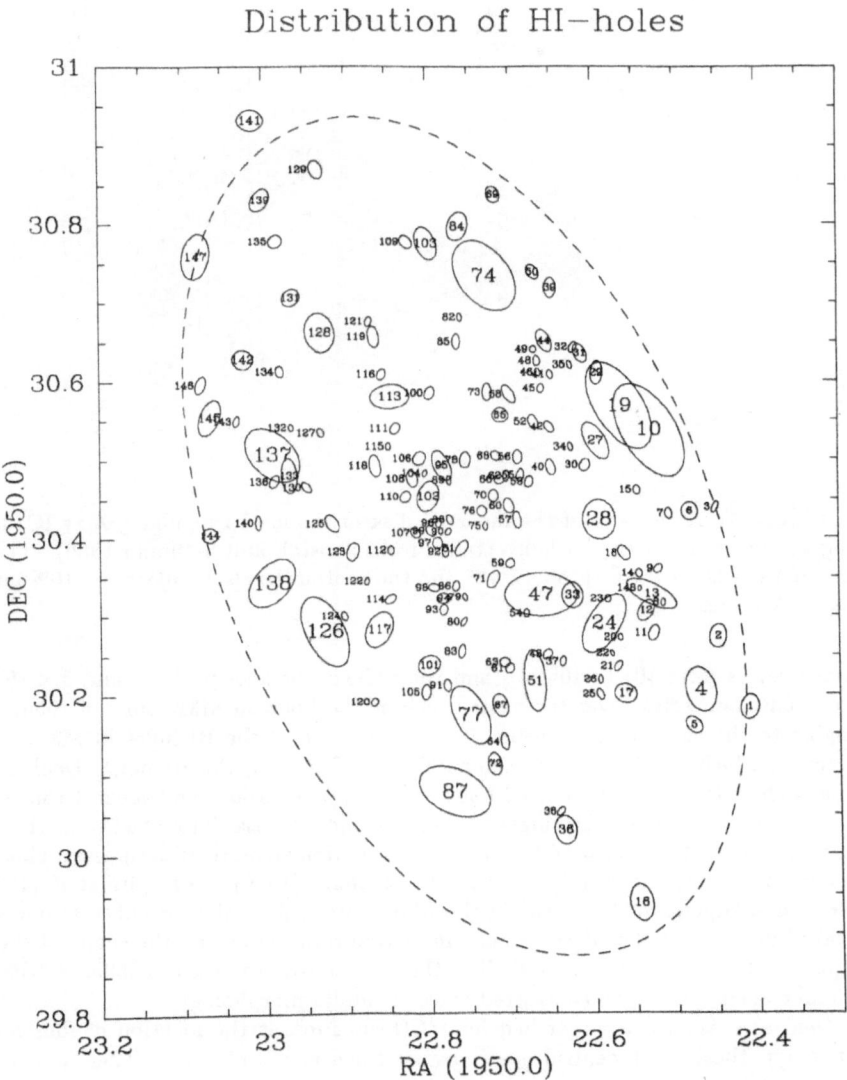

Figure 6: Spatial distribution of the HI holes in M33 as outlined by Deul and den Hartog (1989). The scale for this picture is the same as for the integrated surface brightness map shown in Figure 5. Note the concentration of small holes in the inner galaxy. The larger holes are generally interarm cavities.

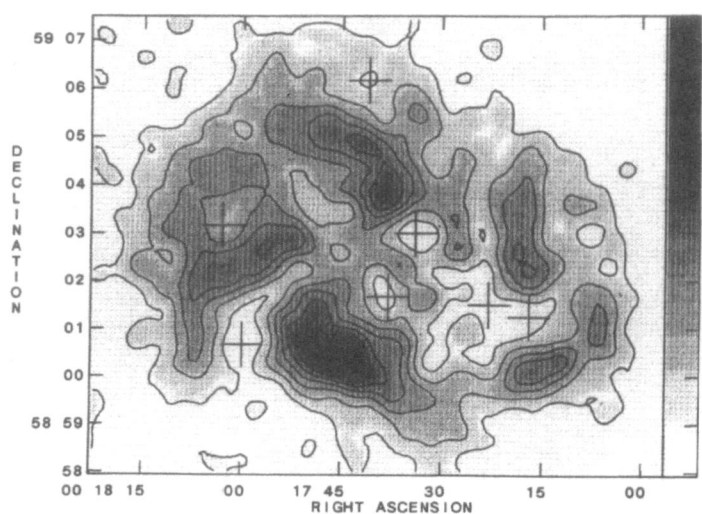

Figure 7: A close-up of the inner HI disk of the local irregular galaxy IC10, identifying the positions of the HI holes (taken from Shostak and Skillman 1989). The peak column density is 4.1×10^{21} atoms cm^{-2} and the contour levels are placed at 10% intervals of the peak value.

up mass ranges from 10^3 to 10^5 M$_\odot$ and their kinematic ages peak around 5×10^6 year. The estimated energies cover the same range as the holes in M31, none of them clearly belonging to the class of supershells. The distribution of the HI holes in M33 is shown in Figure 6 which can be directly compared with Figure 5, the HI map. Deul and den Hartog confirm the correlation of HI holes with OB associations and suggest that smaller HI holes tend to coincide with single HII regions and OB associations whereas the larger, and presumably older holes tend to show star formation along their periphery. This would be in agreement with the description given by Meaburn (1980) and Dopita *et al.* (1985) for the holes and supergiant Hα shells in the LMC, although in the case of M33 one should also take into account the effects which differential rotation has on the shape of the holes as explained by Palouš *et al.* (1989). The three giant HII regions, NGC595, NGC604 and IC133 all coincide with well-contrasted holes of similar morphology.

Deul and den Hartog searched for ^{12}CO emission at the location of four holes to test the hypothesis that neutral gas is transformed into molecules. Their upper limits confirm that there is less material at the locations they sampled, unless the conversion factor between the measured ^{12}CO surface brightness and H$_2$ column density is smaller than 0.3 times the Galactic conversion factor. More sensitive ^{12}CO observations and at a higher spatial resolution would settle this question once and for all. In brief, the results on M33 generally agree very well with what has been found in M31, the LMC and the Galaxy.

5.4 IC10 AND IC1613

IC10 is a Local Group galaxy at a distance of 1.3 Mpc which has been mapped in HI by

Shostak and Skillman (1989, and references therein). Its HI distribution is quite unusual and the authors propose that the outer HI envelope is still in a collapse phase. In the inner disk seven prominent holes are visible (see Figure 7). Due to the patchy nature the holes are somewhat less easily defined than, for example, those in M31 or M33. The derived linear sizes range from 200 to 800 pc. The estimated energies which are required to produce them are 2–300×10^{51} erg, agreeing quite well with what we know about similar features in other galaxies. Interestingly, in addition to the HI holes, Skillman finds an area measuring about 230 pc in diameter which stands out in non-thermal radiation and which coincides with the most massive HI concentration. He interprets this region as a superbubble in an early stage of its evolution and predicts that eventually this bubble might evolve to a supergiant shell with a diameter of the order of 1000 pc.

Another galaxy worth mentioning is IC1613 which was imaged in Hα by Meaburn et al. (1988) and in HI by Lake and Skillman (1989). As in, for example, the LMC a multitude of Hα filaments is visible which link up to form giant and supergiant Hα shells. The HI maps, although made at a moderate resolution, show several HI holes. It would be interesting to reobserve this galaxy at a higher resolution and to compare the HI morphology with the Hα structures.

5.5 M101 AND NGC 6946

Until now I have mainly dealt with those features in the ISM of nearby galaxies which can be understood in terms of the evolution of a group of massive stars and their influence on their environment. But, at several instances it was pointed out that some of the structures might in fact be too large or require too much energy to be explained this way. Van der Hulst and Sancisi (1988) recently reobserved M101 with the WSRT (see Figure 8). They noticed in their HI maps two regions in which the HI is displaced to very high velocities with respect to the normal disk rotation. An example is shown in Figure 9. Simple energy considerations would imply kinetic energies of over 10^{54} erg. There are no obvious signs of young stars which could be held responsible for creating these giant structures. Similar features, though with lower energies, have been found by Kamphuis et al. (1989) in another system, NGC6946, but in this case it is not yet quite clear if a supernova origin is ruled out by the data.

The explanation which van der Hulst and Sancisi propose for their observations of M101 is the infall of high velocity clouds. Support for this interpretation can be found in the Galaxy. As argued by Tenorio-Tagle and Bodenheimer (1988) it is very difficult to account for the supershells found by Heiles (1979) by multiple supernova explosions. The reason is that once a shell grows large enough to break through the disk, all the energy from subsequent supernova events will be channeled to the halo. On the other hand, infall of a cloud with a velocity of 100 km s^{-1} and a mass of 10^5 M$_\odot$ corresponds to the deposition of some 10^{52} erg in the disk of a galaxy. As shown by Tenorio-Tagle et al. (1987) the conversion efficiency is in the region of 15–25%, much higher than the few percent of the total energy output of an OB association which is converted to kinetic energy. In other words, HVC infall can be an efficient mechanism to create large scale remnants. Another point of interest is that the infall model explains why many of the shells show one hemisphere only, as was noticed by Heiles (1979). Tenorio-Tagle et al. show that extrapolation of such a partial shell to a complete bubble overestimates their energy tremendously.

58

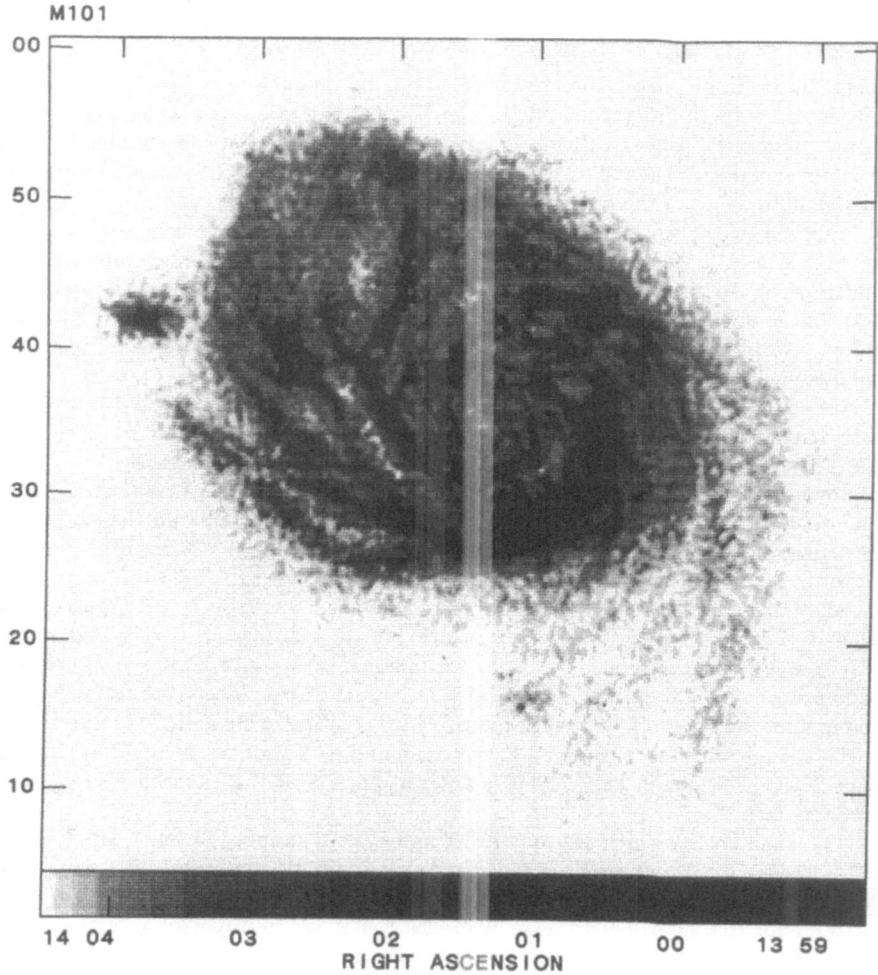

Figure 8: Total HI surface brightness map of M101. At the assumed distance of 7.2 Mpc, 1 arcmin corresponds to 2 kpc. The most prominent area where deviating HI gas is found is located at 14^h02^m and $54°44'$. (Image courtesy of J. Kamphuis).

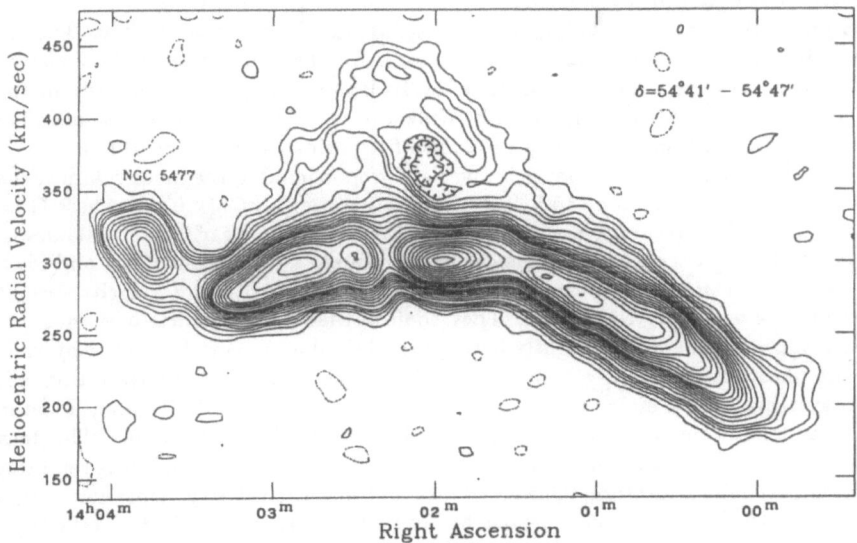

Figure 9: Position-velocity diagram averaging the emission in a declination strip some 7 arcmin wide and centred on the hole mentioned in Fig. 8. Local minima are indicated by tick marks. Contours levels are $-2, 2, 4 \ldots, 10, 15, \ldots, 50, 60, \ldots, 100, 120, \ldots, 280$ K. Figure taken from van der Hulst and Sancisi (1988).

From the large scale surveys of Hulsbosch and Wakker (1988) and Bajaja *et al.* (1985) it is clear that the sky is filled with HVCs. Looking at those clouds which have the highest velocities (VHVCs), Mirabel and Morras (1984) and Mirabel (1989) claim that they represent high velocity *inflow* of material toward the Galaxy. This does not settle the question if the VHVCs are infalling intergalactic matter or tidal debris from the Magellanic Stream, or if the material is indirectly connected to star formation. It could be hot matter which was expelled from the disk after an HI shell has broken through and which is condensing out and raining down on the disk, as described in galactic fountain and chimney models (Bregman 1980, Corbelli and Salpeter 1988, Norman and Ikeuchi 1989). Additional observations might eventually provide an answer.

5.6 EDGE-ON GALAXIES

In order to study the HI distribution of galaxies one usually chooses systems which are relatively face-on. However, to obtain information on the z-distribution one has to turn to edge-on galaxies. Rupen (1989) has successfully resolved the HI layer in two edge-on systems, NGC 4565 and NGC 891. His VLA data extend the work on the three-dimensional structure of disks in spiral galaxies which was recently reviewed by van der Kruit (1988). Figure 10 shows a set of channel maps of NGC 891. Both galaxies show a flaring of the HI layer when moving to larger distances from the centre. Rupen finds for NGC 4565 values for the one sigma scale height, $h(r)$, ranging from 120 pc for the inner disk to about 300 pc at around 16 kpc at which radius the warp sets in. Corresponding values for NGC 891 are 105 pc for the centre and 235 pc at the last measured point near 11 kpc. The values at small radii agree comfortably with those quoted for the inner disks of the Galaxy and M31. These observations also confirm that the HI layers at large galactocentric distances flare. For the Galaxy this was discussed by Kulkarni *et al.* (1982). In the case of M31 this could only be inferred on the basis of the model fits (Brinks and Burton 1984).

The scale height of the gas layer is inversely proportional to the mass volume density, $\rho(r, z)$, in the disk at z=0, and can be written as $h(r) = \sigma_g/[4\pi G\rho(r,0)]^{1/2}$ where G is the gravitational constant (see e.g., Kulkarni *et al.* 1982). From studies of face-on galaxies it has been shown that σ_g is constant and ranges from 7–10 km s^{-1} (van der Kruit and Shostak 1984). Hanson *et al.* (1989) observed with improved sensitivity and velocity resolution the nearby face-on galaxy NGC 1058. They confirm that σ_g is constant over most of the disk at a value of 5.7 km s^{-1}, slightly lower than the value previously quoted by van der Kruit and Shostak. Assuming a constant velocity dispersion, Rupen's observations of the edge-on systems yield a direct measurement of the mass volume density and, if the mass distribution follows the old star light, the M/L value as a function of radius. The data on both galaxies are consistent with a constant $M/L \approx 3$ M$_\odot$/L$_\odot$. Although it is tempting to assume that all galaxies feature a flaring HI disk and that the velocity dispersion of the gas perpendicular to the disk is constant, one has to be cautious. Bottema *et al.* (1986) found at least one (dwarf) galaxy, NGC 5023, in which the scale height of the HI gas appears constant.

Interestingly the maps of NGC 891 show numerous wiggles which Rupen ascribes to non-thermal processes. It would be interesting to see to what extent these wiggles are associated with the features which are seen in Hα and which were presented at this meeting by Dettmar *et al.* (1989) and Kulkarni *et al.* (1989). Both groups indicate that there are filaments in their Hα maps which originate in star forming regions in the disk and which stick out of the plane. Are we seeing here evidence for the disk-halo interaction as proposed by the galactic fountain and chimney models?

The last galaxy which I would like to mention is the edge-on spiral NGC 3079 which was observed with the VLA in HI by Irwin and Seaquist (1989). This galaxy not only displays remarkable activity in the centre but it also contains half a dozen HI shells in its disk. Their dimensions and energy requirements as derived by Irwin and Seaquist place them in the catagory of Heiles supershells. Again, if these features are caused by the collective effects of several star forming regions, it is tempting to interpret these observations in terms of chimney models in which hot material from a bubbling disk flows into the halo.

6. Forward Look

I hope that the above has indeed illustrated that studying the ISM in nearby galaxies via the 21-cm line of HI is not only feasible but very exciting and rewarding. Moreover, in order to put to the test the multitude of models which exist for the overall structure of the ISM and for the detailed evolution of an OB association, this kind of observations is indispensable. At present beam averaged values for several parameters describing the state of the neutral gas can be determined, typically sampling volumes with a diameter

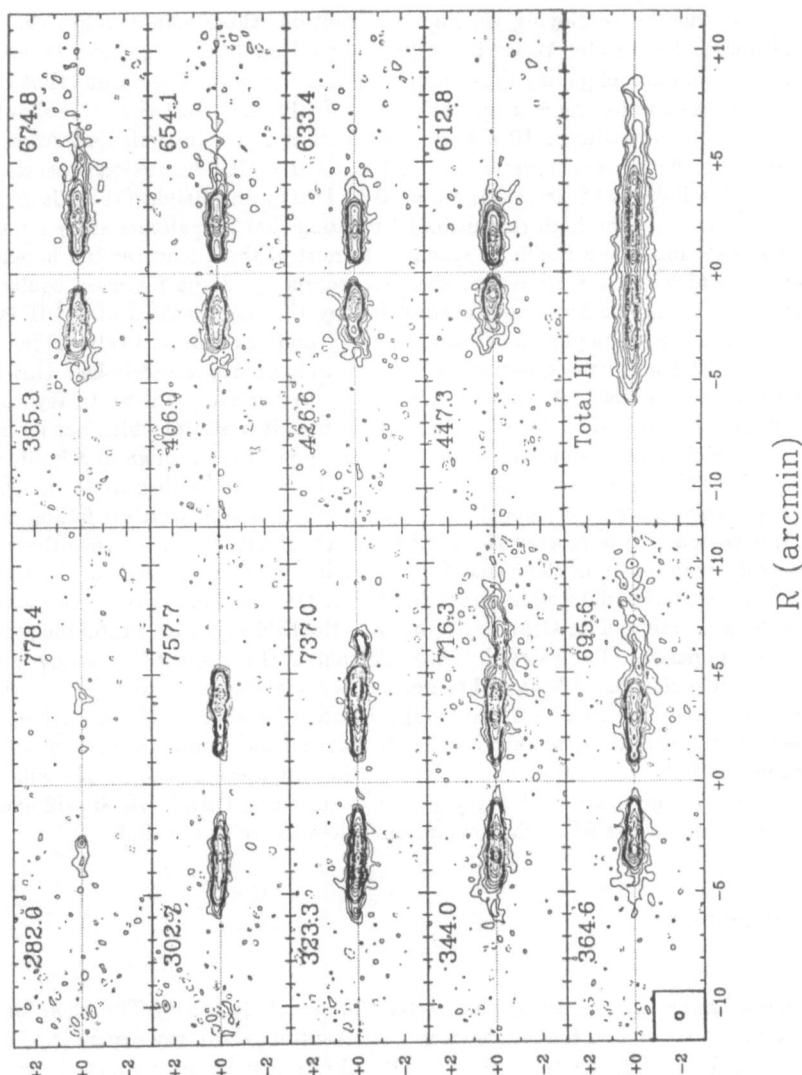

Figure 10: Series of HI channel maps at 20 arcsec resolution of NGC 891. The maps have been rotated to put the major axis horizontal, north pointing to the left. Each panel shows the two channel maps which are symmetrically placed with respect to the systemic velocity (530 km s^{-1}). The heliocentric velocities are indicated on top. Contour levels are plotted at $-10, -8, \ldots, -2, 2, 4, \ldots, 10, 15, \ldots, 45$ mJy beam^{-1}. The bottom right hand panel shows a map of the integrated HI distribution with contour levels at $0.1, 0.2, 0.5, 1.0, \ldots, 4.0$ Jy beam^{-1} km s^{-1}. (Map courtesy of M. Rupen).

of a few hundred parsec. Values for the measured velocity dispersion, volume density and spin temperature are generally in line with values derived for the local Galaxy. There are indications that by employing yet higher resolutions these values might change. As mentioned above, Braun and Walterbos (1989) show that peak brightness temperatures might vary as a function of galaxy type and position in a galaxy. They argue that the peak brightness is a measure for the star formation rate. If this is correct then this should show up in their VLA high resolution HI survey of a dozen or so nearby galaxies. Another way to throw light on the intrinsic temperature of the gas is by HI absorption observations. A new survey, extending the observations reported by Dickey and Brinks (1988) is on its way.

It is evident from the high resolution HI imaging that all galaxies show a multitude of filaments, arcs, loops and shells. It seems as if most of these features can be explained by the effects which OB stars have on their environment. Some features, equivalent to Heiles' supershells and perhaps best exemplified by the observations of M101 (van der Hulst and Sancisi 1988) require much more energy and their most likely explanation is infall of material. One of the questions which can now be addressed is how this infall is related to the presence of high velocity clouds. A further step will be to try to decide where these clouds come from; are they tidal debris or is it material which has its origin in the disk and which has been ejected by supernovae upon breakthrough of a bubble blown by an evolving OB association. Alternatively, the HVCs could be disk material which has been shot into the halo by the impact of a small companion and which is falling back.

In this review I have restricted myself to HI observations which trace the cool and warm neutral medium. It is clear that for a complete picture one would like to include Hα and radio continuum data which trace the WIM. One such project, which looks at the smaller scale interaction of an OB association with the ISM and which combines radio and optical data is underway (Brinks et al. 1989). The aim of this project is to study in detail a number of OB associations in M31 and to provide constraints for the various models which have been proposed and to derive the energy balance for these regions. Other obvious components which should be added in order to gain a more complete understanding are the cool dust and the molecular phase whereas the last piece of the puzzle will have to come from the new generation of X-ray satellites such as ROSAT which will eventually supply information on the HIM, thus completing the picture of the ISM.

Acknowledgements

I would like to thank the organizing committee and especially Harley Thronson for having invited me to Wyoming and for reimbursing a substantial part of my expenses. The other part, not less substantial, was covered by the Royal Greenwich Observatory. I am grateful to all those who have supplied me with photographic material and verbal input which was used in the preparation of this manuscript, notably Erik Deul, Jurjen Kamphuis, Michael Rupen, Ed Salpeter, Thijs van der Hulst, and Jacqueline van Gorkom. It is a pleasure to thank Robert Braun, John Dickey, Carl Heiles, Colin Norman, Daniel Puche, Evan Skillman, Guillermo Tenorio-Tagle, Thijs van der Hulst, and David Westpfahl for their extensive comments on an earlier version of this manuscript.

References

Abbott, D.C. 1982, *Ap. J.*, **263**, 723.

Bajaja, E., Cappa de Nicolau, C.E., Cersosimo, J.C., Loiseau, N., Martín, M.C., Morras, R., Olano, C.A., and Pöppel, W.G.L. 1985, *Ap. J. Suppl.*, **58**, 143.

Bottema, R., Shostak, G.S, and van der Kruit, P.C. 1986, *Astr. Ap.*, **167**, 34.

Bouchet, P., Lequeux, J., Maurice, E., Prévot, L., and Prévot-Burnichon, M.L. 1985, *Astr. Ap.*, **149**, 330.

Braun, R. 1989, *Ap. J. Suppl.*, submitted.

Braun, R., and Walterbos, R.A.M. 1989, in *The Interstellar Medium in External Galaxies, (Poster Session)*, eds. D.J. Hollenbach and H.A. Thronson, Jr., *NASA Conf. Proc.*, (Washington: NASA), *in press*.

Bregman, J.N. 1980, *Ap. J.*, **236**, 577.

Brinks, E., and Bajaja, E. 1986, *Astr. Ap.*, **169**, 14.

Brinks, E., Braun, R., and Unger, S.W. 1989 in *IAU Colloquium 120, Structure and Dynamics of the Interstellar Medium*, eds. G. Tenorio-Tagle, M. Moles, and J. Melnick, *Lecture Notes in Physics*, (New York: Springer-Verlag) *in press*.

Brinks, E., and Burton, W.B. 1984, *Astr. Ap.*, **141**, 195.

Brinks, E., and Shane, W.W. 1984, *Astr. Ap. Suppl.*, **55**, 179.

Clark, B.G. 1965, *Ap. J.*, **142**, 1398.

Cohen, R.S., Dame, T.M., Garay, G., Montani, J., Rubio, M., and Thaddeus, P. 1988, *Ap. J. (Letters)*, **331**, L95.

Corbelli, E., and Salpeter, E.E. 1988, *Ap. J.*, **326**, 551.

Corbelli, E., and Salpeter, E.E. 1989, in *The Interstellar Medium in External Galaxies, (Poster Session)*, eds. D.J. Hollenbach and H.A. Thronson, Jr., *NASA Conf. Proc.*, (Washington: NASA), *in press*.

Cowie, L.L. 1987, in *Interstellar Processes*, eds. D.J. Hollenbach and H.A. Thronson, Jr. (Dordrecht: Reidel), p. 245.

Cox, D.P. 1981, *Ap. J.*, **245**, 534.

Cox, D.P, and Smith, B.W. 1974, *Ap. J. (Letters)*, **189**, L105.

Crovisier, J. 1981, *Astr. Ap.*, **94**, 162.

Dettmar, R.-J., Keppel, J., Roberts, M.S., and Gallagher, J.S., III 1989, in *The Interstellar Medium in External Galaxies, (Poster Session)*, eds. D.J. Hollenbach and H.A. Thronson, Jr., *NASA Conf. Proc.*, (Washington: NASA), *in press*.

Deul, E.R. 1988, *Ph. D. Thesis*, Leiden Observatory.

Deul, E.R., and den Hartog, R.H. 1989, *Astr. Ap.*, in press.

Deul, E.R., and van der Hulst, J.M. 1987, *Astr. Ap. Suppl.*, **67**, 509.

Dickey, J.M., and Brinks, E. 1988, *M.N.R.A.S.*, **233**, 781.

Dopita, M.A., Mathewson, D.S., and Ford, V.L. 1985, *Ap. J.*, **297**, 599.

Field, G.B., Goldsmith, D.W., and Habing, H.J. 1969, *Ap. J. (Letters)*, **155**, L149.

Giovanelli, R., and Haynes, M.P. 1988, in *Galactic and Extragalactic Radio Astronomy, 2nd edition*, eds. G.L. Verschuur and K.I. Kellermann, (New York: Springer-Verlag), p. 522.

Hanson, M.M., Dickey, J.M., and Helou, G. 1989, in *The Interstellar Medium in External Galaxies, (Poster Session)*, eds. D.J. Hollenbach and H.A. Thronson, Jr., *NASA Conf. Proc.*, (Washington: NASA), *in press*.

Heiles, C. 1979, *Ap. J.*, **229**, 533.

Heiles, C. 1984, *Ap. J. Suppl.*, **55**, 585.

Heiles, C. 1987, *Ap. J.*, **315**, 555.

Heiles, C. 1989, in *IAU Colloquium 120, Structure and Dynamics of the Interstellar Medium*, eds. G. Tenorio-Tagle, M. Moles, and J. Melnick, *Lecture Notes in Physics*, (New York: Springer-Verlag) *in press*.

Hindman, J.V. 1967, *Aust. J. Phys.*, **20**, 147.

Huchtmeier, W.K., and Richter, O.-G. 1988, *Astr. Ap.*, **203**, 237.

Hulsbosch, A.N.M., and Wakker, B.P. 1988, *Astr. Ap. Suppl.*, **75**, 191.

Ikeuchi, S., Habe, A., and Tanaka, Y.D. 1984, *M.N.R.A.S.*, **207**, 909.

Irwin, J.A., and Seaquist, E.R. 1989, *Ap. J.*, in press.

Kamphuis, J., van der Hulst, T., and Sancisi, R. 1989, in *The Interstellar Medium in External Galaxies, (Poster Session)*, eds. D.J. Hollenbach and H.A. Thronson, Jr., *NASA Conf. Proc.*, (Washington: NASA), *in press*.

Kerr, F.J., Hindman, J.V., and Robinson, B.J. 1954, *Aust. J. Phys.*, **7**, 297.

Knapp, G.R., Turner, E.L., and Cunniffe, P.E. 1985 *A. J.*, **90**, 454.

Koornneef, J. 1982, *Astr. Ap.*, **107**, 247.

Kraan-Korteweg, R.C., and Tammann, G.A. 1979, *Astron. Nachr.*, **300**, 181.

Kulkarni, S. R., Blitz, L., and Heiles, C. 1982, *Ap. J. (Letters)*, **259**, L63.

Kulkarni, S. R., and Heiles, C. 1987, in *Interstellar Processes*, eds. D.J. Hollenbach and H.A. Thronson, Jr. (Dordrecht: Reidel), p. 87.

Kulkarni, S. R., and Heiles, C. 1988, in *Galactic and Extragalactic Radio Astronomy, 2nd edition*, eds. G.L. Verschuur and K.I. Kellermann, (New York: Springer-Verlag), p. 95.

Kulkarni, S.R., Rand, R.J., and Hester, J.J. 1989, in *The Interstellar Medium in External Galaxies, (Poster Session)*, eds. D.J. Hollenbach and H.A. Thronson, Jr., *NASA Conf. Proc.*, (Washington: NASA), *in press*.

Lake, G., and Skillman, E.D. 1989, *A. J.*, **98**, 1274.

Lockman, F.J. 1984, *Ap. J.*, **283**, 90.

Martin, N., Maurice, E., and Lequeux, J. 1989, *Astr. Ap.*, **215**, 219.

Mathewson, D.S, Ford, V.L., and Visvanathan, N. 1988, *Ap. J.*, **333**, 617.

McGee, R.X., and Milton, J.A. 1966, *Aust. J. Phys.*, **19**, 343.

McKee, C.F., and Ostriker, J.P. 1977, *Ap. J.*, **218**, 148.

Meaburn, J. 1980, *M.N.R.A.S.*, **192**, 365.

Meaburn, J., Clayton, C.A., and Whitehead, M.J. 1988, *M.N.R.A.S.*, **235**, 479.

Meaburn, J., Marston, A.P., McGee, R.X., and Newton, L.M. 1987, *M.N.R.A.S.*, **225**, 591.

Mihalas, D., and Binney, J. 1981, *Galactic Astronomy*, (San Francisco: Freeman).

Mirabel, I.F., and Morras, R. 1984, *Ap. J.*, **279**, 86.

Mirabel, I.F. 1989 in *IAU Colloquium 120, Structure and Dynamics of the Interstellar Medium*, eds. G. Tenorio-Tagle, M. Moles, and J. Melnick, *Lecture Notes in Physics*, (New York: Springer-Verlag) *in press*.

Norman, C.A., and Ikeuchi, S. 1989, *Ap. J.* **345**, 372.

Palouš, J., Franco, J., and Tenorio-Tagle, G. 1989, *Astr. Ap.* submitted.

Perley, R.A., Schwab, F.R., and Bridle, A.H. eds. 1989, *Synthesis Imaging in Radio Astronomy, Astr. Soc. Pac. Conf. Series*, **6**.

Reynolds, R.J. 1984, *Ap. J.*, **282**, 191.

Rohlfs, K., Kreitschmann, J., Siegman, B.C., and Feitzinger, J.V. 1984, *Astr. Ap.*, **137**, 343.

Rupen, M.P. 1989, *A.J.*, submitted.

Schneider, S.E., Helou, G., Salpeter, E.E., and Terzian, Y. 1983, *Ap. J. (Letters)*, **273**, L1.

Schneider, S.E., Skrutskie, M.F., Hacking, P.B., Young, J.S., Dickman, R.L., Claussen, M.J., Salpeter, E.E., Houck, J.R., Terzian, Y., Lewis, B.M., and Shure, M.A. 1989, *A. J.*, **97**, 666.

Shostak, G.S., and Skillman, E.D. 1989, *Astr. Ap.*, **214**, 33.

Shull, J.M. 1987, in *Interstellar Processes*, eds. D.J. Hollenbach and H.A. Thronson, Jr. (Dordrecht: Reidel), p. 225.

Shull, J.M., and Van Steenberg, M.E. 1985, *Ap. J.*, **294**, 599.

Skillman, E.D. 1988, in *IAU Colloquium 101, Supernova Remnants and the Interstellar Medium*, eds. R.S. Roger and T.L. Landecker, (Cambridge University Press, Cambridge), p. 465.

Spitzer, L., Jr. 1978, *Physical Processes in the Interstellar Medium*, (New York: Wiley and Sons).

Tenorio-Tagle, G., and Bodenheimer, P. 1988, *Ann. Rev. Astr. Ap.*, **26**, 145.

Tenorio-Tagle, G., Franco, J., Bodenheimer, P., and Rozyczka, M. 1987, *Astr. Ap.*, **179**, 219.

van der Hulst, T., and Sancisi, R. 1988, *A. J.*, **95**, 1354.

van der Kruit, P.C. 1988, *Astr. Ap.*, **192**, 117.

van der Kruit, P.C., and Shostak, G.S. 1984, *Astr. Ap.*, **134**, 258.

Wardle, M., and Knapp, G.R. 1986, *A. J.*, **91**, 23.

Wild, J.P. 1952, *Ap. J.*, **115**, 206.

Wolfire, M.G., Tielens, A.G.G.M., and Hollenbach, D.J. 1989, in *The Interstellar Medium in External Galaxies, (Poster Session)*, eds. D.J. Hollenbach and H.A. Thronson, Jr., *NASA Conf. Proc.*, (Washington: NASA), *in press*.

Molecular Clouds in Spiral Galaxies

Judith S. Young
Department of Physics and Astronomy
University of Massachusetts
Amherst, MA 01003

ABSTRACT. Observations of the 2.6mm CO emission in more than 200 galaxies are used to deduce the molecular content, the yield in young stars per unit molecular gas mass, and the H_2/HI ratio as a function of radius in individual galaxies and globally as a function of morphological type, luminosity, and environment. New evidence indicates that the molecular masses for spiral galaxies are accurate to better than a factor of 2, that the ratio of molecular to atomic gas mass decreases by a factor of ~20 among spiral galaxies from type S0/a to Sd/m, and that the mean yield of high mass stars per unit mass of molecular gas in spiral galaxies is constant as a function of type.

1. Introduction

Studies of molecular clouds are essential to our understanding of galactic evolution, both on small and large scales. On the individual cloud scale, this involves understanding why stars form where they form, and the interaction between young stars and their parent molecular clouds. On the galaxy-wide scale, the comparison of molecular cloud distributions with other tracers in a galaxy provides an understanding of the large scale processes which influence cloud and star formation as a function of galaxy type, luminosity, and environment.

The distribution and abundance of molecular clouds in spiral galaxies are critical in determining both the morphology and evolution of the galactic disk; it is within the giant molecular clouds that interstellar gas is cycled into the next generation of stars, and the most massive of these young stars produce a major part of the galactic luminosity. In order to fully understand the star formation histories of galaxies, and thereby their evolution, it is necessary to synthesize the details of the distributions of stars, gas, and dust in nearby galaxies, in conjunction with statistical studies of large samples of

H. A. Thronson, Jr. and J. M. Shull (eds.), The Interstellar Medium in Galaxies, 67–120.
© 1990 *Kluwer Academic Publishers.*

galaxies. Since the discovery of the molecular component of the interstellar medium, CO observations have uncovered the details of the cloud distribution in the Milky Way, of hundreds of galaxies from the Local Group to the Virgo cluster, and in luminous galaxies with recession velocities as high as 24,000 km s^{-1}.

In this review, I concentrate on the molecular content of spiral galaxies, probed through observations of the CO J=1→0 transition at λ2.6mm (115 GHz) with angular resolutions of 7" to 90". There have been a large number of molecular studies of galaxies conducted recently, including both detailed analyses of nearby galaxies and comparisons of the global properties of selected samples of galaxies. These two approaches are complimentary, and both are necessary to improve our understanding of the large scale physical processes which govern star formation and molecular cloud evolution. Tables 1 and 2 list the individual galaxies whose CO radial distributions have been measured and the statistical studies of larger samples, respectively.

Based on the above studies, CO fluxes and luminosities have been derived in over 200 galaxies. This data base is now sufficiently large to begin to statistically address the global H_2 content of galaxies in relation to other components of the disk, specifically the atomic gas content and star formation, as a function of morphological type, luminosity, and environment. The questions which we will address in this review are:

1) What is the range of H_2 masses in galaxies, and is there a dependence on galaxy type, luminosity, or environment?
2) What is the ratio L_*/M_{gas}, or the star formation efficiency (SFE = yield of young stars per unit mass of molecular gas), with radius in individual galaxies and from galaxy to galaxy?
3) What is the range of $M(H_2)/M(HI)$ in galaxies, and is there a dependence on galaxy type, luminosity, or environment?
4) What are the effects of spiral arms on the $\sigma(H_2)/\sigma(HI)$ ratio and the SFE in galaxies of a range of spiral arm amplitudes?

The answers to these and related questions depend on a multiwavelength analysis of galaxies. Specifically, the CO observations are used to deduce the masses and distributions of molecular hydrogen in galaxies, which are then compared with other tracers in the disk. In this review, I will first discuss CO as a tracer of H_2 mass in spiral galaxies (§2); there is now evidence, based on a new study of the gas/dust ratio in galaxies (Devereux and Young 1989a), that the global H_2 mass derivations for spirals are accurate to better than a factor of 2, including uncertainties arising from the CO→H_2 conversion. The other topics covered in this review include the global gas content and star formation properties of galaxies (§3), the distributions of gas and star formation within galaxies, (§4), spiral structure (§4), molecular properties of galaxies in the Local Group (§4), the star formation efficiency (§5), the effects of environment (§6), and nuclear gas concentrations (§7).

Table 1

CO Studies of Individual Galaxies

Galaxy	Type[a]	Telescope[b]	Resolution (arcsec)	Reference
NGC 55	Sc	SEST	44	Dettmar and Heithausen (1989)
NGC 224 (M31)	SbI-II	BTL	100	Stark (1979)
		OSO	33	Boulanger et al. (1984)
		BTL	100	Ryden and Stark (1986)
		NRO	15	Ichikawa et al. (1987)
		NRO	15	Lada et al. (1988)
		OVRO	7	Vogel et al. (1987)
NGC 253	Sc(s)	NRAO	67	Rickard et al. (1977)
		FCRAO	45	Scoville et al. (1985)
		OVRO	7	Canzian et al. (1988)
NGC 598	Sc(s)II-III	FCRAO	45	Young and Scoville (1982c)
NGC 891	Sb on edge	FCRAO	45	Solomon (1982)
		NRO	15	Sofue (1987)
		FCRAO	45	Sanders and Young (1989)
NGC 1068	Sb(rs)II	FCRAO	45	Scoville et al. (1983)
		OVRO	7	Myers and Scoville (1987)
		IRAM	21	Planesas et al. (1988)
		NRO	15	Kaneko et al. (1989)
NGC 1097	RSBbc(rs)I-II	IRAM	21	Gerin et al. (1988)
NGC 1275	E pec	IRAM	21	Lazareff et al. (1989)
NGC 1365	SBb(s)I	OSO	33	Sandquist et al. (1988)
NGC 1569	SmIV	FCRAO	45	Young et al. (1984a)
		NRAO	60	Hunter et al. (1988)

Table 1 (continued)
CO Studies of Individual Galaxies

Galaxy	Type[a]	Telescope[b]	Resolution (arcsec)	Reference
NGC 2146	SbII pec	FCRAO	45	Young et al. (1986a)
		BIMA	7	Jackson and Ho (1988)
		OVRO	7	Young et al. (1988b)
NGC 2403	Sc(s)III	FCRAO	45	Young and Scoville (1982c)
NGC 2623	Merger	IRAM	12	Casoli et al. (1988)
NGC 2841	Sb	FCRAO	45	Young and Scoville (1982b)
NGC 3034 (M82)	Amorphous	NRAO	67	Rickard et al. (1977)
		FCRAO	45	Young and Scoville (1984)
		OSO	33	Olofsson and Rydbeck (1984)
		NRO	15	Nakai et al. (1987)
		OVRO	7	Lo et al. (1987)
		BIMA	7	Carlstrom (1987)
NGC 3077	Amorphous	OSO	33	Becker et al. (1989)
NGC 3079	Sc pec	FCRAO	45	Young et al. (1986a)
		OVRO	7	Young et al. (1988a)
NGC 3147	Sb(s)I.8	FCRAO	45	Young et al. (1986a)
NGC 3593	Sa pec	NRAO	60	Hunter et al. (1988)
NGC 3623	Sa(s)II	FCRAO	45	Young et al. (1983)
NGC 3627	Sb(s)II.2	FCRAO	45	Young et al. (1983)
NGC 3628	Sbc	FCRAO	45	Young et al. (1983)
NGC 3738	SdIII	FCRAO	45	Tacconi and Young (1985)
NGC 4419	SBab	FCRAO	45	Kenney et al. (1989)

Table 1 (continued)
CO Studies of Individual Galaxies

Galaxy	Type[a]	Telescope[b]	Resolution (arcsec)	Reference
Virgo Cluster	Spirals	FCRAO	45	Kenney and Young (1986,88,89)
		BTL	100	Stark et al. (1986)
NGC 4438	Sb (tides)	IRAM	21	Combes et al. (1988)
NGC 4449	SmIV	FCRAO	45	Tacconi and Young (1985)
NGC 4485/90	Pair	NRAO	60	Thronson et al. (1989a)
NGC 4565	Sb	BTL	100	Richmond and Knapp (1986)
NGC 4736	RSab(s)	FCRAO	45	Garman and Young (1986)
NGC 5005	Sb(s)II	FCRAO	45	Young et al. (1986a)
NGC 5055	Sbc(s)II-III	OSO	30	Johansson and Booth (1987)
NGC 5194	Sbc(s)I-II	NRAO	67	Rickard and Palmer (1981)
		FCRAO	45	Scoville and Young (1983)
		OSO	33	Rydbeck et al. (1985)
		BIMA	7	Lo et al. (1987)
		OVRO	7	Vogel et al. (1988)
		IRAM	12	Guelin et al. (1989)
		OVRO	7	Rand and Kulkarni (1989)
		FCRAO	45	Lord and Young (1989)
NGC 5195	SB0 pec	FCRAO	45	Lord and Young (1989)
		NRAO	60	Sage (1989)
NGC 5236 (M83)	SBc(s)II	NRAO	67	Combes et al. (1978)
		FCRAO	45	Lord (1987)
		NRO	15	Handa et al. (1988)
NGC 5457 (M101)	Sc(s)I	NRAO	67	Solomon et al. (1983)

Table 1 (continued)
CO Studies of Individual Galaxies

Galaxy	Type[a]	Telescope[b]	Resolution (arcsec)	Reference
NGC 6946	Sc(s)II	NRAO	67	Morris and Lo (1978)
		NRAO	67	Rickard and Palmer (1981)
		FCRAO	45	Young and Scoville (1982a)
		OVRO	7	Ball et al. (1985)
		FCRAO	45	Tacconi and Young (1986,89)
		IRAM	21	Weliachew et al. (1988)
		NRO	15	Sofue et al. (1988a)
NGC 7331	Sb(rs)I-II	FCRAO	45	Young and Scoville (1982b)
NGC 7469	Sab pec	NRAO	60	Heckman et al. (1986)
NGC 7479	SBbc(s)I-II	FCRAO	45	Young et al. (1986a)
Arp 220	Merger	IRAM	12	Casoli et al. (1988)
		OVRO	5	Scoville et al. (1986b)
IC 10	Im	NRO	15	Ohta et al. (1988)
IC 342	SAB(rs)cd	NRAO	67	Morris and Lo (1978)
		NRAO	67	Rickard and Palmer (1981)
		FCRAO	45	Young and Scoville (1982a)
		OVRO	7	Lo et al. (1984)
		OSO	33	Wiklind et al. (1987)
		NRO	15	Hayashi et al. (1987)
Maffei II	Sb	NRAO	67	Rickard et al. (1977)
		IRAM	21	Weliachew et al. (1988)
Mrk 231	Merger	FCRAO	45	Sanders et al. (1987)
Mrk 1014	Quasar	NRAO	60	Sanders et al. (1988b)
VII Zw 31		FCRAO	45	Sage and Solomon (1987)
		FCRAO	45	Goldsmith and Young (1989)

Table 1 (continued)
CO Studies of Individual Galaxies

Galaxy	Type[a]	Telescope[b]	Resolution (arcsec)	Reference
Zw 1	Quasar	IRAM	21	Barvainis et al. (1989)
IRAS 12112+0305		NRAO	67	Mirabel et al. (1988a)
LMC and SMC	SB(s)m		120	Israel et al. (1986)
			530	Cohen et al. (1988)

Notes to Table 1:

[a] Galaxy type from *Revised Shapley Ames Catalogue* (Sandage and Tammann 1981).

[b] Key for telescopes:
BTL = Bell Telephone Laboratories
CSO = Caltech Submillimeter Observatory
FCRAO = Five College Radio Astronomy Observatory
IRAM = Inst. Radio Astronomie Millimetrique
NRAO = National Radio Astronomy Observatory
NRO = Nobeyama Radio Observatory
OSO = Onsala Space Observatory
OVRO = Owens Valley Radio Observatory

Table 2
Published Surveys of CO in Galaxies

Galaxy Class	Sample Size	No. Detected	Telescope[a]	Reference
Ellipticals	1	1	IRAM	Huchtmeier et al. (1988)
IR Bright				
Centers	20	10	FCRAO	Young et al. (1984b)
Maps	14	14	FCRAO	Young et al. (1986a)
High Luminosity	15	15	FCRAO	Sanders et al. (1986)
High Luminosity	3	3	OVRO	Sanders et al. (1988a)
Ultra-high Lum.	5	4	SEST	Mirabel et al. (1988b)
High Luminosity	4	1	OVRO	Scoville et al. (1989)
Global Properties	124	108	FCRAO	Young et al. (1989a)
Irregulars				
Magellanic	6	1	NRAO	Elmegreen et al. (1980)
Star-forming	1	1	FCRAO	Young et al. (1984a)
Star-forming	3	3	FCRAO	Tacconi and Young (1985)
Dwarf Irr's	15	6	FCRAO	Tacconi and Young (1987)
Blue Compact Galaxies	12	0	IRAM	Arnault et al. (1988)
Isolated and Interacting				
Maps	26	26	FCRAO	Young et al. (1986b)
Maps	93	90	FCRAO/NRAO	Solomon and Sage (1988)
Radio Bright	21	20	FCRAO/NRAO	Sanders and Mirabel (1985)
Seyferts	9	2	NRAO/BTL	Bieging et al. (1981)
	9	0	NRAO	Wilson et al. (1979)
	43	18	NRAO	Heckman et al. (1989)

Table 2 (continued)
Published Surveys of CO in Galaxies

Galaxy Class	Sample Size	No. Detected	Telescope[a]	Reference
SO's				
Centers	11	5	NRAO	Sage and Wrobel (1989)
SO-Sa	20	11	FCRAO	Thronson et al. (1989a)
Spirals				
Early Types	24	22	OSO,SEST, IRAM	Wiklind and Henkel (1989)
Gas-Rich	29	5	NRAO	Rickard et al. (1977)
Nearby	81	5	NRAO	Rowan-Robinson et al. (1980)
Range of ...				
Arm Types	29	20	BTL	Stark et al. (1987)
Hubble Types	19	6	BTL	Verter (1983,87,88)
SAB Galaxies	23	5	NRAO	Elmegreen & Elmegreen (1982)
Sc's	9	9	FCRAO	Young and Scoville (1982b)
Starbursts	42	9	FCRAO	Jackson et al. (1989)
Virgo Cluster				
Centers	25	18	FCRAO	Young et al. (1985)
Major Axis Maps	42	34	FCRAO	Kenney and Young (1986,88,89)
Total CO	47	25	BTL	Stark et al. (1986)
^{13}CO Observations				
Centers	8	5	BTL	Encrenaz et al. (1979)
Disks	6	6	NRAO	Rickard and Blitz (1985)
Disks	6	6	FCRAO	Young and Sanders (1986)

Notes to Table 2:

[a] Telescopes are as in Table 1.

2. Molecular Mass Determinations

A number of papers have been written on the use of CO as a tracer of the mass of molecular hydrogen (see for example Young and Scoville 1982; Liszt 1982; Sanders, Solomon, and Scoville 1984; Bloemen et al. 1986; Dickman, Snell and Schloerb 1986; Young et al. 1986a; Scoville and Sanders 1987; Maloney 1988; Maloney and Black 1988; Dickman 1988; Elmegreen 1989). In spite of the high optical depth of the 2.6mm CO $J=1\rightarrow0$ line, the emission is shown to trace the molecular gas mass of giant molecular clouds (GMCs). In this section we briefly review the theoretical, empirical, and observational basis for the use of CO as a tracer of H_2.

2.1 Theoretical Basis

Although molecular hydrogen is the dominant component of the interstellar medium over the inner disk of the Milky Way galaxy (Scoville and Solomon 1975; Gordon and Burton 1976), the H_2 molecule has no permanent electric dipole moment and the quadropole rotational transitions lie in the infrared. It is therefore necessary to utilize indirect techniques to estimate the H_2 abundance in cool, dense clouds; because of its brightness and widespread distribution, ^{12}CO is commonly used as a tracer of molecular hydrogen gas.

Using the CO integrated intensity, I_{CO}, which is given by the integral of the brightness temperature over the telescope beam $(I_{CO}=\int T \, dv)$, Dickman, Snell and Schloerb (1986) and Scoville and Sanders (1987) have shown that the CO luminosity of a cloud can be defined as

$$L_{CO} = d^2 \int I_{CO} \, dv \qquad (1)$$

where d is the distance to the cloud. For a cloud of radius R and linewidth ΔV, the CO luminosity is given by

$$L_{CO} = \pi R^2 \, T_{CO} \, \Delta V \qquad (2)$$

where T_{CO} is the peak brightness temperature in the CO line. For a cloud of mass M in virial equilibrium, $\Delta V = \sqrt{GM/R}$ and

$$L_{CO} = \sqrt{3\pi G/4\rho} \; T_{CO} \, M. \qquad (3)$$

Thus, the CO luminosity is proportional to the molecular mass and to the ratio $T_{CO}/\rho^{0.5}$.

To the extent that the ratio $T/\sqrt{\rho}$ does not vary in the mean from galaxy to galaxy, equation (3) indicates that the CO luminosity is directly proportional to the mass of molecular clouds.

Estimates of cloud temperatures and densities have been derived from recent multi-transition studies of CO in M82. Comparison of the $J=3\rightarrow2$, $2\rightarrow1$, and $1\rightarrow0$ observations (Wild et al. 1989; Turner, Martin, and Ho 1989) indicate the presence of several cloud populations -- a cool one ($T\sim10$-$20K$) with low density $[n(H_2)<10^3$ $cm^{-3}]$, and a warm one ($T\sim75K$) with high density $[n(H_2)\sim5\times10^4$ $cm^{-3}]$. In spite of the markedly different properties in these two cloud components, the difference in $T/\sqrt{\rho}$ between these two populations is less than a factor of 2. Therefore, *even in M82*, which is one of the most extreme galaxies in our part of the universe, the assumption of a constant conversion between CO luminosity and H_2 mass leads to a global molecular gas mass that is accurate to within a factor of 2.

Maloney (1988), Maloney and Black (1989) and Elmegreen (1989) have considered the theoretical basis for the derivation of H_2 masses from CO luminosities also considering the effects of metallicity variations. Maloney and Black (1989) discuss the effects of very low metallicity with regard to the H_2 masses in irregular galaxies. They suggest that a reduction in the sheilding of the gas by the dust will lead to smaller molecular cores for clouds located in low metallicity regions. While there has been much discussion of the $CO\rightarrow H_2$ conversion in irregular galaxies (cf. Elmegreen, Elmegreen and Morris 1980; Young, Gallagher, and Hunter 1984; Tacconi and Young 1987; Israel 1988), much work remains to be done to establish the molecular gas content of these systems. Elmegreen (1989) concludes that for galaxies similar to the Milky Way, the derivation of H_2 masses from L_{CO} will be reliable.

2.2 The $CO\rightarrow H_2$ Conversion in the Milky Way and Nearby Galaxies

2.2.1 The Galaxy

There have been a number of recent observational attempts to empirically determine the constant of proportionality for deriving molecular masses from CO luminosities (Frerking, Langer, and Wilson 1982; Young and Scoville 1982a; Liszt 1982; LeBrun et al. 1983; Sanders, Solomon and Scoville 1984; Bloemen et al. 1984, 1986; Scoville et al. 1986; Solomon et al. 1987). The four general techniques which have been applied include (1) the correlation of ^{12}CO in nearbly clouds with visual extinction (A_v), (2) the correlation of ^{13}CO, which is less optically thick than ^{12}CO, versus A_v, (3) the derivation of virial masses from the ^{12}CO and ^{13}CO cloud sizes and line widths, and (4) the comparison of the distribution of γ-rays with the total column density of nucleons. The derived ratios of $N(H_2)/I_{CO}$ cover the range $(1.8$-$4.8)\times10^{20}$ cm^{-2} $[K(T_R)$ km $s^{-1}]^{-1}$. It is interesting that different studies using different techniques have yielded similar values for the conversion of CO integrated intensities to H_2 column densities. The reasonably good agreement among these independent analyses suggests that global H_2 mass estimates for spiral galaxies are probably reliable to a factor of 2. Throughout this work we adopt a constant value of $N(H_2)/I_{CO} = 2.8 \times 10^{20}$ cm^{-2} [$K(T_R)$ km s^{-1}]$^{-1}$ (Bloemen et al. 1986). As discussed in §3 below, Devereux and Young (1989a) find that the small scatter in the observed gas/dust ratio in spiral disks is consistent with the 30% measurement uncertainty in the molecular mass determinations, including uncertainties from galaxy to galaxy in the CO→H_2 conversion.

It is noteworthy that the CO →H_2 conversion factor has been derived in the Milky Way between 2 and 10 kpc (Bloemen et al. 1986), and found to show no significant large-scale variations over this region. Certainly within the Galaxy, variations in metallicity and cloud properties do not seem to affect the H_2 mass derivations.

Figure 1 shows the correlation between the virial masses and CO luminosities of molecular clouds in the Milky Way (Scoville et al. 1987; Solomon et al. 1987). The best fit to these data for clouds with mass between 10^5 and 2×10^6 $M_\odot$ yields a CO→H_2 proportionality which is consistent with the determination of Bloemen et al. (1986).

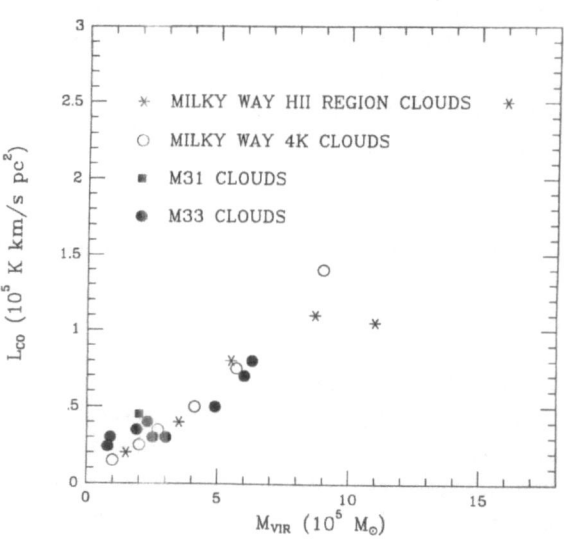

Figure 1. Comparison of CO luminosities and virial masses of molecular clouds in the Milky Way (Scoville et al. 1987), in M31 (Vogel et al. 1987), and in M33 (Wilson and Scoville 1989). In the Milky Way, the CO luminosity is closely correlated with the virial masses of the clouds, both with and without high mass star formation. This linear proportionality justifies the use of CO as a tracer of the mass of H_2; the best fit to these data for clouds with mass between 10^5 and 2×10^6 $M_\odot$ yields a constant of proportionality of 3.6×10^{20} H_2 cm^{-2} (K km s^{-1})$^{-1}$. The similarity of the clouds in M31 and M33 to the Milky Way justifies the use of the the same CO→H_2 proportionality in the external galaxies.

Most of the data used in the above studies is presented in units of T_R^* (Kutner and Ulich 1981), which does not correct for the coupling between the source and the telescope beam. Since giant molecular clouds have large angular sizes, the coupling efficiency, η_c, is close to unity, and $T_R = T_R^* \, \eta_c^{-1} \approx T_R^*$. However, the external galaxies studied in CO have smaller angular sizes and coupling efficiencies should be corrected for when deriving molecular masses or the masses will be underestimated.

2.2.2 Nearby Galaxies

One technique to observationally test the reliability of H_2 masses in external galaxies which are derived using the $CO \rightarrow H_2$ proportionality for the Milky Way, is to measure the sizes and CO luminosities of individual molecular clouds in other galaxies. If those clouds follow the same relation as the clouds for the Milky Way, then H_2 masses will not be systematically biased. In order to measure the sizes of individual clouds, it is crucial to have a spatial resolution considerably better than 40 pc, the "typical" size of a GMC (Scoville and Sanders 1987). The external galaxies which have been studied on such small spatial scales are M31 and M33, for which the CO observations have been made with ~7" (~20 pc) resolution (Vogel et al. 1987; Wilson and Scoville 1989). Moderate resolution studies of the LMC and SMC (2.0' or 33 pc -- Israel et al. 1986; 8.8' or 140 pc -- Cohen et al. 1988) have also been made, but in these studies the individual clouds have not been resolved, and do not therefore provide a contrast to the CO luminosity-virial mass relation for Milky Way molecular clouds. The CO luminosities and virial masses for the M31 and M33 molecular clouds (Vogel et al. 1987; Wilson and Scoville 1989) are plotted in Figure 1 along with those for the Milky Way. Here it can be seen that these extragalactic molecular clouds are similar to those in the Milky Way, despite the fact that they represent regions of lower metallicity (by a factor of 2 relative to the sun -- Pagel and Edmunds 1981) and are found in galaxies ranging from type Sb (M31) to Scd (M33).

From a study of LMC molecular clouds, Cohen et al. (1987) have suggested that the molecular clouds in that galaxy have 6 times more mass per unit CO luminosity than molecular clouds in the Milky Way. It must be borne in mind, however, that their angular resolution of 140 pc is insufficient to resolve individual molecular cloud sizes and linewidths. The linewidth observed over a region considerably larger than an individual cloud will reflect not only the mass in the cloud, but also the *total* mass (stellar + gaseous) in the region. Cohen et al. find that clouds of a given CO luminosity in the LMC have 6 times larger linewidths than clouds in the Milky Way; however, the coarse resolution of the CO observations can cause an overestimate of the cloud virial mass. In the limit that an entire galaxy is observed in CO or HI, the *dynamical* mass of the galaxy is estimated from the linewidth and the size of the region sampled, not just the gas mass.

Obviously, the linewidth samples more than simply the mass of gas when individual clouds are not resolved.

2.2.3 ^{13}CO Observations of Galaxies

An observational test of the accuracy of molecular mass determinations is the comparison of the CO radial distribution in a galaxy with that of the more optically thin ^{13}CO line. The first ^{13}CO observations of galaxies were made in 8 systems (Encrenaz et al. 1979), with subsequent observations in a small number of other galaxies (Young and Scoville 1982a, 1984; Stark and Carlson 1984; Rydbeck et al. 1985). Defining the ratio of ^{12}CO to ^{13}CO integrated intensities to be $R = I(^{12}CO)/I(^{13}CO)$, Encrenaz et al. found that R has a value of ~12, or roughly twice the value typical of GMCs in the molecular annulus of our Galaxy (Solomon, Scoville, and Sanders 1979; Stark 1983).

Rickard and Blitz (1985) and Young and Sanders (1986) have measured the ^{12}CO and ^{13}CO emission in the disks of nearby luminous spiral galaxies. Globally, the value of R was found to range between 6 and 20 for individual galaxies, with low values typical of the Milky Way and NGC 891 (Sanders and Young 1989). The highest value of R is found in the center of M82 (Stark and Carlson 1984; Young and Scoville 1984), where the ratio of integrated intensities is 20. Within individual galaxies, Rickard and Blitz and Young and Sanders found that R varies by a factor of 2 on the scale of 1 to 2 kpc. These variations may be related to the differences in the distributions of hot and cold cloud cores, such as in the Milky Way (Solomon, Sanders, and Rivolo 1985).

From observations of the distributions of ^{12}CO and ^{13}CO in 6 galaxies, Young and Sanders find that the shapes of the two azimuthally averaged distributions are similar, as shown in Figure 2. This result indicates that the surface density distribution of H_2 is traced by observations of both ^{12}CO and the weaker ^{13}CO. In this paper, the H_2 masses and distributions are inferred from ^{12}CO observations, with an adopted CO→H_2 constant of proportionality as described in §2.2.1.

2.2.4 CO Total Flux Determinations and Uncertainties

Ideally, we would like to know the CO flux (in telescope-independent units) for all galaxies in the sky. This means it is necessary to correct for the source-beam coupling, since different telescopes will observe different signals from the same source, as the size of the source changes relative to the beam. In order to determine the global CO flux of a galaxy, therefore, knowledge of the beam pattern and assumptions about the intrinsic distribution of CO emission are necessary. Several different techniques have now been employed for determining global CO fluxes of galaxies; in this section we compare the global CO fluxes derived from the different methods in use, and evaluate the associated uncertainties.

Based on single CO observations of the centers of galaxies, Verter (1987) derived global CO fluxes from an assumed exponential form for the CO distribution (cf. Young and Scoville 1982a; Scoville and Young 1983; Solomon et al. 1983), and adopted the scale length appropriate to the blue light distribution in the galaxy. This exponential was then integrated out to a radius of $\frac{2}{3}R_{25}$ (where $R_{25}=D_{25}/2$; de Vaucouleurs, de Vaucouleurs, and Corwin 1976).

Stark et al. (1986) have taken the more general approach of deriving the CO flux from the model distribution which best fits the observations when convolved with the appropriate beam pattern. In this study, the CO fluxes were extrapolated out to R_{25} based on observations at the center of each galaxy, with two additional points on the major axis and two on the minor axis. Kenney and Young (1988a) and Young et al. (1989a) have used a procedure similar to Stark et al., and derived CO fluxes from the models which best match the observed distributions of integrated intensities when convolved with the appropriate gaussian. The principal difference between this method and that of Stark et al. is that Kenney and Young observe the CO distribution along the major axis out to where the emission falls below 1 K km s^{-1}, determine the flux within the region observed, and make no extrapolations. Both of these methods correct for the source-beam coupling.

Another method of determining the total CO emission has been used (Solomon and Sage 1988), although this method does not correct for the source-beam coupling. In this case, the observed emission is simply integrated over the region observed.

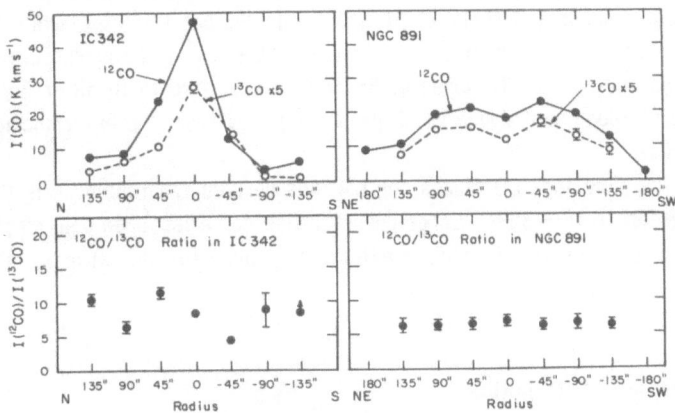

Figure 2. ^{12}CO and ^{13}CO distributions (upper panels) in the face-on Sc galaxy IC342 (Young and Sanders 1986) and in the edge-on Sb galaxy NGC 891 (Sanders and Young 1989). The solid dots represent the ^{12}CO distributions and the open circles represent the ^{13}CO distributions scaled up by a factor of 5. The center of each galaxy is plotted in the center of the panel. The lower panels show the ratio of ^{12}CO/^{13}CO integrated intensities at each position.

Uncertainties in the CO fluxes arise from a number of sources, most notably (1) the uncertainties in the individual integrated intensities due to calibration, baseline-fitting, and signal-to-noise, (2) the undersampling of emission, and (3) the quality of the fit of the model to the observed integrated intensities. Kenney and Young (1988a) have estimated the fractional uncertainty contributed by each of the above components to the overall CO flux uncertainty. They find individual integrated intensity uncertainties to be ~15%, undersampling uncertainties to be typically ~20%, and quality of fit uncertainties to be ~20%, leading to an overall CO flux uncertainty of ~32%.

For the 58 galaxies in common between the Solomon and Sage (1988) study and those of Kenney and Young (1988a) and Young et al. (1989a), the dispersion in the flux measurements is ±0.17 dex, which is consistent with a 35% error in the two sets of fluxes. This result confirms that the uncertainties estimated by Kenney and Young (1988a) and Young et al. (1989a) are realistic, and of order 35%.

There is considerably more scatter in the comparison of Verter's and Stark et al.'s CO fluxes with those of Kenney and Young, Young et al., and Solomon and Sage. Taking the 35% uncertainty in the latter fluxes derived above, the ±0.30 dex dispersion in comparing Verter and Stark's data with the other studies leads to a 150% error in Stark et al. and Verter's fluxes. Verter (1987) points out that her CO fluxes may be uncertain by a factor of 3-4, but are more likely to be in error by a factor of 1.3-1.5, consistent with the above analysis. We note that for 9 galaxies in the sample of Young et al. (1989a), Verter has used the *same data* and derived fluxes which are on average 2.4 times higher than those of Young et al. This difference is due primarily to the fact that Verter adopts an exponential scale length and integrates out to $2/3R_{25}$, while Young et al. *fit* the data for the scale length, and derive the flux out to the maximum radius observed. The dispersion in the CO fluxes of Stark et al. relative to those of Solomon and Sage, Kenney and Young, and Young et al. probably arises from the extrapolation over the entire optical disk.

In summary, we find that CO fluxes for galaxies derived from major axis CO observations are accurate to ~35%. Large uncertainties can arise when only single positions are observed, and when the flux is extrapolated over the optical disk; these fluxes should be used with caution.

3. Global Properties of Galaxies

The principal goal of CO studies of galaxies is to understand galaxy evolution by elucidating the relationship between molecular clouds, atomic gas, and star formation from galaxy to galaxy as a function of galaxy type, luminosity, and environment. In order to achieve this goal, it is necessary to observe the molecular cloud distributions and content in several hundred galaxies, because of the broad parameter space which galaxies occupy.

In this section we review the global H_2, HI, IR, Hα, and dust content of galaxies in which CO observations have been made. Extensive results have been obtainted recently from the Five College Radio Astronomy Observatory (FCRAO) Extragalactic CO Survey. The galaxies whose CO distributions have been observed along the major axis as part of this Survey are either (1) brighter than $B_T{}^\circ = 12$ mag. in the blue, or (2) brighter than 20 Jy at 100 μm (from coadded IRAS data) (Kenney and Young 1988b; Young et al. 1989a).

The principal parameters with respect to which the global properties will be discussed below are morphological type, galaxy luminosity or mass, and environment. In investigating global galaxy properties, it is important to analyze *both* the comparison of absolute luminosities and masses as well as the ratios of these quantities. In comparisons of the global quantities, it is essential to determine the slope, intercept, and scatter in a correlation. If the slope of a correlation is unity, and the scatter is small, then one could conclude that galaxies are found with a wide range of masses, and that small galaxies are just scaled down versions of large galaxies. If, on the other hand, the correlation of two related quantities indicates that the slope is not unity and the scatter is not small, then there is some physical process to be understood. Only when the scatter in the absolute quantities is known can the scatter in the normalized quantities be understood. And only when the slope of a correlation between two quantities is known can the scatter in the normalized quantities be predicted.

To illustrate the point that galaxies have a wide range of absolute sizes, Figure 3 shows CCD images for 6 galaxies -- NGC 253, 520, 660, 2146, IC 342 and M82 -- as if they were all at the same distance. In this Figure it is apparent that M82 and NGC 253 are tiny galaxies, while NGC 520, 660, and 2146 are enormous.

An appreciation of the uncertainties in the H_2 mass determinations is critical to the interpretation of the yield of young stars per unit mass of molecular gas and to the H_2/HI ratio within and among galaxies. For this reason, I begin below with a discussion of a recent study which addresses the accuracy of molecular mass determinations through an investigation of the gas/dust ratio in spiral galaxies.

3.1 The Gas-to-Dust Ratio in Spiral Galaxies

One of the treasures which astronomers have received as a result of the successful completion of the IRAS mission is the measurement of infrared luminosities, dust temperatures, and dust masses for complete samples galaxies across the sky. Since the infrared emission is predominantly thermal dust emission heated primarily by young stars (Telesco and Harper 1980; Rieke et al. 1980; Telesco 1988; Rice et al. 1989; Devereux and Young 1989), the infrared luminosities can be used to indicate the star formation rate (SFR). Until recently, the best correlation which had been found with the IRAS data was that between the global infrared and radio continuum luminosities in spirals, extending over 6 orders of magnitude with little scatter (Dickey and Salpeter 1984; Helou et al.

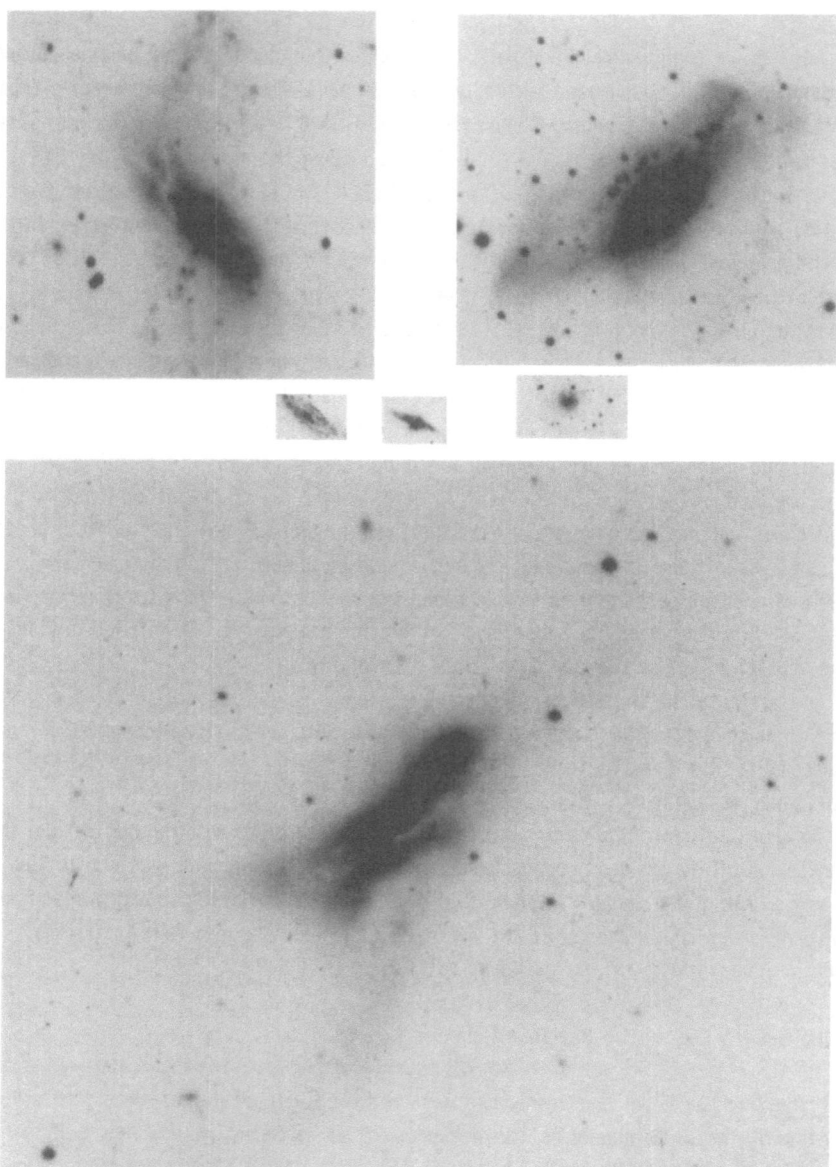

<u>Figure 3</u>. Red CCD images (Hα + continuum) of 6 galaxies all reproduced at the same *absolute* scale. The images were obtained with the #1-0.9 m telescope of KPNO. The galaxies illustrated are as follows: NGC 660 (top left), NGC 2146 (top right), NGC 253 (left center), M82 (middle), IC 342 (right center), and NGC 520 (bottom). It is clear that M82 and NGC 253 are both quite small galaxies.

1985; Devereux and Eales 1989).

Of all the global correlations which have been presented involving the molecular content of galaxies, the best one found thus far is that between molecular masses and IRAS-derived dust masses (Young et al. 1986a, 1989a; Stark et al. 1986). Young et al. (1989a) also find that the comparison of the dust masses with HI masses for the same galaxies exhibits considerably more scatter. Comparison of molecular gas masses with the dust masses that are required to produce the infrared emission measured by IRAS lead to gas to dust ratios that are typically 570±50 rather than the value of ~150 which is widely used for the Galaxy (Spitzer 1978; Hildebrand 1983; Draine and Lee 1984). The inclusion of atomic gas in spiral galaxies only accentuates the discrepancy. One plausible explanation for the high gas/dust ratio is that the bulk of the dust mass is cold ($T_{dust} < 30K$) and radiating beyond 100μm. Henceforth, the dust detected by IRAS will be referred to as "warm dust."

In a recent study of the gas/dust ratio in 58 galaxies whose H_2 *and* HI distributions are known, Devereux and Young (1989a) have considered not only the molecular gas, but also that fraction of the atomic gas associated with the inner disk, i.e. the area of the galaxies with most of the star formation and hence infrared emission. They find that the correlation between the warm dust mass and the molecular gas mass is as significant as the correlation between warm dust mass and the total H_2+HI mass, although the dispersion in both correlations is larger than the measurement uncertainty. The dispersion can be dramatically reduced, however, if one includes the molecular gas and only that fraction of the atomic gas that is located in the inner disks ($R < R_{25}/2$) of the galaxies. It is therefore concluded that the warm dust detected by IRAS is located in the inner regions of spiral disks.

Excluding the extended HI component in spiral galaxies reduces the dispersion in the gas/warm-dust ratio to a level that is consistent with the 30% measurement uncertainties estimated for the global gas mass determinations (see Figure 4). Consequently, the gas/warm-dust ratio is well determined, with a value of 1080±70. No significant variations in the ratio are found as a function of morphological type, dust temperature, or whether the HI or H_2 is the dominant phase of the ISM. If the true gas to dust ratio in spiral galaxies is the same as the value of ~150 measured within the Galaxy, the high value that is found from the IRAS data indicates that 80-90% of the dust mass in spiral galaxies is radiating at λ > 100 μm and has a dust temperature colder than ~30K.

Additionally, there is a very important conclusion which has been drawn from the above study with regard to the accuracy of molecular mass determinations, including both measurement error *and* galaxy to galaxy variations in the CO→H_2 conversion in spiral galaxies. Devereux and Young (1989a) point out that the small dispersion in the inner disk gas to warm dust ratio (Figure 4d) can be attributed entirely to a 30% uncertainty

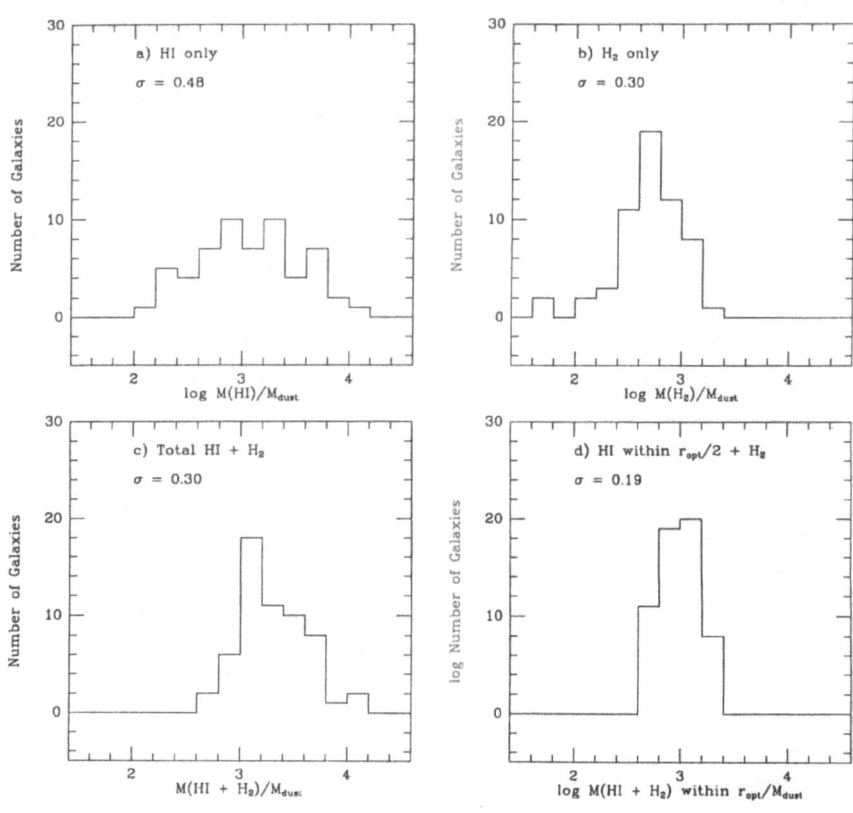

Figure 4. Histograms of the gas/dust ratio in 58 galaxies for which the distributions of molecular and atomic gas are known (Devereux and Young 1989a). The individual panels represent the ratios of (a) HI/dust, (b) H_2/dust, (c) total (HI+H_2) gas/dust, and (d) inner disk ($R<R_{opt}/2$) gas/dust. The value of σ in each panel represents the dispersion in the observed gas/dust ratio.

in each of the H_2 and HI masses, and a 10% uncertainty in the dust masses. Even if all of the scatter in the correlation were introduced by uncertainties in H_2 masses, and specifically in the $CO \rightarrow H_2$ proportionality, Devereux and Young find less than a factor of 2 uncertainty in H_2 masses, even though such diverse galaxies as M82, NGC 6946, and M33 were included in the sample. Thus, molecular mass determinations using a constant $CO \rightarrow H_2$ proportionality are supported theoretically, empirically, and from ^{13}CO observations (see §II), and now in a sample of galaxies where the inner disk gas to dust ratio is shown to exhibit very little scatter. It is therefore concluded that the method of determining global molecular gas masses for spiral galaxies is reliable, because the dispersion in the correlation of global H_2 masses with independently determined HI and dust masses can be attributed to the measurement errors.

3.2 Total H_2 Masses

The masses of molecular gas which are derived for galaxies cover a broad range, from more than 10^{10} $M_\odot$ for the most massive spirals (Sanders et al. 1986; Young et al. 1986a) to less than 10^6 $M_\odot$ for some low mass dwarfs (Tacconi and Young 1987). The range of total HI masses found in these same galaxies is similar (Young et al. 1989a).

3.3 Ratio of Molecular to Atomic Gas as a Function of Morphological Type

In order to gain an understanding of the global processes which influence molecular cloud formation in disk galaxies, it is necessary to determine the relative amounts of atomic and molecular gas as a function of position in galaxies and from galaxy to galaxy. While atomic gas studies of galaxies have been underway for more than 30 years, the relative youth of the field of extragalactic molecular studies has meant that knowledge of the relative amounts of molecular and atomic gas in galaxies has been limited until recently to the relatively small number of galaxies observed in CO.

Ever since the pioneering work in the 1950's and 1960's, it has been recognized that there is a morphological type dependence to the total atomic gas content of galaxies. In particular, Roberts (1969) showed that the HI mass to blue luminosity ratio, $M(HI)/L_B$, increases by a factor of ~5 among spiral galaxies from types Sa through Scd. Do late type spirals have more gas than early types when the molecular gas content is included?

Young and Knezek (1989) have analyzed the global ratio of molecular to atomic gas mass for 150 of the galaxies in the FCRAO Extragalactic CO Survey for which HI masses were available in the literature (Huchtmeier et al. 1983). Within this sample, they confirm the previously found trend for the mean ratio of total HI mass to blue luminosity to increase with type by a factor of 5, and they find that the mean ratio of H_2 mass to blue luminosity is roughly constant for types Sa-Sc, with a decrease of a factor of >3 for types Scd-Sdm. The combination of these two effects is that *the mean value of the ratio of molecular to atomic gas decreases smoothly by a factor of ~20 as a function of morphological type for types Sa-Sd.* The mean $M(H_2)/M(HI)$ ratio is 4.0±1.9 for S0/Sa galaxies, and 0.2±0.1 for Sd/Sm galaxies.

The $M(H_2)/M(HI)$ ratio is shown in Figure 5 for the spiral galaxies from the FCRAO Extragalactic CO Survey and for the early type galaxies from Thronson et al. (1989). While there is considerable scatter in the ratio of molecular to atomic gas within a given type, there is a clear trend in that the $M(H_2)/M(HI)$ ratio changes smoothly with type. A volume limited subset of the galaxies more nearby than the Virgo cluster also displays a similar decrease in the $M(H_2)/M(HI)$ ratio. In an earlier study, Verter (1987) concluded that the CO/HI flux ratio peaks for intermediate morphological types, but her study was based on many fewer galaxies.

For spiral galaxies in the Young and Knezek study, the mean ratio of total ISM gas mass to optical area is 17 $M_{\odot}/pc^2$ with only a factor of 2 increase in the mean gas surface density of spiral galaxies from early to late morphological type. Thus, the dominant effect is that the *phase* of the gas in the cool ISM varies systematically along the Hubble sequence. One possibility is that gas becomes molecular at a lower critical column density in early than in late-type spiral galaxies.

3.4 Gas Mass Fractions in Spiral Galaxies

The fractional gas content as a function of morphological type was also investigated for the 150 galaxies described above. Using the dynamical results of Rubin et al. (1985) -- where the ratio M_{dyn}/L_B is 6.2 $M_{\odot}/L_{\odot}$ for Sa galaxies, 4.5 $M_{\odot}/L_{\odot}$ for Sb galaxies, and 2.6 $M_{\odot}/L_{\odot}$ for Sc galaxies -- in conjunction with the blue luminosities from RC2, the dynamical masses have been computed for the spiral galaxies for which H_2 and HI masses are known. The mean ratio of the total neutral gas mass $[=M(H_2)+M(HI)]$ to the dynamical mass, M_{gas}/M_{dyn}, ranges from 4% for Sa galaxies to 25% for Scd galaxies, as shown in Figure 6. This indicates that the early type galaxies have processed a considerably larger fraction of their mass into stars than late type galaxies. It is rather remarkable, then, that the present mean gas surface densities for early type galaxies are only a factor of two lower than in late types.

3.5 CO-Radio Continuum Comparisons

Recent observational results suggest a close connection between non-thermal radio continuum emission and star formation in spiral galaxies, based on the small dispersion in the correlations of the global 1.49 GHz emission with Hα emission (Kennicutt 1983) and far infrared continuum emission (Dickey and Salpeter 1984; Helou et al. 1985; Devereux and Eales 1989). Because IRAS data provide no spatial resolution on the star formation

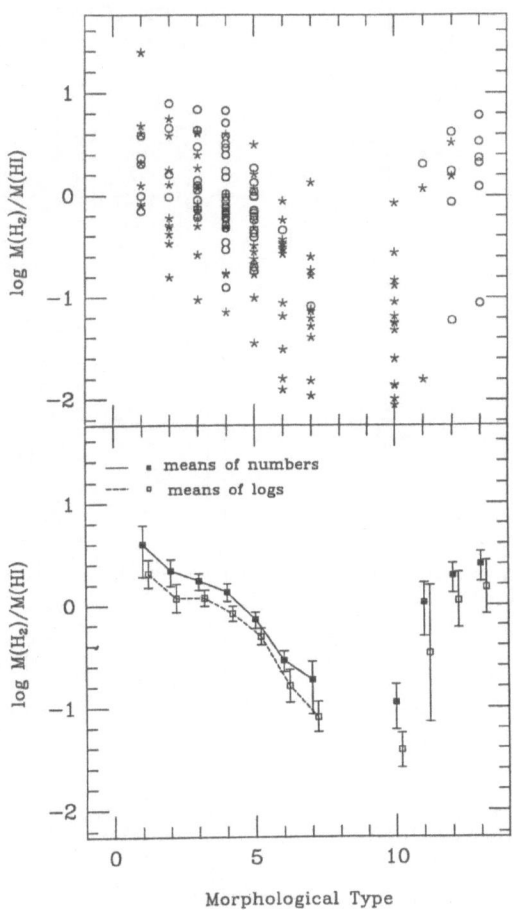

Figure 5. Ratio of molecular to atomic gas mass as a function of morphological type (upper panel) for 170 galaxies (Young and Knezek 1989). Open circles represent galaxies with $L_{total} = (L_{IR} + L_B) > 6 \times 10^{10} \ L_\odot$, while stars indicate lower luminosity galaxies. We have combined RC2 type 0 with type 1, and types 8 and 9 with type 7. Category 10 includes dwarf and irregular galaxies, category 11 includes I0 galaxies (i.e. M82), category 12 represents interacting galaxy pairs, and category 13 represents merger remnants.

The lower panel represents the mean molecular to atomic gas mass ratio as a function of morphological type. Types are the same as in the upper panel. The solid line represents the mean value for each type, while the dashed line represents the mean of the logs. Error bars represent the uncertainty in the mean.

90

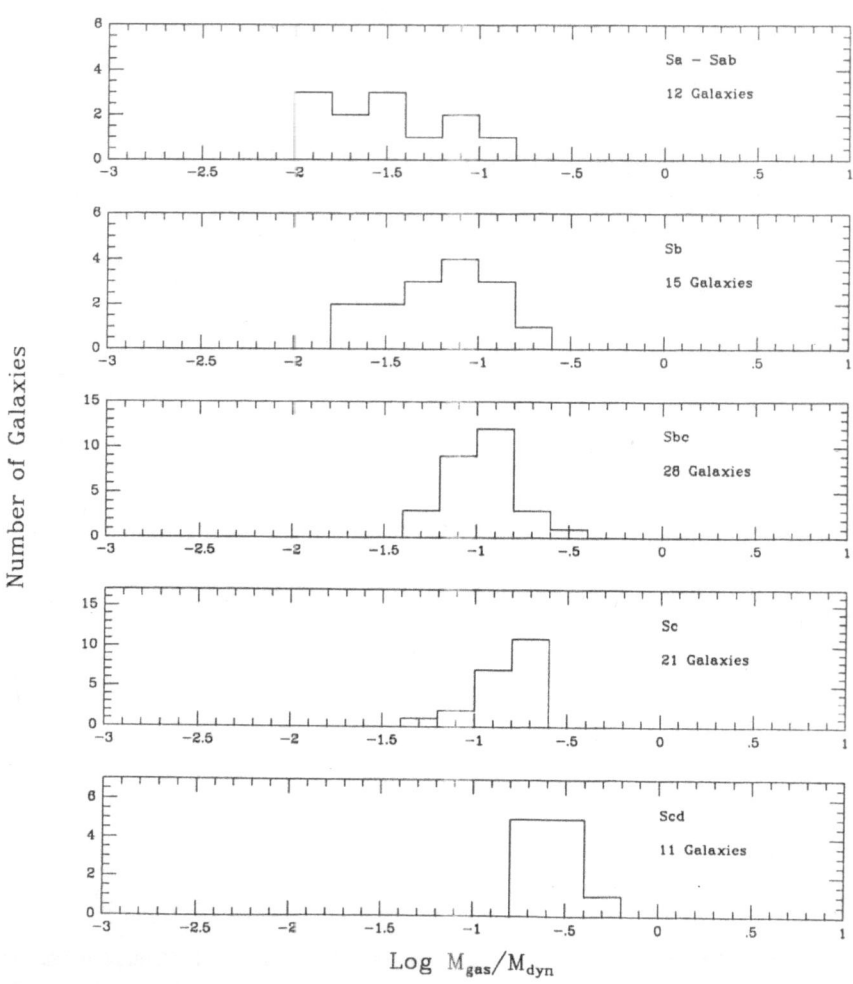

Figure 6. Ratio of total interstellar gas mass (HI+H₂) to dynamical mass for 150 galaxies. Dynamical masses were derived using the results of Rubin et al. (1985) for the ratio M_{dyn}/L_B as a function of morphological type.

rate within galaxies, comparison of the CO and radio continuum distributions in galaxies can in principal provide information on the yield of young stars per unit mass of molecular gas within galaxies.

Studies of the relationship between CO and radio continuum emission in spiral disks have been conducted by Israel and Rowan-Robinson (1984), Adler et al. (1989) and Devereux and Young (1989c). From a comparison of CO and radio distributions in 13 spiral disks, Israel and Rowan-Robinson (1984) found that Sb and Sc galaxies had similar ratios of radio to CO flux, which they interpreted in terms of similar efficiencies of OB star production in molecular clouds. Devereux and Young (1989c) have compared the CO radial distributions in 65 galaxies from the FCRAO Extragalactic CO Survey with the 1.49 GHz radio continuum distributions (Condon 1987). Although the radial distributions of radio continuum and CO emission both decrease by as much as a factor of 100 with increasing radius, the ratio of radio emission to CO emission does not vary by more than a factor of 3 for the majority of spirals in the sample. Devereux and Young (1989c) suggest that this is consistent with star formation efficiencies which do not systematically increase or decrease with radius. This argues against a global dynamical mechanism, such as a spiral density wave, for being the dominant mechanism triggering disk star formation for the majority of spirals in this sample, and supports the view that star formation in disks is a local process.

There are several unusual galaxies in which the ratio of radio to CO emission decreases by more than an order of magnitude (~10% of the galaxies in the Devereux and Young sample), for which large radially decreasing gradients in the star formation efficiency are present. Similarly large radial gradients in the Hα to CO ratio have been noted in M82 and NGC 2146 (Young 1988; Allen and Young 1989).

3.6 CO and Blue Light in Galaxies

Another important measure of the past star formation which has occurred in a disk galaxy is provided by the blue light. There are several problems with blue light as a tracer of star formation, however. First, it is difficult to isolate the contribution to the total blue light which is made by young stars (cf. Searle, Sargent, and Bagnuolo 1973), although approximately one half of the blue light is contributed by stars younger than a few billion years in disk galaxies. Second, extinction can alter both the distribution of blue light and the total flux. Because of the widespread availability of blue magnitudes for galaxies, especially prior to the availability of IRAS data, a number of comparisons of H$_2$ masses and blue luminosities have been made.

The CO distributions exhibit the same radial behavior as the exponential profiles of blue luminosity in several high luminosity, late type spiral galaxies (Young and Scoville 1982a; Solomon et al. 1983; Scoville and Young 1983; Kenney and Young 1989). In addition, the central 5 kpc blue luminosity was found to be proportional to the central 5 kpc H_2 mass for 19 late-type spiral galaxies in the Virgo cluster and in the field (Young and Scoville 1982c; Young, Scoville and Brady 1985). These results apply to measurements of the blue luminosities and H_2 masses in the *same regions* from galaxy to galaxy.

The correlation between *global* measurements of the blue luminosity and H_2 mass has been investigated by Rengarajan and Verma (1986), Stark et al. (1986), Young (1987), Sage and Solomon (1989), and Young, Xie, Kenney and Rice (1989a). These investigators find that the blue luminosity is proportional to the 0.6±0.2 power of the H_2 mass, which is considerably shallower than the unity slope of the correlation between IR luminosity and H_2 mass (see §5.2). Both Sage and Solomon (1989) and Young et al. (1989a) suggest that extinction of the blue light in luminous galaxies, where high central column densities of gas and dust are found, is at least in part responsible for the non unity slope. Additionally, there is little dependence of the $L_B/M(H_2)$ ratio on morphological type among spiral galaxies.

4. Profiles of Molecular and Atomic Gas in Spiral Galaxies

The radial distributions of molecular gas have been measured in ~100 galaxies (see Table 1). In this section, we review the results of measurements of the *azimuthally averaged* CO radial distributions in galaxies as a function of morphological type. This discussion refers to observations at 45"-60" resolution, which in absolute terms represents a resolution ranging from ~1 to 4 kpc.

4.1 Unbarred Galaxies

4.1.1 Radial Distributions

There are over 30 Sc galaxies whose CO radial distributions have been measured along the major axis. Figure 7 shows the CO integrated intensity distributions for a small sample of Sc galaxies, where all observations have been corrected for inclination. Also shown in Figure 7 is the Milky Way CO distribution at 1 kpc resolution (Sanders, Solomon and Scoville 1984), illustrating the central peak, the absence of gas between 1 and 4 kpc, and the molecular annulus between 4 and 8 kpc. None of the face-on Sc galaxies have CO radial distributions which resemble that in the Milky Way. While the dominant feature in the Sc galaxy CO radial distributions is the central peak and intensity decrease with radius, it should also be pointed out that there is a scatter of a factor of 2 in the CO integrated intensities at a particular radius, relative to the mean value measured at that radius (Rickard 1982; Young and Scoville 1982a; Solomon et al. 1983; Scoville and Young

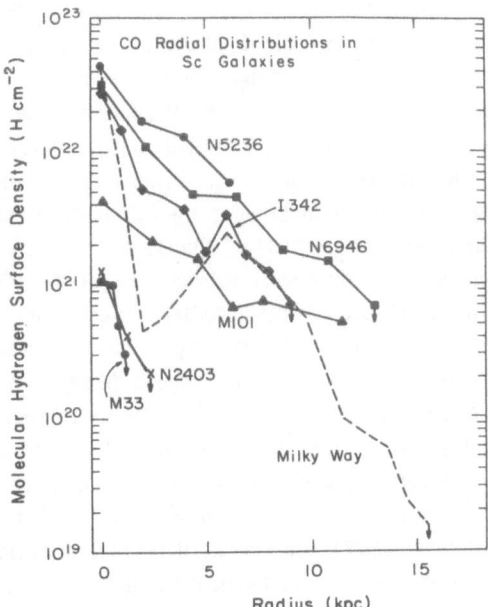

Figure 7. Radial distributions of CO integrated intensity, corrected to face-on, for 6 relatively face-on Sc galaxies (solid lines) and the Milky Way (dashed line -- Sanders, Solomon, and Scoville 1984). Galaxy types and references for the CO distributions are found in Table 1.

1983; Tacconi and Young 1986; Kenney and Young 1988a; Lord and Young 1989).

From a comparison of the H_2 and HI distributions in luminous, face-on spirals, it is apparent that the azimuthally averaged distribution of H_2 is markedly different from that of HI (Morris and Rickard 1982; Young and Scoville 1982a; Scoville and Young 1983; Kenney and Young 1989). The central peaks in the H_2 distributions for Sc galaxies are contrasted with central depressions in the HI distributions and relatively constant HI surface densities across the optical disk at a value of $\sim 10^{21}$ cm $^{-2}$ (cf. Rogstad, Shostak, and Rots 1973). In NGC 6946, for example, the ratio of H_2 to HI surface densities decreases from a central value of 30 to a value of approximately 1 at a radius of 10 kpc (Tacconi and Young 1986). In the luminous Sc galaxies, the molecular gas more than fills in the central holes present in the atomic gas distributions such that the overall gas distribution (H_2+HI) is similar to the exponential light distribution.

The total number of Sb and Sbc galaxies in which the CO distributions have been measured is over 25. Among the Sb galaxies which are at distances of less than 50 Mpc, less than half have been found to exhibit central CO depressions. These galaxies include M31 (Stark 1979), NGC 2841 and 7331 (Young and Scoville 1982b), NGC 1068 (Scoville, Young, and Lucy 1983; Myers and Scoville 1987), NGC 891 (Sofue et al. 1985; Sanders and Young 1989), NGC 4216 (Kenney and Young 1988a), and NGC 488, 2336, 3147, and IC 356 (Young et al. 1989b). Figure 8 illustrates the CO radial distributions for the Sb galaxies with and without central CO depressions. There is no obvious correlation of the presence or absence of the central CO depression with other galaxy properties. In the Sb galaxies for which both H_2 and HI distributions are available in the literature, it is apparent that the molecular surface density can exceed the atomic surface density in the inner disk *even* at the location of the CO minimum (Young and Scoville 1982b).

Major axis observations of the CO distributions in early type spiral galaxies have been published for only a small number of systems. For NGC 4736, the CO radial distribution exhibits a central depression (Garman and Young 1986) similar to those in the Sb/Sbc galaxies. All of the remaining Sa galaxies observed, however, exhibit CO radial distributions which peak in the center and decrease as a function of radius, like the Sc galaxies and the majority of the Sb's. The H_2 surface densities in the centers of the early type galaxies are as high as those found in the centers of the Sc galaxies. The HI distributions for Sa galaxies, on the other hand, have surface densities which peak at 1-4 $M_\odot$ pc^{-2}, rather than in the range 5-10 $M_\odot$ pc^{-2} found for Sc galaxies (Wevers et al. 1986; van Driel 1987). The lower HI surface densities in the early type galaxies are partially responsible for the H_2/HI ratio being higher in the early types.

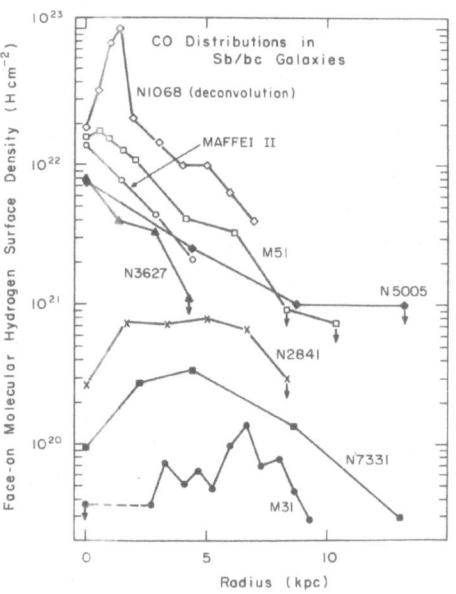

Figure 8. Radial distributions of CO integrated intensity, corrected to face-on, for 8 Sb/Sbc galaxies which are not edge-on. Galaxy types and references for the CO distributions are found in Table 1. The CO distribution for M31 has been scaled down by a factor of 5. Five of these galaxies exhibit central CO depressions.

4.1.2 CO Extents in Galaxies

For the 100 galaxies in the FCRAO Extragalactic CO Survey for which CO emission was detected in 3 or more positions, the CO isophotal diameters have been determined. This isophotal diameter, D_{CO}, is taken to be the diameter at which the face-on CO integrated intensity falls to 1.5 K(T_R) km s^{-1} (see Kenney and Young 1988a). Figure 9 shows a plot of the ratio of CO to optical diameters (D_{CO}/D_{25}) as a function of morphological type. Overall, the mean ratio of CO to optical diameters has a value of 0.5±0.2, with a tendency for the CO diameters in the early type spirals to be slightly smaller than those in the late type spirals.

4.2 Galaxies with CO Bars, and CO in Barred Spiral Galaxies

High resolution CO observations of IC 342 and NGC 6946 indicate that each of these galaxies has a CO bar in the central arcminute (Lo et al. 1984; Ball et al. 1985). Optically these galaxies do not have strong bar morphologies, while the central CO structures have distinctly elongated appearances. These observations, along with the indication that NGC 253 has a bar (Scoville et al. 1985), support the suggestion that the bar serves as the mechanism feeding gas into the center of the galaxy, thus providing the interstellar matter out of which a central burst of star formation may arise (refs).

The best examples of barred spiral galaxies in which the distribution of molecular matter has been mapped are NGC 5236 (M83) (Combes et al. 1978; Lord 1987), NGC 1097 (Gerin et al. 1988), and NGC 1365 (Sandquist et al. 1988). In these galaxies, the CO emission is stronger along the bar rather than perpendicular to it. Thus, these galaxies provide excellent examples of the CO distribution following the large scale structural features in a galaxy. Certainly, more studies of the distribution of molecular clouds in barred spiral galaxies and the star formation efficiency on and off the bar will be useful in determining the roles of bars in the centers of galaxies.

4.3 Spiral Structure in Molecular Cloud Distributions

Since the earliest observations of molecular clouds in galaxies, there has been great interest in determining the extent to which molecular clouds are correlated with spiral arms. Since it is generally accepted that stars form in molecular clouds, the optical patterns seen in galaxies were thought to possibly reflect a spiral pattern in the underlying distribution of molecular mass. Specifically, the spiral density wave believed responsible for the spiral pattern might provide a mechanism for molecular cloud growth through gravitational instabilities or by increasing the incidence of cloud-cloud collisions and coalescence. Secondly, if the density wave in a galaxy were responsible for

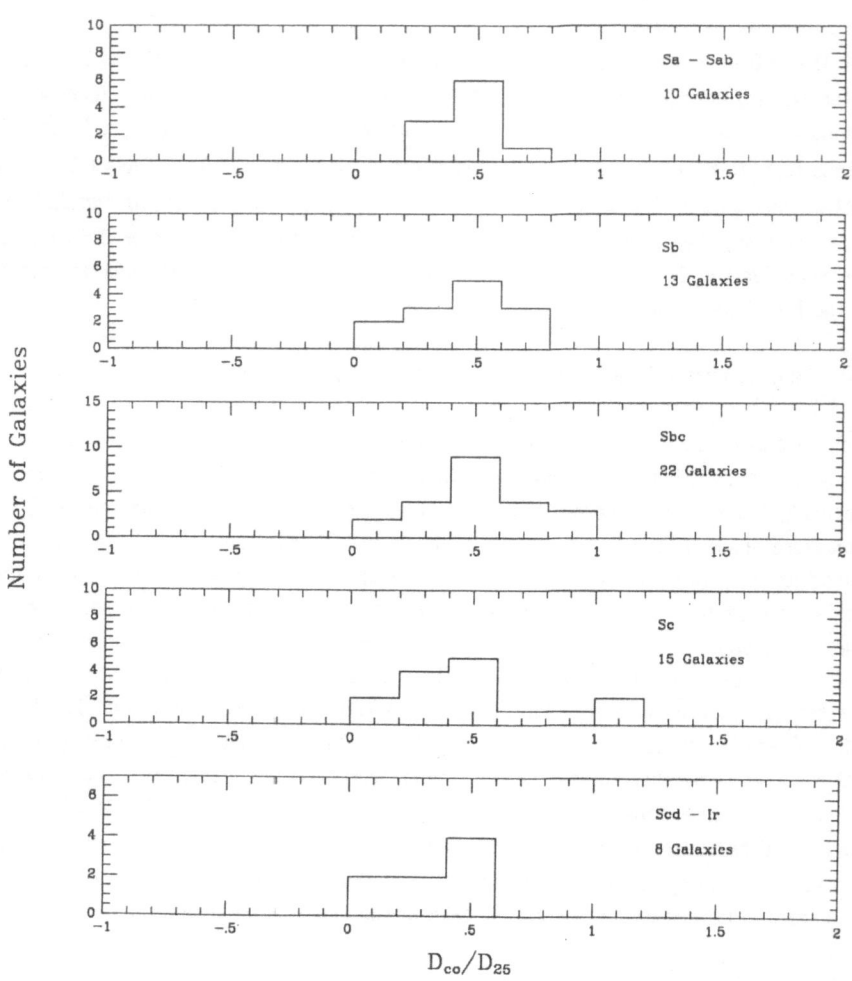

<u>Figure 9,</u> Ratio of CO diameter (out to where the CO integrated intensity falls to 1.5 $K(T_R)$ km s^{-1}) to optical diameter (D_{25}) as a function of morphological type.

triggering star formation through action of density wave shock fronts or cloud collisions, then the subsequent star formation might disrupt molecular clouds and thus produce dissociated gas in the interarm regions of spiral galaxies.

Despite these expectations, the results of numerous studies of molecular clouds in spiral galaxies have shown that emission is widespread and not solely confined to spiral arms. While the initial results, based on observations of galaxies employing 45" to 66" resolution (Rickard and Palmer 1981; Young and Scoville 1982a; Scoville and Young 1983), may have appeared to be the effect of insufficient angular resolution, more recent higher resolution studies have confirmed that ~75% of the molecular emission in galaxies is not confined to spiral arms.

Perhaps *the* best example of a galaxy in which to investigate the extent to which molecular clouds in galaxies are confined to spiral arms is M51. In this galaxy, at a distance of 9.6 Mpc, one arcsecond corresponds to 50 pc. Studies of CO emission in M51 at 33" resolution, i.e. 1.6 kpc (Rydbeck, Hjalmarson, and Rydbeck 1985), show a CO enhancement of 20% in the arms. The contrast is most apparent after subtracting the underlying exponential distribution, which contributes 75% of the emission.

Aperture synthesis mapping of the CO in M51 at 7" resolution using the Owens Valley Interferometer has recently been done by Vogel, Kulkarni, and Scoville (1988) and Rand and Kulkarni (1989). The observations show that ~25% of the emission is concentrated in ridges which are coincident with the dust lanes. In these studies, over 70% of the total emission is "resolved out", and thus represents the extended underlying distribution. Finally, high resolution (~11") single dish observations which detect all of the flux (Guelin et al. 1989) show that molecular clouds are located both in the arm and interarm regions. In their study, the arm-interarm contrast, estimated after accounting for the underlying exponential distribution, is 3-5:1.

Why, then, are the spiral arms so apparent in galaxies, if the underlying distribution of star forming material is relatively smooth? The answer may lie in the *efficiency* with which stars form in the arm regions in spiral galaxies. In order to ascertain whether the star formation efficiency is higher in spiral arms, it is necessary study the azimuthal structure in galaxies, as described in §5.5.

4.4. Local Group Galaxies

4.4.1 M31

M31 is one of the prime galaxies in which to study the detailed distribution of molecular clouds due to its proximity. Because M31 covers such a large area on the sky, only selected regions have been observed in CO, whether to determine the CO radial distribution, to investigate the relationship between CO and the spiral arms, or to resolve individual GMCs. The first investigation of CO in M31 consisted of observations at 90" resolution along the minor axis (Stark 1979). These observations revealed an absence of molecular clouds in the center, and an annulus of emission coincident with the optical arms. Along the minor axis, the inferred H_2 surface density is everywhere comparable to or lower than that of HI (Brinks 1985).

Numerous investigations of the concentration of molecular clouds in spiral arms have been conducted for M31 (Combes et al. 1977a,b; Boulanger 1984; Ryden and Stark 1986; Lada et al. 1988). In the outer disk, the CO emission is concentrated in an annulus which is coincident with HII regions and optical arms. The observed contrast between the molecular surface density on and off the arms, over a linear scale of about 300 pc, is 7±4 to 1, as inferred from Figure 4 of Ryden and Stark (1986). Ryden and Stark also find kinematic evidence of streaming across the spiral arms.

In a recent CO study, in which individual GMCs in M31 were partially resolved, Lada et al. (1988) inferred sizes, densities, and masses similar to those of GMCs in the solar neighborhood. Comparison with HI observations led Lada et al. to conclude that the GMC-HI complexes have total gas (H_2+HI) masses on the order of 10^6 $M_\odot$, of which ~30% is atomic. They suggest that the hydrogen column density threshold for H_2 formation is similar to that estimated for the Milky Way.

4.4.2 The Magellanic Clouds

The Large and Small Magellanic Clouds are sufficiently nearby that they provide the opportunity to achieve high spatial resolution (1'=23 pc) relative to the sizes of the molecular clouds within them. Early CO observations of the LMC have been reported by Israel (1984) and Israel et al. (1986). The most comprehensive study is that of Cohen et al. (1988), who obtained a fully sampled CO map at 8.8' resolution (200 pc) of the central 6°x6° in the LMC. Cohen et al. detected CO emission over approximately 10% of the region studied, with emission dominated by a molecular cloud complex extending several kpc south of 30 Doradus. Extended high velocity CO emission was also seen in this region, possibly arising from gas accelerated by stellar winds and supernovae. Cohen et al. also derived a CO luminosity-linewidth relation for the LMC molecular clouds, although the spatial resolution of 200 pc for their observations is insufficient to resolve individual GMCs.

5. Star Formation Rates and Efficiencies in Galaxies

5.1 Star Formation Tracers in Galaxies

The rate at which the ISM in a galaxy evolves is probably determined in large part by the rate of star formation within molecular clouds. The most widely available measures of the star formation activity in galaxies are global Hα fluxes (Kennicutt and Kent 1983; Bushouse 1986; Kennicutt et al. 1987; Allen and Young 1989) and IR fluxes from the IRAS survey (Lonsdale et al. 1985). While the Hα emission traces the young massive stars, the disadvantages of Hα are several. First, the emission suffers extinction, and will therefore provide a lower limit to the SFR. Second, the emission traces only high mass stars, so that assumptions about the initial mass function (IMF) must be made in order to deduce the overall SFR. On the other hand, the principal difficulty with the infrared emission as a tracer of the SFR is that there are dust heating sources in addition to young stars, so that the IR luminosity provides an upper limit to the SFR. The infrared emission is also sensitive the high mass end of the IMF.

The comparison of IR and Hα luminosities in galaxies has been addressed by Lonsdale and Helou (1987) and by Devereux and Young (1989b). Devereux and Young find that the Hα and far-IR luminosities can both be attributed to young high mass stars since the ratio of Hα to bolometric luminosity is similar to that expected from HII regions around O and B stars, with 1 magnitude of extinction. Figure 10 shows the comparison of the IR and Hα luminosities for the galaxies in the Hα survey of Kennicutt and Kent (1983), from which it is apparent that the IR and Hα luminosities can be

Figure 10. Comparison of Hα emission line and the IRAS 40-120 μm luminosities for spiral galaxies (Devereux and Young 1989b) using the Hα data of Kennicutt and Kent (1983). The points are coded by morphological type withstars for Sa's, open circles for Sb's, and filled circles for Sc's and later. Extrapolating the IRAS 40-120 μm luminosity to 1000 μm (L_{Bol}) and correcting the observed Hα emission line luminosity by 1 magnitude of extinction translates the data points by an amount equivalent to that illustrated by the vector R. The diagonal lines show the ratio of Hα to bolometric luminosity expected for HII regions powered by stars of masses 40, 20, 16, and 6 $M_\odot$.

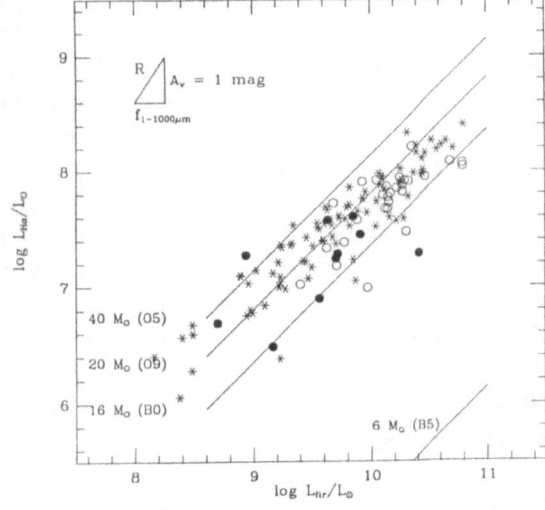

generated by stars more massive than 6 $M_\odot$.

Because of the sensitivity and all-sky coverage of IRAS, and because of extinction reducing the Hα luminosities, the IRAS data provide the most reliable and abundant measures of the high mass star formation rates in spiral galaxies presently available.

5.2 Star Formation Rate Per Unit Molecular Mass

5.2.1 Comparisons of IR Luminosities and H_2 Masses

In this section, IR luminosities are used to trace global SFR's in galaxies. It is the comparison of the IR luminosity with the mass of molecular gas that enables us to deduce the star formation efficiency (SFE=yield of young massive stars per unit mass of molecular gas) from galaxy to galaxy or as a function of radius in individual galaxies.

A large number of CO-IR comparisons have now been reported in the literature (Rickard and Harvey 1984; Young et al. 1984; Sanders and Mirabel 1985; Young et al. 1986a, 1986b; Sanders et al. 1986, 1987; Tacconi and Young 1987; Solomon and Sage 1988; Kenney and Young 1988b; Young et al. 1989a). All of the above correlations exhibit several orders of magnitude of scatter, which is much more than that expected on the basis of measurement uncertainties. Several studies have shown that one order of magnitude of this scatter arises from variations in the star formation efficiency which are related to galaxian environment, as described below.

Figure 11 shows a plot of the infrared (IR) luminosities and CO luminosities for 150 galaxies from several studies (Stark 1979; Young et al. 1986a, 1986b; Sanders et al. 1986; Kenney and Young 1988a; Young et al. 1989a; Thronson et al. 1989). The CO luminosities for all of these galaxies were derived from CO fluxes using the approach of Kenney and Young (1988a), while the infrared luminosities are based on coadded IRAS data and computed following the method of Lonsdale et al. (1985). The galaxies shown in Figure 11 are coded by dust temperature, from which it is clear that dust temperature variations from galaxy to galaxy are responsible for some of the scatter. There are two very simple reasons for this. First, as shown in §3.1, the ratio of molecular gas mass to the warm dust mass is relatively constant from galaxy to galaxy. Second, the IR luminosity has a strong temperature dependence (T^5 for a λ^{-1} emissivity law), while the CO luminosity has only a weak one (see eq. 3). Were it not for *both* of these characteristics, the separation according to dust temperature shown in Figure 11 would not be apparent. Such a separation by dust temperature is <u>not</u> found for the comparison of IR luminosities and HI masses, from which it is concluded that the dust emission in these galaxies is more closely tied to the molecular than to the total atomic gas content. Kenney and Young (1988a) and Devereux and Young (1989a) have shown that the IR emission in galaxies is more closely tied to the *inner disk* gas (molecular plus atomic) than to the total gas content, indicating that it is the *location* and not the phase of the gas which is important.

Among spiral galaxies of type Sbc and later, considerable variation is observed in the ratio $L_{IR}/M(H_2)$. Those galaxies with high yields of massive stars per unit molecular gas mass are also found to have high $H\alpha$ emission line equivalent widths, while galaxies with low values of $L_{IR}/M(H_2)$ have low equivalent widths (Young et al. 1989a). The $H\alpha$ equivalent width ($EWH\alpha$) is the $H\alpha$ line flux normalized by the red continuum flux; to the extent that the bulge to disk ratio is similar for the late type spirals, the variations in $EWH\alpha$ among late type spirals reflect variations in the SFR. Thus, the correlation of $EWH\alpha$ with $L_{IR}/M(H_2)$ indicates that late type spiral galaxies which are presently forming many high mass stars are doing so through efficient conversion of gas into stars.

<u>Figure 11.</u> Comparison of total CO and IR luminosities for 150 galaxies from several studies (Stark 1979; Young et al. 1986a,1986b; Sanders et al. 1986a; Kenney and Young 1988a; Young et al. 1989a; Thronson et al. 1989a). The CO luminosities have all been derived in the same way following the method described in Kenney and Young (1988a); the IR luminosities are from coadded IRAS Survey data. Points are coded by dust temperature, with stars for $T_{dust} > 40$ K, filled circles for T_{dust} between 30 and 40 K, and open circles for $T_{dust} < 30$ K. Total H_2 masses are given by the scale at the top. The dashed lines labeled a, b, and c represent values of $L_{IR}/M(H_2)$ of 100, 10, and 1 $L_\odot/M_\odot$, respectively.

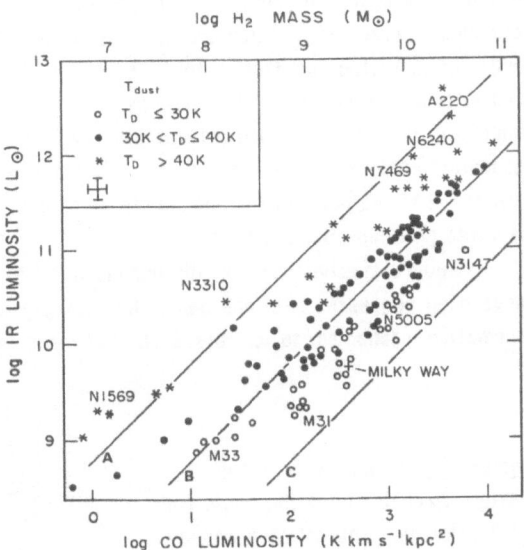

5.2.2 The Star Formation Efficiency as a Function of Environment

In the studies of Sanders and Mirabel (1985) and Young et al. (1986a), it was recognized that high values of the ratio $L_{IR}/M(H_2)$, or high SFE's, were present in interacting galaxies. This led to several studies of the efficiency of star formation in interacting/merging galaxies versus isolated galaxies (Young et al. 1986b; Solomon and Sage 1988), in which it was found that isolated galaxies have a mean $L_{IR}/M(H_2)$ of 12 ± 3 $L_\odot/M_\odot$ while interacting galaxies have a mean value of 78 ± 14 $L_\odot/M_\odot$. If the infrared luminosity measures the rate of high mass star formation in a galaxy, then these results imply that the perturbations due to galaxy-galaxy interactions result in more efficient production of massive stars in these systems than in their isolated counterparts. Solomon and Sage (1988) also showed that only the most closely interacting systems had a noticeable enhancement in the SFE relative to isolated galaxies.

Figure 12 shows a plot of the ratio $L_{IR}/M(H_2)$ versus dust temperature for 150 galaxies, where the points are coded by the galaxy's environment (isolated, group/cluster, interacting/merging galaxies). Here the effects of galaxy size are normalized out, and intrinsic variations in the SFE from galaxy to galaxy are apparent. While the most strongly interacting galaxies are responsible for many of the high SFE's, there is an order of magnitude scatter in the SFE observed among the isolated galaxies and among the interacting galaxies. This scatter, which is considerably larger than the measurement uncertainty in the SFE, may reflect time evolution of the IR luminosity and H_2 content associated with bursts of star formation in the interacting galaxies (Young et al. 1986a, 1986b), and the possibility that bursts of star formation control the evolution even in isolated galaxies (Scalo 1987).

Several mechanisms for the enhancement in the SFE in interacting/merging galaxies have been suggested. On one hand, the galaxy-galaxy interaction may cause a new star formation mechanism to operate such that more stars form per unit molecular mass.

Figure 12. Comparison of the ratio $L_{IR}/M(H_2)$ with S_{60}/S_{100} for 150 galaxies. The galaxies coded by environment as follows: open circles for isolated galaxies, triangles for Virgo galaxies, stars for interacting/merging galaxies, and open squares for other galaxies.

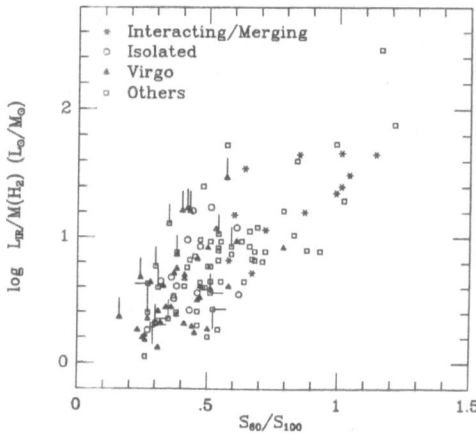

Alternatively, the physical process which causes high mass stars to form in merging/interacting galaxies may be no different than that in isolated galaxies, but the interaction may enhance its effectiveness. The explanation which we favor is the latter, where the interaction causes a star formation mechanism already present to operate at an increased level, as opposed to invoking a new mechanism.

One process which probably becomes important during galaxy-galaxy interactions is that of cloud-cloud collisions. Noguchi and Ishibashi (1986) and Olsen and Kwan (1989) have made numerical simulations of galaxy-galaxy interactions including gas clouds as well as stars. They find that the cloud-cloud collision rate increases as bridges and tails develop during violent galaxy-galaxy encounters. In the simulations of Noguchi and Ishibashi, the cloud collision rate reaches a maximum value of an order of magnitude greater than the pre-encounter value 3×10^8 years after closest approach of the perturber. If cloud-cloud collisions are responsible for high mass star formation in galaxies (Scoville, Sanders, and Clemens 1986), then the enhanced rate of cloud-cloud collisions in interacting galaxies should result in an increase in the SFR per unit H_2 mass.

Because of the low spatial resolution of the IRAS detectors, the FIR luminosity derived from the 60 and 100 µm flux densities represents the global FIR emission only; it does not contain information on the distribution of star formation within a galaxy. In order to determine the regions within galaxies which give rise to the most efficient star formation, it is necessary to compare the CO distributions in galaxies with the distribution of star formation deduced, for example, from Hα CCD imaging.

5.2.3 The Star Formation Efficiency as a Function of Morphological Type

In order to examine the morphological type dependence of the SFE, a number of investigators have determined the $L_{IR}/M(H_2)$ ratio for early and late type spirals. Rengarajan and Verma (1986), Young et al. (1989a), Thronson et al. (1989), and Wiklind and Henkel (1989) all find that early and late type spirals have similar global SFE's. Allen, Young, Kenney, and Kleinmann (1989) find similar mean ratios of $L_{IR}/M(H_2)$ for types Sa-Scd; the mean value is 5 $L_\odot/M_\odot$. Figure 13 illustrates the type dependence of the SFE from the ratio $L_{IR}/M(H_2)$. The similarity in the yield of high mass stars per unit mass of molecular gas as a function of morphological type among spirals indicates that the global SFE's in spiral disks do not depend on morphological type.

5.3 Gas Depletion Times

If the star formation rates and gas supplies in galaxies are known, it is possible to compute the length of time for which the present SFR can be sustained. Assuming that the IR luminosities of galaxies are produced by O, B, and A stars while on the main sequence, and assuming that the stellar energy is generated by processing 13% of the

104

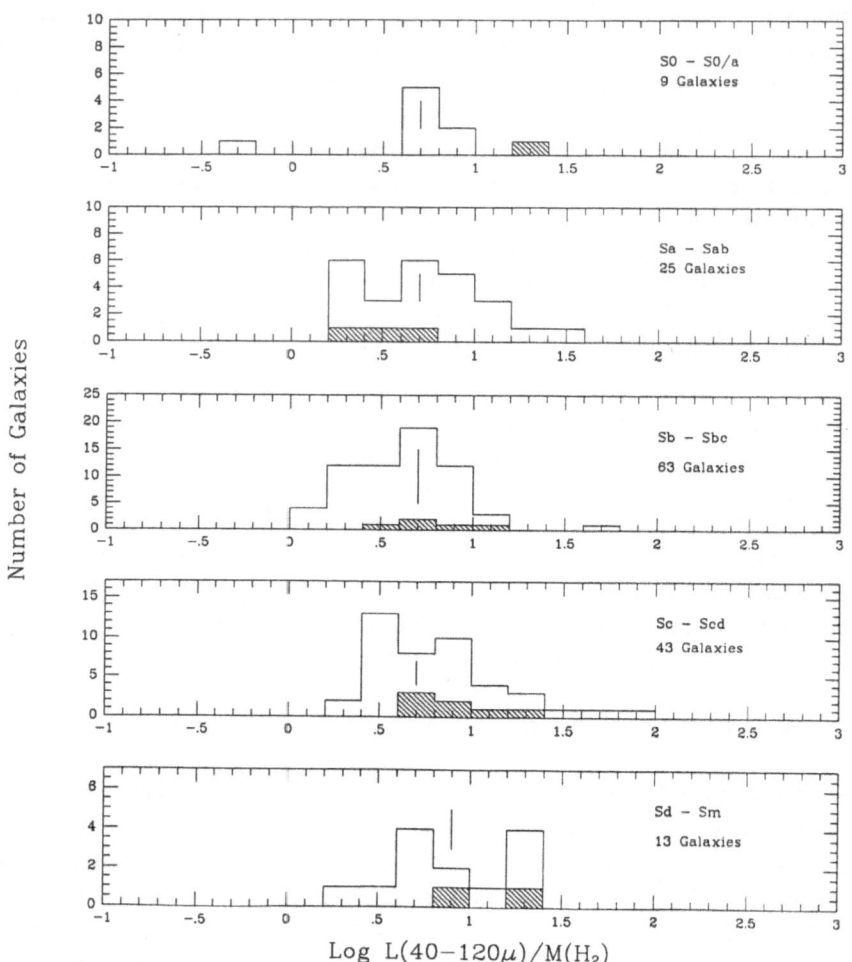

<u>Figure 13</u>. Histograms of the ratio $L_{IR}/M(H_2)$ according to morphological type among spiral galaxies. Hatched areas represent lower limits to the ratio. The vertical tic mark in each panel represents the median value of $L_{IR}/M(H_2)$ for that type.

stellar mass through the CNO cycle, Scoville and Young (1983) and Sandage (1986) have shown that the SFR's are given by

$$\dot{M}_{O,B,A} = 7.7 \times 10^{-11} \, L_{O,B,A}/L_\odot \quad [M_\odot \, yr^{-1}] \quad (4)$$

where $L_{O,B,A}$ is the luminosity in young stars, and $\dot{M}_{O,B,A}$ is the SFR.

Gas depletion times are given by $M_{gas}/\dot{M}_{O,B,A}$, where M_{gas} is the mass of gas available for star formation. For the galaxies studied by Young et al. (1986a, 1989a) and Solomon and Sage (1988), global depletion times range from 10^8 years to almost 10^{10} years. These depletion times do not take into account the formation of low mass stars or the recycling of gas by high mass stars. One implicit assumption which enters this calculation is that the present SFR is maintained into the future. If the episodes of intense star formation in interacting galaxies are relatively short-lived compared to 10^8 years, then the gas depletion timescales for these objects could be longer. If the star formation rate *per unit mass of gas* is constant, so that the SFR decreases as does the gas supply, the computed gas depletion times are somewhat longer (Lord and Young 1989), and in fact are the e-folding times for consuming the gas. If the IMF is weighted toward the high mass end, then the gas supply will last considerably longer due to recycling. In any case, galaxies appear to be in the latter portions of their lives.

5.4 The Star Formation History of an Individual Galaxy: NGC 6946

A complimentary approach to understanding galaxy evolution, as opposed to comparing global galaxy properties, is to compare the detailed distributions within the disks of individual galaxies. From a comparison of the CO distribution with past (blue light) and present (Hα or IR) tracers of star formation, we can deduce the star formation history of a galaxy. NGC 6946 and M51 are two nearby well-studied nearby face-on late-type spiral galaxies which are ideal for comparing the distributions of young stars and gas within both the disks and the spiral arms. NGC 6946 is a luminous example of spiral type Scd; it is quite isolated, with no significant companions within 1 Mpc.

In NGC 6946, observations have been made of the distributions of Hα, HI, CO, radio continuum, blue light, and far-infrared continuum (for references see Tacconi and Young 1986). In both NGC 6946 and M51, the azimuthally averaged distributions -- FIR, blue light, Hα, radio continuum, and CO -- show the same radial falloff, which is much steeper than that of the atomic gas (Young and Scoville 1982a; Scoville and Young 1983; Solomon et al. 1983; Tacconi and Young 1986; Kenney and Young 1989; Lord and Young 1989), as shown in Figure 14.

The fact that the azimuthally averaged CO, blue light, Hα, FIR and radio continuum distributions all show similar radial gradients when sampled at ~2 kpc resolution is

106

Figure 14. Comparison of the
radial distributions of CO (H_2), HI,
Hα, blue and radio continuum in
NGC 6946 from Tacconi and Young
(1986). All intensity scales are
relative except that for the HI,
which is plotted relative to H_2,
assuming the H_2 surface densities
are derived as in §2.2.1.

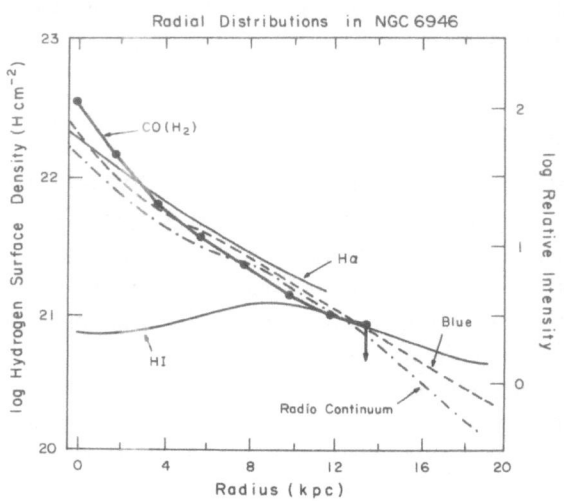

significant in terms of galaxy evolution. If the blue light measures the star formation
rate integrated over the last $\sim 2 \times 10^9$ years (cf. Searle, Sargent, and Bagnuolo 1973;
Gallagher, Hunter, and Tutukov 1984; Sandage 1986), and the Hα flux measures the rate
of formation of high mass stars, the fact that the blue/CO and Hα/CO ratios are
constant as a function of radius indicates that *both the present-day formation rate for
high mass stars and the long-term integrated formation rate for intermediate mass
stars are proportional to the available supply of molecular gas,* i.e. *the star formation
efficiency is constant.*

In contrast to the similar radial profiles for tracers of past, present, and future
star formation in NGC 6946, the H_2 and HI distributions are entirely different, as
discussed above (see §4.4.1). If the ratio of H_2/HI surface densities is a measure of
the efficiency with which molecular clouds form, the radial decrease in the H_2/HI ratio
indicates that molecular cloud formation proceeds most efficiently or that the clouds have
longer lifetimes in the centers of luminous galaxies.

Schmidt (1959) has parameterized the large scale star formation rate in galaxies
as depending on the gas density to the power n, SFR $\propto \rho^n$. Based on a comparison of
the HI and stellar scale heights and distributions in the Milky Way, Schmidt concluded that
n has a value of ~ 2. However, since 1959, the discovery of the molecular component of
the interstellar medium, and the observation that star formation is linked to the molecular
gas, leads to the parameterization of the star formation process in terms of two
separate steps -- star formation from molecular clouds, and molecular cloud formation
from atomic gas. Many authors have made this distinction between star and cloud
formation (cf. Elmegreen 1979; Seiden 1983; Tacconi and Young 1986; Wyse 1986;

Wyse and Silk 1989; although see also Kennicutt 1989), as the observational evidence suggests.

We can then write SFR $\propto \rho(H_2)^n$ and CFR $\propto \rho(HI)^m$, where CFR is the molecular cloud formation rate, and $\rho(H_2)$ and $\rho(HI)$ are the large-scale molecular and atomic gas densities. The one-to-one correspondence between the CO and Hα surface density distributions in NGC 6946 indicates that the exponent n has a value of 1 in that galaxy, and that this value does not change with radius. In contrast, the decreasing $\sigma(H_2)/\sigma(HI)$ ratio with radius indicates that the value of the exponent m decreases with radius, so that molecular cloud formation is least efficient in the outer parts of the galaxy. Tacconi and Young (1986) suggest that this may be due to an increase in the HI scale height and a corresponding decrease in the HI volume density in the outer parts of the galaxy. Furthermore, they suggest that the optical edges of galaxies may reflect the edges of the molecular disks, where the efficiency of molecular cloud formation from atomic gas is reduced. However, once formed, the molecular clouds in NGC 6946 appear to form stars in proportion to the available mass of molecular gas.

Several models for the star formation efficiency within galactic disks have been proposed (Gusten and Mezger 1982; Dopita 1985). Lord and Young (1989) show that the density-wave stellar-production model of Gusten and Mezger predicts that the SFE should depend on $[\Omega(R)-\Omega_p]$, where $\Omega(R)$ is the angular velocity and Ω_p is the pattern speed. The quantity $\Omega(R)-\Omega_p$ decreases as a function of radius unlike the constant SFE observed as a function of radius. Lord and Young (1989) also show that the model dependent on the pressure of the ISM (Dopita 1985) predicts a SFE which is proportional to the stellar surface density, which also decreases with radius unlike the observations. Thus, these models for the SFE in galactic disks do not explain the observation that the SFE is constant in spiral disks, as shown in Figure 15.

5.5 The Star Formation Efficiency On and Off the Spiral Arms

Lord and Young (1989) and Tacconi and Young (1989) have investigated the SFE on and off the spiral arms in M51 and NGC 6946, respectively, through a comparison of CO and Hα emission, both sampled at 45" resolution. These studies do not address the arm/interarm contrast in the molecular cloud distribution, but do address the SFE on and off the arms at 45" resolution. Figure 16 shows the distributions of I-band light, Hα/H2, H2/HI, and Hα/blue as a function of azimuth in NGC 6946 at a radius of 1.5' (Tacconi and Young 1989). At this radius, the Hα arms at 45" resolution in NGC 6946 stand out as factor of ~4 enhancements relative to interarm locations; in CO at this resolution, the spiral arms are much less apparent. Thus, at 45" resolution, the ratio of ionized to molecular gas in NGC 6946, or the SFE, is enhanced by more than a factor of 2 on the spiral arms relative to the interarm regions. Lord and Young (1989) find a similar result for the enhancement in the SFE on the arms of M51. Tacconi and Young

(1989) have also investigated radial variations in the star formation efficiency on and off the spiral arms. They find that the arm-interarm contrast in Hα/H$_2$ increases with radius to a value which exceeds 10 beyond a radius of 2'.

The results of these studies indicate, therefore, that *the yield in young massive stars per unit mass of H$_2$ in spiral arms of luminous galaxies is enhanced relative to the interarms regions by at least a factor of 2.* Elevated Hα/CO ratios at particular locations in a galaxy indicate that there is a non-linear dependence of the star formation rate on the gas surface density. One possible mechanism which can explain the enhanced SFE is that of cloud-cloud collisions, which should depend on the square of the cloud number density. Lord and Young (1989) show that their observations of the Hα/CO ratio on and off the spiral arms are consistent with the SFR depending on the square of the H$_2$ surface density on the spiral arms, although their observations do not tightly constrain the value of the exponent.

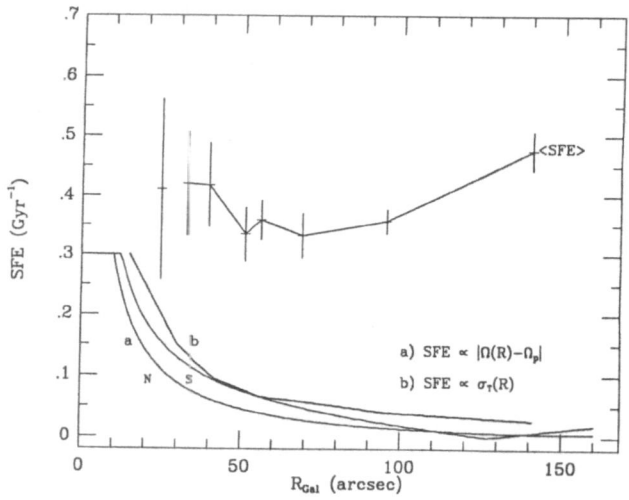

Figure 15. The observed and modeled star formation rate (SFR) per unit gas surface density (σ_p) as a function of galactic radius in M51 from Lord and Young (1989). The upper curve shows the current value of <SFR>/<σ_p> averaged over each radial bin, where the SFR is extrapolated from the massive SFR using an extended Miller-Scalo IMF (cf. Kennicutt 1983). Curves a, (N and S), give the normalized angular velocity of material passing through the spiral density wave using the North and South rotation curves of Goad, De Veny, and Goad (1979) and the pattern speed from Tully (1974). Curve b shows the total mass distribution with radius, an important parameter in the SFR ~ σ_{gas} × σ_{total} model of Dopita (1985).

R = 1′.50

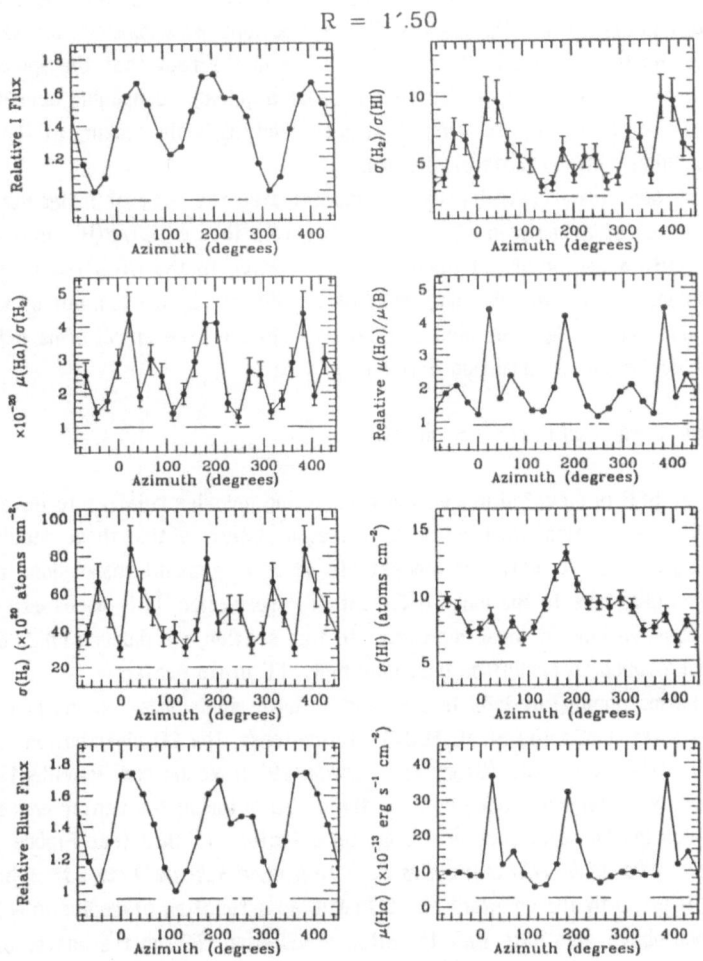

Figure 16. Azimuthal distributions of I-band, Hα, H₂, Hα/H₂, H₂/HI, and Hα/blue at a radius of 1.5′ in the disk of the Sc galaxy NGC 6946 from Tacconi and Young (1989), all derived at 45″ resolution. Azimuths are defined from north through east, with north at 0°. The solid horizontal lines in the lower portion of each panel indicate the locations of the I-band spiral arms before smoothing the data to 45″ resolution. The Hα/H₂ ratio as a function of azimuth shows enhancements in the spiral arm regions.

How is it, then, that the azimuthally averaged Hα and CO radial distributions are similar in NGC 6946 and M51, while the Hα and CO distributions *as a function of* azimuth show enhancements on the spiral arms? The answer lies in the fact that the spiral arms consist of a small fraction of the surface area of a galaxy; averaging azimuthally averages out the effects of the arms in both Hα and CO, while looking at the azimuthal structure highlights the effects of the arms.

Tacconi and Young (1989) have also investigated the ratio of molecular to atomic gas as a function of azimuth in NGC 6946. They find that $\sigma(H_2)/\sigma(HI)$ is highest on the spiral arms, with a factor of ≤ 10 enhancement relative to the $\sigma(H_2)/\sigma(HI)$ ratio in the interarm regions. If the atomic component of the ISM is the medium out of which molecular clouds form, then the high $\sigma(H_2)/\sigma(HI)$ ratio on the spiral arms indicates an enhanced molecular cloud formation efficiency there.

5.6 The Star Formation Efficiency in M82

Given that the SFE is elevated in merging/interacting galaxies relative to isolated galaxies (see §5.2.2), one question which naturally arises is *where* within these peculiar galaxies the efficiency of star formation is highest. In order to pinpoint the regions *within* galaxies which give rise to the most efficient star formation, it is necessary to compare Hα and CO observations in these systems. In this section, we discuss M82, a galaxy which has been found to exhibit an unusually high SFE in its center.

The Hα morphology of M82 is well-known, with emission extending primarily along the minor axis (cf. McCarthy et al. 1987). In contrast, the CO distribution is more extended along the major axis (Stark and Carlson 1984; Young and Scoville 1984). While the CO integrated intensities drop by a factor of 50 between the center and a radius of 3', the Hα flux in the same aperture drops by a factor of 2000 (see Figure 17). Thus, *the observed ratio of Hα/CO decreases by a factor of 400 with radius!* This result is in sharp contrast with the constant Hα/CO ratio as a function of radius in NGC 6946 and M51 described above. We note that the intense radiation field in the center of M82 will heat the gas to some extent. However, as pointed out in §2.1, the ratio $T/\sqrt{p}$ for the hot dense gas in M82 is within a factor of 2 of the ratio for the cooler molecular component, so that the H$_2$ mass estimates will be reasonably accurate.

The high central ratio of ionized to molecular gas in M82 suggests a high efficiency of high mass star formation in that region. Figure 17 shows the SFE as a function of radius in M82. If the present star formation rates are maintained relative to the present gas supply, the gas depletion time in the center of the galaxy will be less than 10^9 years, while the depletion time in the outer parts will be close to 10^{11} years. Thus, the consumption and ionization of gas through star formation may produce a large number of stars and a gas distribution with a central depression. Such a scenario may also occur

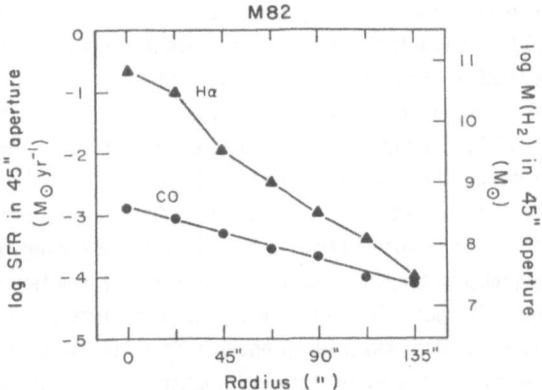

<u>Figure 17</u>. Radial distributions of Hα, CO, and Hα/CO along the major axis of M82. This galaxy shows a dramatic decrease in the Hα/CO ratio as a function of radius, in marked contrast to that seen in NGC 6946 and M51.

early in the evolution of an early type spiral when the burst of star formation occurs which forms the bulge of the galaxy.

What is the cause of the enhanced central SFE in M82? It is noteworthy that the mass of H_2 in the central 45" is ~20% of the dynamical mass in the same region. Given such a large central gas concentration, one expects a high rate of cloud-cloud collisions. If cloud-cloud collisions lead to the formation of high mass stars, as discussed in relation to interacting/merging galaxies, then one expects a high star formation efficiency in regions of high gas concentration, where the cloud-cloud velocity dispersion is high. The mechanisms for producing such a high central gas concentration include gas flow along bars in barred spiral galaxies, and galaxy-galaxy interactions which cause the gas clouds to lose angular momentum and fall to the center.

6. Summary

6.1 The Star Formation Efficiency

In summary, within galaxies at 2-4 kpc resolution, the SFE is found to be constant as a function of radius in the disks of NGC 6946, M51, and numerous luminous late-type Virgo spirals (Scoville and Young 1983; DeGioia-Eastwood et al. 1984; Tacconi and Young 1986; Lord and Young 1989; Kenney and Young 1989). In contrast, the SFE is found to be elevated both on the spiral arms in NGC 6946 and M51 relative to the interarm regions (Tacconi 1987; Lord 1987), and in the center of M82 relative to the disk (Young 1988) where there is a strong concentration of molecular gas (Lo et al. 1987; Nakai et al. 1987). From galaxy to galaxy, the mean value of the global SFE shows no variation with morphological type (Rengarajan and Verma 1986; Young et al. 1989a; Thronson et al. 1989; Wiklind and Henkel 1989). In contrast, the SFE is elevated by a factor of ~7 in interacting/merging galaxies relative to isolated galaxies (Young et al. 1986; Solomon and Sage 1988).

These observations are consistent with star formation in the disk of a galaxy being a local process, in which the rate of star formation depends on the available supply of molecular gas. The SFE is independent of location in the disk of a late type spiral and independent of morphology because star formation is inherently a small scale process. The exceptions to this scenario, where the SFE is enhanced, occur in regions of high gas concentration -- in spiral arms, in the centers of peculiar galaxies, and globally in interacting/merging galaxies. It has been suggested that the enhanced rate of cloud-cloud collisions which should occur in these regions of high molecular gas concentration may lead to an enhanced rate of production of high mass stars per unit molecular gas mass (Scoville et al. 1986). Such a situation could arise either through the heating of the gas during collisions, in which case the first stars to form subsequently would be high mass stars (Silk 1987; Scoville et al. 1986), or through cloud growth as a result of collisions, to the point where self-gravity overcomes cloud support (Shu, Adams, and Lisano 1987).

6.2 The H_2/HI Ratio

Within luminous Sc galaxies at 2-4 kpc resolution, the ratio of molecular to atomic gas surface densities decreases by a factor of ~30 as a function of radius (Morris and Rickard 982; Young and Scoville 1982a; Scoville and Young 1983; Solomon et al. 1983; Tacconi and Young 1986). At a particular radius in NGC 6946, the $\sigma(H_2)/\sigma(HI)$ ratio is higher on the spiral arms than in the interarm regions, and the arm/interarm contrast in the $\sigma(H_2)/\sigma(HI)$ ratio is found to increase with radius (Tacconi and Young 1989). Globally, the ratio $M(H_2)/M(HI)$ decreases by a factor of 20 with morphological type from Sa→Sd (Young and Knezek 1989), and the ratio is higher in high than in low luminosity Sc galaxies (Young et al. 1982c). All of the regions with high H_2/HI ratios have one feature in common -- they occur in regions of higher total mass density, or deeper gravitational potential, whether it is the center versus the disk of an Sc galaxy, the arm versus interarm in an Sc galaxy, or an Sa versus an Sc galaxy.

In order to interpret the above observations, it is necessary to establish whether the HI distribution and H_2/HI ratio in a galaxy reflect the photo-destruction of H_2 by the UV radiation field (cf. Shaya and Federman 1987), or the efficiency of molecular cloud formation (cf. Tacconi and Young 1986). In order to differentiate between these two possibilities, it is necessary to compare the distributions of H_2, HI, and Hα within the disks of galaxies. In particular, the constant Hα/CO ratio with radius in NGC 6946 implies that the UV flux per unit H_2 should also be constant. This should lead to the dissociation of the same fraction of H_2 at all radii, in which case the HI distribution should closely resemble that of H_2. However, the HI distributions in galaxies bear no resemblance to the H_2 distributions, so that dissociation of H_2 probably does not explain the majority of HI in H_2-dominated regions of galaxies.

It is more likely, therefore, that the H_2/HI ratio at 2-4 kpc resolution represents the efficiency of H_2 cloud formation (Wyse 1986; Tacconi and Young 1986; Wyse and Silk 1989; Wang 1989). The variation in the H_2/HI ratio with morphological type, then, indicates that the efficiency of molecular cloud formation is higher in early- than late-type spirals; it is plausible that the H_2/HI ratio depends on morphology because gravity on the large scale is involved in molecular cloud formation. One scenario which is consistent with the observations is that the massive, centrally concentrated galaxies are able to achieve a molecular-dominated ISM through the collection of more gas in the central potential. This gas may then form molecular clouds when a critical column density is exceeded. It is possible that this critical column density is lower in early- than in late-type spirals, a situation which could arise if the velocity dispersion in the gas is lower in the early types.

6.3 Effects of Galactic Environment on Molecular Gas in Galaxies

A complete review of the effects of environment on the cool phase of the ISM in galaxies is presented elsewhere in this volume (Kenney 1989). Brief mention of the effects of galaxian environment on molecular gas in galaxies is made below.

The Virgo cluster provides a laboratory for studying the cycling of gas between the atomic and molecular phases, and the role of each in star formation. It is now well known that the atomic gas content of many Virgo spirals is low by factors ranging from 2 to 10, compared with more isolated galaxies of the same type and optical size (Giovanelli and Haynes 1983; van Gorkom and Kotanyi 1985; Warmels 1986). To determine the fate of the molecular gas in galaxies in a cluster environment, several surveys of CO emission in Virgo spirals have been conducted (Kenney and Young 1986, 1988a, 1988b, 1989; Stark et al. 1986). From these studies, it has been concluded that the H_2 is not deficient; some process has removed the low density atomic gas and left the denser molecular gas relatively unscathed.

Galaxy-galaxy interactions can dramatically change the gas distributions in galaxies, either through the enhanced conversion of HI into H_2, the redistribution of H_2 during the interaction, or both. High resolution CO observations indicate that some luminous merging/interacting galaxies have strongly centrally concentrated gas distributions, with as much as $10^{10} M_\odot$ of H_2 in the central few kpc (Scoville et al. 1986; Sanders et al.

1988). Modeling of the effects of galaxy-galaxy interactions on the distribution of molecular clouds (Olsen and Kwan 1989) indicates that close interactions can lead to large-scale redistributions of the gas, with the end result that the gas is concentrated in the center of the galaxy. Finally, galaxy-galaxy interactions may lead to enhanced conversions of gas into young stars, since the $L_{IR}/M(H_2)$ ratios are found to be higher in interacting/merging galaxies than in isolated galaxies (Young et al. 1986b; Solomon and Sage 1988).

I would like to gratefully acknowledge the support of the FCRAO and the generous allottment of observing time over the years, as well as the contributions toward the observations by a numer of faculty, postdos, and students -- L. Allen, M. Claussen, N. Devereux, Y.L. Huang, J. Kenney, P. Knezek, L. Tacconi, L. Tacconi-Garman, P. Schloerb, S. Schneider, P. Viscuso, and S. Xie. I would also like to thank N. Devereux, L. Allen, and N. Scoville for helpful comments and discussions, as well as help with the figures from N. Devereux.

7. References

Adler, D.S., Allen, R., and Lo, K.Y. 1989 in the Second Teton Conference on *The ISM in Galaxies* (eds. M. Shull and H. Thronson).

Adler, D., and Liszt, H. 1989, Ap.J., **339**, 836.

Allen, L., and Young, J.S. 1989, in the Second Teton Conference on *The ISM in Galaxies* (eds. M. Shull and H. Thronson).

Allen, L., Young, J.S., Kenney, J., and Kleinmann, S.G. 1989, in preparation.

Arnault, P., Casoli, F., Combes, F., and Kunth, F. 1988, Astr.Ap., **205**, 41.

Ball, R., Sargent, A., Scoville, N., Lo, F., and Scott, S. 1985, Ap.J.(Letters), **298**, L21.

Barvainis, R., Alloin, D., and Antonucci, R. 1989, Ap.J.(Letters), **337**, L69.

Bash, F., Davis, J., Jaffe, D., Wall, W., and Sutton, E. 1988, preprint.

Becker, R., Schilke, P., and Henkel, C. 1989, Astr.Ap., **211**, L19.

Bieging, J., Blitz, L., Lada, C., and Stark, A. 1981, Ap.J., **247**, 443.

Bloemen, J.B.G.L. 1986, Astr.Ap., **154**, 25.

Boulanger, F., Bystedt, J., Casoli, F., and Combes, F. 1984, Astr.Ap., **140**, L5.

Brand, J., Routerloot, J., Becker, R., and Stirpe, G. 1989, Astr.Ap., **211**, 315.

Brinks, E. 1985, Ph.D. Thesis, Sterrewacht Leiden.

Bushouse, H. 1986, Ph.D. Thesis, Univ. of Illinois.

Canzian, B., Mundy, L., and Scoville, N. 1988, Ap.J., **333**, 157.

Carlstrom, J. 1987, in *Galactic and Extragalactic Star Formation* (eds. M. Fich and R. Pudritz).

Casoli, F., et al. 1988, Astr.Ap., **192**, L17.

Cohen, R.S., et al. 1988, Ap.J.(Letters), **331**, L95.

Combes, F., Dupraz, C., Casoli, F., and Pagani, L. 1988, Astr.Ap., **203**, L9.

Combes, F., Encrenaz, P., Lucas, R., and Weliachew, L. 1977a, Astr.Ap., **55**, 311.

----- 1977b, Astr.Ap., **61**, L7.

----- 1978, Astr.Ap., **63**, L13.

Condon, J. 1987, Ap.J.Suppl., **65**, 485.

de Jong, T., et al. 1984, Ap.J.(Letters), **278**, L67.

DeGioia-Eastwood, K., Grasdalen, G.L., Strom, S.E., and Strom, K.M. 1984, Ap.J., **278**, 564.

Dettmar, R.-J., and Heithausen, A. 1989, Ap.J.(Letters), **344**, L61.

de Vaucouleurs, G., de Vaucouleurs, A., and Corwin, H.G. 1976, Second Reference Catalogue of Bright Galaxies (Austin: Univ. of Texas Press) (RC2).

Devereux, N., and Eales, S. 1989, Ap.J., **340**, 708.

Devereux, N., and Young, J.S. 1989a, Ap.J., submitted.

----- 1989b, Ap.J.(Letters), in press.

----- 1989c, in the Second Teton Conference on *The ISM in Galaxies* (eds. M. Shull and H. Thronson).

Dickey, J., and Salpeter, E. 1984, Ap.J., **284**, 461.

Dickman, R.L. 1988, in *Molecular Clouds in the Milky Way and External Galaxies* (eds. R. Dickman, R. Snell, and J. Young).

Dickman, R.L., Snell, R.L., and Schloerb, F.P. 1986, Ap.J., **309**, 326.

Doi, M., Ishizuki, S., Sofue, Y., Nakai, N., and Handa, T. 1988, preprint.

Dopita, M.A. 1985, Ap.J.(Letters), **295**, L5.

Draine, B.T., and Lee, H.M. 1984, Ap.J., **285**, 89.

Elmegreen, B.G. 1979, Ap.J., **231**, 372.

----- 1989, Ap.J, **338**, 178.

Elmegreen, B., and Elmegreen, D. 1982, A.J., **87**, 626.

Elmegreen, B., Elmegreen, D., and Morris, M. 1980, Ap.J., **240**, 455.

Encrenaz, P., Stark, A., Combes, F., and Wilson, R. 1979, Astr.Ap., **78**, L1.

Gallagher, J.S., Hunter, D.A., and Tutukov, A.V. 1984, Ap.J., **284**, 544.

Garman, L. and Young, J.S. 1986, Astr.Ap., **154**, 8.

Gerin, M., Nakai, N., and COmbes, F. 1988, Astr.Ap., **203**, 44.

Giovanelli, R., and Haynes, M.P. 1985, Ap.J., **292**, 404.

Goad, J., De Veny, J., and Goad, L. 1979, Ap.J.Suppl., **39**, 439.

Goldsmith, P., and Young, J. 1989, Ap.J., **341**, 718.

Gordon, M.A., and Burton, W.B. 1976, Ap.J., **208**, 346.

Guelin, M., Garcia-Burillo, S., Blundell, R., Cernicharo, J., Despois, D., and Steppe, H. 1988, in *Highlights of Astronomy*, Vol. 8.

Gusten, R., and Mezger, P. 1982, *Vistas in Astronomy*, **26**, 159.

Handa, T., Nakai, N., Sofue, Y., Hayashi, M., and Fujimoto, M. 1988, preprint.

Hayashi, M., et al. 1987, in IAU No. 115 *Star Forming Regions* (eds. M. Peimbert and J. Jugaku).

Heckman, T., Blitz, L., Wilson, A., Armus, L., and Miley, G. 1989, Ap.J., **342**, 735.

Heckman, T., Beckwith, S., Blitz, L., Skrutskie, M., and Wilson, A. 1986, Ap.J., **305**, 157.

Helou, G., Soifer, B.T., and Rowan-Robinson, M. 1985, Ap.J.(Letters), **305**, L15.

Hildebrand, R.H. 1983, Quart.J.R.A.S., **24**, 267.

Huchtmeier, W., Bregman, J., Terlebvich, R., and Tueben, P. 1988, Astr.Ap., **198**, 33.

Huchtmeier, W., Richter, O.-G., Bohnenstengel, H.-D., and Hauschildt, M. 1983, *A General Catalog of HI Observations of External Galaxies,* ESO Preprint No. 250.

Hunter, D.A., Thronson, H., Casey, S., and Harper, D. 1988, preprint.

Ichikawa, T., Nakano, M., and Tanaka, Y. 1987, in IAU No. 115 *Star Forming Regions* (eds. M. Peimbert and J. Jugaku).

Israel, F. 1988, in *Millimeter and Submillimeter Astronomy,* eds. R. Wolstencroft and W. Burton (Dordrecht: Kluwer), p. 281.

Israel, F., DeGraauw, T., Van de Stadt, H., and DeVries, C. 1986, Ap.J., **303**, 186.

Israel, F., and Rowan-Robinson, M. 1984, Ap.J., **283**, 81.

Jackson, J., and Ho, P.T.P. 1988, Ap.J.(Letters), **324**, L5.

Jackson, J., Snell, R., Ho, P., and Barrett, A. 1989, Ap.J., **337**, 680.

Johansson, L., and Booth, R. 1987, in IAU No. 115 *Star Forming Regions* (eds. M. Peimbert and J. Jugaku).

Joint IRAS Science Working Group. 1985, IRAS Point Source Catalogue (Washington, D.C.: U.S. Government Printing Office).

Kaneko, N., et al. 1989, Ap.J., **337**, 691.

Kenney, J. 1987, Ph.D. Thesis, University of Mass., Amherst.

Kenney, J. and Young, J. 1986, Ap.J.(Letters), **301**, L13.

----- 1988a, Ap.J.Suppl., **66**, 261.

----- 1988b, Ap.J., **326**, 588.

----- 1989, Ap.J., submitted.

Kenney, J., Young, J., Rubin, V., and Ford, W. 1989, Ap.J., in press.

Kennicutt, R.C. 1983, Ap.J., **272**, 54.

----- 1989, Ap.J., **344**, 685.

Kennicutt, R.C. Jr. and Kent, S.M. 1983, A.J., **88**, 1094.

Kennicutt, R.C. Jr., et al. 1987, A.J., **93**, 1011.

Lada, C.J., Margulis, M., Sofue, Y., Nakai, N., Handa, T. 1988, Ap.J., **328**, 143.

Lazareff, B., Castets, A., Kim, D., and Jura, M. 1989, Ap.J.(Letters), **336**, L13.

Liszt, H.S. 1982, Ap.J., **262**, 198.

Lo, K.Y., et al. 1984, Ap.J.(Letters), **282**, L59.

Lo, K.Y., Ball, R., Masson, C., Phillips, T., Scott, S., and Woody, D. 1987, Ap.J.(Letters), **317**, L63.

Lo, K.Y., et al. 1987, in IAU No. 115 *Star Forming Regions* (eds. M. Peimbert and J. Jugaku).

Lonsdale, C., and Helou, G. 1987, Ap.J., **314**, 513.

Lonsdale, C.J., Helou, G., Good, J.C., and Rice, W.L. 1985, "Catalogued Galaxies and Quasars Observed in the IRAS Survey" (Washington, D.C.: U.S. Government Printing Office).

Lord, S. 1987, Ph.D. Thesis, University of Massachusetts.

Lord, S., and Young, J.S. 1989, Ap.J., submitted.

Maloney, P. 1988, Ap.J., **334**, 761.

Maloney, P., and Black, J. 1988, Ap.J., **325**, 389.

McCarthy, P., Heckman, T., and van Breugel, W. 1987, A.J., **93**, 264.

Mirabel, F., Booth, R., Garay, G., Johansson, L., and Sanders, D. 1988b, Astr.Ap., **206**. L20.

Mirabel, F., Kazes, I., and Sanders, D. 1988a, Ap.J.(Letters), **324**, L59.

Myers, S.T., and Scoville, N.Z. 1987, Ap.J.(Letters), **312**, L39.

Morris, M., and Lo, K.Y. 1978, Ap.J., **223**, 803.

Morris, M., and Rickard, L.J. 1982, Ann.Rev.Ast.Ap., **20**, 517.

Nakai, N., et al. 1987, P.A.S.J., **40**, 653.

Nilson, P. 1973, *Uppsala General Catalogue of Galaxies* (Uppsala) (UGC).

Noguchi, M., and Ishibashi, S. 1986, M.N.R.A.S., **219**, 305.

Ohta, K., Sasaki, W., and Saito, M. 1988, P.A.S.J., **39**, 685.

Olofsson, H., and Rydbeck, G. 1984, Astr.Ap., **136**, 17.

Olsen, K., and Kwan, J. 1989, Ap.J., in press.

Pagel, B.E.J., and Edmunds, M.G. 1981, Ann.Rev.Astr.Ap., **19**, 77.

Planesas, P., Gomez-Gonzales, J., and Martin-Pintado, J. 1988, preprint.

Rand, R., and Kulkarni, S. 1989 in the Second Teton Conference on *The ISM in Galaxies* (eds. M. Shull and H. Thronson).

Rengarajan, T.N., and Verma, R.P. 1987, in IAU No. 115 *Star Forming Regions* (eds. M. Peimbert and J. Jugaku).

Rice, W., Boulanger, F., Viallefond, F., Soifer, B.T., and Freedman, W. 1989, preprint.

Richmond, M., and Knapp, G. 1986, A.J., **91**, 517.

Rickard, L.J. 1982, in *Greenbank Workshop on Extragalactic Molecules* , p. 1.

Rickard, L.J., and Blitz, L. 1985, Ap.J.(Letters), **292**, L57.

Rickard, L.J., and Harvey, P. 1984, Ap.J.(Letters), **268**, L7.

Rickard, L.J., and Palmer, P. 1981, Astr.Ap., **102**, L13.

Rickard, L., Turner, B., and Palmer, P. 1977, Ap.J.(Letters), **218**, L51.

Rieke, G.H., Lebofsky, M.J., Thompson, R., Low, F., and Tokunaga, A. 1980, Ap.J., **238**, 24.

Roberts, M.S. 1969, A.J., **74**, 859.

Rogstad, D.H., Shostak, S., and Rots, A. 1973, Astr.Ap., **22**, 111.

Rowan-Robinson, M., Phillips, T., and White, G. 1980, Astr.Ap., **82**, 381.

Rubin, V.C., Burstein, D., Ford, W.K., and Thonnard, N. 1985, Ap.J., **289**, 81.

Ryden, B.S., and Stark, A.A. 1986, Ap.J., 305, 823.

Rydbeck, G., Hjalmarson, A., and Rydbeck, O.E.H. 1985, Astr.Ap., 144, 282.

Sage, L. 1989, Ap.J., 344, 200.

Sage, L. and Solomon, P.M. 1987, Ap.J.(Letters), 321, L103.

----- 1989, Ap.J., 342, L15.

Sage, L., and Wrobel, J. 1989, Ap.J., 344, 204.

Sandage, A. 1986, Astr.Ap., 161, 89.

Sandage, A., and Tammann, G. 1981, The Revised Shapley-Ames Catalogue (Washington: Carnegie Inst.) (RSA).

Sanders, D., Scoville, N.Z., Sargent, A.I., and Soifer, B.T. 1988a, Ap.J.(Letters), 324, L55.

Sanders, D.B., and Mirabel, F. 1985, Ap.J.(Letters), 298, L31.

Sanders, D., Scoville, N., and Soifer, B.T. 1988b, Ap.J., 335, L1.

Sanders, D.B., Solomon, P.M., and Scoville, N.Z. 1984, Ap.J., 276, 182.

Sanders, D.B., and Young, J.S. 1989, Ap.J., submitted.

Sanders, D., Young, J., Scoville, N., Soifer, T., and Danielson, G. 1987, Ap.J.(Letters), 312, L5.

Sanders, D.B., et al. 1986a, Ap.J.(Letters), 305, L45.

Sandquist, A., Elfhag, T., and Jorsater, S. 1988, Astr.Ap., 201, 223.

Scalo, J. 1987, in Galactic and Extragalactic Star Formation (eds. M. Fich and R. Pudritz).

Scoville, N.Z. and Good, J. 1989, Ap.J., 339, 149.

Scoville, N.Z. and Sanders, D.B. 1987, in Interstellar Proceses, eds. D. Hollenbach and H. Thronson (Dordrecht: Reidel), p. 21.

Scoville, N.Z., Sanders, D.B., and Clemens, D. 1986a, Ap.J.(Letters), 310, L77.

Scoville. N.Z., et al. 1986b, Ap.J., 311, L47.

Scoville, N., Sanders, D., Sargent, A., Soifer, T., and Tinney, C. 1989, Ap.J.etters), 345, L25.

Scoville, N.Z., Soifer, B.T., Neugebauer, G., Young, J.S., Mathews, K., and Yerka, J. 1985, Ap.J., 289, 129.

Scoville, N.Z., and Solomon, P.M. 1975, Ap.J., 199, L105.

Scoville, N.Z., and Young, J.S. 1983, Ap.J., 265, 148.

Scoville, N.Z., Young, J.S., and Lucy, L. 1983, Ap.J., 270, 443.

Scoville, N., Yun, M., Clemens, D., Sanders, D., Waller, W. 1987, Ap.J.(Suppl)., 63, 821.

Searle, L., Sargent, W.L.W., and Bagnuolo, W. 1973, Ap.J., 179, 427.

Seiden, P. 1983, Ap.J., 266, 555.

Shaya, E., and Federman, S. 1987, Ap.J., 319, 76.

Shostak, G.S. 1978, Ast.Ap., 68, 321.

Shu, F., Adams, and Lisano, S. 1987, Ann.Rev.Astr.Ap., 25, 23.

Silk, J. 1987, in Galactic and Extragalactic Star Formation (eds. M. Fich and R. Pudritz).

Smith, J. 1982, Ap.J., **261**, 463.

Smith, J., Harper, D.A., and Lowenstein, R.F. 1984, in Airborne Astronomy Symposium, ed. Thronson and Erickson, p. 277.

Sofue, Y. 1987, in *Galactic and Extragalactic Star Formation* (eds. M. Fich and R. Pudritz).

Sofue, Y., Doi, M., Ishizuki, N., Nakai, N., and Handa, T. 1988a, P.A.S.J., **40**, 511.

Sofue, Y., Doi, M., Krause, M., Nakai, N., and Handa, Y. 1988b, preprint.

Soifer, B.T., et al. 1984, Ap.J.(Letters), **278**, L71.

Solomon, P.M. 1982, in *Greenbank Workshop on Extragalactic Molecules* .

Solomon, P., Barrett, J., Sanders, D.B., and de Zafra, R. 1983, Ap.J.(Letters), **266**, L103.

Solomon, P.M., Rivolo, R., Barrett, J., and Yahil, A. 1987, Ap.J., **319**, 730.

Solomon, P.M., and Sage, L. 1988, Ap.J., **334**, 613.

Solomon, P.M., Sanders, D.B., and Rivolo, R. 1985, Ap.J., **292**, L19.

Solomon, P.M., Scoville, N.Z., and Sanders, D.B. 1979, Ap.J.(Letters), **232**, L89.

Spitzer, L. 1978, in *Physical Processes in the Interstellar Medium* (Wiley-Interscience), p. 162.

Stark, A.A. 1979, Ph.D. Thesis, Princeton University.

----- 1983, in *Kinematics, Dynamics, and Structure of the Milky Way,* ed. W. Shuter (Dordrecht: Reidel), p. 127.

Stark, A.A., and Carlson, E. 1984, Ap.J., **279**, 122.

Stark, A., Elmegreen, B., and Chance, D. 1987, Ap.J., **322**, 64.

Stark, A.A., et al. 1986, Ap.J., **310**, 660.

Tacconi, L. 1987, Ph.D. Thesis, University of Massachusetts.

Tacconi, L., and Young, J.S. 1985, Ap.J., **290**, 602.

----- 1986, Ap.J., **308**, 600.

----- 1987, Ap.J., **322**, 681.

----- 1989a, Ap.J., in press.

----- 1989b, Ap.J., submitted.

Telesco, C.M. 1988, Ann.Rev.Astr.Ap., **26**, 343.

Telesco, C.M., and Harper, D.A. 1980, Ap.J., **235**, 392.

Thronson, H., et al. 1989b, Ap.J., **344**, 747.

Thronson, H., Hunter, D., Casey, S., Latter, W., and Harper, D. 1989a, **339**, 803.

Thronson, H., et al. 1988, Ap.J., **334**, 605.

Tully, B. 1974, Ap.J.Suppl., **27**, 449.

Turner, J., Martin, B., and Ho, P.T.P. 1989, preprint.

van Driel, W. 1987, Ph.D. Thesis, University of Groningen.

van Gorkom, J., and Kotanyi, K. 1985, in ESO Workshop on the Virgo Cluster of Galaxies, ed. Richter and Bingelli. p. 51.

Verter, F. 1983, Ph.D. thesis, Princeton University.

----- 1987, Ap.J.Suppl., **64**, 555.

----- 1988, Ap.J.(Suppl.), **68**, 129.

Vogel, S., Boulanger, F., and Ball, R. 1987, Ap.J.(Letters), **316**, 243.

Vogel, S., Kulkarni, S., and Scoville, N. 1988, Nature, **334**, 402.

Walterbos, R.M. 1987, in Galactic and Extragalactic Star Formation, ed. Fich and Pudritz.

Wang, Z. 1989, in the Second Teton Conference on *The ISM in Galaxies* (eds. M. Shull and H. Thronson).

Warmels, R.H. 1986, Ph.D. Thesis, Groningen.

Weliachew, L., Casoli, F., and Combes, F. 1988, Astr.Ap., **199**, 353.

Wevers, B., van der Kruit, P., and Allen, R. 1986, Astr.Ap.Suppl., **55**, 505.

Wiklind, T., and Henkel, C. 1989, preprint.

Wiklind, T., Rydbeck, G. Hjalmarson, A., and Rydbech, O. 1987, in IAU No. 115 *Star Forming Regions* (eds. M. Peimbert and J. Jugaku).

Wild, W., et al. 1989, in the Second Teton Conference on *The ISM in Galaxies* (eds. M. Shull and H. Thronson).

Wilson, C., and Scoville, N. 1989, preprint.

Wilson, T., Fricke, K., and Biermann, P. 1979, Astr.Ap., **79**, 245.

Wyse, R. 1986, Ap.J.(Letters), **311**, L41.

Wyse, R. and Silk, J. 1989, Ap.J., **339**, 700.

Young, J.S. 1987, in IAU No. 115 *Star Forming Regions* (eds. M. Peimbert and J. Jugaku).

----- 1988, in *Molecular Clouds in the Milky Way and External Galaxies* (eds. R. Dickman, R. Snell, and J. Young).

Young, J.S., Claussen, M., and Scoville, N. 1988a, Ap.J., **324**, 115.

Young, J.S., Gallagher, J.G., and Hunter, D.A. 1984a, Ap.J., **276**, 476.

Young, J.S., Kenney, J., Lord, S., and Schloerb, F.P. 1984b, Ap.J.(Letters), **287**, L65.

Young, J.S., Claussen, M., Kleinmann, S.G., Rubin, V., and Scoville, N. 1988b, Ap.J.(Letters), **331**, L81.

Young, J.S., and Knezek, P. 1989, Ap.J.(Letters), in press (Dec. 15).

Young, J.S., et al. 1986b, Ap.J.(Letters), **311**, L17.

----- 1989b, in preparation.

Young J.S., and Sanders, D.B. 1986, Ap.J., **302**, 680.

Young, J.S., Schloerb, F.P., Kenney, J, and Lord, S. 1986a, Ap.J., **304**, 443.

Young, J.S., and Scoville, N.Z. 1982a, Ap.J., **258**, 467.

----- 1982b, Ap.J.(Letters), **260**, L11.

----- 1982c, Ap.J.(Letters), **260**, L41.

----- 1984, Ap.J., **287**, 153.

Young, J.S., Scoville, N.Z., and Brady, E. 1985, Ap.J., **288**, 487.

Young, J.S., Tacconi, L., and Scoville, N. 1983, Ap.J., **269**, 136.

Young, J.S., Xie, S., Kenney, J., and Rice, W.L. 1989a, Ap.J.Suppl., **70**, 699.

INTERSTELLAR DUST IN GALAXIES

Michael Rowan-Robinson

Astronomy Unit, Queen Mary College
Mile End Rd, London E1 4NS

ABSTRACT. Current ideas about the nature of interstellar dust in galaxies are reviewed, with a strong emphasis on the nature of the very small grain component needed to explain the mid-infrared diffuse emission and unidentified infrared features. Models for the infrared spectra of galaxies are reviewed and the evidence that most of the radiation in star-forming regions is being absorbed by a high visible-uv optical depth of dust is summarised. The evidence for destruction of very small grains in regions of high radiation intensity is discussed.

A new model for interstellar grains in galaxies is presented, based on a revised version of the model of Rowan-Robinson (1986) and is compared to observed far infrared colour-colour diagrams and to far infrared spectra of galaxies which have been mapped at 800 μm by Hughes et al (1989). Work on far infrared and submillimetre mapping of galaxies is reviewed, as also is recent work on infrared emission from ellipticals and lenticulars. The determination of dust mass in galaxies is briefly discussed.

1. INTRODUCTION

My task in reviewing interstellar dust in galaxies is greatly simplified by the appearance of several excellent review articles on this area during the past year or so. Although each covers only a specific aspect of the subject, together they comprise a good introduction to our current knowledge.

A comprehensive review of infrared emission from our Galaxy, with much historical background, has been given by Cox and Mezger (1989). They emphasize that the results from IRAS have led to a major reappraisal of estimates of the fraction of the infrared emission from our Galaxy which comes from interstellar dust illuminated by the interstellar radiation field, as opposed to regions of massive star formation. The latter are now believed to contribute only about 10% of the total infrared emission from the Galaxy. Boulanger and Perault (1988) have given an authoritative discussion of the infrared emission observed by IRAS from the different components of diffuse emission from our Galaxy, and the correlations between them, which must be

H. A. Thronson, Jr. and J. M. Shull (eds.), The Interstellar Medium in Galaxies, 121–150.
© 1990 *Kluwer Academic Publishers.*

the starting point for any analysis of the interstellar dust in normal galaxies. A general review of the IRAS view of the extragalactic sky has been given by Soifer et al (1987). Telesco (1988) has reviewed enhanced star formation and infrared emission in the centres of galaxies, with a strong emphasis on imaging and spectroscopic data derived from ground-based studies. Helou (1988) has reviewed the far infrared emission from Galactic and extragalactic dust seen by IRAS, emphasizing the similarity in the range of far infrared colours seen in external galaxies and in reflection nebulae in our Galaxy. Roche (1988) has given an interesting summary of the results from near and middle infrared spectroscopy of galaxies. Puget and Leger (1989) have given a very thorough review of the evidence for small grains and large aromatic molecules in the interstellar medium of our own and other galaxies. Finally Draine (1988a) has reviewed interstellar extinction in the infrared.

In this review I shall concentrate on those areas where major controversy exists and where significant progress may be expected in the next few years. The topics I have selected are grain models, first attempts to explain the infrared spectra of IRAS galaxies, the destruction of the very small grain component, a new picture of interstellar dust in galaxies, results from far infrared and submillimetre mapping of galaxies, determination of the dust mass in galaxies, and dust in ellipticals and lenticulars.

2. GRAIN MODELS

Classical grain models consisting of silicate and carbon grains of radius 0.01-0.1 μ, for example those of Mathis et al (1977), Draine and Lee (1984), Rowan-Robinson (1986), Tielens and Allamandola (1987), are successful in accounting for the observed visible and ultraviolet extinction curve and the emission longward of 60 μ. However there is not yet a concensus on the grain properties longward of 300 μ, as emphasized by Draine (1988a). I will discuss this further in section 5 below. The observations which the classical grain model definitely can not account for are (i) excess diffuse emission from the Milky Way at 2-20 μ (Price 1981, Boulanger et al 1985), (ii) 2-20 μ emission from reflection nebulae with colour temperature approximately independent of distance from the star (Sellgren 1984) and (iii) the broad features at 3.3, 6.2, 7.7, 8.6 and 11.3 μ seen ubiquitously in emission (Gillett et al 1973).

Current models for these three phenomena all involve the non-equilibrium response of very small particles to absorption of an ultraviolet photon (Greenberg 1968, Duley 1973, Allen and Robinson 1975, Purcell 1976, Andriesse 1978, Sellgren 1984, Draine and Anderson 1985). The main contenders are:
(A) Polycyclic aromatic hydrocarbons (PAH), which can be thought of as hydrogenated graphite platelets consisting of about 50 atoms (Platt 1956, Donn 1968, Leger and Puget 1984, Allamandola et al 1985, Puget and Leger 1989). To account for the full range of observed phenomena, Puget and Leger (1989) have to include also a very small carbonaceous grain (VSG) component with radii in the range 0.0015-0.01 μ. Fig 1a shows how some particular examples of PAHs can give at least

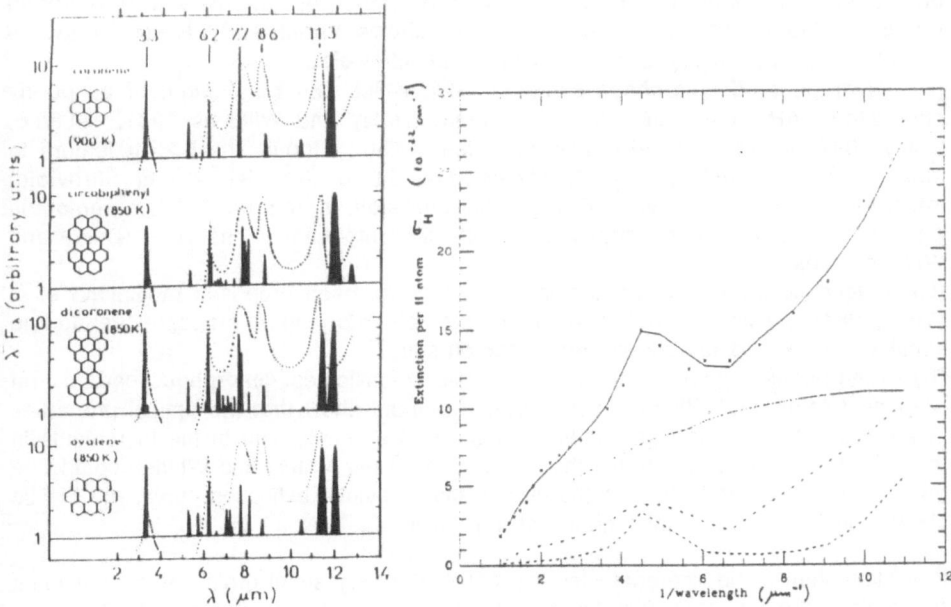

Fig 1:   (a) Emission spectra of several PAHs calculated from their laboratory absorption spectra, compared with observations of the reflection nebula NGC2023. (b) Fit to the interstellar extinction curve.(Puget and Leger 1989)

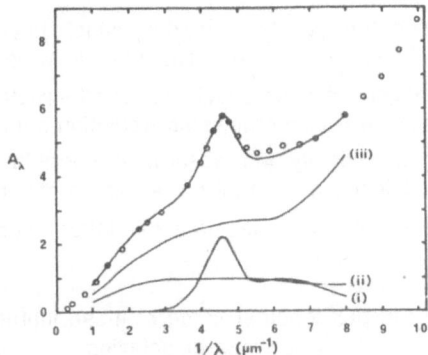

Fig 2: Fit by Jones et al (1987) to the interstellar extinction curve in the visible and ultraviolet.

qualitative agreement in the wavelengths of (most of) the 3-12 μ broad-band features (Puget and Leger 1989). Fig 1b shows Puget and Leget's fit to the interstellar extinction curve in the visible and ultraviolet.

(B) Hydrogenated amorphous carbon (HAC), which can be thought of as poorly connected PAH islands in a larger structure (Duley and Williams 1981, 1988a,b, Duley 1987, Jones et al 1987, Williams 1989). They attribute the 0.22 μ feature to small silicate particles. Fig 2 shows their fit to the visible and ultraviolet interstellar extinction curve. Broad-band emission in the 0.6-0.9 μm region is attributed to luminescence from a diamond-like component in the HAC (Duley and Williams 1988b).

(C) Quenched carbonaceous composite (QCC) has been proposed by Sakato et al (1983,1984). This material is made in the laboratory in a process intended to simulate the expanding atmospheres of carbon stars.

(D) Amorphous aggregates of small particles of silicates, amorphous carbon and graphite (Mathis and Whiffen 1989). These authors show that the optical properties of an aggregate can be significantly different from a simple sum of the ingredients in the aggregate. Fig 3a illustrates the appearance of the Mathis and Whiffen composite grains, Fig 3b shows their fit to the visible and ultraviolet extinction curve and Fig 3c shows the properties of their grains at 1-1000 μ.

It is clear that in the aggregate models (B-D), the very small grain component must retain its thermodynamic identity in order to explain the phenomena (i-iii) above. From the point of view of understanding infrared emission from dust, it may therefore be academic whether the very small grain component is integrated into a larger structure or not, since this integration must be so weak as to leave the specific properties of the component intact. Puget and Leger (1989) in fact query whether aggregate grain models can localize the energy of an incident photon for the several seconds required for infrared emission.

Draine (1988b) has reviewed the variety of models which have been put forward specifically to explain the 0.2175 μ feature. The models which he considers are graphite, nongraphitic carbonaceous solids, OH^- on small silicate grains, PAH, small MgO or CaO particles, dessicated microorganisms, radiation-damaged SiO_2, charge transfer on Si, Fe or Mg, and finally the absorption edge in silicate grains. He concludes that only two are consistent with all the available observations, graphite or OH^- on small silicate grains, and he notes that the latter hypothesis is less well developed than the graphite hypothesis.

In section 5 below I shall try to pull together some of these ingredients into a simple but comprehensive picture for interstellar dust in galaxies.

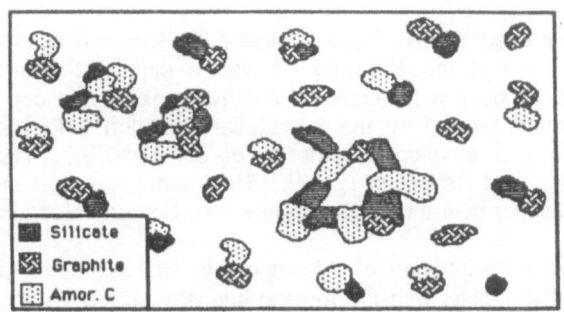

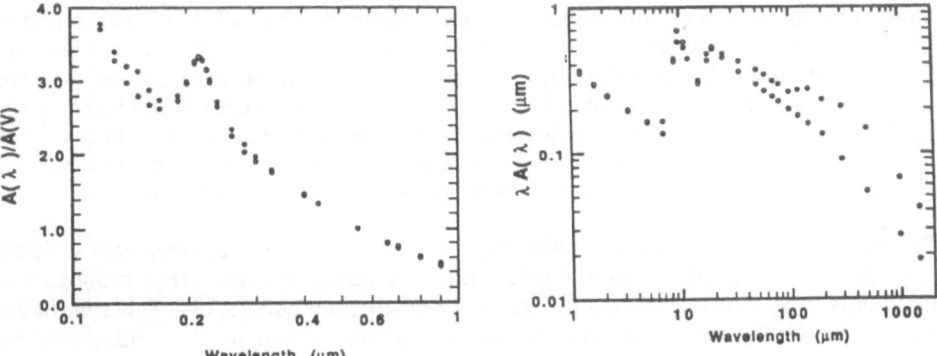

Fig 3: (a) Schematic picture of the Mathis & Whiffen (1989) grain model.
(b)   Their fit to the ultraviolet and visual interstellar extinction curve.
(c) The same for infrared wavelengths. (filled circles:calculated, open circles: observations)

3. FIRST ATTEMPTS TO EXPLAIN THE INFRARED SPECTRA OF IRAS GALAXIES

Models for IRAS galaxy spectra have been reviewed by Rowan-Robinson (1987 a,b) and Helou (1988). The first model proposed was a simple 2-component model consisting of warm (50 K) dust in molecular clouds/HII regions and cool (20 K) dust in the interstellar medium heated by the interstellar radiation field (de Jong et al 1984). This model has been developed further by de Jong and Brink (1987) and has been criticized by Eales and Devereux (1989). The model is rather similar to that proposed by Cox and Mezger over a number of years (see Cox and Mezger 1989).

Helou (1986) proposed an extension of this model in which the warm component becomes a one-parameter family, with the heating intensity as the parameter. As the intensity increases from that found in the solar neighbourhood to the much higher value found in star-forming regions, the dust temperature increases from 20 to 50 K. More recently, Helou (1988) emphasizes the similar range of IRAS colours found in galaxies and in Galactic sources. Fig 4a shows log{ S(60)/S(100)} versus log{ S(12)/S(25)} for IRAS galaxies and Fig 4b shows the same diagram for Galactic sources. The sequence of colours found in the reflection nebulosity surrounding ζ Per by Boulanger et al (1988) with increasing distance from the star is also shown. This appears to be telling support for Helou's hypothesis that the variation of colour is simply due to variation of the heating intensity experienced by the grains.

Rowan-Robinson and Crawford (1986,1989) have also used the analogy with Galactic sources to derive a rather different model for IRAS galaxy spectra. They propose that the galaxy spectra are a mixture of three components, the general disc emission of the galaxy consisting of reradiation of the interstellar radiation field absorbed by interstellar grains (Fig 5a), a component present in Seyferts peaking at 25 μm due to dust in the narrow-line region, and a starburst component with a spectrum similar to that for Galactic compact HII regions. Their models for the latter (Crawford and Rowan-Robinson 1987) are optically thick at visible and ultraviolet wavelengths, with $A_V \approx 20$ (they are optically thin in the far infrared, of course). Fig 5b compares their starburst model spectrum with the Telesco et al (1984) spectrum of the NGC1068 starburst component and with the average spectrum for Galactic compact HII regions/regions of massive star formation derived by Rowan-Robinson (1979). Confirmation of the fact that most of the massive star formation in galaxy starbursts takes place at high visible-uv optical depth comes from a comparison of the 60 μ luminosity of a large sample of IRAS galaxies with their Hα luminosity (Leech et al 1988, Fig 5c). Ratios of these luminosities range from 200-4000, compared with 30-100 for the nearby normal galaxies studied by Persson and Helou (1987). The Hα/Hβ ratios for these IRAS galaxies indicate values for A_V of only a few, so the bulk of the far infrared radiation must come from stars whose visible light is heavily extinguished, while the Hα radiation must come from near the surface of the star-forming volume (Leech et al 1989). Further evidence for high visual extinction comes from the Brackett-alpha and -gamma observations of Kawara et al (1989) for a sample of starburst galaxies. From these they infer values for A_V in the range 7-

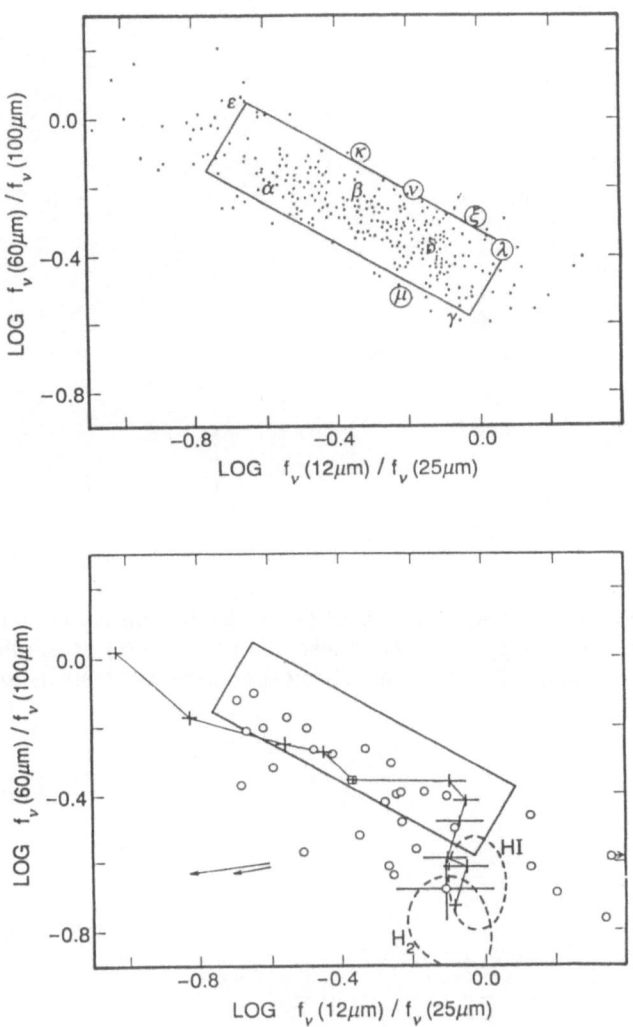

Fig 4: IRAS colour-colour diagrams for (a) galaxies, (b) Galactic star-forming regions (Helou 1988). In (b) the crosses denote data for ξ Per.

128

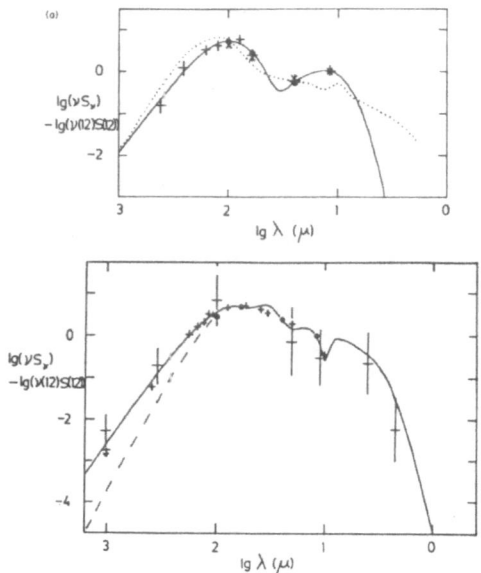

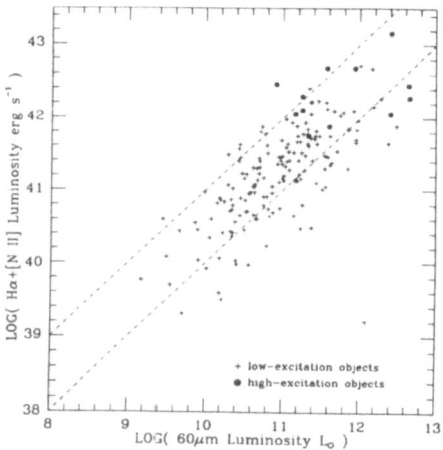

Fig 5: Models by Rowan-Robinson & Crawford (1989) for (a) the cirrus and (b) the starburst components in galaxy spectra. The broken curve in Fig (b) shows the effect of changing the wavelength at which the grain absorption efficiency steepens to 80 μm.

Fig 5c: H-alpha luminosity versus infrared luminosity for sample of IRAS galaxies (Leech et al 1988). The broken lines correspond to L(60μm)/L(H-alpha) = 400 and 4000.

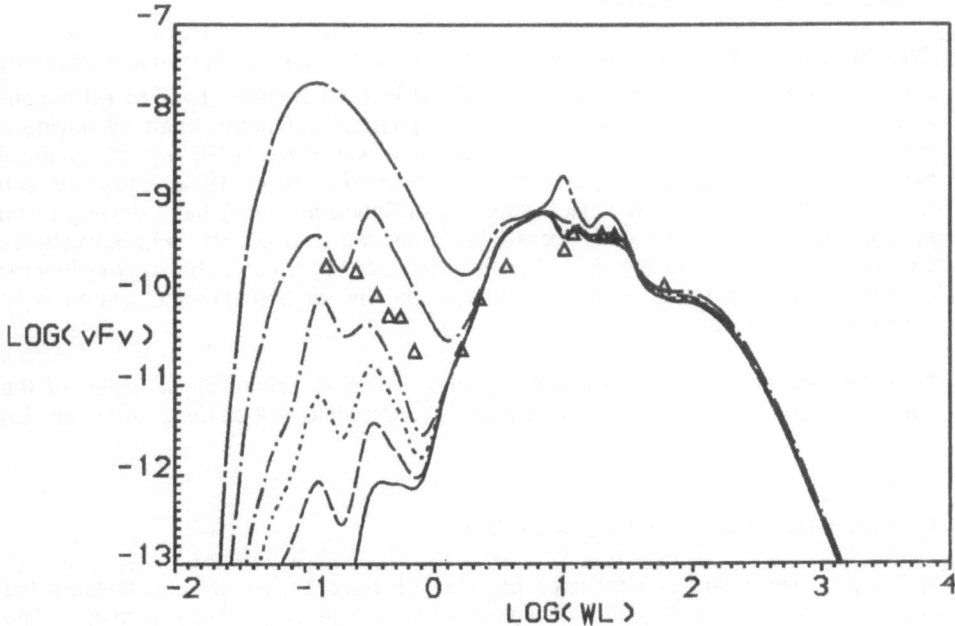

Fig 6: Sequence of flared disc models for the narrow-line region of NGC4151, as a function of the viewing angle, from face-on (top) to edge-on (bottom) (Efstathiou & Rowan-Robinson 1989)

33. These values are in agreement with those inferred from the depth of the 10 μm silicate feature in these galaxies.

Once we are dealing with dust clouds with $A_V \gg 1$, then the illumination geometry becomes of critical importance for models of the infrared spectra. Evolved HII regions in our Galaxy show strong deviations from spherical geometry, often displaying a blister geometry, although it is possible that for young compact HII regions spherical symmetry is a reasonable approximation (Rowan-Robinson 1982, Crawford and Rowan-Robinson 1987). Efstathiou and Rowan-Robinson (1989) have developed an accurate radiative transfer code for axially symmetric dust clouds. Fig 6 illustrates the crucial importance of the aspect angle when viewing a non spherically-symmetric system. Leisawitz (1989) has also studied the role of non-spherical geometry in star-forming regions.

An improved model for IRAS galaxy spectra, which is essentially a fusion of the approaches of Helou and of Rowan-Robinson and Crawford, will be described in section 5.

4. THE DESTRUCTION OF VERY SMALL GRAINS

In the past two years several lines of evidence have begun to point towards the destruction of very small grains in regions of very high uv radiation intensity. The most direct evidence comes from infrared spectroscopy. Roche (1988) and Desert and Dennefeld (1988) have shown that the broad 3-12 μ features attributed to very small grains are absent in the spectra of many Seyfert galaxies (Fig 7a). Destruction of very small grains is also presumably the reason that Rowan-Robinson and Crawford (1989) found that the disc component was very weak or absent in many Seyferts (Fig 7b).

Reasonably direct evidence for the destruction of very small grains in a high radiation intensity comes from the decline in the ratio of S(12)/S(100) near hot stars. Ryter et al (1987) showed this effect for σ Sco and Boulanger et al (1988) showed it for ξ Per.

Telesco et al (1989) argue that a similar effect is seen in the centre of M82. Fig 8a shows the increase in S(25)/S(12) with increasing uv intensity found by Telesco et al for M82 superposed on the curve derived from Boulanger et al's observations of ξ Per. However the spectrum of the emission from outside the nucleus of M82 (and of the integrated emission from the galaxy) is very similar to that for the NGC1068 starburst, and for compact Galactic HII regions, shown in Fig 5b, and one would normally assume that the bulk of this emission arises in regions where the visible and ultraviolet optical depth is $\gg 1$. The 10 μm emission from such a cloud does not arise from very small grains. Fig 8b shows the integrated spectrum of M82 compared to the optically thick starburst model of Rowan-Robinson and Crawford (1989): the agreement is good. Also shown is the shape of the spectrum of the central region of

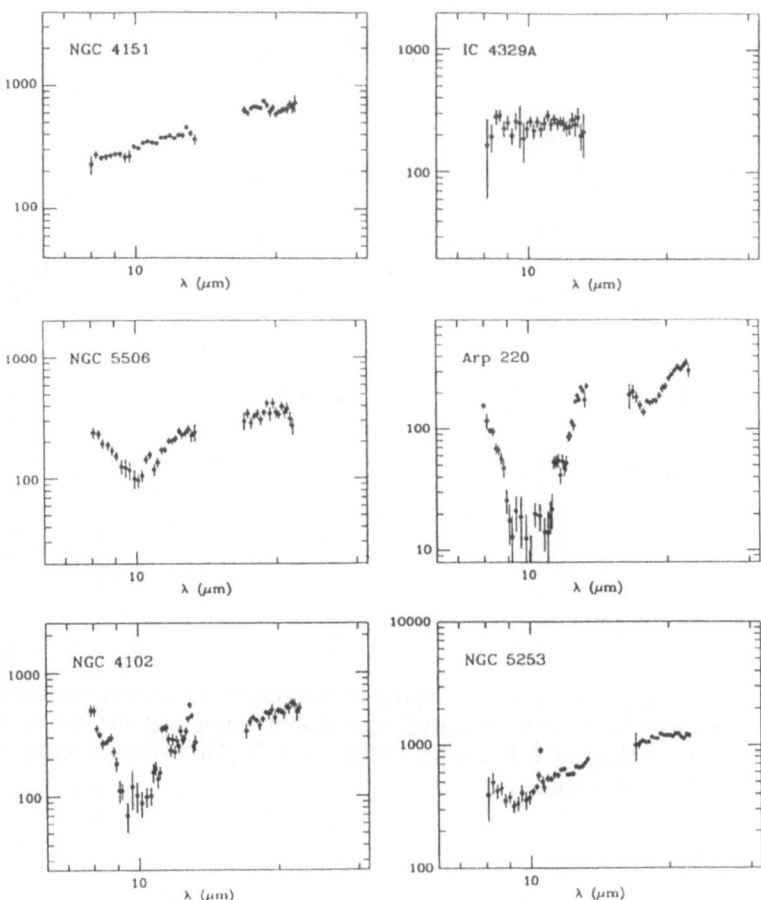

Fig 7: (a)   8-13μm and 17-22μm spectra of six galaxy nuclei.   Note that the unidentified ir features are completely absent from the Seyferts   NGC4151 and IC4329A.

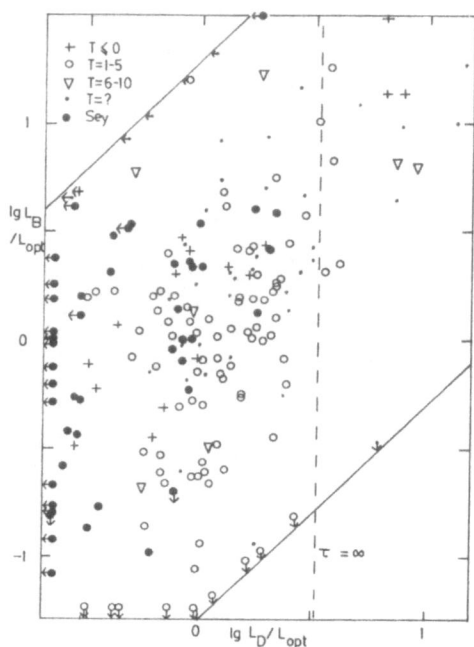

Fig 7(b)   Ratio of infrared luminosity in starburst component to optical luminosity, versus ratio of infrared luminosity in cirrus component to optical luminosity for IRAS galaxies (Rowan-Robinson & Crawford 1989).   The Seyferts (filled circles) are deficient in the cirrus component.

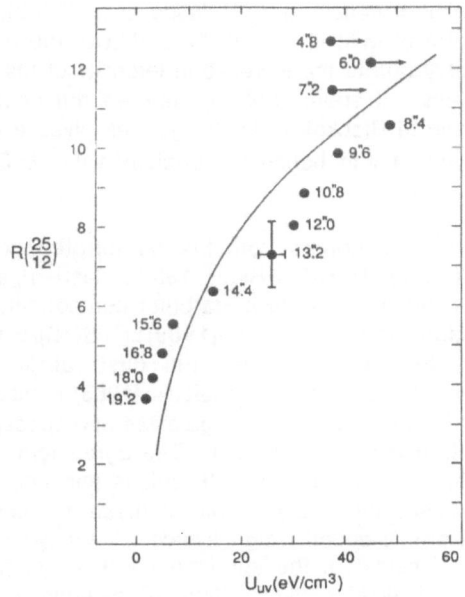

Fig 8a: Variation of 25/12 μm colour ratio with intensity of radiation field in centre of M82 (filled circles, Telesco et al 1989)) compared with relation found in ξ Per by Boulanger et al (1988).

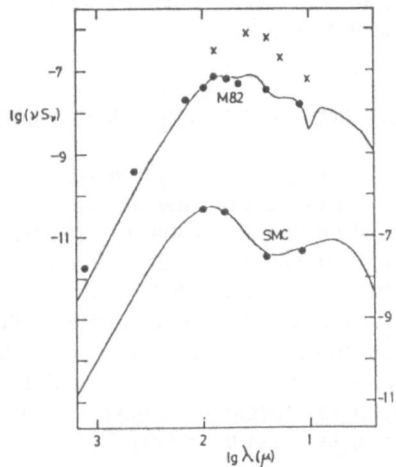

Fig 8b. Top: Integrated spectrum of M82 (filled circles) compared with starburst model (data from Telesco 1988, Smith et al 1989a). The crosses (arbitrary vertical scale) show the relative shape of the spectrum of the core of M82. Bottom: Integrated spectrum of the SMC compared with cirrus model (X=30) in which abundance of 5 Å grains has been reduced by 2/3rds.

M82, derived from the colours measured by Telesco et al (1989). The change in spectrum towards the centre of M82 is essentially a shift of the emission peak from 80 μm to 60 μm, presumably due to the increase in intensity of the radiation from the starburst towards to nucleus. It seems unlikely that we are seeing emission from optically thin dust (the ratio of Brackett-alpha to -gamma gives a value for A_V of 14 for M82 (Kawara et al 1989)) and hence the analogy with ξ Per appears to be spurious.

Similarly unconvincing evidence comes from the far infrared colours of galaxies (Pajot et al 1986, Gosh & Drapatz 1987, Helou 1988). Here again the problem is confusion with the role of the optically thick starburst component, for which, in the model of Rowan-Robinson and Crawford (1989), S(12)/S(60) =0.04 , but radiative transfer effects in normal 0.01-0.1 μm dust rather than small grain depletion is the cause. Fig 4b above showed Helou's (1988) compilation of the IRAS colours of compact Galactic HII regions and of galaxies superposed on the range of colours seen in ξ Per by Boulanger et al (1988). The agreement is good, but in my view this is fortuitous in the case of Galactic HII regions and galaxies dominated by starbursts since in most cases the optical depth in these sources is high and the analogy with ξ Per therefore of doubtful significance. If the 60/25 μm colour ratio, ignored by Helou, is also considered, the agreement with ξ Per is less impressive. However the case of the Small Magellanic Cloud (Schwering 1988) is convincing because the spectrum of this galaxy does indeed look like cirrus in which the smallest grain component is depleted (see Fig 8b).

In an interesting development, Leene and Cox (1987) have found that the 0.22 μ feature is also suppressed in regions of high radiation intensity, which suggests that this feature is associated with the very small carbonaceous grains responsible for the broad features and diffuse emission at 2-20 μ .

5. TOWARDS A NEW PICTURE OF INTERSTELLAR DUST IN GALAXIES

If we concentrate first on the 'cirrus' component in galaxies, the reradiation by interstellar dust of the energy absorbed from the interstellar radiation field, then it is clear that a satisfactory model involves a number of ingredients. Firstly a multiple (or aggregate ?) grain model is required to account for the interstellar extinction curve and it must incorporate very small grains and/or PAH. Secondly the model must allow for the fact that there is a range of heating intensities within galaxies and from galaxy to galaxy. For our Galaxy and a few other nearby galaxies we may hope to study how the observed spectrum varies with heating intensity. For more distant galaxies for which we have only the integrated spectrum we have to make do, for the moment, with a characteristic heating intensity. Let me define $X = I/I_{isrf}$, where I is the intensity in the region under consideration and I_{isrf} is the intensity of the interstellar radiation field in the solar neighbourhood, which I assume to be as characterised by Mathis et al (1983). Finally we may have to allow for the fact that for $X >$ some critical value, the very small grain component starts to be destroyed.

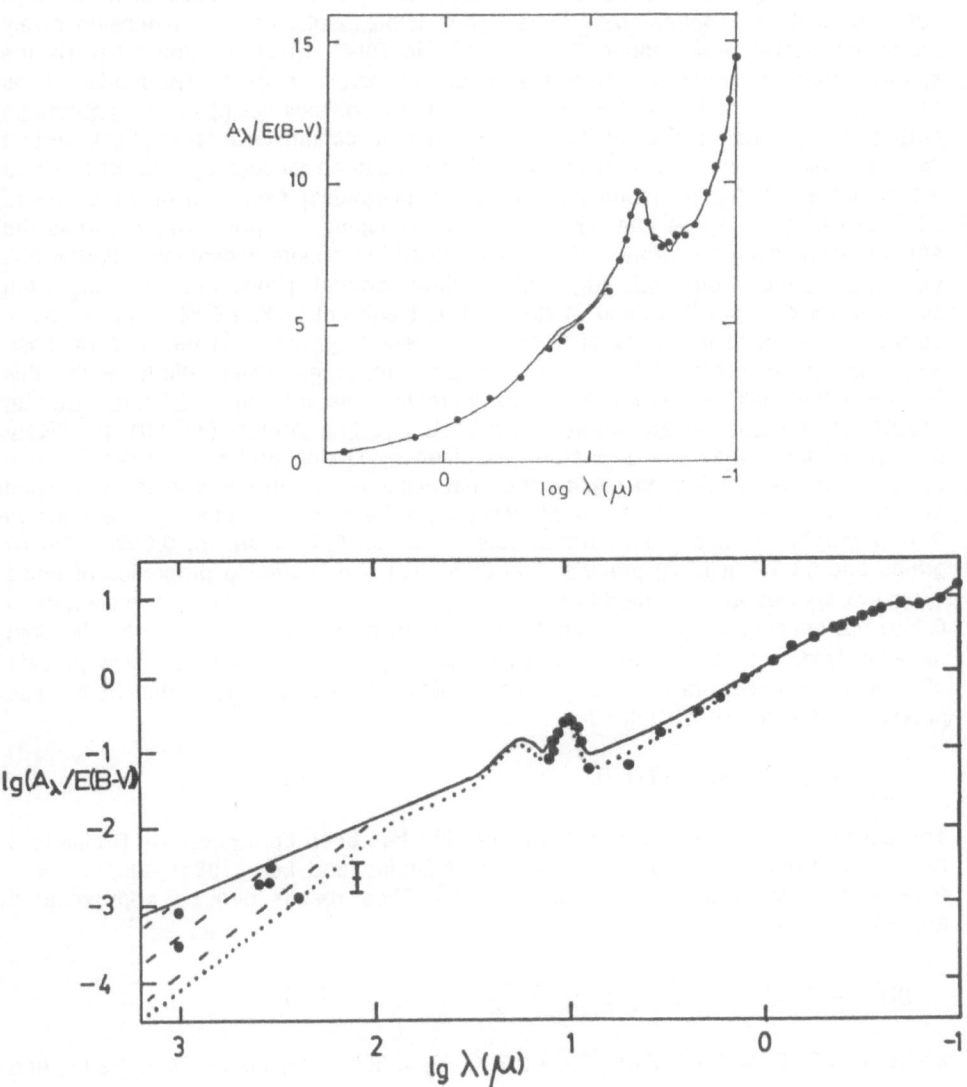

Fig 9: Fit to interstellar extinction curve for model described in section 5
(a) at visible and ultraviolet wavelengths (upper curve: Rowan-Robinson 1986,
lower curve: revised model (differs only near 0.4 μm) (b) in the infrared (solid and
broken curves: Rowan-Robinson 1986, dotted curve: revised model). References for
observations are given in Rowan-Robinson (1986).

Models which satisfy the first two of these requirements were presented by Draine and Anderson (1985). Bernard and Desert (1989) have given some details of work which satisfies all three requirements. Here I give some results from an extension of my earlier interstellar grain model (Rowan-Robinson 1986), which is intended to be the simplest possible model that fits all the present observational data. The model retains the 6 grain types of the earlier work, with some modifications: (i) 0.1 μ amorphous carbon grains, their optical properties derived from circumstellar dust shells around carbon stars. The absorption efficiency of these has been reduced by a factor of 1.5 at wavelengths > 0.4 μ to improve the fit to the interstellar extinction curve at 5-9 μ, while retaining the same total extinction at wavelengths < 1 μm. This also has the effect of increasing the visible and ultraviolet albedo to a more acceptable value of 0.7. (ii) 0.1 μ amorphous silicate grains, their optical properties derived from circumstellar dust shells around M stars. (It is worth noting that 50% of the mass of carbon and 80% of the mass of silicon in interstellar grains is in the form of these larger amorphous grains. We see them being manufactured in situ. We know that this is where the bulk of interstellar grains were last made.) (iii) 0.03 μ graphite grains, (iv) 0.03 μ silicate grains, (v) 0.01 μ graphite grains, (vi) 0.01 μ silicate grains, all four types with properties as given by Draine and Lee (1984). These components are required to explain the interstellar extinction curve in the ultraviolet and the 0.22 μ feature. The main difference from the earlier model is that the mass in 0.01 μ graphite grains is now redistributed between 0.01 μ grains, 0.002 μ (20 Å) grains and 0.0005 μ (5 Å) grains. The absorption and scattering properties of these latter two species are assumed to be the same as the 0.01 μ grains at wavelengths > 0.1 μ, but because they are so small they will not be in equilibrium with the incident radiation field. Instead we have to assume that they have a certain probability p(T) dT of having a temperature between T and T + dT. The emission spectrum from these grains then has to be calculated from

$$I_\nu = \int Q_\nu \, B_\nu(T) \, p(T) \, dT \quad . \qquad (1)$$

The calculation of p(T) is a complex matter but has been carried out by Draine and Anderson (1985) for the grain properties of Draine and Lee (1984) adopted here (see also Guhathakurta and Draine 1989). Their results can be approximated analytically as

$$p(T) = k \, T^{-b} \quad \text{for } T_1 \le T \le T_2 , \qquad (2)$$

where b = 2.75, and k = 6.68, T_1 = 2.7 K, T_2 = 500 K, for the a= 20 Å grains, and k = 0.168, T_1 = 2.7 K, T_2 = 80 K, for the a = 5 Å grains. Here I am assuming that the very small grains emit the bulk of their radiation as a continuum. Roche (1988) estimates that galaxies emit 1% of their energy in the form of unidentified features and as only 10% of the energy of galaxies is emitted at 2-20 μ, we can infer that only about 10% of the radiation from very small grains emerges as the unidentified features. It will be relatively simple to incorporate these features into the calculation in future.

TABLE 1: PARAMETERS FOR GRAIN MODEL

type	B_0	abundance % cosmic ($\times 10^{-4}$)		grain temperatures		
				X=1	X=10	X=500
0.1μ amor Si	1.36	0.257	77	14.5	21.4	42.6
0.03μ Si	1.10	0.063	19	16.8	24.5	47.0
0.01μ Si	0.52	0.010	3	17.5	25.4	48.5
0.1μ amor C	0.45	1.02	20	15.3	22.6	46.0
0.03μ C	0.65	0.45	9	19.3	28.7	60.3
0.01μ C	1.0	0.24	4.7	19.7	29.0	61.7
0.002μ C	0.8	0.039	0.8			
0.0005μ C	0.8	0.010	0.2			

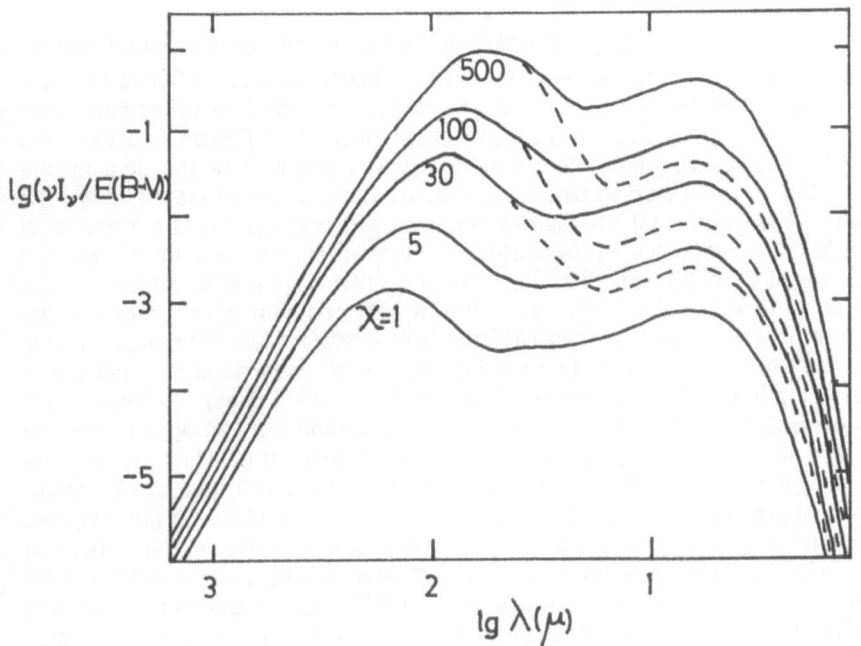

Fig 10: Predicted infrared emissivity of interstellar grains (ergs/cm^2/s/mag) as a function of the intensity of the radiation field. For X ≥30, the effect of a 90% destruction of the smallest grains (5 Å) is shown as broken curves.

In the earlier calculation (Rowan-Robinson 1986), I considered values of the wavelength at which the absrprtion efficiency of the 0.1 μ grains steepened from Q_ν $\propto \nu$ to $Q_\nu \propto \nu^2$ of 100, 316 and 1000 μ. In the present model I take this wavelength to be 80 μ, which is still consistent with the IRAS data for circumstellar dust shells (with the possible exception of IRC+10216, Rowan-Robinson et al 1986) and gives an acceptable fit to the data for high latitude dust clouds in our Galaxy. The possibility that $Q_\nu \propto \nu$ to a wavelength significantly larger than 100 μm is now completely ruled out by observations in our Galaxy and other galaxies. Fig 9a shows the fit to the interstellar extinction curve at visible and ultraviolet wavelengths. Fig 9b shows the overall fit at 0.1 -1000 μ. Fig 10 shows the predicted emission spectra for interstellar dust in the infrared for a range of heating intensities (the temperatures of the different grain components are given in Table 1). For $X \geq 30$, the effect of 90% depletion of the 5 Å grains is also illustrated. Fig 11a shows the predicted emissivity for grains immersed in the local interstellar radiation field compared with obervations of high latitude clouds. The agreement with observations is excellent both in the shape of the spectrum and in the absolute value of the emissivity. The 12-100 μm emissivity of the isolated cloud observed by Herter et al (1989) also agrees with that predicted in Fig 11a.

Fig 11b shows the corresponding prediction and observations for the central regions of the Galaxy ($l < 30^\circ$), where the intensity of the radiation field corresponds to $X = 5$: the fit is also satisfactory. Fig 12 shows the predicted IRAS colour-colour diagrams for $X = 1$-500 (colour-corrected as in Appendix A of Rowan-Robinson and Crawford 1989) compared with observations. Figs 12a and b show the data for the unresolved IRAS galaxies studied by Rowan-Robinson and Crawford (1989). Figs 12c and d show data for resolved IRAS galaxies mapped by Rice et al(1988) and Young et al (1989). Galaxies with $\log\{S(60)/S(25)\} \leq 0.5$ need the additional Seyfert component peaking at 25 μm. For the resolved galaxies, most of which can be explained as pure disc (cirrus) emission, it can be seen that a range of heating intensities are present, from $X = 1$ for NGC205 to $X = 30$ for M33. The vast majority of the galaxies whose colours are shown in Figs 12 can be understood as a mixture of Seyfert (S) + starburst (B) + one of the cirrus models (curved lines). There are 3 classes of exception to this. (a) Two galaxies, NGC1569 and Arp 220 appear to lie on the locus of a highly extinguished starburst model. (b) Several galaxies, notably the SMC, lie to the right of the cirrus curve in the 25-60-100 μm colour-colour diagram and above and to the right of the cirrus curve in the 12-25-60 μm diagram, consistent with the effect of destruction of very small grains at high heating intensities. However not all galaxies with high heating intensity show evidence for very small grain destruction. M33, which like the SMC has a spectrum consistent with $X = 30$, appears to have a normal abundance of very small grains. It is also worth noting that there appear to be no galaxies in which the abundance of 5 Å grains is reduced by more than a factor of 10 compared to the solar neighbourhood. The possibility that reduction in the carbon abundance in galaxies (but not the silicon) is the cause of the anomalous colours needs to be explored, especially for the SMC. (c) Several galaxies, for example M31, have 25-60-100 μm colours consistent with cirrus but have very low values of $S(25)/S(12)$, implying excess radiation at 12

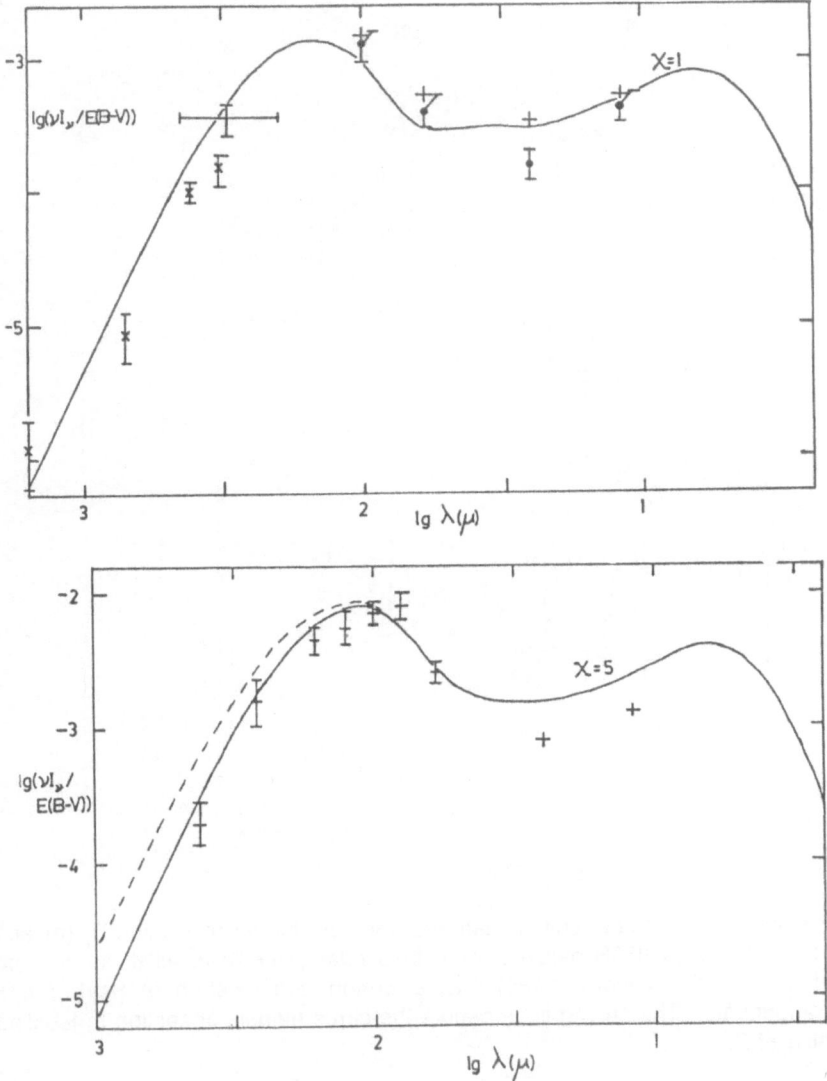

Fig 11: Predicted emissivity, compared with observations (a) towards the Galactic pole (X=1, data from Boulanger & Perault 1989, Halpern et al 1988, Fabbri et al 1988, assumed E(B-V)=0.05), (b) towards the central regions of the Galaxy (X=5, data from Beichman 1987 and refs therein, assumed E(B-V)=6.1). The broken curve shows the effect of assuming the wavelength at which the absoption efficiency of the grains steepens is > 1mm.

140

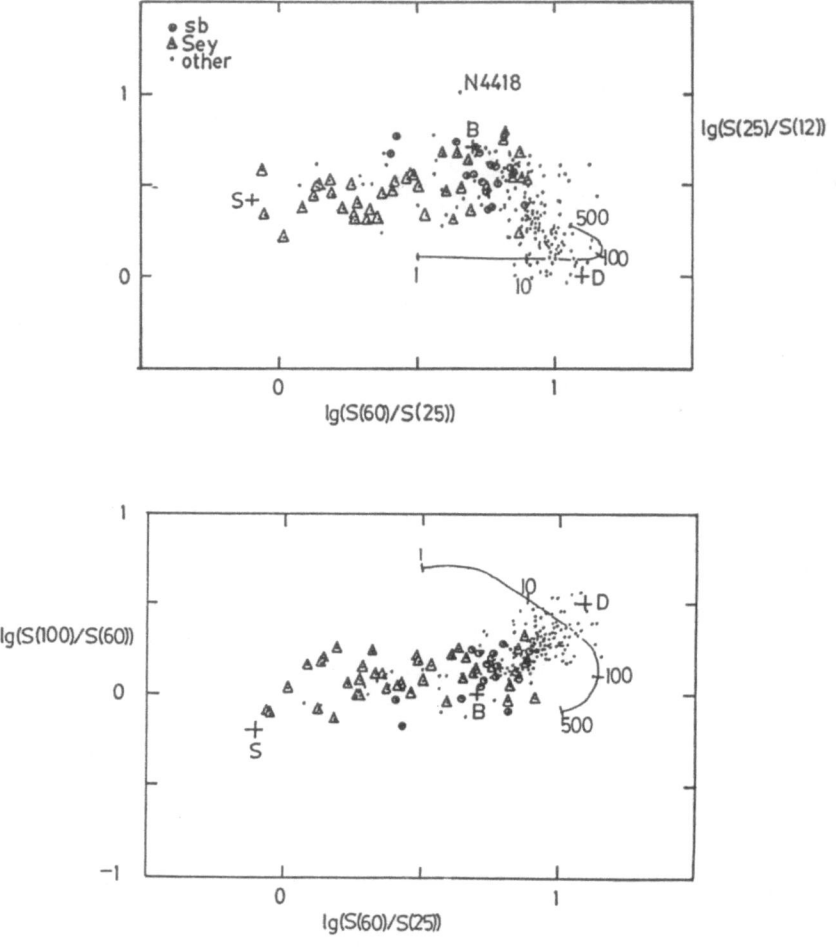

Fig 12: Predicted IRAS colour-colour diagrams for dust model of section 5. (a) and (b) Data for unresolved IRAS galaxies with good quality fluxes in all 4 bands from Rowan-Robinson and Crawford (1989). S, B denote the location of Seyfert and starburst components. The curved lines denote the cirrus models of section 5, labelled with the value of X.

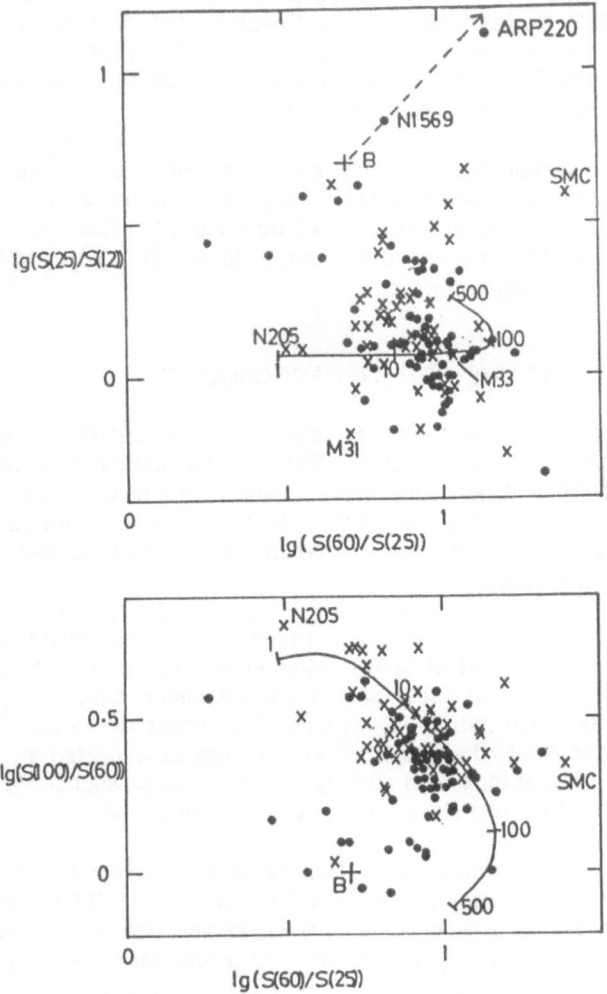

Fig 12 (c) and (d). Data for resolved IRAS galaxies with coadded fluxes in all four bands from Rice et al (1988, crosses) and Young et al (1989, filled circles). Only galaxies with fluxes brighter than 0.4 Jy in all four bands were included. The broken line in Fig 12c shows the effect of reddening ($A_V = 40$) on the starburst component.

μm. Possible explanations of this are a strong contribution from circumstellar dust shells (Soifer et al 1986, Rowan-Robinson and Chester 1987: though for M31 the spatial distribution of the 12 μm radiation does not differ from that at longer wavelengths) or an unusually strong contribution from PAH/very small grains. 8-13 μm spectroscopy of these galaxies would be very valuable.

Discrepant colours occasionally result from poorly determined fluxes, especially at 25 μm where not all IRAS detectors were functioning. For this reason galaxies with fluxes lesss than 0.4 Jy in any band were omitted from Fig 12. However in general experience suggests that IRAS colours are accurate to 0.1 in $\log_{10}$ and that any discrepancy greater than this has a real cause.

6. FAR INFRARED AND SUBMILLIMETRE MAPPING OF GALAXIES

Prior to the launch of IRAS rather little information on the spatial extent of far infrared emission in galaxies was available. Some of the earlier work has been reviewed by Telesco (1988). One of the most significant pre-IRAS studies was by Smith (1982), who produced a 170 μ map of the disk of M51, which showed that the bulk of the far infrared emission in M51 is produced by dust associated with the diffuse gas in the disk of the galaxy.

IRAS extended data is still under active study by several groups. Detailed maps have been produced of M31 (Habing et al 1986, Soifer et al 1987 and Walterbos and Schwering 1987), M33 (Rice et al 1989) and of the Magellanic Clouds (Schwering 1988). Rice et al (1988) have published coadded IRAS maps for all galaxies with optical extent greater than 8'. Higher resolution images may be expected for many of these galaxies from the use of maximum entropy and other deconvolution techniques now under active study at IPAC and elsewhere (eg Canterna et al 1989).

Subsequent studies have for the most part concentrated on wavelengths longer than 100 μ. Stark et al (1988) have mapped 4 Virgo spirals at 160 and 350 μ and shown that that there is no evidence for a grain component whose emission peaks beyond 200 μ, a prediction of grain models with emissivity $Q_\nu \propto \nu$ at wavelenths > 100 μ. Eales and Wynn-Williams (1989) have measured 350 and 450 μ fluxes at locations centred on several galaxies with a 100" beam. 160 μm maps have been published of NGC4449 (Thronson et al 1987), NGC4214 (Tnronson et al 1988), NGC 4485 and 4490 (Thronson et al 1989a) and NGC1569 and 3593 (Hunter et al 1989). Thronson et al (1989b) have published maps of IC10 at 95 and 160 μ and given fluxes for several other galaxies. Eckart et al (1989) have mapped Centaurus A at 50 and 100 μ, and have separated the cirrus and starburst components. Smith et al(1989a) have mapped M82 at 450 μm, Smith et al (1989b) have mapped M83 at 100 μm and Engargiola and Harper (1989) have mapped NGC6946 at 100, 160 and 250 μ. Hughes et al (1989) have mapped 8 IRAS galaxies at 800 μm with JCMT and given some 450 and 1100 μm data for some of them. The importance of the longer wavelengths is that the most reliable estimates of dust mass can be obtained at these wavelengths. Fig 13 shows far infrared spectra of selected galaxies from this latter study, with theoretical fits derived from the models described in section 5. Fig 14a-d

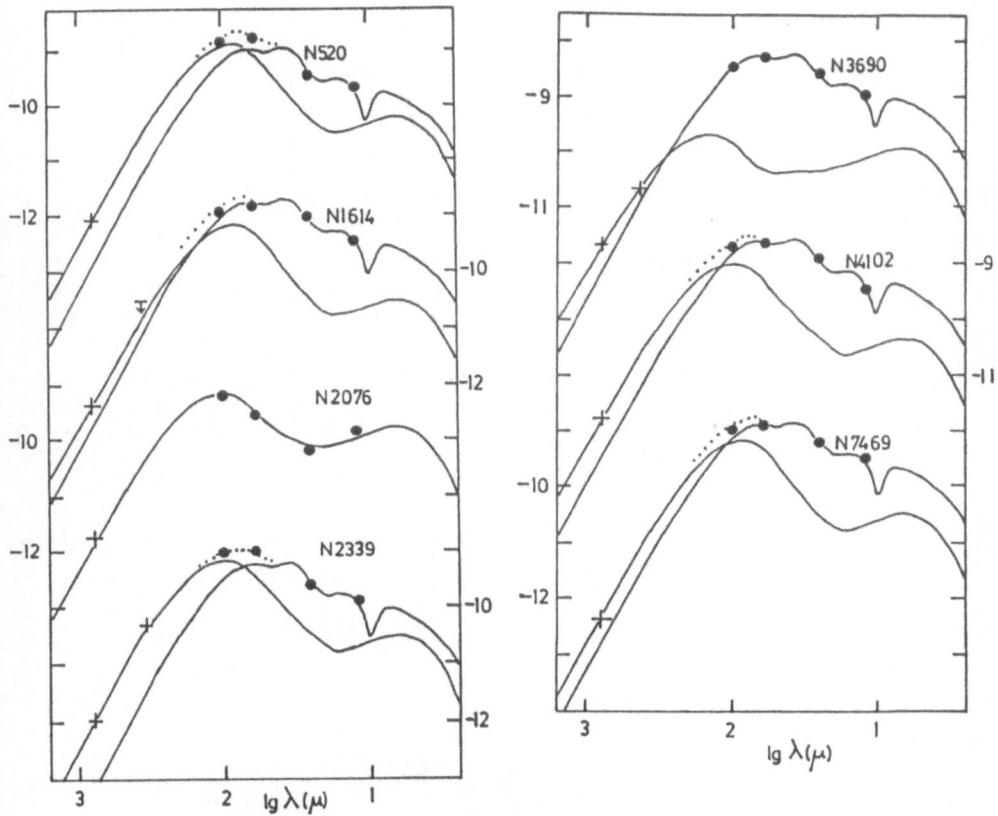

Fig 13: Far infrared and submillimetre spectra of galaxies mapped by Hughes et al (1989) at 800 μm, compared with models of section 5. For NGC 2076 a pure cirrus model, with depletion of the smallest grains, is satisfactory. For the other galaxies both cirrus and starburst components are required (the dotted curve indicates the total predicted flux). Parameters for the models are given in Table 2.

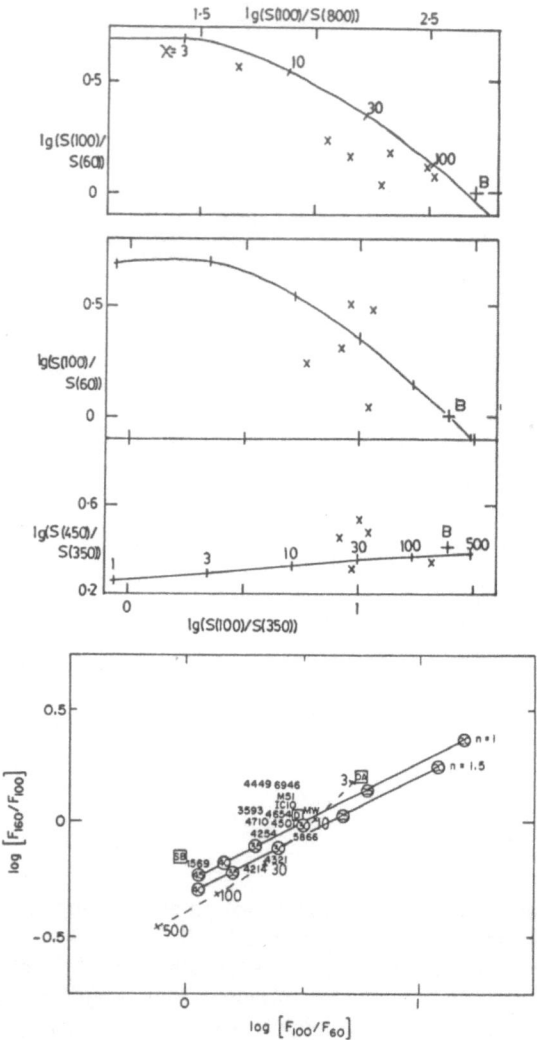

Fig 14: Far infrared colour-colur diagrams, compared with predictions of interstellar grain model of section 5. (a) 100/60 versus 100/800 (data from Hughes et al 1989), (b) 100/60 and (c) 450/350 versus 100/350 (data from Eales et al 1989). (d) 160/100 versus 100/60 (data from Thronson et al 1989b: solid curves, $\nu^n B_\nu$ (T) fits; broken curve, cirrus model of section 5).

show colour-colour diagrams derived from the work of Thronson et al (1989b), Hughes et al (1989) and Eales and Wynn-Williams (1989) compared with the predictions of the models of section 5.

7. DETERMINATION OF DUST MASS IN GALAXIES

Hildebrand (1983) gave a prescription for deriving dust masses in galaxies from far infrared data which has been widely used. Young et al (1989) have used the Hildebrand prescription to conclude that the average gas-to-dust ratio in galaxies is 1200. Draine (1989) has given a discussion of the derivation of dust masses which emphasizes some of the difficulties. He emphasizes that here is considerable disagreement about the grain opacity at long wavelengths, though this disagreement is somewhat exaggerated by illustrating the most extreme of the models discussed by Rowan-Robinson (1986) in which $Q_{y} \propto \nu$ all the way to 1 mm. Draine shows that if only IRAS observations are available of galaxies, then the derivation of dust mass is very uncertain, since several rather different models could in principle be fitted to the same observations. However provided a significant proportion of the 12-100 μm emission from a galaxy is due to cirrus, and fluxes are available in all four IRAS bands, a good separation into cirrus and starburst components can be made, and reasonable estimates of dust mass derived. Observations at long wavelength (>300 μm) are very valuable in tying down the value of X , the radiation field intensity, and are essential if the 12-100 μm spectrum is dominated by a starburst.

Table 2 gives dust masses derived from the study of Hughes et al (1989) based on the grain model of Rowan-Robinson (1989) described in section 5 above. Comparison of the dust mass in the cirrus component with the neutral hydrogen masses given by Young et al (1989) shows normal gas-to-dust ratios for this model in most cases. However since in many cases the neutral hydrogen in a galaxy extends well beyond the optical image, whereas the bulk of the infrared emission is generally located within the optical image, there may be a tendency to underestimate the total dust mass from far infrared observations. Dust in the outer parts of a galaxy, illuminated with a starlight intensity much lower than in the central regions, may contribute only a very small fraction of the total infrared flux. Sensitive observations at long wavelengths with large beam-throws will be needed to characterize such dust.

It is important when modelling the far infrared emission from dust in galaxies to take account of the fact that several grain components are present, at different temperatures. Calculations based on the assumption of a single composite grain model and a single temperature are unlikely to yield accurate results. However the cirrus models of Fig 10 can be approximately fitted at long wavelengths with a $\nu^{2}B$ (T) curve, with the values of T as given in Table 3 for different X . The validity of this fit is for $\lambda > 1700/T$ μm. Also given for these models are the values of $\log\{S(100)/S(60)\}$ and $\log \{M_{d}/S(100\mu m) D^{2}\}$, $\log \{M_{d}/S(800\mu m) D^{2}\}$. Note that whereas $M_{d}/S_{100} D^{2} \propto X$, $M_{d}/S_{800} D^{2} \propto X^{0.3}$, so much more accurate dust masses can be obtained if long wavelength observations are available.

TABLE 2: PARAMETERS FOR GALAXIES MAPPED BY HUGHES ET AL (1989)

| galaxy | distance (Mpc) (H=50) | cirrus model | | | | starburst model |
		X	depletion of 5 Å grains	log M_d(C)	log M(HI)	log M_d(SB)
NGC520	45.4	30	90%	7.25	10.10	5.94
NGC1614	92.9	30	90%	7.57	9.88	6.83
NGC2076	48.4	10	50%	7.73		-
NGC2339	46.7	10	90%	7.50	10.05	5.75
NGC3690	62.1	1	-	8.34	<9.73	6.97
NGC4102	19.7	10	90%	6.97	9.02	5.60
NGC7469	102.0	30	90%	7.63	9.90	6.80

TABLE 3: CIRRUS MODEL PARAMETERS FOR DUST MASS DETERMINATION

X=	1	3	5	10	20	30	50	100	200	500	SB
log{S(100)/S(60)}	0.69	0.70	0.64	0.54	0.44	0.34	0.24	0.13	0.03	-0.11	0.0
$T(\nu^2 B_\nu)$ [a]	16	19	21	24	27	29	31	34	37	43	40
log(M_d/S_{100} D^2) [b]	4.08	3.42	3.16	2.83	2.60	2.37	2.17	1.93	1.67	1.44	1.33
log (M_d/S_{800} D^2) [b]	4.99	4.84	4.78	4.70	4.62	4.57	4.52	4.43	4.34	4.19	

a   valid for $\lambda > 1700/T$ μm
b   solar masses/(Jy Mpc^2)

8. DUST IN ELLIPTICALS AND LENTICULARS

There has been a growing realization that ellipticals and lenticulars have a significant interstellar medium, and that interesting amounts of star formation take place there. There has been a decade of work on HI emission from ellipticals and more recently CO observations in several cases (see eg the reviews by Wardle & Knapp (1986) and Schweizer (1987)). Although the majority of the galaxies detected by IRAS are spirals, quite a number of ellipticals and lenticulars were detected (eg Jura 1986, Jura et al 1987, Knapp et al 1989). Thronson and Bally (1987) have studied the IRAS colour-colour diagrams for these galaxies and conclude that they occupy the same region of the diagrams as spiral galaxies (and, for that matter, star forming regions in our Galaxy). About 2/3rds of the sample they studied have the colours characteristic of cirrus and 1/3rd those of dusty regions surrounding young stars. The star formation rate they derive (0.1-1 $M_\odot$ /year) is comparable to the mass-loss rate for evolved stars in these galaxies, but mergers and gas infall may also contribute significantly. Bally and Thronson (1989) studied the IRAS data for a sample of 74 S0 galaxies which had known single-dish radio fluxes. 30% were detected in all 4 IRAS bands and 80% were detected in at least one band. The galaxies divided into those which followed the infrared-radio relation for spirals, for which the radio emission is presumably due to normal star formation, and those with excess radio emission, presumably due to an active nucleus and jets or lobes. A small number showed a slight excess of infrared to radio. Similar conclusions were reached by Walsh et al (1989). Knapp et al (1989) report that 2/3rds of a sample of several hundred SOs are detected by IRAS at 60 and 100 μm.

Thronson et al (1989c) examined the IRAS data for 150 lenticular and elliptical 'shell' galaxies (Malin and Carter 1983), which are believed to be the result of low velocity mergers. Although some of the galaxies showed evidence for enhanced star formation, the majority did not and they concluded that either (1) the merging galaxies are almost always E or S0 with only modest amounts of interstellar gas, or (2) the time-scale for creation and maintenance of the shell is longer than the time-scale for the starburst event, or (3) the formation of a shell structure requires a mass difference between the galaxies of a factor 10-100, so only a small fraction of the i.s.m. is heated or participates in star formation.

Walsh and Knapp (1989) find that the ellipticals detected by IRAS tend preferentially to be those with dust lanes visible in the optical. However the infrared properties are not strongly dependent on the visible dust content. They also find a slightly enhanced 100 μm detection rate for ellipticals with shells, boxy isophotes or inner discs, all of which are evidence of a recent merger, a result which is not necessarily inconsistent with that of Thronson et al (1989b).

It is unfortunate that the infrared sources associated with early type galaxies are almost all rather weak, so that there is little immediate prospect of detection at wavelengths > 300 μm, and hence of accurate dust mass determinations.

REFERENCES

Allamandola, L.J., Tielens, A.G., & Barker, J.R., 1985, Ap.J. 290, L25
Allen, M., & Robinson, G.W., 1975, Ap.J. 195, 81
Andriesse, C.D.,1978, A.A. 66, 169
Bally, J., & Thronson, H.A.,Jr, 1989, A.J. 97, 69
Beichman, C.A., 1987, A.R.A.A. 25
Bernard, J.P, & Desert, X., 1989, this volume
Boulanger, F., Baud, B., & van Albada, G.D., 1985, A.A. 144, L9
Boulanger, F., Beichman, C., Desert, F.X., Helou, G., Perault, M., & Ryter, C., 1988,
    Ap.J. 332, 328
Boulanger, F., & Perault, M., 1988, Ap.J. 330, 964
Canterna, R., Hackwell, J.A., & Grasdalen, G.L., 1989, this volume
Crawford, J., & Rowan-Robinson, M., 1986, MNRAS 221, 923
Cox, P., & Mezger, P.G., 1989, Astron.Astrophys.Review
Desert, F.X., and Dennefeld, M., 1988, A.A. 206, 227
Donn, B., 1968, Ap.J. 152, L129
Draine, B.T., 1988a, 22nd ESLAB Symposium, Infrared Spectroscopy in   Astronomy
Draine, B.T., 1988b, in IAU Symposium 135,   Interstellar Dust, eds
    L.J.Allamandola & A.G.G.M.Tielens (Reidel)
Draine, B.T., 1989, this volume
Draine, B.T., & Anderson, N., 1985, Ap.J. 292, 494
Draine, B.T., & Lee, H.M., 1984, Ap.J. 285, 89
Duley, W.M.,1973, Nature Phys.Sci. 244, 57
Duley, W.W., 1987, MNRAS 229, 203
Duley, W.W., & Williams, D.A., 1981, MNRAS 196, 269
Duley, W.W., & Williams, D.A., 1988a, MNRAS 231, 969
Duley, W.W., & Williams, D.A., 1988b, MNRAS 230, 1p
Eckart, A., Cameron, M., Rothermel, H., Wild, W., Zinnecker, H., Olberg, M.,
    Rydbeck, G., & Wiklind, T., 1989, this volume
Eales, S.A., and Devereux, N., 1989, this volume
Eales, S.A., Wynn-Williams, G., & Duncan, W.D., 1989, Ap.J. 339, 859
Efstathiou, A., & Rowan-Robinson, M., 1989, MNRAS (submitted)
Engargiola, G., & Harper, D.A., 1989, this volume
Fabbri, R., Guidi, I., Natale, V., & Ventura, G., 1988, preprint
Gillett, F.C., Forrest, W.J., & Merrill, K.M., 1973. Ap.J. 183, 87
Gosh, S.K., & Drapatz, S., 1987, A.A.
Greenberg, J.M., 1968, in Stars and stellar Systems, Vol 7, ed. Middlehurst et al
    (Chicago Univ.Press), p.221
Guhathakutra, P., & Draine, B.T., 1989, Ap.J. (in press)
Habing, H.J., et al, 1984, Ap.J. 278, L59
Halpern, M., Benford, R., Meyer, S., Muehlner, D., & Weiss, R., 1988, Ap.J. 332,
    596
Helou, G., 1986, Ap.J. 311, L33
Helou, G., 1988, in IAU Symposium 135, Interstellar Dust in Galaxies
Herter, T., Shupe, D.L., & Chernoff, D.F., 1989, Ap.J. (in press)
Hildebrand, R.H., 1983, QJRAS 24, 267

Hughes, J., Rowan-Robinson, M., Lawrence, A., & Crawford, J., 1989 (in preparation)

Hunter, D.A., Thronson, H.A.Jr, Casey, S., & Harper, D.A., 1989, Ap.J. 341, 697

de Jong, T., et al, 1984, Ap.J. 278, L67

de Jong, T., & Brink, K., 1987, in Star Formation in Galaxies, ed. C.L.Persson, p.323

Jones, A.P., Duley, W.W., & Williams, D.A., 1987, MNRAS 229, 213

Jura, M., 1986, Ap.J. 306, 483

Jura, M., Kim, D.W., Knapp, G.R., & Guhathahurta, P., 1987, Ap.J. 312, L11

Knapp, G.R., Guhathakurta, P., Kim, D.-W., & Jura, M., 1989, Ap.J. Supp. (in press)

Leech,K.J., Lawrence, A., Rowan-Robinson, M., Walker, D., & Penston, M.V., 1988, MNRAS 231, 977

Leech, K.J., Penston, M.V., Terlevich, R., Lawrence, A., Rowan-Robinson, M., & Crawford, J., 1989, MNRAS 240, 349

Leene & Cox, 1987, A.A. 174, L1

Leger, A., & Puget, J.L., 1984, A.A. 128, 212

Leisawitz, D., 1989, this volume

Malin, D.F., & Carter, D., 1983, Ap.J. 274, 534

Mathis, J.S., Rumpl, W., & Nordsieck, K.H., 1977, Ap.J. 217, 425

Mathis, J.S., Mezger, P.G., & Panagia, N., 1983, A.A. 128, 212

Mathis, J.S., & Whiffen, G., 1989, Ap.J. 341, 808

Pajot, F., Boisse, P., Gispert, R., Lamarre, J.M., Puget, J.-L., & Serra, G., 1986 , A.A. 157, 393

Persson, C., & Helou, G., 1987, Ap.J. 314, 513

Platt, J.R., 1956, Ap.J. 123, 486

Price, S.D., 1981, A.J. 86, 193

Puget, J.L., & Leger, A., 1989, A.R.A.A. 27

Purcell, E.M., 1976, Ap.J. 206, 685

Rice, W., Lonsdale, C.J., Soifer, B.T., Neugebauer, G., Kopan, E.L., Lloyd, L.A., de Jong, T., & Habing, H.J.,1988, Ap.J. Supp

Rice, W., Boulanger, F., Viallefond, F., Soifer, B.T., & Freedman, W.L., 1989, Ap.J. (in press)

Roche, P.F., 1988, in 22nd ESLAB Symposium, Infrared Spectroscopy in Astronomy

Rowan-Robinson, M., 1979, Ap.J. 234, 111

Rowan-Robinson, M., 1982, in Submillimeter Astronomy, ed. P.Phillips & J.Beckman (CUP), p.47

Rowan-Robinson, M., 1986, MNRAS 219, 737

Rowan-Robinson, M., 1987a, in Star Formation in Galaxies, ed. C.Persson, p.133

Rowan-Robinson, M., 1987b, in Starbursts and Galaxy Evolution, eds T.X.Thuan, T.Montmerle & J.T.T.Van (Edition Frontieres) p.235

Rowan-Robinson, M., & Chester, T., 1987, Ap.J. 313, 413

Rowan-Robinson, M., Lock, T.D., Walker, D.W., & Harris, S., 1986, MNRAS 222, 273

Rowan-Robinson, M., & Crawford, J., 1986, in Light on Dark Matter, ed. F.P.Israel (Reidel) p.421

Rowan-Robinson, M., & Crawford, J., 1989, MNRAS 238, 523

150

Ryter et al, 1987, A.A. 186, 312

Sakato, A., Wada, S., Tanabe, T., & Onaka, T., 1983, Nature 301, 493

Sakato, A., Wada, S., Tanabe, T., & Onaka, T., 1984, Ap.J. 287, L51

Schweizer, F., 1987, in IAU Symposium 127, Structure and Dynamics of   Elliptical
  Galaxies (Reidel)

Schwering, P., 1988, Ph.D. thesis, Univ. of Leiden

Sellgren, K., 1984, Ap.J. 277, 623

Soifer, B.T., Houck, J.R., & Neugebauer, G., 1987, A.R.A.A. 25, 187

Soifer, B.T., Rice, W.L., Mould, J.R., Gillett, F.C., Rowan-Robinson, M., &   Habing,
  H.J.,1986, Ap.J. 304, 651

Smith, J., 1982, Ap.J. Ap.J. 261, 463

Smith, P.A., Brand, P.W.J.L., Puxley, P.J., Mountain, C.M., Gear, W.K., & Nakai,   N.,
  1989a, this volume

Smith, B.J., Lester, D.F., & Harvey, P.M., 1989b, this volume

Stark, A.A., Davidson, J.A., Harper, D.A., Pernic, R., Loewenstain, R., & Casey,   S.,
  1989, Ap.J. (in press)

Telesco,, 1988, A.R.A.A. 26, 343

Telesco, C.M., Becklin, E.E., & Wynn-Williams, G., 1984, Ap.J. 282, 427

Telesco, C.M., Decher, R., & Joy, M., 1989, Ap.J. (in press)

Thronson, H.A., Jr, & Bally, J., 1987, Ap.J. 319, L63

Thronson, H.A.Jr, Hunter, D.A., Telesco, C.M., Harper, D.A., & Decher, R., 1987,
  Ap.J. 317, 180

Thronson, H.A.Jr, Hunter, D.A., Telesco, C.M., Greenhouse, M., & Harper, D.A., 1988,
  Ap.J. 334, 605

Thronson, H.A. Jr, Hunter, D.A., Casey, S., Latter, W.B., & Harper, D.A., 1989a, Ap.J.
  339, 803

Thronson, H.A.Jr, Hunter, D.A., Casey, S., & Harper, D.A., 1989b, Ap.J. (in   press)

Thronson, H.A., Jr, Bally, J., & Hacking, P., 1989b, A.J. 97, 363

Tielens, A.G.G.M., & Allamandola, L.J., 1987, in Interstellar Processes, eds
  D.Hollenbach & H.A.Thronson Jr (Reidel) p.397

Walsh, D.E.P., Knapp, G.R., Wrobel, J.M., & Kim, D.-W., 1989, Ap.J. 337, 209

Walsh, D., & Knapp, J., 1989, this volume

Walterbos, R.A.M., & Schwering, P.B.W., 1987, A.A.Supp. 180, 27

Wardle, M., & Knapp, G.R., 1986, A.J. 91, 23

Williams, D.A., 1989, preprint

Young, J.S., Xie, S., Kenney, J.D.P., & Rice, W.L., 1989, Ap.J. Supp. (in press)

The Effects of Environment on the Cool Phase
of the Interstellar Medium in Galaxies

Jeffrey D. P. Kenney
Owens Valley Radio Observatory
California Institute of Technology

ABSTRACT. The atomic and molecular gas properties of many galaxies are altered by external influences. Ionization by the cosmic ultraviolet background radiation can account for sharp cutoffs observed in the HI disks of spirals. Tidal interactions between galaxies strip outer disk gas and cause gas clouds to collide more frequently, leading to large radial inflows of gas and enhanced rates of star formation. And in many clusters, interactions between gas in the intracluster medium (ICM) and gas in a galaxy's interstellar medium (ISM) lead to stripping of outer disk gas, and may perturb the remaining ISM sufficiently to trigger star formation.

1. Introduction

As we learn more about galaxies, we find increasingly that their evolution is influenced by their environment. Many of the peculiar galaxies known in the local universe owe much of their optical appearance to stellar orbits perturbed by gravitational interactions between galaxies (Toomre and Toomre 1972). Yet the gaseous components of galaxies are more fragile than the stellar components, being susceptible to processes other than gravity. Since the distribution and dynamics of gas controls both the rate of star formation and the feeding of active nuclei, understanding the environmental influences on a galaxy's gaseous reservoir is essential for a complete understanding of galaxy morphology and evolution.

The interstellar media of galaxies are known to be affected by three general outside influences: gravity from other galaxies, gas from other galaxies or an intergalactic medium, and cosmic ultraviolet photons. In this review I give an overview of the external processes known to affect the atomic and molecular gas properties of galaxies. Brief discussions will be given on tidal encounters and the ionization of HI by the cosmic UV background. A more extensive discussion will be given on the environmental effects on galaxies within clusters. The article by Noguchi in this volume discusses the interstellar medium during galaxy collisions.

H. A. Thronson, Jr. and J. M. Shull (eds.), The Interstellar Medium in Galaxies, 151–180.
© 1990 *Kluwer Academic Publishers.*

2. Ionization of the Neutral ISM by the Cosmic UV Background

The ultraviolet radiation field in intergalactic space ionizes the atoms at the periphery of galaxies, creating a galaxy-sized, enveloping HII region. Twenty years ago, Sunyaev (1969) argued that the universe could not be closed by hot gas in an intergalactic medium, since HI column densities of $\sim 10^{20}$ cm^{-2} were commonly observed in galaxies. The detection of neutral hydrogen at that column density is inconsistent with the bremsstrahlung radiation which would be produced in the UV by hot ($T \geq 10^5$ K) intergalactic gas at the critical density ($n \sim 10^{-5}$ cm^{-3}). Today, although HI surface densities of a few times $\sim 10^{19}$ cm^{-2} have been measured in many galaxies, gas with $\sigma_{HI} \leq 3 \times 10^{18}$ cm^{-2} in the outer disks of galaxies is relatively rare (Briggs et al. 1980). The lack of low column density HI extensions to disk galaxies may be the result of ionization by the cosmic UV radiation field.

Extragalactic photons undoubtedly contribute to the ionization of halo gas above the plane of the Milky Way in the vicinity of the solar neighborhood. The best evidence for this is the abrupt increase in the ionization state of the gas at ~ 1 kpc above the galactic plane (Savage 1988). But it is difficult to unambiguously distinguish cosmic UV photons from Milky Way UV photons when there are Milky Way UV sources so nearby. The presence of the cosmic UV background can be more directly measured at the outer edges of galaxies, far away from any internal ionizing sources. The ionization by cosmic photons will cause a cutoff in the HI column density in the outer disk, which should be observable given sufficient sensitivity.

Sharp HI cutoffs which may be the result of external ionization have been observed in at least two galaxies. Figure 1 shows the HI surface density versus radius in the outlying regions of M33 from sensitive Arecibo observations by Corbelli, Schneider, and Salpeter (1989). In the inner galaxy (not shown) the HI surface density changes slowly with radius, but at a distance of 18 kpc from the center of M33, the HI surface density suddenly drops by more than an order of magnitude within 1 kpc. Very deep imaging with the VLA has revealed a similarly sharp truncation in the HI disk of the normal spiral NGC 3198 (van Gorkom et al. 1989). In both cases, the HI surface density drops from a few times 10^{19} cm^{-2} to a few times 10^{18} cm^{-2} within ~ 1-2 kpc, a rate of change greater than any disk scale length.

If this "missing" atomic gas has been ionized by extragalactic UV photons, then the HI column density where the falloff begins and the steepness of the falloff constrain the intensity and spectral index of the background radiation (Silk and Sunyaev 1976). The falloff in HI surface density is not expected to be as sharp as that in HII regions around hot stars since the cosmic UV background has a harder spectrum (Silk 1971). Compared to stellar HII regions, the cosmic background has many more soft x-ray photons, which penetrate to a greater depth than Lyman limit photons. Helium is photoionized by soft x-ray photons, and its UV recombination radiation ionizes HI at greater depths than primary Lyman limit photons. An HI cutoff at $\sim 3 \times 10^{19}$ cm^{-2} can be produced by a UV intensity of $\sim 4 \times 10^{-23}$ erg cm^{-2} s^{-1} Hz^{-1} sr^{-1} at the Lyman limit if the spectral index is ~ 1.5 (Maloney 1989). Galaxies at large redshifts should have their HI disks truncated at even higher column densities, since the extragalactic UV radiation field is believed to be at least 10 times stronger at redshifts z=1.7-3.8 (Bajtlik, Duncan, and Ostriker 1988).

The nature of the HI falloff also depends upon the scale height of the gas layer, the degree of clumpiness, and the gas temperature (Maloney 1989), so that the physical properties of gas in outlying disk regions can be constrained by modelling. Maloney finds that the sharpness of the HI edge in NGC 3198 is difficult to explain unless the gas density decreases rapidly over a short range in radius, meaning that the HI disk must flare significantly. Of course, the steep radial falloffs in HI cannot at present be unambiguously ascribed to ionization. Ionized gas has not yet been observed from HI cutoff regions. In the case of M33, the steep falloff is based on only 1 radial cut.

Observations of sharp HI edges in more galaxies and detailed modelling of the HI-HII transition zone are needed to improve our understanding of gas properties in both the halo and outer disk, as well as the nature of the ultraviolet radiation field. The origin of the UV background is currently uncertain, at least at high redshifts. While the presence of highly ionized species in QSO absorption line systems indicates that the UV background has an AGN-like spectrum (Steidel and Sargent 1989), there are not enough quasars known to account for the necessary luminosity (Lin and Phinney 1989). Knowledge of the intensity and spectral index of the cosmic background in the local universe may provide clues toward understanding the universe during the epoch of galaxy formation.

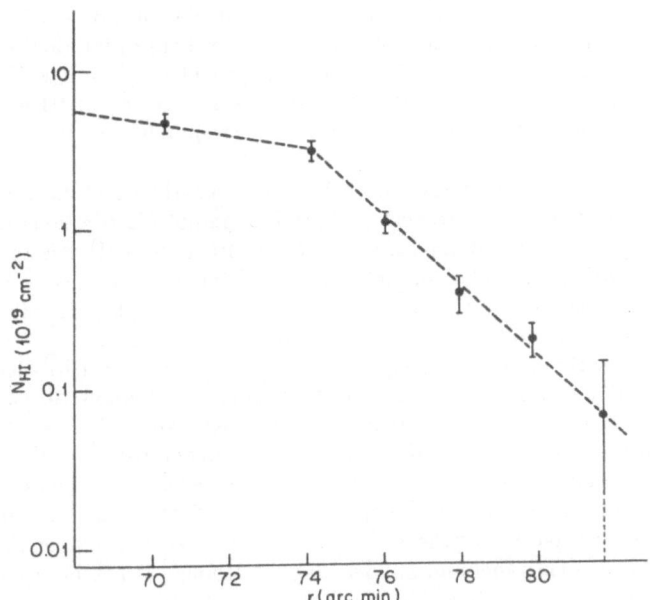

Figure 1 HI column density vs. radius in the outer disk of M33. Although the HI column density changes relatively slowly with radius in the inner disk, at R=18 kpc it drops sharply by more than an order of magnitude within 1 kpc (4'). This sharp cutoff could be the result of ionization by the extragalactic ultraviolet radiation field. Figure is from Corbelli, Schneider, and Salpeter (1989).

3. Galaxy-Galaxy Interactions

3.1 INTRODUCTION

Far from being just carried along for the ride with the stars in a galaxy-galaxy collision, gas actually determines many of the essential features of galactic interactions. In this section I will briefly review the various effects on the ISM which are known, or proposed, to occur when one galaxy experiences a transient gravitational encounter or merger with another. These include stripping of gas by tidal forces and ram pressure, enhanced collision rates of gas clouds leading to enhanced massive star formation and radial gas flows, and large scale conversion of atomic to molecular gas. Most of the important effects in galaxy-galaxy interactions are the by-product of tidal forces. Although gravity does not discriminate between gas and stars, the *consequences* of a gravitational interaction are very different for gas and stars since gas clouds collide and dissipate angular momentum and energy.

3.2 OBSERVED GAS PROPERTIES IN GALAXY GROUPS

Tidal interactions are especially important in groups of galaxies, where the velocity dispersion among the member galaxies is comparable to the random and rotational velocities within galaxies. In such an environment, the most obvious consequence of tidal interactions is the stripping of stars and gas from the outer disks of interacting galaxies. A good example of a tidal encounter in a loose group is provided by the Leo Triplet. Mapping of HI (Haynes, Giovanelli and Roberts 1979) reveals a large tail of HI near the spiral NGC 3628. The morphology of the HI tail and its correspondence with a stellar tail provide direct evidence for a tidal interaction.

Even though gravity removes both stars and gas, the effect on the HI is greater than on the stars in at least one respect. The initial radial distributions of HI and stars in galaxies are typically very different, with a larger fraction of HI existing in the outer disk as compared with stars. Thus outer disk tidal stripping may make a galaxy moderately HI-poor (i.e., lower its M_{HI}/L_B or M_{HI}/D_{opt}^2 ratio), although probably not very *gas-poor* when the other components of the ISM are taken into account.

Galaxies within *compact* groups are observed to be more HI-deficient than those in *loose* groups by an average of a factor of 2 (Williams and Rood 1987). The high rate of galaxy-galaxy collisions and mergers in these compact groups (Barnes 1989) probably lead to their lack of HI. The cause of HI-deficiency in compact groups could be outer disk tidal stripping, an enhanced conversion of HI into H_2, or even the depletion of gas due to an increased rate of star formation. Comparisons of CO with HI emission in highly luminous merging systems suggest that the global H_2/HI ratio becomes higher than normal during the merger process (Mirabel and Sanders 1989), although it has been questioned whether the relationship between CO and H_2 is the same in these peculiar galaxies as it is in the Milky Way (Maloney and Black 1988). This uncertainty aside, the large CO/HI ratios (Mirabel and Sanders 1989) and the large amounts of molecular gas inferred to exist in the central regions of the most luminous mergers (Scoville *et al.* 1989), and the HI deficiency of galaxies in compact groups (Williams and Rood 1987) all suggest that atomic gas is converted into molecular gas in strong tidal interactions and mergers.

Although gravitational interactions are the dominant environmental effect in groups, ICM-ISM interactions (e.g., ram pressure) may be dominant in clusters. Yet ram pressure stripping doesn't happen only in clusters. There is one known case of outflowing gas from an active galactic nucleus stripping the neutral hydrogen from a nearby group member. In the NGC 3079 group, the dwarf S0 galaxy NGC 3073 has an HI tail which is aligned with and pointing away from the nucleus of the large galaxy NGC 3079 (Irwin *et al.* 1987), which is known to be active. A tidal origin for the HI tail seems unlikely due to the lack of a corresponding stellar tail. The density and velocity of NGC 3079's outflowing wind at the location of NGC 3073 are unknown, but a velocity of $\sim 10^3$ km s^{-1} and a density of a few times 10^{-5} cm^{-3} are reasonable estimates based on the emission properties of the NGC 3079 radio loops (Irwin *et al.* 1987). These densities and velocities are comparable with those in the regions of clusters which contain HI-deficient galaxies, and would result in a ram pressure force sufficient to strip the HI in NGC 3073. It is not known how frequent an occurrence this is, but it could be important in those groups which have galaxies with active nuclei, and would presumably be more common at high redshifts when the nuclear activity of galaxies was greater.

3.3 DISTANT TIDAL INTERACTIONS

Distant weak encounters are by far the most common type of tidal interaction, so it is important to understand whether they affect the dynamics of the ISM. The orbits of stars and gas clouds become appreciably altered in a tidal encounter when the velocity perturbation is larger than their normal velocity dispersion. Icke (1985) has argued that even in distant encounters when the velocity perturbation is *smaller* than the velocity dispersion of stars, the gaseous component can be shocked if the velocity perturbation is greater than the sound speed in the gas. Since the stellar velocity dispersions in the disks of spiral galaxies are typically 10-80 km s^{-1} (Mihalas and Binney 1981) and the sound speed in diffuse atomic hydrogen is ~ 1 km s^{-1} (and less in molecular gas), there is a regime where gas will be affected but not the stars. In Icke's simple treatment of the ISM as a single fluid phase with constant density, the shocks can lead to significant radial transport of the gas. However, in a clumpy ISM, the relevant sound speed is the velocity dispersion between gas clouds, which is ~ 4-13 km s^{-1} in the Milky Way (Dickey, Salpeter, and Terzian 1978; Clemens 1985), significantly greater than the sound speed within the gas itself. Tidal interactions will not affect a clumpy ISM until the velocity perturbation is comparable to the normal cloud velocity dispersion. Since most of the ISM mass in typical spiral galaxies exists in clumps, perturbations less than ~ 10 km s^{-1} should not significantly affect the ISM. This is consistent with the lack of direct observational evidence for radial transport of gas due to weak tidal interactions.

3.4 CLOSE TIDAL INTERACTIONS AND MERGERS

When one galaxy passes another, the time-varying gravitational field perturbs the orbits of the gas and stars. If the magnitude of the velocity perturbation to the gas clouds in a tidal encounter exceeds their normal velocity dispersion, then the clouds may collide more frequently. There are several consequences of an enhanced cloud-cloud collision rate, including the dissipation of both energy and angular momentum, the formation and/or disruption of giant molecular clouds, and the triggering of massive star formation.

Observational evidence for radial transport of gas toward the nucleus can be found in merging systems. In several cases, gas surface densities of over 1000 M$_\odot$ pc^{-2} in the central 1-2 kpc are inferred from CO observations. If the CO line is indeed a good tracer of H_2 mass in these systems, then the gas comprises 30% or more of the dynamical mass in the central regions of some mergers. (Sanders *et al.* 1988; Scoville *et al.* 1989). Since such large concentrations of gas are not observed in non-interacting systems, this argues that the interaction has driven gas toward the nucleus. The larger than normal radial gradients in the gas surface density and star formation rate should consequently produce a large chemical abundance gradient in the merger remnant. Such gradients are observed in elliptical galaxies, which is consistent with the hypothesis that giant ellipticals are indeed merger remnants (Schweizer 1986; Barnes 1989). Recently it has been suggested that the deposition of a large gas mass in the nucleus will trigger a nuclear starburst and eventually lead to the growth of a massive black hole (Norman and Scoville 1988).

Cloud-cloud collisions are believed to play an important role in our galaxy, so it is important to understand what happens to an ISM if the cloud-cloud collision rate is increased dramatically by a tidal interaction. Cloud-cloud collisions may explain the formation of giant molecular clouds and the observed cloud mass spectrum in the Milky Way (Scoville and Hersh 1979; Kwan 1979; Kwan and Valdes 1987; but see Elmegreen 1989b for arguments which suggest that other processes are also important), and have also been proposed to explain the enhanced massive star formation rate in the spiral arms of the Milky Way (Scoville, Sanders, and Clemens 1986). Large $H\alpha$ equivalent widths and large far-infrared luminosities due to elevated massive star formation rates in both the nuclei and disks are a well-documented feature of tidally disturbed galaxies (Lonsdale, Persson and Matthews 1984; Keel *et al.* 1985; Joseph and Wright 1985; Bushouse 1986; Bushouse 1987; Kennicutt *et al.* 1987), and this may be due to an increased rate of cloud collisions. Simple 2-dimensional modelling of cloud orbits finds that the cloud-cloud collision rate can increase by an order of magnitude in close tidal encounters (Noguchi and Ishibashi 1986), which is similar to the global increase in the massive star formation efficiency of strongly interacting systems (Young *et al.* 1986).

The cloud-cloud collision rate in a galaxy-galaxy collision will depend on the radial distance from the galactic center, the impact parameter of the collision, the relative velocity, inclination, mass ratio, gas content, and spin angle of the galaxies. Colliding clouds can coalesce or disrupt each other, depending upon their masses, velocities, and relative spins (Lattanzio and Henriksen 1988). In simulations of collisions between gas-rich and gas-poor galaxies, Olson and Kwan (1989) have separately tracked coalescent and disruptive cloud collisions. Since the rate of disruptive collisions increases much more dramatically than coalescent collisions in strong interactions, Olson and Kwan argue that the elevated massive star formation rates in strongly interacting systems are due to disruptive rather than coalescing collisions. If so, this might be consistent with massive star formation triggered at the interface of colliding clouds (e.g., Scoville, Sanders, and Clemens 1986), and would be inconsistent with models in which cloud agglomeration is required to produce high mass star formation (e.g., Shu 1987).

Since collisions dissipate energy and angular momentum, large amounts of gas can be driven toward the nucleus in galaxy-galaxy interactions. Several recent simulations of tidal encounters of a gas-rich with a gas-free galaxy have produced inward transport of gas, but in different ways. In the N-body simulations of transient tidal encounters by Noguchi (1988), significant radial gas motion occurs only as the indirect result of a tidal encounter. The perturbation produces a stellar bar which eventually causes gas to flow inward.

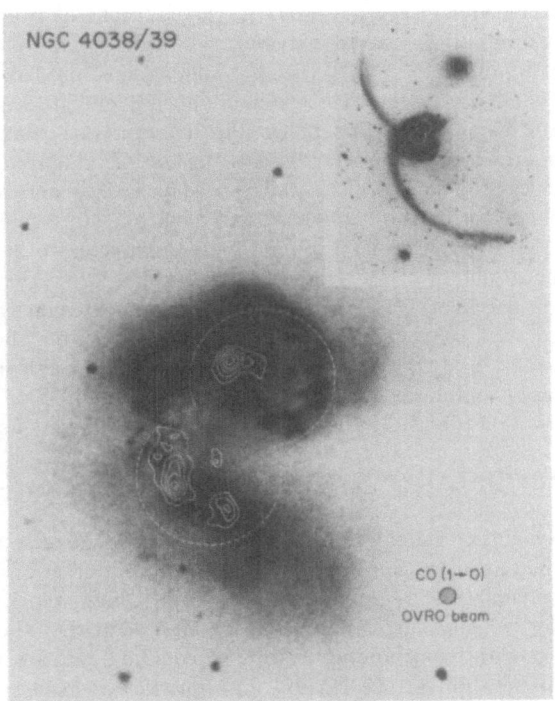

Figure 2 CO emission (contours) from "The Antennæ" (NGC 4038/39) superposed on an optical photograph of the merger system. Dotted lines indicate HPBW of primary beam. Of the 3 large gas concentrations detected, 2 are located at the nuclei of NGC 4038 and NGC 4039, and the largest complex is located near the region of overlap between the 2 galaxies. Inset shows larger scale, higher contrast optical image with interferometer fields. Figure from Stanford *et al.* (1989).

In their simulations of a transient tidal encounters, Olson and Kwan (1989) produce significant radial inflow due to dissipation from cloud-cloud collisions without the need for stellar bars. A hybrid N-body/hydrodynamics method was used by Hernquist (1989) to model the merger of a small gasless galaxy with a large gas-rich disk galaxy. An instability produced by hydrodynamical effects which is *initially* driven by gas dissipation forms a large gas concentration which eventually sinks to the nucleus. But the angular momentum lost by the gas is gained by the disk stars, which suggests that tidal torquing and dynamical friction, not gas viscosity, are the means by which gas loses angular momentum. It seems clear that large amounts of gas can be driven inwards in strong tidal interactions, but the physical mechanism by which the gas loses angular momentum is not yet established.

The different behavior of gas and stars in the collisions of two gas-rich galaxies is illustrated nicely by maps of the spatial distribution of gas. In high resolution CO maps of "The Antennae" (NGC 4038/NGC 4039) (Stanford *et al.* 1989) and Arp 299 (NGC 3690/IC 694) (Sargent *et al.* 1987) galaxy systems which are both in the process of merging, strong CO emission is detected in the region of overlap between the galaxies. These regions are also seen to be sites of vigorous massive star formation. Figure 2 shows the OVRO interferometer CO map of "The Antennae" superposed on an optical image. Concentrations of dense gas are detected from the two nuclei, and from one large complex in the overlap region with $\sim 10^9$ M$_\odot$. A large concentration of gas is expected to be produced near the collision interface by a pile-up of the viscous cloud systems while the collisionless stellar systems pass through one another. The velocity field of the extranuclear complex in "The Antennae" (Stanford *et al.* 1989) suggests that such a gas pile-up may be partly responsible for its origin. An extension of the northern tail of NGC 4039 through the inner disk would pass through this large complex, which suggests that gas compression resulting from the gravitational perturbation might also be partially responsible for its formation.

3.5 HIGH VELOCITY GALAXY-GALAXY COLLISIONS

The highest velocity galaxy-galaxy collisions occur in rich clusters, where the velocity dispersions are ~ 1000 km s^{-1}. Since the effect of a tidal interaction in the impulse approximation is proportional to the interaction time t$\sim$v^{-1}, high velocity encounters cause less damage than low velocity encounters. At such high velocities, the collisional kinetic energy exceeds the gravitational binding energy, so colliding galaxies will not merge. In a well-known early paper, Spitzer and Baade (1951) predicted that head-on, high-velocity collisions would leave the stellar disks relatively unscathed, but that the ISMs would coalesce and form one gaseous remnant which the stellar disks leave behind. But it is now known that tidal disruption to at least portions of stellar disks can be significant in a head-on, high-velocity collision (Combes *et al.* 1988).

The fate of gas at a particular place in the disk will depend, among other things, on its location relative to the two nuclei at closest approach and the column densities of gas in both galaxies. Gas near the nucleus of its parent galaxy is less likely to be stripped, because of the large gravitational binding force, and because the column density of gas near the nucleus is likely to be larger than that of the colliding column density of gas (since gas densities generally peak in the nuclei, and collisions are rarely head-on).

When lower density gas from one galaxy collides with higher density gas clouds from another, a high density cloud may be bodily stripped by ram pressure, in much the same way that ICM gas is thought to strip HI clouds from cluster galaxies. Differences are that gas densities within a galaxy can be several orders of magnitude larger than ICM gas densities, but the collision timescale is shorter than the ISM-ICM interaction timescale. Thus even giant molecular clouds can be stripped in a galaxy-galaxy collision, if the interaction lasts long enough for it to achieve escape velocity.

One possible example of low density gas in a galaxy halo stripping higher density atomic gas from a colliding dwarf can be found in the Virgo cluster. This is the explanation proposed by Sancisi, Thonnard, and Ekers (1987) for the HI cloud observed between the giant Virgo cluster elliptical NGC 4472 and the dwarf irregular UGC 7636, although a tidal explanation cannot be ruled out. No HI is detected at the position of the dwarf, which is very HI-poor for its type. X-ray emission from NGC 4472 reveals the presence of a halo with density $n \sim 10^{-3}$ cm^{-3}. The two galaxies differ in radial velocity by ~ 700 km s^{-1}, thus conditions are more than sufficient to cause ram pressure stripping. The radial velocity of the HI cloud is 200 km s^{-1} greater than that of the dwarf and 500 km s^{-1} less than that of NGC 4472, indicating that it may have been accreted by the giant elliptical.

The galaxy which presently exhibits the heaviest environmental damage in the nearby Virgo cluster is NGC 4438. Its highly disturbed stellar disk (Arp 1966) is in all likelihood due to a collision with the nearby galaxy NGC 4435. The interaction has been modelled by Combes et al. (1988), who find that the closest approach occurred $\sim 10^8$ years ago at a nuclear separation of ~ 4 kpc with a relative velocity of ~ 900 km s^{-1}. Thus much of the present morphology of NGC 4438 is due to a rare, nearly head-on high-velocity collision. Many observations have shown that the remaining ISM in NGC 4438 is highly disturbed. HI is displaced to the west of the nucleus (Hibbard and van Gorkom 1989) and x-ray (Kotanyi, van Gorkom, and Ekers 1983), radio continuum (Hummel et al. 1983), and CO (Combes et al. 1988) emission have all been detected both near the nucleus, and displaced to the west. Several well-defined Hα+[NII] filaments originate from the nucleus and extend westward, suggestive of either ongoing shock heating or cooling filaments (Kenney et al. 1990c).

It's not entirely clear which environmental effects are responsible for the different features of NGC 4438's anatomy. Although NGC 4438 has evidently suffered a high-velocity galactic collision, it is also moving at over 1000 km s^{-1} through the center of the Virgo cluster, and is therefore subject to large ICM-ISM interaction forces. The peculiarities of its ISM have been attributed to both ongoing ICM-ISM stripping (Kotanyi et al. 1983) and tidal effects (Combes et al. 1988). Since a high velocity galaxy-galaxy collision is required to explain the disrupted stellar disk, it may be that the disturbed ISM is also due to the collision. If the colliding galaxy NGC 4435 was originally gas-rich, then the resulting collision of the ISMs at 900 km s^{-1} could produce many features of the highly disturbed ISM which is observed today (e.g., Harwit et al. 1987).

3.6 THE ACCRETION OF SMALL GALAXIES AND INTERGALACTIC GAS

A growing number of galaxies show evidence that all or part of their interstellar medium has been recently acquired by the capture of small galaxies or intergalactic gas. As opposed to the more dramatic mergers of similar mass galaxies, this type of object is not associated with large-scale bursts of star formation, and has often been identified by unusual HI properties. By supplying a fresh supply of gas, such events prove that galaxies do not evolve as closed-box systems and give us glimpses of the processes which occurred during galaxy formation. The capture of small objects is a source of energy input to the ISM and can noticeably affect the morphology of localized areas of the disk.

In the large nearly face-on spiral M101, van der Hulst and Sancisi (1988) have discovered high-velocity HI gas in 2 regions of the disk. The gas in these 2 complexes have motions as large as 150 km s^{-1} perpendicular to the disk, and is all redshifted. Figure 3 shows the HI distribution as a function of right ascension and velocity at a declination which cuts through the largest high-velocity complex. This figure shows that the high-velocity gas is connected to the disk gas, and is located in regions where the HI column density at the local disk velocity is small. The lack of blushifted gas, the physical connection of the high-velocity gas with the "normal" disk gas, the hole in the "normal" disk gas,

160

the total mass (10^7-10^8 M$_\odot$) and kinetic energy (10^{55} ergs) involved all suggest that the high-velocity gas is the result of an intergalactic gas cloud or small galaxy that crashed into M101's disk. This event seems to have altered the structure of the local spiral arms.

The signatures of gaseous accretion are most obvious in early type galaxies, where even a small mass of accreted gas can dominate the mass of the ISM. The most certain examples of accretions are the polar ring galaxies (e.g., van Gorkom, Schechter, and Kristian 1987) and systems with counterrotating gas and stars (Galetta 1987; Schweizer, van Gorkom, and Seitzer 1989). Polar rings, which are composed of rings of gas and/or stars oriented nearly perpendicular to the main disk, are thought to exist only in S0 galaxies since these galaxies have little indigenous interstellar matter to dissipate the kinetic energy of the gas in the acquired galaxy. Likewise, counterrotating gas can only occur in a galaxy which has almost no ISM of its own.

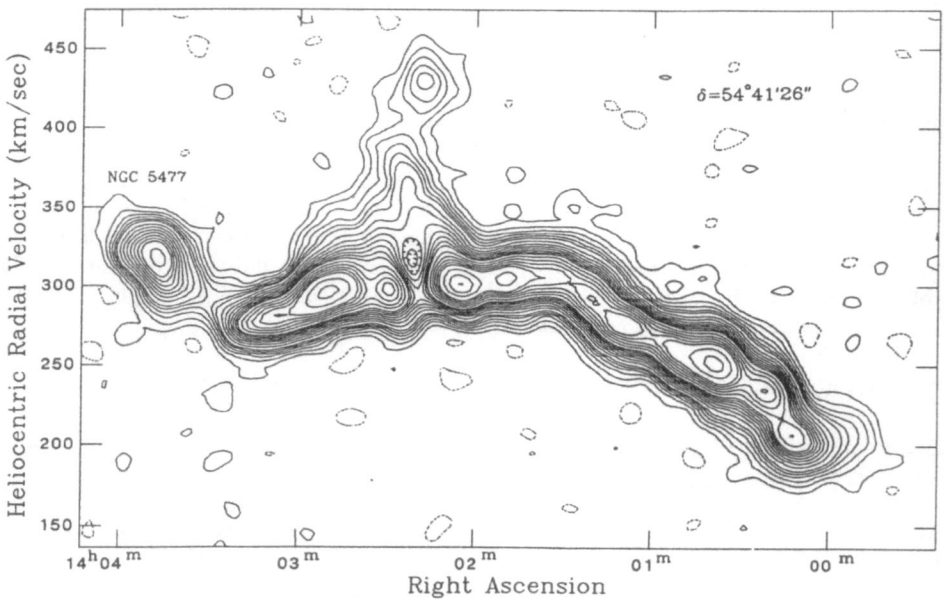

Figure 3 The distribution of HI emission as a function of right ascension and velocity at a declination which cuts through a large high-velocity HI complex in M101. The lack of blue-shifted gas, the physical connection with the rest of the disk, and the HI hole at the normal disk velocity all suggest that a small galaxy or intergalactic gas cloud has collided with M101's disk. Figure from van der Hulst and Sancisi (1988).

The distribution of HI in outer rings which are inclined to the optical disk has been taken as evidence that the gas in many early type galaxies has been accreted (van Driel 1987). Using statistical arguments, Wardle and Knapp (1986) have argued that the HI in a large fraction of S0 galaxies has been accreted. The accreted gas is generally presumed to originate from the gravitational capture of an entire small galaxy or the outer disk of larger galaxy. Schneider (1989) suggests that the outer HI ring of the Sab galaxy M96 is fed by the unusual intergalactic HI ring in the M96 group. The challenge is to estimate how frequently these events occur, and in so doing determine the mass accretion rate as a function of time for a typical galaxy.

4. Interactions Between Interstellar and Intracluster Gas

4.1 THE PHENOMENON OF HI DEFICIENCY

The phenomenon of HI deficiency in Virgo cluster galaxies was first discovered seventeen years ago by Davies and Lewis (1973), who compared the ratio $M(HI)/L_B$ for samples of Virgo and non-cluster galaxies. The interpretation of these early results as indicating gas depletion was questioned (Bottinelli and Gouguenheim 1974; Huchtmeier et al. 1976), since the type and luminosity dependence of HI emission in "normal" galaxies was not yet established, and it was not clear that the low $M(HI)/L_B$ ratios could be ascribed to the environment. Studies over the last decade have now shown convincingly that many cluster galaxies are deficient in HI as compared with isolated galaxies of the same morphological type and size (Chamaraux, Balkowski, and Gerard 1980; Giovanardi et al. 1983; Giovanelli and Haynes 1983; Giovanelli and Haynes 1985; Haynes and Giovanelli 1986; Chamaraux, Balkowski, and Fontanelli 1986; Huchtmeier and Richter 1989).

It order to assess the normality of HI emission in cluster galaxies, it is first necessary to establish the HI properties of galaxies which are relatively isolated but otherwise similar to the cluster galaxies. Extensive HI studies of isolated galaxies have been carried out by Haynes and Giovanelli (1984, 1986), who find that the HI mass of an isolated spiral galaxy is well-correlated with its optical diameter. Figure 4a shows the relationship between HI mass and optical diameter for a sample of isolated galaxies from Haynes and Giovanelli (1986). The scatter in this relation is a factor of ~1.5, which is comparable to the difference between Sab galaxies and Sc galaxies. The use of the optical diameter, which is a disk quantity, minimizes the difference between morphological types, and is therefore preferred to either optical luminosity or total mass, both of which reflect the presence of the bulge. The relatively good correlation between HI mass and optical diameter allows one to define an expected HI mass for a given optical diameter (and morphological type). This has been used to define an HI deficiency parameter, given by:

$$\text{Def}_{HI} = \log M_{HI}(\text{expected}) - \log M_{HI}(\text{observed}). \tag{1}$$

Figure 4b shows the HI masses of a sample of low-mass Virgo cluster galaxies, plotted against their optical diameters. The line is the best fit from the isolated galaxies, which defines the expected HI mass. While there are some Virgo cluster galaxies which are deficient in HI by more than a factor of 10 ($\text{Def}_{HI} = 1.0$), others have normal HI properties. And while some clusters contain large numbers of HI-deficient galaxies, others contain mostly HI-normal galaxies. Ongoing investigations focus on two principle issues. What

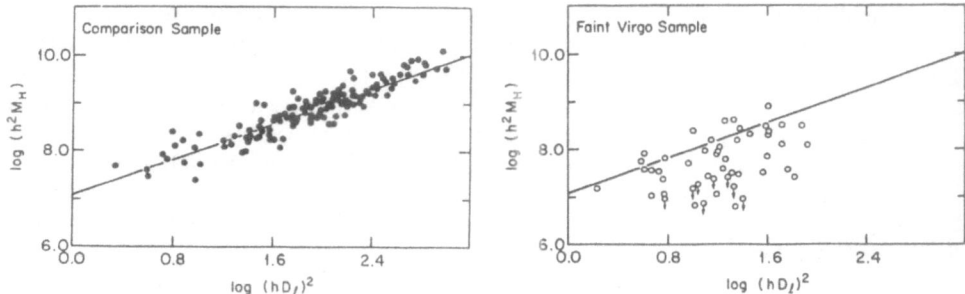

Figure 4 The relationship between HI mass and optical linear diameter for nearby Sc-Irr galaxies in relatively isolated environments (left), and faint Sa-Irr Virgo cluster galaxies (right). The solid line in each panel indicates a least squares fit to the comparison sample. Many of the Virgo galaxies have weak HI emission for their optical diameter. Figure is from Haynes and Giovanelli (1986).

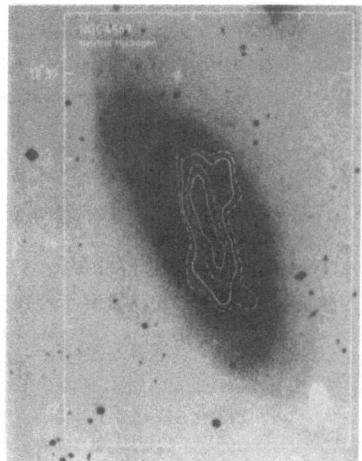

Figure 5 HI map of the Virgo cluster spiral NGC 4569 superposed on optical image. The extent of HI emission is much less than the optical extent in this very HI-deficient galaxy. Figure from Warmels (1988a).

is the cause of the HI deficiency? And what is the effect on galaxy evolution? As the nearest cluster containing HI-deficient galaxies, the Virgo cluster has been studied in far greater detail than any other cluster. Consequently, much of the discussion which follows in §4.2-4.9 focuses on Virgo.

4.2 THE SPATIAL DISTRIBUTION OF HI IN HI-DEFICIENT SPIRALS

Insight into the cause of HI-deficiency has been provided by high resolution mapping of HI in Virgo spirals. Some 25 spirals have been mapped at the VLA (van Gorkom and Kotanyi 1985; Cayatte et al. 1989) and 36 have been mapped at Westerbork in either one or two dimensions (Warmels 1988a, 1988b). Figure 5 shows Warmels' HI contour map for the massive Sab galaxy NGC 4569 superposed on its optical image. NGC 4569 is near the center of the Virgo cluster, is moving at over 1000 km s^{-1} with respect to the rest of the cluster, and has 10 times less HI than a typical isolated galaxy of the same morphological type and mass. The radial extent of atomic gas is sharply reduced as compared with the radial distribution of starlight, implying that the removal mechanism operates selectively on gas. Examination of the HI radial distributions in Virgo spirals reveals that it is mainly the outer disks which are HI-poor (Warmels 1986, Kenney and Young 1989), consistent with an external depletion mechanism, as opposed to the effects of internal winds. The HI distributions in many HI-deficient spirals show pronounced asymmetries (Warmels 1988c; Cayatte et al. 1989). Such gas asymmetries are expected when galaxy motions are involved in the gas removal process (e.g., through ram pressure stripping as opposed to thermal evaporation).

Figure 6 shows a map of the Virgo cluster, with the 25 VLA HI maps put in their correct positions, but magnified by a factor of 5 (Cayatte et al. 1989). Note that the galaxies closest to the cluster center tend to have the smallest HI disks. This diagram suggests what more detailed analysis proves, namely that the HI-poorest galaxies tend to be in the regions of the highest ICM density near the cluster center (Magri et al. 1988). There are both HI-normal galaxies near the cluster center and HI-deficient galaxies away from the cluster center. These can be understood by projection effects as well as by orbits which carry some stripped galaxies from the cluster center to the outskirts.

4.3 HOT GAS IN THE INTRACLUSTER MEDIUM

There are several pieces of evidence indicating that an interaction between hot gas in the intracluster medium (ICM) and the interstellar media (ISM) of the galaxies are predominantly responsible for HI deficiency in clusters. The x-ray emission that has been detected in many clusters is thermal bremsstrahlung radiation from hot gas in the intracluster medium (Forman, Jones and Tucker 1985). The densities (typically $n\sim10^{-3}$-10^{-5} cm^{-3}) and temperatures (typically $T\sim10^{7}$-10^{8} K) derived for this ICM gas are sufficiently high to significantly affect the interstellar media of the cluster members, especially when the galaxies move at speeds in excess of 1000 km s^{-1}. In Virgo, the strongest x-ray emission is centered on the giant elliptical M87. From fitting the Einstein satellite x-ray surface brightness profile around M87, under the assumption that the gas is in hydrostatic equilibrium, Fabricant and Gorenstein (1983) estimate that the ICM density 1° from M87 is 4×10^{-4} cm^{-3}, and a simple extrapolation suggests that the density drops by roughly an order of magnitude at a radial distance of 5°. New and sensitive observations with the

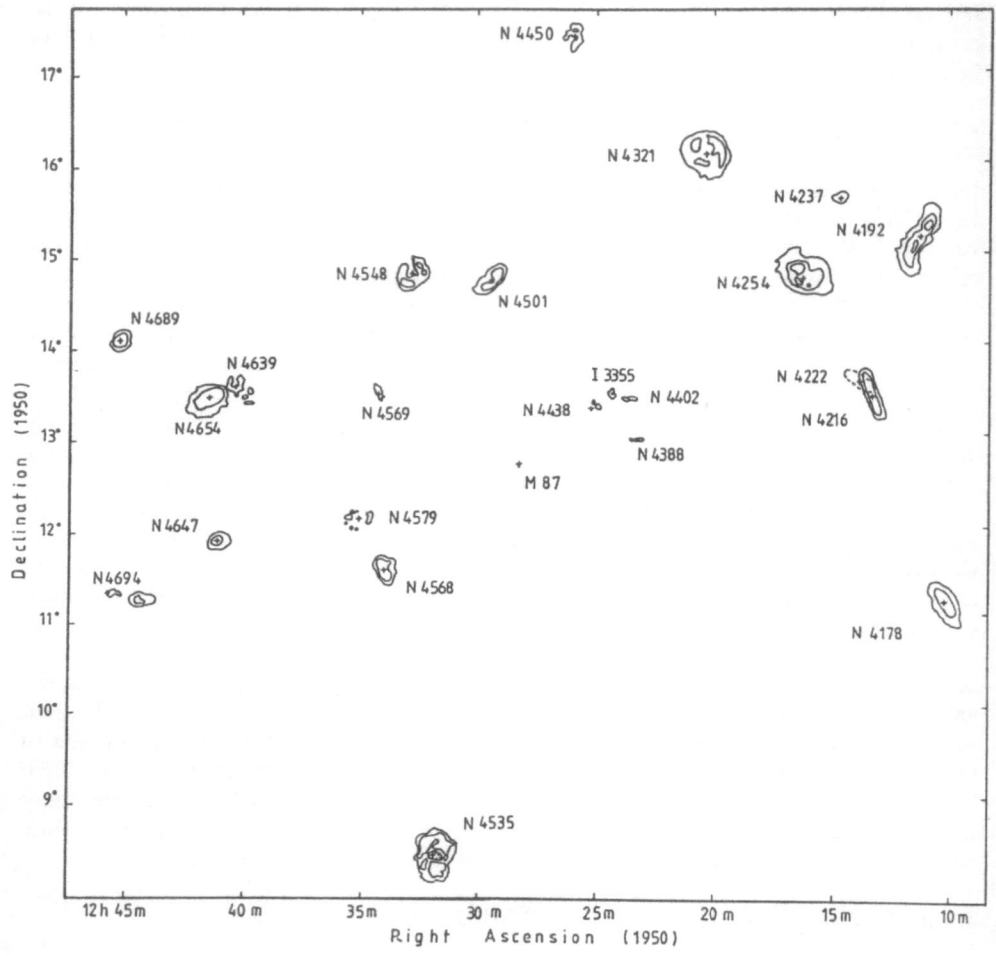

Figure 6 HI maps of 25 Virgo spirals placed at the correct cluster location, but magnified by a factor of 5 with respect to the scale in right ascension and declination. From this figure it may be seen that the galaxies with the smallest HI disks (with respect to their optical disks) tend to be near the cluster center. Figure is from Cayatte *et al.* (1989).

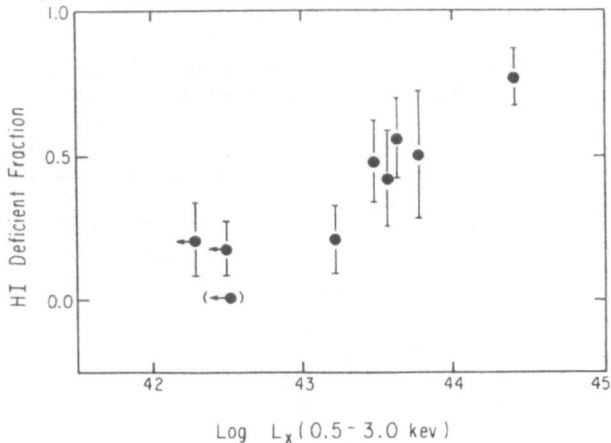

Figure 7 Relationship between cluster x-ray luminosity (in 0.5-3.0 keV range) and the fraction of cluster galaxies which are HI-deficient by more than a factor of 3. This relationship suggests that the ICM, which produces the x-ray emission, is responsible for HI deficiency. Figure taken from Giovanelli and Haynes (1985).

Ginga satellite reveal that x-ray emission (at 1.5-4.5 keV) extends out to a radius of $\sim6°$ from M87 (Takano *et al.* 1989), which is spatially coincident with the region where most of the HI-deficient galaxies are found (Haynes and Giovanelli 1986).

There is further evidence that the ICM is somehow responsible for HI deficiency. Figure 7 shows the x-ray luminosity of 9 clusters plotted against the fraction of the cluster galaxies which are HI-deficient by more than a factor of 3, from Giovanelli and Haynes (1985). The good correlation indicates that clusters with the densest intracluster media contain the most HI-deficient galaxies. This correlation by itself does not prove that an ICM-ISM interaction causes HI deficiency, since there are other parameters (e.g., cluster velocity dispersion) which correlate with x-ray luminosity, but it is strongly suggestive.

4.4 THE HI DEFICIENCIES OF DIFFERENT MORPHOLOGICAL TYPES

Different morphological types exhibit different degrees of HI deficiency. Table 1 lists the mean HI deficiency for different morphological types in Virgo, compiled from Chamaraux, Balkowski, and Fontanelli (1986) and Hoffman, Helou, and Salpeter (1988). Here it can be seen that HI deficiency increases systematically along the Hubble sequence from Sc to Sa galaxies. Three types of explanations have been proposed for this and all are probably relevant. One explanation is that early type disk galaxies have their HI distributed in an outer ring (van Driel 1987) where it is less tightly bound and easier to strip. The effect of selective stripping of outer disk gas is to remove a larger fraction of an Sa's HI reservoir

Table 1

HI Deficiency of Virgo Cluster Galaxies as a Function of Morphological Type

Type	HI Deficiency	Reference
S0	≥ 1.00	1
S0/a	≥ 1.00	1
Sa	1.34 ± 0.13	1
Sab	0.72 ± 0.17	1
Sb	0.51 ± 0.09	1
Sbc	0.42 ± 0.12	1
Sc	0.17 ± 0.10	1,2
Sm	0.20 ± 0.14	1,2
dI	0.27 ± 0.12	2

Explanation of Table 1: Central column lists the mean HI deficiency for Virgo galaxies of the morphological type listed in first column. See equation 1 in §4.1 for the definition of HI deficiency. References: (1) Chamaraux, Balkowski, and Fontanelli (1986); (2) Hoffman, Helou, and Salpeter (1988).

than an Sc's. A second explanation is that early type spirals are closer to the cluster center, on average, than late type galaxies (Dressler 1980), and therefore are subject to stronger ICM-ISM interactions.

A third and compelling explanation is that gas-rich and gas-poor spirals have systematically different types of orbits within clusters (Dressler 1986). Figure 8, from Dressler (1986), shows the dispersion in radial velocity versus the distance from the cluster center for different classes of cluster galaxies. As shown in the bottom panel of Figure 8, the HI-poor spirals (mostly early types) from 5 "deficient" clusters have velocity dispersions which decrease with increasing radius. The trend appears to be in the opposite sense for the HI-rich spirals (mostly late types), although this conclusion is weakened by the small overlap in cluster radius in which significant numbers of both types of galaxies are found. The top panel in Figure 8 shows an even more pronounced kinematic difference between HI-poor Sa-Sb galaxies in 5 "deficient" clusters as compared to all types of galaxies (but mostly late types) in 4 "non-deficient" clusters. These trends suggest that gas-poor early type spirals tend to have radial orbits, while late type spirals and gas-rich early type spirals may have more isotropic (including circular) orbits. Independent evidence that the orbits

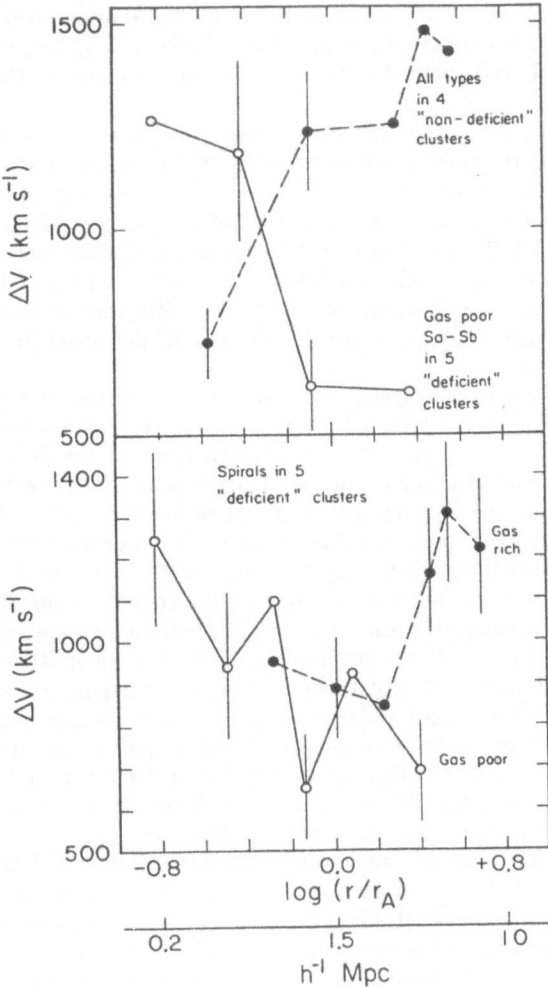

Figure 8 Relationship between velocity dispersion and projected radius (normalized by Abell radius r_A) for galaxies in 9 clusters. In the bottom panel, the HI-poor spirals in 5 "deficient" clusters (i.e., those with many HI-deficient galaxies) have velocity dispersions which decrease with increasing radius. The HI-rich spirals seem to show the opposite trend. In the top panel, an even larger difference is observed between all types of galaxies (but mostly late types) in "non-deficient" clusters and gas-poor Sa-Sb galaxies in "deficient" clusters. Relationships indicate that galaxy orbits vary within and among clusters, and suggest that the most HI-deficient galaxies are on nearly radial orbits which carry them through the cluster center. Figure is from Dressler (1986).

vary systematically within clusters has been found by O'Dea, Sarazin, and Owen (1987). From a comparison of the orientation of narrow-angle tail radio sources with the projected distance from the cluster center, they find that galaxies in regular clusters tend to have predominantly radial orbits near the cluster center, and predominantly circular orbits far from the center.

Galaxies on radial orbits plunge close to the cluster center, where the ICM density is highest, and therefore experience stronger ICM-ISM interactions than galaxies on more circular orbits. Part of the large velocity dispersion for late type spirals which are far from cluster centers is probably due to the inclusion of infalling galaxies which have not yet dynamically relaxed (Tully and Shaya 1984). Compared with the early type spirals, many of the late types are recent arrivals which have not yet had a passage through the cluster center. Different cluster locations, different orbits and different amounts of time spent in clusters all help to explain the systematic difference in HI-deficiency for early and late type spirals.

It has been something of a mystery that most dwarf irregular (dI) galaxies in Virgo are only moderately HI-deficient. Although there are some very HI-deficient dwarfs (Vigroux et al. 1986), Table 1 shows that the mean HI deficiency of the Sc and dwarf irregular galaxies are quite similar (Hoffman et al. 1985; Haynes and Giovanelli 1986; Magri et al. 1988; Hoffman, Helou, and Salpeter 1988). Being at the low end of the mass spectrum, dwarf galaxies should be more susceptible to most gas removal processes than their high mass counterparts, since the gas is less tightly bound to them. While this has been taken as evidence that the gas removal process operating in clusters behaves qualitatively differently from ram pressure stripping, this conclusion seems premature. Many of the faintest dwarf irregulars remain undetected in HI, making a statement of their HI deficiency uncertain (Hoffman, Helou, and Salpeter 1988). Few dwarf irregulars populate the central part of the cluster, which is instead inhabited by large numbers of gas-poor dwarf ellipticals (Hoffman, Helou, and Salpeter 1988). It has been proposed that stripping turns dwarf irregulars into dwarf ellipticals (dE's), but it is difficult to explain the higher central surface brightness of dE's by this mechanism (for discussions on the differences between dE's and dI's, see Ichikawa et al. 1988 and Gallagher and Hunter 1989).

In order to understand the evolution of cluster spiral and dwarf irregulars, it is necessary to measure both the atomic and molecular gas content of the galaxies. As described in §4.5, the total gas $(HI+H_2)$ deficiency of large Virgo spirals is only a small fraction of the HI deficiency, while for low mass spirals and dwarfs, the total gas deficiency is probably comparable to the HI deficiency. If so, then dwarf irregular galaxies are more gas-deficient than large spirals, in keeping with standard models of gas removal.

4.5 MOLECULAR GAS IN HI-DEFICIENT SPIRALS

Molecular gas comprises a significant fraction of the interstellar gas mass in large spirals, and is the medium in which stars form. Therefore, the evolutionary impact of stripping in cluster galaxies is critically dependent upon the fate of both atomic and molecular gas in the stripping event. At present, the only cluster containing HI-deficient galaxies that has been well-surveyed in CO is the nearby Virgo cluster. $CO(J=1\rightarrow0)$ has been measured in over 50 Virgo cluster spirals by Kenney and Young (1986, 1989) and Stark et al. (1986). Both groups find that the HI-deficient spirals have high CO/HI ratios, thus molecular gas is not as deficient as atomic gas. At least three factors may contribute to the survivability of molecular gas. The radial distribution of CO is typically more centrally peaked than HI, so that a larger fraction of the molecular gas component resides deep in the galaxy's potential well where it is difficult to remove by any means. Also, the gas in giant molecular clouds is trapped locally by the cloud's own self-gravity. Finally, the surface density of molecular gas is much greater than atomic gas, making it less susceptible to removal by an external pressure.

Figure 9, from Kenney and Young (1989), allows one to make a quantitative estimate of the total gas deficiency. The horizontal axis in this figure is the HI deficiency parameter (see §4.1), while the vertical axis is the HI+H_2 mass, normalized by the optical area of the galaxy. Plotted in this plane are all Virgo spirals brighter than 12th magnitude in the B passband. The curves A, B, and C represent 3 possible evolutionary paths for the gas in Virgo spirals. Curve A represents a constant total gas mass, which would occur if HI were converted into H_2 as a galaxy becomes HI-deficient. Curve B represents an H_2 mass which is independent of the HI mass, a condition synonymous with "normal" CO emission. Curve C represents an H_2-deficiency (and total gas deficiency) equal to the HI-deficiency. Curve B provides the closest match to the data for both the Sa-Sb and Sbc-Sm galaxies, which implies that most HI-deficient Virgo spirals have normal CO emission. A least squares fit to the points in Figure 9c indicates that large Virgo Sc galaxies which are HI-deficient by a factor of 10 are gas-deficient in the mean by only a factor of 2-3, due to the survival of molecular gas (Kenney and Young 1989). Figure 9b shows that the total gas deficiency is similar but less certain for the Sa-Sb galaxies, since there is more scatter and there are fewer early types in the cluster with normal HI emission.

The overall gas deficiency may be comparable to the HI deficiency in smaller spirals and dwarf galaxies. The CO luminosity per unit area or mass drops sharply for Virgo spirals fainter than $B_T^o \leq 12$ (Kenney and Young 1988), and it is not yet clear how much of this is due to a higher CO-H_2 proportionality factor for low mass galaxies, and how much is due to a smaller molecular gas fraction (M_{H_2}/M_{HI+H_2}). There is evidence that both effects are important. A comparison of the virial masses and CO luminosities of molecular clouds suggests that the CO-H_2 proportionality factor is $\sim 5 - 6$ times higher in the Large Magellanic Cloud than in the Milky Way (Cohen et al. 1988; Johannsen 1989), suggesting that a systematically different proportionality factor may be part of the explanation.

There is also both empirical and theoretical evidence that the global molecular gas fraction decreases with decreasing galaxy mass. If the virial technique used to estimate the H_2 mass in the LMC is correct (Cohen et al. 1988), then the H_2 mass of the LMC is only ~20% of its HI mass, as compared with the roughly equal amounts of HI and H_2 seen in the Milky Way (Scoville and Sanders 1987). In a comparison of the HI and H_2 properties of the centers of Sc galaxies of varying mass, Young and Scoville (1982) found that the H_2 surface density increased with the mass surface density, whereas the HI surface density was roughly the same in all galaxies. Although the mean HI surface density of galaxies does not vary *strongly* with mass (Rogstad and Shostak 1972), there are some low mass Virgo spirals with larger HI surface densities than are found in any high mass spirals (Kenney and Young 1988). This indicates that gas does not enter and remain in the molecular phase as readily in low mass galaxies as in high mass galaxies. In several ways, low mass galaxies are similar to the outer disks of large spirals, which are HI-dominated (Wang 1989). Applying the virial theorem and molecular self-shielding to gravitationally-bound clouds, Elmegreen (1989a) argues that large clouds in low mass galaxies should contain most of their mass in HI, since both the cloud boundary pressures and dust-to-gas ratios are probably lower in

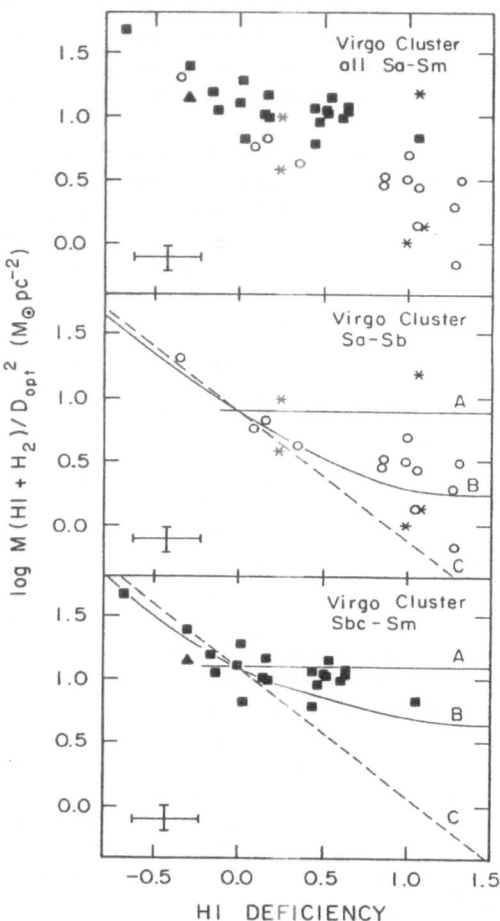

Figure 9 Relationship between total gas (HI+H$_2$) deficiency and HI deficiency in bright Virgo cluster galaxies. a) The log of total gas mass normalized by optical diameter squared M(HI+H$_2$)/D$_{opt}^2$ vs. HI deficiency for all Virgo cluster Sa-Sm galaxies with B$_T^o \geq 12$. The different symbols represent different morphological types (star = Sa, circle = Sab-Sb, square = Sbc-Sd, triangle = Sm). b) Same as Figure 9a, but with only the Sa-Sb galaxies plotted. The curves labelled A, B, and C represent 3 possible gas evolutionary tracks, which are described in the text. c) Same as Figure 9b, but for Sbc-Sm galaxies. Curve B (H$_2$ "normal") provides the best overall fit to the Virgo galaxies, although the fit is uncertain for the early types. Figure is from Kenney and Young (1989).

low-mass galaxies than in high-mass galaxies. If small galaxies are indeed HI-dominated, then the total gas deficiency is close to the HI-deficiency in low mass galaxies, and small galaxies have greater *total gas deficiencies* than large galaxies.

Stripped galaxies have lost most of their HI from the outer disks, but appear to have lower HI surface densities even in their inner disks. Virgo spirals which are globally HI-deficient by a factor of 10 have 2-3 times less HI within their inner disks (Kenney and Young 1989). Since the inner disks of large spirals are H_2-dominated, it is not presently known whether the inner disk HI has been removed from the galaxies, or has entered the molecular phase. The selective stripping of low column density HI gas is predicted by ram pressure stripping models and might be expected to occur with other removal processes. Alternatively, the lower rates of massive star formation observed in HI-deficient Virgo spirals (see §4.8) must result in less H_2-dissociating radiation, and could lead to smaller HI envelopes around molecular clouds (Shaya and Federman 1987).

4.6 GAS REMOVAL MECHANISMS IN CLUSTERS

Several mechanisms have been proposed to explain the removal of HI from cluster galaxies. In this section I review the evidence for and against each model.

As discussed in §3.2, tidal interactions strip both stars and gas from the outer disks, and can result in mild HI deficiency since $M(HI)/L_{stars}$ increases with increasing radius. However, tidal interactions do not appear capable of producing the bulk of HI deficiency in clusters. Tidal effects cannot explain HI disks which are shrunken with respect to the optical disks (Giovanelli and Haynes 1983; van Gorkom and Kotanyi 1985; Warmels 1988c; Cayatte *et al.* 1989).

Since tidal damage decreases with increased velocity, tidal stripping is less effective in clusters than in groups, because the velocity dispersions are higher in clusters. While the evidence for shrunken HI disks indicates that tidal interactions are not the primary cause of HI deficiency, they may be an important secondary cause. There do exist clumps within clusters with local velocity dispersions much less than that of the cluster as a whole (Binggeli, Tammann, and Sandage 1987; Dressler and Schectman 1988). Compared to groups, clusters have a larger galaxy density and a larger relative velocity of galaxies, which results in a higher *rate* of interactions. For these reasons, galaxy-galaxy interactions are believed to be an important dynamical influence on galaxies within clusters (e.g., Miller and Smith 1982). And although high galaxy velocities imply less tidal damage for encounters with large impact parameters, the high velocities can produce enormous damage to the ISM for the relatively rare encounters with small impact parameters. One such case is discussed in §3.5.

Ram pressure stripping occurs when the pressure of the ICM gas, due to a galaxy's motion through the ICM, exceeds the gravitational force binding ISM gas to the galaxy, as expressed by the condition

$$\rho v^2 \geq 2\pi G \sigma_{tot}(r)\sigma_{gas}(r) \tag{2}$$

where ρ is the ICM gas density, v is the component of a galaxy's velocity relative to the cluster that is perpendicular to the disk, and $\sigma_{tot}(r)$ and $\sigma_{gas}(r)$ are the total and gas mass surface densities in the disk, each of which are functions of the distance from the galaxy center r (Gunn and Gott 1972). "Diffuse" HI gas with surface densities of a few $M_\odot$ pc^{-2} can be stripped from the disk of a medium-sized spiral in a typical encounter, whereas giant molecular clouds with $\sigma_{gas} \geq 100$ M$_\odot$ pc^{-2} will not be stripped from anywhere in a galaxy, even with an extreme galaxy velocity and an extreme ICM density (Kritsuk 1983; Kenney and Young 1986, 1989). The observations of shrunken HI disks with respect to optical disks and asymmetric HI distributions are consistent with ram pressure stripping. The predicted strong dependence upon galaxy velocity has not been observed (Giovanelli and Haynes 1983; Magri *et al.* 1988), although the projection effects on velocity and cluster location, the unknown inclination of galaxies with respect to their motion through the ICM, and the existence of different types of orbits within clusters (Dressler 1986) add a good deal of scatter to any such correlation. Therefore it's not clear that this is a problem.

Significant mass loss rates could also result from thermal conduction and the consequent evaporation of interstellar gas in contact with a hot intracluster medium (Cowie and Songaila 1977). Thermal evaporation is highly sensitive to ICM gas temperature and less sensitive to ICM gas density. Although only crude estimates of ICM gas temperatures exist in the clusters studied, there appears to be no correlation between temperature and the fraction of galaxies which are HI deficient (Magri *et al.* 1988). Another problem with this theory is that the effectiveness of evaporation in real galaxies may be limited by magnetic fields, which reduce thermal conductivity.

Several effects arising from the flow of a hot ICM gas past a galaxy may lead to gas loss. Nulsen (1982) has analyzed the effects of viscosity, thermal conduction, and turbulence in such a flow. In the case where the flow past the galaxy is not turbulent, then gas is lost due to both viscous drag and thermal conduction, with thermal conduction always the dominant mass loss mechanism. If the flow is turbulent, which it may be if the ICM has even a weak and tangled magnetic field, then a Kelvin-Helmholtz instability sets in which can result in significant mass loss (which Nulsen terms "turbulent viscous stripping"). Viscous stripping removes gas from the leading edge of a galaxy, and so is consistent with the observation of asymmetrical HI distributions in Virgo spirals (Warmels 1988c). For any of these transport processes, the mass loss rate will usually exceed the total mass of hot gas through which a galaxy has passed. Since large galaxies (with large surface areas) pass through more ICM gas than small galaxies, the larger ones should lose more gas than the smaller ones, and all galaxies should lose an amount proportional to their area (which is roughly proportional to mass). Therefore the turbulent viscous stripping model predicts that the average gas deficiency should be roughly the same for big and small galaxies. Haynes and Giovanelli (1986) find that the amount of HI missing from Virgo spirals is comparable, on average, with the amount of hot gas the galaxy has passed through, seemingly consistent with turbulent viscous stripping. But until the molecular gas content of small galaxies is determined (see §4.5), it will not be known if the observations are truly consistent with turbulent viscous stripping.

At this time, the empirical evidence does not strongly favor either ram pressure stripping or turbulent viscous stripping. One shortcoming of the turbulent viscous stripping model is that it's not clear how closely the mass loss rates calculated for the simple case of a uniform, single component ISM apply to a clumpy, multi-phase ISM. This complication aside, any of the transport processes supplement ram pressure stripping while being independent of it. They may be more effective than ram pressure stripping when the galaxy moves edge-on or slowly through the ICM.

4.7 PERTURBATIONS TO GAS WHICH IS NOT STRIPPED

Gas clouds which are not removed from "stripped" galaxies may be perturbed in some way by an ICM-ISM interaction. Since the ICM-ISM interaction is strong enough to remove some gas clouds, it must also perturb those clouds which have barely avoided removal. Many authors have suggested that star formation is enhanced when an ISM is perturbed during a galaxy's first encounter with an ICM, and this topic will be discussed in the next section (§4.8). Here we discuss one example of a cluster galaxy which displays a strongly asymmetric molecular gas distribution that may be the result of an ICM-ISM encounter.

Although most Virgo spirals appear to have "normal" CO emission, there is one galaxy known to have peculiar CO properties. The CO distribution in NGC 4419 is strongly asymmetric, with 3-4 times more emission on one side of the galaxy than the other (Kenney et al. 1990a). NGC 4419 is an Sa galaxy located only 2° from the cluster center, and moving at an extreme radial velocity of 1300 km s^{-1} with respect to the cluster. Its CO asymmetry, large HI deficiency, and the large molecular to atomic gas mass ratio of $\sim$15 all indicate that NGC 4419 is currently experiencing a strong ISM-ICM interaction. Since NGC 4419 is rushing through the ICM at a large inclination angle, interstellar clouds on the approaching side of the nucleus will have a velocity with respect to the ICM $\sim$ 400 km s^{-1} greater than those on the receding side, and the ram pressure force (which goes as v^2) will be $\sim$ 2 times greater on the approaching side. While ram pressure of the predicted magnitude is insufficient to strip molecular clouds from large galaxies, the force may be strong enough to significantly perturb the orbits of clouds which are not stripped.

An analytical treatment of the interaction of the ICM with a cloudy ISM by Kritsuk (1983) shows that perturbed clouds generally lose angular momentum and fall toward the nucleus, and that an asymmetric ISM should result from a strong edge-on interaction. However, it is not presently clear whether this type of interaction can produce such a strongly asymmetric distribution in the dense molecular component. The lifetime of the asymmetry presents a puzzle – differential rotation should destroy the observed asymmetry in less than 10^8 years if the gas is in circular rotation. While GMCs might gradually achieve elongated orbits after 10^9 years of ICM ram pressure, they are too massive to be significantly perturbed in less than 10^8 years. The present CO data do not resolve the velocity field sufficiently to determine whether elongated orbits are possible. Further modelling of the interactions between the ICM and a cloudy, evolving multi-phase ISM are needed, especially to see what happens to the gas which is perturbed but not stripped.

4.8 ALTERED RATES OF STAR FORMATION IN CLUSTER GALAXIES

Suggestions can be found in the literature that ICM-ISM interactions trigger star formation as well as strip gas. Gunn and Dressler (e.g., Gunn 1989) have recently proposed that the enhanced fraction of "active" galaxies (i.e., those with blue colors and strong emission lines) observed in high-redshift clusters may be due to a burst of star formation initiated by the galaxy's first encounter with a dense ICM. While this hypothesis has several merits (Evrard 1989), it will be difficult to observationally establish that an ICM-ISM interaction is the trigger for the activity in such distant galaxies. We must instead look to more nearby clusters for such evidence. The evidence for ICM-induced star formation in nearby clusters is suggestive but not yet compelling.

In Virgo spirals which are presently HI-deficient, global star formation rates have been *reduced*. Kennicutt (1983) first showed that Hα equivalent widths (measures of the present day massive star formation rates normalized to the past average star formation rates), are systematically lower in Virgo spirals within 6° of the cluster center than both non-cluster and outer Virgo cluster spirals. Large Virgo spirals which are HI-deficient by a factor of 10 have both massive star formation rates (as traced by either Hα line or far-infrared continuum emission) and total gas ($HI+H_2$) contents which are lower by factors of 2-3 (Kenney and Young 1989). The star formation rate in Virgo spirals is reduced by an amount comparable to the amount of gas lost, but this close agreement may be somewhat misleading, since gas is lost mostly from the outer disk where star formation is relatively inefficient (see Kennicutt 1989 and references therein).

In Virgo, there exist some galaxies with enhanced star formation over part of the disk, and asymmetries in their gas distributions suggestive of some kind of interaction. For example, the large Sc galaxies NGC 4254 and NGC 4654 each have compressed HI contours and enhanced massive star formation on one side of the disk (Cayatte *et al.* 1989). Each are ~3° from M87 and each have normal HI masses. The asymmetries in NGC 4654 may be due to a tidal encounter with the nearby spiral NGC 4639, but there exists no obvious perturber for NGC 4254. NGC 3312 in the Hydra I cluster also has compressed HI contours, enhanced optical and radio continuum emission on its eastern side (McMahon *et al.* 1989), and faint optical filaments located on the opposite side (Gallagher 1978). Its properties are consistent with enhanced star formation due to an ICM-ISM interaction, but they might also be the result of a tidal interaction with NGC 3314 (McMahon *et al.* 1989).

In the more distant and less well-studied Coma, Cancer, and Abell 1367 clusters, there appears to be no correlation between HI deficiency and Hα equivalent width (Kennicutt, Bothun, and Schommer 1984). This may be partly due to the survival of molecular gas, but could also reflect enhanced star formation triggered by ICM-ISM interactions in some of the HI-deficient galaxies. Among clusters with $z \leq 0.1$, the Coma cluster should have the strongest interactions since it is has the highest ICM gas densities (several times that of Virgo). Most of the best evidence for ICM-induced star formation among nearby clusters comes from Coma.

From a radio continuum survey of the Coma Supercluster, Gavazzi and Jaffe (1986) find that the cluster spirals tend to have systematically stronger radio continuum emission, which could be due to enhanced rates of star formation. They also identify three Coma cluster galaxies with unusually strong radio continuum and Hα emission, as well as asymmetries in their radio and optical distributions (Gavazzi and Jaffe 1985). These galaxies contain normal amounts of HI, and are outside the region of the cluster where x-ray emission has been detected.

Bothun and Dressler (1986) describe several HI-poor Coma cluster galaxies with very blue colors, relatively strong Hα emission (from ongoing massive star formation) and unusually strong Balmer absorption lines (from a large population of A-F stars). These properties indicate that these galaxies are in the waning stages of relatively recent starbursts. Bothun and Dressler postulate that an ICM-ISM interaction triggered star formation and began to strip HI at roughly the same time. Their observations are consistent with the strongest burst of star formation preceeding most of the gas loss.

If ICM-ISM interactions do trigger massive star formation, then it seems to occur as galaxies are encountering the dense part of the ICM for the first time and before significant amounts of gas are stripped, producing an asymmetric enhancement in star formation. The relative timing of the star formation burst and the gas loss indicates that astration may be partly responsible for HI deficiency. However, astration cannot be the general cause of HI deficiency in clusters, since the massive star formation rates are not sufficiently enhanced in most HI-deficient clusters.

By what physical mechanism might an ISM-ICM interaction induce star formation? The magnetic and turbulent forces which support gravitationally-bound molecular clouds (Myers and Goodman 1988; Elmegreen 1989a) are generally much stronger than the ram pressure forces experienced during an ISM-ICM interaction in a cluster like Virgo (Kenney and Young 1989). This makes it unlikely that the interaction will accelerate the collapse of pre-existing molecular clouds and in this manner increase the star formation rate, at least in clusters with central ICM gas densities of $\leq 10^{-4}$ cm^{-3}. On the other hand, ram pressure forces easily exceed the forces which support low density HI clouds and are expected to be dynamically important for HI clouds which are not stripped. Therefore, if star formation is enhanced in Virgo-type clusters as the result of an ICM-ISM interaction, then it could result from compressing HI to form molecular clouds, or by increasing the cloud-cloud collision rate, but not by compressing pre-existing molecular clouds.

In the centers of rich clusters with ICM gas densities of $\geq 10^{-3}$ cm^{-3}, ICM pressures which are 10-100 times greater than Milky Way ISM pressures can be experienced (Evrard 1989). These high pressures are sufficient to influence the internal dynamics of even gravitationally-bound molecular clouds, which have magnetic and turbulent energy densities also in the range of 10-100 times typical ISM pressures. Thus, star formation triggered by squeezing molecular clouds might occur in these dense clusters. This may explain why the center of Coma shows more clear-cut evidence for ICM-induced star formation than Virgo: Coma has a peak ICM pressure which is comparable to or exceeds the internal equivalent pressure of molecular clouds, whereas Virgo does not.

4.9 THE EVOLUTION OF STRIPPED CLUSTER SPIRALS

The phenomenon of gas removal in cluster galaxies has led to much discussion over whether S0's in clusters could be stripped spirals. Either environmental effects or initial conditions (or both) must account for the preferred location of S0 galaxies in high density environments (Dressler 1980). There are several facts about S0's that stripping cannot explain, including the presence of S0's in non-cluster environments, and larger bulge velocity dispersions for S0's than spirals (Dressler and Sandage 1983). Yet these facts still allow the possibility that spirals inside clusters with a dense ICM turn into S0's more rapidly than spirals outside of dense clusters. The survival of molecular gas in the inner disks of large spirals makes this morphological transformation more difficult, but probably does not preclude it. The existence of high CO/HI ratios in the inner disks of virtually all the HI-deficient Virgo spirals argues that the molecular gas reservoirs in the inner disks do not respond to the HI removal for at least a cluster crossing time, or $\sim 2 \times 10^9$ years (Kenney and Young 1986; 1989). The outer disks, however, are where most of the gas is lost. The lifetime of H_2 in outer disks is probably much shorter than that in inner disks (Scoville and Hersh 1979), so that even if outer disk GMCs are not stripped, they may quickly evolve into lower density HI which is stripped.

Certainly the gas-poor outer disks stop forming stars earlier than they would otherwise, and in this sense stripped outer disks undergo accelerated evolution with respect to both unstripped galaxies and their own inner disks. It still remains to be determined whether this effect is large enough to significantly accelerate *morphological evolution*. (All

S0 galaxies looked something like spirals when their disks were most actively forming stars, so in this sense all S0 galaxies have undergone morphological evolution). If accelerated morphological evolution occurs because of stripping, then the induced change must be roughly one Hubble type or less, since stripping does not change bulge luminosities. The galaxies which are most susceptible to this effect are the early types, since they are more HI-deficient on average, and tend to move on the more dangerous radial orbits through the cluster centers. There are several Virgo S0-Sa galaxies which are extremely HI-poor, yet contain significant CO and Hα emission from their central regions (Kenney et al. 1990b). The high CO/HI ratios in these systems imply global $M(H_2)/M(HI)$ ratios of 15-30, which are the highest known among galaxies. These galaxies exhibit the expected aftereffects of strong ISM-ICM interactions, have outer disks which have experienced accelerated evolution, and are perhaps the best candidates for galaxies whose morphological evolution has been accelerated by ICM-ISM stripping.

5. Conclusions

Galactic evolution is determined primarily by the rate at which gas turns into stars. Since the physical state, the dynamics, and even the quantity of gas are affected by many environmental factors, external influences on interstellar media play a large role in galactic evolution. Some of the processes discussed in this paper speed up the evolution of galaxies, either by removing gas or accelerating the rate at which it turns into stars. Other processes may be considered to slow down evolution, either by resupplying gas to otherwise dormant galaxies, or by increasing the energy of the ISM and thereby impeding the contraction of gas clouds.

Most of the empirical evidence for environmental effects is from nearby galaxies. Many of these effects are expected to be greater at larger redshifts, where galaxies are more gas-rich, are closer together, are emitting more ultraviolet radiation, and are entering clusters for the first time. Compared to the environmental effects on the gaseous reservoirs of galaxies in the early universe, those in the present universe may seem tame indeed.

Thanks go to Jaqueline van Gorkom and Phil Maloney for sending data and preprints in advance of publication. I am indebted to Alan Dressler, Steve Lord, Wal Sargent, Nick Scoville, Chuck Steidel for critical readings of various parts of the manuscript, and Joanna Papa for her careful scrutiny of the final draft. I'd also like to thank Harley Thronson for an excellent conference and for the money to get there.

6. References

Arp, H. 1966, *Atlas of Peculiar Galaxies* (Pasadena: California Institute of Technology).
Bajtlik, S., Duncan, R. C., and Ostriker, J. P. 1988, *Ap.J.*, **327**, 570.
Barnes, J. 1989, *Nature*, **338**, 123.

Binggeli, B., Tammann, G. A., and Sandage, A. 1987, *Astr.J.*, **94**, 251.
Bothun, G. D., and Dressler, A. 1986, *Ap.J.*, **301**, 57.
Bottinelli, L., and Gouguenheim, L. 1974, *Astr.Ap.*, **36**, 461.
Briggs, F. H., Wolfe, A. M., Krumm, N., and Salpeter, E. E. 1980, *Ap.J.*, **238**, 510.
Bushouse, H. A. 1986, *Astr.J.*, **91**, 255.
Bushouse, H. A. 1987, *Ap.J.*, **320**, 49.
Cayatte, V., van Gorkom, J. H., Balkowski, C., and Kotanyi, C. 1989, *Astr.J.*, submitted.
Chamaraux, P., Balkowski, C., and Gerard, E. 1980, *Astr.Ap.*, **83**, 38.
Chamaraux, P., Balkowski, C., and Fontanelli, P. 1986, *Astr.Ap.*, **165**, 15.
Clemens, D. P. 1985, *Ap.J.*, **295**, 422.
Cohen, R. S., Dame, T. M., Garay, G., Montani, J., Rubio, M., and Thaddeus, P. 1988, *Ap.J.(Letters)*, **331**, L95.
Combes, F., Dupraz, C., Casoli, F., and Pagani, L. 1988, *Astr.Ap.*, **203**, L9.
Corbelli, E., Schneider, S. E., and Salpeter, E. E. 1989, *Astr.J.*, **97**, 390.
Cowie, L. L., and Songaila, A. 1977, *Nature*, **266**, 501.
Davies, R. D., and Lewis, B. M. 1973, *M.N.R.A.S.*, **165**, 231.
Dickey, J. M., Salpeter, E. E., and Terzian, Y. 1978, *Ap.J.Suppl.*, **36**, 77.
Dressler, A. 1980, *Ap.J.*, **236**, 351.
Dressler, A. 1986, *Ap.J.*, **301**, 35.
Dressler, A., and Sandage, A. 1983, *Ap.J.*, **265**, 664.
Dressler, A., and Schectman, S. 1988, *Astr.J.*, **95**, 985.
Elmegreen, B. G. 1989a, *Ap.J.*, **338**, 178.
Elmegreen, B. G. 1989b, *Ap.J.(Letters)*, submitted.
Evrard, A. E. 1989, in the summary of poster papers for this conference, eds. D. J. Hollenbach and H. A. Thronson, Jr. (Washington: NASA), in press.
Fabricant, D., and Gorenstein, P. 1983, *Ap.J.*, **267**, 535.
Federman, S. R., Glassgold, A. E., and Kwan, J. 1979, *Ap.J.*, **227**, 466.
Forman, W., Jones, C., and Tucker, W. 1985, *Ap.J.*, **293**, 102.
Gallagher, J. S. 1978, *Ap.J.*, **223**, 386.
Gallagher, J. S., and Hunter, D. A. 1989, *Astr.J.*, (Sept).
Galletta, G. 1987, *Ap.J.*, **318**, 531.
Gavazzi, G., and Jaffe, W. 1985, *Ap.J.*, **294**, L89.
Gavazzi, G., and Jaffe, W. 1986, *Ap.J.*, **310**, 53.
Giovanardi, C., Helou, G., Salpeter, E. E., and Krumm, N. 1983, *Ap.J.*, **35**, 51.
Giovanelli, R., and Haynes, M. P. 1983, *Astr.J.*, **88**, 881.
Giovanelli, R., and Haynes, M. P. 1985, *Ap.J.*, **292**, 404.
Gunn, J. E., and Gott, J. R. 1972, *Ap.J.*, **176**, 1.
Gunn, J. E. 1989, in *The Epoch of Galaxy Formation*, eds. Ellis, *et al.* (Dordrecht: Kluwer), in press.
Harwit, M., Houck, J. R., Soifer, B. T., and Palumbo, G. G. C. 1987, *Ap.J.*, **315**, 28.
Haynes, M. P., and Giovanelli, R. 1984, *Astr.J.*, **89**, 758.
Haynes, M. P., and Giovanelli, R. 1986, *Ap.J.*, **306**, 466.
Haynes, M. P., Giovanelli, R., and Roberts, M. S. 1979, *Ap.J.*, **229**, 83.
Hernquist, L. 1989, *Nature*, in press.
Hibbard, J. E., and van Gorkom, J. H. 1989, in the summary of poster papers for this conference, eds. D. J. Hollenbach and H. A. Thronson, Jr. (Washington: NASA), in press.

Hoffman, G. L., Helou, G., Salpeter, E. E., and Sandage, A. 1985, *Ap.J.(Letters)*, **287**, L15.

Hoffman, G. L., Helou, G., and Salpeter, E. E. 1988, *Ap.J.*, **324**, 75.

Hummel, E., van Gorkom, J. H., and Kotanyi, C. G. 1983, *Ap.J.(Letters)*, **267**, L5.

Huchtmeier, W. K., and Richter, O.-G. 1989, *Astr.Ap.*, **210**, 1.

Huchtmeier, W. K., Tammann, G. A., and Wendker, H. J. 1976, *Astr.Ap.*, **46**, 381.

Ichikawa, S., Okamura, S., Kodaira, K., and Wakamatsu, K. 1988, *Astr.J.*, **96**, 62.

Icke, V. 1985, *Astr.Ap.*, **144**, 115.

Irwin, J. A., Seaquist, E. R., Taylor, A. R., and Duric, N. 1987, *Ap.J.(Letters)*, **313**, L91.

Johansson, S. 1989, in *Kona Symposium on Submillimetre and Millimetre Astronomy*, ed. A. Webster, in press.

Joseph, R. D., and Wright, G. S. 1985, *M.N.R.A.S.*, **214**, 87.

Keel, W. C., Kennicutt, R. C., Hummel, E., and van der Hulst, J. M. 1985, *A.J.*, **90**, 708.

Kenney, J. D., and Young, J. S. 1986, *Ap.J.(Letters)*, **301**, L13.

Kenney, J. D., and Young, J. S. 1988, *Ap.J.*, **326**, 588.

Kenney, J. D. P., and Young, J. S. 1989, *Ap.J.*, **344**, 171.

Kenney, J. D. P., Young, J. S., Hasegawa, T., and Nakai, N. 1990a, *Ap.J.*, (in press).

Kenney, J. D. P., Young, J. S., Rubin, V. C., and Ford, W. K., Jr. 1990b, in preparation.

Kenney, J. D. P., Young, J. S., Rubin, V. C., and Ford, W. K., Jr. 1990c, in preparation.

Kennicutt, R. C. 1983, *A.J.*, **88**, 483.

Kennicutt, R. C. 1989, *Ap.J.*, in press.

Kennicutt, R. C., Bothun, G. D., and Schommer, R. A. 1984, *A.J.*, **89**, 1279.

Kennicutt, R. C., Keel, W. C., van der Hulst, J. M., Hummel, E., and Roettinger, K. A. 1987, *A.J.*, **93**, 1011.

Kotanyi, C., van Gorkom, J. H., and Ekers, R. D. 1983, *Ap.J.(Letters)*, **273**, L7.

Kritsuk, A. G. 1983, *Astrofizika*, **19**, 263.

Kwan, J. 1979, *Ap.J.*, **229**, 567.

Kwan, J., and Valdes, F. 1987, *Ap.J.*, **315**, 92.

Lattanzio, J. C., and Henriksen, R. N. 1988, *M.N.R.A.S.*, **232**, 565.

Lin, Z., and Phinney, E. S. 1989, preprint.

Lonsdale, C. J., Persson, S. E., and Matthews, K. 1984, *Ap.J.*, **287**, 95.

McMahon, P. M., Richter, O.-G., van Gorkon, J. H., and Ferguson, H. C. 1989, in the summary of poster papers for this conference, eds. D. J. Hollenbach and H. A. Thronson, Jr. (Washington: NASA), in press.

Magri, C., Haynes, M. P., Forman, W., Jones, C., and Giovanelli, R. 1988, *Ap.J.*, **333**, 136.

Maloney, P. 1989, in the summary of poster papers for this conference, eds. D. J. Hollenbach and H. A. Thronson, Jr. (Washington: NASA), in press.

Maloney, P., and Black, J. H. 1988, *Ap.J.*, **325**, 389.

Mihalas, D., and Binney, J. 1981, in *Galactic Astronomy: Structure and Kinematics* (San Fransisco: Freeman), p. 421.

Miller, R. H., and Smith, B. F. 1982, *Ap.J.*, **253**, 58.

Mirabel, I. F., and Sanders, D. B. 1989, *Ap.J.(Letters)*, **340**, L53.

Myers, P. C., and Goodman, A. A. 1988, *Ap.J.(Letters)*, **326**, L27.

Noguchi, M. 1988, *Astr.Ap.*, **203**, 259.

Noguchi, M., and Ishibashi, S. 1986, *M.N.R.A.S.*, **219**, 305.

Norman, C., and Scoville, N. Z. 1988, *Ap.J.*, **332**, 163.

Nulsen, P. E. J. 1982, *M.N.R.A.S.*, **198**, 1007.

O'Dea, C. P., Sarazin, C. L., and Owen, F. N. 1987, *Ap.J.*, **316**, 113.

Olson, K. M., and Kwan, J. 1989, *Ap.J.*, in press.

Rogstad, D. H., and Shostak, G. S. 1972, *Ap.J.*, **176**, 315.

Sancisi, R., Thonnard, N., and Ekers, R. D. 1987, *Ap.J.(Letters)*, **315**, L39.

Sanders, D. B., Scoville, N. Z., Sargent, A. I., and Soifer, B. T. 1988, *Ap.J.(Letters)*, **324**, L55.

Sargent, A. I., Sanders, D. B., Scoville, N. Z., and Soifer, B. T. 1987, *Ap.J.(Letters)*, **312**, L35.

Savage, B. D. 1988, in *QSO Absorption Lines: Probing the Universe*, eds. J. C. Blades, D. Turnshek, and C. A. Norman (Cambridge: Cambridge University Press), p. 195.

Schneider, S. E. 1989, *Ap.J.*, **343**, 94.

Schweizer, F. 1986, *Science*, **231**, 227.

Schweizer, F., van Gorkom, J. H., and Seitzer, P. 1989, *Ap.J.*, **338**, 770.

Scoville, N. Z., Sanders, D. B., Sargent, A. I., Soifer, B. T., and Tinney, C. G. 1989, *Ap.J.(Letters)*, submitted.

Scoville, N. Z., and Hersh, K. 1979, *Ap.J.*, **229**, 578.

Scoville, N. Z., and Sanders, D. B. 1987, in *Interstellar Processes*, eds. D. J. Hollenbach and H. A. Thronson, Jr. (Dordrecht: Reidel), p. 21.

Scoville, N. Z., Sanders, D. B., and Clemens, D. 1986, *Ap.J.(Letters)*, **310**, L77.

Shaya, E. J., and Federman, S. R. 1987, *Ap.J.*, **319**, 76.

Shu, F. 1987, in *Star Formation in Galaxies*, ed. C. J. Lonsdale Persson (Washington: NASA), p. 743.

Silk, J. 1971, *Astr.Ap.*, **12**, 421.

Silk, J., and Sunyaev, R. A. 1976, *Nature*, **260**, 508.

Spitzer, L., and Baade, W. 1951, *Ap.J.*, **113**, 413.

Stanford, S. A., Sargent, A. I., Sanders, D. B., and Scoville, N. Z. 1989, *Ap.J.*, in press.

Stark, A. A., Knapp, G. R., Bally, J., Wilson, R. W., Penzias, A. A., and Rowe, H. E. 1986, *Ap.J.*, **310**, 660.

Steidel, C. C., and Sargent, W. L. W. 1989, *Ap.J.(Letters)*, **343**, L33.

Sunyaev, R. A. 1969, *Astrophys.Let.*, **3**, 33.

Takano, S., Awaki, H., Koyama, K., Kunieda, H., Tawara, Y., Yamauchi, S., Makishima, K., and Ohashi, T. 1989, *Nature*, **340**, 289.

Toomre, A., and Toomre, J. 1972, *Ap.J.*, **178**, 623.

Tully, R. B., and Shaya, E. 1984, *Ap.J.*, **281**, 31.

van der Hulst, T., and Sancisi, R. 1988, *Astr.J.*, **95**, 1354.

van Driel, W. 1987, Ph.D. thesis, Groningen.

van Gorkom, J. H., and Kotanyi, C. 1985, in *ESO Workshop on the Virgo Cluster of Galaxies*, eds. O.-G. Richter and B. Binggeli (Garching: ESO), p. 61.

van Gorkom, J. H., van Albada, T. S., Cornwell, T., and Sancisi, R. 1989, in preparation.

van Gorkom, J. H., Schechter, P. L., and Kristian, J. 1987, *Ap.J.*, **314**, 457.

Vigroux, L., Thuan, T. X., Vader, J. P., and Lachieze-Rey, M. 1986, *Astr.J.*, **91**, 70.

Wang, Z. 1989, *Ap.J.*, submitted.

Wardle, M., and Knapp, G. R. 1986, *Astr.J.*, **91**, 23.

Warmels, R. H. 1986, Ph.D. thesis, Groningen.

Warmels, R. H. 1988a, *Astr.Ap.Suppl.*, **72**, 19.

Warmels, R. H. 1988b, *Astr.Ap.Suppl.*, **72**, 57.

Warmels, R. H. 1988c, *Astr.Ap.Suppl.*, **72**, 427.
Williams, B. A., and Rood, H. J. 1987, *Ap.J.Supp.*, **63**, 265.
Young, J. S., and Scoville, N. Z. 1982, *Ap.J.*, **260**, L11.
Young, J. S., Kenney, J. D., Tacconi, L., Claussen, M. J., Huang, Y.-L., Tacconi-Garman, L., Xie, S., and Schloerb, F. P. 1986, *Ap.J.(Letters)*, **311**, L17.

The Diffuse Interstellar Medium

Donald P. Cox
University of Wisconsin-Madison

ABSTRACT. This article reviews the last 20 years of the struggle to understand the diffuse ISM, highlighting recent changes of fundamental aspects. It then provides a pictorial description consistent with those developments. Various caveats are discussed and a number of areas for future investigation identified. It ends with a hopeful glance toward understanding ISM phase segregation.

1. Background

A great deal of attention in interstellar medium studies is directed toward understanding denser regions where activity associated with star formation takes place. Nevertheless, there remains a modest level of effort toward understanding the diffuse interstellar medium, where the boundary conditions for the denser regions are established. This paper provides a brief (and somewhat parochial) survey of developments in this activity over the last few years, and proposes a new picture of the significant aspects of the medium.

At the time of the great ISM model of George Field, Donald Goldsmith, and Harm Habing (1969 FGH), it was widely believed that the vast majority of interstellar space was occupied by a "warm" diffuse component ($n \sim 0.1$ cm^{-3}, $T \sim 10^4$ K) with a low level of partial ionization. The FGH model invoked a fairly thick blanket of this intercloud medium so that its weight would provide the pressure required to hold together the interstellar clouds, found mainly at rather low z. A very straightforward phase diagram for the medium was constructed, based on heating of both cloud and intercloud material by (hypothesized) low energy cosmic rays, transferring the majority of supernova power. This was apparently the first "supernova dominated" ISM model, but one in which the explosion effects were diffuse in nature, providing widespread cosmic ray heating. (Since that time, observations whose details I have forgotten have set an upper limit on CR heating and ionization rates in clouds, one that is too low for the model to be appropriate.)

H. A. Thronson, Jr. and J. M. Shull (eds.), The Interstellar Medium in Galaxies, 181–200.
© 1990 Kluwer Academic Publishers.

Although the heating mechanism behind the FGH model was incorrect, certain aspects of the picture have received new attention recently, as we shall see.

In 1974, Barry Smith and I (Cox and Smith 1974) drew attention to the long lifetime of bubbles of hot gas generated by supernovae, and estimated the filling factor of such bubbles to be perhaps 10% of interstellar space. We also pointed out the nonlinear nature of the problem and suggested that bubble rejuvenation by subsequent SNe might cause growth of a tunnel network through the ISM, with hot gas occupying a considerably larger proportion of the volume. This point of view was encouraged at that time by recent observations of the soft x-ray background (SXRB) and the common occurence of O^{+5} ions in the interstellar absorption spectra of hot stars.

In 1977 this picture was pushed to its logical extreme by Chris McKee and Jerry Ostriker (McKee and Ostriker, 1977, MO) who pointed out that with an intercloud density ~ 0.1 cm^{-3}, the supernova remnants would in one generation expand to fill all space. It was believed that an individual remnant consisted at late time, of a dense shell of cold gas surrounding a bubble of hot gas. Thus, one generation produced a froth of bubbles, with the former intercloud material converted to dense gas at the bubble intersections.

The numbers entering this calculation are sufficiently firm that for a long time I regarded the conclusion as airtight. I came to refer to it as The Porosity Imperative.

MO went on to search for an ISM structure for which the initial conditions encountered by SNe and the final state in which they left it were the same. They found such a structure, with plausible mechanisms and rates to assure energy and mass conservation during the complex redistribution, and which made the configuration thermally stable. Their result is the well known and highly durable McKee and Ostriker Model for the ISM. For over a decade, many if not most observations and modeling of interstellar behavior have used the MO model as the context for their interpretations.

In barest form, the MO Model's picture of the ISM is that clouds consist of small dense cores surrounded by layers of warm neutral (WNM) and ionized (WIM) zones. The total cloud structure occupies perhaps 30% of interstellar space while the remaining 70% contains a rather hot $(T \sim 3 \times 10^5$ K) low density coronal component with comparable thermal pressure.

In this model, specific regions recently reheated by SN occurences find themselves with an even hotter coronal component, within which thermal evaporation of the immersed clouds is active. It was thought that the SXRB showed the Sun to be inside such a recently reheated region, since the x-rays indicated a local temperature for the hot gas of 10^6 K. The origin of the narrow interstellar O^{+5} absorption features was thought to be in evaporative boundaries of the clouds. The high abundance of O^{+5} which would be present in the coronal component itself was thought to have evaded detection owing to its expected broader velocity profile and the uncertain continua of the hot stars needed to observe around 1000 Å.

Over the next several years a variety of problems surfaced when the MO picture was compared with the observational situation. Many of these could be remedied in one way or another by modifying the picture somewhat, grouping clouds into complexes, for example, to remedy the difficulty that the vast numbers expected were not found near the Sun. This helped also with an associated difficulty that the Lyman continuum mean free path of 12 pc was absurdly short when the model used O star radiation to maintain the ionized layer on cloud peripheries. The fact that the O^{+5} column density in cloud evaporation boundaries had been seriously underestimated could be compensated by restricting active evaporation to clouds in unusually hot regions. The criticism was made that thermal evaporation would be quenched by tangential magnetic fields on cloud boundaries, fields swept into that configuration (as with the Earth's magnetosphere) by remnant expansion itself. But mass transfer from clouds to hot gas might take place anyway via a process called hydrodynamic stripping. There was considerable carping at the model yet there appeared to be no reasonable alternative. The porosity imperative was simply too strong an underpinning.

During this period it was quickly realized that hot gas in the ISM was buoyant (Jones, 1973) and that, if it opened directly to the space above the galactic disk, would rapidly rise into the halo or coronal region of the Galaxy. Thus were born or reborn the concepts of galactic fountains (Shapiro and Field, 1976) and galactic winds (Bregman, 1978). Considerable attention was devoted to the dynamics of such situations (Mathews and Baker, 1971; Chevalier and Oegerle, 1979; Bregman, 1980) and turned to ways in which such fountains could be the source for high velocity clouds, and for populations of coronal ions being found at high galactic latitude. They were also interesting in that a Galactic fountain provided a potentially powerful mechanism for radial redistribution of recent products of nucleosynthesis. The Proceedings of the 1985 NRAO Workshop on Gaseous Haloes of Galaxies provides a good view of the ideas and participants in much of this discussion.

Following the dramatic HI surveys of Heiles and Habing (1974), Colomb et al. (1980), and Cleary et. al. (1979), and the ensuing discoveries of HI shells and supershells, much effort was devoted to correlated supernovae and winds of massive stars, in essence the explosions of OB associations. Supernova effects were segregated, with the more uniform Type I's left to take responsibility for ISM properties in quiescent regions. The Type II's were then lumped into associations and used to generate immense blisters in the interstellar gas disk. Concepts such as "breakout" and "blowout" were invoked to describe the degree to which the blisters succeeded in penetrating the entire disk and in spewing their vertical accumulation of material (and/or hot gas from their interiors) into the lower halo. The source for galactic fountain material became conceptually localized in the successful breakouts. The subject is well covered in a recent review by Tenorio-Tagle and Bodenheimer (1988).

By 1982 I was ready to give up. I had been unable to locate the fatal flaw in the porosity imperative. On a boat in the Venice harbor, with Chris McKee, I offered to work together to patch up what I regarded as the difficulties of the MO model, and to move toward a greater truth and beauty in the process. Thanks largely to Chris's boundless optimism and enthusiasm, he was apparently unable to take my depression very seriously. The proposed collaboration never materialized and by 1985 the fatal flaw seemed to have revealed itself.

As usual in astrophysics, the way out of a difficulty is to invoke the poorly understood magnetic field. Although the approach tends to be one of last resort, the tendency to avoid it is well described by Sofue et al. (1986) in their definition of "pangravitationalism". One tends to ignor the field so long as one can get away with it. Gravity is used as the only fundamental force, with gas pressure the only operative against compression.

What I discovered in 1985 was that although it was to some extent reasonable to neglect the magnetic field effects on supernova evolution when one is concerned with models of observed SNR properties, it is folly to neglect it during the later development of remnants in the low density intercloud medium. Magnetic pressure rapidly becomes the dominant contributor to the bulk modulus of the gas in the shell. When one follows the evolution of a remnant in the intercloud medium to an age in exess of 10^6 years, when overlap of remnants becomes probable, the resulting configuration is not a thin dense shell of swept up intercloud gas enclosing a 100 pc radius bubble of hot gas. It is instead a very thick wall of intercloud material at very nearly its original density, with a small bubble of hot gas well isolated deep in the interior. Except for the generation of the bubble, the final state is very little different from the initial one. Remnants do overlap, but only at a time when they are already rather weak disturbances, and when their bubbles occupy only a small fraction of the volume.

The spell of the porosity imperative was broken.

This led to a series of conference papers exploring various aspects of this new insight (Cox 1986; Cox and Snowden 1986; Cox 1988; Cox and Slavin, 1989; Cox 1989). But early this year, Jon Slavin built a coupled hydro, ionization, and radiation code. When he ran the first ever (1D) models of the complete evolution of an SNR, through final disappearance of the hot bubble, it became absolutely clear that the new direction was sure to be fruitful. The magnetic pressure was included in an approximate fashion, assuming a tangential field but negligible tension. And there before us lay the development of the thick blanket of scarcely compressed intercloud material, with its long lived bubble nestled deep inside. What's more, we found that over the bubble lifetime its population of O VI, N V, C IV, and Si IV were apparently sufficient to provide the observed quantities of interstellar high ions. There was no longer any need to invoke a general hot gas component of interstellar gas and thermal evaporation of clouds to explain the presence of those ions.

With the passages of the porosity imperative and the need for cloud evaporation (and the high intercloud temperatures and cooperative field geometry needed to provide it), the only remaining observational evidence for the existence of a general phase of high temperature material would seem to be the soft x-ray background. Yet oddly enough, details of the X-ray observations directly contradict the assumption that 10^6 K gas is common in the interstellar medium. (In the MO picture, recall, 10^6 K gas was not common. It was present only in recently reheated regions, while most of the hot component was too cool for appreciable X-ray emission.) The discussion of this point has a long and somewhat quarrelsome history and I refer the reader to the Cox and Reynolds (1987) review of the Local ISM for details. My view, and I think that of most active students of the X-ray background, is that the observed X-rays below 1/4 KeV originate almost entirely in a local region varying in extent from 30 to 200 pc from the Sun. Within this region, the Local Bubble, 10^6 K gas is the overwhelmingly dominant occupier of volume, yet there is no indication that the hot gas continues beyond this confined region. There are other localized regions of hot gas outside our own, particularly the volume bounded by Loop I, and the Orion-Eridanus "Hot Spot". But the existence of these regions too serves to emphasize that X-ray emitting gas is a localized phenomenon deriving from major localized disruptions of the interstellar medium.

Thus the SXRB actually shows only that the Sun is within a large bubble of hot gas. It does not directly suggest that there is a pervasive coronal component.

X-ray observations have provided a second and much more damning piece of evidence that has strongly shaped my own convictions. The halos, or coronae of spiral galaxies are not the strong emitters of X-rays that afficionados of galactic fountains would expect. This is true of the Milky Way where the 100 pc or so of Local Bubble provides more X-ray surface brightness than the subsequent several kpc of possible corona (Cox, 1981). It is similarly true of other edge-on and face-on spirals that were studied with the Einstein X-ray satellite (refs in Cox and McCammon, 1986). Very little of the total anticipated SN power is radiated in X-rays by galactic coronae, winds, or fountains. One is led to invent reasons for fountains being uniformly too cool for X-ray emission, or to doubting their existence altogether. I currently find myself in the latter frame of mind.

With this background in hand, we turn to an exposition of the major features of the diffuse interstellar medium that follow from a fresh look at the data and theories.

2. The Interstellar Pressure and Its Components

This section discusses the vertical distribution of total ISM pressure and its subdivision into cosmic ray, magnetic and kinetic components. The results are that the total midplane pressure is <u>at least</u> 3.5×10^{-12} dyn cm^{-2} (p/k $\gtrsim$ 25,000 cm^{-3} K), of which there are roughly equal contributions from the three components listed above. The contribution from strictly thermal pressure, however, is almost negligible. At z ~ 1 kpc, the total pressure is down only to about a third of the midplane value, so that the pressure components are comparable to those at midplane in older ISM work. Several confirming lines of investigation concur that the diffuse ISM has both this higher pressure and much greater overall thickness than commonly appreciated.

2.1 THE WEIGHT OF THE ISM

Approximating the local situation as one of vertical hydrostatics, the pressure at height z is the weight of the overlying ISM plus any halo pressure that may be present. See Badhwar and Stevens (1977), Bloemen (1987), Spitzer (1990), or the forthcoming Boulares and Cox (1990) for more complete discussions. There have been three recent developments affecting one's best estimate of the results. These include:

(1) The work of Lockman (1984) and Lockman, Hobbs, and Shull (1986) which followed a low density component of HI, the WNM, to several hundred parsecs off the plane;

(2) The work of Reynolds (1989) which used pulsar dispersion measures of high z pulsars to trace an ionized component, the WIM, to over 1 kpc off the plane, with a measure of total column density provided by DMs to pulsars in globular clusters; and

(3) Recent studies (refs. in Boulares and Cox, 1990) which have tended to favor a near plane gravitational field that has little or no contribution from unobserved (dark) matter.

The first two developments added to our mass estimate for the ISM, while the latter reduced the weight per unit mass. Certainly we have not reached the final word in these developments, but it appears that the total weight between z=0 and 5 kpc is at least 3.5×10^{-12} dyn cm^{-2}, with a third of that being above 1 kpc. This is higher than early estimates concentrating on cloud material, chiefly because the high z components experience high gravity and add very substantially to the weight. The weight of the intercloud components alone provides 2.5×10^{-12} dyn cm^{-2} at midplane.)

2.2 THE MIDPLANE PRESSURE CONTRIBUTIONS

This minimum midplane pressure, corresponding to p/k = 25,000 cm^{-3} K, is much larger than the sum of its parts, if one constrains the contributions of the sum to estimates made a decade ago. On the other hand, recent improvements in measurements of both cosmic ray and magnetic field contributions have provided significant increases in both. In addition, the uncertainties in the velocity dispersions of the very diffuse HI and HII components (e.g. Spitzer 1990) allow the possibility that the kinetic pressure could also be significantly larger.

Holzer (1989) quotes a cosmic ray pressure of 1.3×10^{-12} dyn cm^{-2}. The increase over past estimates apparently derives from measurements by spacecraft in the outer solar system indicating that solar modulation had been underestimated.

Rand and Kulkarni (1989) studied pulsar rotation measures for 200 pulsars in the solar neighborhood. They found that the magnetic field had a uniform component of about 1.6 μG, a random component of 5μG, and a fluctuation cell size of 50 pc. The corresponding magnetic pressure was 1.3×10^{-12} dyn cm^{-2}. Curiously, however, the authors were uneasy with their results since they did not believe that the magnetic pressure would be so large compared with the, in their minds, well established cosmic ray pressure, the latter quoted by them as only 0.4×10^{-12} dyn cm^{-2} from Spitzer (1978).

This behavior is not unique to these authors. Measurements of cosmic rays, magnetic fields, and velocity dispersions are each moving away from conventional wisdom and toward higher pressure, consistent with the conclusions from hydrostatics. But each change, taken alone, appears to be an anomaly and leads to uneasiness. As a friend of mine noted, it seems almost as though Simplicio has kept a finger in our business. (Your job is to figure out whether he's writing this article.)

Within current uncertainties, the cosmic ray, magnetic field, and kinetic pressure contributions to the midplane pressure are equal, each about 1.2×10^{-12} dyn cm^{-2}. All three could be somewhat higher. In further contrast to conventional wisdom, it appears that the magnetic contribution is dominated by the random component, rather than the random and ordered fields being comparable. (The cell size found for the random component, 50 pc, may leave the field rather orderly on the smaller scales encountered in many ISM problems.)

Pulsar rotation measures do not provide the only samples of the magnetic field strength, but in the past they have tended to give consistently lower field strength than other techniques. With the recently improved pulsar sample, this is apparently no longer the case.

Some form of equipartition seems still to be alive and well, but it is apparently no longer inclusive of thermal pressure. The known contributions to kinetic pressure are dominated by bulk motions of the gas. The cosmic rays, magnetic field, and volume averaged kinetic terms each correspond to $p/k \sim 8000$ cm^{-3} K, while the thermal pressure within clouds is believed to be only about 3000 cm^{-3} K. Combined with a cloud filling factor of only about 2% of the disk (n/n ~ 1/40), this is negligible. Similarly the diffuse WNM and WIM have together $\langle n + n_e \rangle \sim 0.2$ cm^{-3}, T ~ 8000 K for $p/k \sim 1600$ cm^{-3} K. Only a substantial contribution from unseen hot gas could make the thermal contribution nonnegligible. I doubt that it does. As Charlie Goebel puts it, the thermal pool is the sink for energy dumped from the other forms, the only one capable of appreciable radiation. The energy flow could well depend on the thermal energy density being lower.

2.3 PRESSURE CONTRIBUTIONS AT 1 kpc

The total pressure at 1 kpc is roughly one third its midplane value. The mass density of that height, however, is rather low, perhaps corresponding to n ~ 0.02 cm^{-3}. In order for the kinetic pressure to be comparable to the magnetic and cosmic ray terms, the 1D velocity dispersion of the gas would have to be about 30 km s^{-1}, the magnetic field 3μG, and the cosmic ray pressure 0.4×10^{-12} dyn cm^{-2}. The high required velocity dispersion is not at odds with any observations of which I am aware. The persistence of cosmic rays and magnetic field to this height are confirmed independently from the low average density within the CR trapping volume, the high galactic synchrotron surface brightness toward the poles, and rotation measures of extragalactic radio sources.

Progress on understanding the diffuse ISM, particularly phenomena occuring at high z, will be made only when the very thick distribution of pressure and mass is included in the analyses.

2.4 EXCUSES FOR EQUIPARTITION

What velocity dispersion might one expect for a gas in a high magnetic field environment? The complex mixture of components at low z makes this question difficult, but above the cloud layer there may be a straightforward answer. Alfven waves are generated by disturbances in the disk and may propagate rather violently into the lower density material at higher z. Such waves are difficult to damp until they are nonlinear, at which point if the gas temperature is rather low, supersonic convergences in the flow must become common, yielding highly dissipative shocks. One might therefore expect the velocity field to accumulate in such waves until the dispersion velocity is comparable to the Alfven velocity, making the kinetic pressure comparable to the magnetic pressure.

What cosmic ray pressure should be expected? We tend to think in terms of CR diffusion being responsible for their escape. But there appears to be little understanding of how the diffusion coefficient depends on CR energy density in such a way that it would lead to equipartition. In fact, simple theories seem to suggest (to me at least) that a higher energy density in CR will decrease the diffusion coefficient, making it more difficult for the particles to leave. This picture surely breaks down once the CR population accumulates to have a pressure comparable to that of the confining magnetic field. In that case, diffusion occurs within the trapping volume, but with a safety valve that opens occasionally to release a portion of the prisoner population. This regulation naturally leads to cosmic ray pressure comparable to that of the magnetic field (Kraushaar,1963).

Finally, what magnetic field pressure should be expected? The field generation probably takes place via some sort of dynamo action. Such dynamos' ability to generate field is generally estimated by assuming the action of random velocity fields and calculating their effects, when combined with differential rotation, to amplify existing fields. But experience shows that when dynamos are loaded, their cranks become difficult to turn. Very likely the field strength builds up until it begins to interfere with velocity perturbations responsible for the generation. If so, the field pressure is naturally expected to be comparable to the energy densities in the disturbances responsible for the dynamo. This might well be the reason why the field strength is just right to alter meaningfully the evolution of old supernova remnants.

This series of excuses yields the relationships

$$P_k \sim P_B, \quad P_{CR} \sim P_B, \quad P_B \sim P_{SNR}.$$

The existence of the rough equalities has long been known, but this particular logic ties all forms to the supernovae through their turning of the dynamo crank.

There are further elements which could play significant roles, among them that the cosmic rays are probably accelerated by the supernovae, the waves associated with P_k probably derive from the supernovae, and the cosmic ray escape timescale, about 2×10^7 years, is very similar to the dynamical timescale of the hydrostatic disk distribution. These all should probably enter meaningfully in a comprehensive theory.

For newcomers to this game, the notion of an effective SNR pressure may require some clarification. (Its first exposition appeared in the MO model.) The basic idea is that evolving remnants at late times have some fraction f of the initial explosion energy E_o still within them. This provides an effective pressure fE_o/V, which decreases as the volume V increases. The evolution of f(V) depends on the model one has for the ambient medium. The expansion velocity is found from $p \sim \rho v^2$ implying

$$v \sim (fE_o/(\rho V))^{1/2}.$$

In familiar terms, v is roughly sonic. At any time, the age of a remnant is $t \sim (1/3) R/v$ and the time integrated volume occupation is

$$\int_0^t V \, dt' \sim \frac{1}{2} V(t) \, t.$$

For a population of supernova remnants with rate S per unit volume, the fraction of volume occupied by remnants younger than t (neglecting overlap effects) is

$$Q \sim S \int_0^t V(t') dt' \sim \frac{1}{2} SVt \sim \frac{1}{6} SVR/v.$$

The effective pressure of the supernova population is evaluated by finding the average energy density within the remnant population younger than the t for which Q = 1, at which point the population fills the space. The effective pressure is then

$$P_{SNR} \sim \frac{3}{2} \int_0^{t(Q=1)} S \, (fE_0)dt' = \frac{3}{2} \, S\overline{fE}_0 t(Q=1)$$

while

$$1 \sim Q = \frac{1}{2} \, SV(t)t$$

implying $P_{SNR} \sim 3\overline{fE}_0/V(Q=1)$. Using the substitutions for v, t, and assuming spherical remnants, one finds

$$P_{SNR} \sim S^{6/11} \rho^{3/11} (fE_0)^{8/11}.$$

As an example, the estimates $fE_0 \sim 3 \times 10^{50}$ ergs, $\rho \sim \rho_{ICM} \sim 2 \times 10^{-25}$ gm cm^{-3}, and $S \sim 4 \times 10^{-14}$ SN pc^{-3} yr^{-1} $\sim 4 \times 10^{-77}$ cm^{-3} s^{-1} yield

$$P_{SNR} \sim 2 \times 10^{-12} \text{ dyn cm}^{-2},$$

comparable to the controlled pressure components at midplane.

An underlying assumption is that the medium dissipates this energy locally, or disperses it to higher z at the same rate it is supplied.

3. A Picture of the Intercloud Medium

It is my impression from discussions with Shri Kulkarni that the observed "Warm Neutral Medium" consists of two parts, one fairly dense and cool, perhaps $\gtrsim 15$ cm^{-3} and $\lesssim 400$ K that is associated with clouds, and another much less dense and warm, perhaps 0.1 cm^{-3} and 8000 K, which is distributed independently from the clouds. The latter independence is quite clear from, for example, Lockman's work, following this low density component far higher in z than the cloud population. As a rough approximation to those results, following Bloemen (1987), I adopt

$$n_I(z) \simeq 0.1 \text{ cm}^{-3} \exp(-|z|/400 \text{ pc}).$$

for the mean density distribution of neutral intercloud gas.

Similarly there is an ionized intercloud component (WIM) studied by Reynolds, among others, with a distribution

$$n_{II}(z) \simeq 0.025 \text{ cm}^{-3} \exp(-|z|/1500 \text{ pc})$$

Apart from small bubbles of hot gas generated by Type I supernovae and large bubbles created by exploding OB associations, I propose to interpret the sum of $\overline{n}_I$ and $\overline{n}_{II}$ as a relatively homogeneous intercloud distribution of material.

As discussed in the introduction, the work of Cox and Slavin (1989) has shown that this low density distribution would not immediately be disrupted by the Type I supernovae. The bubbles generated by the supernovae were in fact estimated to fill only about 12% of interstellar space. In addition, the highly disturbed regions around OB associations can tentatively be regarded as isolated inflamations in the medium, between which the conditions are nearly quiescent and under local control. On the other hand, thermal pressure within the intercloud gas is low, having $p/k \sim 800\text{-}3000$ cm^{-3} K, and acts so slowly in restoring disruptions that exact homogeneity is certainly lacking. Similarly, the disturbances by SNe are ever present, compressing things a bit here, shuffling them about there, also contributing to widespread density variation at a moderate level.

In the midplane, then, the intercloud component has mean neutral density about 0.1 cm^{-3} and mean ionized density 0.025 cm^{-3}. For various reasons, however, it is unlikely that these two are completely intermixed. In particular, the diffuse Hα background is several times brighter than could be provided by the uniform distribution of n_e. In addition, Reynolds has searched for forbidden lines of neutrals in brighter regions of the diffuse background and interpreted their absence as being due to nearly complete ionization within the emitting region.

My own opinion is that the neutral regions are at least partially ionized, perhaps having $n_e/n \sim 0.1$ and therefore contributing about 40% of the mean electron density. At the same time, there are almost certainly regions of similar total density, fully ionized by UV from 0 stars, as well as ionized boundaries of denser regions which may be contributing somewhat to the seemingly diffuse Hα. Investigation of this is still active, however, and details may change.

At high z, the overlap of intercloud ionization cones from OB associations may provide the apparently observed rapid increase in ionization fraction with z. This too is under study, building on the early work of Trinh Thuan (1975). (See Cox, 1989, for further discussion.)

In the galactic midplane, the intercloud medium probably has an rms 1D velocity dispersion no greater than 20 km s^{-1} so that its mean dynamic pressure is
$$P_d = \rho \sigma^2 < 10^{-12} \text{ dyn cm}^{-2},$$
while its thermal pressures in neutral and ionized portions are roughly .16 and 0.3×10^{-12} dyn cm^{-2} respectively. It is threaded by a magnetic field of about 5μG, pressure $p_B \approx 1.2 \times 10^{-12}$ dyn cm^{-2}, and permeated by cosmic rays with a similar pressure. The magnetic field may be rather orderly on scales smaller than 50 pc, but possibly not entirely orderly since CR induced waves might be present in abundance at scales of order gyroradii, about 2×10^{-7} pc at 1 GeV, and wave reverberation among the cloud population could introduce parsec sized irregularities.

This configuration exists in such a thick layer compared to the cloud distribution that the name "intercloud medium" misrepresents it. Ron Reynolds and I have suggested instead "extracloud medium" and "extracloud layer" but a name with no reference whatever to clouds would be better. (I hope the search fares better than the one I attempted before settling on Local Fluff as the name for the circumsolar portion. If not, it may well end up Galactic Fluff.)

The effective thickness of the layer as a whole appears from the pressure distribution to be about 3 kpc although most of the mass is more concentrated than that. The velocity dispersion of the gas may increase sufficiently rapidly with z to keep the kinetic pressure nonnegligible.

One interesting characterization of the intercloud component goes by the name supernova recurrence timescale. It is the length of time between major disturbances by supernovae.

In Slavin's model (Cox and Slavin, 1989) for SNR evolution in $n \sim 0.1$ cm^{-3} with $B \sim 5\mu G$, the radius of the outer shock was approximately
$$R \simeq 104 \text{pc } (t/10^6 \text{ years})^{1/2},$$
while a supernova rate of $S \sim 4 \times 10^{-14}$ pc^{-3} yr^{-1} was implied by the mean interstellar O^{+5} density. (This estimate is entirely consistant with Type I rate statistics – see Heiles, 1987.) As a consequence, the criterion for $Q = 1$ is

$$1 = Q = S \int_0^t V dt \simeq \frac{2}{5} SV(t) t$$

which yields $t(Q = 1) \simeq 2.8 \times 10^6$ years and overlap radius $R(Q = 1) \simeq 174$pc, at which point the (weak) shock velocity reaches the Alfven speed, about 30 km s^{-1}. Thus there is a timescale for SNR overlap of about 3×10^6 years.

More useful in many cases is the time between somewhat more severe encounters of a parcel with a supernova shock. In this model, for example, assuming an initial temperature of 8000 K, the post shock temperature is more than twice that only for shock radii less than 110 pc. At this epoch, the age is 1.12 Myr, the compression factor is 1.63, and the shock velocity is 48 km s^{-1}. Most of the enthalpy is stored in the compressed field.

Since each SNR hits parcels at least this hard within 110 pc, 5.6×10^6 pc^3, the mean time between such encounters for a random parcel is

$$t_{SNR} = (SV)^{-1} \sim 4.5 \times 10^6 \text{ years.}$$

There are probably two significant heating mechanisms for the WNM. Diffuse clouds apparently are heated at a rate

$$h \sim 10^{-25} \text{ erg s}^{-1}(\text{H atom})^{-1},$$

possibly by photoelectric ejection from grains by starlight. This same rate may occur in the WNM. In addition, the late dissipation of a significant fraction of Type I SNR energy by large scale wave damping should provide

$$H \sim 10^{-26} \text{ erg cm}^{-3} \text{ s}^{-1}.$$

Since the region has $n \sim 0.1$ to 0.2 cm^{-3}, the "starlight" and SNR contributions are comparable.

A total heating rate of perhaps $H \sim 2 \times 10^{-26} \text{ erg cm}^{-3} \text{ s}^{-1}$, when $n \sim n_H \sim 0.1$ and $n_e \sim 0.01$, requires an extremely high cooling coefficient judged by

$$H/n^2 \sim 2 \times 10^{-24} \text{ erg cm}^3 \text{ s}^{-1}$$

or

$$H/n_e n \sim 2 \times 10^{-23} \text{erg cm}^3 \text{ s}^{-1}.$$

For comparison, the cooling rate due to C^+ excitation near 10^4 K seems to be (from cooling timescale in Kulkarni & Heiles, 1988) only

$$C_{C+}/n_e n \sim 5 \times 10^{-25} \text{ erg cm}^3 \text{ s}^{-1}.$$

It is my impression that the enormous rate of intercloud heating can be dissipated only via collisional excitation and ionization of hydrogen. For $T \sim 10^4$K, I estimate the excitation cooling rate to be

$$C_H/ \; n_e n \sim 10^{-[18 \; + \; 5.14/T_4]} erg \; cm^3 \; s^{-1}$$

(where $T_4 = T/10^4$ K), which would balance H with T $\approx$ 11,000 K.

Perhaps we can find some interesting aliases of this cooling though most is probably converted to two photon emission and/or absorbed by dust. One natural consequence is a collisional contribution to the diffuse Hα. From Osterbrock (1989, AGN[2]) I estimate the ratio of collisional excitations of n = 3 to that of n = 2 to be $0.44 \times 10^{-0.95/T_4}$. With $T_4 = 1.1$, this is 0.06. For the required total cooling rate, this ratio yields a diffuse Hα production rate of 7×10^{-17} photons cm^{-3} s^{-1}, which would resemble recombination from $n_e^2 = 5.4 \times 10^{-4}$ cm^{-6}, $n_e \simeq 0.023$ cm^{-3}. Though several times smaller than the observed diffuse Hα production rate, this could help smooth the spatial distribution somewhat.

With 10^{-25} ergs s^{-1} per atom as the characteristic heating and cooling rates, the timescale for thermal equilibration is only 70T years, or about 7×10^5 years for the WNM. This is much faster than all other timescales. On the other hand, the strong variation of the cooling rate with temperature is important. If heating is occasionally absent, a much longer time ($\sim 10^7$ years) seems to be required to move the temperature significantly lower. Conversely, intermittent high temperatures are lowered much faster. (In such regions the collisional Hα signature is also stronger.)

One final aspect that may be of some importance is that the mechanical (SNR) contribution to heating may be shared in a region somewhat thicker than the SN distribution, lowering the midplane heating which we have been considering.

4. Caveats

There are three areas of weakness that I know of in the picture I've been discussing. The first is that although I'm reasonably convinced that supernovae will not easily disrupt the intercloud medium, I'm not absolutely sure that in time they cannot or will not do so. Sufficiently large cavities or systems of tunnels may also be stable (or even unstable toward further growth).

There are only two excuses I have for not worrying more about this, one being that there is no evidence for a pervasive hot phase while there is abundant evidence for the warm intercloud gas. The other is that with all earlier arguments for the hot phase now inactivated, I prefer to begin anew with the simplest picture consistent with the observations we have. This may turn out to have been a mistake.

The second caveat is related; it is my embarrassment over our non-Copernican location inside a region of hot gas, the Local Bubble. I do not know why this bubble exists, although I have worked on it a great deal. (One hears rumors that perhaps Harold Weaver knows the answer, but I haven't seen him in some time.)

The bubble contains approximately 1 SN worth of thermal energy, yet cannot have been excavated by a single supernova. I favor a picture invented by Innes and Hartquist (1984), further elaborated in Cox and Reynolds (1987), that a cluster of explosions generated the cavity, perhaps 10^7 years ago. But if so, where are the residual longer lived members of the cluster? We will know more when spectra exist for the SXRB, but that may not lessen my embarrassment.

The third place in which I pretend to be cavalier has to do with the distributions high off the plane. It may be very unrealistic to think of conditions being primarily a function of z. Quite possibly the upper disk structure is filled with loops and plumes and great frothing irregularities. Nevertheless, we can be reasonably confident from the cosmic ray path length distribution and mean age that there is connectedness between the high z volume and the midplane. Cosmic rays not only reach these great heights, they return to tell the story.

From discussions with Dick Edgar and Blair Savage, I have the impression that interstellar Ti II is found mostly in the WNM while Al III traces the WIM. It would be interesting to compare the profiles of these ions along a line of sight to a distant high z star, particularly in a direction for which differential rotation easily translates to distance. It might be possible to map the space filling of WNM and WIM along such a sight line, explore them for mutual exclusion and for voids in the intercloud gas, measuring directly the degree of inhomogeneity.

5. New Directions and Reassessments

There are very many investigations needing doing in this new context. The hydrostatics has been pretty well covered by Boulares and Cox (1990). The rotational dynamics and internal waves of the extracloud layer are being considered by Marco Marto. The properties of SN remnants at high z are being investigated by Slavin. The 2 D aspects of magnetized SNR evolution are being undertaken by Mike Norman. The high z ionization pattern that would be generated by the known O star population is being studied by Warren Miller.

It is important to assess the degree to which the high ions at high z can be contributed by the higher members of the SNR population, whether porosity increases or decreases with z, whether a weak fountain could be driven from the outer parts of the layer.

Cosmic Ray acceleration by supernova remnants should be revisited to see whether old remnants with high magnetic pressure can contribute appreciably, and whether they cause too much diffuse acceleration to be consistent with the decreasing particle path length with increasing energy.

I discussed this a bit with Ian Axford in 1986 who seemed confident that acceleration took place preferentially in smaller younger remnants, leading me to think immediately of the Cygnus Loop. Thus far, there are tantalyzing possibilities for seeing acceleration effects in the observables, but no smoking gun (Boulares and Cox, 1988).

Since almost everyone who works with cosmic rays carries around an estimate of the fraction of SN power that is required by cosmic rays, it's probably worth noting that if remnants of Type I SNe are responsible, then the efficiency could be uncomfortably high. Our tentative estimate of the volume rate, based on 5×10^{50} ergs per SN is $S \sim 4 \times 10^{-14}$ SN pc^{-3} yr^{-1} or 4×10^{-77} SN cm^{-3} s^{-1}. The scale height has been estimated to be 300 pc (Heiles,1987). That corresponds to a surface power density in Type I SNe of 4×10^{-5} erg cm^{-2} s^{-1}. The corresponding estimate of the cosmic ray power density is (with the increased CR energy density) about 2×10^{-5} erg cm^{-2} s^{-1}. There is enough uncertainty that the actual required efficiency is not well determined. Still, the notion that it could be 0.5 should be something of a surprise.

Calculations of OB association "explosions" to create superbubbles must also be redone, including the realistic high z mass, pressure, and gravity distributions before we can have much confidence in the meaningfulness of the results. My expectations are that there will be few if any blowouts, that small superbubbles will be smaller than previously estimated (because of significant external pressure) with less dense and less noticeable shells. Conversely, larger associations may make larger bubbles because the greater pressure scale height tends to prevent depressurization of the interior by blowout.

6. Steps Toward a Successful Phase Segregation Model

A long held goal, and one which the FGH model addressed nicely, is to be able to understand the reasons behind the segregation of interstellar matter into its component phases. One first attempts to understand the phases, intercloud, diffuse cloud, and molecular cloud, and then the mechanisms for mass exchange among them.

In my view, molecular clouds are truly bizarre entities. They depend on gravity, opacity, and internal star formation for their existence. José Franco and I have suggested that in galactic systems lacking the coincidence that optical depth one and self gravity arrive at the same column density, the evolution of massive clouds could be quite different. In particular, we suggested that if the opacity were less, the cloud would evolve directly to a globular cluster rather than stabilizing in a molecular phase. (The continued formation of GCs in the LMC was considered a prime example.)

A number of papers have been written on mechanisms by which a slow rate of star formation in a molecular cloud can give the cloud as a whole a quasistable existence. For present purposes we shall suppose we understand how this works, and that ultimately a large cloud will self destruct by making OB stars. The residual gas is returned to more diffuse phases.

Diffuse clouds, on the other hand, appear to be truly stable entities. They have heating cooling balance and are confined by external pressure. The heating mechanism is not well understood, but its magnitude is known to be about 10^{-25} ergs s^{-1} (H atom)$^{-1}$, possibly due to starlight on dust as I mentioned in section 3. With that heating rate and C+ cooling, one can calculate the n(T) for which there is energy balance. This in turn determines the thermal pressure $P_c(T) = n(T)KT$.

In Cox (1988) I presented a phase diagram for the clouds, with $P_c(T)$ plotted against n(T). The pressure has a weak minimum, varying only by a factor of two over the range $n \approx 15$ cm^{-3}, T ~ 400 K and $n \approx 150$ cm^{-3}, T $\approx$ 40 K. At these two extremes $p/k = nT \approx 6000$ cm^{-3} K while the minimum, $p/k \approx 3000$ cm^{-3} K is found at T ~ 100 K, n ~ 30 cm^{-3}.

Since the cloud thermal pressures are considerably less than the system pressure, or even the magnetic contribution with $p/k \sim 8000$ cm^{-3} K, the clouds can easily be confined by their surroundings. They will probably have very slightly lower magnetic fields than the intercloud component, since they have higher thermal pressure. (In any particular case, this is sensitive to the cloud's dynamical environment as well.)

Whatever the cause of the heating mechanism, it satisfies a very strange criterion. It provides just the right amount of heating so that the diffuse clouds will have thermal pressures that can be comfortably confined. If the heating were several times larger or the C+ abundance several times smaller, clouds would be driven back into the intercloud medium by their own internal pressures. There would probably be some similar catastrophe if the heating were lower, perhaps the intercloud thermal pressure forcing the clouds into a very cold dense state, with non-negligible self gravity.

This is a great mystery. The heating mechanism, if photoelectric ejection, depends sensitively on the diffuse UV flux, the grain abundance and details of the grain properties. The ambient pressure is determined by the surface density of intercloud gas and the supernova power which inflates it to give it weight. How both of these are connected to provide a very comfortable pressure equilibrium has been very difficult to imagine.

But I've thought of a way. The key piece of information is that it has been very difficult for quite clever and dedicated astropysicists to calculate the photoelectric heating rate. What that may mean is that grain environment, with its many destruction, growth, annealing and fragmention processes may make the photoelectric heating rate a strong function of a parcel's recent history.

Under that circumstance, it is quite possible that the value of the heating rate found in diffuse clouds is determined by the maximum h allowed for cloud condensation out of the intercloud component. We would then get what we see.

Perhaps the following scenario could develop into a useful model. When material is in molecular clouds, it develops a grain population containing fragile members, able to provide large photoelectric heating if placed in an optically thin environment. Molecular cloud destruction moves material into the intercloud phase where photoelectric heating is initially quite large. Supernovae and the total intercloud surface density control the vertical distribution of mass, the pressure, and the intercloud density. With time, grain modification in this environment reduces h, the heating coefficient.

Supernova shocks sweeping through the ICM become radiative when their velocities are about 150 km s^{-1}, from which one can estimate the maximum compression factor allowed by the magnetic field, and therefore the highest density brought about by the remnants. (For v_c = 150 km s^{-1}, n_0 = 0.1 cm^{-3}, B_0 = 5μG, the maximum density is about 7 cm^{-3}.) If at that density the heating rate is such that the equilibrium thermal pressure is low enough, the gas might not fully reexpand to join the intercloud phase, instead being left in a stable cloud component.

The density 7 cm^{-3} is quite low for the h observed in diffuse clouds, but would be acceptable for half that heating rate. Once the gas reaches the cloud phase, however, dust alteration sets in to raise h very gradually with time.

Notice that if h is sufficiently sensitive to dust properties, the magnitude of the UV flux matters hardly at all. It just has to be there.

There is a great deal more involved in making this into a model. So far, it appears that the intercloud mass content would be determined by the molecular cloud destruction rate and the dust processing timescale. But the latter is also very sensitive to environment. If the intercloud surface density were raised a little, perhaps a factor of three, the SNR heating could not keep the intercloud temperature at 10^4 K. The system would go into wholesale cooling, permeated by many shocks as waves damped. The dust destruction rate could rise dramatically, stable clouds condense out, and the intercloud medium return to 10^4 K with a much reduced surface density. (Something of this sort could happen on spiral arm traversal.) Thus the dust processing timescale is probably also regulated by the medium (making it also very difficult to calculate). This regulates the intercloud surface density as well, forcing it to lie in a regime which is close to wholesale condensation and to marginal h for cloud stability.

With this picture, the intercloud surface density and all its other properties become functions only of the supernova rate.

A next step would be to investigate the cloud population for a criterion that would drive their collection into the molecular component. A straightforward estimate of the effective Jeans' Mass implies that the clouds should gather when their average spatial density is comparable to that of the stellar component, again consistent with observation.

The stellar distribution with its density, gravity, and Type I supernova rate thus controls the intercloud medium properties, the diffuse cloud heating rate and pressure, and the volume filling factor of diffuse clouds. Molecular clouds get what is left over.

Can this be? Get to Work.

For further information on the diffuse ISM, I refer the reader to the pair of beautiful review articles by Kulkari and Heiles (1987,1988).

This work was supported in part by the National Aeronautics and Space Administration under Grant NAG5-629 to the University of Wisconsin. I would like to express my appreciation for the heroic efforts of M. Brunner, S. Christenson, A. Abad, and J. Joers in the preparation of this manuscript.

7. References

Badhwar, G. D. and Stevens, S. A. 1977, Ap. J, 212,494.
Bloemen, J.B.G.M. 1987, Ap. J., 322,694.
Boulares, A., and Cox, D. P. 1988, Ap. J., 333,198.
Boulares, A. and Cox, D. P. 1990, to be submitted to Ap. J.
Bregman, J. N. 1978, Ap. J., 224, 768.
Bregman, J. N. 1980, Ap. J., 236, 577.
Chevalier, R. A. and Oegerle, W. R. 1979, Ap. J., 227,398.
Cleary, M. N., Heiles, C., and Haslam, C.G.T. 1979, Astron. Astrophys. Suppl., 36,95.
Colomb, F. R., Poppel, W.G.L., and Heiles, C. 1980, Astron. Astrophys. Suppl., 40,47.
Cox, D. P. 1981, Ap. J., 245,534.

Cox, D. P. 1986 in Workshop on Model Nebulae, Observ. de Meudon, ed. Pequinot, D. (Paris Observatory) p. 11.

Cox, D. P. 1988, in Supernova Remnants and the Interstellar Medium, eds. R. S. Roger and T. L. Landecker (Cambridge) p. 73.

Cox, D. P. 1989 in Proc. of IAU Colloq. 120, Structure and Dynamics of the Interstellar Medium, eds. Tenorio-Tagle, G., Moles, M., and Melnick, J. (Berlin, Springer-Verlag) in press.

Cox, D. P., and McCammon, D. 1986, Ap. J., 304,657.

Cox, D. P. and Reynolds, R. J. 1987, Ann Rev. Astron. and Astrophys., 25,303.

Cox, D. P. and Slavin, J. D. 1989 in EUV Astronomy, eds. R. F. Malina and S. Bowyer (Pergomon) in press.

Cox, D. P. and Smith, B. W. 1974, Ap. J. (Lett), 189,L105.

Cox, D. P. and Snowden, S. L. 1986, Adv. Space Res., 6,97.

Field, G. B., Goldsmith, D. W. and Habing, H. J. 1969, Ap. J. (Letters), 155,L149.

Heiles, C. 1987, Ap. J., 315,555.

Heiles, C. and Habing, H. J. 1974, Astron. Astrophys. Suppl., 14,1.

Holzer, T. E. 1989, in Ann. Rev. Astron. and Astrophys., 27,199.

Innes, D. E. and Hartquist, T. W. 1984, M.N.R.A.S., 209,7.

Jones, E. M. 1973, Ap. J., 182,559.

Kraushaar, W. L. 1963, in Proc. Int. Conf. on Cosmic Rays at Jaipur, 3,379.

Kulkarni, S. R. and Heiles, C. 1988, in Galactic and Extragalactic Radio Astronomy, ed. Kellerman, K. I. and Verschuur, G. L., 2nd ed., (New York:Springer Verlag) p. 95.

Kulkarni, S. R. and Heiles, C. 1987, in Interstellar Processes, ed. Hollenbach, D. J. and Thronson, H. A., Jr., (Dordrecht:Reidel) p. 87.

Lockman, F. J. 1984, Ap. J., 283,90.

Lockman, F. J., Hobbs, L. M., and Shull, J. M. 1986, Ap. J., 301,380.

Mathews, W. G., and Baker, J. C. 1971, Ap. J., 170,241.

McKee, C. F. and Ostriker, J. P. 1977, Ap. J., 218,148.

Osterbrock, D. E. 1989, Astrophysics of Gaseous Nebulae and Active Galactic Nuclei, (University Science Books:Mill Valley)

Rand, R. J. and Kulkarni, S. R. 1989, Ap. J., 343,760.

Reynolds, R. J. 1989, Ap. J. (Lett.), 339,L29.

Shapiro, P. R. and Field, G. B. 1976, Ap. J., 205, 762.

Sofue, Y., Fujimoto, M., and Wielebinski, R. 1986, Ann. Rev. Astron. Astrophys., 24,459.

Spitzer, L., Jr. 1978, Physical Processes in the Interstellar Medium (New York:John Wiley and Sons)

Spitzer, L., Jr. 1990, Ann. Rev. Astron. and Astrophys., 28, in press.

Tenorio-Tagle, G. and Bodenheimer, P. 1988, Ann. Rev. Astron. and Astrophys., 26, 145.

Thuan, T. X. 1975, Ap. J., 198,307.

Cooling Flows and X-Ray Emission in Early-Type Galaxies

Craig L. Sarazin
Department of Astronomy
University of Virginia
P. O. Box 3818 University Station
Charlottesville, Virginia 22903 U.S.A.

ABSTRACT. Early-type galaxies are luminous sources of X-ray radiation, and contain large amounts of hot, interstellar gas. In the brighter X-ray galaxies, the inferred masses of hot gas are consistent with those expected given the present rates of stellar mass loss. The required rates of heating of the gas are also roughly consistent with those expected from the motions of gas losing stars and supernovae. The X-ray observations suggest a lower rate of Type Ia supernovae than has been previously thought to apply. In the brightest X-ray galaxies, the cooling times in the gas are short, which suggests that the gas forms steady-state cooling flows. Cooling flow models explain most of the properties of the brighter X-ray galaxies, including their luminosities, the X-ray–optical correlation, their temperatures, and their surface brightness profiles. Homogeneous cooling flow models, in which all the hot gas flows into the central regions of the galaxy, have surface brightness profiles which are too centrally peaked, and which decline too rapidly in the outer parts. On the other hand, if the gas is inhomogeneous and much of the gas cools below X-ray emitting temperature near the point where it is injected, good agreement with the observed X-ray surface brightness profiles is found. Although the optical and X-ray luminosities of early-type galaxies are strongly correlated, there is a large dispersion in this correlation. One suggestion is that this dispersion indicates that early-type galaxies are not generally in steady-state, and that the fainter galaxies are in an earlier hydrodynamical phase in which most of the heating of the gas goes into the kinetic energy in a wind or partial wind, or into the thermal energy of subsonic inflation. Alternatively, the fainter galaxies may have lost much of their gas to ram pressure ablation. At a given optical luminosity, galaxies which are brighter in X-rays are found in less dense regions, which supports this last suggestion.

1. Introduction

One of the most exciting and least anticipated discoveries of the *Einstein* X-ray Observatory was that normal elliptical and S0 galaxies are generally strong X-ray sources (Forman *et al.* 1979; Nulsen, Stewart, and Fabian 1984; Forman, Jones, and Tucker 1985; Trinchieri and Fabbiano 1985). The X-ray emission indicates that these galaxies contain extensive atmospheres of hot, diffuse interstellar gas. Prior to this discovery, it was generally be-

H. A. Thronson, Jr. and J. M. Shull (eds.), The Interstellar Medium in Galaxies, 201–238.
© 1990 *Kluwer Academic Publishers.*

lieved that early-type galaxies were gas-poor systems. The amounts of cool gas observed to be present in these galaxies are generally much less than the amounts expected from stellar mass loss (Sandage 1957; Faber and Gallagher 1976; Knapp 1988). For example, an elliptical with a blue luminosity of $\sim 10^{11}\ L_\odot$ would be expected to accumulate $\sim 10^{10}\ M_\odot$ of gas from stellar mass loss at the present rates over a Hubble time (Faber and Gallagher 1976). The lack of cool gas or obvious evidence for star formation in early-type galaxies led to the theory that supernova-powered galactic winds might drive almost all of the interstellar gas out of these galaxies (Mathews and Baker 1971; Faber and Gallagher 1976; Bregman 1978; Bailey 1980; MacDonald and Bailey 1981; White and Chevalier 1983). We now know that this gas is actually present in many elliptical galaxies. It is hot (with a temperature $T \sim 10^7$ K), and escaped earlier detection because most of its emission is the X-ray band.

The first elliptical galaxy to be detected in X-rays was M87; in fact, this was the first extragalactic X-ray source to be discovered, and one of the earliest identified X-ray sources of any kind (Byram, Chubb, and Friedman 1966; Bradt et al. 1967). The X-ray spectrum of M87 established that the emission was thermal emission from hot, diffuse gas (Serlemitsos et al. 1977). However, in the case of this and other cluster central galaxies, it is uncertain whether to attribute the hot gas to the galaxy or to the surrounding cluster. Because M87 is much more luminous in X-rays than other elliptical galaxies with similar optical luminosities in the Virgo cluster, it is likely that much of the gas that surrounds M87 may arise from the cluster rather than from M87 alone. Luminous X-ray emission is detected around many other cluster central elliptical galaxies.

X-ray emission from more "normal" (not cluster central) ellipticals was first detected in an early *Einstein* Observatory survey of the Virgo cluster (Forman et al. 1979). Five normal ellipticals were detected at that time, and all were much fainter in X-rays than M87. Several of these galaxies were far enough from the center of the Virgo cluster X-ray source (M87) that it was unlikely that the emission was due to intracluster gas. Following this initial detection, many other ellipticals and S0s have been detected as X-ray sources with the *Einstein* and EXOSAT Observatories (Kriss et al. 1980; Biermann, Kronberg, and Madore 1982; Bechtold et al. 1983; Biermann and Kronberg 1983; Kriss, Cioffi, and Canizares 1983; Nulsen, Stewart, and Fabian 1984; Forman, Jones, and Tucker 1984, 1985; Dressel and Wilson 1985; Trinchieri and Fabbiano 1985; Mason and Rosen 1985; Canizares et al. 1986; Killeen, Bicknell, and Carter 1986; Thomas et al. 1986; Trinchieri, Fabbiano, and Canizares 1986; Canizares, Fabbiano, and Trinchieri 1987; Killeen and Bicknell 1988). These include several relatively isolated ellipticals (*e.g.*, Nulsen et al. 1984). The ubiquity of the X-ray emission suggests that it connected with the galaxies themselves and not with their environment. However, most ellipticals and S0s are found in relatively dense environments, and it is always somewhat uncertain whether to attribute the hot gas halo of an early-type galaxy to the galaxy itself, or to the surrounding group, poor cluster, or subcluster. For example, one of the brightest normal X-ray ellipticals is NGC 4472. This galaxy is near the center of a subcluster of galaxies in the southern Virgo cluster, and it has been argued that the gaseous halo of this galaxy is accreted intergalactic gas rather than gas from the galaxy itself (*e.g.*, Thomas 1986).

This paper will review the X-ray properties of normal early-type galaxies, and our limited theoretical understanding of the physics of the hot interstellar medium in these galaxies. There are a number of other excellent reviews of the X-ray properties of early-type galaxies and related systems, which were a great help in preparing this paper. Reviews of the X-ray properties of ellipticals and S0s include Long and Van Speybroeck (1983), and Fabbiano (1986a, 1989). The last paper, an *Annual Reviews* paper on X-ray emission from normal galaxies, is very comprehensive. As will be discussed below, the hot gas in ellipticals and S0s is similar in many ways to the hot intracluster medium. In particular, the gas in ellipticals may form "cooling flows" similar to those observed around the central galaxies in many clusters of galaxies. These cooling flows are reviewed extensively in Fabian, Nulsen, and Canizares (1984), Sarazin (1986a, 1988), and Fabian (1988). The latter is the Proceedings of a NATO workshop on cooling flows. Most of these reviews on clusters and cooling flows also discuss normal elliptical galaxies.

The X-ray properties of early-type galaxies are summarized in §2. Some simple arguments about the physical state of the gas are given in §3. §4 presents steady-state cooling flow models for these galaxies, and their time-dependent evolution is discussed in §5. Early-type galaxies of a given optical luminosity are observed to have a wide range of X-ray luminosities; the origin of this dispersion is discussed in §6. Some suggestions for further work are summarized in §7.

2. X-ray Emission

2.1 OBSERVED PROPERTIES

2.1.1 X-ray Luminosities. At present, X-ray fluxes or interesting upper limits exist for $\sim$100 elliptical and S0 galaxies; the largest surveys include Forman *et al.* (1985), Trinchieri and Fabbiano (1985), and Canizares *et al.* (1987). These galaxies have X-ray luminosities which range over $L_X \approx 10^{39} - 10^{42}$ erg s^{-1}. (All comparisons to observations in this paper assume a Hubble constant $H_o = 50$ km/s/Mpc, and a distance to the center of the Virgo cluster of 25 Mpc.) As mentioned above, the central galaxies in groups or clusters often are much more luminous. There appears to be a strong correlation between the X-ray and blue optical luminosities of early-type galaxies, with $L_X \propto L_B^{1.6-2.3}$. Figure 1 shows the X-ray and optical luminosities of the early-type galaxies in the survey of Canizares *et al.* (1987). The filled circles are detections, while the inverted triangles are upper limits on L_X. The best-fit correlation (including the upper limits using maximum-likelihood survival statistics) is

$$\left(\frac{L_X}{10^{41} \text{ ergs s}^{-1}} \right) = 1.22 \left(\frac{L_B}{10^{11} \, L_\odot} \right)^{1.77} . \tag{1}$$

Better distance estimates are available for a subsample of elliptical galaxies from the Seven Samurai survey (Faber *et al.* 1989). When these are combined with the X-ray fluxes and limits from Canizares *et al.* (1987), a steeper correlation is found (White and Sarazin 1989)

$$\left(\frac{L_X}{10^{41} \text{ ergs s}^{-1}} \right) = 0.94 \left(\frac{L_B}{10^{11} \, L_\odot} \right)^{2.27} . \tag{2}$$

It is particularly important for the interpretation of the X-ray emission that the X-ray–optical correlation is considerably steeper than linear. One should also note that there is considerable dispersion about this correlation; this point will be discussed further in §6.

2.1.2 X-ray Surface Brightness. The spatial distribution of the X-ray emission has been studied for a much smaller sample of ellipticals for which the observations have adequate

204

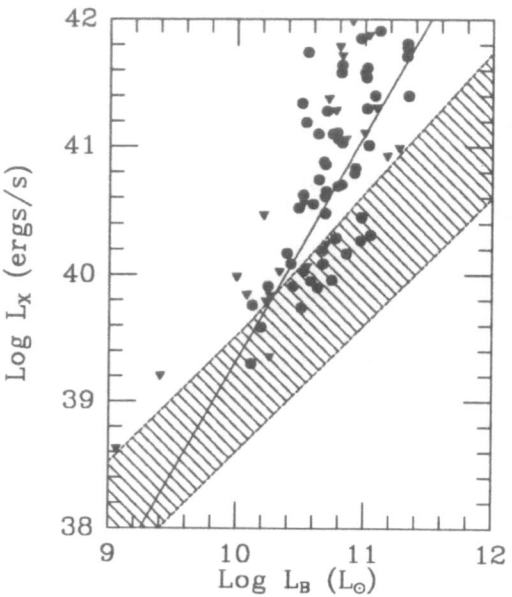

Figure 1. The correlation of X-ray and optical luminosities of early-type galaxies from Canizares *et al.* (1987). The filled circles are detections, and the inverted triangles are upper limits. The solid line is the best-fit correlation (equation 1). The hatched area gives the range of estimated stellar X-ray luminosities from Forman *et al.* (1985), Canizares *et al.* (1987), and Fabbiano *et al.* (1989).

signal to noise. The X-ray emission in these sources extends out to radii of $R_X \sim 50$ kpc (Forman *et al.* 1985; Trinchieri and Fabbiano 1985; Thomas *et al.* 1986; Trinchieri *et al.* 1986). While this radius is considerably larger than the effective radius of these large elliptical galaxies, it is similar to the largest radii at which optical light is detected. Of course, both the observed optical and X-ray extents of galaxies are determined largely by the background. Figure 2 shows the *Einstein* Observatory IPC X-ray image of the Virgo elliptical NGC 4472 from Fabbiano (1989). Typically, the X-ray isophotes are circular in the inner parts of the images, and become irregular in the outer parts. In several objects, the outer X-ray isophotes are significantly distorted and non-circular. As can be seen in Figure 2, the X-ray emission in NGC 4472 extends further to south than to the north. A more dramatic case is M86 (Figure 3), where the X-ray emission forms a plume extending to the north of the galaxy (Forman and Jones 1982).

The azimuthally-averaged surface brightness profiles of elliptical have been fit using the "beta" model which was developed for clusters of galaxies:

$$I_X(r) = I_{X_o} \left[1 + \left(\frac{r}{a_X}\right)^2\right]^{-3\beta + 1/2}, \tag{3}$$

where I_{X_o} is the central surface brightness, r is the projected radius, and a_X is the core radius of the X-ray emission. In Table 1, values of the X-ray core radius a_X, the parameter

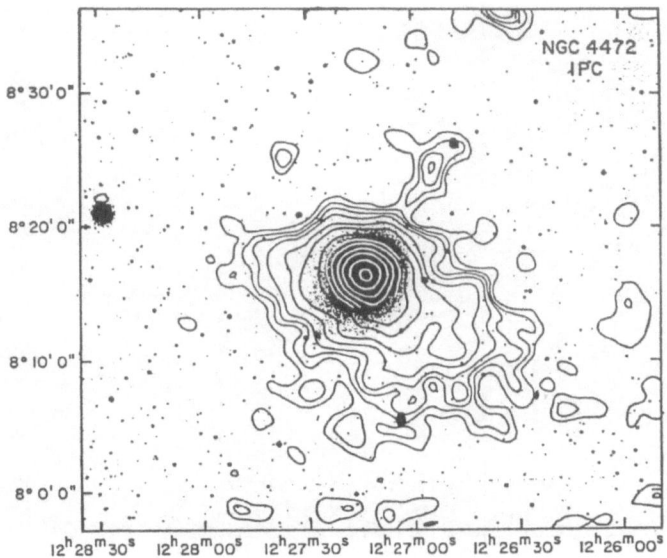

Figure 2. The *Einstein* Observatory IPC X-ray image of the elliptical galaxy NGC 4472 is shown as contours superimposed on the Palomar Sky Survey optical image (Fabbiano 1989). Note the north-south asymmetry in the outer portions of the X-ray image.

β, and the maximum radius of detected X-ray emission R_X are given for several bright X-ray galaxies from Forman *et al.* (1985; FJT) and Trinchieri *et al.* (1986; TFC). For the brighter ellipticals, values of $\beta \approx 1/2$ and $a_X \sim 2$ kpc are typically found. This implies that in the outer parts of ellipticals, the X-ray surface brightness falls off roughly as $I_X \propto r^{-2}$. This is similar to the decline in the optical surface brightness of ellipticals. For a Hubble or a King law, the optical surface brightness declines as $I_B \propto r^{-2}$ in the outer parts (King 1978). Trinchieri *et al.* (1986) find that the X-ray surface brightness follows the optical surface brightness fairly well in the three early-type galaxies with the best images. Figure 4 shows the X-ray and optical surface brightness profiles of NGC 4472, which illustrates the correspondence. The scale in the optical surface brightness is arbitrary.

2.1.3 X-ray Spectra. At present, X-ray spectral information is available only for the brighter X-ray galaxies. These early-type galaxies have fairly soft X-ray spectra; thermal fits to the *Einstein* Observatory IPC spectra lead to temperature of $T_X \sim 0.5 - 2$ keV (Forman *et al.* 1985; Trinchieri *et al.* 1986). In one galaxy (NGC 4472), there were a sufficient number of counts in the IPC image to divide the spectra into three radial bins, and the spectral temperatures were found to be roughly constant (Forman *et al.* 1985). The spectra from the IPC are limited by the poor spectral resolution and the small bandwidth of this instrument. Recently, higher resolution spectra extending to higher photon energies have be obtained with the *Ginga* X-ray Observatory for two of the brightest ellipticals, NGC 4472 and NGC 4636 (Ohashi 1990). These spectra are fit by thermal emission with a

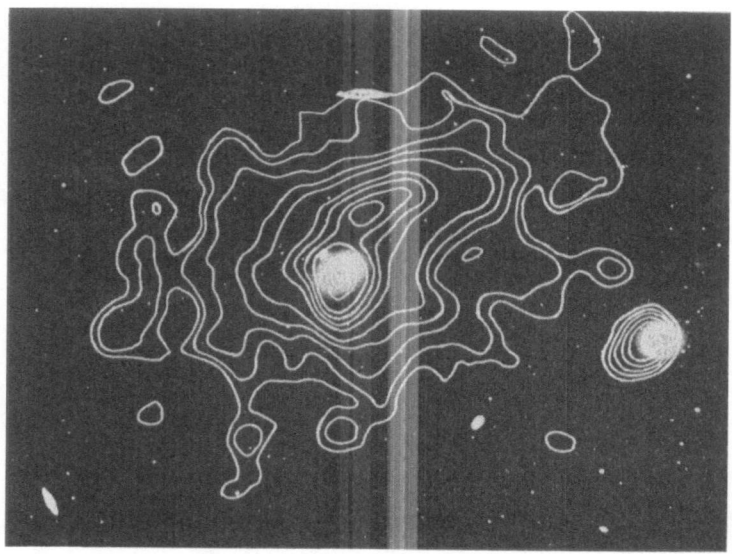

Figure 3. The *Einstein* Observatory IPC X-ray image of the elliptical galaxy M86 (NGC 4406) in the Virgo cluster. The image is shown as contours superimposed on the Palomar Sky Survey optical image (Forman and Jones 1982). Note the X-ray "plume" extending to the north.

relatively low temperature $T_X \approx 1.9$ keV. The 7 keV iron line complex is not detected, which leads to an upper limit of roughly twice the solar abundance of iron. On the other hand, the S0 galaxy NGC 3998 apparently had a harder, power-law X-ray spectrum.

2.2 THERMAL VS. STELLAR EMISSION

Two X-ray emission mechanisms have been suggested which might play a major role in early-type galaxies. First, these galaxies almost certainly contain stellar X-ray sources (probably mainly compact binaries) of the sort that are found in our own galaxy and in nearby spiral galaxies. Second, the emission might be due to hot, interstellar gas. In the brighter X-ray ellipticals and S0s, there are compelling arguments that the emission is thermal emission by diffuse hot gas (Forman *et al.* 1985; Trinchieri and Fabbiano 1985). First, the spectra are well-fit by thermal emission from gas at $T \approx 10^7$ K. The spectra are not fit as well by a non-thermal power-law. The X-ray spectra of the galactic bulge binary X-ray sources, which are of the type which might be expected to make up the bulk of the stellar X-ray emission of elliptical and S0s, are generally much harder. The X-ray spectra of spiral galaxies, which are believed to be dominated by stellar X-ray sources, are also much harder than those of early-type galaxies (Fabbiano and Trinchieri 1987).

Second, the asymmetries in the outer portions of the X-ray images of ellipticals (*e.g.*, Figure 2) would be difficult to understand if the X-ray emission were stellar. Similar

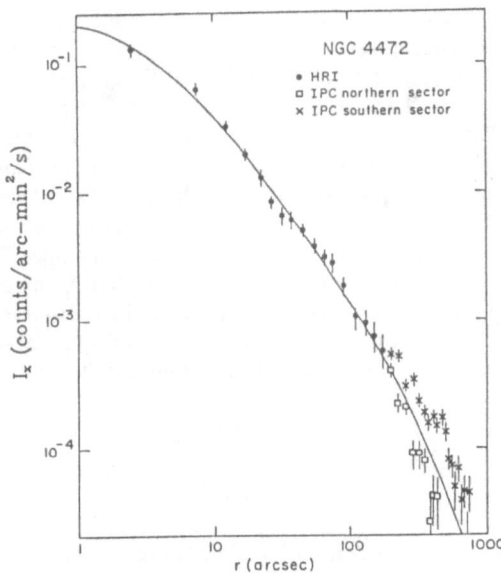

Figure 4. The *Einstein* Observatory X-ray surface brightness profile for the elliptical galaxy NGC 4472 in the Virgo cluster (Trinchieri *et al.* 1986). The filled circles at smaller radii are from the HRI image, while the crosses and open squares at larger radii are from the IPC. Because of the north-south asymmetry in the outer portions of the X-ray image, the northern and southern portions of the IPC image are shown separately. The solid curve is the blue optical surface brightness of the galaxy from King (1978); this curve has been shifted vertically to match the X-ray surface brightness.

asymmetries are not generally seen in the stellar optical emission. On the other hand, these asymmetries can be understood as the hydrodynamical interactions of the hot interstellar gas in ellipticals and S0s with any surrounding intergalactic medium. For example, an extreme case of such an asymmetry is the plume in M86 (Figure 3). M86 has a high radial velocity relative to M87 and to the mean velocity of galaxies near the center of the Virgo cluster, and is located in a region of relatively high intracluster gas density. The X-ray morphology appears to be well-explained by the ram pressure stripping of the hot interstellar gas in this galaxy by intracluster gas from the Virgo cluster (Forman *et al.* 1979; Fabian, Schwarz, and Forman 1980; Forman *et al.* 1984; Takeda, Nulsen, and Fabian 1984; Nulsen and Carter 1987; White *et al.* 1989).

Third, most elliptical and S0 galaxies have rather similar stellar colors and spectra, suggesting that they have similar stellar populations (Sandage and Visvanathan 1978). If the X-ray emission were stellar in origin, one would expect that the optical and X-ray luminosities would vary in proportion to one another (Forman *et al.* 1985; Trinchieri and Fabbiano 1985). The nonlinear X-ray–optical correlation which is observed (equations 1 and 2) would appear to be inconsistent with the hypothesis that stellar X-ray emission dominates in all ellipticals. As will be discussed below, the observed X-ray–optical correlation can be understood if the X-ray emission is thermal emission from hot interstellar gas which has been ejected by stars.

Finally, the X-ray luminosities of the brighter X-ray galaxies are too large to be plausibly produced by stellar X-ray sources (Forman *et al.* 1985; Trinchieri and Fabbiano 1985). The details of the formation and stellar evolution of the bulge X-rays sources is poorly understood, so the most secure way to infer their population in ellipticals is by scaling from other galaxies. Several estimates have been made in this way; unfortunately, they disagree by a factor of about ten. Forman *et al.* (1985) scaled the stellar X-ray emission from Cen A, and argued that this was actually an upper limit. Trinchieri and Fabbiano (1985) and Canizares *et al.* (1987) scaled the stellar X-ray emission from that of the bulge of M31, where the X-ray emission is resolved into discrete sources. Fabbiano, Gioia, and Trinchieri (1989) used the X-ray–optical correlation of late-type spiral galaxies. These different determinations give the stellar X-ray luminosity of

$$\left[\frac{L_X(stellar)}{10^{41} \text{ ergs s}^{-1}} \right] \approx \left\{ \begin{matrix} 0.42 \\ 0.33 \\ 0.04 \end{matrix} \right\} \left(\frac{L_B}{10^{11} L_\odot} \right), \qquad (4)$$

where the top value in the brackets is from Fabbiano *et al.* (1989), the middle value is from Canizares *et al.* (1987), and the bottom value is from Forman *et al.* (1985). The Fabbiano *et al.* stellar X-ray–optical correlation is actually a little steeper than the linear relationship shown in equation (4). The range of stellar X-ray luminosities from these three estimates is shown as a hatched region in Figure 1.

It is clear from this plot that the brighter X-ray sources cannot be due to stellar X-ray sources. This plot also emphasizes the point made earlier, that the stellar X-ray luminosity should vary in a nearly linear fashion with the optical luminosity. The brighter X-ray ellipticals and S0s include all of those galaxies having good X-ray images and reasonable X-ray spectra. As noted before, the images and spectra of these galaxies are inconsistent with stellar X-ray emission, but consistent with thermal emission by diffuse gas. Thus, it seems fairly certain that the bulk of the X-ray emission in the brighter X-ray galaxies is due to thermal emission by hot gas.

However, the situation is much less clear for the fainter X-ray galaxies. Note that there are quite a number of detected galaxies with $10 \lesssim \log L_B \lesssim 11$ and $\log L_X \lesssim 40.5$ which are within or near the hatched region in Figure 1. Thus, if the higher of the estimates for the stellar X-ray emission is correct, these galaxies are dominated by stellar X-ray sources (Canizares *et al.* 1987; Fabbiano *et al.* 1989). If the lower of the estimates applies, then all of the detected galaxies brighter than $L_B \gtrsim 10^{10} L_\odot$ are dominated by thermal X-ray emission by hot gas (Forman *et al.* 1985). While there are interesting arguments given in support of each of these positions, it is probably fair to regard the range of the estimates as an indication of our ignorance of the population of bulge X-ray sources in early-type galaxies. At present, the most direct way to resolve this issue would be if X-ray spectra could be obtained for the fainter X-ray galaxies (those in the shaded region of Figure 1). If the emission from these galaxies is dominated by stellar sources, the spectra should be hard ($T_X \sim 10$ keV). If the emission is due to diffuse interstellar gas, the spectra should be soft ($T_X \sim 1$ keV), similar to the observed spectra of the brighter X-ray galaxies.

2.3 GAS PROPERTIES

Assuming that the X-ray emission in the brighter X-ray galaxies is thermal emission from a diffuse plasma with solar abundances, one can use the X-ray properties to infer the properties of the gas. In this section, the gas will be assumed to be spherically symmetric and homogeneous; in §4.2, evidence that the gas may be clumped will be presented. Then, the gas in any region can be characterized by a single temperature $T(r)$ and a single gas density $n_p(r)$, which are functions only of the radius r. (Here, n_p is the number density of hydrogen nuclei.)

The X-ray spectra indicate that the average temperature in the gas in the brighter X-ray galaxies is $T \approx 10^7$ K (Forman et al. 1985; Trinchieri et al. 1986). As mentioned above, there is evidence that the temperature is very roughly constant in NGC 4472, but in general, little is known about the spatial variation of T.

There are several techniques by which one can derive the distribution of gas densities from the X-ray surface brightness profile, if some assumption is made about the gas temperature variation. The *Einstein* Observatory IPC was not particularly sensitive to the temperature of the gas, so the gas density can be determined fairly accurately even if the gas temperature is poorly known. The gas density can be directly deconvolved from the X-ray surface brightness (Thomas 1986; Thomas et al. 1986). Alternatively, one can fit the observed X-ray surface brightness to some parametric form, such as equation (3). If one assumes that the gas is isothermal, equation (3) can be inverted to give the gas density as a function of position

$$n_p(r) = n_o \left[1 + \left(\frac{r}{a_X} \right)^2 \right]^{-3\beta/2} , \qquad (5)$$

where n_o is the central value of the number density of hydrogen nuclei. In Table 1, values of the central density n_o, the X-ray core radius a_X, the parameter β, and the maximum radius of detected X-ray emission R_X are given for several bright X-ray galaxies from Forman et al. (1985; FJT) and Trinchieri et al. (1986; TFC). For the brighter ellipticals, values of $\beta \approx 1/2$ are typical. This implies that the density gas in the outer regions of the galaxy declines roughly as $n_p \propto r^{-3/2}$. The central gas densities in the best studied cases exceed $0.01 \, \text{cm}^{-3}$ and approach $0.1 \, \text{cm}^{-3}$.

Table 1 also gives the total mass of gas M_{gas} (out to a radius of R_X) inferred from the X-ray surface brightness profiles. If the gas density varies as $n_p \propto r^{-3/2}$ at large radii, then the gas mass diverges as $M_{gas} \propto r^{3/2}$. The outer portions of the images of early-type galaxies tend to merge into the background, making the total extents and masses of the gas fairly uncertain. The gas masses in large ellipticals are $\sim 10^{10} \, M_\odot$. Dividing the gas mass by the total blue optical luminosity of the galaxies in Table 1 gives a mean ratio of

$$\left(\frac{M_{gas}}{M_\odot} \right) \approx 0.16 \left(\frac{L_B}{L_\odot} \right) , \qquad (6)$$

with a considerable range in the ratio (0.01–0.8). If a stellar mass to light ratio of $M_*/L_B \approx 10 \, M_\odot/L_\odot$ is assumed, then this implies that the gas mass to stellar mass ratio is typically $\sim 2\%$. This applies mainly to the brighter X-ray galaxies; for the galaxies near the shaded area in Figure 1, the fraction of the gas mass is much smaller.

TABLE 1
Gas Distributions in Early-Type Galaxies

Galaxy	Source	β	a_X (kpc)	n_o (cm^{-3})	R_X (kpc)	M_{gas} ($10^{10}\,M_\odot$)	R_{cool} (kpc)
NGC 315	FJT	≤ 0.70	≤ 5.4	≥ 0.0105	110	6.47	97
NGC 720	TFC	0.45	≤ 6.4	≥ 0.0024	85	1.20	38
NGC 1316	FJT	0.50	1.4	0.0336	34	0.327	41
NGC 1332	FJT	0.60	≤ 2.8	≥ 0.0151	36	0.247	30
NGC 1395	FJT	≤ 0.50	≤ 6.5	≥ 0.0030	98	1.96	46
	TFC	0.50	≤ 5.9	≥ 0.0168	77	4.70	108
NGC 2563	FJT	≤ 1.00	≤ 8.8	≥ 0.0045	80	3.02	83
NGC 4374	FJT	0.50	2.0	0.0173	15	0.0784	37
NGC 4382	FJT	≤ 0.45	≤ 4.0	≥ 0.0030	48	0.305	28
	TFC	≥ 0.40			34	0.082	16
NGC 4406	FJT	0.45	3.0	0.0131	88	2.54	63
NGC 4472	FJT	0.50	1.5	0.0530	80	2.09	59
	TFC	0.50	0.5	0.5240	57	2.40	91
NGC 4594	FJT	≤ 0.65	≤ 9.3	≥ 0.0024	32	0.325	54
NGC 4636	FJT	0.45	1.6	0.0293	44	0.773	61
	TFC	0.50	1.2	0.0622	38	0.57	53
NGC 4649	FJT	0.50	≤ 2.0	≥ 0.0282	36	0.506	52
	TFC	0.60	1.0	0.0995	36	0.27	30
NGC 5128	FJT	0.50			20	0.0664	24

3. Physical State of the Gas

In this sections, a number of simple arguments concerning the origin, dynamics, and energetics of the hot gas in early-type galaxies will be discussed. Comparisons to more detailed models follow in §§4 and 5.

3.1 ORIGIN OF THE GAS — STELLAR MASS LOSS

Early-type galaxies with X-ray emission occur in a range of environments; there are a number of rather bright galaxies which are relatively isolated (Nulsen *et al.* 1984). In fact, I will argue below that bright X-ray ellipticals and S0s occur more often in sparse environments (§6.0). This suggests that the source of the X-ray emitting gas in most cases is intrinsic to the galaxy, and that the gas has not been accreted from some ambient medium. (On the other hand, the cluster dominant elliptical are accreting ambient gas, and some of the "normal" ellipticals may also be accreting gas from surrounding groups or poor clusters.)

The obvious source for the gas in early-type galaxies is then stellar mass loss within the galaxy. The present rate of stellar mass loss in early-type galaxies is determined primarily from the difference by the initial and final (*i.e.*, white dwarf) masses of stars currently undergoing evolution off the main sequence in galaxy, and the rate of death of such stars (Renzini and Buzzoni 1986). Since these stars contribute much of the luminosity of the galaxy, their death rate is proportional to the optical luminosity. This argument is fairly independent of the detailed mechanism for stellar mass loss, as long as the initial and final masses of presently evolving stars are are known, and as long as these stars contribute much of the optical luminosity. As a result, the rate of mass loss is probably known to better than a factor of two. If $\dot{M}_*$ is the total rate of stellar mass loss, then stellar evolution calculations give

$$\frac{\dot{M}_*}{L_B} \approx 1.5 \times 10^{-11} \, M_\odot \, \text{yr}^{-1} \, L_\odot^{-1} \tag{7}$$

(Faber and Gallagher 1976; Tinsley 1980; Renzini and Buzzoni 1986). The rate of gas injection in a large elliptical with $L_B = 10^{11} \, L_\odot$ is then $\dot{M}_* \approx 1.5 \, M_\odot \, \text{yr}^{-1}$. Although the mass loss rate is given most directly in terms of the stellar optical luminosity, it is sometimes useful to give the rate in terms of the stellar mass. It is conventional to define the stellar mass loss parameter $\alpha_* \equiv \dot{M}_*/M_*$. Assuming a stellar mass to light ratio of $(M/L_B)_* = 10 M_\odot/L_\odot$, one then finds

$$\alpha_* = \left(\frac{\dot{M}_*}{L_B}\right) (M/L_B)_*^{-1} \approx 1.5 \times 10^{-12} \, \text{yr}^{-1} = 4.7 \times 10^{-20} \, \text{s}^{-1}. \tag{8}$$

From equation (7), the total amount of gas M_{inj} injected over the lifetime t_o of the galaxy is

$$\left(\frac{M_{inj}}{M_\odot}\right) \approx 0.15 \left(\frac{L_B}{L_\odot}\right) \left(\frac{t_o}{10^{10} \, \text{yr}}\right) \left[\frac{\langle \alpha_*(t) \rangle}{\alpha_*(t_o)}\right], \tag{9}$$

where $\langle \alpha_*(t) \rangle$ is the average value of the stellar mass loss rate over the age of the galaxy, and $\alpha_*(t_o)$ is the present value. Comparing with equation (6), we see that the present rates of gas loss are capable of generating the observed amounts of hot gas in the brighter X-ray ellipticals and S0s over a plausible lifetime for the galaxies.

There are several caveats that need to be given concerning this agreement of the observed and predicted gas masses. First, models for the stellar evolution of elliptical galaxies predict that $\alpha_*(t)$ was much higher in the past, when these galaxies contained more massive stars. These models suggest that $\langle \alpha_*(t) \rangle \approx (3 - 10)\alpha_*(t_o)$ (Renzini and Buzzoni 1986; Mathews 1989) for the era following the formation of the galaxy. The amount of intracluster medium in rich clusters of galaxies may indicate that the total rates of stellar gas loss (including the epoch of galaxy formation) are even larger. In the richest clusters, the total mass of intracluster gas exceeds the mass of the stars in the galaxies by a factor of perhaps five (e.g., Sarazin 1988). This gas has an iron abundance which is roughly half of the solar value. This suggests that a very large amount of gas was ejected from stars and subsequently lost from the galaxies into the intracluster medium; the total mass of ejected gas may be comparable to the mass of all the stars in the galaxies. This is $\sim 10^2$ times the present rate of gas loss by stars. Thus, early-type galaxies appear to have lost substantial amounts of gas in the past.

In any case, the agreement of the present stellar mass lost rate and the amount of hot gas really only applies to the brightest X-ray galaxies. In many galaxies, particularly the faint X-ray galaxies in the hatched band of Figure 1, the masses of hot gas are apparently much smaller than expected from even the present rates of stellar mass loss. Another caveat, mentioned previously, is that large ellipticals near the centers of groups or poor clusters may be accreting gas (Thomas 1986). Finally, it is not obvious that all the gas resulting from stellar mass loss will be heated to X-ray emitting temperatures (White and Chevalier 1983; Thomas 1986; Thomas *et al.* 1986; Mathews 1989).

3.2 DYNAMICS OF THE GAS — COOLING FLOWS

The following arguments concerning the dynamics of the X-ray gas in early-type galaxies apply only to the brighter X-ray galaxies for which detailed X-ray surface brightness profiles and spectra are available. Fainter X-ray galaxies may not have the same dynamical properties. These arguments also assume that the gas is spherically symmetric; in the inner parts of the galaxy, this is supported by the observed X-ray images. However, it is likely that the gas is actually inhomogeneous at small scales, due to thermal instabilities and convection. Given spherical symmetry, the gas is either static, flowing radially inward, or flowing radially out at each radius.

Prior to the discovery of the X-ray emission from elliptical galaxies, they were generally thought to possess transonic global galactic winds (Mathews and Baker 1971; Faber and Gallagher 1976; Bregman 1978; Bailey 1980; MacDonald and Bailey 1981; White and Chevalier 1983). This picture can now be ruled out, at least for the brightest ellipticals (Nulsen *et al.* 1984; Forman *et al.* 1985; Sarazin 1986b; Thomas 1986; Fabian, Arnaud, and Thomas 1987; Loewenstein and Mathews 1987; Sarazin and White 1987). If they did possess such winds, the X-ray luminosities of ellipticals would be much smaller than those observed (by a factor of 10^{3-4}; White and Chevalier 1983). It is easy to see why this is the case. Unless the sonic point in the flow is very far out (which requires a very fine balancing of the heating rate in the gas), the gas will leave the galaxy at roughly the sound speed or higher. The sound–crossing time for an elliptical galaxy is roughly $\sim 10^8 (R/50\,\mathrm{kpc})/(T/10^7\,\mathrm{K})$ yr, where R is the galaxy radius and T is the gas temperature. Since this is $\sim 1\%$ of the age of the galaxy assumed in §3.1 above, the galaxy would contain only about 1% of the gas generated by stellar mass loss during its lifetime. At least for the brighter X-ray galaxies, the observed X-ray luminosities imply gas masses which would be generated by the present rates of stellar mass loss over essentially the entire age of the galaxy (see equations 6 and 9 above). If the galaxies had galactic winds, their gas masses would be $\sim 10^{-2}$ times smaller; since the thermal X-ray emissivity scales with the square of the gas density, their X-ray luminosities would be $\sim 10^{-4}$ times smaller than those observed. Another problem with transonic wind models for the X-ray emission in early-type galaxies is that the gas density profiles would be expected to follow $n_p \propto r^{-2}$, which is steeper than the observed profiles.

Another possible dynamical structure for the gas in ellipticals is a "partial wind" (MacDonald and Bailey 1981). In such a model, radiative losses might suppress an outflow in the inner parts of the galaxy, but a wind would occur in the outer, lower density, less tightly bound regions. This model may apply to the fainter X-ray galaxies. It probably cannot apply to the observed regions of the brighter X-ray galaxies, however. The problem is that the X-ray surface brightness profiles of these galaxies extend out as far as the optical surface brightness profiles (Trinchieri *et al.* 1986). The outer, wind portion of such a partial wind will be transonic, and, by the arguments given above, this region would have an undetectably faint X-ray surface brightness. From the X-ray observations, it would seem that bright X-ray galaxies have partial winds, they start at radii beyond the effective radii of the optical emission.

The possibility that the gas forms a static atmosphere (Forman *et al.* 1985) seems unlikely given the very short cooling times in the gas. In the brighter X-ray ellipticals and S0s with good surface brightness profiles (Table 1), the central cooling time in the gas is generally less than 10^8 years (Trinchieri et al. 1986). Even in the outer regions, the cooling times are shorter than the likely ages of the galaxies ($t_o \approx 10^{10}$ years). In Table 1, the last column gives the radius R_{cool} at which the integrated isobaric cooling time in the gas reaches 10^{10} years. This may be compared to R_X, the maximum radius of the detected X-ray emission. The integrated isobaric cooling time is defined as

$$T_c \equiv \frac{5}{2\rho\theta} \int_0^\theta \frac{\theta \, d\theta}{\Lambda(\theta)}, \tag{10}$$

where $\theta \equiv kT/(\mu m_p)$, μ is the mean mass per particle in the gas, ρ is the gas mass density, and $\rho^2 \Lambda(\theta)$ is the cooling rate per unit volume in the gas. A gas temperature of 10^7 K was assumed in calculating the cooling times. In general, the cooling time in the gas reaches 10^{10} years at or beyond the outermost radius of the observed X-ray emission. Thus, unless the gas is reheated, it will cool. In fact, it is unlikely that any physically plausible reheating mechanism will be able to keep the bulk of the mass of the gas hot, because gas at these temperatures ($T \sim 10^7$ K) cools very rapidly.

In the brighter X-ray ellipticals and S0s, the cooling time (equation 10) is shorter than the age but longer than the dynamical time scale (the sound crossing time) over essentially all of the galaxy. A system satisfying these limits is often referred to as a "cooling flow" (Fabian et al. 1984; Nulsen et al. 1984; White and Chevalier 1984; Sarazin 1986b, 1987a; Thomas 1986; Thomas et al. 1986; Canizares 1987; Fabian et al. 1987). Because the cooling time is shorter than the sound crossing time, the gas will remain nearly hydrostatic as it begins to cool. As the gas cools, its density will increase in order that the gas pressure continue to support the weight of the overlying gas. As the density of any given element of gas increases, its volume will decrease. If the gas is homogeneous and spherically symmetric, it must slump slower inward in the galaxy. Thus, homogeneous cooling flows will be characterized by slow (subsonic) inflow. However, it is important to emphasize that the directly observable property of the gas is its radiative cooling, which produces the X-ray which we observe. The fact that the cooling time is short argues that there will be a steady-state established between the sources of mass and energy in the gas, and the loss of the energy and mass in hot gas through radiative cooling.

3.3 HEATING OF THE GAS

There are several indications that the gas in early-type galaxies undergoes heating at the present epoch. Because the cooling times in the gas are short, the gas would not have remained hot had it been heated in the remote past. The gas is "hotter" (has a higher velocity dispersion) than the stars (Fabbiano 1986a, 1989; Killeen and Bicknell 1988). This is established both by comparing the observed (but poorly known) gas temperatures with the velocity dispersions of the stars in the galaxy, and by comparing the gas density distribution with that of the stars. The gas has a much flatter density distribution, indicating that it is hotter than the stars.

Because the X-ray emission appears to be a fairly ubiquitous property of early-type galaxies, the heating of the gas to X-ray emitting temperatures is probably not generally either an environmental effect or one associated with unusual activity (such as strong radio emission) in the galaxy. There are two heating mechanisms for the gas which are associated with the injection of the gas itself, and therefore should always occur to some extent. First, a small fraction of the gas may be ejected with large kinetic energy by supernovae. Second, even in the more quiescent ejection of gas by giant stars, planetary nebulae, and so on, the gas-losing star will generally be moving at a high velocity (given by the stellar velocity dispersion of the galaxy) relative to the hot gas or the gas ejected by neighboring stars. Collisions between the ejected gas and either gas ejected by neighboring stars or ambient hot gas may thermalize the kinetic energy of motion of the gas (White and Chevalier 1983; Mathews 1989).

The mass loss can be divided into mass loss by supernovae, with specific rate α_{SN}, and more quiescent stellar mass loss, with specific rate α_S, so that $\alpha_* = \alpha_S + \alpha_{SN}$. Because both of these mechanisms give an amount of heat which is proportional to the amount of gas injected, it is useful to define the energy per unit injected gas mass as $3kT_{inj}/2\mu m_p$, where

$$T_{inj} = T_S + T_{SN}. \tag{11}$$

Here, T_S and T_{SN} give the heating due to the motions of the gas-losing stars and to supernovae, respectively.

3.3.1 *Supernova Heating.* Only Type Ia supernova are observed in elliptical and S0 galaxies. It is conventional to give the supernova rate as a function of the blue luminosity; the rate of supernovae is r_{SN} per $10^{10} L_\odot$ (in blue luminosity) per century. Tammann (1982) gives $r_{SN} = 0.22$, but more recent surveys give much lower values. van den Bergh, McClure, and Evans (1987) and Evans, van den Bergh, and McClure (1989) find $r_{SN} \approx 0.075$. Then, the heating rate per unit injected gas mass due to supernovae is

$$T_{SN} = 3.0 \times 10^7 \text{ K} \left(\frac{r_{SN}}{0.22}\right) \left(\frac{E_{SN}}{8 \times 10^{50} \text{ ergs}}\right) \left[\frac{(M/L_B)_*}{10 M_\odot/L_\odot}\right]^{-1} \left(\frac{\alpha_*}{4.7 \times 10^{-20} \text{ s}^{-1}}\right)^{-1}, \tag{12}$$

where E_{SN} is the average kinetic energy of each supernova ejection. Obviously, supernovae are capable of heating the gas to temperatures comparable to those observed.

It is worth noting that although the supernova heating rate per unit mass T_{SN} and the gas loss rate parameter α_* depend on the stellar mass to light ratio $(M/L_B)_*$ (equations 8 and 12), this is only because they are normalized to the galaxy mass. The directly measurable quantity is the galaxy optical luminosity, and the supernova heating rate and gas loss rate are independent of the mass to light ratio when expressed in terms of the optical luminosity.

3.3.2 *Heating by Stellar Motions.* Next, consider the heating of the gas due to the motions of the gas-losing stars. Gas which is lost from stars initially has the same energy per unit mass as the stars, assuming the velocity of ejection is small compared to the stellar velocity dispersion in the galaxy. This energy is assumed to be thermalized by encounters with other gas. The resulting heating is then

$$T_S = \frac{\mu m_p \sigma_*^2}{k} = 6.8 \times 10^6 \text{ K} \left(\frac{\sigma_*}{300 \text{ km s}^{-1}}\right)^2, \tag{13}$$

where σ_* is the one-dimensional stellar velocity dispersion.

Comparing equations (12) and (13), it is clear that supernova heating would dominate in all ellipticals if the supernova rate is as high as that given by Tammann (1982). On the other hand, if the rate is as low as the more recent value of Evans *et al.* (1989), then heating by stellar motions will be important, at least for the larger ellipticals.

3.3.3 *Heating by Infall and Adiabatic Compression.* In §3.2, arguments were given which suggested that the hot gas in at least the brighter X-ray galaxies forms a cooling flow. If the hot interstellar gas does flow subsonically inward in a galaxy, there is additional heating of the gas due to infall in the galactic potential and adiabatic compression of the gas (Nulsen *et al.* 1984). Detailed models, such as those described in §4 below, are needed to accurately assess the importance of this heating term. However, the following crude arguments are helpful. For a subsonic flow, the gas is nearly hydrostatic and the virial theorem suggests that roughly one–half of the increase in the gravitational potential energy of the inflowing gas appears as binding energy and roughly one–half is radiated away. Thus, the heating per unit mass of gas due to infall and compression is roughly $\langle \Delta \phi \rangle / 2$, where ϕ is the gravitational potential of the galaxy and $\langle \Delta \phi \rangle$ is the average change in the potential between the point where the gas is injected into the galaxy and the point where it stops flowing or cools below X-ray emitting temperatures.

Since the stellar velocity dispersion measures the depth of the potential well in an elliptical galaxy, this source of heating is also proportional to σ_*^2. Both the motions of the stars in a galaxy, and the infall and compression of the gas are results of the gravitational field of the galaxy. In a global sense, one can combine the effects of infall and compressional heating with the heating by stellar motions, and describe the total effect as "gravitational heating." Globally, the gravitational heating rate should be given by equation (13), with the numerical coefficient increased by some factor. There is some evidence that elliptical galaxies have "heavy halos" containing much of the mass but producing little observable radiation (see §7.0). If these heavy halos cause the velocity dispersion in the galaxy to be approximately constant, then the gravitational potential may be given approximately by the isothermal sphere potential $\phi = 2\sigma_*^2 \ln(r)$. Then, the the heating per unit mass of gas due to infall and compression is roughly $\sigma_*^2 \ln(r_{in}/r_{out})$, where r_{in}/r_{out} is the average ratio of the radius at which gas is injected into the galaxy to the radius at which it stops flowing or drops below X-ray emitting temperatures. In the detailed model calculations (§4), one finds that r_{in} is somewhat smaller than the total optical radius of the galaxy (or somewhat larger than the effective radius), and r_{out} is somewhat larger than the core radius of the galaxy. This implies that the heating rate due to stellar motions is augmented by a factor of at most ~ 3.

3.4 SIMPLE ESTIMATE OF X-RAY LUMINOSITY

Based on the argument given above, it is possible to give a simple estimate for the X-ray luminosity of early-type galaxies (Nulsen *et al.* 1984; Sarazin 1986b). This estimate requires the following assumptions. First, one assumes that elliptical and S0 galaxies are isolated, in the sense that all of the gas and thermal energy in the interstellar gas originates within the galaxy, and that the gas remains bound to the galaxy. Second, one assumes that the gas originates in normal stellar mass loss, as argued in §3.1. Third, the heating of the gas is assumed to be due to either gravitational heating (stellar motions, and infall and compression) or supernovae (as argued in §3.3). Fourth, one assumes that there is a steady-state balance between the rates of heating and cooling in the gas (as discussed in §3.2). The main cooling is radiative cooling. A fifth assumption is that the gas temperature is high enough ($T \gtrsim 3 \times 10^6$ K) that essentially all of the radiative cooling occurs in the X-ray band. These assumptions imply that the X-ray luminosity equals the total heating rate of the gas. Thus, a simple estimate of L_X is given by

$$L_X \approx \left(\frac{3kT_{inj}}{2\mu m_p} \right) \dot{M}_*. \tag{14}$$

It is useful to consider separately the contribution to the X-ray luminosity from supernova and gravitational heating.

Combining equations (7), (8), (12), and (14), the total heating rate due to supernovae is found to be

$$L_{SN} = 5.5 \times 10^{41} \text{ ergs s}^{-1} \left(\frac{r_{SN}}{0.22} \right) \left(\frac{E_{SN}}{8 \times 10^{50} \text{ ergs}} \right) \left(\frac{L_B}{10^{11} L_\odot} \right). \tag{15}$$

Since most early-type galaxies have similar colors and spectra, one would expect that the first two factors in parentheses equation (15) would not vary much from galaxy to galaxy. These factors depend only on the stellar population of the galaxy. Given that early-type galaxies have similar stellar populations, the supernova heating rate should just depend on the number of stars in the galaxy (or the optical luminosity L_B).

Figure 5 compares the X-ray luminosity predicted by equation (15) with the observed X-ray luminosities of early-type galaxies from Canizares et $al.$ (1987). The Tammann (1982) supernova rate ($r_{SN} = 0.22$) and a reasonable value for the Type I supernova energy ($E_{SN} = 8 \times 10^{50}$ ergs) are assumed. Comparison of the predicted supernova luminosities (equation 15) with the observed correlation between the optical and X-ray luminosities of early-type galaxies reveals two problems. First, the predicted X-ray luminosities tend to be too large, particularly for the lower luminosity galaxies (but not for those with the highest luminosities). Second, the predicted correlation between X-ray and optical luminosities is $L_X \propto L_B$, which is much flatter than that observed. This suggests that either the supernova rates in early-type galaxies are lower than those given by Tammann, or most of the supernova luminosity does not appear in the X-ray band. Recently, van den Bergh et $al.$ (1987) and Evans et $al.$ (1989) have argued that the Type Ia supernova rate in early-type galaxies is ~ 4 times smaller than the Tammann (1982) rate.

Next, consider the luminosity due to heating of the gas from the motions of the gas-losing stars. Since the velocity dispersion (the galaxy gravitational potential) increases with the mass of the galaxy, this gives a heating rate per unit mass of gas which increases with the optical luminosity of the galaxy, resulting in an X-ray–to–optical luminosity correlation which is steeper than a linear correlation. Brighter ellipticals have larger stellar velocity dispersions (Faber and Jackson 1976). While Faber and Jackson found $L_B \propto \sigma_*^4$, more recent work suggests that the slope is flatter for lower-luminosity ellipticals (Tonry 1981; Davies et $al.$ 1983), becoming $L_B \propto \sigma_*^{2.4}$ at low optical luminosities. The observed velocity dispersions are mainly measured for the centers of the galaxies. For the present discussion, we will assume that the velocity dispersion is essentially constant throughout the galaxy; this requires that elliptical galaxies have heavy halos of optically dark matter. If the "Faber-Jackson" relation for early-type galaxies is assumed to be $L_B \approx 10^{11} L_\odot (\sigma_*/300 \text{ km s}^{-1})^{2.4-4}$ the total heating rate due to stellar motions is (equations 7, 13, and 14)

$$L_S = 1.3 \times 10^{41} \text{ ergs s}^{-1} \left(\frac{L_B}{10^{11} L_\odot} \right)^{1.83-1.5}. \tag{16}$$

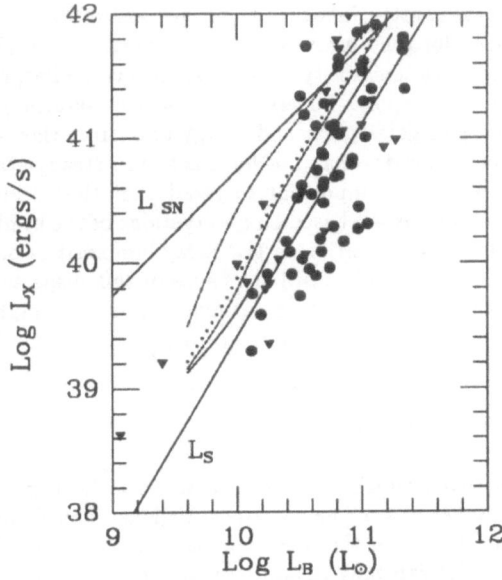

Figure 5. A comparison between the observed X-ray luminosity of early-type galaxies from Canizares *et al.* (1987), and those predicted by models. The filled circles are detections, and the inverted triangles are upper limits. The solid line marked L_{SN} is the expected supernova luminosity (equation 15). The solid line marked L_S is the predicted total heating rate due to stellar motions (equation [16] with the exponent set to 1.7). The shorter curves give the X-ray luminosities of the steady-state cooling flow models of Sarazin and White (1987). The solid curve shows models without either heavy halos or supernova heating. The dotted line shows models with supernova heating but without heavy halos. The dashed line is models with heavy halos but without supernova heating. The dash-dot line is for models with both supernova heating and heavy halos.

The range in the exponent of the L_B term corresponds to the range in the exponent of the L_B vs. σ_* relation. The variation of L_S with L_B is shown in Figure 5 (assuming the exponent in equation [16] is 1.7). Equation (16) is a reasonable fit to both the form and the magnitude of the observed X-ray-to-optical luminosity correlation for larger values of the exponent (Nulsen *et al.* 1984; Sarazin 1986b).

As discussed above (§3.3.3), inflow and adiabatic compression can increase the total gravitational heating rate up to roughly three times the value in equation (16). This would make the gravitational heating rate larger than the typical X-ray luminosities of early-type galaxies, but smaller than the largest luminosities.

Thus, the primary conclusion from these simple estimates of the X-ray luminosities of early-type galaxies is that the motions of the gas-losing stars are capable of heating the gas sufficiently to produce the observed X-ray emission. Supernova would provide too much energy for most galaxies unless the supernova rates are lower than previously thought ($r_{SN} \lesssim 0.07$). Also, supernova heating would be expected to give a roughly linear relation between X-ray and optical luminosity, which is not observed. If ellipticals and S0s have heavy halos of optically dark matter, inflow and compression in the galactic potential would also give too much heating for the observed X-ray luminosities of typical galaxies. There are several worries about these luminosity estimates, however. First, one or several of the required assumptions may not apply. For example, most early-type galaxies may not be isolated; they may be losing or gaining gas and energy to or from their environment (see §6 below). Alternatively, most galaxies may not have achieved steady-state between heating and cooling (see §5 below). It is important to recall that the observational arguments underlying these two assumptions are based on observations of the brightest X-ray galaxies only. Also, there is a very large dispersion in the X-ray luminosities of early-type galaxies of a given optical luminosity (see §6). Until the cause of this dispersion is understood, it is difficult to know whether models for these galaxies should attempt to fit the medium X-ray luminosities, or the maximum X-ray luminosities.

4. Steady-State Cooling Flow Models

In §3.2, arguments were given which suggest that at least the brighter X-ray ellipticals and S0s are in steady-state, except in the outermost regions. The X-ray isophotes of these galaxies are quite round, at least in the inner regions. This suggests that the interstellar gas in the brighter X-ray ellipticals may be approximated by spherically symmetric, steady-state hydrodynamical models. As noted in §3.2, if the gas cools steadily and is homogeneous and spherically symmetric, it will form a cooling inflow. One dimensional, steady-state cooling flow models for ellipticals have been calculated by Bailey (1980), MacDonald and Bailey (1981), White and Chevalier (1984), Thomas (1986), Thomas et al. (1986), Sarazin and White (1987, 1988), Vedder, Trester, and Canizares (1988), and Sarazin and Ashe (1989). These models include a wide range of galaxy optical luminosities, and a wide range of assumptions about the stellar and mass distributions of the galaxies, the supernova rates, the effects of the environment on the galaxy, and the cooling and loss of the hot gas. On the other hand, these simple steady-state models neglect several important physical effects. First, these models generally do not include thermal conduction. Second, the angular momentum of the inflowing gas cannot be included in such spherically symmetric models. Third, time-dependent effects must become important in the outer regions of the flow even in the brightest X-ray galaxies, and may be important throughout the flow in the fainter galaxies. Time dependent hydrodynamical models are discussed in §5 below. While fully time–dependent calculations are certainly needed, it is difficult to perform such calculations with sufficient resolution and for a sufficiently wide range of models to permit detailed comparisons with observations. The spherical, steady-state models involve only ordinary differential equations, and it is feasible to calculate highly resolved models for many galaxy parameters. Fourth, gas in early-type galaxies is probably very inhomogeneous, and these models can only include this in an approximate fashion.

In most of the models, the only sources of the gas are assumed to be stellar mass loss, and possibly inflow from the exterior. The only heating mechanisms are those discussed in §3.3 (supernovae and the motions of gas losing stars). The only cooling mechanism is radiative cooling. Then, the hydrodynamical equations for such a galactic cooling flow are

$$\frac{1}{r^2}\frac{d}{dr}\left(\rho v r^2\right) = \alpha_* \rho_* - \dot{\rho}, \qquad (17)$$

$$\rho v \frac{dv}{dr} + \frac{dP}{dr} + \frac{GM}{r^2}\rho + \alpha\rho_* v = 0, \tag{18}$$

$$\rho v \left(\frac{3}{2}\frac{d\theta}{dr} - \frac{\theta}{\rho}\frac{d\rho}{dr} \right) = -\Lambda\rho^2 + \alpha\rho_* \left[\frac{v^2}{2} + \frac{3}{2}\left(\theta_{inj} - \frac{5}{3}\theta \right) \right]. \tag{19}$$

Here r, v, P, ρ, and ρ_* are the radius, gas velocity, gas pressure, gas density, and stellar density, respectively. The temperature parameter $\theta \equiv kT/(\mu m_p)$, where $\mu \approx 0.63$ is the mean particle mass in units of the proton mass m_p. M is the total galaxy mass and $\Lambda(T)$ is the cooling function, defined so that $\Lambda\rho^2$ is the cooling rate per unit volume in the gas. The energy per unit mass injected into the gas is $3\theta_{inj}/2 = (3/2)(kT_{inj})/(\mu m_p)$, and the pressure $P = \rho\theta$. The term $-\dot\rho$ in equation (17) allows for the removal of gas from the flow (see §4.2 below).

One also needs to specify the boundary conditions on the solution. If one wishes to treat the galaxies as being isolated, the required outer boundary condition is that there be no inflow of gas from the exterior of the galaxy (Sarazin and White 1987; Sarazin and Ashe 1989). The rate of inflow of gas at any radius is given by

$$\dot M(r) \equiv -4\pi r^2 \rho v. \tag{20}$$

Thus, an outer boundary condition for isolated galaxies would be that $\dot M(r) \to 0$ at the outer boundary of the galaxy. Substituting this condition into the energy equation (19) shows that the outer boundary of the galaxy will then be a singular point of this equation. One can show that there are generally no solutions unless the right hand side of equation (19) vanishes at the outer boundary of the galaxy (Sarazin and White 1987; Sarazin and Ashe 1989). This gives a second outer boundary condition. Alternatively, part of the gas in the galaxy may originate in the exterior of the galaxy. In this case, suitable outer boundary conditions are to specify the mass inflow rate $\dot M$ and the pressure at the outer boundary (Thomas 1986; Vedder et al. 1988). Except for models with a very large rate of loss of gas, the solutions of equations (17)–(19) will have an inner sonic radius. The requirement that the solutions remain regular at this point gives a suitable inner boundary condition (White and Chevalier 1984). For models with a very large rate of gas loss, a suitable inner boundary condition (necessary if the flows are to be steady) is that the mass inflow rate approach zero as the radius approaches zero (Sarazin and Ashe 1989).

4.1 MODELS WITHOUT GAS LOSS

First, models without gas loss ($\dot\rho = 0$) will be considered. A grid of models in which the galaxies are treated as isolated were calculated by Sarazin and White (1987). Figure 6 shows the results for four models for an elliptical galaxy with an optical luminosity of $L_B = 10^{11} L_\odot$. The figures show the variation of the gas temperature T, the gas proton density n_p, inflow velocity v, and gas pressure P (actually, $P/k \equiv nT$ is shown, where n is the total particle density) as a function of radius. In order to assess the effects of varying the mass profiles of ellipticals and their supernova rates, models were calculated with a constant mass to light ratio equal to that of the stellar population (no "dark matter") and with a heavy halo such that the velocity dispersion was constant. Similarly, models were calculated with the full Tammann supernova rate, and without any supernova heating. The solid curves show models without either heavy halos or supernova heating. The dotted lines show models with supernova heating but without heavy halos. The dashed lines are

220

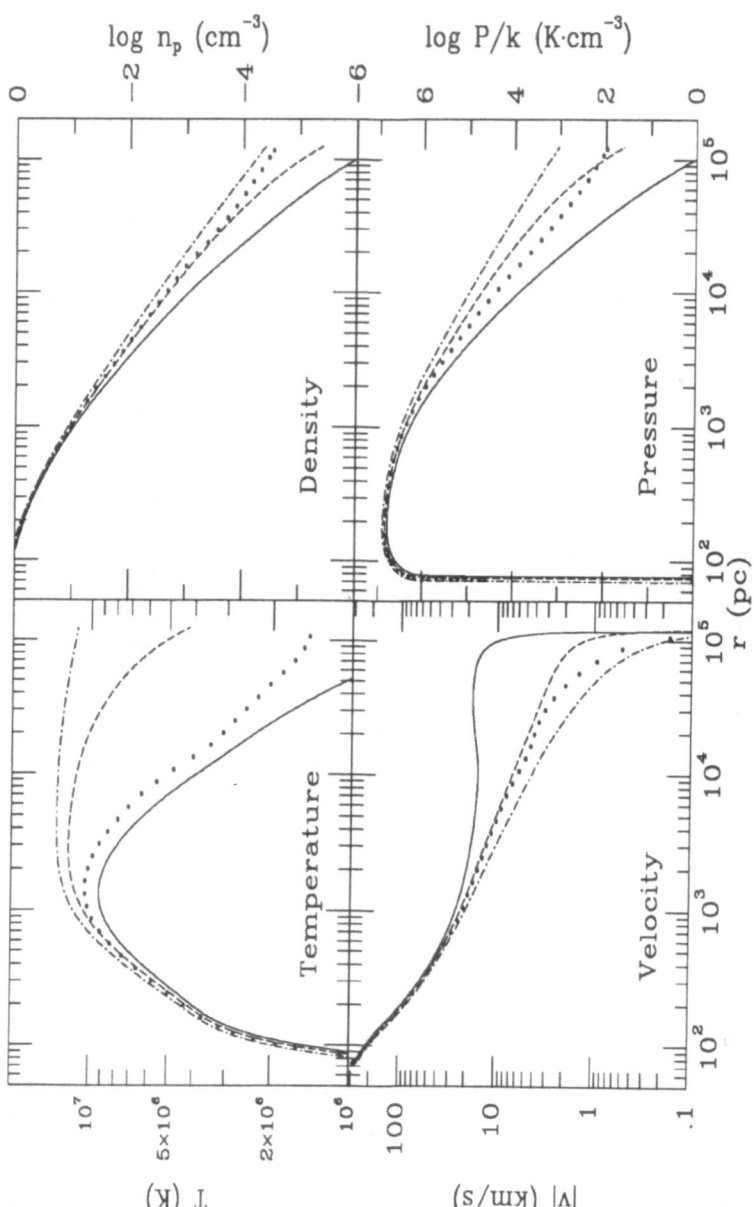

Figure 6. The variation of the gas temperature T, gas proton density n_p, inflow velocity v, and gas pressure P as a function of radius, from the steady-state cooling flow models of Sarazin and White (1987). The models are for an elliptical galaxy with $L_B = 10^{11} L_\odot$. The solid curves show models without either heavy halos or supernova heating. The dotted lines show models with supernova heating but without heavy halos. The dashed lines are models with heavy halos but without supernova heating. The dash-dot lines are for models with both supernova heating and heavy halos.

models with heavy halos but without supernova heating. The dash-dot lines are for models with both supernova heating and heavy halos. Because of the high rate heating (equation 12), the models with supernovae may be capable of maintaining global winds if they did not start with a high ambient gas density (White and Chevalier 1984). These cooling flow solutions then correspond to "smothered winds" or "radiatively induced infall" and may be unstable to wind formation through the networking of supernova remnants (Cox and Smith 1974).

The flows are very slow and subsonic over all of the galaxy except near the center (Figure 6). All of the models involve transonic inflow in the center. As gas accumulates in the flow from stellar mass loss, the inflow rate $\dot{M}$ increases, and higher inflow velocities are needed to carry this gas at smaller radii. As the same time, as the gas flows inward, it cools which reduces its sound speed. The sonic radii are typically $r_s \sim 100$ pc, which is smaller than the resolution of any of the existing X-ray images. Within the sonic radius, the gas temperature and pressure rapidly drop to essentially zero. It is easy to understand why this occurs (Sarazin and White 1987; Soker and Sarazin 1988). First, within the sonic radius the inflow is supersonic, and conditions downstream can have no influence on the properties of the gas at the sonic radius. Thus, the drop in pressure doesn't lower the gas density and pressure or alter the position at the sonic radius. Second, at the sonic radius in a cooling flow, the cooling time and inflow time are comparable. At the temperatures of interest ($T \sim 10^6 - 10^7$ K), as the temperature decreases cooling accelerates rapidly, and the cooling time gets much shorter. On the other hand, within the sonic radius, the gas velocity is roughly constant, so the flow time does not decrease rapidly. As a result, the gas cools almost completely before it flows in very far from the sonic radius. The drop in temperature and pressure within the sonic radius is shown in Figure 6.

The temperature profiles in the models flatten out or reach a maximum outside of the sonic radius; the maximum temperature is typically $0.75\mu m_p \sigma_*^2$, where σ_* is the central stellar velocity dispersion (Figure 6). In the models with heavy halos, the gas is fairly isothermal beyond $r \sim 1$ kpc. In the models without heavy halos, the gas temperature drops rapidly in the outer parts. The temperature profiles are most strongly affected by the gravitational potential of the galaxy, and less by the rate of heating (the injection temperature). Because the gas is very subsonic over most of the galaxy, it is nearly in hydrostatic equilibrium, and this condition roughly determines the temperature profile (Sarazin and White 1987; Vedder *et al.* 1988).

The density profiles of the models with heavy halos and significant supernova heating are roughly power law in the outer parts, with $n_p \propto r^{-1.7}$. This is similar to but slightly steeper than the observed profiles. The other models have steeper profiles. The density profiles are affected most strongly by the heating rate (the injection temperature). They correspond to densities which are slightly smaller than those which would give a local balance between heating and cooling, $\Lambda(T)\rho^2 \approx 3\alpha_*\rho_*\theta_{inj}/2$, as suggested by Sarazin (1986b).

The X-ray luminosities of these models are compared to the observed values in Figure 5. The stellar X-ray luminosities predicted by Canizares *et al.* (1987; the central value in equation 4) have been added to the gaseous emission. The notation is the same as in Figure 6. It is clear that the models generally have luminosities which are higher than the average of the observed elliptical galaxies. The models tend to lie near the upper envelope of the observed X-ray luminosities. In particular, the models with both high rates of supernova heating and massive halos are the most X-ray luminous, and their luminosities exceed the bulk of the data. The models with supernova heating but without heavy halos (dotted curve) and the models without supernova heating but with heavy halos (dashed curve) have similar luminosities. This is contrary to the expectations (§3.4) that the supernova heated models would be more luminous, and would have $L_X \propto L_B$ (equation 15). That this is not borne out by detailed models is the product of several factors. First, infall and adiabatic compression increase L_S in equation (16) by several times, making it similar to L_{SN} for the brighter galaxies. Second, for the fainter galaxies, $L_{SN} \gg L_S$ (equations 15 and 16), but the gas in the supernova heated models is sufficiently cool that most of the radiation is in the UV rather than X-ray band, particularly in the outer regions (see Figure 6). The models give an X-ray–optical correlation with about the correct slope.

The result that model galaxies with the predicted rates of stellar mass loss and the predicted heating rates give X-ray luminosities which are similar to those of the brighter X-ray galaxies, but larger than those of typical galaxies agrees with the result of deconvolutions by Thomas (1986) and Thomas et al. (1986). They suggest that the amounts of stellar mass loss which are heated to X-ray emitting temperatures may be smaller than the total rates of mass loss suggested by stellar evolution theory. Thomas et al. argue that cooling outflows occur in some galaxies, as well.

Vedder et al. (1988) have constructed models including the effects of a high external pressure. They find that the pressure modifies the temperature profiles dramatically, producing a sharp rise in the temperature in the outer regions of the galaxy. In the outer regions, the interstellar gas is confined by the external pressure, rather than by the gravity of the galaxy. On the other hand, the external pressure doesn't have a very strong effect on the X-ray luminosity of the models. As argued in §3.4, the luminosities of steady-state models are determined mainly by the rate of energy injection in the galaxy. The external pressure doesn't modify this, so the X-ray luminosity is not strongly affected. An external medium could affect the X-ray luminosity if it provides a significant amount of energy through the inflow of hot gas or the inflow of heat from thermal conduction.

A major problem with the cooling flow models without gas loss concerns their X-ray surface brightness profiles. In the outer regions, these models give X-ray profiles which are somewhat more steeply decreasing than the observed profiles. A more serious problem occurs with the central surface brightness. Figure 7 compares the observed surface brightness profile of NGC 4472 (same data as in Figure 4) with the predictions of four models from Sarazin and White (1987, 1988). The models have strongly peaked X-ray surface brightness profiles with central surface brightnesses which are much larger than those observed. Similar results were found by Thomas (1986), Thomas et al. (1986), and Vedder et al. (1988). It is easy to understand why this occurs. The temperature profiles in these models (Figure 6) show that the gas temperature is either constant or increases inward to roughly $r \sim 1$ kpc. Thus, outside of this point infall and adiabatic compression are at least as important as cooling, and the enthalpy in the gas does not decrease as the gas flows inward, despite cooling. However, within the sonic radius, the gas cools very rapidly. In a cooling flow model without gas loss, all of the gas flows into this point, and thus a nonnegligible fraction of the thermal energy in the gas throughout the entire galaxy is radiated within this region. Since this inner region subtends at most a few seconds of arc even for a large, nearby elliptical, this region is basically unresolved even in the *Einstein* HRI. The observed X-ray surface brightness profiles of the elliptical galaxies studied with the HRI do not show this strong central concentration.

The result that cooling flow models without gas loss give surface brightness profiles

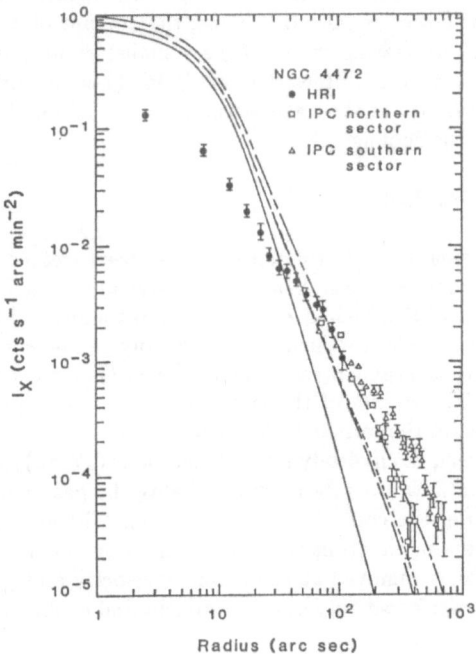

Figure 7. The observed and model X-ray surface brightness profiles for NGC 4472. The data points are from Trinchieri *et al.* (1986), and are the same data as plotted in Figure 4. The curves give the predicted X-ray surface brightness profiles of models (Sarazin and White 1988), using the same notation as that in Figures 5 and 6.

which are too centrally condensed is independent of the stellar mass loss rate, the detailed nature of the heating source of the gas, or the detailed form of the optical distribution of the galaxy. Thus, if the basic picture of cooling flows in ellipticals is correct, one must conclude that the bulk of the cooling gas *does not* flow into the center of the galaxy. There are other arguments which support this. If all of the gas released by stellar mass loss were to flow into the center of elliptical galaxies, the central velocity dispersion, mass to light ratio, and optical luminosity profile would be affected in ways which are not observed (Mathews 1988). There are several physical mechanisms which might keep the gas from flowing into the center of a galaxy. First, if the inflowing gas has angular momentum and viscosity is weak, centrifugal forces might cause the inflow to stagnate and form a disk. Against this suggestion is the fact that the inner contours of the observed X-ray surface brightness profiles of ellipticals are quite circular (Forman *et al.* 1985; Trinchieri *et al.* 1986). Second, some heating process might stop the inflow. Another possible explanation is provided by the thermal instability of the cooling gas (Fabian and Nulsen 1977; Mathews and Bregman 1978; Balbus 1986, 1988; Nulsen 1986). If the gas in the galaxy is sufficiently inhomogeneous, regions of the gas having different densities will cool at different rates, and the denser regions will cool below X-ray emitting temperatures before the lower density gas has cooled much or flowed in very far. Thus, the amount of X-ray emitting, inflowing gas would decrease with decreasing radius. A very similar result has been found for cluster cooling flows (Fabian *et al.* 1984; Stewart *et al.* 1984; Thomas, Fabian, and Nulsen 1987; White and Sarazin 1987a,b). In the next section, we will adopt this mechanism of loss of gas due to inhomogeneous cooling.

4.2 MODELS WITH GAS LOSS

The reduction in the amount of inflowing, hot gas has been treated in two ways. First, the loss of hot gas has been treated as a sink in the hydrodynamical equations of a homogeneous hot phase of gas (Thomas 1986; Vedder *et al.* 1988; Sarazin and Ashe 1989). Second, multi-phase models have been made in which the gas at any radius consists of many different density components with the same pressure and velocity (Fabian *et al.* 1984; Thomas 1986; Thomas *et al.* 1986). The results of the two approaches are in good general agreement; this section will emphasize the former technique.

In the first approach, the hydrodynamical equations (17–19) for a steady-state, spherically symmetric cooling flow are only modified slightly by gas loss. The only difference is the mass sink term $-\dot{\rho}$ in the continuity equation (17). The absence of terms containing the rate of gas loss in the momentum and energy equations (18) and (19) is based on the assumption that the gas is removed at the ambient velocity and pressure. Then, no momentum or entropy is transferred between the ambient and cooling gas (White and Sarazin 1987a).

Two forms have been assumed for the gas loss rate. Vedder *et al.* (1988) and Thomas (1986) assumed that the rate of removal of gas was proportional to the density of stars in the galaxy $\dot{\rho} \propto \rho_*$. A concern with this assumption is that one would expect that the rate for any physical process in the gas should depend on the properties of the gas, and not on the stars. Once the stars have injected the gas into ellipticals, they should have no further effect on the gas. However, the gas loss is due to cooling, so it might be reasonable to expect that the rate of removal of would increase with the cooling rate of the gas. Sarazin and Ashe (1989) assumed that the rate of gas loss was proportional to the cooling rate of the gas. Similar models have been given for cooling flows in clusters of galaxies by White and Sarazin (1987a,c). Sarazin and Ashe assume

$$\dot{\rho} = q\frac{\rho}{t_c}, \qquad (21)$$

where q is a dimensionless parameter assumed to be of order unity. The isobaric cooling time t_c is defined to be

$$t_c \equiv \frac{5}{2}\frac{\theta}{\rho\Lambda}. \qquad (22)$$

While the fact that the gas must cool to be removed from the flow suggests that equation (21) might give the mass loss rate to the right order of magnitude, there is no real reason why q might not depend on position either explicitly or implicitly through the gas properties. The two approaches to the calculation of the gas loss lead to results which are in general qualitative agreement.

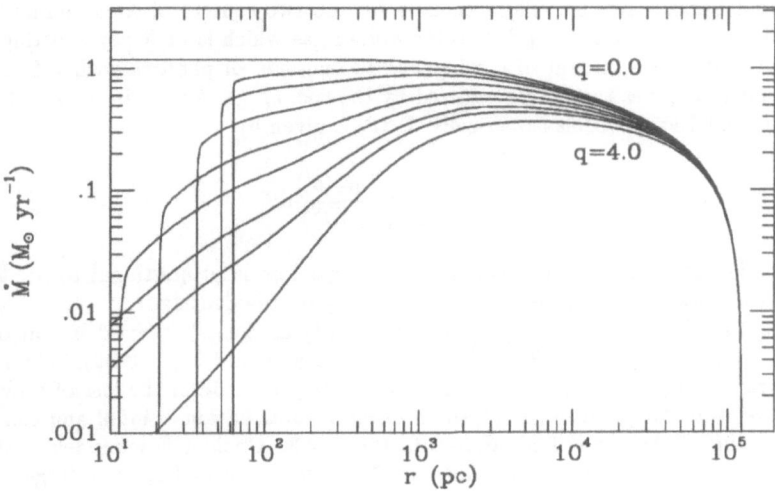

Figure 8. The variation of the mass inflow rate $\dot{M}$ versus radius in the steady-state cooling flow models with gas loss of Sarazin and Ashe (1989). The models range from a model with no gas loss ($q = 0$) to a model with very rapid gas loss ($q = 4$ in equation 21).

In models with gas loss, the diffuse gas is hotter and of lower density than the gas in homogeneous models. The sonic radius is reduced in the gas loss models. If the gas loss is very efficient, there is no sonic radius and the mass flow rate $\dot{M} \to 0$ as the center of the galaxy is approached ($r \to 0$). As expected, the rate of inflow of gas is reduced at essentially all radii as the rate of gas loss is increased. As a result, much less hot gas flows into the center of the galaxy. Figure 8 gives the mass inflow rate $\dot{M}$ as a function of radius for models with gas loss (and one model without gas loss) from Sarazin and Ashe (1989). For the model without gas loss ($q = 0$), $\dot{M}$ approaches a constant at the galaxy center, since there are neither significant sources nor sinks of gas within the central region. For the models with gas loss, $\dot{M}$ reaches a maximum and decreases at smaller radii where gas loss dominates. As q increases, this maximum occurs at larger radii, and at essentially all radii, $\dot{M}$ is smaller. For the models with gas loss, $\dot{M}$ drops rapidly to zero within the sonic radius. As noted above, the gas cools catastrophically in this region. White and Sarazin (1987a) show that the relationship between cooling and gas loss means that the mass flux $\dot{M}$ and entropy S in the gas vary together, with $\dot{M} \propto S^{3q/5}$. Inside the sonic radius, the temperature and entropy in the gas drop to zero (Figure 6), so the mass flux also goes to zero unless $q = 0$.

In these models with gas loss, the gas is removed by radiative cooling. Since this radiative cooling involves the emission of X-rays, the removal of gas does not, in itself, lower the amount of X-ray emission. It is important that the X-ray emission produced by the cooling gas be included. The resulting X-ray emissivity at a photon frequency ν of cooling flow gas including gas loss is given by

$$\epsilon_\nu(r) = \rho^2(r)\Lambda_\nu(\theta) + \dot{\rho}(r)\Gamma_\nu(\theta) \tag{23}$$

(White and Sarazin 1987b). This accounts for the two sources of X-ray emission in the cooling flows. The first term is due to the diffuse gas which is at X-ray emitting temperatures. Λ_ν gives the amount of cooling due to emission of photons with a frequency ν. The second term gives the X-rays emitted by the cooling gas which drops out of the flow. Assuming that this gas cools isobarically, $\Gamma_\nu(\theta)$ is given by

$$\Gamma_\nu(\theta) = \frac{5}{2} \int_0^\theta \frac{\Lambda_\nu(\theta')}{\Lambda(\theta')} d\theta'. \tag{24}$$

In the models of Sarazin and Ashe, the rate of gas loss is proportional to ρ^2, and both terms in the X-ray emissivity depend quadratically on the density.

The loss of hot gas lowers the X-ray luminosity of the galaxy models, but only by a factor of $\lesssim 3$ (Thomas 1986; Vedder et al. 1988; Sarazin and Ashe 1989). The reduction in the X-ray luminosity of gas loss models occurs only because of the loss of gravitational potential energy. As noted in §3.3.3, the maximum contribution of infall and compression to the heating of the gas is limited to a factor of $\lesssim 3$. Cooling flow models without gas loss gave X-ray luminosities which were roughly a factor of at least two larger than the average of those observed (Figure 5 and §4.1). Cooling flow models with efficient gas loss have X-ray luminosities in better agreement with the average of those observed.

The integrated X-ray spectra of models with and without gas loss are rather similar (Sarazin and Ashe 1989). The distinction between the models with and without gas loss is not in how much gas is cooling, but rather in where the gas is cooling. As a result, a major difference is found in the X-ray spectra of the outer portions ($r \gtrsim 3$ kpc) of the models. In this region, the gas loss models have much stronger soft X-ray line emission than the homogeneous models, because gas cools below 10^6 K in all parts of the gas loss models, but only at the centers of the models without gas loss. This difference suggests that the X-ray spectra of the outer regions of elliptical galaxies be used as a test to distinguish between the two classes of models.

The reduction in the amount of hot gas flowing into the center of the models brings the models into much better agreement with the observed X-ray surface brightness profiles of elliptical galaxies (Thomas 1986; Thomas et al. 1986; Vedder et al. 1988; Sarazin and Ashe 1989). A major problem with the models without gas loss was that their X-ray surface brightness profiles were too centrally peaked. The models with very efficient gas loss have peak X-ray surface brightnesses which are more than an order of magnitude smaller than those of the models without gas loss.

The X-ray surface brightness profiles in the outer regions of the models with gas loss and heavy halo mass distributions also agree well with the observed profiles. There is a simple argument which explains this (Sarazin 1986b). The argument is easiest if one considers the extreme form of a cooling flow model with gas loss. In this extreme "local cooling" model, one assumes that gas loss is so efficient that all of the gas cools before it can flow inward any significant distance. Then, in steady-state the local rates of heating and cooling must balance. Thus,

$$\Lambda(T)\rho^2 \approx \alpha_* \rho_* \frac{3kT_{inj}}{2\mu m_p}. \tag{25}$$

The cooling function Λ depends on the temperature, but the dependence is weaker than the quadratic dependence of the cooling rate on density. Thus, for a constant value of α_*, the gas density should vary crudely as $\rho \propto (\rho_* T_{inj})^{1/2}$. If supernova heating is dominant and/or the galactic velocity dispersion is independent of radius (as in a heavy halo model), then the gas injection temperature T_{inj} is constant, and the only variable factor on the right-hand-side of equation (25) is the stellar density ρ_*. In elliptical galaxies and stellar bulges, the stellar density varies crudely as $\rho_* \propto r^{-3}$ (the outer portions of a Hubble law). Then, the gas density goes as $\rho \propto r^{-3/2}$. This agrees with the gas profiles derived from the X-ray observations of elliptical galaxies. To compare to the X-ray surface brightness profiles, let $\Lambda_X(T)$ be the cooling function for emission within the X-ray band, so the $\Lambda_X(T)\rho^2$ is the X-ray emissivity. Then, integrating equation (25) along a line-of-sight through the galaxy gives

$$I_X(r) \approx \alpha_* \frac{3k}{2\mu m_p} \langle T_{inj} \rangle (M/L_B)_* \left\langle \frac{\Lambda_X}{\Lambda} \right\rangle I_B(r), \qquad (26)$$

where I_X is the X-ray surface brightness, I_B is the blue optical surface brightness, and the averages are along the line-of-sight at r. The factor $\langle \Lambda_X / \Lambda \rangle$ is a bolometric correction for the fraction of the gas emission is the X-ray band. Thus, if the T_{inj} is nearly constant as a function of radius, and if the gas is everywhere hot enough ($T \gg 10^6$ K) so that all of the emission is in the X-ray band, then equation (26) predicts that

$$I_X \propto I_B. \qquad (27)$$

Again, T_{inj} is roughly constant if supernova heating is dominant and/or if the stellar velocity dispersion is independent of position. These results are in good agreement with the observed X-ray images of early-type galaxies (Forman et al. 1985; Trinchieri et al. 1986).

5. Evolution of the Gas in Early-Type Galaxies

While steady-state models may explain many of the properties of the hot gas in the brighter X-ray early-type galaxies, time-dependent effects must be of interest in many galaxies. First, in the fainter X-ray galaxies the gas density is probably too low to allow cooling to balance heating; the cooling time is longer than the age of the galaxy. Even if the cooling time is short at the center of a galaxy, cooling may not be effective in the outer regions. As the gas density decreases with increasing radius, there must be some point in the outer portions of the galaxy where the cooling time is longer than the age. In addition to the finite age of galaxies, time-dependence is introduced by the fact that the rate of stellar mass loss and supernova heating probably vary with time. The present state of the gas in galaxies may be affected by the results of the earlier evolution. For example, cooling flow solutions with high supernova heating rates ("smothered winds" or "radiatively induced infall"; see §4.1) are only possible if the galaxy retains a very high density of interstellar gas. This may not be true because of the earlier evolution of the galaxy. As another example, it has been suggested ellipticals may have had outflows at early times which were retained within the heavy halos around the galaxies (Mathews and Loewenstein 1986; Loewenstein and Mathews 1987). Finally, the studies of time-dependent evolution of early-type galaxies are needed to predict their X-ray properties at high redshift, and to understand the history of star formation and nucleosynthesis.

Since the discovery of the X-ray emission from early-type galaxies, a number of detailed, spherically symmetric, time-dependent hydrodynamical simulations of elliptical galaxies have been made (Mathews and Loewenstein 1986; Hattori, Habe, and Ikeuchi 1987; Loewenstein and Mathews 1987; Umemura and Ikeuchi 1988; D'Ercole et al. 1989; David, Forman, and Jones 1990). In addition to the basic model for the stellar and mass distributions of present day galaxies, these models require that many other quantities be specified. These studies differ significantly in their conclusions, largely because they make different assumptions about these quantities. First, one needs to specify the initial conditions on the models; most of these models assume that elliptical galaxies form in a single short burst of star formation. Most of the models follow the evolution only after this burst (the models of David et al. [1990] are an exception). Most of the models assume that the initial gas density is very low. The rates of stellar mass loss and supernova heating must be specified as a function of time. All of these models assume that the stellar mass loss rate decreases with time. Hattori et al. (1987) assume that $\alpha_* \propto t^{-1}$. The other models calculate the mass loss rate from a model for the initial mass function of the stars (MacDonald and Bailey 1981; Renzini and Buzzoni 1986). Mathews and Loewenstein (1986) use an expression with varies roughly as $\alpha_* \propto t^{-5/4}$ at late times; Loewenstein and Mathews (1987) find a similar variation, but normalize their rate to a present day mass loss rate which is much smaller than that expected from stellar evolution calculations (D'Ercole et al. 1989). Hattori et al. (1987) include gas loss terms (§4.2), and Loewenstein and Mathews (1987) calculate a model with thermal conduction.

The time variation of the supernova heating rate is a very important parameter in these evolution models. With the exception of David et al. (1990), all of the calculations start well after the initial burst of star formation, when the Type II and Type Ib supernova rate is assumed to be very low. The sole contribution to supernova heating is then from Type Ia supernovae. Unfortunately, the progenitors and evolution of Type Ia supernovae are poorly understood. As a result, many different forms for the variation of the supernova heating rate have been assumed. Hattori et al. (1987) assumed that the Type Ia rate and energy were independent of time. This leads to a total heating rate which is independent of time. Hattori et al. also assumed that the stellar mass loss rate was much higher in the past. Loewenstein and Mathews (1987) assume that the Type Ia supernova rate is slightly higher in the past, but decreases more slowly with time than the stellar mass loss rate. In both cases, this means that the supernova heating rate per unit mass (or the resulting injection temperature T_{SN}) was much *lower* in the past. Mathews and Loewenstein (1986) assume that the Type Ia supernova rate is proportional to the stellar mass loss rate $\alpha_{SN} \propto \alpha_*$. Then, the supernova heating rate increases in the past, but the heating rate per unit mass and the resulting injection temperature T_{SN} are *constant*. Based on arguments about stellar evolution in binary stars, and the variation of the Type I supernova rate with the color of the parent galaxy (Renzini 1990), D'Ercole et al. (1989) assume a supernova rate which decreases with time faster than the rate of stellar mass loss. This implies that the heating rate per unit mass and the injection temperature T_{SN} was much *higher* in the past. David et al. (1990) include Type II supernovae, and therefore have a supernova heating rate and injection temperature which are much higher during the early stages of the evolution of the galaxies. Unfortunately, the variations of the injection temperature with time is a factor which is of great importance in determining how the hot gas atmospheres of elliptical galaxies evolve. The range of assumptions made by these authors is probably indicative of our lack of understanding to the evolution of Type Ia supernovae.

Although these studies of the evolution of ellipticals contain a wealth of detailed information, it is possible to crudely characterize the dynamical state of the gas at any given time as being in one of four phases. First, if the injection temperature of the gas is higher than a critical value set by the depth of the galactic gravitation potential (and the initial gas density is not so high that cooling is important), the gas forms a *transonic wind*. In this case, the X-ray luminosity is much smaller than the rate of heat injection, and most of the injected energy appears as kinetic energy of the wind. Second, if the injection temperature is to low to unbind the gas, but the cooling time in the center of the galaxy is longer than the age, the gas undergoes *subsonic inflation*. The heating of the gas is insufficient to unbind it from the galaxy, and most of the injected energy goes into increasing the pressure in the gas and causing it to inflate slowly. Third, if the injection temperature is too low to unbind the gas, and the cooling time is shorter than the age at the center of the galaxy but longer than the age in the outer regions, the gas forms a *partial wind*. In the inner regions where cooling is effective, the gas basically forms a steady-state cooling flow, but forms a subsonic or transonic wind in the outer regions. Finally, if the cooling time is short throughout the galaxy, it forms a steady-state cooling flow. In this phase, the X-ray luminosity of the galaxy is essentially equal to the total rate of heating of the gas. In the subsonic inflation and partial wind phases, the X-ray luminosity is smaller than the heating rate, but higher than the same galaxy would have if the gas formed a transonic wind.

Figure 9 shows the variation of the X-ray luminosity in three published models for elliptical galaxies with an optical luminosity of $L_B \approx 5 \times 10^{10} L_\odot$. The model labeled ML is from Mathews and Loewenstein (1986); it is their model B. The model labeled LM is Loewenstein and Mathews (1987) model B2. The model labeled D is a D'Ercole et al. (1989) model (specifically, it is the model with $\delta = 100$, $\gamma = 0.03$, and $(M/L_B)_* = 8$ in their notation).

In the models where the injection temperature increases with time but the total heating rate does not (Hattori et al. 1987; Loewenstein and Mathews 1987), the gas initially undergoes a partial wind or subsonic inflations, and eventually forms a cooling flow, a partial wind, or a supersonic wind. In any case, the X-ray luminosity generally decreases with time, possibly after a brief early period of increasing luminosity. An example is model LM in Figure 9. The models with a constant injection temperature are very similar (Mathews and Loewenstein 1986). Model ML in Figure 9 is an example of this.

In the models with an injection temperature which decreases with time (D'Ercole et al. 1989; David et al. 1990), the galaxies initially have transonic winds. During this period, the X-ray luminosity decreases with time. Then, the gas undergoes either a partial wind or subsonic inflation, during which time the X-ray luminosity generally increases. During both of these two phases, the X-ray luminosity is less than the total heating rate. Finally, the gas forms a steady-state cooling flow, and the X-ray luminosity is essentially equal to the total heating rate. The curve D in Figure 9 is an example of this behavior. The two points in the curve where $(d \log L_X / dt)$ changes sign mark the transitions from a supersonic wind to a subsonic inflation, and from a subsonic inflation to a cooling flow.

The evolution of the gas in these models also depends strongly on the mass profile of the galaxy. Galaxies with heavy halos are much more likely to retain their gas and eventually form cooling flows (Mathews and Loewenstein 1986; Hattori et al. 1987; Loewenstein and Mathews 1987; David et al. 1990).

230

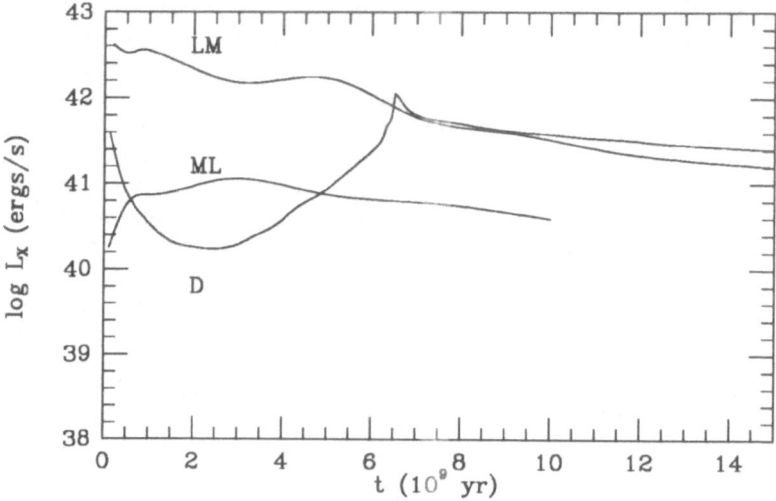

Figure 9. The variation of the X-ray luminosity with time in three evolution models. The curve labeled ML is from Mathews and Loewenstein (1986); it is their model B. It is a model in which the injection temperature increases with time. The model labeled LM is Loewenstein and Mathews (1987) model B2. In this model, the injection temperature remains constant. The model labeled D is a D'Ercole *et al.* (1989) model (specifically, it is the model with $\delta = 100$, $\gamma = 0.03$, and $(M/L_B)_* = 8$ in their notation). The injection temperature in this model decreases with time.

6. Dispersion in X-Ray Luminosities of Early-Type Galaxies

Although there is a strong correlation between the optical and X-ray luminosities of early-type galaxies, galaxies of a given optical luminosity exhibit a very significant range in in their X-ray luminosities (Figure 1). This dispersion in X-ray luminosity is much greater than the dispersion in many other properties of early-type galaxies (for a given L_B), such as their sizes, velocity dispersions, colors, and metallicities. In particular, the X-ray luminosities of early-type galaxies with blue luminosities of $L_B \approx 10^{11}$ $L_\odot$ range over a factor of roughly fifty.

In §3.4, a simple estimate for the X-ray luminosity of ellipticals was given. This estimate assumed that ellipticals were effectively isolated, so that all of the gas and energy in the interstellar gas was due to stellar mass loss within the galaxy, and was retained by the galaxy. Another assumption was that the galaxies were in steady-state, with heating and radiative cooling in global balance. If the gas is hot enough, all of the radiative cooling will occur in the X-ray band. The heating processes were assumed to be the motions of gas losing stars and supernovae. Then, if all early-type galaxies of a given optical luminosity have similar stellar populations and similar stellar velocity dispersions (Sandage and Visvanathan 1978), this simple argument predicts that they should all have similar X-ray luminosities. The fact that they do not indicates that at least one of these assumptions is wrong.

There are several possible sources of this dispersion in the X-ray luminosities. One possibility is that the scatter results from experimental error in either the fluxes of the galaxies or in their distances. The scatter is probably too large to be due to errors in fluxes, although these errors probably do contribute. Errors in distances also contribute significantly; White and Sarazin (1989) find that using the Seven Samurai distances (Faber et al. 1989) reduces the dispersion in $\log L_X$ by about 25%. Another possibility is that there may be a considerable scatter in the intrinsic properties of galaxies (i.e., stellar mass loss rates, supernova rates, velocity dispersions, ...) of a given optical luminosity (Djorgovski and de Carvalho 1990). This might produce a dispersion in the rates of heating of the interstellar gas. It is possible that the a multivariate correlation with the correct set of optical parameters (the "fundamental plane" parameters) might greatly reduce the dispersion is L_X (Djorgovski and de Carvalho 1990).

Another possibility is that E/S0s may not generally be in steady-state, and that galaxies with differing X-ray luminosities but similar optical luminosities are in different stages of their hydrodynamical evolution (Hattori et al. 1987; Loewenstein and Mathews 1987; D'Ercole et al. 1989). For example, the lowest X-ray luminosity early-type galaxies might have supersonic winds, intermediate luminosity galaxies might be undergoing partial winds or subsonic inflation, and only the highest luminosity galaxies may have steady-state cooling flows (D'Ercole et al. 1989). With this suggestion, one still needs to understand why early-type galaxies with similar optical luminosities have different effective hydrodynamical ages. Either their actual ages vary over a large range (say 5 to 15 billion years), or their hydrodynamical evolution is strongly affected by their environment.

Canizares et al. (1987), Sarazin and White (1988), and White and Sarazin (1989) suggest that the X-ray emission of early-type galaxies is affected by their environment. For example, many ellipticals and S0s may have lost much of their gas via ram-pressure stripping. It would not be surprising if E/S0s were subject to ram-pressure stripping. First of all, spiral galaxies in dense environment are stripped of their H I (see the paper by Young in this volume). Early-type galaxies are generally found in denser environment than spirals, and early-type galaxies should be easier to strip. The gas in E/S0s has lower column densities than the H I in spirals, and is not concentrated in a disk. Also, the X-ray isophotes of several E/S0s show distortions in their outer regions which may be the result of ram-pressure ablation; the most prominent example is the "plume" in M86 (Figure 3), which is generally interpreted as being due to ram-pressure ablation (Forman and Jones 1982).

One feature of the dispersion in X-ray luminosities of ellipticals is that it extends very crudely from a lower limit which is consistent with purely stellar X-ray emission (Figure 1) to an upper limit which is consistent with the highest X-ray luminosities in steady-state cooling flow models (Figure 5). This suggests that the brightest X-ray galaxies have retained all of their hot gas and that the input of heat is converted directly to X-ray emission, and that the fainter galaxies have less hot gas and less of the heat input goes into X-ray emission. This is consistent with the dispersion being due to hydrodynamical evolution; in the fainter galaxies, the hot gas may be lost in winds, and the heating may be converted into kinetic energy (or into increasing thermal energy in subsonic inflation models). Alternatively, the pattern in the X-ray luminosities is consistent with ram-pressure stripping being important; the fainter galaxies may have lost much of their hot gas to the ambient medium. One way to test between these two hypotheses is to observe the X-ray surface brightness profiles of the fainter X-ray galaxies. If these galaxies are deficient because of winds or inflation, they should have more extended X-ray emission than the X-ray bright galaxies. On the other hand, if they are faint because of stripping, they should have truncated X-ray surface brightness profiles.

The uncertainty in the source of the L_X dispersion makes it difficult to compare theoretical models to the X-ray observations. If the scatter in L_X is due to intrinsic properties, models using average optical properties should reproduce the average X-ray luminosity of early-type galaxies of a given optical luminosity. On the other hand, if the scatter is induced by varying degrees of ram-pressure stripping, theoretical models should provide an upper envelope for L_X as a function of L_B. This difference would be significant; for example, the supernova rate inferred from these respective approaches would differ by at least a factor of ~ 3 (cf. Sarazin and White 1988).

White and Sarazin (1989) have studied to the origin of this dispersion in the X-ray luminosities of early-type galaxies by performing multivariate analyses. They searched for correlations of $\log L_X$ with other galaxy properties, and for each correlation, the standard deviation of the residual $\delta(\log L_X)$ was found. The object of this study was to find galaxy properties which were strongly correlated with the X-ray luminosity. A property is "strongly correlated" with $\log L_X$ if the standard deviation of the residual $\delta(\log L_X)$ is significantly reduced. For each of the early-type galaxies, the X-ray fluxes were taken from Canizares et $al.$ (1987). For many of the galaxies, only an upper limit to the X-ray flux is known, leading to an upper limit for L_X. In order to properly take these upper limits into account, "survival statistic" techniques were used. Maximum-likelihood linear regressions were done. The form of the linear regression was

$$\log L_X = c_0 + m_1 p_1 + m_2 p_2 + R, \tag{28}$$

where p_1 and (in the case of multivariate regressions) p_2 are two properties of the galaxies, m_1 and m_2 are the slopes of the correlations, c_0 is the constant, and the residual R is a random variable drawn from a normal distribution with a standard deviation $\sigma(\log L_X)$. A few of the correlations from White and Sarazin (1989) are listed in Table 2. The first column gives the source of the distance to the galaxy; "CFT" stands for Canizares et $al.$ (1987), while "7Sam" stands for the Seven Samurai distances (Faber et $al.$ 1989). Columns 2 and 3 give the first independent variable p_1 in the correlation, and its coefficient m_1 in the linear regression (equation 28). For the cases with multivariate correlations, this information is repeated for the second independent variable p_2 in columns 4 and 5. Column 6 gives the constant c_0 in the linear regression, and column 7 gives the number of galaxies in the sample. In the eighth column, the value of χ^2 gives the cumulative probability that such a tend in the data could have arisen randomly from uncorrelated data. Finally, the values of the dispersion $\sigma(\log L_X)$ are given in column 9.

Table 2 shows that a significant source of the dispersion in the X-ray luminosities are errors in the distances to galaxies. When the "Seven Samurai" (Faber et $al.$ 1989) distances are used, the dispersion in $\log L_X$ is substantially reduced. Table 2 contains correlations of $\log L_X$ with intrinsic galaxy properties. In addition to the well-known correlation of L_X with L_B, correlations of L_X with galaxy color ($U - V$ or $B - V$), metallicity (measured by Mg_2), effective radius, velocity dispersion σ_*, H I content, and far-infrared luminosity L_{IR} were tested. The latter two measures of the correlation between the hot gas and cool gas contents of E/S0s showed no correlation. There was a weak correlation of L_X with color and metallicity. The sense of the correlation was that redder, more metal-rich galaxies (of a fixed optical luminosity) had slightly higher X-ray luminosities. This might be explained

TABLE 2
Early-Type Galaxy X-ray–Optical Correlations

Distance Source	Property 1 p_1	m_1	Property 2 p_2	m_2	Constant c_0	No.	χ^2	Dispersion $\sigma(\log L_X)$
CFT	$\log L_B$	1.77±0.20			21.65±2.10	81	34.16	0.51
7Sam	$\log L_B$	2.27±0.24			15.97±2.61	47	40.25	0.37
CFT	$\log L_B$	1.44±0.22	$(U-V)$	2.61±0.66	18.93±2.37	61	26.19	0.47
7Sam	$\log L_B$	2.17±0.23	Mg_2	7.08±3.54	14.86±2.63	47	40.41	0.35
CFT	$\log L_B$	1.33±0.26	$\log \sigma_*$	1.91±0.79	21.80±2.21	57	28.68	0.51
CFT	$\log(L_B\sigma_*^2)$	1.19±0.13			22.13±2.09	57	28.36	0.51
CFT	$\log L_{IR}$	0.64±0.35			15.13±6.97	30	1.83	0.61
CFT	$\log L_B$	1.43±0.22	$\log L_{IR}$	−0.04±0.15	26.62±5.19	22	12.61	0.49
7Sam	SB_e	0.52±0.09	$\log \sigma_*$	4.58±0.70	18.82±2.41	47	40.60	0.39

if stellar mass loss rates increase somewhat with stellar metallicity (Greggio 1990). There was also a weak correlation with the velocity dispersion of ellipticals. If the motions of gas losing stars dominate the heating of the gas in ellipticals, then the X-ray luminosity might be expected to vary in proportion to $L_B\sigma_*^2$ (equations 13 and 16). This is consistent with the observed correlation. Yet, none of the correlations tested by White and Sarazin significantly reduced the dispersion in the residual X-ray luminosities (by more than about 10%). Djorgovski and de Carvalho (1990) argue that X-ray luminosity is strongly correlated with the elliptical galaxy "fundamental parameters", the effective optical surface brightness SB_e and the stellar velocity dispersion σ_*. As shown in Table 2, White and Sarazin do not confirm this correlation. Djorgovski and de Carvalho also suggest that X-ray luminosity is strongly correlated with the photometric parameter a_4, which measures the "boxyness" of the optical isophotes of the galaxy. This might indicate that X-ray emission is affected by merging.

In order to assess the effect of environment on the X-ray luminosity of E/S0s, White and Sarazin (1989) looked for a correlation of L_X with the number of neighboring galaxies in various galaxy catalogs. The "neighbors" were defined either in projection (close to the X-ray E/S0 on the sky) or in angle and in recession velocity (to reduce the effect of projections). A fairly strong correlation between X-ray luminosity and the density of the environment was found. Collectively, galaxies with low X-ray luminosities (for a given L_B) tend to be in denser environments than galaxies with higher X-ray luminosities: galaxies with low L_X have ~50% more neighbors between 1 and 5 degrees than do galaxies with high L_X (for a given L_B).

This suggests that the dispersion in the X-ray luminosity of early-type galaxies of a given optical luminosity is caused by the environment of the galaxy. The fainter X-ray galaxies are found in denser locations. This is the sense of the correlation expected if ram-pressure stripping affects the amount of gas in galaxies in dense environments. Detailed calculations indicate that ram-pressure should be important in the environments in which early-type galaxies are located (White and Sarazin 1989).

7. Conclusions

Early-type galaxies are luminous sources of X-ray emission. This emission indicates that these galaxies contain significant amounts of hot interstellar gas, and are not the gas-poor systems they were previously thought to be. In the brighter X-ray galaxies, the amounts of hot gas observed are consistent with those expected given the present rates of stellar mass loss. The required rates of heating of the gas are consistent with those expected from the motions of gas losing stars and supernovae. The X-ray observations are generally more consistent with a lower rate of Type I supernovae than was previously thought (Tammann 1982). In the brightest X-ray galaxies, the cooling times in the gas are shorter than the Hubble time essentially throughout the observed X-ray image. This suggests that the gas in these galaxies forms a steady-state cooling flow. Cooling flow models explain most of the properties of the brighter X-ray galaxies, including their luminosities, the X-ray–optical correlation, their temperatures, and their surface brightness profiles. Homogeneous cooling flow models, in which all the hot gas flows into the central regions of the galaxy, have surface brightness profiles which are too centrally peaked, and which decline too rapidly in the outer parts. On the other hand, if the gas is inhomogeneous and blobs of gas cool below X-ray emitting temperature near the position where they start, good agreement with the observed X-ray surface brightness profiles is found.

Although the optical and X-ray luminosities of early-type galaxies are strongly correlated, there is a large dispersion in this correlation. One suggestion is that this dispersion indicates that early-type galaxies are not generally in steady-state, and that the fainter galaxies are in an earlier hydrodynamical phase in which most of the heating of the gas goes into the kinetic energy in a wind or partial wind, or into the thermal energy of subsonic inflation. Alternatively, the fainter galaxies may have lost much of their gas to ram pressure ablation. At a given optical luminosity, the brightest X-ray galaxies are found in lower density regions, which supports this last suggestion. There are several new X-ray observations which would greatly clarify this situation. First, X-ray spectra of the early-type galaxies of a given optical luminosity with the smallest X-ray luminosities (those in the hatched region in Figure 1) are needed to determine whether the X-ray emission of these galaxies is from diffuse hot gas or from stellar X-ray sources. X-ray surface brightness profiles and images of these galaxies would also be useful. If the emission is due to hot gas, the image should show distortions due to the environment in the outer parts. If the emission is stellar, the X-ray images should accurately follow the optical images. If the fainter galaxies are experiencing winds or subsonic inflation, the X-ray emission should be more extended than the optical emission. If they have been stripped of their gas by ram pressure, the X-ray emission will be less extended than the optical emission.

The X-ray observations of early-type galaxies suggest that these galaxies have heavy halos of optically dark material which contain much of the galaxy mass (Forman et al. 1985; Fabian et al. 1986; Mathews and Loewenstein 1986; Thomas 1986; Hattori et al. 1987; Loewenstein and Mathews 1987; Sarazin 1987b; Sarazin and White 1987, 1988; Killeen and Bicknell 1988). Unfortunately, the mass profiles of ellipticals and S0s are very uncertain, because of the lack of temperature profiles for the gas (Trinchieri et al. 1986; Canizares et al. 1987; Vedder et al. 1988; Fabbiano 1989). Spatially and spectrally resolved observations of early-type galaxies are needed to determine their masses, and would help to understand the role of ambient pressure, supernova heating, thermal conduction, and gas loss processes as well (e.g., Vedder et al. 1988; Sarazin and Ashe 1989).

If steady-state cooling flows are present in at least the brighter X-ray emitting early-type galaxies, where does the cooling gas go? Both the small amounts of cooler gas and their lack of correlation with the X-ray luminosities (Jura 1986; Jura *et al.* 1987; Bregman, Roberts, and Giovanelli 1988; Huchtmeier *et al.* 1988; Kim *et al.* 1988; Knapp 1988; Bally and Thronson 1989; Table 2) indicate that the gas is not stored as cool gas. The most likely suggestion is that the cooling gas forms stars. Many elliptical galaxies have strong UV emission from a very hot stellar population (Burstein *et al.* 1988), and it is possible that this emission is due to ongoing massive star formation (Kjaergaard 1987; Rocca-Volmerange 1989). However, there are several arguments that suggest that the UV emission is due to evolved, old stars (*e.g.*, Burstein *et al.* 1988). It may be that star formation in ellipticals favors low mass stars (Jura 1977), as has been suggested for cluster cooling flows (Fabian, Nulsen, and Canizares 1982; Sarazin O'Connell 1983).

8. Acknowledgments

I would like to thank Harley Thronson, Mike Shull, and the other organizers of this meeting for their efforts, which made this a very enjoyable meeting. I would also like to thank Pepi Fabbiano and Bill Forman for permission to reproduce several of the figures from their publications. This work was supported in part by NASA Astrophysical Theory Program Grant NAGW-764.

9. References

Bailey, M. E. 1980, *M.N.R.A.S.*, **191**, 195.

Balbus, S. A. 1986, *Ap. J. (Letters)*, **303**, L79.

Balbus, S. A. 1988, *Ap. J.*, **328**, 395.

Bally, J., and Thronson, H. A. 1989, *A. J.*, **97**, 69.

Bechtold, J., Forman, W., Giacconi, R., Jones, C., Schwarz, J., Tucker, W., and Van Speybroeck, L. 1983, *Ap. J.*, **256**, 26.

Biermann, P., and Kronberg, P. 1983, *Ap. J. (Letters)*, **268**, L69.

Biermann, P., Kronberg, P. P., and Madore, B. F. 1982, *Ap. J. (Letters)*, **256**, L37.

Biermann, P., Kronberg, P. P., and Schmutzler, T. 1989, *Astr. Ap.*, **208**, 22.

Bradt, H., Mayer, W., Naranan, S., Rappaport, S., and Spuda, G. 1967, *Ap. J. (Letters)*, **161**, L1.

Bregman, J. N. 1978, *Ap. J.*, **224**, 768.

Bregman, J. N., Roberts, M. S., and Giovanelli, R. 1988, *Ap. J. (Letters)*, **330**, L93.

Burstein, D., Bertola, F., Buson, L. M., Faber, S. M., and Lauer, T. R. 1988, *Ap. J.*, **328**, 440.

Byram, E. T., Chubb, T. A., and Friedman, H. 1966, *Science*, **152**, 66.

Canizares, C. R. 1987, in *Proceedings of IAU Symposium 117: Dark Matter in the Universe*, ed. by G. Knapp and J. Kormendy, p. 167.

Canizares, C. R., Donahue, M., Trinchieri, G., Stewart, G., and McGlynn, T. 1986, *Ap. J.*, **304**, 312.

Canizares, C. R., Fabbiano, G., and Trinchieri, G. 1987, *Ap. J.*, **312**, 503.

Cox, D. P. and Smith, B. W. 1974, *Ap. J. (Letters)*, **189**, L105.

David, L. P., Forman, W., and Jones, C. 1990, in *Proceedings of the Wyoming Conference on the Interstellar Medium of External Galaxies*, ed. by H. A. Thronson and J. M. Shull (Dordrecht: Kluwer), in press.

Davies, R. L., Efstathiou, G., Fall, S. M., Illingworth, G., and Schechter, P. L. 1983, *Ap. J.*, **266**, 41.

D'Ercole, A., Renzini, A., Ciotti, L., and Pellegrini, S. 1989, *Ap. J. (Letters)*, **341**, L9.

Djorgovski, S. G., and de Carvalho, R. R. 1990, in *Window on Galaxies*, ed. by G. Fabbiano, J. S. Gallagher, and A. Renzini, in press.

Dressel, L. L., and Wilson, A. S. 1985, *Ap. J.*, **291**, 668.

Evans, R., van den Bergh, S., and McClure, R. D. 1989, preprint.

Fabbiano, G. 1986a, *Pub. A.S.P.*, **98**, 525.

Fabbiano, G. 1986b, in *Proceedings of the Greenbank Workshop on Gaseous Halos around Galaxies*, ed. by J. Bregman and F. Lockman (Greenbank: NRAO), p. 203.

Fabbiano, G. 1989, *Ann. Rev. Astr. Ap.*, **27**, in press.

Fabbiano, G., Gioia, I. M., and Trinchieri, G. 1989, preprint.

Fabbiano, G., Klein, U., Trinchieri, G., and Wielebinski, R. 1987, *Ap. J.*, **312**, 111.

Fabbiano, G., and Trinchieri, G. 1987, *Ap. J.*, **315**, 46.

Faber, S. M., and Gallagher, J. S. 1976, *Ap. J.*, **204**, 365.

Faber, S. M., and Jackson, R. E. 1976, *Ap. J.*, **204**, 668.

Faber, S. M., Wegner, G., Burstein, D., Davies, R. L., Dressler, A., Lynden-Bell, D., and Terlevich, R. J. 1989, *Ap. J. Suppl.*, **69**, 763.

Fabian, A. C. 1981, in *The Structure and Evolution of Normal Galaxies*, ed. by S. M. Fall and D. Lynden-Bell (Dordrecht: Reidel), p. 181.

Fabian, A. C. 1988, *Cooling Flows in Clusters and Galaxies*, (Dordrecht: Reidel).

Fabian, A. C., Arnaud, K. A., and Thomas, P. A. 1987, in *Proceedings of IAU Symposium 117: Dark Matter in the Universe*, ed. by G. Knapp and J. Kormendy, p. 201.

Fabian, A. C., and Nulsen, P. E. J. 1977, *M.N.R.A.S.*, **180**, 479.

Fabian, A. C., Nulsen, P. E. J., and Canizares, C. R. 1982, *M.N.R.A.S.*, **201**, 933.

Fabian, A. C., Nulsen, P. E. J., and Canizares, C. R. 1984, *Nature*, **310**, 733.

Fabian, A. C., Schwarz, J., and Forman, W. 1980, *M.N.R.A.S.*, **192**, 135.

Fabian, A. C., and Thomas, P. A. 1987, in *Proceedings of IAU Symposium 127: Structure and Dynamics of Elliptical Galaxies*, ed. by T. de Zeeuw (Dordrecht: Reidel), p. 155.

Fabian, A. C., Thomas, P. A., Fall, S. M., and White, R. E. III 1986, *M.N.R.A.S.*, **211**, 1049.

Forman, W., and Jones, C. 1982, *Ann. Rev. Astr. Ap.*, **20**, 547.

Forman, W., Jones, C., and Tucker, W. 1984, in *Clusters and Groups of Galaxies*, ed. by F. Meudirossian, G. Guiricin, and M. Mezzetti (Dordrecht: Reidel), p. 297.

237

Forman, W., Jones, C., and Tucker, W. 1985, *Ap. J.*, **293**, 102.
Forman, W., Schwarz, J., Jones, C., Liller, W., and Fabian, A. C., 1979, *Ap. J. (Letters)*, **234**, L27.
Greggio, L. 1990, in *Window on Galaxies*, ed. by G. Fabbiano, J. S. Gallagher, and A. Renzini, in press.
Hattori, M., Habe, A., and Ikeuchi, S. 1987, *Prog. Theor. Phys.*, **78**, 1099.
Huchtmeier, W. K., Bregman, J. N., Hogg, D. E., and Roberts, M. S. 1988, *Astr. Ap.*, **198**, L17.
Jura, M. 1977, *Ap. J.*, **212**, 634.
Jura, M. 1986, *Ap. J.*, **303**, 327.
Jura, M., Kim, D. W., Knapp, G. R., and Guhathakurta, P. 1987, *Ap. J. (Letters)*, **312**, L11.
Killeen, N. E., and Bicknell, G. V. 1988, *Ap. J.*, **325**, 165.
Killeen, N. E., Bicknell, G. V., and Carter, D. 1986, *Ap. J.*, **309**, 45.
Kim, D. W., Guhathakurta, P., van Gorkom, J. H., Jura, M., and Knapp, G. R. 1988, *Ap. J.*, **330**, 684.
King, I. R. 1978, *Ap. J.*, **222**, 1.
Kjaergaard, P 1987, *Astr. Ap.*, **176**, 210.
Knapp, G. R. 1988, in *Proceedings of the NATO Advanced Study Workshop: Cooling Flows in Galaxies and Clusters*, ed. by A. Fabian, (Dordrecht: Reidel), p. 93.
Kriss, G. A., Canizares, C. E., McClintock, J. E., and Feigelson, E. D. 1980, *Ap. J. (Letters)*, **235**, L61.
Kriss, G. A., Cioffi, D. F., and Canizares, C. R. 1983, *Ap. J.*, **272**, 439.
Loewenstein, M., Fabian, A. C., and Nulsen, P. E. J. 1987, *M.N.R.A.S.*, **229**, 129.
Loewenstein, M., and Mathews, W. G. 1987, *Ap. J.*, **319**, 614.
Long, K. S., and Van Speybroeck, L. P. 1983, in *Accretion Driven X-ray Sources*, ed. W. Lewin and E. van den Heuvel (Cambridge: Cambridge University press), p. 117.
MacDonald, J., and Bailey, M. E. 1981, *M.N.R.A.S.*, **197**, 995.
Mason, K. O., and Rosen, S. R. 1985, *Space Sci. Rev.*, **40**, 675.
Mathews, W. G. 1988, *A. J.*, **95**, 1047.
Mathews, W. G. 1989, *A. J.*, **97**, 42.
Mathews, W. G., and Baker, J. 1971, *Ap. J.*, **170**, 241.
Mathews, W. G., and Bregman, J. N. 1978, *Ap. J.*, **224**, 308.
Mathews, W. G., and Loewenstein, M. 1986, *Ap. J. (Letters)*, **306**, L7.
Nulsen, P. E. J. 1986, *M.N.R.A.S.*, **221**, 377.
Nulsen, P. E. J., and Carter, D. 1987, *M.N.R.A.S.*, **225**, 939.
Nulsen, P. E., Stewart, G. C., and Fabian, A. C. 1984, *M.N.R.A.S.*, **208**, 185.
Ohashi, T. 1990, in *Window on Galaxies*, ed. by G. Fabbiano, J. S. Gallagher, and A. Renzini, in press.
Renzini, A. 1990, in *Window on Galaxies*, ed. by G. Fabbiano, J. S. Gallagher, and A. Renzini, in press.
Renzini, A., and Buzzoni, A. 1986, in *Spectral Evolution is Galaxies*, ed. C. Chiosi and A. Renzini (Dordrecht: Reidel), p. 195.
Rocca-Volmerange, B. 1989, *M.N.R.A.S.*, **236**, 47.
Sandage, A. 1957, *Ap. J.*, **125**, 422.

238

Sandage, A., and Visvanathan, N. 1978, *Ap. J.*, **223**, 707.

Sarazin, C. L. 1986a, *Rev. Mod. Phys.*, **58**, 1.

Sarazin, C. L. 1986b, in *Proceedings of the Greenbank Workshop on Gaseous Halos around Galaxies*, ed. by J. Bregman and F. Lockman (Greenbank: NRAO), p. 223.

Sarazin, C. L. 1987a, in *Proceedings of IAU Symposium 117: Dark Matter in the Universe*, ed. by J. Kormendy and G. Knapp (Dordrecht: Reidel), p. 183.

Sarazin, C. L. 1987b, in *Proceedings of IAU Symposium 127: Structure and Dynamics of Elliptical Galaxies*, ed. by T. de Zeeuw (Dordrecht: Reidel), p. 179.

Sarazin, C. L. 1988, *X-ray Emission from Clusters of Galaxies*, (Cambridge: Cambridge University Press).

Sarazin, C. L., and Ashe, G. A. 1989, *Ap. J.*, **345**, in press.

Sarazin, C. L., and O'Connell, R. W. 1983, *Ap. J.*, **268**, 552.

Sarazin, C. L., and White, R. E. III 1987, *Ap. J.*, **320**, 32.

Sarazin, C. L., and White, R. E. III 1988, *Ap. J.*, **331**, 102.

Serlemitsos, P., Smith, B., Boldt, E., Holt, S., and Swank, J. 1977, *Ap. J. (Letters)*, **211**, L63.

Soker, N. and Sarazin, C. L. 1988, *Ap. J.*, **327**, 66.

Stewart, G. C., Canizares, C. R., Fabian, A. C., and Nulsen, P. E. J. 1984, *Ap. J.*, **278**, 536.

Takeda, H., Nulsen, P. E. J., and Fabian, A. C. 1987, *M.N.R.A.S.*, **208**, 261.

Tammann, G. A. 1982, in *Supernovae: A Survey of Current Research*, ed. by M. J. Rees and R. J. Stoneham (Dordrecht: Reidel), p. 371.

Thomas, P. A. 1986, *M.N.R.A.S.*, **220**, 949.

Thomas, P. A., Fabian, A. C., Arnaud, K. A., Forman, W., and Jones, C. 1986, *M.N.R.A.S.*, **222**, 655.

Thomas, P. A., Fabian, A. C., and Nulsen, P. E. J. 1987, *M.N.R.A.S.*, **228**, 973.

Tinsley, B. M. 1980, *Fund. Cosmic Phys.*, **5**, 287.

Tonry, J. 1981, *Ap. J. (Letters)*, **251**, L1.

Trinchieri, G. 1986, in *Proceedings of the Greenbank Workshop on Gaseous Halos around Galaxies*, ed. by J. Bregman and F. Lockman, p. 215.

Trinchieri, G., and Fabbiano, G. 1985, *Ap. J.*, **296**, 447.

Trinchieri, G., Fabbiano, G., and Canizares, C. R. 1986, *Ap. J.*, **310**, 637.

Umemura, M., and Ikeuchi, S. 1988, *Ap. J.*, **319**, 163.

van den Bergh, S., McClure, R. D., and Evans, R. 1987, *Ap. J.*, **323**, 44.

Vedder, P. W., Trester, J. J., and Canizares, C. R. 1988, *Ap. J.*, **332**, 725.

Walsh, D. E., Knapp, G. R., Wrobel, J. M., and Kim, D. W. 1989, *Ap. J.*, **337**, 209.

White, D. A., Fabian, A. C., Forman, W., Jones, C., and Stern, C. 1989, preprint.

White, R. E. III, and Chevalier, R. A. 1983, *Ap. J.*, **275**, 69.

White, R. E. III, and Chevalier, R. A. 1984, *Ap. J.*, **280**, 561.

White, R. E. III, and Sarazin, C. L. 1987a, *Ap. J.*, **318**, 612.

White, R. E. III, and Sarazin, C. L. 1987b, *Ap. J.*, **318**, 621.

White, R. E. III, and Sarazin, C. L. 1987c, *Ap. J.*, **318**, 629.

White, R. E. III, and Sarazin, C. L. 1988, *Ap. J.*, **335**, 688.

White, R. E. III, and Sarazin, C. L. 1989, preprint.

The Interstellar Medium in Active Galaxies

Julian H. Krolik
Johns Hopkins University

ABSTRACT. Simple estimates based on the observed ionizing luminosities of active galactic nuclei suggest that they should have profound influence on the interstellar media of their host galaxies. Surprisingly, they do not. The simple estimates fail because they do not adequately account for the complicated exchange of mass between the stellar population and the interstellar medium, and because they do not include any obscuration by dusty material close to the nucleus. Recent observational work has demonstrated the existence of obscuration close to the nucleus; recent theoretical work has clarified the physical properties of this material. Because the obscuration is not complete, the ionizing luminosity does generate extended emission line radiation from the host galaxy's interstellar medium along certain preferred directions. Future work will use these extended emission line regions as diagnostics of interstellar conditions in the inner parts of active nuclei. In this regard, several speculations are presented about the dynamics of this gas and what role it may play in fuelling the central nucleus.

1. Some Simple Estimates

When thinking about the interstellar media of active galaxies, presumably what we wish to learn is how the interstellar gas in these galaxies is different from that in ordinary galaxies. The active nucleus can alter the state of its host's interstellar gas, but the state of the host's interstellar gas can also influence the character of nuclear activity. Both effects can lead to correlations between nuclear activity and properties of the host's interstellar medium, but, of course, with the opposite sense of causality. Historically, it has been much easier to analyze how the nucleus should affect the host's interstellar medium than the other way around, so much of this review will take that point of view. However, it is important to bear in mind that considering the question from the opposite direction may be at least as important. In particular, it is possible (see §5.2) that the character of the host's interstellar medium may determine whether activity occurs at all.

The most obvious way in which an active nucleus can affect its host galaxy's interstellar medium is by ionizing it. The Strømgren sphere associated with a quasar inside a garden-variety spiral galaxy could in principle be very large:

$$R_s \simeq 27 L_{ion,46}^{1/3} n^{-2/3} \text{kpc},\qquad(1)$$

239

H. A. Thronson, Jr. and J. M. Shull (eds.), The Interstellar Medium in Galaxies, 239–255.
© 1990 *Kluwer Academic Publishers.*

where the ionizing luminosity (the luminosity between 1 and 1000 Rydbergs) is normalized to 10^{46}erg s^{-1}, and the interstellar gas is assumed to have a uniform density of $n$cm^{-3}.

A similar, but slightly more sophisticated, estimate arises from consideration of the ionization balance for elements other than H. Define the ionization parameter $\Xi \equiv J_{ion}/(nkTc)$, where J_{ion} is the mean intensity of ionizing radiation and T is the gas temperature. Then, more or less independent of the shape of the ionizing spectrum (at least within the range of different spectral shapes observed in active nuclei) and likewise more or less independent of the elemental composition (so long as H dominates), a warm ($\sim 10^4$K) photoionized equilibrium can be maintained only if $\Xi \leq \Xi_c^* \simeq 10$ (Krolik, et al. 1981). If we again imagine a quasar inside a spiral galaxy, the typical Ξ is very large compared to Ξ_c^*:

$$\Xi \simeq 2 \times 10^3 L_{ion,46} r_{kpc}^{-2} p_4^{-1}, \tag{2}$$

where the distance to the nucleus r is scaled in kiloparsecs and p is the pressure due to H nuclei in units of K cm^{-3}. Indeed, $\Xi > \Xi_c^*$ out to $r \simeq 14 L_{ion,46}^{1/2} p_4^{-1/2}$kpc. On the basis of this estimate we would predict that essentially the whole interstellar medium of a galaxy housing a quasar would be completely ionized and very hot.

If fully-stripped plasma is hydrostatic and has time to come to thermal equilibrium with a radiation bath, its equilibrium temperature is the Compton temperature $T_c \equiv \int d\nu J_\nu h\nu/(4J)$. For typical active nuclei, $T_c \sim 3 \times 10^7$K, with at least a factor of 3 dispersion over the population. But the escape temperature T_{grav} in a spiral galaxy is generally only $\sim 4 \times 10^6$K, so there would be no reason for this gas to stay bound to the galaxy. In this situation the temperature of the gas is limited by a balance between continued Compton heating and adiabatic expansion cooling (Balbus and McKee 1982):

$$T_{ch} \simeq 2 \times 10^6 L_{46}^{2/3} \left(\frac{T_c}{3 \times 10^7 \text{K}} \right)^{2/3} r_{kpc}^{-2/3} K, \tag{3}$$

where now the relevant radiation to figure in the luminosity is the complete spectrum, not just the ionizing part. Since $T_{ch} \sim T_{grav}$ here, we would expect a substantial wind to blow, stealing mass from the interstellar medium at a very high rate (Begelman 1985):

$$\dot{M} \sim 130 L_{46} M_\odot \text{ yr}^{-1}. \tag{4}$$

Subtler effects could also help destroy the interstellar medium of the host galaxy. As Chang, et al. (1987) pointed out, radiation pressure on dust grains in the interstellar gas could overwhelm gravity:

$$\frac{g_{rad}}{g_{grav}} \simeq 2 L_{opt,46} r_{kpc}^{-1} \left(\frac{N}{3 \times 10^{21} \text{cm}^{-2}} \right)^{-1}, \tag{5}$$

where now the relevant luminosity is the range in the optical and ultraviolet effectively scattered by the dust grains, N is the column density of H atoms in the interstellar medium measured in the plane of the galaxy, and a normal dust/gas ratio has been assumed.

The picture, then, which existed in most people's minds as of a few years ago was that active nuclei, especially at the high end of the luminosity distribution, would have drastic effects on their host galaxies' interstellar media. The gas would be ionized, heated above escape temperature, and pushed out by radiation pressure at a rate far greater than it could be replenished by stellar processes.

2. The Actual Interstellar Media of Local Active Galaxies: An Observational Interlude

The only problem with this very clear picture is that it doesn't seem to describe the active galaxies we see. Figure 1 is reproduced from Wilson, *et al.* (1985), and shows how, even within 1 kpc from the nucleus of the narrow line X-ray galaxy NGC 5506, there remain large quantities of warm ($\sim 10^4$K) interstellar gas.

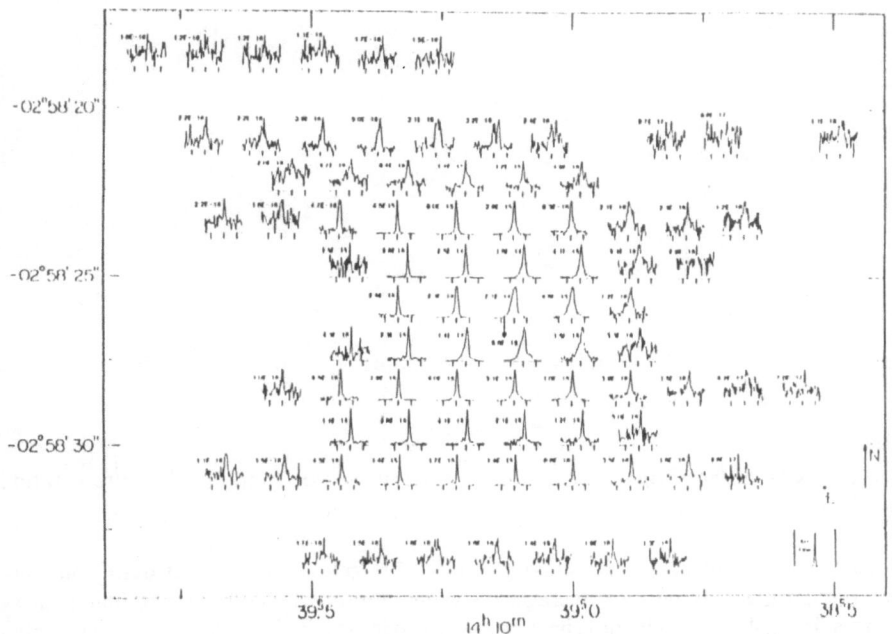

Figure 1. A map of the [OIII]5007 line profiles in NGC 5506. The distance between vertical ticks is $220h^{-1}$pc, where h is the usual abbreviation for the Hubble constant in units of 100km s^{-1} Mpc^{-1}.

Cool gas also seems to have no problem surviving in active galaxies. Figure 2 (from Meixner, *et al.* 1989) shows that substantial molecular gas (represented by CO) as well as considerable photoionized gas (attested to by Hα) exists very close to the nucleus of the type 1 Seyfert galaxy NGC 3227. Nearly all the CO is contained within $500h^{-1}$pc of the nucleus, and half lies within $50h^{-1}$pc. Using the conventional H$_2$/CO conversion (while acknowledging its considerable uncertainty), $2 \times 10^9 M_\odot$ of H$_2$ is associated with the observed CO. This is an order of magnitude greater than the quantity of molecular gas this close to the center of our own Galaxy. In another type 1 Seyfert galaxy, NGC 7469, there is so much CO emission that the associated H$_2$ is about a third of the dynamical mass within 1 kpc of the galactic center (Meixner, *et al.* 1989).

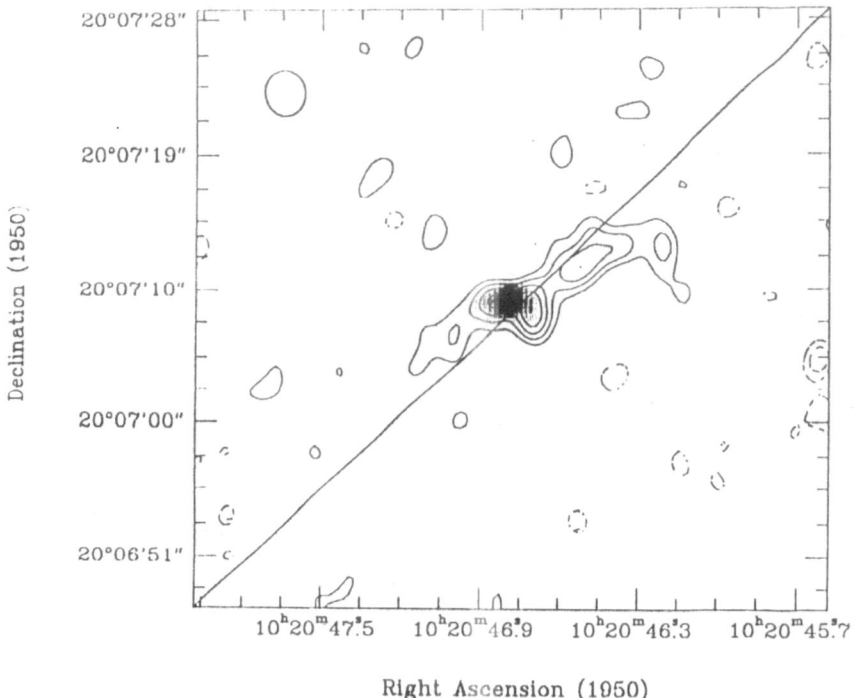

Figure 2. CO surface brightness contours superimposed on a gray-scale representation of the Hα surface brightness in NGC 3227.

Such large quantities of molecular gas in Seyfert galaxies are in fact quite common. Type 1 Seyfert galaxies have, on average, the same ratio of CO luminosity to blue light as ordinary spiral galaxies, while in type 2 Seyfert galaxies this ratio is twice its normal value (Heckman 1989). We must conclude, then, that, contrary to expectations, Seyfert galaxies have at least as much cool gas as ordinary spirals.

3. How Can the Simple Estimates Have Been So Wrong?

Clearly, at least one major consideration was omitted from the simple estimates. In fact, there were at least two.

3.1 THE MASS BUDGET

The first omission was a sufficiently close look at the mass budget of the interstellar medium, particularly the mass exchange between stars and gas. Because nuclear activity may well be short-lived in any one galaxy (or at least episodic), it is important to specify the initial quantity of gas in the interstellar medium. Particularly at high redshift, where the most luminous quasars are found, there is good reason to think the initial quantity of gas may be considerably larger than in local spiral galaxies. All the mass which ultimately ends up in stars in a spiral galaxy existed in gaseous form in the disk initially, though the ratio of gas mass to stellar mass depends on how fast the galaxy accumulates mass relative to its rate of gas depletion by star formation. Even at a loss rate of $100 M_\odot$ yr^{-1}, an initial inventory of $10^{10} M_\odot$ could survive a plausible quasar lifetime of 10^8 yr.

Moreover, although in the long-run star formation is a net loss of mass to the gas phase because at least some of the mass gets locked up in stable remnants, a large portion of the mass used initially to make stars is ultimately returned to the interstellar medium, and at some delay. Thus, depending on the star-formation history and the stellar initial mass function, for comparatively short times the net flow of mass between stars and gas could have *either* sign and (in extreme cases) a magnitude of up to $100 M_\odot$ yr^{-1}. Simply changing from a case with no current star formation and a Population II stellar mix to one with no current star formation and predominantly Population I stars alters the mass return rate by an order of magnitude, and substantially diminishes the region of interstellar medium stripped by an active nucleus (Shanbhag and Kembhavi 1988).

Adding yet another layer of complexity, there can be radial flows of gas both into and out of the regions heated by the nuclear radiation. For example, the estimate of equation 2 took the pressure in the interstellar medium throughout the galaxy to be comparable to the pressure in the midplane of our own Galaxy at the solar circle. In real galaxies, substantial radial and vertical pressure gradients exist; for example, the pressure a few parsecs from our own Galactic Center is $\sim 3 \times 10^7$ K cm^{-3} (Genzel and Townes 1987), while the vertical pressure scale height is generally ~ 100 pc. Consequently, it would be quite possible for Ξ in the mid-plane of a Seyfert galaxy to be less than Ξ_c^* even though several scale heights up the gas pressure is low enough for a wind to take off. In that case one could easily imagine replenishment by radial inflow in the midplane even while losses occur at higher altitude.

3.2 SHADOWING

The second major omission was the possibility of obscuration close to the nucleus. Although there had been several foreshadowings of this idea (Osterbrock 1978, Lawrence and Elvis 1982), the first clear demonstration that substantial obscuration exists close to the nuclei of active galaxies was by Antonucci and Miller (1985). Prior to their work, Seyfert galaxies had been classified primarily by the relative strength of their broad (*i.e.*, several thousand km s^{-1} wide) and narrow (*i.e.*, several hundred km s^{-1} wide) emission lines, type 1 galaxies having strong broad lines, type 2 having essentially none. In addition, there are secondary correlations involving the strength of [OIII]5007 compared to the Balmer lines (large in type 2, small in type 1), the fraction of the bolometric luminosity radiated in X-rays (substantial in type 1, negligible in type 2), and the radio to (non-stellar) optical luminosity ratio (a factor of 30 greater on average in type 2's than type 1's). Upon obtaining the *polarization* spectrum of the archetypical type 2 Seyfert galaxy NGC 1068, Antonucci and Miller discovered that in polarized light NGC 1068 had all the optical characteristics of a type 1 galaxy. Moreover, after a careful subtraction of starlight from the total flux spectrum, they showed that the fractional linear polarization was quite high ($\simeq 15\%$), and independent of wavelength, including the broad components of the emission lines. Their explanation of these facts—and none other has yet been proposed—is that NGC 1068 possesses a type 1 nucleus hidden by a very optically thick torus, and the only way we can see any indication of this type 1 nucleus is by the small fraction of its light which electron scatters off a warm ($< 10^6$ K) plasma filling the hole of the torus and extending some ways above its top. Because this scattering region is not circularly symmetric on the sky, the reflected light is linearly polarized.

Since their discovery, considerable supporting evidence for this picture has emerged. Miller and Goodrich (1989) have measured the polarization spectra of a number of other type 2 Seyfert galaxies, with similar results. Krolik and Begelman (1986) pointed out that this picture also explains two of the other type distinctions: The larger ratio of radio to non-stellar optical luminosity found in type 2 Seyfert galaxies is simply due to the radio emission being generated at radii far outside the obscuration, while the near absence of X-rays in type 2 Seyfert's can be explained if the torus is optically thick to Compton scattering. Edelson, *et al.* (1987) showed that the infrared spectra of all type 2 Seyfert galaxies seem to be thermal, whereas this is true for only some type 1 galaxies, from which Krolik and Begelman (1988) inferred that all type 2 galaxies have dusty obscuring tori which reradiate a large part of the nuclear luminosity in the infrared, while only some type 1 galaxies have such tori.

The evidence which most directly shows that obscuration of the nucleus prevents its ionizing photons from harming much of the host galaxy's interstellar medium is emission line imaging. Numerous studies (Corbin, *et al.* 1988; Pogge 1989; Haniff, *et al.* 1988) have now shown that in a great many Seyfert galaxies, particularly type 2's, there is an extended conical zone of high ionization emission line production, while in the majority of the solid angle around the nucleus little that cannot be attributed stars appears. The total line production in the extended high ionization emission line zone in type 2 Seyfert's is in general much greater than would be predicted on the basis of the directly observed non-stellar spectrum, suggesting that the ionizing luminosity seen from those directions is much greater than we are able to see (Wilson, *et al.* 1988; Pogge 1989).

Recently this paradigm of nuclear obscuration has been extended to include radio galaxies. Examples have been found in which photoionized gas extends along the radio axis of radio galaxies, but is clearly not directly associated with the radio emission (van Bruegel, *et al.* 1985; McCarthy 1988; Tadhunter, *et al.* 1988; Baum, *et al.* 1988). The most natural interpretation of this geometry is that the ionizing luminosity which drives the line emission escapes along the radio axis, but very little emerges in our line of sight. Barthel (1989) has suggested that this arrangement is generic to radio galaxies, and that if we viewed them along the radio axis, we would call them quasars rather than radio galaxies. In support of this argument, he points out that among all identified 3C sources with redshifts between 0.5 and 1, 30% are quasars, while the remainder, which show no nuclear optical emission at all, are called radio galaxies. This population ratio is easily consistent with an inner torus which obscures the majority of the solid angle around an active nucleus. The redshift distributions of the two subsets are similar, but the linear sizes of the radio lobes in the quasars are systematically smaller than for the radio galaxies, just as would be expected if they are seen more nearly on-axis.

The proof of this proposition would be, of course, if the infrared luminosity of the radio galaxies were comparable to the bolometric luminosity of the quasars. Unfortunately, these active galaxies are too far away for IRAS to be able to do much more than detect the brightest of them. The detected fractions of the 3C radio galaxies and quasars are similar, consistent with this picture, but stronger evidence is necessary to confirm (or destroy) this idea. If it is correct, we will have achieved a considerable unification of the various subclasses of active galactic nuclei all on the basis of toroidal obscuration and viewing angle.

4. The Nature of the Obscuring Material

The preceding section demonstrated from an observational point of view how important obscuration is in determining (*i.e.*, limiting) the influence of an active nucleus on its host galaxy. We could simply proceed on to discuss what happens in those regions with an unimpeded view of the nucleus, but the obscuring matter is of interest in its own right. For a start, it is interesting from the point of view of AGN studies because it clearly plays an important role in determining how an active nucleus appears to the outside world. In fact, it is likely that the obscuring material may also be an important waystation for the accretion fuel on its way in toward the central black hole (Krolik and Begelman 1988), but that is another story. However, it also has some intrinsic interest in this context, for it is the first interstellar matter encountered by the nuclear radiation, and its inner portion must surely be the portion of the host galaxy's interstellar medium most strongly affected by the nucleus. This section will be devoted to reviewing recent progress in understanding the physical state of the obscuring matter.

4.1 LOCATION OF THE OBSCURING MATERIAL

The first question to be considered is the location of the obscuring material. Antonucci and Miller (1985) found that the broad emission lines in NGC 1068 are polarized exactly the same as the continuum, but the narrow emission lines are quite different. In the scattering picture, this means that the solid angle through which photons can escape from the broad line region is essentially the same as for the continuum, while the narrow line region is outside the region of obscuration altogether. Thus, the sizes of the two line emitting regions are bounds for the scale of the obscuring matter.

The best-developed method for estimating the sizes of emission line regions is through photoionization modelling. Using this technique, one typically finds that broad line regions are $\sim 0.05 L_{ion,44}^{1/2} pc$ across, while the narrow line region is $\sim 10^3 \times$ larger. Here the ionizing luminosity of the nucleus has been scaled to $10^{44} erg\ s^{-1}$, a typical value for a Seyfert galaxy. The estimate of the broad line region size can be tested and refined by monitoring line and continuum flux variation (Blandford and McKee 1982), but results to date are still controversial (Peterson 1988, Edelson and Krolik 1988). Thus, a likely distance from the nucleus to the obscuring matter is somewhere in the neighborhood of several parsecs.

This is a particularly interesting location, for it is here that the gravitational potential due to the stars in the galaxy becomes comparable to the gravitational potential due to the central nucleus. In terms of the (unknown) ratio of the total luminosity to the Eddington luminosity, the radius at which the two contributions to the potential would be equal is

$$r_{eq} \simeq 2.5 \frac{L_{44}}{L/L_E} \left(\frac{v_*}{250 km\ s^{-1}} \right)^{-2} pc, \qquad (6)$$

where v_* is the characteristic orbital speed due just to the stellar gravity. Outside r_{eq} the orbital frequency is a relatively slow function of r, while inside it is Keplerian, $\propto r^{-3/2}$. This change in the rate of shear may have important consequences for the dynamics of the obscuring material (Krolik and Begelman 1988).

It is also worth pointing out in this context that the orientation of the torus (as indicated by extended high ionization line emission) appears to be entirely uncorrelated with the axes of the galaxy on large scales (Morris, *et al.* 1984; Baldwin, *et al.* 1987;

Antonucci and Miller 1985; Pogge 1989), although it *is* well-correlated with the kiloparsec-scale radio axis (Antonucci and Miller 1985; Wilson and Heckman 1985; Unger, *et al.* 1987). We do not know whether the lack of correlation with the galactic axes comes about because the material came from a source outside the galaxy, or because it was subjected to forces directed out of the orbital plane at some earlier point in its career (see §5.2). Curiously, the molecular torus a few parsecs from the center of our own Galaxy is also tilted with respect to the galactic plane (Genzel and Townes 1987). In a similar vein, the correlation that does exist with the radio axes may be due to a direct dynamical connection (*e.g.*, the radio jet punches a hole in the obscuring material, creating the toroidal geometry), or a more indirect relationship (*e.g.*, the accretion fuel shares the same direction of net angular momentum as the obscuring material, spins up the black hole, and the black hole orientation determines the axis of the radio jet).

4.2 OPACITY AND GEOMETRY OF THE OBSCURING TORUS

Next we consider the nature of the opacity responsible for blocking the nuclear radiation. For photons from the mid-infrared through the extreme ultraviolet, the dominant opacity source is probably dust. Dust is, of course, very efficient at scattering and absorbing optical and ultraviolet photons, but its presence is made even more plausible by the large thermal infrared luminosities of most type 2 Seyfert galaxies (Edelson, *et al.* 1986). Unfortunately, in nearly every example, these galaxies are unresolved in the 10μ band where the flux peaks, so variability is the only way to separate the nucleus from the host galaxy. However, in the one case bright enough and near enough to work on (NGC 1068 again), the contributions from the nucleus and the rest of the galaxy are roughly comparable (Tresch-Fienberg *et al.* 1987). Thus, dust seems at least a good working hypothesis.

From $\sim$ 50eV up to $\sim$ 10keV photoionization dominates the opacity of astrophysical gases; above 10keV, Compton scattering is the principal contributor. At such high energies, the Compton scattering cross section per electron is independent of whether the electrons are bound to atoms or free, so the requirement of a significant Compton depth in order to stop X-rays places no constraint on the physical state of the obscuring matter. It is worth noting that if the dust/gas ratio is the same as in the local interstellar medium, a Compton optical depth $\sim$ 1 is accompanied by $\sim$ 1000mag extinction in the V band (Krolik and Begelman 1986)!

We infer the geometric thickness of the obscuring torus by comparing the numbers of type 1 and type 2 Seyfert galaxies—if all Seyfert galaxies had tori, then the opening (solid) angle of the torus would be $4\pi\times$ the fraction of type 1 galaxies in a sample unbiased with respect to Seyfert type. Since it is possible that there are some type 1 Seyfert galaxies with no tori at all, the best measure of the torus's opening angle is the fraction of type 1 galaxies in the Seyfert sample produced by the extended CfA redshift survey (Huchra and Burg 1989) culled to include only those with thermal IR spectra as determined by IRAS (Edelson, *et al.* 1986). That fraction is $\simeq 1/4$ (Krolik and Begelman 1988), so we conclude that the torus is quite thick geometrically as well as optically.

As was pointed out in the previous subsection, v_*, the characteristic orbital speed in the neighborhood of the obscuring material, is probably $\gtrsim$ 100km s^{-1}. In order for the torus to be geometrically thick, its material must have random speeds in the vertical direction which are almost this great. If these random speeds are the thermal motions of atoms, the temperature would be $\sim 10^6$K and it would be impossible for dust grains to survive. Therefore, the material must instead be clumped into clouds whose internal temperatures are low enough to permit dust grains to exist, but which are moving vertically with highly supersonic speeds.

A cartoon of the region of obscuration (called "the intermediate zone" by Krolik and Begelman 1986) is shown in Figure 3.

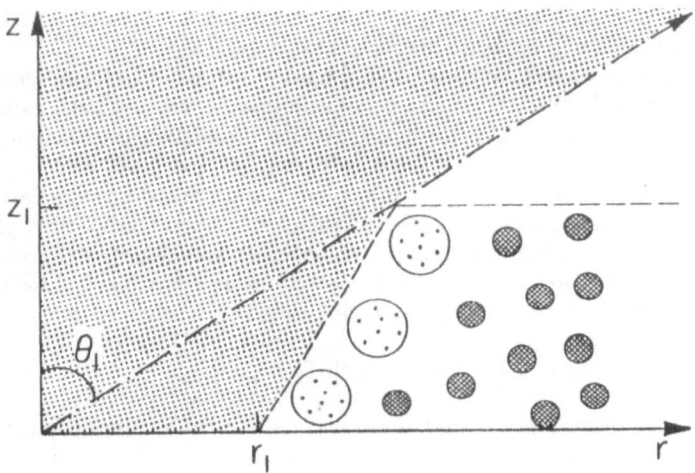

Figure 3.   A schematic representation of the intermediate zone. The actual structure is symmetric with respect to rotations around the vertical axis. When viewed in the equatorial plane, this would be called a type 2 Seyfert galaxy; when viewed along the axis, it would be seen as a type 1. The "central engine" is at the origin, the shadowed clouds of the obscuring torus are the cross-hatched circles, evaporating clouds are the larger circles with interior dots, and the warm reflecting plasma is the shaded area filling the central hole of the torus and extending some ways above its top.

4.3 THE WARM REFLECTING PLASMA

Krolik and Begelman (1986) showed that the plasma responsible for reflecting a small portion of the nuclear light into our line of sight could be understood as a Compton-heated wind. At the inner edge of the torus, $\Xi > \Xi_c^*$, so material is rapidly photoionized and heated. It ultimately achieves a temperature $\sim 10^6$K as it expands upward through the central hole. This picture predicts a Thomson optical depth through the plasma ~ 0.1, in rough agreement with what is observed, and a total mass loss rate of $\sim 0.5 M_\odot$ yr^{-1}.

The following year, Krolik and Kallman (1987) noted that this model made a very striking prediction: type 2 Seyfert galaxies should be very weak continuum X-ray sources ($L_x \sim (10^{-3} - 10^{-2})L_{bol}$), but they should be strong emitters of the Fe Kα line. Independent of everything but the Fe abundance, the equivalent width of this line should be $\simeq$ 1keV. Their argument was that the Fe K-edge opacity is always comparable to the Thomson opacity (provided the Fe isn't totally stripped), so that $\tau_{FeK} \sim \tau_T \sim 0.1$. Consequently, roughly equal numbers of photons at the Fe K edge are scattered and absorbed, with a fixed fraction of those absorbed producing Kα photons by fluorescence. Moreover, the ionization conditions required in the Krolik and Begelman (1986) model for the reflecting plasma imply that Fe is sufficiently highly-ionized (between FeXVII and FeXXIII) that the energy of the Kα line can be distinguished from that of cold iron.

Fortunately, experimental test followed closely upon prediction. *Ginga* observations of NGC 1068 have now shown that this is exactly what happens. The equivalent width of Fe Kα is 1.3 keV, and its energy is closely consistent with the predicted ionization state (Koyama 1989).

These observations thus give strong backing to the picture of the intermediate zone developed thus far: an obscuring torus optically thick enough to stop hard X-rays; plasma on the toroidal axis that is highly, but not totally ionized; and reflection by electron scattering. In addition, they give us a measurement of the Fe abundance close to the center of this Seyfert galaxy—a few times greater than solar.

4.4 PHYSICAL STATE OF THE OBSCURING MATTER

We have already been able to infer a number of properties of the obscuring torus from negative evidence—the absence of any photons coming directly from the nucleus in type 2 Seyfert galaxies. It would be helpful if *positive* diagnostics could be found as well. To that end, Krolik and Lepp (1989) have studied the physical state of the obscuring material, calculating its ionization state and temperature, and predicting emission line strengths.

They found that the cool, dusty matter of the torus is in a rather unusual state. From the inference that it is thick enough to reflect hard X-rays, it immediately follows that the only heat source in the bulk of the torus is the fraction of hard X-rays able to penetrate that far. Soft X-rays and ultraviolet photons are stopped in a thin layer at the inner edge of the torus. Similarly, assuming a normal dust/gas ratio implies a very large extinction, and significant opacity even deep into the infrared: only wavelengths longer than $50 - 100\mu$ can be assured of leaving the torus.

The energy flow starts with electron recoil upon scattering a hard X-ray. These fast electrons create large numbers of electron secondaries with energies a few times the ionization potential of H_2. Some of the energy in the secondaries goes into exciting bound-bound electronic transitions in H_2 and He, but the photons resulting from these excitations face such a large extinction in order to escape the torus that essentially all their energy is converted into a thermal IR continuum. When the secondaries fall below the electronic transition threshold, they give up the rest of their energy by exciting vibrational transitions in H_2 and, to a lesser degree, rotational transitions in CO. Because the dust opacity is so great, only those H_2 vibrational line photons created within $\sim 10^{22}$cm^{-2} of the torus's outside surface can escape; the rest are also absorbed by dust grains and reradiated in the IR continuum. Thus, the principal line photons escaping from most of the volume of the torus come from CO rotational transitions.

Even though the torus shields itself from irradiation by ultraviolet and soft X-ray photons from the nucleus, the heating and ionization rates are still quite high by comparison with ordinary interstellar molecular clouds. Typical temperatures are ~ 1000K, while the typical ionized fraction is $\sim 10^{-3}$. Because the temperature is so high, the CO rotational ladder is populated up to $J \simeq 60$, and most of the CO luminosity comes out in the highest J lines. In addition, because the column density is so great, the $^{13}C^{16}O$ and $^{12}C^{18}O$ lines are optically thick up to $J \simeq 20$, and as bright as the corresponding $^{12}C^{16}O$ lines.

As Table 1 shows, many Seyfert galaxies have now been detected in the 2μ vibrational lines of H_2. Some of this flux should be due to the outer layers of the obscuring torus, but it is probably a minority contributor. Compared to the intrinsic X-ray luminosity, Krolik and Lepp (1989) predict the total H_2 luminosity to be in the range $10^{-5} - 10^{-4}$, with type 1 Seyfert galaxies relatively brighter in H_2 lines because we see a warmer surface of the torus. The brightest several individual lines each carry about 8 – 10% of the total power. In addition, because the density in these clouds is probably $\sim 10^7 cm^{-3}$, the H_2 level populations are very well thermalized, and the ratios between lines corresponding to different vibrational quantum number transitions should have the usual thermal ratios. These predictions can be compared to the H_2/L_x ratios compiled by Kawara, et al. (1989). The single brightest 2μ line in their 7 Seyfert galaxies has a luminosity $\sim 3 \times 10^{-4}L_x$, or about 30× brighter than Krolik and Lepp (1989) predict from the torus. It is possible that the dust/gas ratio is smaller in the torus than in the local interstellar medium: the escaping H_2 luminosity rises in direct proportion to the column density of gas per unit dust extinction. It is likewise possible that the very hard X-ray band (around 100 keV) is stronger in these galaxies than would be predicted on the basis of a $\nu^{-0.7}$ extrapolation from the Einstein band fluxes; that, too, would increase the H_2/L_x ratio (particularly since the X-ray luminosity is usually measured near 1 keV). If these effects, or some other, could combine to make up the factor of 30, the nearly constant ratio between the H_2 vibrational luminosity and the X-ray luminosity would be readily explained; otherwise, there must be some regulatory mechanism tying the radiation losses of molecular clouds out in the host galaxy to the X-ray luminosity of the nucleus.

Krolik and Lepp's calculations also predict a new *negative* diagnostic of the obscuring torus: free-free optical depth at radio frequencies. Because the ionized fraction and atomic density are both relatively high, the free-free optical depth should be substantial:

$$\tau_{ff} \simeq 0.1 \frac{p_{10} N_{24}}{\nu_{10}^2}, \qquad (7)$$

where the pressure is referred to 10^{10} K cm^{-3}, the column density N of the torus is in units of 10^{24}cm^{-2} and the radio frequency ν is in units of 10GHz. An immediate prediction, then, is that there should be no very compact low frequency radio components in type 2 Seyfert galaxies. This should be testable in the near future.

5. The Narrow Line Region

In times past, the narrow line region was thought of as another bit of exotica attached to the active nucleus. In fact, it is probably more productive to think of it as that portion of the host galaxy's interstellar medium irradiated by the nucleus. The division between resolved and extended high ionization line emission can now be seen as completely artificial:

Table 1

H_2 MOLECULES IN TYPE 1 SEYFERT GALAXIES AND QUASARS

Galaxy	Hot H_2
I Zw 1	< 4.38
Mrk 1014	< 4.87
NGC 1275	4.14
NGC 1365	3.25
NGC 1566	2.40
3C 120	< 3.60
Mrk 618	< 3.79
Akn 120	< 3.51
IRAS 0518-25	3.98
NGC 2992	< 2.82
Mrk 1239	< 3.40
ESO 438 -G9	3.03
NGC 3783	2.66
NGC 4151	2.35
3C 273	4.90
IRAS 1334+24	4.88
NGC 5506	2.48
Mrk 841	< 3.70
NGC 7469	4.09
NGC 7582	2.34

All figures are logarithms of the mass in solar masses. Hot H_2 is derived from the luminosity in 2μ emission lines. The data are taken from the compilation in Kawara, *et al.* (1989).

Pogge (1989) has shown that in the nearer Seyfert galaxies approximately half the total emission is resolved. Moreover, the detailed spectroscopic maps collected by Baldwin *et al.* (1987), Wilson *et al.* (1985), and Cecil (1989) show that the highly-ionized regions are also the ones which are dynamically disturbed. When HST flies, the border between nuclear and extended will change again, demonstrating clearly that such distinctions are purely instrumental. Its resolution will allow us to study regions as close to the nucleus as $5h^{-1}$pc in the nearest Seyfert galaxies, while 100pc will be resolved out to redshifts $\simeq 0.07h$.

Given this somewhat different point of view, narrow line region studies undergo a change in purpose. No longer a pursuit whose goal is simply to find the physical state of the stuff responsible for the light, this subject acquires two new directions: to serve as a "dentist's mirror", allowing us to see into the active nucleus from directions other than our own lie of sight; and as a diagnostic of interstellar conditions which can tell us both about how active nuclei alter their host galaxies, and possibly how the host furnishes accretion fuel to the nucleus.

5.1 PHYSICAL CONDITIONS

Because considerable effort has already been expended on analyzing the physical conditions in the narrow line region, we already know a good deal about this portion of the interstellar media of galaxies with active nuclei. The most striking contrasts between this gas and ordinary interstellar gas are its pressure and its ionization state. From the relative strengths of [OII]3727, [OIII]5007, and [OIII]4363 we infer a pressure $\sim 10^8$K cm^{-3}: $\sim 10^4 \times$ greater than in the local interstellar medium. The strength of CIII]1909 compared to CIV 1549 suggests $\Xi \sim 10^{-1}$ and a distance from the nucleus $\sim 100 L_{ion,44}^{1/2}$pc. In truth, the line profiles are sufficiently different from one another that one-zone models such as this are not particularly faithful to the data (Wilson and Heckman 1985), but these estimates at least serve to give a rough indication of conditions.

Given a normalization from the total line luminosity, it is possible to infer a volume filling factor for this line-emitting matter:

$$ f \sim 8 \times 10^{-5} \left(\frac{L[\mathrm{H}\beta]/L_{ion}}{10^{-3}} \right) L_{ion,44}^{-1/2} \left(\frac{\Xi}{0.1} \right)^{3/2} p_8^{-1/2} \tag{8} $$

It is clear the material is strongly clumped. Outside the clumps there may be hotter, more rarefied material of comparable pressure (Krolik and Vrtilek 1984, Filippenko and Halpern 1984). If there were no such material, the clumps would freely expand and self-destruct, but that would introduce new material into the inter-clump space. By definition, the narrow line region has an unimpeded view of the nucleus, so this newly-injected gas would rapidly rise to the characteristic temperature (*cf.* equation 3) $T_{ch} \simeq 4 \times 10^5 L_{44}^{1/3} (T_c/3 \times 10^7 K)^{2/3} p_8^{1/3} (\Xi/0.1)^{1/3}$K. T_{ch} is defined assuming an outflow which is transonic; the corresponding equilibrium density in the region would be $n_h \sim 500 (L[\mathrm{H}\beta]/10^{-3} L_{ion}) L_{44}^{-1/6} (\Xi/0.1)^{5/6} p_8^{5/6} N_{cl,21}^{-1}$cm^{-3}, where we have normalized the cloud column density N_{cl} to 10^{21} cm^{-2}because Halpern and Steiner (1983) and Ferland and Netzer (1983) have found that the ratio of [OI]6300 to [OIII]5007 is best reproduced by clouds of that thickness. The expected intercloud pressure is then:

$$ p_h \sim 2 \times 10^8 \left(\frac{L[\mathrm{H}\beta]/L_{ion}}{10^{-3}} \right) L_{44}^{1/6} N_{cl,21}^{-1} (\Xi/0.1)^{7/6} p_8^{7/6}\mathrm{K \ cm^{-3}}. \tag{9} $$

In other words, if there were no pressure-confining medium in the narrow line region, the resulting cloud destruction would result in the formation of one. However, unless there is an additional source of material, the cloud lifetime produced by this pressure confinement cannot be too much greater than the free expansion time. This conclusion should be quite broadly applicable, for it depends on only the 1/6 power of luminosity.

5.2 DYNAMICS

The study of interstellar dynamics in normal galaxies amounts to analyzing small deviations from circular orbits (*cf.* Balbus's review in this volume). These deviations might be caused by spiral arms, or on smaller scales, by expanding supernova remnants or HII regions. In the narrow line region of active galaxies, the deviations from circular orbits are not small. Although there is evidence that the magnitude of the mean gas speed in the narrow line region of a particular active galaxy is correlated with the orbital velocity at large radius in that galaxy, the relationship between the gas velocity and the *local* orbital speed is less clear (Whittle 1989). In addition, somewhere there must be material flowing inward at a reasonable velocity in order to feed the active nucleus.

Nothing is simple about this problem. At least three forces act on the narrow line gas: the galaxy's gravity, radiation pressure from the central nucleus, and gas pressure gradients:

$$g_{grav} \sim 3 \times 10^{-7} \left(\frac{v_*}{100 \text{km s}^{-1}} \right)^2 \left(\frac{r}{100 \text{pc}} \right)^{-1} \text{cm s}^{-2}$$

$$g_{rad} \sim 6 \times 10^{-7} p_8 \text{cm s}^{-2}$$

$$g_{press} \sim 3 \times 10^{-7} T_6 \left(\frac{r}{100 pc} \right)^{-1} \text{cm s}^{-2} \tag{10}$$

In addition, axisymmetry, usually quite a good assumption in spiral galaxies, is badly broken here. The fact that the nuclear radiation only escapes in certain preferred directions means that the inner interstellar medium is only significantly heated in these directions; strong azimuthal pressure gradients result.

Beyond this, as was discussed briefly in §2, there are a number of cases in which so much gas is accumulated near the center of the galaxy that it accounts for a significant part of the gravity. Large amplitude bars are then likely because the soft equation of state of the gas keeps it very flat. This is the complementary case to the one argued by Ostriker and Peebles (1973); they called for spherical dark halos in order to stabilize flat galaxies against bars, whereas here bars may be necessary in order to account for the non-circular gas orbits. Inside corotation, gas is driven efficiently inward by bars, while outside corotation gas is driven equally efficiently outward (Schwarz 1985). Thus, these bars may share responsibility for the galaxy's nuclear activity by driving the accretion flow through this region (Shlosman, *et al.* 1989; Hernquist 1989; Norman 1988; Noguchi 1988).

By providing a new preferred plane at small radii, they may also explain the tilted tori discussed in §4.1 (Tohline and Osterbrock 1982). However, a great deal remains to be done to understand this situation better: none of the calculations to date has included gas self-gravity, which is essential to generating the bars; all simulational results are contaminated at some level by artificial viscosity; and the gas equation of state is generally approximated extremely crudely. New computational techniques are likely to allow rapid improvement in this situation but results in hand are still little more than suggestive (Hernquist 1989).

In addition to torques due to gravitational forces, there is another mechanism which may contribute to driving radial flows in the inner parts of active galaxies, but has not yet received much attention: magnetic braking. First developed in the context of star formation (Mouschovias and Paleologou 1979, 1980; Königl 1987), it is a mechanism which removes angular momentum from a gas cloud linked by poloidal magnetic field lines to external material rotating more slowly. The field lines attempt to rotate with the angular rate of the cloud in which they are anchored, but flux-freezing in the external material makes them drag behind. A torque density of magnitude $r^2 B_r B_\phi$ is then exerted on the external material, taking angular momentum from the inner cloud. In the time required for an Alfvén wave to travel outward from the cloud through material having moment of inertia comparable to the cloud, it loses roughly half its initial angular momentum.

The inner regions of host galaxies to active nuclei are in fact likely places for this mechanism to act. Gas accumulates there, drawing in the field lines which initially threaded it at larger radii. The equilibrium rotation rate declines rapidly outward, both as a result of the shape of the stellar potential and due to the significant contribution to the gravity made by the gas. If the Alfvén speed is a modest fraction of the orbital speed, this mechanism may drive radial flows as rapidly as gravitational torquing (Krolik and Meiksin 1989).

6. Summary

A few years ago it was generally thought that the interstellar media of galaxies with active nuclei would be mostly distinguished by their absence. At the very least, it was expected that the ionizing luminosity coming out of an active nucleus would ensure that any interstellar gas in the host was very hot and highly ionized. In fact, most active galaxies (or at least the nearby low luminosity ones which can be observed) seem to contain considerable quantities of gas, and large quantities of cool molecular gas can be found quite close to their nuclei.

Two reasons stand out as the principal explanations for the failure of the early predictions. Subtleties in the mass budget of the interstellar medium were ignored, and the possibility of thick obscuration close to the nucleus was overlooked. Predictions that the interstellar medium would be depleted did not take into account the possibility that the initial inventory of gas would be large enough to survive the lifetime of the active nucleus, nor did they take careful account of the varieties of star-gas interchange possible for different star formation histories and stellar populations. The first hard evidence for near-nuclear obscuration appeared four years ago in the spectropolarimetry of Antonucci and Miller (1985), but since then there has been a flood of observations finding signs of it in many active galaxies. Where it is present, of course, the active nucleus can only irradiate the host galaxy along special directions.

Rapid progress has been made in understanding the nature of that portion of the interstellar medium most affected by the nucleus, namely the near-nuclear obscuring material. Its dusty molecular clouds are probably warmer and more ionized than ordinary interstellar molecular gas, and should eventually be detectable in H_2 vibrational emission lines and high-J CO emission lines. The electron density inside these clouds is high enough to present considerable free-free opacity to radio frequencies below a few GHz, so that the obscuring material is effective against low-frequency radio emission as well as higher frequency infrared, optical, ultraviolet, and X-ray photons. Inside the toroidal obscuring material there is warm plasma whose highly-ionized Fe $K\alpha$ emission line photons have now been detected.

Moving out in the galaxy, one next encounters the narrow emission line region. Here ordinary interstellar gas is subjected to irradiation by the nuclear continuum (in those places where the nucleus is visible), and normal photoionization diagnostics can be employed. What is abnormal about this region is its high pressure and unusual dynamics, both having to do with the presence of the active nucleus. Although the pressure is probably a result of heating by the nuclear radiation, the dynamics may be both cause and effect: radial flows through the narrow line region may mark the paths of accretion fuel heading towards the nucleus. Because so little is yet known about the forces affecting this gas, the employment of spatially-resolved spectroscopy (*e.g.*, the scanning Fabry-Perot techniques of Cecil 1989) on this region promises to teach us a great deal, while HST will enable less comprehensive studies at an order of magnitude finer resolution. The new IR arrays should also be extremely valuable for penetrating regions of moderate obscuration.

On the theoretical side, the next several years' work is likely to focus on the dynamics of interstellar gas in the inner parts of active galaxies. With the advent of merged hydrodynamic and N-body simulation codes, it should be possible to explore the genesis and development of non-axisymmetric gravitational potentials in enough detail to understand what role, if any, they play in the inner regions of active galactic nuclei. At the same time, exploratory work on other dynamical elements—radiation pressure, magnetic fields, *etc.* —should elucidate what influence they may have.

7. References

Antonucci, R.R.J. and Miller, J.S. 1985, *Ap. J.* **297**, 621.

Balbus, S.A. and McKee, C.F. 1982, *Ap. J.* **252**, 529.

Baldwin, J.A., Wilson, A.S., and Whittle, M. 1987, *Ap. J.* **319**, 84.

Baum, S.A., Heckman, T.M., Bridle, A., van Bruegel, W., and Miley, G. 1988, *Ap. J. Suppl.* **68**, 643.

Begelman, M.C. 1985, *Ap. J.* **297**, 492.

Blandford, R.D. and McKee, C.F. 1982, *Ap. J.* **255**, 419.

Cecil, G. 1989 in *The Interstellar Medium in External Galaxies*, H.A. Thronson and J.M. Shull, eds. (Kluwer: Dordrecht)

Chang, C.A., Schiano, A.V.R., and Wolfe, A.M. 1987, *Ap. J.* **322**, 180.

Corbin, M.R., Baldwin, J.A., and Wilson, A.S. 1988, *Ap. J.* **334**, 584.

Davidson, K. and Netzer, H. 1979, *Revs. Mod. Phys.* **51**, 715.

Edelson, R.A. and Krolik, J.H. 1988, *Ap. J.* **333**, 646.

Edelson, R.A., Malkan, M.A., and Rieke, G. 1986, *Ap. J.* **321**, 238.

Ferland, G.J. 1981, *Ap. J.* **249**, 17.

Ferland, G.J. and Netzer, H. 1983, *Ap. J.* **264**, 105.

Filippenko, A. and Halpern, J. 1984, *Ap. J.* **285**, 458.

Genzel, R. and Townes, C. 1987, *Ann. Rev. Astron. Astrop.* **25**, 377.

Halpern, J.P. and Steiner, J.E. 1983, *Ap. J. Lett.* **269**, L37.

Haniff, C.A., Wilson, A.S., and Ward, M.J. 1988, *Ap. J.* **334**, 104.

Heckman, T.M. 1989, in *IAU Symposium 134: Active Galactic Nuclei*, D.E. Osterbrock and J.S. Miller, eds. (Kluwer: Dordrecht) p.359

Heckman, T.M., *et al.* 1989

Hernquist, L. 1989, to appear in *Annals of the New York Academy of Sciences: the 14th Texas Symposium on Relativistic Astrophysics*

Huchra, J.P. and Burg, R., 1989, preprint

Kawara, K., Nishida, M., and Gregory, B. 1989, preprint

Königl, A. 1987, *Ap. J.* **320**, 726.

Koyama, K. 1989, in *IAU Symposium 134: Active Galactic Nuclei*, D.E. Osterbrock and J.S. Miller, eds. (Kluwer: Dordrecht) p. 167

Krolik, J.H. and Begelman, M.C. 1986, *Ap. J. Lett.* **308**, L55.

Krolik, J.H. and Begelman, M.C. 1988, *Ap. J.* **329**, 702.

Krolik, J.H. and Kallman, T.R. 1987, *Ap. J. Lett.* **320**, L5.

Krolik, J.H. and Lepp, S.L. 1989, *Ap. J.* in press.

Krolik, J.H., McKee, C.F., and Tarter, C.B. 1981, *Ap. J.* **249**, 422.
Krolik, J.H. and Meiksin, A. 1989, in preparation.
Krolik, J.H. and Vrtilek, J.M. 1984, *Ap. J.* **279**, 521.
Lawrence, A. and Elvis, M. 1982, *Ap. J.* **256**, 410.
McCarthy, P.J. 1988, University of California at Berkeley Ph.D. thesis
Meixner, M., Puchalsky, R., Blitz, L., Wright, M.C.H. 1989, preprint
Miller, J.S. and Goodrich, R. 1989, in preparation (quoted in Miller, J.S. 1989, in *IAU Symposium 134: Active Galactic Nuclei*, D.E. Osterbrock and J.S. Miller, eds. (Kluwer: Dordrecht) p. 273
Morris, S., Ward, M.J., Whittle, M., Wilson, A.S., and Taylor, K. 1984, *M.N.R.A.S.* **216**, 193.
Mouschovias, T. and Paleologou, E.V., 1979, *Ap. J.* **230**, 204.
Mouschovias, T. and Paleologou, E.V., 1980, *Ap. J.* **237**, 877.
Noguchi, M. 1988, *Astr. Ap.* **203**, 259.
Norman, C.A. 1988, in *Comets to Cosmology*, A. Lawrence, ed. (Springer-Verlag: Berlin)
Osterbrock, D.E. 1978, *Proc. Natl. Acad. Sci. USA* **75**, 540.
Ostriker, J. and Peebles, P.J.E. 1973, *Ap. J.* **186**, 467.
Peterson, B. 1988, *P.A.S.P.* **100**, 18.
Pogge, R. 1989, *Ap. J.* in press.
Schwarz, M.P. 1985, *M.N.R.A.S.* **212**, 677.
Shanbhag, S. and Kembhavi, A. 1988, *Ap. J.* **334**, 34.
Shlosman, I., Frank, J., and Begelman, M.C. 1989, *Nature* **338**, 45.
Tadhunter, C.N., Fosbury, R.A.E., di Serego Alighieri, S., Bland, J., Danziger, I.J., Goss, W.M., McAdam, W.B., and Snijders, M.A.J. 1988, *M.N.R.A.S.* **235**, 405.
Tohline, J. and Osterbrock, D.E. 1982, *Ap. J. Lett.* **252**, L49.
Tresch-Fienberg, R., Fazio, G.G., Gezari, D.Y., Hoffman, W.F., Lamb, G.M., Shu, P.K., and McCreight, C.R. 1987, *Ap. J.* **312**, 542.
Unger, S.W., Pedlar, A., Axon, D.J., Whittle, M., Meurs, E.J.A., and Ward, M.J. 1987, *M.N.R.A.S.* **228**, 671.
van Bruegel, W., Miley, G., Heckman, T., Butcher, H., and Bridle, A. 1985, *Ap. J.* **290**, 496.
Whittle, M. 1989, in *IAU Symposium 134: Active Galactic Nuclei*, D.E. Osterbrock and J.S. Miller, eds. (Kluwer: Dordrecht) p.349
Wilson, A.S., Baldwin, J.A., and Ulvestad, J.S. 1985 **291**, 627.
Wilson, A.S. and Heckman, T.M. 1985 in *Astrophysics of Active Galaxies and Quasi-Stellar Objects*, J.S. Miller, ed. (University Science Books: Mill Valley) p. 39.
Wilson, A.S., Ward, M.J., and Haniff, C.A. 1988, *Ap. J.* **334**, 121.

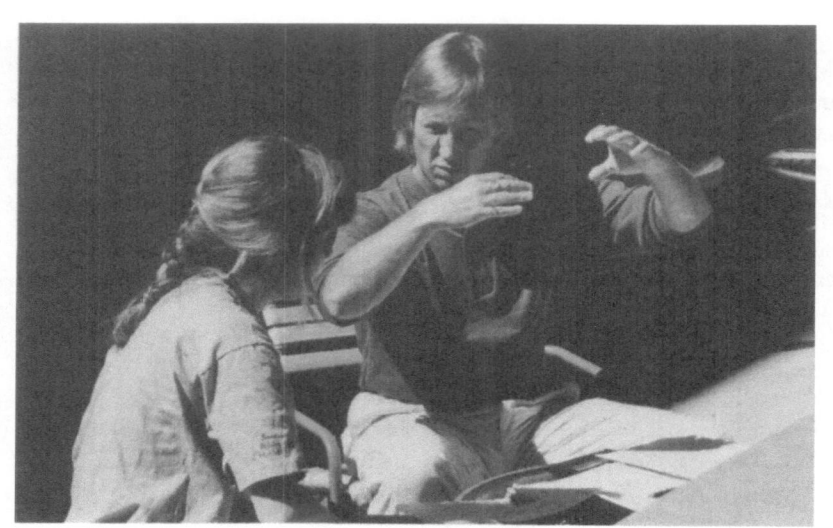

Abundances in Extragalactic H II Regions

Harriet L. Dinerstein
University of Texas at Austin

ABSTRACT. Giant H II regions can be observed out to large distances. They trace recent star formation and, through the analysis of their chemical composition, previous star formation activity. Systematic effects such as metallicity gradients across the disks of spirals and the correlation of metallicity with total galaxy mass offer important constraints on galaxy evolution. The helium abundance in regions where the gas has experienced little stellar nuclear processing provides an important test of cosmological theories. Some of the current uncertainties in interpreting nebular spectra are likely to be resolved by the next generation of ground-based, airborne, and spaceborne telescopes.

1. Introduction

1.1. H II REGIONS IN THE CONTEXT OF THE ISM

The original two-phase model of the interstellar medium was the H II/H I region dichotomy introduced by Strömgren (1939). He showed that photoionized gas near hot stars is segregated into physically distinct volumes, separated from their neutral environment by sharp boundaries. The study of ionized (formerly "gaseous") nebulae differs in many ways from that of most of the other components of the ISM, for both historical and physical reasons. Because H II regions are the only form of interstellar material which emits strongly in the optical spectral region, there is a much longer and richer history of observations and theory for them than for the other thermal phases of interstellar matter. (The discovery of nebular emission lines dates back to the mid-1860's.) In contrast, most of our information about the other phases depends on recently developed technologies such as radio, infrared, and X-ray astronomy.

Optical observations of H II regions provide fairly complete information about their elemental composition. From their spectra, abundances relative to hydrogen can be estimated for nearly all of the most common elements, particularly He, N, O, Ne, Ar, and S. (Note that oxygen alone constitutes nearly 50% by mass of the elements heavier than helium.) Furthermore, ionized nebulae are remarkably efficient machines for converting ultraviolet continuum energy from OB stars, originally diluted over wide bandpasses, into a few narrow, intense, optically-thin emission lines. The intrinsic emissivities of these lines are easy to calculate in principle, although they are sensitive to the local thermodynamic state of the gas (n_e and T_e). On the other hand, the thermal parameters can also be

H. A. Thronson, Jr. and J. M. Shull (eds.), The Interstellar Medium in Galaxies, 257–285.
© 1990 *Kluwer Academic Publishers.*

determined from the spectra, using diagnostic line-intensity ratios. In this way, H II regions can be used to measure element abundances in the (present-day) gas of distant galaxies. The sample of extragalactic H II regions studied so far has metal abundances ranging from about .02 to several times solar. This is a useful complement to studies of our own Galaxy, which contains no severely metal-deficient H II regions (except for a handful of planetary nebulae formed by stars of the halo population). In contrast, for many H II regions in the outskirts of late-type spirals and in some dwarf irregular galaxies, the process of metal enrichment by stellar nucleosynthesis is still in its early stages, providing a "window" on the early chemical evolution of galaxies. The flip side of this coin is that these low-metallicity H II regions are also presumed to have experienced only a small degree of alteration in their helium abundances due to stellar activity. Therefore, their present He/H ratios should be nearly the same as the primordial value, providing valuable tests for cosmological theories.

1.2. ENVIRONMENTS AND SYSTEMATICS

Although a comprehensive review of all the properties of extragalactic H II regions is beyond the scope of this paper, a brief discussion is needed to provide a context for the subject of abundances. The various "categories" of extragalactic H II regions are essentially lists of their environments. These include: (1) disk H II regions in spiral and irregular galaxies; (2) gassy dwarf irregular galaxies with spectra which are heavily dominated by H II regions; and (3) nuclear and near-nuclear regions sometimes called "starburst" or "hotspot" H II regions (e.g. Kennicutt, Keel, and Blaha 1989). In the present review I will concentrate on the first two categories, for which the best abundance data are available. Regions in the third group tend to have relatively strong stellar continua and to be fairly metal-rich, which make it to difficult to obtain accurate measurements of the emission lines from which abundances are determined. On the other hand, members of the first two categories are universally regarded as members of the same family. H II regions in nearby galaxies have been well-catalogued; atlases are available for the LMC, SMC, and a large number of other galaxies (Hodge and Wright 1967, 1977; Hodge and Kennicutt 1983). The star-forming dwarf irregulars are usually found by spectroscopic surveys for emission-line galaxies (see, for example, the review by Kinman 1984).

The statistical properties of the H II region populations in spiral and irregular galaxies were most recently addressed by Kennicutt (1988) and Kennicutt, Edgar, and Hodge (1989). They find that late-type galaxies have both intrinsically higher-luminosity first-ranked H II regions, and larger total numbers of H II regions after normalization by galaxy size, than do early-type spirals. Within a galaxy, the differential luminosity function of the H II regions is roughly power-law, $N \propto L^{-2 \pm 0.5}$, although some low-luminosity irregulars have an exceptional supergiant complex, and Sa-Sb galaxies are deficient in luminous regions. While the positive correlation between the luminosity of the brightest H II region and that of the parent galaxy can be understood as chiefly a sample-size effect, the dependence on morphological type is a real and separate factor. Typical large galaxies contain hundreds of optically detectable H II regions. We note here that, of all the regions detected and cataloged in $H\alpha$ or $H\beta$, it is usually the nearest and the most luminous ("giant") H II regions for which abundances are derived.

The dimensions of these "giant" extragalactic H II regions are typically 100-200 pc. They are ionized by clusters of 10^{1-4} OB stars and contain 10^{3-5} $M_\odot$ of ionized gas. Some of the best-studied regions are the 30 Dor complex in the LMC, NGC 604 in M33, and NGC 5461 and 5471 in M101. Certain selection effects must be present. Necessarily poorer spatial resolution contributes to a tendency to identify larger regions in more distant galaxies. This effect is illustrated by Israel, Goss, and Allen (1975), who compare large-beam radio measurements with optical images of the same H II regions in M101; at better resolution these regions break up into groups or chains of smaller clumps. Likewise, H II regions in dwarf irregulars are also found to have complex structure when closely examined (e.g. Hodge, Lee, and Kennicutt 1989; Davidson, Kinman, and Friedman 1989). In more distant galaxies, we will always be looking at more heterogeneous volumes; for example, a typical aperture size (4") for spectrophotometric studies corresponds to 1 pc at 50 kpc (the LMC) and 2 kpc at 100 Mpc.

1.3. INTERNAL STRUCTURE

The morphology of many giant extragalactic H II regions can be characterized to first order as a "core-halo" structure, on the basis of both optical and radio-continuum data (see, for example, the multi-wavelength study of NGC 604 by Israel *et al.* 1982). The cores are composed of dense material, often in several distinct clumps, close to the ionizing stars. The diffuse, lower-density envelopes are presumably ionized by photons escaping from the inner regions and represent the radiation-bounded edges of the Strömgren volume. Most giant extragalactic H II regions are believed to be essentially radiation-bounded (e.g. McCall, Rybski, and Shields 1985). In addition, the denser regions themselves are inhomogeneous, as seen in the recent detailed studies of NGC 5471 by Skillman (1985), and of NGC 604 by Diaz *et al.* (1987). That there are also inhomogeneities on smaller spatial scales is shown by the discrepancy between (rms) n_e values derived from recombination emission and local values determined from density-sensitive line ratios. The dense clumps are embedded in a much lower-density medium, with typical clump volume filling factors of .01 - .1 (e.g. Kennicutt 1984; McCall, Rybski, and Shields 1985). The interclump material is often treated as a vacuum in nebular models, because it does not contribute significantly to the optical emission lines.

A good deal of recent work has focused on the velocity structure of the emission lines. Giant extragalactic H II regions display supersonic velocities, which appear to correlate with Hβ luminosity. Terlevich and Melnick (1981) interpret the line-widths as virial and therefore usable for determining the local gravitational field; they also find a secondary dependence on metallicity. An alternative interpretation of the origin of the line-widths is that they are a result of stellar winds from the exciting stars, and possibly also from embedded supernova remnants (e.g. Dopita 1981; Skillman 1985). For nearby regions, it is possible to actually identify the stars which may be responsible for driving the high-velocity gas, as in a recent kinematic model of NGC 604 by Clayton (1988).

1.4. IONIZING CLUSTERS

Luminous extragalactic H II regions are ionized by OB associations. For nearby regions, the members of the stellar cluster can be distinguished individually and HR diagrams constructed (see papers in De Loore, Willis, and Laskarides 1986). The nebular ionization structure and emitted spectrum will evolve as the cluster ages and the UV radiation field diminishes and softens. Exploratory models of this kind have been calculated (e.g. Melnick, Terlevich, and Eggleton 1985; Kennicutt and Chu 1988). The process has been inverted, using the nebular spectrum to infer the age of the star cluster or "burst" (Lequeux *et al.* 1981; Copetti, Pastoriza, and Dottori 1985). A more controversial issue is whether the stellar initial mass function varies with metallicity, as proposed by Terlevich (1985).

Over the last decade it has become apparent that Wolf-Rayet stars are often present in extragalactic H II regions. Wolf-Rayet features have been seen in M 33 by D'Odorico and

Rosa (1981) and Conti and Massey (1981), and in many other regions as well (see Rosa and D'Odorico 1986 for a recent review and further references.) The frequency of Wolf-Rayet stars is higher for higher-metallicity regions (Maeder, Lequeux, and Azzopardi 1980), a point to which we return below. Wolf-Rayet stars are important in our context because they furnish metal-rich outflows which are capable of altering the chemical composition of their gaseous environment (Chiosi and Maeder 1986). Giant H II regions are also known hosts of Type II supernovae (Richter and Rosa 1984). Winds and supernovae from the massive stars can contaminate (or enrich) the local gas in H II regions in He, C, O, and other species. Evidence for such local enrichments has been sought and perhaps seen in some regions (Skillman 1985; Pagel, Terlevich, and Melnick 1986).

2. Methods of Abundance Determination

2.1. THE "CONVENTIONAL" (DIRECT) METHOD

2.1.1. Basic Steps. In this review, I will emphasize the techniques of abundance determinations and their advantages and pitfalls, rather than the latest set of results, in the hope that a critical discussion of methods will have greater longevity than a set of numbers subject to future revision. The standard method of deriving nebular abundances from emission lines is explained in great detail in references on the physics of gaseous nebulae (Aller 1984; Osterbrock 1989). The procedure is summarized by the following flow-chart.

Observed Emission Line Intensities

⇓   *{Correction for Extinction using Hydrogen Recombination Decrement}*

Reddening-Corrected Line Intensities

⇓   *{Diagnostic Line Ratios, e.g. [O III] for T_e, [S II] for n_e}*

Local Physical Conditions (n_e, T_e)

⇓   *{Ionic Level Populations and Calculated Line Emissivities}*

Ionic Abundance Ratios (O^+/H^+, O^{++}/H^+, etc.)

⇓   *{Correction for Unobserved Ions}*

Elemental Abundance Ratios (O/H, etc.)

2.1.2. Choice of T_e. A key species for nearly all nebular abundance determinations is the O^{++} ion. Figure 1 shows the energy-level structure and major transitions for the ground configuration. The intensity ratio [O III] 4363Å/5007Å (and/or 4959Å) has long been the chief diagnostic of the gas temperature in nebulae. Recent improvements in the calculated collision strengths (Mendoza 1983; Osterbrock 1989) cause systematic revisions by up to factors of two in the derived abundances (e.g. Zamorano and Rego 1985).

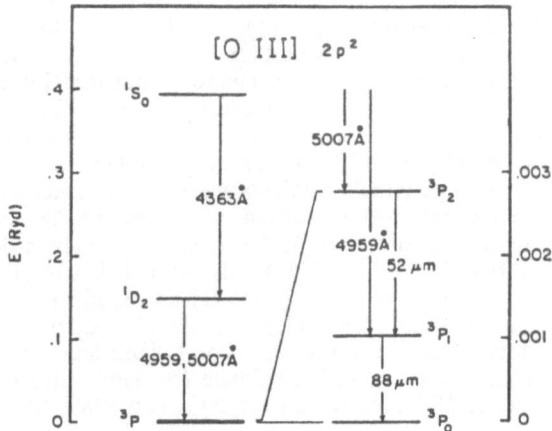

Figure 1. The energy-level diagram for the ground configuration of the O^{++} ion, with the most important transitions indicated, from Dinerstein, Lester, and Werner (1985).

A more subtle effect arises because the derived temperature is necessarily weighted by the temperature dependences of the lines. Both 4363Å, and to a lesser and different degree 5007Å, are more strongly emitted by hotter gas. For a non-isothermal nebula, this results in a tendency to overestimate T_e and underestimate O^{++}/H^+. A procedure for correcting for these effects was developed by Peimbert (1967; also see Rubin 1969), who characterized the temperature distribution in terms of a mean value T_0 and an rms "fluctuation" parameter t^2 which serves as an indicator of the amplitude of the temperature spread within the region; these variations may be either microscopic or macroscopic. In principle, measurement of two line ratios with known functional dependences on temperature yields a simultaneous solution for both T_0 and t^2, and allows one to correct the line emissivities used to derive ionic abundances. The magnitude of these corrections for reasonable values, $t^2 = .00$, .035, and .055, can be seen in Table 10 of Peimbert and Torres-Peimbert's study of Orion (1977). For N, O, Ne, and S, the elemental abundances increase by roughly .25 and .45 dex respectively (factors of about 2 and 3) if $t^2 = .035$ or .055. The corrections to ionic abundance ratios such as C/O and N/O are substantially smaller (Zuckerman and Aller 1986). It is notable that the abundances relative to hydrogen derived assuming $t^2 = .00$ are systematically lower than stellar abundances, while a modest nonzero t^2 brings them into agreement with the solar and stellar abundance scale. In principle, values for t^2 are best determined from two temperature-diagnostic line ratios which arise from the same volume, e.g. from the same ion. Such a solution has been done recently for planetary nebulae using the far-infrared lines of [O III] in combination with the optical lines (Dinerstein, Lester, and Werner 1985), who find evidence for non-zero values. Nevertheless, the correction for non-isothermality is rarely made for extragalactic H II regions, primarily because it is difficult to measure the necessary line ratios.

2.1.3. Correction for Unobserved Ions. Perhaps the most difficult step in the abundance solution is to correct for ions which may be common in the nebulae but do not give rise to conveniently observable emission lines. The magnitude of this correction varies greatly from element to element. Oxygen is the best case, since there are intrinsically strong optical lines for the two generally most abundant ions, O^+ and O^{++}. A set of formulae defining "ionization correction factors" (i_{cf}'s) which convert ionic into elemental abundance ratios was introduced by Peimbert and Costero (1969). These formulae are based on near-coincidences between the ionization potentials (I.P.'s) of various ions, and assume that the fractional ionic populations are the same as those of ions with similar I.P.'s. For example, the I.P. for N^+ is 29.6 eV, while that of O^+ is 35.1 eV, so one assumes that $(N/H) = (N^+/H^+) \times i_{cf}(N) = (N^+/H^+) \times (O^+ + O^{++})/O^+$, or $(N/O) = (N^+/O^+)$. Likewise, the ionization potentials of O^+ and S^{++} are similar, suggesting a parallel correction formula for S^{+3}. These formulae are still often used, although it is clear that the I.P.'s do not tell the whole story; in some cases other factors, such as recombination coefficients and charge-exchange reactions, can significantly change the ionization balance (for sulfur, for example: see Natta, Panagia, and Preite-Martinez 1980; Dinerstein 1980; Garnett 1989a,b).

2.2. NEBULAR MODELS

Another approach to interpreting nebular spectra is to construct models of the nebular ionization and thermal structure. One can either tailor a model to a specific region and optimize the match to the observed line intensities (e.g. Dinerstein and Shields 1986), or utilize grids of models covering an appropriate range of nebular parameters (e.g. Stasinska 1978; Dufour *et al.* 1980; Mathis 1982; Evans and Dopita 1985). In the latter method, ionic abundances are basically derived using the direct method, but the models are used to estimate the electron temperatures and ionization correction factors for individual ions.

The ingredients of a model are: (1) T_*, the effective temperature of the ionizing stars (or, more generally, the spectral energy distribution of the ionizing radiation, taken from model stellar atmospheres); (2) a set of elemental abundances such as (O/H), which determine the local cooling efficiency, and hence the gas temperature; and (3) the gas density, often parametrized in terms of the "ionization parameter". The ionization parameter is given by $U = Q(H^0)/[4\pi r^2_S nc] \propto [Q(H^0)nf^2]^{1/3}$, where $Q(H^0)$ is the number of H-ionizing photons, r_S is the Strömgren radius, n is the gas density within the clumps, and f is the volume filling factor of the clumps (e.g. Shields 1986; note that other authors may use slightly different definitions of U). The ionization parameter is essentially the local ratio of Lyman-continuum photons to gas density, which determines the degree of ionization at any particular location within the nebula (a point stressed by Mathis 1985). The abundances of the other elements are often assumed to scale with oxygen, which is itself the main driver of the gas cooling, through its strong emission lines. As the abundances vary, the nebular structure and emergent spectrum may change drastically (e.g. Stasinska 1980). A further refinement is to allow for the fact that presumably the abundances in the atmospheres of the ionizing stars are the same as those in the nebular gas (Balick and Sneden 1976; Shields and Searle 1978; Skillman 1989).

Most models, especially those designed for extragalactic H II regions, assume a spherical geometry but allow for clumping of the gas (with filling factors of .01 - .001 as discussed above). Some workers have suggested that dust within the nebula softens the ionizing radiation field (Sarazin 1977; Herter, Helfer, and Pipher 1983), but others find that the effect of dust is unimportant (Mathis 1986); the result depends the assumed grain optical properties, which are poorly known.

2.3. "BRIGHT-LINE" METHODS

Many extragalactic nebulae are too faint to permit measurement of weak lines, such as [O III] 4363Å, which are needed for the analysis methods described above. In these cases, it is still possible to obtain estimates of the nebular abundances, using intensity ratios among the handful of brightest ionic emission lines. This is referred to as the "empirical" or "semi-empirical" approach, or, more descriptively, as the "bright-line" method. The physical basis lies primarily in the energy-level structure of the main coolant for most H II regions, the O^{++} ion. The rate of energy lost via a collisionally excited line is proportional to the factor $\exp(-\chi/kT)$, where χ is the energy of the level from which the line arises. The exponential term varies steeply near $T_e = 10^4$ K, a typical value for H II regions, for the strong optical lines 4959, 5007Å. However, it is nearly constant for the 52 and 88 μm lines which arise from fine structure in the ^3P ground term (see Figure 1). This leads to a behavior which may initially seem counterintuitive: as O/H decreases, the cooling in all of the [O III] lines drops and T_e rises; as a result, 4959 and 5007Å brighten dramatically. On the other hand, as O/H increases, T_e drops and the optical lines weaken greatly; in this case, most of the cooling occurs through the infrared lines. This basic mechanism was recognized by Searle (1971) and further developed by Shields (1974). The other optical lines also have fairly strong temperature dependences, but in some cases, such as [O II] 3727Å, the ion does not have fine-structure in the ground term providing an alternate outlet for radiative energy losses. On the other hand, the (generally) most abundant ion of sulfur, S^{++}, has an energy-level structure similar to that of O^{++}, with strong lines at 9069, 9532Å analogous to 4959, 5007Å, and infrared lines at 18, 35 μm.

In H II regions where both can be measured directly, the electron temperature does correlate with O/H (see, for example, Figure 1 in Pagel 1986). The line ratio ([O II] 3727Å+ [O III] 4959,5007Å)/Hβ was calibrated and employed as an abundance indicator for extragalactic H II regions by Pagel *et al.* (1979), and has since been recalibrated several times (e.g. Edmunds and Pagel 1984; Dopita and Evans 1986). Permutations of these line ratios and combinations with lines of [N II], [S II], and [S III] have been considered by McCall, Rybski, and Shields (1985), Mathis (1985), and others. It is clearly preferable to use the ratio ([O II]+[O III])/Hβ rather than [O III]/Hβ, although the latter is easy to measure because the lines are adjacent in wavelength (e.g. Zaritsky, Elston, and Hill 1989), because [O III]/Hβ depends on the degree of ionization. Unfortunately, once the abundance approaches solar, the radiative cooling becomes so efficient that the weak, highly temperature sensitive line [O III] 4363Å is essentially unobservable (in distant H II regions). Thus, the high-abundance end of this relation is intrinsically difficult to calibrate, and relies heavily on nebular models. This calibration is one of the main sources of disagreement in the abundance values derived by different authors.

It is also important to recognize that this simple inverse relation between abundance and [O III] line strengths breaks down at very low metallicities, where 4959, 5007Å weaken in proportion to the O/H abundance because this ion no longer dominates the cooling. This effect produces another "branch" in the ([O II]+[O III])/Hβ diagram, for which the line strengths vary in direct proportion to abundance. Since the "turn-around" point falls near (O/H) = 10^{-4}, the most metal-poor (and hence, often the most interesting!) H II regions fall in a part of the diagram where the line strength-abundance relationship is double-valued. This fact must be kept in mind when applying the bright-line method to samples of nebulae which might include very low metallicity regions.

3. Abundance Gradients in Galactic Disks

3.1. THE PHENOMENON

It was long ago noticed that certain H II region emission-line ratios, such as [O III]/Hβ, vary across the disks of nearby spiral galaxies (Aller 1942). The interpretation of this variation in terms of a metallicity trend was introduced by Searle (1971), in a paper that laid the groundwork for the entire field of abundance gradients. It was soon followed up by further observational studies (Smith 1975; Jensen, Strom, and Strom 1976) and a more rigorous analysis involving the construction of realistic nebular models (Shields 1974; Shields and Searle 1978). From the start it was recognized that there was a need for a "second parameter" in addition to the O/H ratio, to explain an observed systematic increase in O++/O+ with decreasing O/H. Shields and Tinsley (1976) suggested that this secondary effect results from a tendency for the effective temperatures of the ionizing stars to be hotter for lower O/H, and interpreted it as a metallicity-dependent truncation of the top end of the initial mass function (i.e. that the formation of very massive stars is inhibited by higher metallicity). Some form of the idea of a Z-dependent IMF is still the most popular interpretation of the "excitation" trend (e.g. Vilchez and Pagel 1988; Campbell 1988), but it is also the case that a similar effect can arise from systematic variations in the nebular geometry and/or filling factor (Mathis 1985; Dopita and Evans 1986).

3.2. TRENDS WITH GALAXY TYPE

An extensive body of literature has been amassed on the subject of abundance gradients in galaxies over the past two decades. Results prior to 1981 were reviewed by Pagel and Edmunds (1981; also see the review by Shields 1990). More recent major studies involving large numbers of galaxies include those of Webster and Smith (1983) and McCall, Rybski, and Shields (1985). Not surprisingly, many workers have focused on large, nearby galaxies with many observable H II regions, such as M 33 (Kwitter and Aller 1981; Vilchez et al. 1988) and M 101 (Evans 1986; Torres-Peimbert, Peimbert, and Fierro 1989). The gradients are usually expressed as a logarithmic fit to some 5-10 regions per galaxy, and have a magnitude of about $\Delta\log(O/H)/\Delta R = -.08$ ($\pm .03$) dex/kpc. This is similar to the values derived for the solar-neighborhood metallicity gradient in the Milky Way galaxy (see Section 3.4. below). It is possible that the gradients may steepen in the inner parts of galactic disks. However, this is difficult to prove, both because the H II region samples are often small, and more fundamentally because these are generally the most metal-rich regions, for which [O III] 4363Å is unobservable and therefore the derived abundances are heavily model-dependent.

The steepest abundance gradients were initially seen in late-type spiral galaxies (types Sb-Scd). Irregulars and barred spirals tend to have weak or zero radial gradients (Webster and Smith 1983). Early type spirals are harder to study because their H II regions are intrinsically fainter (see Section 1.2 above), but recent studies of M 81 (Sab) show it to have an O/H gradient similar to those of M 33 and M 101 (Stauffer and Bothun 1984; Garnett and Shields 1987). There is at present no convincing evidence that the O/H gradient depends on morphological type among spiral galaxies. However, there *is* evidence for a good correlation between mean O/H abundance and the overall galaxy mass or luminosity. This can be seen, for example, in Figure 2 (from Garnett and Shields 1987). This trend resembles the correlation of stellar metallicity with galaxy mass, and probably has its roots in the fundamental processes of galaxy formation and evolution (Larson 1976). The extension of this correlation to low-mass galaxies is discussed below.

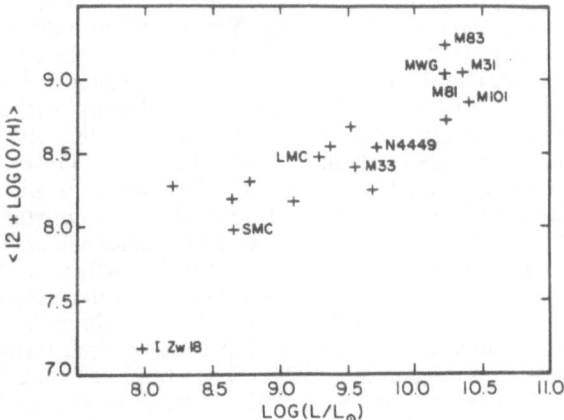

Figure 2. The relationship of mean oxygen abundance to total galaxy luminosity. Reprinted from Garnett and Shields (1987).

3.3. ELEMENTS OTHER THAN OXYGEN

3.3.1. Nitrogen. Along with the trend in [O III]/Hβ, a similar radial trend was noted for the ratio [N II]/Hα, which decreases with increasing distance from the centers of spiral galaxies. Although part of this trend is due to the generally lower degree of ionization in the outer H II regions, there also must be a real variation in abundance (e.g. Shields 1974). Unlike oxygen, for nitrogen one usually can measure the singly-ionized state only; unfortunately, N^{++} has no strong optical lines. As discussed in Section 2.1.3., the nitrogen abundance is basically derived from [N II]/[O II]. Figure 3, from Vilchez *et al.* (1988), shows the O/H and N/H gradients for M 33. The relative behavior of O and N is often displayed by plotting N/O vs. O/H. Some studies find that N/O varies almost as steeply as O/H, which has special significance in the context of chemical evolution models (see Section 3.4.), but others claim that N/O varies only slightly or is constant across the disks of galaxies such as M 101, M 33, M 81, and M 83 (references given above). There also appear to be variations in N/O at a given O/H from galaxy to galaxy (e.g. Webster and Smith 1983). Some of these variations may be an artifact of the analysis, especially since N^{+} contains only a small fraction of the nitrogen for the lowest-abundance, most highly ionized regions. For such regions, the ionization correction factors (Section 2.1.3.) are very large, and the uncertainties in the ionization structure translate into large uncertainties in the elemental abundance of nitrogen. Nevertheless, there is accumulating evidence that nitrogen has a more complicated behavior than does oxygen, with N/O being roughly constant at low values of O/H and increasing at higher O/H (e.g. Pagel 1985; Torres-Peimbert, Peimbert, and Fierro 1989). Measurements of N/O in metal-poor dwarf irregular galaxies are an important ingredient in this argument (see Section 4.5.).

3.3.2. Sulfur. Another element with bright optical emission lines is sulfur, but it suffers from a similar problem as does nitrogen. In most studies to date, sulfur abundances have been determined from [S II] 6717, 6731Å. Comparison of these lines with [N II], [O II]

266

yields values for S/N and S/O. Unfortunately, the ionization potential of S+ is lower than those of N+ or O+, so that one should really measure S++ as well, in order to estimate the elemental abundance ratio (e.g. Hawley and Grandi 1977). In fact, the S+ ion is present outside the Strömgren sphere, and the [S II] emission comes primarily from the H+/H° transition zone (Shields 1974). As a result, abundances derived from [S II] alone are very sensitive to the procedure used to correct for the ionization structure. Discrepant results have been reported by different authors for the same H II regions. For example, Evans (1986) reports that S/O increases as O/H decreases in M101, while for the same galaxy, Torres-Peimbert, Peimbert, and Fierro (1989) find S/O to be constant. Figure 4 reproduces the N/O and S/O vs. O/H variations reported by the latter authors for M 101.

Progress in determinations of sulfur abundances probably will require much more extensive measurements of S++. This ion can be sampled either via the weak [S III] 6312Å line, which is the temperature-sensitive analog to [O III] 4363Å, or through the strong lines at 9069, 9532Å. It is preferable to use the latter both because of their intrinsic strengths and their gentler dependence on temperature. These lines fall in a spectral region only recently opened up by CCD detectors, and potentially suffer from absorption by telluric water vapor, but are beginning to be observed (Vilchez and Pagel 1988; Garnett 1989a,b) and probably offer the best prospect for improved sulfur abundances. When comparing results on sulfur abundances from different studies one should be aware of the lines observed and the method of analysis employed. Because of its fine-structure in ground term, S++ can behave as a "thermostat" in a similar way as O++, and is the major coolant in low-ionization regions such as those in the inner parts of galaxy disks (Mathis 1985).

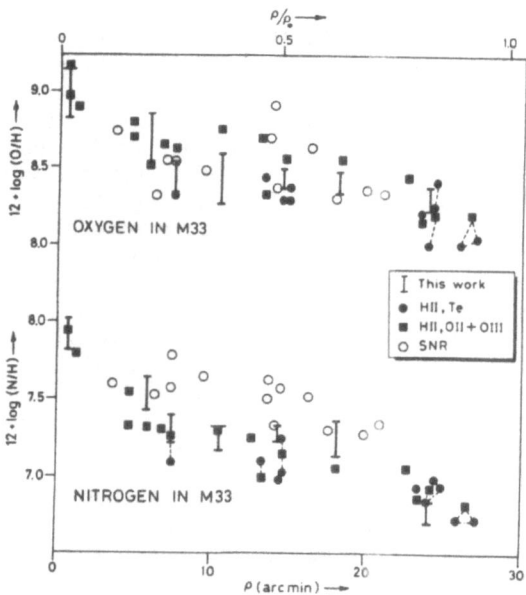

Figure 3. The O/H and N/H gradients in M 33, reprinted from Vilchez et al. (1988).

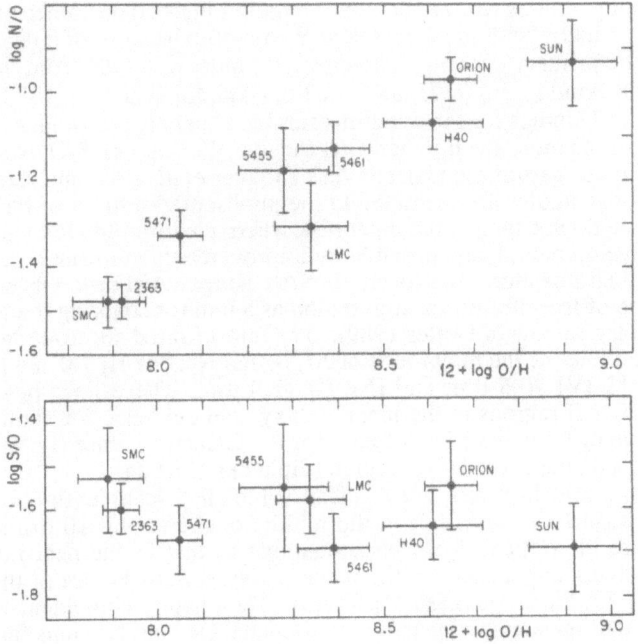

Figure 4. The relative abundance ratios N/O and S/O vs. O/H in M 101, from Torres-Peimbert, Peimbert, and Fierro (1989).

3.3.3. Neon and Argon. Although they do have emission lines in the optical spectral region, abundance determinations for neon and argon in extragalactic H II regions are much rarer than for the elements discussed above. For Ne, the abundances are usually derived from [Ne III] 3869, 3889Å. Although Ne+ is the majority ion in most H II regions, except for those ionized by the very hottest stars, the assumption that the Ne++/Ne fraction is similar to O++/O is borne out by nebular models. Argon is a more difficult case. The strongest optical line is [Ar III] 7136Å; [Ar IV] 4711, 4740Å are weak lines and blended with other species, and [Ar II] 7.0 μm falls in the infrared. Attempts to find simple formulas for argon based on ionization potentials have not been very successful, so most recent studies based their ionization corrections on nebular models. The results show fairly large scatter, but most studies conclude that the observations are consistent with constant values for Ne/O and Ar/O (e.g. Torres-Peimbert, Peimbert, and Fierro 1989).

3.4. COMPARISON WITH THE GALACTIC ABUNDANCE GRADIENT

Because of interstellar extinction, one can use the same techniques as for extragalactic H II regions only for the part of our Galaxy outside a galactocentric distance of about 7 kpc (for $R_G = 10$ kpc). Studies such as those of Hawley (1978) found gradients similar to those in other spirals, $\Delta\log(O/H)/\Delta R = -.04$ to $-.06$ dex/kpc and $\Delta\log(N/H)/\Delta R = -.10$ dex/kpc.

Determination of abundances in the inner galaxy requires the use of other techniques, such as measuring electron temperatures from radio recombination lines. The values of T_e are found to increase systematically with increasing R_G, presumably because of a decreasing abundance of oxygen, the primary coolant. The inferred gradient in O/H from the most recent major survey of this kind is $\Delta\log(O/H)/\Delta R = -.07$ dex/kpc (Shaver et al. 1983).

The results from optical studies for the other measurable elements are similar to those for other galaxies: N/H varies more steeply than O/H (hence, N/O varies); S/O, Ne/O, and Ar/O do not vary in the outer part of the Galactic disk (Shaver et al. 1983 and references therein). Again, the optical studies are restricted to the unobscured portion of the Milky Way galaxy, and therefore do not sample the inner disk where the inferred O/H values are high. A more recent development, made possible by improvements in infrared detectors and the availability of a stratospheric platform, NASA's Kuiper Airborne Observatory (KAO), is the exploration of the infrared spectral region as a tool for studying the galactic abundance gradient (see the review of Herter 1989). The mid-infrared spectral region (5-30 μm) contains emission lines of the major ions of Ar, S, and Ne: [Ar II] 7.0 and [Ar III] 9.0 μm; [S III] 18 and [S IV] 10.5 μm; and [Ne II] 12.8 μm. These lines have been measured in a number of H II regions in the inner Galaxy, and evidence for abundances elevated by factors of two or three have been found for the Galactic Center (Lester et al. 1981), and for H II regions in the 5 kpc "ring" region (Pipher et al. 1984).

However, even these mid-infrared lines suffer somewhat from extinction. In particular, the [Ar III] and [S IV] lines fall in the middle of the strong 10 μm silicate absorption feature, where the optical depth is comparable to that in the near-infrared. Another approach to studying abundances in the inner galaxy is to make use of the fine-structure lines of [O III] 52, 88 μm and [N III] 57 μm. By a happy coincidence, these lines from the abundant and (presumably) usually co-extensive O++ and N++ ions fall close together in wavelength and have fairly similar dependences on the electron density. The line emissivities are also essentially independent of the electron temperature. Measurements of these three lines therefore yield a relatively accurate value for the N/O ratio (Watson et al. 1981; Lester et al. 1983). A survey of about a dozen galactic H II regions in these lines yielded strong evidence that N/O in the Galactic Center and 5 kpc "ring" is elevated by a factor of 2 or 3 as compared to the solar neighborhood (Lester et al. 1987). There remain some unsettled questions regarding N/O determinations from the far-infrared lines, including possible ionization structure effects in H II regions ionized by very cool stars (e.g. Rubin et al. 1988), and a systematic discrepancy between values derived from the infrared lines and those derived optically from [N II]/[O II] (Simpson et al. 1986). Nevertheless, this is clearly an area where further progress can be expected.

3.5. COMPARISON WITH CHEMICAL EVOLUTION MODELS

The recognition of significant variations in the gas composition within and among galaxies, along with parallel results on the stellar populations, inspired the development of chemical evolution models which attempt to explain these patterns (see reviews by Audouze and Tinsley 1976; Tinsley 1980; Clayton 1986; Pagel 1987). The so-called "simple model" postulates a closed system of gas and stars, which self-enriches in metals as generations of stars age, die, and seed the ambient gas in the heavy elements (Searle and Sargent 1972). This model also makes the approximations that the stellar lifetimes and timescale for complete mixing of nucleosynthetic products are negligible in comparison to the timescale on which the metallicity evolves ("instantaneous recycling"). The simple model makes a specific prediction regarding the metallicity and system properties: $Z = y \ln(M_{tot}/M_{gas})$. In this equation, Z is the metal abundance, y is the fraction of the stellar mass converted to heavy elements (the "yield"), and $M_{tot} = M_{gas} + M_{stars}$. Although this model is most appropriate for the low-mass galaxies discussed in Section 4, it can also be applied to large disk galaxies if concentric radii are treated as independent zones. However, it does not explain the observed gradients, so modifications such as radial flows, matter exchange with an outside reservoir (infall and outflow), or a variable stellar initial mass function, have been proposed as modifications to the model (e.g. Güsten and Mezger 1982; Lacey and Fall 1985; Matteucci and Francois 1989; Dopita 1990).

The relative abundances of nitrogen and oxygen are of particular interest, since they are synthesized in different astrophysical sites. Oxygen is synthesized in massive stars and distributed into the interstellar medium by Type II supernovae, while the origin of nitrogen is more problematical. A distinction is frequently made between "primary" nucleosynthetic products, which can be synthesized directly from H and He in Population III stars, and "secondary" products, which require a "seed" heavy nucleus to be initially present in the star where its synthesis occurs. By this definition, oxygen is a primary species. Nitrogen is secondary when made as a by-product of CNO-cycle hydrogen burning. According to the simple closed-box model, the abundance of a secondary species is quadratic, so that if N is secondary and O primary, then $(N/H) \propto (O/H)^2$, or $(N/O) \propto (O/H)$. The N/O ratio does appear to approach this behavior, for H II regions with moderately high O/H values in M 101 (Torres-Peimbert, Peimbert, and Fierro 1989) and in the Milky Way (Lester *et al.* 1987). However, below a certain values of O/H, it appears that N/O is constant; these low-metallicity H II regions occur mostly in low-mass galaxies (see Section 4.5.1.). Thus, it is becoming clear that nitrogen is not purely a secondary nucleosynthetic product. Indeed, N may be produced within intermediate-mass stars by an effectively primary process, if C synthesized within the star by the triple-alpha reaction is later subjected to the CN cycle (Alloin *et al.* 1979; Renzini and Voli 1981). Nitrogen made by this process would be primary, but there might be a time-delay in building up its abundance relative to the nuclear products of supernovae, because of the longer lifetimes of the source stars (Edmunds and Pagel 1978; Serrano and Peimbert 1983; Diaz and Tosi 1986).

The other elements measured in extragalactic H II regions, S, Ne, and Ar, are not likely to be dominated by secondary processes. They might still, however, vary differently than oxgyen, if they were produced in stars of different mass ranges and the initial mass function varied or the timescales for enrichment differed substantially. There are known variations in the abundance ratios of certain elements. For example the fact that the iron-group is deficient relative to oxygen in Population II stars is thought to reflect an origin for the former chiefly in Type I supernovae, which originate in long-lived progenitors, as opposed to synthesis of oxygen in massive stars and Type II supernovae. For a recent review of this subject see Wheeler, Sneden, and Truran (1989).

4. Abundances in Metal-Poor Dwarf Galaxies

4.1. NOMENCLATURE

Giant H II regions similar to those in spiral galaxy disks are also found in dwarf irregular galaxies. These regions have acquired a variety of aliases, all of which refer to essentially the same type of object, dwarf irregular galaxies with spectra dominated by narrow emission lines. Observers making spectroscopic observations of these objects have called them "isolated extragalactic H II regions" (Sargent and Searle 1970, the discovery paper); "H II region-like galaxies" (French 1980); or "H II Galaxies" (Campbell, Terlevich, and Melnick 1986). Other names refer explicitly to the host galaxies, which are generally irregulars: e.g. "blue compact dwarf galaxies (BCDG's or BCG's)", "star-forming dwarf irregular galaxies". Searle and Sargent (1972) called particular attention to the prototype objects, I Zw 18 and II Zw 40, because of their extraordinarily low elemental abundances of oxygen and neon. From their composition, these authors inferred that the current star formation rate in these galaxies is much greater than the past average rate. They concluded that star formation occurred intermittently rather than continuously in these galaxies, and coined the term "bursts" of star formation to describe this phenomenon. Dwarf irregular galaxies experiencing active star formation, and "starbursts" in general, have received a great deal of attention lately. The proceedings of two conferences cover much of the recent work on the subject (Kunth, Thuan, and Van 1985; Thuan, Montmerle, and Van 1987).

4.2. THE "I ZW 18 PROBLEM" AND THE MASS-METALLICITY RELATION

The very low elemental abundances in these dwarf irregular galaxies imply that these are "young" or "new" galaxies, at least in terms of their net star formation activity and nucleosynthetic evolution. They therefore offer an opportunity to watch the early stages of chemical enrichment, which happened in the distant past in our own galaxy. Furthermore, in these galaxies there also has presumably been relatively little helium produced by stars, so they are attractive objects to study for the determination of the pre-galactic, cosmological component of helium. Both of these factors helped motivate a lively search for the most metal-poor H II regions which could be found.

Surveys such as those of Kinman and Davidson (1981), Kunth and Sargent (1983), and Kunth and Joubert (1985), yielded around 100 galaxies for which abundances could be measured. However, despite extensive searches, primarily using objective-prism techniques (e.g. Kunth, Sargent, and Kowal 1981), for nearly two decades no other galaxies were found with O/H abundances as low as I Zw 18, one of the two original prototypes. In fact, virtually no galaxies were found with O/H < 1/10 solar, and the median value was closer to 1/6 solar, $\log(O/H) + 12 = 8.0$. Kunth and Sargent (1986) considered whether this result could be due to selection effects; this median value happens to fall near the abundance at which the strong optical [O III] lines reach their peak intensity, as discussed in Section 2.3. (Campbell, Terlevich, and Melnick 1986). It has been noted that I Zw 18 itself was *not* found by objective prism work; it was first identified as a blue compact galaxy. Nevertheless, Kunth and Sargent (1986) concluded that the failure to find other "I Zw 18's" was not a result of selection bias, but rather indicated that extremely metal-poor H II regions are intrinsically rare. In order to reconcile this conclusion with the existence of gas-poor dwarf spheroidal galaxies having metallicities far lower than those of I Zw 18, they suggested a scenario of rapid self-enrichment of the gas in giant H II regions by supernovae from the ionizing cluster. If the supernova products remained concentrated within the H II region rather than becoming diluted throughout the entire system, then the gas metallicity would rise to that of I Zw 18, while most of the stars remained metal-poor.

The case of I Zw 18 still remained problematical, however. For an instantaneous starburst, the heavy elements are not released until the massive stars begin to die, several million years later, and by then the ionizing radiation field required to produce a bright H II region will have declined drastically, because these are the very stars that provide most of the UV photons (e.g. Lequeux *et al.* 1979; also see Figure 8 in Kunth 1985). If I Zw 18 is still experiencing its first starburst, the stellar population cannot be coeval; that is, the star formation epoch must have a duration longer than several-million-year timescale (Kunth and Sargent 1986). If C and N were enriched relative to O in I Zw 18, compared to their ratios in other dwarf galaxies, then an even more complex star formation history would be required (Dufour, Garnett, and Shields 1988; Pantelaki 1988).

Recall, however, the mean metallicity-galaxy mass relation discussed above (Figure 2). A similar relation is seen for the dwarf irregulars, along with a correlation between the O/H abundance and the gas mass fraction, as expected for the simple model of galactic chemical evolution (e.g. Lequeux *et al.* 1979; Talent 1980; Kinman and Davidson 1981; Vigroux, Stasinska, and Comte 1987). This suggests an alternate strategy for searching for extremely metal-poor H II regions: namely, choose galaxies known to be of low mass, rather than selecting objects because of their emission lines. This is the approach taken recently by Skillman and his collaborators, who examined a number of dwarf galaxies in the Local Group and found several H II regions with O/H values as low as that of I Zw 18 (Skillman *et al.* 1988b; Skillman, Kennicutt, and Hodge 1989; Skillman, Terlevich, and Melnick 1989). The success of this strategy is apparent from the histogram of O/H values for their sample as compared to earlier samples (Figure 5). There still appears to be a threshold value near the O/H abundance of I Zw 18, suggesting that the self-enrichment mechanism advocated by Kunth and Sargent (1986) may apply, but this more recent work does fill in the metallicity "gap" between I Zw 18 and the other previously studied dwarf irregular galaxies. These low-metallicity H II regions extend the mass-metallicity relation seen for higher-mass galaxies (Figure 2) down to a regime that overlaps with metal-poor dwarf spheroidal systems that lack a substantial Population II component. There seems to be reasonable agreement between the two sets of objects, allowing for uncertainties introduced by the fact that the stellar "metallicities" refer to the iron-group, while the H II region abundances are determined for oxygen (e.g. Aaronson 1985; Pagel 1987).

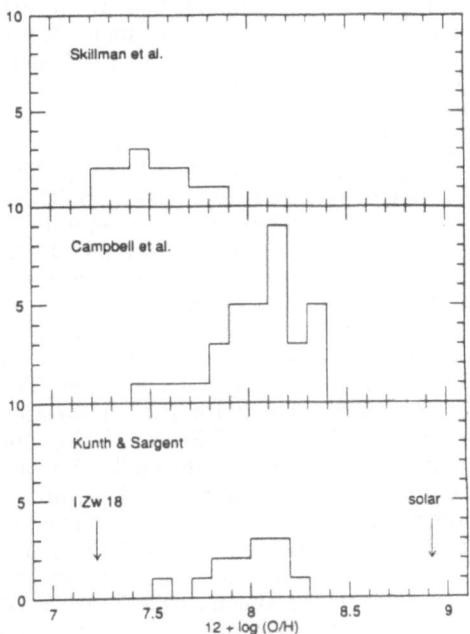

Figure 5. Comparison of the histograms of O/H abundance for different samples of dwarf irregular galaxies, reprinted from Skillman, Terlevich, and Melnick 1989.

4.3. DWARF GALAXIES AS TESTS OF THE SIMPLE MODEL

The simple model of galactic chemical evolution, discussed in Section 3.5, was originally designed for dwarf irregular galaxies like I Zw 18 (Searle and Sargent 1972). Thus it is natural to compare their properties with the predictions of the model, such as the expected linear relation between metal abundance, Z, and $\ln(M_{tot}/M_{gas})$. In order to make such a comparison, it is necessary to measure both M_{tot} and M_{gas}, which is not easy to do for these faint, low-mass systems. Observations are made of the 21 cm H I emission; its intensity yields the mass of interstellar gas; the total gravitational mass can be inferred from the line-widths. Studies of this type have found that the dwarf galaxies tend to fall near the expected relation but with a fairly large scatter (Lequeux et al. 1979; also see Figure 3.3 of Pagel 1987). The fact that I Zw 18 in particular falls very far from this relation, with far too low a gas fraction for its low metallicity, has frequently been noted (Skillman et al. 1988b). The total masses of these galaxies are somewhat uncertain, however, due to the possible unobserved presence of molecular gas, and issues regarding the dynamics of these systems (e.g. Gallagher and Hunter 1984; Skillman et al. 1988a; Hoffman et al. 1989). Furthermore, these small galaxies might not be closed systems, but might experience infall of primordial material or the loss of metal-rich supernovae ejecta which more easily escapes their weak gravitational fields (Larson 1974; Larson and Dinerstein 1975).

4.4. DWARF GALAXIES AND THE EXTRAGALACTIC H II REGION SEQUENCE

The H II regions in dwarf irregular galaxies can be studied using the same methods as used for regions in the disks of spiral galaxies. In general, they seem to follow the same patterns as the disk regions. When directly measurable, their electron temperatures decrease as O/H increases, following the same relationship (e.g. Figure 1 of Terlevich 1985). As expected, the "bright-line" method of abundance analysis also applies to the dwarf irregular H II regions. However, the regions with O/H abundances less than about 1 x 10-4 fall on the lower branch of the O/H vs. line intensity diagram described in Section 2.3., where R_{23} = {[O II] 3727Å + [O III] 4959+5007Å}/Hβ is *directly* proportional, as opposed to inversely proportional, to O/H (Figure 6, reprinted from Skillman 1989).

Another property that H II regions in dwarf galaxies share with those in spirals is the correlation between the degree of ionization (measured from the O++/O+ ratio) and the metallicity. Recall that, in spiral galaxies, the degree of ionization increases, and O/H drops, from the center out to the edge of the disk. For dwarf galaxies, the radial location drops out of the picture, but the ionization fraction still increases as the abundance decreases. The interpretation of this effect is not universally agreed upon. Many authors interpret this effect as being due to hotter ionizing stars, and infer T_*, or "T_{ion}" using nebular models. In this case, one must account for such a variation in the stellar population. Age effects alone are insufficient (e.g. Searle 1971), although they may account for the observed spread of values, which appear to fall below an upper envelope (e.g. Viallefond 1985). Many authors favor metallicity-dependent variations in the stellar initial mass function, either by truncation at the top end, M_{upper} (Shields and Tinsley 1976), or a change in slope (e.g. Terlevich 1985). Another school of thought suggests that the change in ionization fraction is due to a change in U, the ionization parameter, essentially a metallicity-dependent variation in the nebular geometry (Mathis 1985; Dopita and Evans 1986). Such a circumstance could arise, for example, if metal-rich stellar associations produce stronger winds which more efficiently sweep out the ionized gas, yielding a nebula with a lower filling factor. Vilchez and Pagel (1988) attempted to circumvent this ambiguity by constructing a line intensity ratio involving both O++/O+ and S++/S+; they find

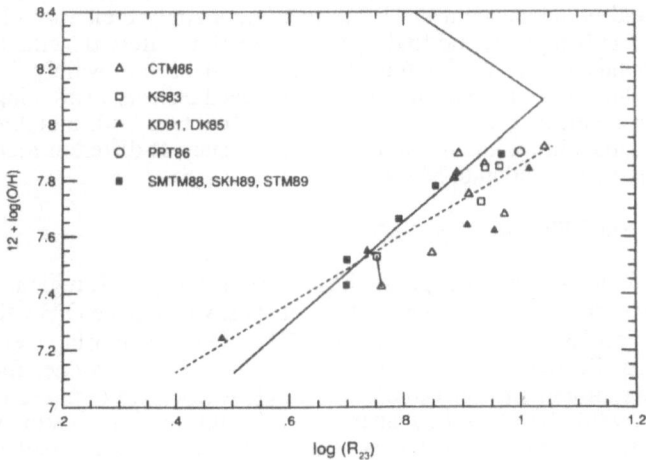

Figure 6. The bright-line ratio vs. abundance relation is shown for low-mass galaxies in the Local Group. The ordinate R_{23} is the line ratio {[O II] 3727Å + [O III] 4959 + 5007Å}/ Hβ. The solid line is the relation of Pagel, Edmunds, and Smith (1980), and the dotted line is a least-squares fit to the data shown here. The reference key is given in Skillman (1989), from which this figure is reprinted.

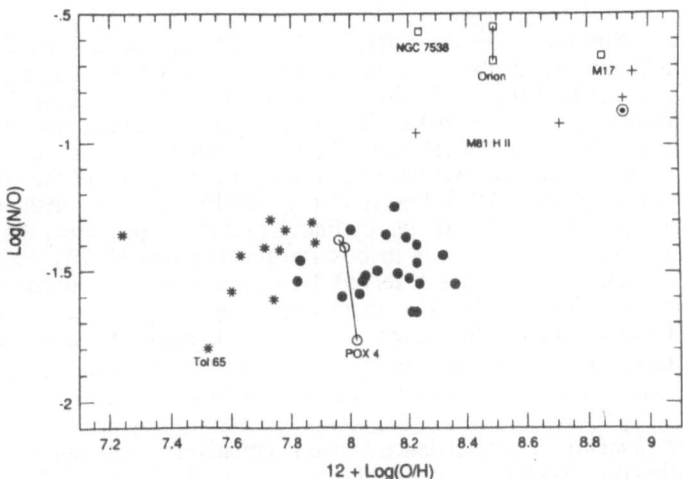

Figure 7. The abundance ratios N/O vs. O/H are shown for a large sample of dwarf irregular galaxies, and also for several relatively metal-rich H II regions (labelled points at upper right). At log (O/H) + 12 < 8.3, (N/O) shows no systematic variation, but scatters around a mean value of about log (N/O) = -1.5. From Garnett (1989b).

that the composite bright-line ratio $\eta = (O^+/O^{++})/(S^+/S^{++})$ depends only on the ionizing-star temperature and is independent of U. These authors apply their formula to a sample of dwarf irregular H II regions, and find that the increase in inferred stellar temperature with decreasing abundance persists. A third line of argument, less widely discussed than the others, is that the change in ionization structure indeed reflects a softening of the radiation field, not as a result of a change in initial mass function with metallicity, but instead because of changes in the *evolution* of the cluster stars at different metallicities. I will return to this issue in Section 6, below.

4.5. ELEMENTS OTHER THAN OXYGEN

4.5.1. Nitrogen. Most spectroscopic studies of dwarf-galaxy H II regions yield abundance values for nitrogen as well as oxygen. These objects, which have O/H values substantially less than solar, do not follow the approximate N/O α O/H relationship seen for more metal-rich regions (see Section 3.3.1. and Figure 4). Rather, the N/O values for dwarf irregular galaxies scatter around a mean value of $\log(N/O) + 12 = -1.5$ (e.g. see reviews by Pagel 1985; Dufour 1986). Some of the scatter may arise from ionization structure effects: these regions are highly ionized, so that N^+ and O^+ contain only a small fraction of their respectively elements. Another source of observational error is the difficulty of obtaining accurate fluxes for [N II] 6548, 6584Å, which are extremely weak in these objects and are situated on the wings of the extremely strong $H\alpha$ line (e.g. Dufour, Garnett, and Shields 1988). Nevertheless, it seems clear that N does not behave like a purely secondary nucleosynthetic product; it must also have a large primary component, which dominates in the most metal-poor regions (e.g. Matteucci and Tosi 1985; Vigroux, Stasinska, and Comte 1987; Garnett 1989c). Results from a recent study of nitrogen abundances in dwarf galaxies is shown as Figure 7 (from Garnett 1989b), which includes, for comparison, the abundances for several regions in the "secondary" domain.

4.5.2. Sulfur. Sulfur abundances determined for H II regions in large disk galaxies were discussed in Section 3.3.2. above. Until recently, relatively few reliable sulfur abundances were available for H II regions in dwarf galaxies other than the Magellanic Clouds (for these, see Dennefeld and Stasinska 1983; Dufour 1986). These regions generally have such high excitation that the [S II] 6717, 6732Å lines are weak and insufficient for measuring accurate total abundances, since a very small fraction of the nebular sulfur is in the S^+ ion. The [S III] 6312Å line is also generally weak, intrinsically temperature-sensitive, and often blended with nearby lines of [O I]. In principle, the best method for determining sulfur abundances is to observe the stronger [S III] 9069, 9532Å lines. Recently, Garnett (1989a, b) has observed these lines in about a dozen dwarf galaxy H II regions, and determined reliable S/O abundances using nebular models to obtain ionization correction factors. The results show no convincing evidence for a variation of S/O with respect to O/H. The mean value is $\log(S/O) + 12 = -1.6$, with a scatter of ± 0.3 dex, probably at least partly due to observational uncertainties.

4.5.3. Other elements: The abundance of neon is relatively easily obtained for metal-poor, high excitation dwarf galaxies, through the strong [Ne III] blue lines. Neon correlates very closely with oxygen, which is expected since it is believed that neon and oxygen are synthesized in the same population of massive stars (Kunth and Sargent 1983; Vigroux, Stasinska, and Comte 1987). Carbon abundances are harder to obtain, since they require measurement of lines in the satellite ultraviolet. At present, with only the relatively small telescope of the International Ultraviolet Explorer (IUE) satellite available to observe these lines, results are available mainly for very nearby galaxies such as the Magellanic Clouds. (This situation will improve dramatically once the much more sensitive Hubble Space Telescope begins collecting data on extragalactic H II regions.) So far, it appears that N/C is constant for any value of C/H, which is interpreted as indicating a mainly primary origin for both N and C (Peimbert 1985; Dufour 1986).

5. Helium Abundances and the Primordial Component

5.1. COSMOLOGICAL MOTIVATION

The universally high abundance of helium in stars and nebulae, $Y \geq 0.24$, is considered to be one of the fundamental pieces of evidence in favor of the "standard" hot big-bang cosmological model. Calculations of nucleosynthesis in the early universe show that the helium to hydrogen abundance ratio is a function of several fundamental cosmological and physical parameters: the baryon to photon number ratio, number of neutrino families, and neutron half-life (e.g. Yang *et al.* 1984). Metal-poor H II regions, where nucleosynthetic activity and enrichment by stars has been minimal, offer extremely attractive sites for attempting to determine the primordial helium fraction Y_P. This realization has led to many attempts to measure helium abundances as precisely as possible for extragalactic H II regions. There are many thorough reviews of these efforts and their results in the astronomical literature; these include the proceedings of the 1983 ESO conference on helium abundances (Shaver, Kunth, and Kjär 1983), and more recent reviews by Shields (1985), Boesgaard and Steigman (1985), and Kunth (1986). In the following discussion, I emphasize the uncertainties and problems in determining Y_P; the interested reader may consult the above references for further details.

A great deal of excitement was generated about ten years ago by the announcement of anomalously low helium values derived from several extragalactic H II regions, $Y_P < 0.22$, by French (1980), French and Miller (1981), Rayo, Peimbert, and Torres-Peimbert (1982). Such low values presented a conundrum for cosmological models. However, the disagreement was significant only if the observational errors were substantially smaller than 10%, prompting further studies to re-examine the question. Kunth and Sargent (1983) studied a sample of about a dozen metal-poor H II regions, and took the average value of measured Y in this sample as being representative of the primordial value, i.e. $Y_P = 0.245$, which was not in conflict with the standard model. Nevertheless, since helium can be presumably be added but not removed by stellar activity, the existence of *any* H II region with a helium fraction less than a cosmologically "allowed" value would seem to disprove the simple big-bang model. The profound consequences of establishing a truly low value of Y in H II regions motivated a number of authors to pursue the issue further.

5.2. DEMANDS ON THE OBSERVATIONS

5.2.1. Corrections to the Line Intensities. In order to definitively establish whether there is a conflict with standard cosmologies, it is necessary to determine the He/H ratio with a precision and accuracy which is unprecedented for astronomical determinations of chemical abundances. Therefore, it becomes essential to consider and correct for every possible source of observational uncertainty, including many effects that can usually be ignored in other contexts. Extensive discussions of the uncertainties involved in determining He/H values in extragalactic H II regions have been given by Davidson and Kinman (1985) and Dinerstein and Shields (1986), in the context of detailed studies of particular objects. Some of the most interesting galaxies (I Zw 18, for example) are so faint that achieving the necessary signal-to-noise in the helium line intensity measurements is a challenge, at least for the present generation of large telescopes (of apertures of 3 to 5 m). Going beyond such standard considerations, several other issues have been raised in the course of pursuing the helium problem. For example, most of the measurements of He/H in the literature were made with IDS (image dissector-scanner) instruments, which have been found to display slight non-linearities in the relationship between counts and flux. This non-linearity, while unimportant under most circumstances, becomes very important in the case of helium. Several different values have been proposed for the magnitude of this effect (e.g. Rosa 1985; Peimbert and Torres-Peimbert 1987); perhaps different individual instruments do indeed have different non-linearities. The issue may become moot, with IDS systems being replaced by CCD's, although the burden of proving the linearity of instruments will remain if the results are to be believed to the percent level.

Another problem affects only He I 5876Å, generally the strongest helium line observed, and therefore given more weight than other measured He I lines. For objects with small positive redshifts (which includes most of the key extragalactic H II regions), 5876Å shifts to the vicinity of the Na I lines at 5889, 5895Å. While telluric emission in the Na I lines can be removed by sky subtraction, it is not so easy to compensate for absorption by Na I in our own Galaxy, an effect which can be at least as large as 10-15% (Davidson and Kinman 1985; Davidson, Kinman, and Friedman 1989). Higher spectral resolution can help somewhat with this problem. Yet another factor which influences the He I line intensities is collisional excitation out of metastable levels. This concern was raised recently by Ferland (1986), who claimed that it could be a large effect; a reassessment by Clegg (1987), using newer calculations of the relevant cross-sections, found the effect on the derived helium abundances to be minor.

In many cases, the entrance aperture for the nebular observations includes not only ionized gas, but also continuum from the ionizing stars (see Section 1.2.). In the spectra of hot stars, the hydrogen and helium lines will be in absorption. Since these observations are generally made with spectral resolutions too low to resolve the narrower emission lines from the underlying absorption features, the emission line intensities will be weakened accordingly. However, unlike the emission decrement, the absorption line decrement is fairly flat; thus, given three or more hydrogen lines, it is possible to solve simultaneously for both interstellar reddening and the strengths of the absorption lines (e.g. Rayo, Peimbert, and Torres-Peimbert 1982; McCall, Rybski, and Shields 1985). In general, the emission equivalent widths of the first few hydrogen recombination lines ($H\alpha$, β, γ) are sufficiently large that this is not a major problem. However, few observers have tried to correct for stellar absorption features underlying the helium lines, which have smaller and less well-known equivalent widths (Dinerstein and Shields 1986).

The problem of correcting for underlying absorption H and He absorption features in the hot star photospheres is not the only problem introduced by the stellar continuum. As discussed in Section 1.4., it is becoming apparent that many extragalactic H II regions contain Wolf-Rayet stars. Such stars produce broad, complex emission features, one of which falls near He I 5876Å. Figure 8 shows this spectral region as well as the region near He II 4686Å, for the dwarf irregular NGC 4861 (Dinerstein and Shields 1986). It is apparent from the figure that, unless one knows the intrinsic shape of the underlying continuum (i.e. whether there is net emission or absorption from the stars), there will be a substantial uncertainty in the strength of the nebular He I 5876Å line. Unfortunately, at present one cannot do much better than to guess at the continuum shape, since the Wolf-Rayet features in extragalactic H II regions display a variety of shapes (D'Odorico, Rosa, and Wampler 1983, and references given above). The best prospect for measuring accurate He I 5876Å lines is to avoid using apertures which contain starlight.

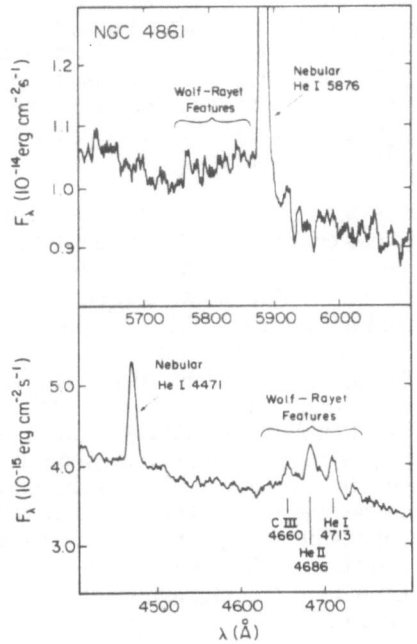

Figure 8. Two segments of the spectrum of the dwarf irregular galaxy NGC 4861 containing Wolf-Rayet emission features are shown. The upper panel shows how these features interfere with the measurement of nebular He I 5876Å. (Figure from Dinerstein and Shields 1986).

5.2.2. The Correction for Neutral Helium.

So far we have discussed only the uncertainties in determining the ionic ratio He+/H+. As with derivations of metal abundances, it is necessary to take into consideration the relative fractions of each element in the observed ions, in order to obtain the elemental abundance ratio. Of the other ions of helium, He++ produces visible recombination lines, primarily 4686Å. This line has been seen in several extragalactic H II regions, but there is some difficulty in distinguishing whether it arises from the nebula or from Wolf-Rayet stars; a true nebular emission line of He II would require the presence of at least some UV radiation from extremely hot stars (e.g. Rayo, Peimbert, and Torres-Peimbert 1982; Dinerstein and Shields 1986). However, even if nebular in origin, He++ represents only a couple of percent of the total helium abundance.

Neutral helium is a potentially much larger component, and, furthermore, it cannot be observed directly. The neutral helium fraction is presumably minimized in nebulae with a high degree of ionization, which, as mentioned above, is usually the case for H II regions with low metal abundances. However, it is still necessary to understand how much neutral helium might be present. Some workers have employed empirical ionization correction formulae for this correction; others have used nebular ionization models. One interesting point is that for $T_* \geq 40,000$ K, the He+ Strömgren sphere actually extends further out than that for H+; therefore the ionization correction factor is less than 1, but only by a few percent at most: He/H ≈ 0.98 x [He+/H+] (Stasinska 1980; Shields and Dinerstein 1986).

On the other hand, it is very difficult to establish that a particular nebular ionization model is a unique solution for an observed set of line intensities. Particularly for distant objects, there is always the possibility that one is observing several spatially distinct regions which are not resolved by the spectrophotometric measurements. In this case, it is possible for there to be a large amount of neutral helium "hidden" inside separate low-ionization nebulae ionized by cooler stars. Such "composite" models have been examined, for example, by Dinerstein and Shields (1986) and Peña (1986), who find that the correction for neutral helium could easily be as great as 10% in this case.

5.2.3. The Correction for Stellar Synthesis. There remains the question of whether or not to attempt to correct for the presumed contribution of helium synthesized by stars. It has been suggested that this contribution should be related linearly to the amount of heavier elements synthesized by the same stars or at least the same population of stars. If such a relationship can be established and the value of $\Delta Y/\Delta Z$ determined, then any measured helium abundance can be extrapolated backwards to obtain Y_P. The problem is that there is no general agreement on the value of this coefficient. Estimates for $\Delta Y/\Delta Z$ have varied from 1.7 (Lequeux *et al.* 1979) to 5.7 (Pagel, Terlevich, and Melnick 1986). Others tend to take a value of around 3 (Rayo, Peimbert, and Torres-Peimbert 1982). Meanwhile, Kunth and Sargent (1983), among others, have argued that there is no such correlation between Y and Z, within the low-metallicity domain. There is also the question of how one determines "Z". Originally it was calculated from O/H, but more recently many authors have suggested calculating the stellar helium contribution from the abundances of N or C, the rationale being that the sites of helium synthesis are also sources N and/or C (e.g. Pagel 1985; Vigroux, Stasinska, and Comte 1987; Steigman, Gallagher, and Schramm 1989; Torres-Peimbert, Peimbert, and Fierro 1989).

5.3. CURRENT STATUS AND FUTURE PROSPECTS

The current situation is that various groups have estimated the value of the primordial helium abundance to fall in the range $0.23 \leq Y_P \leq 0.24$ (Torres-Peimbert, Peimbert, and Fierro 1989; Pagel and Simonson 1989). This is uncomfortably close to the lower limit for the standard cosmological model, but not in actual direct conflict with it (although it does rule out the possible existence of unknown families of neutrinos). However, there is still essentially no decisive proof of the existence of *actual,* as opposed to extrapolated, helium abundances lower than Y = 0.24. There is also no substantial evidence for variations in the primordial abundance from place to place (see Dinerstein and Shields 1986). In view of the caveats discussed in the last section, it seems clear that it is not going to be easy to improve on the current situation. The Hubble Space Telescope will at least provide better opportunities to measure the nebular spectrum without contamination by stellar continuum. It may also help with the correction for neutral helium, because it may be possible to spatially resolve the nebular ionization structure. However, the question of the correction for stellar-synthesized helium will remain. Thus, unless an actual, present-day, helium abundance lower than permitted by the standard cosmology is found, there probably will continue to be controversy about any further inferences regarding the primordial value.

6. Some Outstanding Issues

I would now like to give a personal synopsis of what I see as some of the main results and outstanding problems in the area of chemical abundances in extragalactic H II regions. The first theme that stands out is the *mass-metallicity relation,* which seems to hold all the way from the largest spirals, such as M 81 (Figure 2), down to the smallest Local Group irregulars, such as Gr 8 (see Skillman *et al.* 1988b). This ubiquitous empirical relationship must be accounted for by any convincing model of galactic chemical evolution. It may simply reflect the ease with which hot, metal–rich supernova ejecta can escape from low-mass galaxies, but also implies that there are major similarities in the evolution of all galaxies. A second important issue is that of the *relative evolution of different elements* as compared to oxygen. Probably the clearest case is that of nitrogen, for which there seem to multiple nucleosynthetic sites. Some nitrogen must be made by an effectively "primary" mechanism, either in massive stars via supernovae or Wolf-Rayet star winds, or in intermediate-mass stars by triple-alpha followed by CN-cycle burning and mixing. As the overall metallicity increases and the stellar population evolves, some nitrogen begins to be produced according to the "secondary" formula. The other elements studied so far, chiefly S, Ar, Ne, and C, mostly seem to follow O/H (but note that abundances for the iron-peak elements are not accessible for extragalactic H II regions).

Next, there remain a number of what I will call *"zero-point" questions.* There remains a factor of two or three systematic difference between the nebular and stellar abundance scales, $(O/H)_{Orion} \neq (O/H)_O$. This may either be a true abundance difference, or it may arise from different systematic errors between the two types of analysis. For example, the nominal nebular abundances are always too low if allowance is not made for the presence of temperature inhomogeneities (see 2.1.2.). This issue can be investigated by obtaining multiple diagnostics for the physical parameters, from weak optical lines or infrared measurements, or by spatially resolving the spatial structure. A second problem is the discrepancy between the N/O ratio as determined from optical and infrared techniques, $(N^{++}/O^{++}) \neq (N^{+}/O^{+})$. Possible factors which may contribute to the resolution of this problem include ionization-structure effects, non-collisional excitation of the optical lines (Rubin 1986), or the presence of significant density inhomogeneities (Rubin 1989).

Finally, there is the question of the origin of the *"excitation"-metallicity relation,* the empirical observation that regions with low O/H abundances tend to be more highly ionized. As mentioned above, there are number of different ways in which such a relationship could arise. The first possible cause is a systematic change in the geometry or clumpiness of nebulae as a function of metallicity, producing an effective decrease in the ionization parameter U (see Section 2.2.) as the O/H abundance increases. This possibility has been emphasized by Mathis (1985) and by Dopita and Evans (1986). The second mechanism for lowering the nebular ionization level is by softening the radiation field, which can be accomplished in various ways. Balick and Sneden (1976) pointed out that more metal-rich stars will have deeper ionization edges in their atmospheres. However, calculations using recent model atmospheres stars show that the metal-edge effect in the photospheres is diminished for surface temperatures as high as those in the metal-poor H II regions (Skillman 1989). Dust internal to the nebula might also soften the radiation field, but only for certain assumed optical dust properties, as discussed above. A number of workers have advocated the idea that there is actually a systematic change of some sort in the initial mass function (IMF) with metallicity. Shields and Tinsley (1976) suggested a smaller upper limit to the stellar masses, while Terlevich (1985) and others prefer to invoke a change in the IMF slope. (See Scalo 1986, however, for a critical discussion regarding the lack of direct evidence for a varying IMF.)

As hinted above, there *is* another way to soften the radiation field, without invoking a varying IMF. Stars with different metallicities are predicted to follow different evolutionary tracks. In particular, metal-rich stars tend to have stronger winds, and therefore will spend more time in the Wolf-Rayet (W-R) stage, or else the W-R stage may set in at lower stellar masses. Recent evolutionary models therefore predict that there should be a higher incidence of W-R stars in more metal–rich populations, the trend that is seen in these extragalactic H II regions (Arnault, Kunth, and Schild 1989). Now, since most W-R stars are cooler than main sequence stars of the same mass, the integrated UV radiation field will be effectively "softened" in such regions (e.g. Kunth and Joubert 1985). The presence of W-R stars provides a double "whammy" to the H II region; not only do these stars soften the overall radiation field, but they also provide a natural way to sweep out the local volume and drive down the filling factor, f, and ionization parameter, U (e.g. Dopita and Evans 1986). In lower-metallicity regions, with fewer or no W-R stars, the gas will be less compressed and the ionization parameter larger. It thus seems possible that metallicity-dependent evolution, rather than a metallicity-dependent IMF, can fully account for the empirical ionization-abundance correlation in extragalactic H II regions; this hypothesis ought to be examined further and tested against new observations in the coming years.

7. Future Prospects

I would like to emphasize the diverse strategies and instruments that should be employed in future studies of extragalactic H II regions. Large-aperture, ground-based telescopes will play a key role in obtaining measurements of weak lines (e.g. [O III] 4363Å) in distant, low-luminosity, and high-metallicity nebulae, enabling more reliable "direct" abundance determinations to be made. A good deal more can also be done with such telescopes in studying spatial inhomogeneities (Rosa 1983). Naturally, the Hubble Space Telescope will also be a prime tool for studying extragalactic H II regions. Not only will it yield carbon abundances for many more objects, but the high spatial resolution obtainable with HST will allow better isolation of the nebular spectrum from the stellar continuum, as well as yielding a great deal of information about the properties of the ionizing stars.

Other tools besides optical/UV telescopes should not be overlooked. In particular, the investigation of infrared emission lines from extragalactic H II regions is likely to prove rewarding (e.g. Dinerstein 1986). As of now the surface has barely been scratched, especially for the far-infrared lines; the current largest far-infrared telescope, NASA's KAO, can barely reach the very nearby H II regions in M 82 (Duffy *et al.* 1987) and M 33 (Dinerstein *et al.* 1987). Once instruments such as SOFIA (a 3 m-class airborne telescope) and SIRTF (the Space Infrared Telescope Facility) are in operation, many of the regions which have been the subjects of intense study in the optical will become accessible in the infrared. Finally, one should not overlook the potential contribution of fast contemporary computers in making it practical to undertake realistic calculations of stellar model atmospheres and stellar evolutionary tracks which include non-LTE and metallicity effects and stellar winds (e.g. Oloffson 1989). The prospects seem excellent for answering some of the unsettled questions listed above, in the near future.

Acknowledgments. I would like to thank the meeting organizers for providing support which enabled me to participate in this conference, and Don Garnett and Evan Skillman for many helpful discussions during the preparation of this paper.

8. References

Aaronson, M. 1985, in *Star-forming Dwarf Galaxies* , ed. Kunth, D., Thuan, T.X., and Van, J.T.T. (Gif sur Yvette: Editions Frontieres), p. 125.
Aller, L.H. 1942, *Ap. J.*, **95**, 52.
____. 1984, *Physics of Thermal Gaseous Nebulae* (Dordrecht: Reidel).
Alloin, D., Collin-Souffrin, S., Joly, M., and Vigroux, L. 1979, *Astr. Ap.*, **78**, 200.
Arnault, Ph., Kunth, D., and Schild, H. 1989, *Astr. Ap.*, in press.
Audouze, J., and Tinsley, B.M. 1976, *Ann. Rev. Astr. Ap.*, **14**, 43.
Balick, B., and Sneden, C. 1976, *Ap. J.*, **208**, 336.
Boesgaard, A., and Steigman, G. 1985, *Ann. Rev. Astr. Ap.*, **23**, 319.
Campbell, A. 1988, *Ap. J.*, **335**, 644.
Campbell, A.W., Terlevich, R., and Melnick, J. 1986, *M.N.R.A.S.*, **223**, 811.
Chiosi, C., and Maeder, A. 1986, *Ann. Rev. Astr. Ap.*, **24**, 329.
Clayton, C.A. 1988, *M.N.R.A.S.*, **231**, 191.
Clayton, D.D. 1986, *Pub. A.S.P.*, **98**, 968.
Clegg, R.E.S. 1987, *M.N.R.A.S.*, **229**, 31P.
Conti, P.S., and Massey, P. 1981, *Ap. J.*, **249**, 271.
Copetti, M.V.F., Pastoriza, M.G., and Dottori, H.A. 1985, *Astr. Ap.*, **152**, 427.
Davidson, K., and Kinman, T.D. 1985, *Ap. J. Suppl.*, **58**, 321.
Davidson, K., Kinman, T.D., and Friedman, S.D. 1989, *A. J.*, **97**, 1591.
De Loore, C.W.H., Willis, A.J., and Laskarides, A. 1986, *IAU Symposium 116, Luminous Stars and Associations in Galaxies* (Dordrecht: Reidel).
Dennefeld, M., and Stasinska, G. 1983, *Astr. Ap.*, **118**, 234.
Diaz, A.I., Terlevich, E., Pagel, B.E., Vilchez, J.M., and Edmunds, M.G. 1987, *M.N.R.A.S.*, **226**, 19.
Diaz, A.I., and Tosi, M. 1986, *Astr. Ap.*, **158**, 60.
Dinerstein, H.L. 1980, *Ap. J.*, **237**, 486.
____. 1986, *Pub. A.S.P.*, **86**, 979.
Dinerstein, H.L., Erickson, E.F., Haas, M.R., and Werner, M.W. 1987, *Bull. A.A.S.*, **19**, 1018.
Dinerstein, H.L., Lester, D.F., and Werner, M.W. 1985, *Ap. J.*, **291**, 561.
Dinerstein, H.L., and Shields, G.A. 1986, *Ap. J.*, **311**, 45.
D'Odorico, S., and Rosa, M. 1981, *Ap. J.*, **248**, 1015.
D'Odorico, S., Rosa, M., and Wampler, E.J. 1983, *Astr. Ap. Suppl.*, **53**, 97.
Dopita, M.A. 1981, *Ap. J.*, **246**, 65.
____. 1990, this volume.
Dopita, M.A., and Evans, I.N. 1986, *Ap. J.*, **307**, 431.
Duffy, P.B., Erickson, E.F., Haas, M.R., and Houck, J.R. 1987, *Ap. J.*, **315**, 68.
Dufour, R.J. 1986, *Pub. A.S.P.*, **98**, 1025.
Dufour, R.J., Talbot, R.J., Jensen, E.B., and Shields, G.A. 1980, *Ap. J.*, **236**, 119.
Dufour, R.J., Garnett, D.R., and Shields, G.A. 1988, *Ap. J.*, **332**, 752.
Edmunds, M.G., and Pagel, B.E.J. 1978, *M.N.R.A.S.*, **185**, 77P.
____. 1984, *M.N.R.A.S.*, **211**, 507.
Evans, I.N. 1986, *Ap. J.*, **309**, 544.
Evans, I.N., and Dopita, M.A. 1985, *Ap. J. Suppl.*, **58**, 125.
Ferland, G. J. 1986, *Ap. J. (Letters)*, **310**, L67.
French, H.B. 1980, *Ap. J.*, **240**, 41.
French, H.B., and Miller, J.S. 1981, *Ap. J.*, **248**, 486.
Gallagher, J.S., and Hunter, D.A. 1984, *Ann. Rev. Astr. Ap.*, **22**, 37.
Garnett, D.R. 1989a, *Ap. J.*, **345**, 282.

282

_____. 1989b, Ph.D. thesis, University of Texas at Austin.

_____. 1989c, in preparation.

Garnett, D.R., and Shields, G.A. 1987, *Ap. J.*, **317**, 82.

Güsten, R., and Mezger, P.G. 1982, *Vistas Astr.*, **26**, 159.

Hawley, S.A. 1978, *Ap. J.*, **224**, 417.

Hawley, S.A., and Grandi, S.A. 1977, *Ap. J.*, **217**, 420.

Herter, T. 1989, to appear in *Proceedings of the 22nd ESLAB Symposium, Infrared Spectroscopy in Astronomy,* ed. M. Kessler (ESO).

Herter, T., Helfer, H.L., and Pipher, J.L. 1983, *Astr. Ap. Suppl.*, **51**, 195.

Hodge, P.W., and Kennicutt, R.C. 1983, *An Atlas of H II Regions in 125 Galaxies,* PAPS Document ANJOA88-296-300 (NY:AIP).

Hodge, P.W., Lee, M.G., and Kennicutt, R.C. 1989, *Pub. A.S.P.*, **101**, 640.

Hodge, P.W., and Wright, F.W. 1967, *The Large Magellanic Cloud* (Washington, D.C.: Smithsonian Press).

_____. 1977, *The Small Magellanic Cloud* (Seattle: Univ. Washington Press).

Hoffman, G.L., Helou, G., Salpeter, E.E., and Lewis, B.M. 1989, *Ap. J.*, **339**, 812.

Israel, F.P., Gatley, I., Matthews, K., and Neugebauer, G. 1982, *Astr. Ap.*, **105**, 229.

Israel, F.P., Goss, W.M., and Allen, R.J. 1975, *Astr. Ap.*, **40**, 421.

Jensen, E.B., Strom, K.M., and Strom, S.E. 1976, *Ap. J.*, **209**, 748.

Kennicutt, R.C. 1984, *Ap. J.*, **287**, 116.

_____. 1988, *Ap. J.*, **334**, 144.

Kennicutt, R.C., and Chu, Y.-H. 1988, *A. J.*, **95**, 720.

Kennicutt, R.C., Jr., and Edgar, B.K., and Hodge, P.W. 1989, *Ap. J.*, **337**, 761.

Kennicutt, R.C., Jr., and Keel, W.C., and Blaha, C.A. 1989, *A. J.*, **97**, 1022.

Kinman, T.D. 1984, in *Astronomy with Schmidt-Type Telescopes*, ed. M. Capaccioli (Dordrecht: Reidel), p. 409.

Kinman, T.D., and Davidson, K. 1981, *Ap. J.*, **243**, 127.

Kunth, D. 1985, in *Star-forming Dwarf Galaxies* , ed. Kunth, D., Thuan, T.X., and Van, J.T.T. (Gif sur Yvette: Editions Frontieres), p. 185.

_____. 1986, *Pub. A.S.P.*, **98**, 1025.

Kunth, D., and Joubert, M. 1985, *Astr. Ap.*, **142**, 411.

Kunth, D., and Sargent, W.L.W. 1983, *Ap. J.*, **273**, 81.

_____. 1986, *Ap. J.*, **300**, 496.

Kunth, D., Sargent, W.L.W., and Kowal, C. 1981, *Astr. Ap. Suppl.*, **44**, 229.

Kunth, D., Thuan, T.X., and Van, J.T.T., eds. 1985, *Star-forming Dwarf Galaxies* (Gif sur Yvette: Editions Frontieres).

Kwitter, K.B., and Aller, L.H. 1981, *M.N.R.A.S.*, **195**, 939.

Lacey, C.G., and Fall, S.M. 1985, *Ap. J.*, **290**, 154.

Larson, R.B. 1974, *M.N.R.A.S.*, **169**, 229.

_____. 1976, *M.N.R.A.S.*, **176**, 31.

_____. 1986, *M.N.R.A.S.*, **218**, 409.

Larson, R.B., and Dinerstein, H.L. 1975, *Pub. A.S.P.*, **87**, 911.

Lester, D.F., Bregman, J.D., Witteborn, F.C., Rank, D.M., and Dinerstein, H.L. 1981, *Ap. J.*, **248**, 524.

Lester, D.F., Dinerstein, H.L., Werner, M.W., Watson, D.M., and Genzel, R. L. 1983, *Ap. J.*, **271**, 618.

Lester, D.F., Dinerstein, H.L., Werner, M.W., Watson, D.M., Genzel, R.L., and Storey, J.W.V. 1987, *Ap. J.*, **320**, 573.

Lequeux, J., Maucherat-Joubert, M., Deharveng, J.M., and Kunth, D. 1981, *Astr. Ap.*, **103**, 305.

Lequeux, J., Peimbert, M., Rayo, J.F., Serrano, A., and Torres-Peimbert, S. 1979, *Astr. Ap.*, **80**, 155.

Maeder, A., Lequeux, J., and Azzopardi, M. 1980, *Astr. Ap.*, **90**, L17.

Mathis, J. S. 1982, *Ap. J.*, **261**, 195.

_____. 1985, *Ap. J.*, **291**, 247.

_____. 1986, *Pub. A.S.P.*, **98**, 995.

Matteucci, F., and Tosi, M. 1985, in *Production and Distribution of the C, N, O Elements*, eds. J. Danziger, F. Matteucci, and K. Kjar (Garching: ESO), p. 387.

Matteucci, F., and Francois, P. 1989, *M.N.R.A.S.*, **239**, 885.

Mayor, M. 1979, *Mem.Soc.Astr.Ital.*, **50**, 157.

McCall, M.L., Rybski, P.M., and Shields, G.A. 1985, *Ap. J. Suppl.*, **57**, 1.

Melnick, J., Terlevich, R., and Eggleton, P.P. 1985, *M.N.R.A.S.*, **216**, 255.

Mendoza, C. 1983, in *IAU Symposium 103, Planetary Nebulae*, ed. D. Flower (Dordrecht: Reidel), p. 143.

Natta, A., Panagia, N., and Preite-Martinez, A. 1980, *Ap. J.*, **242**, 596.

Olofsson, K. 1989, *Astr. Ap. Suppl.*, **80**, 317.

Osterbrock, D.E. 1989, *Astrophysics of Gaseous Nebulae and Active Galactic Nuclei* (Mill Valley, CA: University Science Books).

Pagel, B.E.J. 1985, in *Production and Distribution of the C, N,O Elements*, eds. J. Danziger, F. Matteucci, and K. Kjar (Garching: ESO), p. 155.

_____. 1986, *IAU Highlights Astr.*, **7**, 551.

_____. 1987, in *NATO Advanced Study Workshop on The Galaxy*, eds. G. Gilmore and B. Carswell (Dordrecht: Reidel), p. 341.

Pagel, B.E.J., and Edmunds, M.G. 1981, *Ann. Rev. Astr. Ap.*, **19**, 77.

Pagel, B.E.J., Edmunds, M.G., Blackwell, D.E., Chun, M.S., and Smith, G. 1979, *M.N.R.A.S.*, **189**, 95.

Pagel, B.E.J., Edmunds, M.G., and Smith, G. 1980, *M.N.R.A.S.*, **184**, 569.

Pagel, B.E.J., and Simonson, E.A. 1989, *Rev. Mex. Astr. Ap.*, in press.

Pagel, B.E.J., Terlevich, R.J., and Melnick, J. 1986, *Pub. A.S.P.*, **98**, 1005.

Pantelaki, I. 1988, Ph.D. thesis, Rice University.

Peimbert, M. 1967, *Ap. J.*, **150**, 825.

_____. 1985, in*Star-forming Dwarf Galaxies* (Gif sur Yvette: Editions Frontieres), p. 403.

Peimbert, M., and Costero, R. 1969, *Bol. Obs. Tonantzintla y Tacubaya*, **5**, 3.

Peimbert, M., and Torres-Peimbert, S. 1974, *Ap. J.*, **193**, 327.

_____. 1977, *M.N.R.A.S.*, **179**, 217.

_____. 1987, *Rev. Mex. Astr. Ap.*, **14**, 540.

Peña, M. 1986, *Pub. A.S.P.*, **98**, 1061.

Pipher, J.L., Helfer, H.L., Herter, T., Briotta, D.A., Houck, J.R., Willner, S.P., and Jones, B. 1984, *Ap. J.*, **285**, 174.

Rayo, J.F., Peimbert, M., and Torres-Peimbert, S. 1982, *Ap. J.*, **255**, 1.

Renzini, A., and Voli, M. 1981, *Astr. Ap.*, **94**, 175.

Richter, O.-G., and Rosa, M. 1984, *Astr. Ap.*, **140**, L1.

Rosa, M. 1983, *IAU Highlights Astronomy*, **6**, 625.

_____. 1985, *The Messenger*, **39**, 15.

Rosa, M., and D'Odorico, S. 1986, in *IAU Symposium 116, Luminous Stars and Associations in Galaxies*, eds. De Loore, C.W.H., Willis, A.J., and Laskarides, A. (Dordrecht: Reidel), p. 355.

Rubin, R. H. 1969, *Ap. J.*, **155**, 841.

_____. 1986, *Ap. J.*, **309**, 334.

_____. 1989, *Ap. J. Suppl*, , **69**, 897

Rubin, R.H., Simpson, J.P., Erickson, E.F., and Haas, M.R. 1988, *Ap. J.*, **327**, 377.
Sarazin, C.L. 1977, *Ap. J.*, **211**, 772.
Sargent, W.L.W., and Searle, L. 1970, *Ap. J.(Letters)*, **162**, L155.
Scalo, J.M. 1986, in *IAU Symposium 116, Luminous Stars and Associations in Galaxies*, eds. De Loore, C.W.H., Willis, A.J., and Laskarides, A. (Dordrecht: Reidel), p. 451.
Searle, L. 1971, *Ap. J.*, **168**, 327.
Searle, L., and Sargent, W.L.W. 1972, *Ap. J.*, **173**, 25.
Serrano, A., and Peimbert, M. 1983, *Rev. Mex. Astr. Ap.*, **8**, 117.
Shaver, P.A., Kunth, D., and Kjär, K., eds. 1983, *Proc. ESO Workshop on Primordial Helium* (Garching: ESO).
Shaver, P.A., McGee, R.A., Danks, A.C., and Pottasch, S.R. 1983, *M.N.R.A.S.*, **204**, 53.
Shields, G.A. 1974, *Ap. J.*, **193**, 335.
____. 1985, in *Star-forming Dwarf Galaxies* (Gif sur Yvette: Editions Frontieres), p. 197.
____. 1986, in *Proceedings of the Workshop on Model Nebulae*, ed. D. Pequignot (Paris: Observatoire de Meudon), p.225.
____. 1990, *Ann. Rev. Astr. Ap.*, **28**, in press.
Shields, G.A., and Searle, L. 1978, *Ap. J.*, **222**, 821.
Shields, G.A., and Tinsley, B.M. 1976, *Ap. J.*, **203**, 66.
Simpson, J.P., Rubin, R.H., Erickson, E.F., and Haas, M.R. 1986, *Ap. J.*, **311**, 895.
Skillman, E.D. 1985, *Ap. J.*, **290**, 449.
____. 1989, *Ap. J.*, in press.
Skillman, E.D., Bothun, G.D., Murray, M.A., and Warmels, R.H. 1988a, *Astr. Ap.*, **185**, 61.
Skillman, E.D., Melnick, J., Terlevich, R., and Moles, M. 1988b, *Astr. Ap.*, **196**, 31.
Skillman, E.D., Kennicutt, R.C., and Hodge, P.W. 1989, *Ap. J.*, in press.
Skillman, E.D., Terlevich, R., and Melnick, J. 1989, *M.N.R.A.S.*, **240**, 563.
Smith, H.E. 1975, *Ap. J.*, **199**, 591.
Stasinska, G. 1978, *Astr. Ap. Suppl.*, **32**, 429.
____. 1980, *Astr. Ap.*, **85**, 359.
Stauffer, J.R., and Bothun, G.D. 1984, **A. J.**, **89**, 1702.
Steigman, G., Gallagher, J.S., and Schramm, D.N. 1989, *Comm. Astrophys.*, **14**, 97.
Strömgren, B. 1939, *Ap. J.*, **89**, 526.
Talent, D.L. 1980, Ph.D. thesis, Rice University.
Terlevich, R. 1985, in *Star-forming Dwarf Galaxies* , ed. Kunth, D., Thuan, T.X., and Van, J.T.T. (Gif sur Yvette: Editions Frontieres), p. 395.
Terlevich, R., and Melnick, J. 1981, *M.N.R.A.S.*, **195**, 839.
Thuan, T.X., Montmerle, T., and Van, J.T.T., eds. 1987, *Starbursts and Galaxy Evolution* (Gif sur Yvette: Editions Frontieres).
Tinsley, B.M. 1980, *Fund. Cosmic Phys.*, **5**, 287.
Torres-Peimbert, S., Peimbert, M., and Fierro, J. 1989, *Ap. J.*, **345**, 186.
Viallefond, F. 1985, in *Star-forming Dwarf Galaxies* , ed. Kunth, D., Thuan, T.X., and Van, J.T.T. (Gif sur Yvette: Editions Frontieres), p. 207.
Vigroux, L, Stasinska, G., and Comte, G. 1987, *Astr. Ap.*, **172**, 15.
Vilchez, J.M., and Pagel, B.E.J. 1988, *M.N.R.A.S.*, **231**, 257.
Vilchez, J.M., Pagel, B.E.J., Diaz, A.I., Terlevich, E., and Edmunds, M.G. 1988, *M.N.R.A.S.*, **235**, 633.
Watson, D.M., Storey, J.W.V., Townes, C.H., and Haller, E.E. 1981, *Ap. J.*, **250**, 605.
Webster, B.L., and Smith, M.G. 1983, *M.N.R.A.S.*, **204**, 743.

Wheeler, J.C., Sneden, C., and Truran, J.W. 1989, *Ann. Rev. Astr. Ap.*, **27**, 279.
Yang, J., Turner, M.S., Steigman, G., Schramm, D., and Olive, K.A. 1984, *Ap. J.*, **281**, 493.
Zamorano, J., and Rego, M. 1985, *Astr. Ap. Suppl.*, **62**, 173.
Zaritsky, D., Elston, R., and Hill, J.M. 1989, *A. J.*, **97**, 97.
Zuckerman, B., and Aller, L.H. 1986, *Ap. J.*, **301**, 772.

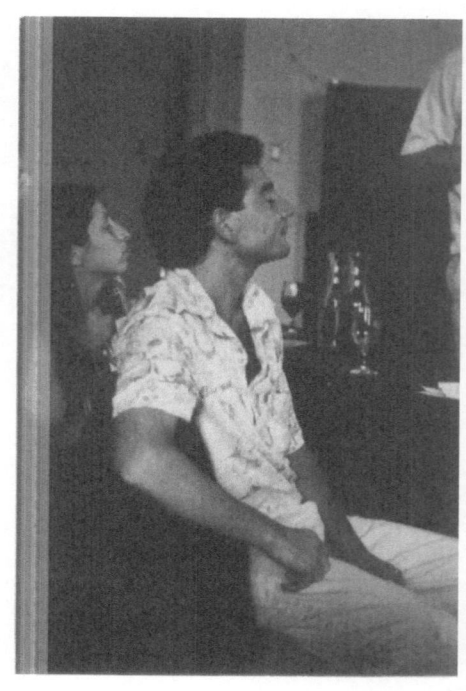

Thermal Phases of the Interstellar Medium in Galaxies

Mitchell C. Begelman
Joint Institute for Laboratory Astrophysics
University of Colorado and National Institute of Standards and Technology
Boulder, CO 80309-0440

ABSTRACT. This review deals with the theory of multiphase media in astrophysical systems. I discuss the basic reasons for the existence of multiple thermal phases, and the fundamental connection between multiphase media and thermal instability. After describing important examples of multiphase media, I examine the interactions among phases, *i.e.*, mass exchange driven by thermal conduction and hydrodynamic ablation. Mass exchange may compete with radiative heating and cooling for control of the thermal state of the hot phase, and may alter the thermal stability properties of the system.

1. Introduction

Astrophysical gases are often highly inhomogeneous, with two or more "thermal phases" coexisting in rough pressure balance with one another. Compared to the scales of typical inhomogeneities, the transitions between neighboring regions of different temperature (and density) can be quite sharp. Such systems are most often modeled as consisting of cold clouds, filaments or sheets embedded in a hotter intercloud medium, although there may be cases in which a model consisting of hot bubbles in a cold matrix is more appropriate. Usually the topology of the phases is highly uncertain, but the conditions which lead to their existence are more reliably established. The temperatures of the phases sometimes differ by orders of magnitude, and are frequently set within rather narrow ranges by the details of atomic and molecular processes or by the spectrum of ambient radiation. It should be stressed that thermal pressure balance may not be exact, *e.g.*, where magnetic fields or cosmic rays supply a significant fraction of the pressure in one or more phases, where self-gravity or turbulent pressure are dynamically important, where ram pressure (associated with differential motion of the phases) provides part of the confinement, in the case of a cool cloud evaporating suprathermally in a hot background (Balbus and McKee 1982), or when there is simply too little time for a system to achieve dynamical equilibrium. Although the concepts of multiphase media are not generally used to describe regions which are wildly out of dynamical equilibrium with their surroundings (such as material behind a propagating shock front), localized pressure fluctuations may be an important means of

H. A. Thronson, Jr. and J. M. Shull (eds.), The Interstellar Medium in Galaxies, 287–304.

transferring material between phases (Wang and Cowie 1988).

This review presents an overview of the theory of thermal phases, with particular attention to their role in the interstellar media of galaxies. In §2 I discuss the basic reasons for the existence of multiphase media, and show the connection between multiple phases and thermal instability. I also give examples of multiphase systems which are important in astrophysics. Since the phases are in physical contact, it is unrealistic to treat them as being isolated from one another. §3 deals with the principal interactions among phases, thermal conduction and ablation. Taking these interactions into account is particularly important if one wishes to understand the temporal evolution of multiphase media; §4 deals with the consequences of mass exchange and with evolutionary models. The "state of the art" is summarized in §5. Much of the original material presented in this review was developed in collaboration with C. F. McKee, and is described in greater detail in Begelman and McKee (1990).

2. Theory of Multiphase Media

2.1 WHY THERMAL PHASES?

The existence of multiple thermal phases is made possible by the flow of energy into and out of a system. Multiple phases do not develop in systems which are thermodynamically isolated from their surroundings. If $\Gamma(n, T, x_j)$ is the heating rate per particle and $\Lambda(n, T, x_j)$ is the cooling function, then the equation of thermal equilibrium may be written

$$n^2\Lambda - n\Gamma \equiv n^2\mathcal{L} = 0, \tag{1}$$

where n is the density of hydrogen nuclei, T is the temperature, and x_j represents the fractional concentrations of various species, $x_j \equiv n_j/n$. The pressure is given by $p = x_t nkT$, where $x_t = \Sigma x_j$ is the number of particles per hydrogen nucleus. An equation analogous to (1) determines the ionization equilibrium. Γ, and sometimes Λ (e.g., in the case of inverse Compton cooling), may also depend on the magnitude of some external heating or ionization agent, which has energy density u_Γ. If u_Γ is held fixed then the solution of the equilibrium equations generates a curve in the $p - n$, $p - V$ (where $V \equiv 1/n$ is the specific density) or $p - T$ plane which separates the heating region ($\Gamma > n\Lambda$) from the cooling region ($\Gamma < n\Lambda$). In general these curves may have complex shapes and be multivalued.

If there are two or more values of n (or, equivalently, of T), which correspond to a given pressure, then a *multiphase equilibrium* is possible: a relatively cool, dense region can coexist with one or more warmer, less dense regions in pressure equilibrium. If this configuration is thermally stable (see §2.2 below), and if there is no mass exchange between phases, then this equilibrium can persist indefinitely. Simple generic examples of multiphase equilibria are shown in Figure 1. In all cases we have assumed that there is a single stable "cloud" phase with a fixed temperature T_{cl}. Figs. 1b and 1d both show systems with two stable phases, while the other panels show systems with only one stable phase. More realistic phase diagrams may show three or more stable phases, e.g., Lepp et al. (1985).

In most cases of astrophysical interest, Γ is linear in u_Γ. If both Λ and Γ are independent of density (as in particle or photon heating and two-body cooling) or depend on density through the ratio n/u_Γ (as in the case of inverse Compton cooling), then both the ionization level and the temperature depend on n and u_Γ only through the combination

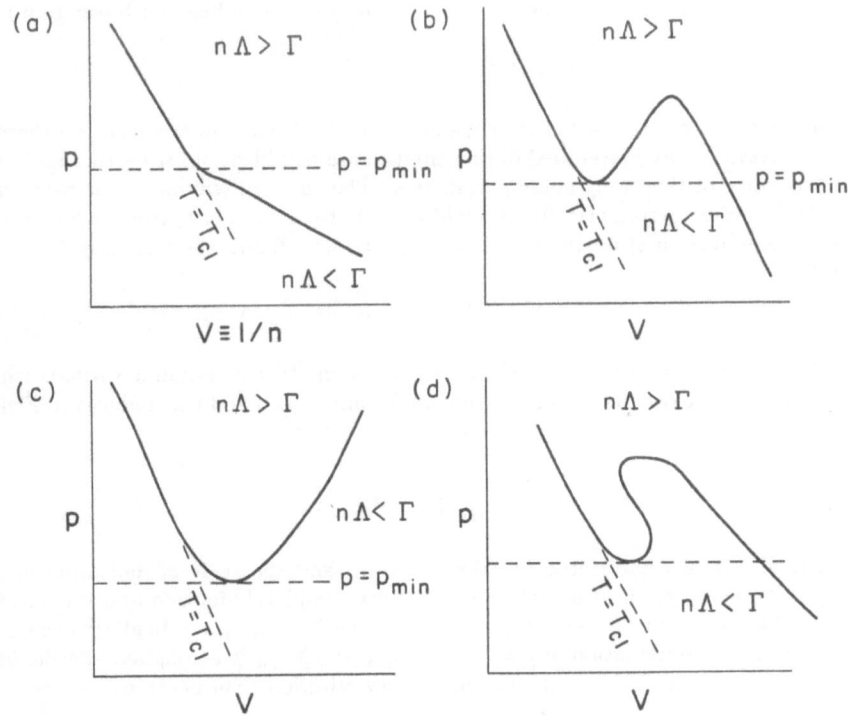

Fig. 1: Radiative thermal equilibrium in the $p - V$ plane. Cooling exceeds heating above the solid line. Clouds exist for pressures exceeding p_{min}; their temperature is fixed at T_{cl}. a) One thermally stable phase. b) Two thermally stable phases separated by a thermally unstable phase. c) Thermally stable clouds in a thermally unstable intercloud medium. d) If $p(V)$ is multivalued (corresponding to gas at a given density having one of several temperatures), the gas may be isochorically as well as isobarically unstable.

n/u_Γ or, equivalently, p/u_Γ. This similarity variable is very useful for characterizing the state of gas heated by cosmic rays (Dalgarno and McCray 1972) or radiation (e.g., Tarter, Tucker and Salpeter 1969; Davidson 1972; Krolik, McKee and Tarter 1981); in various forms it is referred to as the ionization parameter.

Not all thermal phases which are observed in astrophysical systems correspond to stable equilibria. Examples of systems which exhibit long-lived non-equilibrium hot phases, in pressure balance with a stable cold phase, are the three–phase interstellar medium (McKee and Ostriker 1977) and cooling flows in clusters of galaxies (Sarazin 1986, and references therein). The non-equilibrium phases in these systems are in fact thermally unstable, and are observable only because their cooling time scales are extremely long (Spitzer 1956). An accurate analysis of such systems requires the treatment of time-dependence (McKee and Ostriker 1977) and hydrodynamical effects such as buoyancy (Balbus and Soker 1989).

2.2 CONNECTION WITH THERMAL INSTABILITY

There is an intimate connection between the existence of thermal phases and the thermal stability of a system: any system exhibiting multiphase equilibria *must* be thermally unstable over a range of thermodynamic parameters. The thermal stability of astrophysical gases was first studied systematically by Field (1965). His instability criterion was generalized to non-equilibrium systems by Balbus (1986a), who found the following condition for instability:

$$\left[\frac{\partial(n\mathcal{L}/T)}{\partial s}\right]_A < 0. \tag{2}$$

Here s is the entropy per hydrogen nucleus and A is some thermodynamic variable which is held constant during the perturbation. In equilibrium, $\mathcal{L} = 0$ and this reduces to Field's instability criterion

$$\left(\frac{\partial \mathcal{L}}{\partial s}\right)_A < 0 \qquad (\mathcal{L} = 0). \tag{3}$$

In general, s is a complicated function of n, T, and the state of ionization of the gas. However, in many applications the gas is almost completely ionized and the entropy function may be approximated by the expression for an ideal gas, $s \sim \ln pV^{5/3}+\text{const}$. If A is some power law combination of p and V, then $T(\partial s/\partial T)_A$ is a constant specific heat which is positive for cases of interest. The instability criterion then becomes

$$\left[\frac{\partial(n\mathcal{L}/T)}{\partial T}\right]_A < 0. \tag{4}$$

Since the cooling time is proportional to $T/n\mathcal{L}$, this criterion can be rephrased as stating that instability occurs if the cooling time increases with temperature (Balbus 1986a).

If the gas is in equilibrium ($\mathcal{L} = 0$), the instability criterion (4) reduces to

$$\left(\frac{\partial \mathcal{L}}{\partial T}\right)_A < 0 \qquad (\mathcal{L} = 0). \tag{5}$$

Field (1965) showed that for the equilibrium case the isobaric criterion $(\partial \mathcal{L}/\partial T)_p < 0$ is usually the correct one to apply. However, if the system is large enough that the sound crossing time is long compared to the heating or cooling times, then for long wavelengths the isochoric criterion $(\partial \mathcal{L}/\partial T)_V < 0$ is applicable.

The stability criterion (5) may be interpreted geometrically in terms of the equilibrium curve (Figure 1). Typically the cooling region ($n\Lambda > \Gamma$) lies above the heating region because the cooling rate usually increases faster with n and T than does the heating rate. If, on the other hand, the heating region lay above the cooling region, then over much of the curve (wherever $p(V)$ is single–valued) one would have $(\partial \mathcal{L}/\partial T)_V < 0$ and the equilibrium would be isochorically unstable. In this case systems large enough that the sound crossing time is much greater than the heating and cooling times could be unstable even where smaller systems are isobarically stable. This situation does not arise in practice and we therefore assume that the cooling region lies above the heating region in the $p - V$ plane, as shown in Figure 1.

The slope of the equilibrium curve in the $p - V$ plane is directly related to the stability of the system since

$$\left(\frac{dp}{dV} \right)_{\mathcal{L}=0} = - \left(\frac{\partial \mathcal{L}}{\partial T} \right)_p \bigg/ \left(\frac{\partial \mathcal{L}}{\partial T} \right)_V \qquad (6)$$

(Field 1965). For cases in which $p(V)$ is single valued (as in Fig. 1a-c), the condition that the cooling region lie above the heating region implies that the system is isochorically stable, so that the denominator in equation (6) is positive; hence, in this case isobarically stable regions have a negative slope in the $p - V$ plane, whereas unstable regions have a positive slope. The condition for a multiphase equilibrium is that $V(p)$ be a multivalued function, which is equivalent to having $d \ln p / d \ln V$ change sign. Thus, a necessary and sufficient condition for the existence of a multiphase equilibrium is that the system be thermally unstable over a finite range of V. This proves the assertion at the beginning of this section. Fig. 1d illustrates a case in which $p(V)$ is multivalued over a range in V. Such a system can exhibit both isochoric and isobaric instability, where the equilibrium curve has a *negative* slope in the $p - V$ plane.

A system with two stable phases (*e.g.*, Fig. 1b) may be used to illustrate the inevitability of multiple phases under certain circumstances. A characteristic feature of two–phase systems is that the cold phase cannot exist below some minimum pressure p_{min}, while the hot phase cannot exist above some maximum pressure p_{max}. The condition that there be two stable phases implies that $p_{max} > p_{min}$. Now consider a homogeneous system with a density $n_1 < \bar{n} < n_2$, as shown in Fig. 2a. Such a system is clearly unstable in its homogeneous state. However, it is always possible to stabilize the system by making it *inhomogeneous*, while keeping the mean density constant (Fig. 2b). The trick is to put most of the mass in the cold phase, with density $n_c > n_2$, while a small fraction of the matter forms a hot intercloud medium, with density $n_h < n_1$ and temperature T_h. Pressure balance requires $n_c / n_h = T_h / T_{cl}$. If f is the filling factor in cold gas, then the mean density constraint is $\bar{n} = (1 - f)n_h + fn_c$, and f satisfies $T_{cl}/T_h \ll f \ll 1$ if $n_1 \ll \bar{n} \ll n_2$.

2.3 EXAMPLES OF THERMAL PHASES

Field, Goldsmith and Habing (FGH: 1969) produced the first specific model for a two–phase equilibrium of the interstellar medium (ISM), in which radiative cooling is balanced by cosmic ray heating. The two phases in the FGH model include cold clouds ($T \sim 100$ K) and a warm intercloud medium ($T \sim 10^4$ K). Other heating mechanisms which may be important (probably more important than cosmic rays [Spitzer 1978]) include diffuse UV and X-ray flux, photoelectric emission by normal grains (Draine 1978; de Jong 1980; Shull and Woods 1985) or polycyclic aromatic hydrocarbons (PAHs: d'Hendecourt and Leger 1987; Lepp and Dalgarno 1988), mechanical heating (Cox 1979), magnetoacoustic waves (Spitzer 1982; Ikeuchi and Spitzer 1984), and ion-neutral friction (Scalo 1977; Ferrière, Zweibel and Shull 1988). The characteristic temperatures of the warm and cold thermal phases are insensitive to the details of the heating processes; they simply reflect the energies of the resonance and fine-structure lines, respectively, responsible for cooling the gas. Gas at $\sim 10^4$ K may exist in a range of ionization states, and McKee and Ostriker (1977) drew a distinction between the "warm neutral medium" and a "warm (photo)ionized medium" irradiated by UV from hot stars. A molecular phase at ~ 10 K is now known to contain

a)   b)

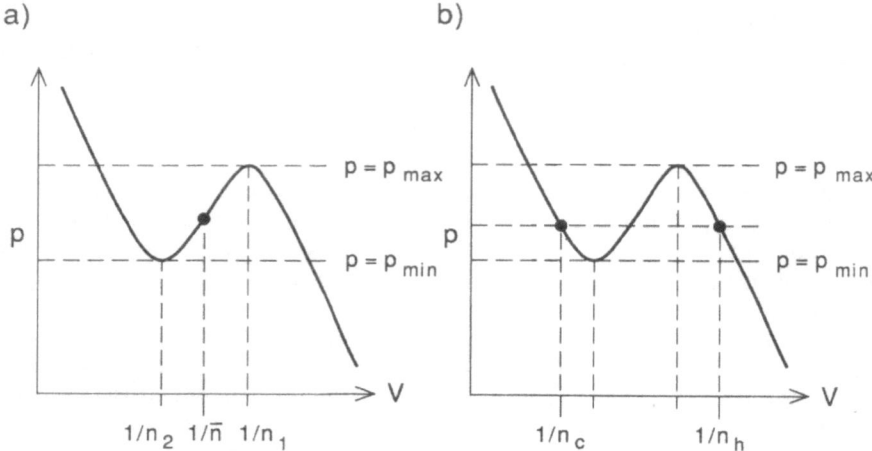

Fig. 2: *Inevitability of thermal instability in a system with a fixed mean density $\bar{n}$ and variable pressure p. Equilibrium curve is identical to that in Fig. 1b. a) Homogeneous state is thermally unstable. b) In the stable two-phase state, most of the mass is in the cold phase with $n_c > n_2$ and a small filling factor.*

most of the mass in the ISM of the Milky Way, but this component appears to form self-gravitating clouds which are out of pressure balance with the rest of the ISM. A phase diagram for these phases is computed by Lepp *et al.* (1985).

The gas which emits the broad emission lines in AGN has also been modeled as part of a stable two–phase medium (McCray 1979; Krolik, McKee, and Tarter 1981 [KMT]; Lepp *et al.* 1985; Krolik 1988). On the basis of observations, the line-emitting gas is inferred to be concentrated in many small clouds which fill a tiny fraction of the volume of the emission line region (Davidson 1972). Compton heating by the observed X–rays provides the minimum level of heating of the hot component of the medium; additional heating due to relativistic particles, radio frequency heating, cloud friction, and shocks may also be important (KMT). Cooling of the hot phase is due to inverse Compton cooling and bremsstrahlung. Clouds at $T \sim 10^4$ K can exist in pressure equilibrium with the hot intercloud medium provided the intercloud heating is strong enough to maintain the temperature above a few times 10^7 K. Recent observations have shown that it is unlikely that Compton heating alone is adequate (Fabian *et al.* 1986), so additional heating is required (Mathews and Ferland 1987). The nature of the two–phase equilibria that can occur are constrained by the mean density of the medium, $\bar{n}$ (KMT): if $\bar{n}$ is sufficiently high, then the gas can either be homogenous and cold or in two phases; if it is sufficiently low, the gas can either be homogeneous and hot or in two phases; and finally, there is usually a range of densities for which the gas *must* be in two phases (*cf.* Fig. 2 and §2.2). KMT showed that unless the temperature of the hot gas in the broad line region is well above 10^8 K, most of the mass is in the hot phase, corresponding to the hot/two–phase case.

Cox and Smith (1974) pointed out that the cooling time of interstellar gas shock-heated by supernova remnants could be longer than the interval between the passage of successive shocks. This suggestion led to the three–phase model of the ISM (McKee and Ostriker 1977), in which most of the volume is occupied by shock-heated gas. This $\sim 10^6$ K gas is an example of a non-equilibrium phase, the possibility of which was foreseen by Spitzer (1956). Because it is produced dynamically, and has a temperature of order the virial temperature of the Galaxy, it has proven very difficult to determine the fate of the hot intercloud medium. It is not at all clear whether it cools radiatively in a region close to the disk (McKee and Ostriker 1977) or is vented into the halo through "chimneys" (McCray and Kafatos 1987; Norman and Ikeuchi 1989), where it undergoes a combination of adiabatic and radiative cooling (the "Galactic fountain": Shapiro and Field 1976; Cox 1981; Wang and Cowie 1988). It is also not known whether the hot gas cools sufficiently in the halo to form clouds which eventually rain down on the disk, remains hot enough to drive a galactic wind, or somehow does both. Finally, the effects of spatial correlations among Type II supernovae (in OB associations) are just beginning to be appreciated (McCray and Kafatos 1987).

Cooling flows in elliptical galaxies and galaxy clusters are also thought to have a nonequilibrium two–phase structure. When the existence of cooling flows was first recognized (Cowie and Binney 1977; Fabian and Nulsen 1977), it was pointed out that the cooling gas should be thermally unstable to the formation of cool ($\sim 10^4$ K) filaments (Fabian and Nulsen 1977; Mathews and Bregman 1978; Cowie, Fabian and Nulsen 1980). Optical emission lines have been observed in the central regions of many cooling flows (Lynds 1970; Heckman 1981; Cowie et al. 1983; Hu, Cowie and Wang 1985; Johnstone, Fabian and Nulsen 1987; Heckman et al. 1989). However, the development of linear thermal instability is severely hampered by buoyancy (Balbus 1988; Balbus and Soker 1989), and it is not clear whether the filaments grow from finite but small perturbations or are advected inward in a highly nonlinear form (Nulsen 1986). Furthermore, the mechanism which excites the emission lines is very uncertain, and may play a role creating and maintaining the multiphase structure. Multiphase models of cooling flows have been studied by Nulsen (1986); Thomas, Fabian and Nulsen (1987); Thomas (1988); and Böhringer and Fabian (1989).

An extreme version of the cooling flow instability has been proposed to account for the masses of protogalaxies (Rees and Ostriker 1977; Silk 1977) and of globular clusters (Fall and Rees 1985). The basic idea of these models is that a self-gravitating gas cloud will fragment only when its cooling time becomes shorter than its free-fall time, and then it will develop a two–phase structure in which just enough material drops out of the hot phase to keep the cooling time roughly comparable to the free-fall time. Characteristic mass scales are determined by the Jeans mass of the cold phase in pressure balance with the hot phase. Triggering of star formation by radio lobes expanding into a protogalactic multiphase medium has been proposed (Rees 1989; Begelman and Cioffi 1989) to account for the observed radio/optical alignments in high-redshift radio galaxies (McCarthy et al. 1987; Chambers, Miley and van Breugel 1987). Cool gas in the multiphase protogalactic environment might give rise to some quasar absorption line systems (Hogan 1987) as well as the extended emission-line "fuzz" around high-redshift quasars (Rees 1988).

3. Interactions Among Phases

3.1 EVAPORATION AND CONDENSATION

Thermal conduction tries to destroy multiphase structure by erasing temperature gradients. Whether this tendency toward homogenization leads to evaporation of clouds or the condensation of hot phase onto existing clouds depends on the cooling function in the hot phase, as well as the sizes and distribution of clouds. The efficiency of thermal conduction also depends on the magnetic connectivity between the phases, which is poorly understood. Although the conductivity perpendicular to a magnetic field line is almost completely suppressed, any connection between the phases, albeit by tangled field lines, is likely to suppress the conductivity only by a factor of a few (Tribble 1989). In a fully ionized cosmic plasma the "classical" coefficient of conductivity is

$$\kappa = 5.6 \times 10^{-7} \phi_c T_e^{5/2} \qquad \text{erg s}^{-1} \text{ K}^{-1} \text{ cm}^{-1} \tag{7}$$

(Spitzer 1962; Draine and Giuliani 1984), where the factor $\phi_c \leq 1$ allows for a reduction in the mean free path due to magnetic fields or turbulence. Equation (7) is appropriate when the electron mean free path is sufficiently short compared to $T/|\nabla T|$ that heat conduction can be treated in the diffusion approximation, $\vec{q} = -\kappa \nabla T$. When the diffusion approximation breaks down, the conductive heat flux first enters the saturated regime (Cowie and McKee 1977), $q_{sat} = 5\phi_s c_s p$ (where $c_s = (p/\rho)^{1/2}$ is the isothermal sound speed in the intercloud medium and ϕ_s is a suppression factor similar to ϕ_c), and eventually the "suprathermal" regime (Balbus and McKee 1982), in which thermal conduction is best treated by a two-fluid approach.

The inhibition of multiphase structure by thermal conduction was first discussed by Field (1965), who found that conduction suppresses thermal instability for wavelengths shorter than a critical value which Begelman and McKee (1990) have generalized and dubbed the *Field length*,

$$\lambda_F \equiv \left(\frac{\kappa T}{n^2 \mathcal{L}_M} \right)^{1/2}, \tag{8}$$

where $\mathcal{L}_M \equiv \text{Max}(\Lambda, \Gamma/n)$. λ_F is the maximum length scale across which thermal conduction can dominate over radiative heating and cooling. Therefore, the thickness of a conductive interface with a radius of curvature r_c is $\sim \text{min}(\lambda_F, r_c)$ (McKee and Cowie 1977). This implies that conduction into clouds with radii smaller than λ_F is unaffected by heating and cooling processes in the surrounding medium. Such "small" clouds always evaporate (Graham and Langer 1973; Cowie and McKee 1977), at a rate given by

$$\dot{M}_{ev} = \frac{16\pi}{25} r_c \frac{\kappa T}{c_s^2} \tag{9}$$

in the classical conduction limit. Clouds with radii larger than λ_F have conductive interfaces whose structures are independent of the cloud size; such interfaces are dominated by the balance between conduction and heating/cooling, and may be treated as plane-parallel.

Steady plane-parallel conduction fronts have been analyzed by Zel'dovich and Pikel'ner (1969), Penston and Brown (1970), and McKee and Begelman (1990). Ballet, Arnaud and Rothenflug (1986) and Böhringer and Hartquist (1987) studied non-equilibrium ionization in steady evaporative flows. Time–dependent mass exchange has been analyzed in one dimension by Doroshkevich and Zel'dovich (1981), by Balbus (1986b), who included magnetic fields, and by Borkowski, Balbus and Fristrom (1989) who also studied the ionization structure. If the hot phase is cooling (and is thermally unstable) then a cooling wave of fixed thickness propagates into the hot gas following an evaporative transient. Doroshkevich and Zel'dovich (and Böhringer and Fabian 1989) used this result to argue that steady-state evaporation solutions are incorrect, i.e., that all clouds embedded in a cooling background medium should condense, not evaporate. However, the evaporative transient lasts until the temperature gradient relaxes to the Field length, and the timescale for this to occur is the cooling time. The evaporative solutions found by Cowie and McKee (1977) persist over a time scale which is short compared to the cooling time, but long compared to the time required to set up the evaporation flow. If the hot phase is thermally stable, then there exists a "saturated vapor pressure" p_{sat} above which "large" clouds condense, and below which they evaporate (Penston and Brown 1970). Zel'dovich and Pikel'ner (1969) devised an approximate method for calculating the evaporation rate when $p \neq p_{sat}$, which was refined and generalized to spherical clouds by McKee and Begelman (1990).

3.2 ABLATION

The motion of clouds with respect to the ambient hot medium leads to Kelvin-Helmholtz and Rayleigh-Taylor instabilities, which can break up the clouds into smaller pieces and accelerate mass exchange between the phases. Both instabilities operate on time scales $t_i \sim (\rho_c/\rho_h)^{1/2} r_c/v$, where v is the relative speed between the cloud and the hot medium and $\rho_c/\rho_h \sim T_h/T_c$ in pressure equilibrium. Most studies have concentrated on the fate of a cloud overtaken by a strong supernova or spiral density-wave shock (Woodward 1976; Nittman, Falle and Gaskell 1982; Heathcote and Brand 1983; McKee 1988; Klein, McKee and Colella 1989). In this case t_i is of the same order as the "cloud-crushing" time, t_{cc}, which is the time scale required for a secondary shock to be driven into a cloud once it is overrun by the main shock (McKee 1988). The cloud destruction process is accelerated by the significant pressure differential between the sides of the cloud and its front and back (Nittman, Falle and Gaskell 1982). The unbalanced forces cause the cloud to "pancake", i.e., to spread sideways, and the increase in cross-section speeds up the momentum deposition which tears apart the cloud. Pressure fluctuations and vorticity generation arising from the interactions of multiple shocks also play an important role in cloud disruption (Klein, McKee and Colella 1989).

The time scale for ablated cloud material to be effectively mixed with the intercloud medium should lie somewhere between t_i and the hydrodynamic drag time, $t_d \sim (\rho_c/\rho_h)r_c/v$. Nulsen (1982), using the longer time scale t_d, estimated that cold gas would be ablated from a cloud at a rate $\dot{M}_{ab} \sim \pi r_c^2 \rho_h v$. If thermal conduction were negligible, the cloud would leave behind a cylindrical "trail" with a radius $\sim r_c$, containing cold material with a mean density $\langle \rho \rangle_{tr} \sim \rho_h$. If the ablated gas is well-mixed with the hot phase downstream of the cloud, as we might expect from a turbulent ablation process, then the global time scale for cooling the hot phase by ablation is simply the time required for the trails to fill space, $t_{ab} \sim r_c/\pi f v$, where f is the filling factor in clouds. t_{ab} is shorter than the cloud disruption time if the clouds contain more mass than the hot phase, and it is longer than the saturated evaporation time by a factor $\sim \mathcal{M}^{-1}$, where $\mathcal{M}$ is the Mach number of cloud motion relative to the hot phase.

For diffuse interstellar clouds moving through the hot phase of the ISM in the Milky

Way, $\mathcal{M} \sim 0.1$. According to the Nulsen (1982) model, ablation from subsonically moving clouds is a less important mechanism for destroying clouds than conduction in the saturated limit, but may be more important than conduction in the classical limit, *i.e.*, for large clouds. For clouds moving nearly sonically, *e.g.*, randomly moving clouds in the spheroidal component of a galaxy, hydrodynamical instabilities are probably the most efficient mechanism for shredding clouds to the point where thermal mixing via conduction is very efficient.

Lateral expansion of the cloud can shorten the hydrodynamic drag time considerably (Nittman, Falle and Gaskell 1982; Klein, McKee and Colella 1989). Klein, McKee and Colella find that the drag time is of order t_i for density contrasts ρ_c/ρ_h as high as 100, but for much larger density contrasts the cloud is torn apart before it slows significantly. These calculations suggest that mixing can occur much more rapidly than predicted by the Nulsen (1982) model. Further numerical simulations capable of following the mixing process with high resolution are clearly needed to test the basic assumptions of any ablation model.

4. Consequences of Mass Exchange

4.1 MASS EXCHANGE VS. RADIATIVE HEATING/COOLING

When the bulk of the energy content and the bulk of the mass content reside in different phases, a relatively small amount of mass or energy transfer between phases can have a large effect on the structure of a multiphase medium. Such a situation is believed to exist in the three–phase model of the ISM (McKee and Ostriker 1977), where the hot phase occupies most of the volume while most of the mass is in cold clouds. Physically, the effect of mass exchange (either by conduction or ablation followed by effective mixing) is to cool the hot phase, since a fixed amount of energy is being distributed among a larger number of particles. Since radiative heating and cooling depend on both density and temperature, mass exchange can affect the radiative evolution of the medium as well. We can illustrate the global consequences of mass and energy exchange between phases by considering a medium with uniform pressure and subsonic motions, in which mass exchange is driven by thermal conduction. The approximate time-dependent equations governing the medium are then

$$\frac{\partial n}{\partial t} + \nabla \cdot (n\vec{v}) = 0 \tag{10}$$

$$\frac{3}{2}\frac{\partial p}{\partial t} + \frac{5}{2}p\nabla \cdot \vec{v} = (n\Gamma - n^2\Lambda) - \nabla \cdot \vec{q}, \tag{11}$$

where $\vec{q}$ represents the conductive heat flux. Note that these equations are appropriate in the single–fluid limit, corresponding to classical or mildly saturated conduction (Cowie and McKee 1977): in the highly saturated suprathermal limit (Balbus and McKee 1982), both the single–fluid approximation and the assumption of pressure balance break down. The evolution of the medium is driven by the terms on the right-hand-side of eq. (11), and may be dominated either by the effects of conduction or by the effects of radiative heating and cooling.

There is a crucial distinction between the heat flux term and the radiative loss term in the energy equation: the energy entering or leaving a volume of radius r due to conduction is proportional to $r^2 q$ ($\propto r$ for classical conduction) whereas that due to heating or cooling is proportional to r^3. Thus, there is a critical length scale which enters the problem, which turns out to be the Field length, λ_F (eq. [8]), when mass exchange is driven by classical conduction.

The temperature structure of the intercloud medium in a system of clouds extending over a region of size R depends on the ratio of the Field length to R. Balbus (1985) developed an elegant electrostatic analogy for an ensemble of clouds in a hot intercloud medium under conditions in which radiative heating and cooling are negligible and the temperature is specified on the boundary of the ensemble. This corresponds to the case $R \ll \lambda_F$. In a steady state, the evaporation rate and the temperature structure in the intercloud medium are then determined by a solution of Laplace's equation with Dirichlet boundary conditions. In the complementary case, $R \gg \lambda_F$, global heat flows are insignificant and the temperature structure of the intercloud gas is determined by a competition between cloud evaporation on the one hand and heating and cooling on the other. Numerically, the Field length is

$$\lambda_F = 2.4 \left(\frac{\phi_c^{1/2} T_6^{7/4}}{n \mathcal{L}_{M-23}^{1/2}} \right) \quad \text{pc}, \tag{12}$$

where $T_6 \equiv T/10^6$ K and $\mathcal{L}_{M-23} \equiv \mathcal{L}_M/(10^{-23} \text{ erg cm}^3 \text{ s}^{-1})$ is normalized to a characteristic value of the radiative cooling rate for astrophysical plasmas.

To quantify the competition between mass exchange and radiative heating/cooling, consider the spatially averaged effect of cloud evaporation on the hot phase. This is meaningful only when the characteristic temperature of the hot phase changes over a length scale which exceeds the mean intercloud separation r_0. This condition is guaranteed to be satisfied when r_0 is smaller than λ_F. When r_0 exceeds λ_F, then mass exchange cannot compete with radiative heating and cooling anyway, so the point is moot. If there are $\mathcal{N}_{cl}(\equiv 3/4\pi r_0^3)$ clouds per unit volume evaporating at a mean rate $\dot{M}_{ev}$ per cloud, then the density of the intercloud medium changes at a rate

$$\dot{n}_{ev} = \mathcal{N}_{cl} \frac{\dot{M}_{ev}}{\mu_H}, \tag{13}$$

where $\dot{M}_{ev}$ is negative for condensation. We may then use $\dot{n}_{ev}$ to define an *effective evaporative cooling rate,*

$$n_h^2 \Lambda_{ev} \equiv \frac{5}{2} p \left(\frac{\dot{n}_{ev}}{n_h} \right) \tag{14}$$

(Begelman and McKee 1990). The evaporative cooling coefficient Λ_{ev} is analogous to the radiative cooling coefficient Λ in that both reduce the specific entropy s, but there are crucial differences between the two: Λ reduces the the *total* entropy of a given volume of intercloud gas, whereas Λ_{ev} increases it; Λ reduces the energy density of the intercloud gas, whereas Λ_{ev} leaves it unchanged. Because of these distinctions, Λ_{ev} should not be included in the net radiative cooling function $\mathcal{L}$. However, the *relative* impact of mass exchange and radiative heating/cooling on the thermal state of the intercloud medium can be expressed by the *radiation/evaporation ratio* (Begelman and McKee 1990),

$$\mathcal{R} \equiv (\mathcal{L}/\Lambda_{ev}). \tag{15}$$

When $\mathcal{R}_M \equiv \mathcal{L}_M/\Lambda_{ev}$ is $\gg 1$ ($\ll 1$), then radiative heating and cooling (mass exchange) determines the thermal state of the intercloud medium.

One can express $\mathcal{R}$ in terms of quantities which characterize the structure of the two–phase medium. Writing the evaporation rate in the form $\dot{M}_{ev} = 4\pi r_c^2 \rho_h c_s F$, where c_s is the isothermal sound speed in the hot phase, we have

$$\mathcal{R} = \frac{2}{15F} \frac{r_0^3}{r_c^2} \frac{n_h^2 \mathcal{L}}{pc_s}. \tag{16}$$

r_0 may be eliminated in favor of the cloud filling factor f (assumed to be $\ll 1$) by substituting r_c/f for r_0^3/r_c^2. In the limit of saturated evaporation $F \sim$ a few (Cowie and McKee 1977), while $F \sim \mathcal{M}/4$ in the Nulsen (1982) model of ablation. In the classical conduction limit, F is twice the "saturation parameter" σ_0' derived by Cowie and McKee (1977), with the result that

$$\mathcal{R}_M = \frac{5}{6} \left(\frac{r_0^3}{r_c \lambda_F^2} \right). \tag{17}$$

Cloud evaporation thus determines the intercloud temperature for $r_c > r_0^3/\lambda_F^2$. The corresponding condition on the filling factor is

$$f > (r_c/\lambda_F)^2 \tag{18}$$

or $f > (r_0/\lambda_F)^6$; for $f \sim 0.03$, as in the ISM, this will be true if $\lambda_F \gtrsim 2r_0$. In terms of the sound-crossing time across r_c (measured in the hot phase), $t_s \sim r_c/2c_s$, and the radiative cooling time in the hot phase, $t_c \sim 5kT_h/2n_h\Lambda(T_h)$, we may express the condition for mass exchange to dominate globally in the form

$$f \gtrsim \frac{t_s}{Ft_c}. \tag{19}$$

Writing (18) in the form $r_0 < r_c^{1/3}\lambda_F^{2/3}$, we see that the intercloud spacing in a conduction-dominated medium must also be smaller than the Field length. This has important consequences for the thermal stability of the hot phase in a conduction-dominated system. Since λ_F is roughly the minimum wavelength which is thermally unstable (Field 1965), potentially unstable regions must contain many clouds. The conduction-modified condition for thermal instability is obtained simply by including Λ_{ev} in Balbus's (1986a) criterion (eq. [4]):

$$\left\{ \frac{\partial}{\partial T_h} \left[\frac{n_h(\Lambda + \Lambda_{ev}) - \Gamma}{T_h} \right] \right\}_A < 0 \tag{20}$$

(Begelman and McKee 1990). $n_h\Lambda_{ev}/T_h$ is generally an increasing function of T_h: for isobaric perturbations $n_h\Lambda_{ev}/T_h \sim q \propto T_h^{7/2} \left(T_h^{1/2} \right)$ for classical (saturated) conduction. Therefore, evaporation has a *stabilizing* influence on the hot phase, and in a conduction-dominated medium thermal instability will be inhibited by the presence of evaporating clouds (Begelman and McKee 1990). This will be true even if the radiative processes place the hot phase in a thermally unstable regime. However, it should be noted that the cooling time scale in the hot phase of a conduction-dominated medium is shorter than the radiative cooling time scale. Therefore, conduction cannot stabilize a hot phase over a time scale which is longer than the time scale for radiative thermal instability. However, it can lead to the hot phase cooling down somewhat before the onset of thermal instability. Since thermal conduction generally becomes less important at low temperatures, such a system may evolve to a state in which evaporative cooling no longer dominates, whereupon thermal instability may occur. Since the Field length is generally a strongly increasing function of temperature, the operation of thermal instability may lead to the production of smaller and more closely spaced clouds than would have formed in the hotter medium.

It is instructive to apply the ideas discussed above to the three–phase model of the ISM (McKee and Ostriker 1977). The three–phase ISM consists of cold HI clouds surrounded by warm HI and HII envelopes, all embedded in a pervasive hot ionized medium (HIM). The physical conditions in the HIM are governed by mass exchange with the clouds and energy injection by supernovae. The model is intrinsically time–dependent: A given element of gas is compressed and heated by SNRs at intervals of about 5×10^5 yr, and this makes its evolution difficult to analyze using concepts developed to treat steady-state or slowly evolving systems. Nonetheless, it is of interest to evaluate the Field length and the radiation/evaporation ratio. Using the fit to the cooling function of Raymond, Cox and Smith (1976) for cosmic abundances, $\Lambda(T) \approx 1.6 \times 10^{-19} T^{-1/2}$ erg cm^3 s^{-1} (10^5 K $< T < 4 \times 10^7$ K), multiplied by an enhancement factor $\beta \equiv 10\beta_1 \simeq 10$ to take account of nonequilibrium ionization and density inhomogeneity near conduction fronts (McKee and Ostriker 1977), we have $\lambda_F = 44(\phi_c^{1/2} T_6^3 / \bar{p}_4 \beta_1^{1/2})$ pc, where $\bar{p}_4 \equiv p/10^4 k$ and $T_6 \equiv T_h/10^6$ K. We can write condition (18) in the form

$$f > 5 \times 10^{-3} \frac{\beta_1}{\phi_c} \left(\frac{r_c}{1 \text{ pc}} \right)^2 \frac{\bar{p}_4^2}{T_6^6}. \tag{21}$$

It is evident that the relative importance of conductive and radiative energy exchange is sensitive to conditions in the hot phase, particularly to T_h, as well as to the filling factor and typical size of clouds. Under the conditions deduced by McKee and Ostriker ($T_6 = 0.45$, $\bar{p}_4 = 0.36$, $\beta_1 = 1$, $r_c = 2.1$ pc, and $f = 0.23$), condition (21) is marginally satisfied. Equivalently, the radiation/evaporation ratio is given by $\mathcal{R} = 0.38$, which implies that evaporative cooling dominates radiative cooling and that the HIM is thermally stable. (The argument that evaporation can stabilize the HIM is originally due to McCray [1986].) However, even a small amount of cloud ablation by hydrodynamic processes (§3.2) or a slight increase in the HIM temperature would lead to a drastic increase in the energetic importance of evaporation. If the typical clouds were sufficiently small that conduction were saturated, i.e., for

$$r_c < 0.5 \phi_c \frac{T_6^3}{\bar{p}_4} \text{ pc}, \tag{22}$$

(Cowie and McKee 1977; Balbus and McKee 1982), the appropriate version of condition (19) would be

$$f > 7 \times 10^{-3} \frac{\beta_1}{\phi_s} \left(\frac{r_c}{1 \text{ pc}} \right) \frac{\bar{p}_4}{T_6^3}. \tag{23}$$

In the McKee–Ostriker picture, each region of HIM is overrun by another supernova remnant before it has time relax to a stationary state with $\mathcal{R} \gtrsim 1$. Stochastic heating by SNR shocks thus leads to discontinuous trajectories in the $p - V$ plane, and heating is balanced by radiative cooling. In the galactic fountain model (Shapiro and Field 1976; Wang and Cowie 1988), the heat is advected into the galactic halo. If clouds are effectively ablated then the accelerated lowering of T_h by evaporation may allow the large tracts of the ISM to cool to a homogeneous state at $T \lesssim 10^4$ K. The intercloud medium remains thermally stable until radiative cooling begins to dominate over evaporative cooling, at $T_h \lesssim 10^5$ K. Since the Field length is a strongly increasing function of T_h, the onset of instability at a reduced temperature may lead to the formation of sub-parsec size clouds.

4.2 EVOLUTION OF MULTIPHASE SYSTEMS

By averaging the equations of mass and energy conservation over a volume $(\mathcal{V})$ which contains many clouds, one can derive equations for the global evolution of the hot intercloud medium in the presence of mass exchange with embedded clouds (Begelman and McKee 1990). The mass of intercloud gas in $\mathcal{V}$ is $\bar{n}_h \mu_H \mathcal{V}$, where $\bar{n}_h$ is the mean density of the intercloud gas in $\mathcal{V}$. Choosing the volume $\mathcal{V}$ to comove with the intercloud gas implies that this mass can change only by cloud evaporation, at a rate $\dot{n}_{ev} \mu_H \mathcal{V}$. Mass conservation for the intercloud gas then becomes

$$\frac{d\bar{n}_h}{dt} + \bar{n}_h \left(\frac{\dot{\mathcal{V}}}{\mathcal{V}} \right) = \dot{n}_{ev}. \tag{24}$$

Integration of equation (11) over $\mathcal{V}$ then implies

$$\frac{3}{2} \mathcal{V} \frac{dp}{dt} + \frac{5}{2} p \dot{\mathcal{V}} = - \int_{\mathcal{V}} n^2 \mathcal{L} dV - \int_S \vec{q} \cdot d\vec{S}, \tag{25}$$

where S is the surface bounding $\mathcal{V}$. By assuming that the characteristic dimension of the averaging volume is large compared to the Field length, we ensure that the conductive heat flux term in equation (25) is negligible compared to the heating and cooling term. The global energy equation then simplifies to

$$\frac{dp}{dt} = -\frac{2}{3} \langle n^2 \mathcal{L} \rangle_{\mathcal{V}} - \frac{5}{3} p \left(\frac{\dot{\mathcal{V}}}{\mathcal{V}} \right), \tag{26}$$

where $\langle \rangle_{\mathcal{V}}$ denotes an average over the volume $\mathcal{V}$. Note that mass exchange does not enter this equation: mass exchange alters the density and temperature of the intercloud gas, but not its pressure.

Additional constraints are required to solve for the evolution of a specific system. Since the cloud evaporation rate depends on the typical cloud size, it is necessary to have an equation for the evolution of r_c in time. There are also likely to be externally imposed constraints on the intercloud medium; Begelman and McKee (1990) considered two limiting cases. In the *isochoric* limit, the comoving volume $\mathcal{V}$ is held constant as the system evolves. If one assumes that the clouds are fixed as well, then the mean density $\bar{n}$ also remains constant. This condition would apply in a system in which the sound crossing time R/c_s is long compared to both the characteristic heating/cooling time and the evaporation time. Such a situation might apply locally within a supersonic accretion flow or wind. Setting $\dot{\mathcal{V}} = 0$ in eqs. (24) and (26), we obtain

$$\frac{dp}{dt} = -\frac{2}{3} n_h^2 \mathcal{L}, \tag{27}$$

and

$$\frac{dn_h}{dt} = \dot{n}_{ev}. \tag{28}$$

In the *isobaric* limit, the intercloud medium can exchange mass with a reservoir in order to maintain a constant pressure, so we set $p(t) = $ constant. Such a situation might apply, for example, if the system were in contact with an X-ray heated wind above an accretion disk (Begelman, McKee and Shields 1983). We then have

$$\frac{dn_h}{dt} = \dot{n}_{ev} + \frac{2}{5}\left(\frac{n_h}{p}\right) n_i^2 \mathcal{L}. \tag{29}$$

The instantaneous state of the intercloud medium can be described by the location of a point in the $p - V$ plane (cf. Fig. 1); $V \equiv n_h^{-1}$ is the specific volume of the intercloud gas. The radiative equilibrium curve $n_h^2 \mathcal{L} = 0$ divides the plane into two regions: above the curve, radiative cooling exceeds the external heating, whereas below the curve the converse is true. The net cooling rate $n_h^2 \mathcal{L}$ may be assumed to be a known function of p and V everywhere on the plane. However, the evaporation rate $\dot{n}_{ev}$ also depends on the distribution of cloud sizes and separations.

The character of a trajectory in the $p - V$ plane is determined by the relative importance of energy exchange and mass exchange, which is expressed quantitatively by the radiation/evaporation ratio $\mathcal{R}$ (eq. [16]). In the isochoric case, the slope of trajectories in the $p - V$ plane is governed by the ratio of equations (27) and (28):

$$\frac{d\ln p}{d\ln V} = \frac{5}{3}\mathcal{R}(p, V). \tag{30}$$

Trajectories are nearly vertical if radiative cooling and heating are dominant ($|\mathcal{R}| \gg 1$), and nearly horizontal if mass exchange is dominant ($|\mathcal{R}| \ll 1$). It is immediately obvious that trajectories must be locally horizontal ($dp/dV = 0$) where they cross the equilibrium curve ($\mathcal{L} = 0$), provided that $\dot{n}_{ev} \neq 0$ at the point of crossing. Points at which $\mathcal{L} = \dot{n}_{ev} = 0$ represent stationary states. The temperature evolves as

$$\frac{d\ln T}{dt} = -\frac{2}{5}\frac{n_h^2 \mathcal{L}}{p}\left(\frac{5}{3} + \frac{1}{\mathcal{R}}\right). \tag{31}$$

Trajectories with $\mathcal{R} = -\frac{3}{5}$ are isothermal. Isobaric trajectories are constrained to be horizontal in the $p - V$ plane. The direction (and rate) of motion is given by equation (29), which may be written in the form

$$\frac{d\ln V}{dt} = -\frac{2}{5}\frac{n_h^2 \mathcal{L}}{p}\left(1 + \frac{1}{\mathcal{R}}\right). \tag{32}$$

Since $T \propto pV \propto V$ in the isobaric case, this equation also describes the temperature evolution of the system. The point $\mathcal{R} = -1$ represents a steady state for the hot phase, although mass continues to be lost (if $\mathcal{L} < 0$) or gained (if $\mathcal{L} > 0$) by clouds in this state (unlike the "true" steady state $\mathcal{L} = \dot{n}_{ev} = 0$ in the isochoric case). For a system dominated by radiative heating or cooling ($|\mathcal{R}| \gg 1$), evolution is leftward above the thermal equilibrium curve and rightward below the curve. In a system dominated by mass exchange ($|\mathcal{R}| \ll 1$), evolution is leftward for evaporation and rightward for condensation. A more detailed discussion of $p - V$ plane trajectories may be found in Begelman and McKee (1990), who considered the specific case of a gas heated by Compton scattering and cooled by bremsstrahlung and the inverse Compton effect. They also discuss the stability properties of evolving systems.

5. Conclusions

Multiple thermal phases are known to exist in many astrophysical systems. The reasons for their existence are understood in general terms, but the detailed properties of specific multiphase systems are poorly known. We can look forward to the further development of multiphase models for the ISM in elliptical galaxies, cooling flows in clusters of galaxies, the intergalactic medium, and protogalactic environments. The structures of multiphase media are sensitive to the rate of mass exchange between phases, which tends to lower the temperature of the hot phase and render it thermally stable. Unfortunately, mass exchange through thermal conduction depends on the topology of conduction fronts and on the magnetic connectivity of the phases, neither of which is understood. However, mixing of the phases may be driven by hydrodynamic instabilities at a rate much faster than that due to thermal conduction. Rapidly improving hydrodynamic codes with a high dynamic range in spatial resolution (*e.g.*, using an adaptive mesh) should clarify some of the physics of the ablation process within the next few years.

ACKNOWLEDGMENTS. Many of the ideas presented above were developed collaboratively with C. F. McKee. Portions of the text borrow heavily (and in some cases verbatim) from Begelman and McKee (1990), which is to be published in *The Astrophysical Journal*. Preparation of this article was supported in part by NSF grant AST88-16140, NASA Astrophysical Theory Center grant NAGW-766, and a grant from the Alfred P. Sloan Foundation.

6. References

Balbus, S. A. 1985, *Ap. J.*, **291**, 518.
——. 1986*a*, *Ap. J. (Letters)*, **303**, L79.
——. 1986*b*, *Ap. J.*, **304**, 787.
——. 1988, *Ap. J.*, **328**, 395.
Balbus, S. A., and McKee, C. F. 1982, *Ap. J.*, **252**, 529.
Balbus, S. A., and Soker, N. 1989, *Ap. J.*, **341**, 611.
Ballet, J., Arnaud, M., and Rothenflug, R. 1986, *Astr. Ap.*, **161**, 12.
Begelman, M. C., and Cioffi, D. F. 1989, *Ap. J. (Letters)*, **345**, L21.
Begelman, M. C., and McKee, C. F. 1990, *Ap. J.*, in press.
Begelman, M. C., McKee, C. F., and Shields, G. A. 1983, *Ap. J.*, **271**, 70.
Böhringer, H., and Fabian, A. C. 1989, *M.N.R.A.S.*, **237**, 1147.
Böhringer, H., and Hartquist, T. W. 1987, *M.N.R.A.S.*, **228**, 915.
Borkowski, K. J., Balbus, S. A., and Fristrom, C. C. 1989, preprint.
Chambers, K. C., Miley, G. K., and van Breugel, W. J. M. 1987, *Nature*, **329**, 604.
Cowie, L. L., and Binney, J. 1977, *Ap. J.*, **215**, 723.
Cowie, L. L., Fabian, A. C., and Nulsen, P. E. J. 1980, *M.N.R.A.S.*, **191**, 399.
Cowie, L. L., Hu, E. M., Jenkins, E., and York, D. 1983, *Ap. J.*, **272**, 29.
Cowie, L. L., and McKee, C. F. 1977, *Ap. J.*, **211**, 135.
Cox, D. P. 1979, *Ap. J.*, **234**, 863.

——. 1981, *Ap. J.*, **245**, 534.

Cox, D. P., and Smith, B. W. 1974, *Ap. J. (Letters)*, **189**, L105.

Dalgarno, A., and McCray, R. A. 1972, *Ann. Rev. Astr. Ap.*, **10**, 375.

Davidson, K. 1972, *Ap. J.*, **171**, 213.

de Jong, T. 1980, *Highlights Astr.*, **5**, 301.

d'Hendecourt, L. B., and Leger, A. 1987, *Astr. Ap.*, **180**, L9.

Doroshkevich, A. G., and Zel'dovich, Ya. B. 1981, *Sov. Phys. J.E.T.P.*, **53**, 405.

Draine, B. T. 1978, *Ap. J. Suppl.*, **36**, 595.

Draine, B. T., and Giuliani, J. L. 1984, *Ap. J.*, **281**, 690.

Fabian, A. C., Guilbert, P., Arnaud, K., Shafer, R., Tennant, A., and Ward, M. 1986, *M.N.R.A.S.*, **218**, 457.

Fabian, A. C., and Nulsen, P. E. J. 1977, *M.N.R.A.S.*, **180**, 479.

Fall, S. M., and Rees, M. J. 1985, *Ap. J.*, **298**, 18.

Ferrière, K. M., Zweibel, E. G., and Shull, J. M. 1988, *Ap. J.*, **332**, 984.

Field, G. B. 1965, *Ap. J.*, **142**, 531.

Field, G. B., Goldsmith, D. W., and Habing, H. J. 1969, *Ap. J. (Letters)*, **155**, L149 (FGH).

Graham, R., and Langer, W. D. 1973, *Ap. J.*, **179**, 469.

Heathcote, S. R., and Brand, P. W. J. L. 1983, *M.N.R.A.S.*, **203**, 67.

Heckman, T. M. 1981, *Ap. J. (Letters)*, **250**, L59.

Heckman, T. M., Baum, S. A., van Breugel, W. J. M., and McCarthy, P. 1989, *Ap. J.*, **338**, 48.

Hogan, C. J. 1987, *Ap. J. (Letters)*, **316**, L59.

Hu, E. M., Cowie, L. L., and Wang, Z. 1985, *Ap. J. Suppl.*, **59**, 447.

Ikeuchi, S., and Spitzer, L. 1984, *Ap. J.*, **283**, 825.

Johnstone, R. M., Fabian, A. C., and Nulsen, P. E. J. 1987, *M.N.R.A.S.*, **224**, 75.

Klein, R. I., McKee, C. F., and Colella, P. 1989, in *Evolution of the Interstellar Medium*, ed. L. Blitz (San Francisco: Astronomical Society of the Pacific), in press.

Krolik, J. H. 1988, *Ap. J.*, **325**, 148.

Krolik, J. H., McKee, C. F., and Tarter, C. B. 1981, *Ap. J.*, **249**, 422 (KMT).

Lepp, S., and Dalgarno, A. 1988, *Ap. J.*, **335**, 769.

Lepp, S., McCray, R., Shull, J. M., Woods, D. T., and Kallman, T. 1985, *Ap. J.*, **288**, 58.

Lynds, R. 1970, *Ap. J. (Letters)*, **159**, L151.

Mathews, W. G., and Bregman, J. N. 1978, *Ap. J.*, **224**, 308.

Mathews, W. G., and Ferland, G. J. 1987, *Ap. J.*, **323**, 456.

McCray, R. 1979, in *Active Galactic Nuclei*, ed. C. Hazard and S. Mitton (Cambridge: Cambridge U. Press), p. 227.

——. 1986, private communication.

McCray, R., and Kafatos, M. C. 1987, *Ap. J.*, **317**, 190.

McCarthy, P. J., van Breugel, W., Spinrad, H., and Djorgovski, S. 1987, *Ap. J. (Letters)*, **321**, L29.

McKee, C. F. 1988, in *Supernova Remnants and the Interstellar Medium, IAU Colloq. 101*, ed. R. S. Roger and I. L. Landecker (Cambridge: Cambridge U. Press), p. 205.

McKee, C. F., and Begelman, M. C. 1990, *Ap. J.*, in press.

McKee, C. F., and Cowie, L. L. 1977, *Ap. J.*, **215**, 213.
McKee, C. F., and Ostriker, J. P. 1977, *Ap. J.*, **218**, 148.
Nittmann, J., Falle, S. A. E. G., and Gaskell, P. H. 1982, *M.N.R.A.S.*, **201**, 833.
Norman, C. A., and Ikeuchi, S. 1989, *Ap. J.*, in press.
Nulsen, P. E. J. 1982, *M.N.R.A.S.*, **198**, 1007.
———. 1986, *M.N.R.A.S.*, **221**, 377.
Penston, M. V., and Brown, F. E. 1970, *M.N.R.A.S.*, **150**, 373.
Raymond, J. C., Cox, D. P., and Smith, B. W. 1976, *Ap. J.*, **204**, 290.
Rees, M. J. 1988, *M.N.R.A.S.*, **231**, 91P.
Rees, M. J. 1989, *M.N.R.A.S.*, **239**, 1P.
Rees, M. J., and Ostriker, J. P. 1977, *M.N.R.A.S.*, **179**, 541.
Sarazin, C. L. 1986, *Rev. Mod. Phys.*, **58**, 1.
Scalo, J. M. 1977, *Ap. J.*, **213**, 705.
Shapiro, P. R., and Field, G. B. 1976, *Ap. J.*, **205**, 762.
Shull, J. M., and Woods, D. T. 1985, *Ap. J.*, **288**, 50.
Silk, J. 1977, *Ap. J.*, **211**, 638.
Spitzer, L. 1956, *Ap. J.*, **124**, 20.
———. 1962, *Physics of Fully Ionized Gases* (New York: Wiley), 2nd edition.
———. 1978, *Physical Processes in the Interstellar Medium* (New York: Wiley).
———. 1982, *Ap. J.*, **262**, 315.
Tarter, C. B., Tucker, W. H., and Salpeter, E.E. 1969, *Ap. J.*, **156**, 943.
Thomas, P. 1988, *M.N.R.A.S.*, **235**, 315.
Thomas, P. A., Fabian, A. C., and Nulsen, P. E. J. 1987, *M.N.R.A.S.*, **228**, 973.
Tribble, P. C. 1989, *M.N.R.A.S.*, **238**, 1247.
Wang, Z., and Cowie, L. L. 1988, *Ap. J.*, **335**, 168.
Woodward, P. 1976, *Ap. J.*, **207**, 464.
Zel'dovich, Ya. B., and Pikel'ner, S. B. 1969, *Sov. Phys. J.E.T.P.*, **29**, 170.
Zweibel, E. G., and Josafatsson, K. 1983, *Ap. J.*, **270**, 511.

Large Scale Interstellar Gasdynamics in Disk Galaxies

Steven A. Balbus
Virginia Institute of Theoretical Astronomy,
Astronomy Department,
University of Virginia

ABSTRACT. The dynamical behavior of interstellar gas in spiral arm potentials and the nature of H I warps in disk galaxies is reviewed. Recent millimeter wave interferometry results of M 51 are discussed, which are supportive in detail of a density wave structure for this galaxy. It is argued that spiral arms, when present, must be an important triggering mechanism for star formation, even though global star formation rates are insensitive to spiral structure. The effect of the unusual velocity shear fields near spiral arms on the development of small disturbances is discussed. We highlight the success of recent investigations treating H I warps as a discrete normal mode of a combined disk-halo system. In this picture, the axes of the halo and the disk are misaligned, and warping modes correspond to what would be a zero frequency trivial tilt of the entire disk in the absence of a halo.

1. Introduction

Faced with the task of reviewing an area as extraordinarily broad as the large-scale gasdynamics in galaxies, and being by nature a rather naive sort of fellow, my initial impulse in response to the organizers' invitation was to begin to prepare a lengthy, encompassing survey of the vast literature of this field. But as the deadline loomed larger and the trickle of preprints I was receiving for inclusion grew into an avalanche, it became apparent that the literature was going to win this confrontation easily. So, at the risk of annoying large numbers of colleagues, I have decided instead to focus rather narrowly on two topics that, in my opinion, have seen some particularly interesting progress recently. I am confident that the resulting compactness at least will appeal to the reader, but I hope that this strategy will prove to be good pedagogy as well.

The first topic covered is the gasdynamics of the interaction of the interstellar medium with spiral arms, with particular attention drawn to the hitherto neglected but unusual velocity shear field. The detailed study of spiral arm dynamics has entered a new observational era as millimeter wave interferometry comes on line (Vogel, Kulkarni, and Scoville 1988). The second topic concerns the puzzling nature of the H I warps, which of late may have been rendered slightly less puzzling (Sparke and Casertano 1988). Reflecting nothing more than my own research interests, spiral structure comprises the majority of this review. But it is quite clear that an understanding of the nature of H I warps is a goal

305

H. A. Thronson, Jr. and J. M. Shull (eds.), The Interstellar Medium in Galaxies, 305–321.

of the highest importance, not only as a fascinating problem in interstellar gasdynamics, but as a powerful new probe of dark haloes.

2. Spiral Structure

The coiled, delicate, filamentary appearance of the spiral patterns seen in many nearby disk galaxies suggested to most early onlookers that spiral arms must be a magnetic phenomenon. A notable exception was B. Lindblad (1963), who argued very early on that a spiral pattern could be gravitationally self-sustaining. But the seminal papers of Lin and Shu (1964, 1966) together with the unpublished PhD thesis of Kalnajs (1965) are generally regarded as marking the beginning of the modern theory of spiral structure.

2.1 DENSITY WAVE DISPERSION FORMULAE AND STABILITY

The idea of Lin and Shu was that spiral structure could be understood as a wave phenomenon primarily in the stellar component of a disk galaxy. They considered very nearly axisymmetric disturbances, so that the shearing of the wave crests was not dynamically significant. Indeed, any "spirality" exhibited by the theory was purely kinematic, not dynamic, in content. (Kalnajs [1965] was an explicitly axisymmetric calculation.) In the usual r, θ disk coordinate system, a nearly axisymmetric plane wave disturbance of the form $\exp(i \int k \, dr + im\theta - i\omega t)$ where k, m, ω represent radial wavenumber, angular wavenumber, and frequency respectively, satisfies a *local* dispersion formula of the form

$$(\omega - m\Omega)^2 = \kappa^2 - 2\pi G\sigma_* |k| \mathcal{F} \tag{2.1a}$$

where Ω is the disk angular velocity, κ the epicyclic frequency, σ_* the disk surface density, and G the gravitational constant. The factor $\mathcal{F}$, the "reduction factor", is a complicated functional of the wavelength and stellar velocity distribution function (see Binney and Tremaine [1987] for the gaussian case), whose precise form need not concern us here, except to note that its value is unity when the stars are dynamically could, and otherwise less than unity. By way of comparison, and for future reference, we list here the corresponding equation for a pure gaseous disk (subscript g) with adiabatic sound speed a,

$$(\omega - m\Omega)^2 = \kappa^2 - 2\pi G\sigma_g |k| + k^2 a^2, \tag{2.1b}$$

and finally, the dispersion formula for a combined rotationally supported star-gas disk (Lin, Yuan, and Shu 1969):

$$\left[(\omega - m\Omega)^2 - \kappa^2\right]^2 + \left[(\omega - m\Omega)^2 - \kappa^2\right]\left[2\pi G|k|\,(\sigma_g + \mathcal{F}\sigma_*) - k^2 a^2\right] - 2\pi G\sigma_* |k| \mathcal{F}k^2 a^2 = 0. \tag{2.1c}$$

where σ_* and σ_g represent stellar and gas surface densities. Disturbances whose dispersion properties are described by any of the formulae (2.1) will be referred to as "density waves".

The first thing we may note about any of these is the possibility of gravitational instability — the disk analogue of the Jeans instability — for sufficiently large surface densities. The simplest case is the gaseous disk (2.1b). One may easily show (Lin and Shu 1968) that if

$$Q \equiv \frac{\kappa a}{\pi G\sigma_g} < 1, \tag{2.2}$$

then a band of wavenumbers will be unstable. If $Q > 1$, large wavenumbers are stabilized by pressure, short wavenumbers by rotation, and there is no intermediate wavenumber range which eludes one or the other. The stability criterion for the stellar disk (2.1a) is harder to derive (Toomre 1964) but just as easily stated: it is simply equation (2.2) written in terms of the stellar density and one-dimensional velocity dispersion, but with a frequency of 3.36 instead of π in the denominator. No simple analytic stability criterion for the star-gas system (2.1c) seems apparent, but in general the tendency for a two-component system is toward mutual destabilization. A simplified version of the dispersion relation has been studied by Jog and Solomon (1984) in which the stars are dynamically modeled as a hot gas. These authors have suggested two-fluid instabilities as the origin of large molecular clouds in the Galaxy.

2.2 RESONANCES

Let us concentrate for the moment on the stellar case (2.1a). Fix the wave frequency ω to be real, and regard $k(r)$ as determined from the dispersion formula. There are three special radii in the disk to note. At the points given by

$$\omega = m\Omega \pm \kappa \qquad (2.3)$$

$k(r)$ must either be zero or infinite, the latter not being obvious from (2.1a) because it depends on the explicit form of $\mathcal{F}$ (Toomre 1977). (Physically, very large k disturbances oscillate at the epicyclic frequency because few stars actively partake of such small wavelength oscillations—they have their own small but finite epicyclic radii to mind—and self-gravity becomes unimportant.) The special locations defined by equation (2.3) are called "Lindblad resonances", and they mark the radii at which the Doppler shifted wave frequency matches the natural epicyclic response frequency. There are in principle two Lindblad resonances: an outer one corresponding to the minus sign, and an inner one associated with the plus sign, which may or may not actually be present in the disk. The third important location is the corotation resonance $\omega = m\Omega$. It lies between the Lindblad resonances, and it plays a particularly important role for the wave transport properties of the disk (Goldreich and Tremaine 1978). For our purposes, the most important point is that density waves of a given frequency can propagate *only* between the Lindblad resonances of that frequency. Additionally, there will be a forbidden zone near corotation in a stable disk.

2.3 MODES AND TRANSIENTS

The initial idea of Lin and Shu was that equation (2.1a) would define $k(r)$ for a "tightly-wrapped", global and discrete spiral mode; i.e. ω real and constant across the face of the disk. But Toomre (1969) pointed out that all that equation (2.1a) really established was that a continuum of WKBJ waves can propagate in a stellar disk according to the prescribed dispersion formula, and he went on to make the elementary but devastating observation that such waves disperse on a time scale associated with their group velocity: some 109 years for waves in the solar neighborhood. *If* a large wavenumber pure spiral mode *could* exist, then equation (2.1a) *would* describe $k(r)$, with ω presumably determined by unspecified boundary conditions. But it is not at all clear that such modes do exist—in fact, for an inviscid gas disk neutrally stable modes do not (Lynden-Bell and Ostriker 1967).

Global mode analyses of model disks do uncover spiral forms (Toomre 1981), but they seem to appear as instabilities—not neutrally stable modes. No one dynamical mechanism seems responsible for global spiral instabilities, and in the past decade and a half or so, investigators have uncovered many surprisingly disparate processes (Erikson 1974, Toomre 1981, Bertin et al. 1989, Sellwood and Lin 1989). Here, we shall look at one in some detail. This is the "swing amplifier" (Toomre 1981), and it is of some importance not only as an approach to the large-scale spiral structure problem (which is not formally in the scope of this review), but also for understanding what happens to interstellar gas upon streaming through a pre-existing spiral arm (which is).

We may begin by considering the fate of a plane wave disturbance in a uniformly shearing background. It is not required that the disturbance be tightly-wound, in other words $m \sim kr \gg 1$. (Such a disturbance, by the way, can certainly be analyzed by WKB methods, although "the WKB approximation" has come to mean only tightly-wound waves ($m/kr \ll 1$) in galactic structure parlance. We will abide by this unfortunate convention.) Obviously, we cannot hope to describe the evolution of such a perturbation as a simple oscillation in the presence of differential rotation. The problem is that shear flow will distort the spacing of the wave crests as a function of time. If, as is normally the case, the underlying flow is such that the angular velocity decreases outward, leading wave crests (i.e. those oriented such that increasing r along the crest corresponds to moving in the direction of rotation) are converted to trailing wave crests. (See fig. [3] from Toomre (1981).) While the analysis of the evolution of such disturbances in a gas disk dates from the classical paper of Goldreich and Lynden-Bell (1965), it was Toomre's key interpretation that the constituent stars (or in the case of a gas disk, fluid elements) in the crest undergo retrograde epicyclic motion which, because of a kinematic match-up with the retrograde shear, results in density enhancements remaining in their compressive phase significantly longer than half of an epicyclic period. The importance of this is that shear does more than just modify wave propagation; it is an agent of destabilization under the right circumstances. "The right circumstances" are achieved considerably more easily if matter stays bunched together longer before entering the expansive phase of an epicyclic oscillation. Indeed, a disk can be stable to axisymmetric wave propagation in accordance with any of our equations (2.1), while being extremely vulnerable to the swing amplifier mechanism (Julian and Toomre 1966).

Note that we speak of an "amplifier", not an "instability". The process we have described will die away (assuming axisymmetric disturbances are stable) once the wave crests have swung to a nearly axisymmetric trailing configuration. A true global instability cannot be caused by swing amplification alone. But it can be an important link in a chain that involves a feedback mechanism (Toomre 1981). An important class of global spiral instabilities seem to be based on this mechanism. For example if trailing WKBJ waves can propagate through the center of the Galaxy, they will emerge as outward propagating waves (e.g. Binney and Tremaine 1987). At the corotation radius of a global mode, the waves are converted from outward propagating leading to inward propagating trailing, with attendant amplification and the start of the feedback cycle again. In this scenario, a combination of wave propagation and local amplification is responsible for the large-scale grand design structure of disk galaxies. But such a view is not yet universally accepted. For example, Sellwood and Lin (1989) have suggested recently that phase space instabilities similar to plasma streaming instabilities may be important, while Bertin et al. (1989) have argued that what amounts to Jeans instability in the interstellar gas is important both to the excitation and maintenance of large scale spiral structure. Our purpose here in concentrating on the swing amplifier is that its very local properties are likely to be quite important for understanding the gasdynamical environment near spiral arms.

2.4 GAS DENSITY ENHANCEMENT AND ITS EFFECT ON STAR FORMATION

The central assumption of this discussion is that a large-scale spiral potential is present in a thin galactic disk, and that the interstellar gas moves relative to the spiral pattern. In no way does this assumption favor one scenario for the origin of spiral structure over another. As we have discussed, even the transient swing-amplifier mechanism, which involves material shear, can in principle find itself at the heart of unstable global mode. But much more persuasive than any theoretical argument, direct observations of the grand design spirals M51 (Vogel, Kulkarni, and Scoville 1988) and M81 (Visser 1980, Kaufman et al. 1989) indicate that gas streaming relative to spiral arms is excellent phenomenology.

The forcing spiral potential may be considered imposed by hot, unresponsive stars, or it may be regarded as a nonlinear wave in a single-fluid system; in either case we shall avoid the complications associated with a two-fluid star-gas system. The potential is small compared to the axisymmetric component of the galaxy. For simplicity, but also because it is a pretty good approximation in many cases, we assume that the forcing potential is tightly-wound. More specifically, the inclination angle i between the spiral arm and the galactocentric circle satisfies $\sin i \ll 1$ (see fig. [1]).

An immediate consequence is that the specific angular momentum of a fluid element is nearly conserved as it passes through the well of the potential. Consider now the history of an unperturbed interstellar gas fluid element on its galactic circumnavigation. Its orbit is very nearly circular; there is some nonclosure of streamlines, but this effect is small (Shu, Milione and Roberts 1973). Even so, upon passing through the region near the minimum of the spiral potential, there is a nonlinear increase in the unperturbed density field, perhaps a factor of several. It is a commonly held notion that the formation of large-scale galactic shocks is inevitable in a smooth fluid if the spiral potential is sufficiently large (Roberts 1969; Shu, Milione, and Roberts 1973), but interestingly self-gravity can eliminate mathematical shocks while retaining large compressions (Lubow, Balbus, and Cowie 1986). Compressions themselves are due primarily to the workings of the Coriolis force, which in turn couples the density increase to a change in the local shear. Of course, the spiral potential has but little effect on the ~ 250 km s^{-1} orbital velocity; the shear is substantially altered because the small change in azimuthal velocity occurs over a small radial extent. The changing shear conditions near the spiral arms are extremely important to an understanding of the growth of density perturbations in this region, but in much of the literature this has either been ignored or misunderstood. We shall examine this in some detail. First however, we focus on density enhancements. A very considerable theoretical and observational effort has been expended in understanding how the enhanced interstellar gas density in the vicinity of spiral arms affects the formation of clouds and stars.

The enhanced density of interstellar gas in the vicinity of spiral arms would seem to be a very natural place for star formation to be triggered and the classical observations of giant H II regions following arms like "beads on a string" would seem to bear this out. However, recently there have appeared in the literature the results of a number of studies pointing out that star formation rates in spiral galaxies seem uncorrelated with Hubble type (Elmegreen and Elmegreen 1986). In fact, irregular galaxies can be hotbeds of star

310

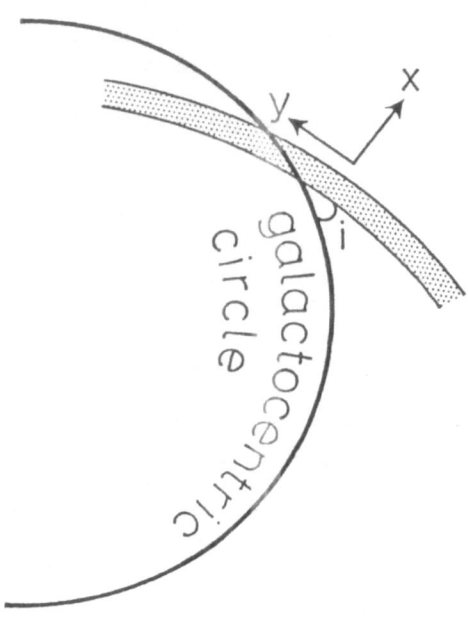

Fig. 1. Tightly wound arm configuration. Stipling indicates spiral arm location (defined by point of maximum density compression), coordinate x is radially outwards, coordinate y is azimuthal in the direction of the rotational velocity. The inclination angle i satisfies $\sin i \ll 1$.

formation (Hunter and Gallagher 1986, 1989), so spiral structure is clearly not a *necessary* ingredient. Larson (1987) has suggested that spiral structure may itself be only incidental to the star formation process, rather than a primary cause.

It is however no easy matter to disentangle cause and effect in this business. In particular, the star formation rates that appear to be uncorrelated from one galaxy to the next are globally determined rates; a particular mechanism may neither be eliminated nor established by this type of observation. One should think perhaps of the nineteenth century laboratory spectroscopist who finds that the emission spectra of all kinds of matter seem quite independent of the type of atom involved, and instead depend only upon the material temperature. Clearly, he thinks, since the emission properties are completely uncorrelated with atomic type, the notion that atoms play a fundamental role in the production of radiation from matter must be greatly suspect!

Of course, what is going on in my slightly loaded example is that the constraints of establishing thermodynamic equilibrium are so powerful that they completely determine the blackbody curve for a given temperature. But this is precisely the point. It seems inevitable that star formation is a highly self-regulating process, and that the global rate of a galaxy is ultimately determined by complicated feedback mechanisms involving energy injection (and possibly mass depletion) that may well not care what the Hubble type is.

To answer the question of whether spiral arms are of fundamental importance in triggering star formation, one must look directly at the "microphysics" of spiral arm gasdynamics, not for statistical correlations.

Just such a study has been done by Vogel, Kulkarni, and Scoville (1988) on the benchmark galaxy M51. Using millimeter interferometric techniques on the CO 1 → 0 transition to obtain the necessary spatial resolution, these authors have obtained surface brightness and kinematic data in good agreement with the basic predictions of density wave theory, including an interesting reversal of the shear near the arm (see below). Moreover, while the CO surface brightness is enhanced a factor of 2–3 near the arm, the number of observed H II regions jumps a factor in excess of 10. Vogel et al. discuss and rule out a number of nonstraightforward interpretations of this result (e.g. star-forming ballistic molecular cloud complexes jumping in number a factor of order 10, while an emission-dominant diffuse molecular component gives the factor ~ 3 surface brightness enhancement), and in the end conclude that there is no viable alternative to the inference that at least in M51, the spiral arm structure is indeed increasing the mass-specific downstream star formation rate. But these observations should not be construed as establishing any sort of a direct connection between the presence of spiral structure in M51 and its *global* star formation rate. And Jay Gallagher's pithy comment at this meeting is a reminder that while the atoms of my analogy certainly are involved with producing radiation spectra, there are also the free electrons not to be overlooked!

Accepting that the passage of the interstellar gas through a spiral potential is likely to trigger condensation processes that ultimately result in star formation, it is not difficult to imagine any number of particular schemes for bringing this about. Density enhancements both speed up collisional processes associated with agglomeration, as well as make it easier to effect large scale gravitational collapse. Cloud formation and destruction processes have been reviewed in Elmegreen's comprehensive review article (1987b) for the first Wyoming Conference, so we present here a most cursory summary of the sorts of ideas that have been suggested.

2.4.1 Magnetic Buoyancy Instability. (Parker 1966) A submerged magnetic field in a gravitationally supported slab is unstable for reasons analogous to the classical Rayleigh-Taylor instability. The increase in amplitude of a galactic magnetic field downstream from a spiral arm leads to a buckling of the field lines perpendicular to the midplane of the disk. Mouschovias, Shu, and Woodward (1974) suggested that OB associations may be formed in the "valleys" of magnetic field lines, while Blitz and Shu (1980) made a similar argument for the formation of giant molecular cloud complexes. Elmegreen (1982) studied the simultaneous effects of self-gravity with the magnetic processes. The characteristic scale length of the instability, some $2\pi \times$ the disk scale height is small (well under 1 kpc) if the disk scale height is set by the cool massive component of the gas, and the details of the gathering process have always remained somewhat vague (Balbus and Cowie 1985), but the general precepts of the Parker instability suggest a way to parcel out chunks of interstellar gas *before* self-gravity becomes an important dynamical factor.

2.4.2 Cloud Coalescence. The idea here is that cloud-cloud collisions become more frequent in the region of heightened interstellar density near a spiral arm, leading to radiative losses in shocks and the growth of large clouds (Hausman 1981; Kwan and Valdes 1983; Scalo and Struck-Marcel 1984; Tomisaka 1986; Elmegreen 1989). At first, it was thought that the smaller clouds would be neutral atomic hydrogen, and that collisions would be an effective way to build up molecular clouds behind shocks (Smith 1980). But it now appears that molecular gas dissociates in spiral arms (Tilanus and Allen 1989), due probably to the activity of newly formed massive stars, and the need to create molecular hydrogen in the arms seems less pressing. The basic agglomeration mechanism would still be present if the small clouds were molecular, of course.

These studies have the virtues of dealing head on with the discrete nature of the interstellar gas, offering a possible insight as to why giant molecular cloud complexes are so clumpy (Blitz and Thadeus 1980) and how they can be supported (Scalo and Pumphrey 1982, Scalo and Struck-Marcell 1984; but see Shu, Adams, and Lizano 1987). But the necessary compromise inherent in this approach is that handling what is in essence a turbulent phenomenon by using "standard clouds" must yield highly model dependent results. Additionally, there probably isn't enough time to assemble the larger molecular complexes seen in spiral arms (Elmegreen 1987a,b, 1989; Rand and Kulkarni 1989) by coalesence alone.

2.4.3. Gravitational Instability. Ultimately, the formation of self-gravitating cloud complexes clearly will involve some form of this instability, and it has been pointed out by a number of investigators (Elmegreen 1979; Cowie 1980, 1981; Balbus and Cowie 1985) that conditions near spiral arms are such that the instability is probably present "right from the start" on scales of about a kiloparsec. Accounting properly for the underlying expansion of the gas flow is important, as perturbation wavelengths are stretched and eventually stabilized by centrifugal forces. Elmegreen (1987a) has pointed out that a magnetic field could transport some angular momentum outwards, which could counteract the stabilization of long wavelength disturbances, but the orbital clock still dictates how much time is available before a fluid element passes through the high density zone. A complete description of local gravitational disturbances requires careful consideration of the peculiar spiral arm shear flow mentioned earlier, and we now turn our attention in this direction.

2.5 THE EVOLUTIONARY EQUATION FOR LOCAL DISTURBANCES

In spite of the very complex microphysics that must characterize cloud-scale interstellar gasdynamics, there is much to be said for adopting a ruthlessly simple constituent fluid model and studying its stability in detail. Important "zeroth order" dynamical processes are thereby isolated, and it becomes more clear just what must be required of the more uncertain thermal, magnetic, and turbulent properties of the gas to explain the development of structure. True to this dictate, our model for the ISM will be extremely crude: a simple polytrope. The notion that an ensemble of interstellar clouds can be treated as a lossy fluid continuum dates at least as far back as Shu (1978), and has been further refined by Cowie (1981). Collisional mean free paths are typically 100–200 pc for H I or small molecular clouds (Spitzer 1978), so apart from a very uncertain cloud viscosity there is reason to expect that the dynamics is not poorly represented by a continuum fluid. The neglect of a magnetic field is a serious omission primarily in so far as the field serves as a viscous couple: an artificial symmetry is introduced into the calculation for the dynamics of a polytropic fluid in the form of a vorticity constant. Restricting density perturbations to evolve along vorticity-conserving paths precludes the possibility that angular momentum transport will

significantly destabilize long wavelength perturbations, a point made in so many words
by Elmegreen (1987a). He found this effect to be more important than the stabilization
brought about by the additional magnetic pressure. But here we shall consider only the
simplest gasdynamics.

What happens to the local shear field in our gas when it is compressed in a spiral
arm potential? We may get to the heart of the matter most quickly by noting that the
ratio of the local epicyclic frequency to gas density κ/σ_g is a constant for a given fluid
element, if the arms are tightly wound. (It is important to appreciate that small radial
streaming motions make the epicyclic frequency a rapidly changing *local* quantity, just like
the density.) Specifically, we have

$$\frac{\kappa^2}{\sigma_g} \equiv 2\Omega \left[\frac{1}{\sigma_g r} \frac{dl}{dr} \right] \tag{2.4}$$

where $l = r^2 \Omega$ is the conserved specific angular momentum. Since Ω itself is hardly changed
by the spiral potential, κ^2/σ_g is essentially a ratio of the differential angular momentum
to fluid element mass, each of which is separately conserved in a Lagrangian sense. We
may define an "unenhanced" epicyclic frequency κ_0 that the gas would have in the absence
of any compression in the arms. If we in turn denote by σ_0 the unenhanced gas surface
density, then we may write

$$\kappa^2 = \frac{\sigma_g}{\sigma_0} \kappa_0^2 \tag{2.5}$$

If the rotation curve of the galaxy is flat in the region of interest, $\kappa_0^2 = 2\Omega^2$. It follows
simply in this case that

$$\frac{d \ln \Omega}{d \ln r} = \frac{\sigma_g}{\sigma_0} - 2, \tag{2.6}$$

or in other words density enhancements in excess of a factor of two will cause the angular
velocity to *increase* outward. (It should be immediately obvious from the above two
equations that compressions that lower the outwardly decreasing shear render the gas *more*
stable to gravitational collapse—due to an increase in the angular momentum gradient—
not, as is often erroneously claimed, less so.) Such an increase has been directly observed
in M51 by Vogel et al. (1988). This reversal of the sense of the shear completely alters
the classical swing amplification mechanism discussed in §2.3, and leads to interesting and
unexpected behavior in the evolution of density perturbations in the interstellar gas.

An analysis of our very local swing amplification is best approached by Lagrangian
techniques (Goldreich and Lynden-Bell 1965). It does no good to consider plane wave
perturbations in fixed Eulerian coordinates; these are not local spatial eigenfunctions. On
the other hand, in coordinates shearing and expanding with the unperturbed background
plane waves, a local "plane wave" analysis is entirely appropriate with the understanding
that the time evolution of the disturbance will generally be neither oscillatory nor exponential. In fact, the point of the analysis is to determine the temporal evolution by solving a
Lagrangian ordinary differential equation. Unfortunately, even to write down the equation
is somewhat complicated, but let us pursue.

We shall consider a gaseous disk (Goldreich and Lynden-Bell 1965; Goldreich and
Tremaine 1978; Julian and Toomre 1966 for the stellar case.) Set up a quasi-cartesian
coordinate system as in fig. [1]. The local Lagrangian approach involves considering a
small patch of fluid as it expands and shears downstream from the arm, but regarding it as

an effectively infinite spatially homogeneous (though temporally changing) background in which a disturbance is initiated. We write k_L for the (constant) Lagrangian wave number of the disturbance (components k_{iL}), and k_E for the "true" Eulerian wavenumber that is distorted by the nonuniform velocity field. An important measure of the distortion is a quantity $T(t)$, which can be defined by:

$$|k_E|^2 = (k_{yL})^2 \left(1 + T^2\right) \tag{2.7}$$

For the case of a uniformly sheared velocity field, T is simply a linear function of time t, but since we will soon need to discuss the case of simultaneously shearing and diverging flow associated with ISM flow near a spiral arm, we leave equation (2.7) in its present general form. Also needed for the case of a slowly expanding (in the x direction of fig. [1]) medium is a volume scale factor $R(t)$, analogous to the cosmological scale factor (Weinberg 1972), but here decidedly anisotropic: no y expansion. This factor simply is a measure of the relative change in volume of a fluid element due to its post compression downstream reexpansion. The surface density σ_g is proportional to $1/R$.

The equation governing the local evolution of density perturbations may be written as follows. The comoving Eulerian density perturbation is denoted by $\delta\sigma_g$. Its evolution is then given by the equation (Goldreich and Tremaine 1978; Balbus 1988):

$$\left[\frac{d^2}{dt^2} + S(t)\right] \frac{\delta\sigma_g/\sigma_g}{(1 + T^2)^{1/2}} = 0 \tag{2.8}$$

where

$$S(t) = (k_{yL}a)^2 (1 + T^2) - 2\pi G\sigma_g|k_{yL}|(1 + T^2)^{1/2} + \frac{\sigma_g}{\sigma_0}\kappa_0^2$$

$$-(1 + T^2)^{1/2}\frac{d^2}{dt^2}(1 + T^2)^{-1/2} + \frac{\sigma_g}{\sigma_0}\frac{\kappa_0^2}{\Omega}\frac{d}{dt}\left(\tan^{-1}T\right). \tag{2.9}$$

Time-dependent flow quantities take on values appropriate to an unperturbed fluid element orbit.

The first three terms in equation (2.9) have been discussed above and are readily interpreted in terms of the dispersion formula (2.1b), and represent the generalization of Jeans-like plane wave behavior in the case when the background is changing due to shear and expansion. The latter two terms are new. The first of these is purely kinematic, as can be seen by its persistent presence in the absence of pressure, gravitational, or Coriolis forces. This term simply measures the changes in relative density due to the shearing and expanding background motions alone. The final term in $S(t)$ is the "Toomre term", and embodies the interplay between the epicyclic oscillations and shearing motions discussed in §2.3, here generalized to allow for an expanding background and variable shear. The term destabilizes if T decreases with time, as it would in a normal disk where the angular velocity decreased outwards, but it is a stabilizing influence if T increases with time. It is tempting to conclude that T will behave this way near spiral arms because of the shear reversal, but one must be careful. Not only does the shear return to its normal outwardly decreasing configuration when the density enhancements drops below a factor of 2, there is another effect of the expansion to consider.

It is not difficult to show that (Balbus 1988):

$$\frac{dT}{dt} \propto \frac{d\ln\Omega}{d\ln r} + \frac{T}{\Omega}\frac{d\ln\sigma_g}{dt}. \tag{2.10}$$

The first term on the right represents the effect of the shear on the density crests of an embedded plane wave and is given by equation (2.6), but the second term shows that more is going on. What is happening is that the *expansion* of the underlying flow also rotates the crests. Leading wave crests are rotated in the retrograde sense of the epicyclic motion, while trailing wave crests are rotated in the opposite sense. This can be easily visualized by imagining that a series of parallel lines has been drawn on a rubber membrane about to be stretched. The stretching causes the lines to reorient themselves along the direction of the expansion velocity regardless of how they are initially drawn, but the reorientation could be either a clockwise or a counterclockwise induced rotation. The upshot of all this grueling shearing and stretching is that the fate of a density perturbation present at the point of maximal compression in an arm depends sensitively on its initial wavenumber direction. A numerical study of the solutions to equation (2.8) (Balbus 1988) shows that two distinct initial wavenumber directions are particularly prone to growth. One direction corresponds to wave crests oriented nearly along a spiral arm, because this direction avoids any (stabilizing) twisting, whereas the other vulnerable direction is basically perpendicular to this! Wave crests that initially stick out from the spiral arm are at first damped as they are wound into a leading configuration due to the effects of prograde shear. But in their subsequent unwinding (due to flow expansion) they come back with vengeance. The diverging flow twists the leading crests in a retrograde manner (consonant with epicyclic motion), stretching them in the process nearly perpendicular to the arm, until at last the expansion stops and normal shear conditions prevail. Depending upon the underlying Q value of the disk, the amplification attained can be quite high: if the azimuthally averaged value of Q is of order 2 or less, and peak density enhancements in the arm about 4, an amplification factor of order 10^2 is possible (Balbus 1988). While it must be remembered that this effect represents only the most local response of which the gas is capable, it is none the less tempting to associate this behavior with the tendency for spiral "spurs" to protrude off the principal arms (Elmegreen 1980).

The fact that two orthogonal wave number directions both exhibit strong, and often comparable growth lends itself to an explanation of one other long-standing observation: the parceling of interstellar gas into clumps and packets along spiral arms at roughly regular intervals. The Parker (1966) instability explained this by using a magnetic field configuration to parcel the gas along the ridge of high density in a spiral arm. Thus, it is of interest to note that the local self-gravitating dynamics of disk gas tends to do this on its own, if we identify spiral arm complexes with the intersection of the growing orthogonal wave crests. Characteristic masses ($\sim 10^5 M_\odot$) and spacings (700 pc) obtained this way are consistent with typical molecular cloud complexes (Stark 1985), but do not seem consistent with the largest observed complexes in M51 (Rand and Kulkarni 1989). Here, something additional seems to be required.

2.6 TWO IMPORTANT PROBLEMS

In closing the discussion of spiral structure and interstellar gasdynamics, I would like to mention two theoretical issues which have received little or no consideration in the literature, but in my opinion are of considerable importance.

2.6.1 Global Instabilities in Spiral Arm Shear Flows. It is a little discouraging to note that despite the rather involved discussion of the previous section, it represents only the most elementary local treatment of the flow. Any sort of behavior that requires, say, an inflection point in the velocity field to be present — e.g. Kelvin-Helmholtz instabilities — would go quite unnoticed in such a local treatment. The observations of Vogel et al. (1988) show quite unambiguously that reversed shear is actually present in the M51 arms and is stable here at least most of the time. But conditions in other galaxies may be quite different, and the formation of growing vortices may occur. These would obviously play an important role in the dissolution of the gaseous arm as well as (perhaps) providing raw material for the largest cloud complexes. Goldreich, Goodman, and Narayan (1986) have elucidated another important type of shearing instability first discovered by Papaloizou and Pringle (1984) involving energy transmission across a corotation point in the flow. The nonlinear resolution of this instability is the development of anticyclonic condensations ("planetesimals") which grow and persist (Blaes and Hawley 1987). The peculiar shear typical of spiral arm zones in disk galaxies would seem to be a prime candidate for these "global" instabilities, and the generally knotted and clumpy appearance of the interstellar gas along spiral ridges is evident. But as yet, there has been no linear nonaxisymmetric study of which this reviewer is aware that considers anything but the very local properties of spiral shear flow.

2.6.2 Chaotic Interstellar Gasdynamics. A tacit assumption of any stability analysis is that there is *some* well-defined equilibrium state of the flow about which one is perturbing. But the results from many fields of nonlinear dynamical studies clearly indicate that this is not something to take for granted. In particular, forced nonlinear periodic systems with feedback often exhibit chaotic behavior (Lichtenberg and Lieberman 1983). The large-scale self-gravitating response of the interstellar medium in a disk galaxy to a forcing potential involves all of these features. Additionally, spiral structure in the disks is extremely varied; we have concentrated here mainly on the well-articulated grand design systems, but they are a minority. Most show an irregular, flocculent spiral pattern in the gas and young stars. All of this suggests that much of the large-scale dynamical interaction between stars and gas in disk galaxies may have an essential element of chaotic structure. Why is this important? For one, it would indicate that grand-design galaxies and messy galaxies may be at heart the same type of system; one just happens to find itself with a somewhat different forcing parameter than the other. For another, the formation of small scale structure like cloud complexes may be understood quite differently in chaotic versus regular systems. In chaotic systems, dynamical structure on all scales may always be present, and the role of some particular instability invoked to explain clumpy structure will be less prominent.

How does one make the notion of chaos well-defined for the large-scale gas response to a forcing spiral potential in a disk galaxy? One simple approach currently being explored (Balbus and Lubow 1990) is to simply expand flow variables in a one-dimensional Fourier series so that, say, the density takes the form

$$\sigma_g(x,t) = \sum_{n=0}^{N} \left(A_n(t) \sin \frac{n\pi x}{L} + B_n(t) \cos \frac{n\pi x}{L} \right) \tag{2.11}$$

where L is an arm-to-arm spacing length and N is some upper limit chosen for convenience. One then follows the Galerkin procedure of plugging the expansions directly into the fully self-gravitating dynamical Roberts (1969) equations and arriving at a series of ordinary differential equations in time for the A_n and B_n. A steady-state equilibrium would correspond to a single point in the $A - B$ phase space, a limit cycle to a closed curve, while the hallmark of chaos would be the wandering of stochastic trajectories either filling the space or confined to an attractor.

3. Warps

A long standing problem of galactic gas dynamics has been to understand the peculiar H I structure seen in the outer regions of disk galaxies, and occasionally in the stellar disks as well. A characteristic integral sign shape in the gaseous disk is the most common manifestation of galactic "warps" (Sancisi 1976). The discovery of the warp in our own Galaxy dates from Kerr (1957). A natural explanation for the presence of this sort of structure is that it arises from ordinary circular gas orbits differentially inclined to the galactic plane. But of course both the the the angular velocity Ω and the vertical oscillation frequency κ_z generally vary with radius (r). The challenge has been to figure out how to make gas persist in its warped state in the face of the relentless tendency of differential precession of the orbits to eliminate coherent structure (Binney and Tremaine 1987).

3.1 FROM WHENCE WARPS?

An early alternative explanation was put forward by Kahn and Woltjer (1959) that the relative motion of our Galaxy through an intergalactic medium in the local group shapes the warp by hydrodynamic processes. These authors recognized the differential precession problem explicitly; their model was an attempt to overcome it. Lynden-Bell (1965) pointed out that "tuning" of the precession problem might occur if self-gravity were included in the dynamics of the disk. This is because the vertical gravitational force is wavelength dependent, and a global mode might therefore be possible with the help of an additional adjustable parameter. The situation is precisely analogous to the spiral structure problem. In fact WKBJ "bending waves" (vertical displacement $\sim e^{i(\int k\, dr + m\theta - \omega t)}$) in the gas satisfy a dispersion formula reminiscent of the stellar density wave equation (2.1a) (Hunter and Toomre 1969):

$$(\omega - m\Omega)^2 = \kappa_z^2 + 2\pi G\sigma_g |k| \qquad (3.1)$$

One merely uses equation (3.1) to *define* $k(r)$ for the correct eigenvalue ω. Notice that self-gravity always raises the effective oscillation frequency (by a varying amount) in a bending wave, whereas it always lowers the epicyclic response in a density wave: there is no gravitational instability for bending waves (Binney and Tremaine 1987). Lynden-Bell found a discrete warp mode in a flattened MacLaurin spheroid. The existence of a *discrete* mode is of paramount importance for the same reason here as it is in the density wave problem. Group velocity dispersion effects will destroy any organized pattern if a continuum of neighboring eigenmodes is present. In an important paper, Hunter and Toomre (1969) showed that any search for a discrete $m = 1$ (the commonly observed azimuthal structure) mode in an isolated, smooth-edged disk is doomed to failure. Such a mode is possible only if the disk is sufficiently sharp-edged: a continuum of WKBJ modes is present if $\int^\infty dr/\sigma_g$ is divergent. Sharp-edged disks can reflect traveling bending waves to set up a modal standing wave pattern.

The importance of the Hunter and Toomre (1969) result became all the more apparent as it was established that warped H I structure in disk galaxies is a common phenomenon

(Sancisi 1976). Not just the occasional oddball exception, warps were turning out to be the rule for spiral galaxies. But while they were emerging as a generic feature of such systems, the one obvious generic feature of disk galaxies — a disk — was not by itself sufficient to understand warps. So, from whence these warps?

3.2 THE MISSING GENERIC: MASSIVE HALOS

With the advent of massive dark haloes (Ostriker and Peebles 1973) came the growing realization that galactic disks could not, for many dynamical problems, be regarded as isolated. This offered a possible route round the Hunter–Toomre roadblock.

Binney (1978, 1981) argued that a triaxial halo (or simply an ordinary bar) could spontaneously excite vertical oscillations at certain resonant radii via the Mathieu instability (e.g. Bender and Orszag 1978). These radii are located where the bar/halo rotation rate happens to equal $\Omega \pm \kappa_z$. But unlike a pure gravitational instability, the Mathieu "instability" is really an overstability. It corresponds to a growth in (when self-gravity is included) *propagating* wave amplitudes, and is not an evanescent phenomenon. Toomre (1983) pointed out the central difficulty here: group velocity dispersal (yet again) carries amplified waves away before the external pumping has driven linear disturbances more than a factor of order unity.

The presence of distant halo matter at radii well beyond the optical disks of a galaxy (M83 is a case in point [Rogstad, Lockhart, and Wright 1974]) can bind H I gas as much as 40 kpc out. At these distances, Tubbs and Sanders (1979) noted that differential precession need not be a problem, provided the halo is not too oblate. Petrou (1980) argued that even oblateness need not necessarily be a problem, if a bit of fine tuning can be arranged. Instead of a local wavenumber adjusting to give an eigenmode, Petrou suggested that the halo oblateness parameter could adjust to give not *differential* precession, but a constant precession rate with radius. But neither of these ideas are very general or encompassing. There are simply too many cases of warps on smaller scales (e.g. our own Galaxy or M31). And in the absence of a convincing dynamical connection, it strains credulity to believe that the oblateness of a massive halo should think twice about making the H I appear presentable.

Genuine progress seems to have come with the notion that the halo and H I disk axes may often be misaligned (Toomre 1983; Dekel and Schlossman 1983; Sparke and Casertano 1988). The idea is that the inner portions of the H I disk tend to follow the stellar disk potential, while the outer portions follow the (misaligned) halo potential. Toomre (1983) gave encouragement to this approach by modeling the halo in a numerical simulation as a distant ring, and Sparke and Casertano (1988) have recently shown that not only do realistic models also achieve persistent warps, but that they are also capable of fitting the observations in some detail (cf. their fit of NGC 4013).

What Sparke and Casertano showed was that a discrete $m = 1$ warping mode does indeed exist for a range of halo oblateness parameters and disk radii in the misaligned disk-halo system. Furthermore, the existence of the discrete mode is *not* sensitive to the details of the disk truncation, although the disk must be finite. The existence of discrete modes can be understood by analogy to the one-dimensional Schrödinger equation,

$$E = V + \frac{\hbar^2 k^2}{2m} \tag{3.2}$$

where E and V represent respectively total and potential energy, and the final term on the right is the usual kinetic energy expression. If $E - V > 0$ everywhere, a continuum spectrum results, whereas if $E - V < 0$ somewhere, a discrete bound state is possible. Similarly, for equation (3.1) discrete states may appear only if

$$(\omega - \Omega)^2 - \kappa_z^2 < 0 \qquad (3.3)$$

somewhere. In practice, this means that the frequencies of the "bound" discrete modes must lie in the range $\Omega \pm \kappa_z$ evaluated at the outer edge of the disk. (Note that the contribution from the halo must be included in κ_z.) A wave with frequency in this range would become evanescent before it reaches the edge. This effective trapping can work like relflecting edge, setting up conditions for a standing wave modal solution.

Actually, for any isolated disk there *is* always one discrete mode: a trivial, nonwarped tilt mode of the entire disk through a constant angle. But in the presence of a nonaligned halo, Sparke and Casertano found that things are no longer so trivial. The tilt mode is endowed with a new status as an active discrete $m = 1$ bending mode whose eigenfreqeuency places it in the continuum gap implied by equation (3.3). To first order in the halo oblateness parameter, the tilt mode frequency changes from zero to

$$\omega_t = \frac{\int_0^\infty \Delta\kappa_z^2 \sigma_g(r) r^3\, dr}{2 \int_0^\infty \Omega(r)\sigma_g(r) r^3\, dr} \qquad (3.4)$$

where $\Delta\kappa_z^2$ is the change in the square of the vertical oscillation frequency due to the departure of the halo from spherical symmetry. Note that the sign of ω_t changes depending upon whether the halo is prolate or oblate indicating respectively prograde or retrograde precession, and that the frequency is not hypersensitive to the details of the disk edge structure. In a more gereral context, this "modified tilt mode" has been suggested by these authors as the underlying dynamical mechanism behind many of the observed warps, and they have modeled the galaxies NGC 4013 and NGC 4065 in some detail. A most important finding is that the existence of a discrete warping mode constrains the halo oblateness parameter from becoming too large in magnitude, with oblate haloes more tightly constrained that prolate ones.

3.3 SUMMARY

The finding of a discrete, persistent mode in the misaligned disk-halo system represents an important advance toward an understanding of galactic warps. It must not be forgotten, however, that as with all bending waves, this eigenmode is neutrally stable. Thus, while persistence is perhaps not the problem it once was for warps, a general excitation mechanism remains elusive. Furthermore, there are cases that continue to defy easy explanation. NGC 4672 displays a warp that bends back toward the optical plane, a configuration plainly incompatible with the modified tilt mode scenario. Tidal interactions (favored by Hunter and Toomre [1969]) are probably involved here and elsewhere. However, if a strong case can be built supporting the modified tilt mode hypothesis as the primary cause for even a few well-observed galactic warps, theorists will have acquired a powerful new tool with which to probe the distribution of dark matter in galactic haloes.

I acknowledge support for this work from NASA grants NAGW-764 and NAGW-1510 and from NSF grant AST 88-20293.

4. References

Balbus, S.A. 1988, *Ap. J.*, **324**, 60.

Balbus, S.A., and Lubow, S.H. 1990, in preparation.

Balbus, S.A., and Cowie, L.L. 1985, *Ap. J.*, **297**, 61.

Bender, C.M., and Orszag, S.A. 1978, *Advanced Mathematical Methods for Scientists and Engineers* (New York: McGraw-Hill)

Bertin, G., Lin, C.C., Lowe, S.A., Thurstans, R.P. 1989, *Ap. J.*, **338**, 78.

Binney, J.J. 1978, *M.N.R.A.S.*, **183**, 501.

Binney, J.J. 1981, *M.N.R.A.S.*, **196**, 455.

Binney, J.J., and Tremaine, S.D. 1987, *Galactic Dynamics* (Princeton: Princeton University Press), pp. 689–693.

Blaes, O.M., and Hawley, J.F. 1987, *Ap. J.*, **326**, 277.

Blitz, L., and Shu, F.H. 1980, *Ap. J.*, **238**, 148.

Blitz, L. and Thadeus, P. 1980, *Ap. J.*, **241**, 676.

Cowie, L.L. 1980, *Ap. J.*, **236**, 868.

Cowie, L.L. 1981, *Ap. J.*, **245**, 66.

Dekel, A., and Schlosman, I. 1983, in *Internal Kinematics and Dynamics of Galaxies, IAU Symposium No. 100*, ed. E. Athanassoula, (Dordrecht: Reidel)

Elmegreen, B.G. 1979, *Ap. J.*, **231**, 372.

Elmegreen, B.G. 1982, *Ap. J.*, **253**, 634.

Elmegreen, B.G. 1987a, *Ap. J.*, **312**, 626.

Elmegreen, B.G. 1987b, in *Interstellar Processes*, eds. D.J. Hollenbach and H.A. Thronson, Jr. (Dordrecht: Reidel), p. 259.

Elmegreen, B.G. 1989, preprint.

Elmegreen, B.G., and Elmegreen, D.M. 1986, *Ap. J.*, **311**, 554.

Elmegreen, D.M. 1980, *Ap. J.*, **242**, 528.

Erickson, S.A. 1974, unpublished Ph.D. thesis, Mass. Inst. of Tech.

Goldreich, P., Goodman, J., and Narayan, R. 1986, *M.N.R.A.S.*, **221**, 339.

Goldreich, P., and Lynden-Bell, D. 1965, *M.N.R.A.S.*, **130**, 125.

Goldreich, P., and Tremaine, S.D. 1978, *Ap. J.*, **233**, 857.

Hausman, M.A. 1981, *Ap. J.*, **245**, 72.

Hunter, C., and Toomre, A. 1969, *Ap. J.*, **155**, 747.

Hunter, D.A., and Gallagher, J.S. 1986, *Publ. Astron. Pacific*, **98**, 5.

Hunter, D.A., and Gallagher, J.S. 1989, *Science*, **243**, 1557.

Jog, C.J., and Solomon, P.M. 1984, *Ap. J.*, **276**, 114.

Julian, W. H., and Toomre, A. 1966, *Ap. J.*, **146**, 810.

Kahn, F.D., and Woltjer, L. 1959, *Ap. J.*, **130**, 705.

Kalnajs, A. 1965, unpublished Ph.D. thesis, Harvard University.

Kaufman, M., Bash, F.N., Hine, B., Rots, A.H., Elmegreen, D.M., and Hodge, P.W. 1989, preprint.

Kerr, F.J. 1957, *Astron. J.*, **62**, 93.

Kwan, J., and Valdes, F. 1983, *Ap. J.*, **271**, 604.

Larson, R.B. 1987, in *Galactic and Extragalactic Star Formation*, eds. R.E. Pudritz and M. Fich, (Dordrecht: Reidel).

Lichtenberg, A.J., and Lieberman, M.A. 1983, *Regular and Stochastic Motion*, (New York:

Springer-Verlag)

Lin, C.C., and Shu, F.H. 1964, *Ap. J.*, **140**, 646.

Lin, C.C., and Shu, F.H. 1966, *Proc. Natl. Acad. Sci. U.S.A.*, **55**, 229.

Lin. C.C., and Shu, F.H. 1968, in *Astrophysics and General Relativity*, Vol. **2**, ed. M. Chretien (New York: Gordon and Breach), p. 235.

Lin, C.C., Yuan, C., and Shu, F.H. 1969, *Ap. J.*, **155**, 721.

Lindblad, B. 1963, *Stockholm Obs. Ann.*, **22**, No. 5.

Lubow, S.H., Balbus, S.A., and Cowie, L.L. 1986, *Ap. J.*, **309**, 496.

Lynden-Bell, D. 1965, *M.N.R.A.S.*, **129**, 299.

Lynden-Bell D., and Ostriker, J.P. 1967, *M.N.R.A.S.*, **136**, 293.

Mouschovias, T. Ch., Shu, F.H., and Woodward, P. 1974, *Astron. Astrophys.*, **33**, 73.

Ostriker, J.P., and Peebles, P.J.E. 1973, *Ap. J.*, **186**, 467.

Papaloizou, J.C.B., and Pringle, J. 1984, *M.N.R.A.S.*, **208**, 721.

Parker, E.N. 1966, *Ap. J.*, **145**, 811.

Petrou, M. 1980, *M.N.R.A.S.*, **191**, 167.

Rand, R. J., and Kulkarni, S.R. 1989, preprint.

Roberts, W.W. 1969, *Ap. J.*, **158**, 123.

Rogstad, D.H., Lockhart, I.A., and Wright, M.C.H. 1974, *Ap. J.*, **193**, 309.

Sancisi, R. 1976, *Astron. Astrophys.*, **53**, 159.

Scalo, J.M., and Pumphrey, W.A. 1982, *Ap. J. (Letters).*, **259**, L29.

Scalo, J.M., and Struck-Marcell, C. 1984, *Ap. J.*, **276**, 60.

Sellwood, J.A., and Lin, D.N.C. 1989, preprint.

Shu, F.H. 1978, in *IAU Symposium 77, Structure and Properties of Nearby Galaxies*, eds. E.M. Berkhuijsen and R. Wieleblinski (Dordrecht: Reidel), p. 139.

Shu, F.H., Adams, F. C., and Lizano, S. 1987, *Ann. Rev. Astron. Astrophys.*, **25**, 23.

Shu, F.H., Milione, V., and Roberts, W.W. 1973, *Ap. J.*, **183**, 819.

Smith, J. 1980, *Ap. J.*, **238**, 842.

Sparke L.S., and Casertano, S. 1988, *M.N.R.A.S.*, **234**, 873.

Spitzer, L. 1978, *Physical Processes in the Interstellar Medium* (New York: Wiley Inter-science).

Stark, A.A. 1985, *Highlights Astr.*, **7**, 507.

Tilanus, R.P.J., and Allen, R.J. 1989, *Ap. J. (Letters).*, **339**, L57.

Tomisaka, K. 1986, *Pub. Astr. Soc. Japan*, **38**, 95.

Toomre, A. 1969, *Ap. J.*, **158**, 899.

Toomre, A. 1977, *Ann. Rev. Astron. Astrophys.*, **15**, 437.

Toomre, A. 1981, in *The Structure and Evolution of Normal Galaxies*, ed. S.M. Fall and D. Lynden-Bell, (Cambridge: Cambridge University Press), p. 111.

Toomre, A. 1983, in *Internal Kinematics and Dynamics of Galaxies, IAU Symposium No. 100*, ed. E. Athanassoula, (Dordrecht: Reidel)

Tubbs, A.D., and Sanders, R.H. 1980, *Ap. J.*, **230**, 736.

Visser, H.C.D. 1980, *Astron. Astrophys.*, **88**, 159.

Vogel, S.N., Kulkarni, S.R., and Scoville, N.Z. 1988, *Nature*, **334**, 402.

Weinberg, S. 1972, *Gravitation and Cosmology* (New York: Wiley Interscience).

Gas during Mergers/Collisions

Masafumi Noguchi
Nobeyama Radio Observatory, National Astronomical Observatory

ABSTRACT. Present status of the theoretical understanding of the gas dynamics in interacting galaxies is described. Recent numerical studies on close encounters, formation of ring galaxies, and mergers are reviewed with special emphasis placed on the gas dynamical results. Generating mechanisms of starbursts and nuclear activity in interacting galaxies are discussed in view of the numerical results. Much attention has been paid to the transient effects caused by galaxy-galaxy interactions. Evolution of tidally-disturbed galaxies is investigated for much longer time span and the permanent aftereffects of interactions on the internal structure of victim galaxies are discussed in detail.

1. Introduction

The interaction of a galaxy with its environment takes various forms. I discuss here the galaxy-galaxy interactions such as galaxy collisions and mergers. Other interesting cases such as galaxy-intergalactic medium or galaxy-galaxy cluster interactions are discussed by Kenney in this volume.

Galaxy-galaxy interactions have long attracted many extra-galactic astronomers in various aspects. Peculiar morphology observed in some galaxies was the first motivation for extensive study of galaxy-galaxy interactions. Numerical simulations by computers have successfully reproduced actual shape of several peculiar galaxies. In recent years, galaxy-galaxy interactions have again become a subject of intensive researches in another context, namely, in relation to various activities in galaxies. Many observations suggest that the star formation process is much enhanced in interacting galaxies compared with isolated galaxies. Although controversial, Seyfert galaxies and quasars may be involved in galaxy-galaxy interactions. At present, much effort (both theoretical and observational) is being made to understand the causal relationship between the galaxy-galaxy interactions and these activities.

Most theoretical studies in the past have treated the stellar constituent of interacting galaxies and placed emphasis on the dynamical aspect. However, the behavior of the interstellar gas in interactions is also an important problem to be studied because the understanding of the activity in interacting galaxies relies on how well we are understanding the behavior of the interstellar gas which is raw material for any activity. Now numerical simulations are beginning to provide some important (though preliminary) results for this problem. Because the gas dynamics in galaxies is not totally but largely governed by the stellar gravitational field, the understanding of gaseous behavior presupposes that of the stellar system. Fortunately, our knowledge about the stellar

H. A. Thronson, Jr. and J. M. Shull (eds.), The Interstellar Medium in Galaxies, 323–347.
© 1990 *Kluwer Academic Publishers.*

dynamical problem in interacting galaxies has also improved appreciably by the aid of numerical simulations in the last decade.

The present article is devoted to summarizing a brief history of the theoretical researches and discussing the recent numerical results. Brief comparison with the currently available observational data will also be presented. For clarity, I take up the different modes of galaxy-galaxy interactions separately. I first discuss the case of close encounters for which the most detailed studies have been carried out to date. After that, ring galaxies and mergers are discussed. In every case, the gas dynamical problem and its possible relationship with the star formation and other activities will be stressed. Recently, the importance of the gaseous self-gravity has often been stressed. This problem is discussed in a separate section. Future prospect will be discussed in the final section. In the most part of the present review, I will take up disk galaxies because they present far more spectacular response than elliptical galaxies because of their internal dynamics and wealthy gaseous content.

2. Close Encounters

We define here a close encounter to be a galaxy-galaxy collision which does not lead to a significant overlapping of two galaxies at the perigalacticon (i.e., the relative orbit is sufficiently non-radial) and does not lead to merger. This case greatly simplifies the numerical treatment because the effect of the perturbing galaxy can be reasonably idealized as tidal force and we can neglect the structure of the perturber in the first approximation.

2.1 PECULIAR MORPHOLOGY - TAILS AND BRIDGES

Toomre and Toomre (1972) have first carried out an extensive numerical experiment for a number of close encounters. Their models are an extended version of the restricted three body problem, with the individual gravitational fields of two galaxies represented by that of a point mass respectively and the disk of each galaxy constructed by a number of test particles circularly rotating around the central mass point. Therefore, two galaxies never merge but continue to execute purely Keplerian motion around each other regardless of the orbital binding energy. It is now well established that a parabolic or elliptic motion leads promptly to merger if the perigalactic distance is smaller than roughly the sum of the radii of two galaxies (e.g., see the summary by Aarseth and Fall 1980). Furthermore, the self gravity of individual galaxies has been neglected, which surely influences the result greatly as seen later.

Despite these shortcomings, Toomre and Toomre (1972) have succeeded in reproducing the peculiar morphology of several galaxy pairs (e.g., "antennae", "mice", M51 system) very well in the sense that the deformation of the outer part, especially the elongated structure such as tails and bridges, has been explained as a tidally-pulled out structure. Test particle simulations have proved to be successful also for the reproduction of the global distribution and kinematics of HI gas around galaxy pairs (e.g., Combes 1978). Therefore, the deformation (at least in the outer part) of galaxies in a pair is considered to have been caused by the tidal force.

Another important finding of Toomre and Toomre (1972) is that the degree of tidal disturbance depends strongly on the inclination between the spin and orbital angular momenta: a prograde encounter brings about the most serious damage while a retrograde one hardly influences the parent galaxy.

2.2 EFFECT ON STAR FORMATION RATE

Since late 1970s, active star formation in interacting galaxies has been noticed by many observers. Among others, Larson and Tinsley (1978) have ascribed a large color scatter observed for the interacting galaxies from *the Atlas of Peculiar Galaxies* (Arp, 1966) to the bursts of star formation induced by tidal interaction. They have concluded that several percent of the stars in the galaxy should have been formed within a period of a few times 10^7 years for the color scatter of the observed amount to be produced. The test-particle simulations, intended for collisionless stellar systems, could not clarify how the star formation process will be affected by galaxy-galaxy interactions. Needless to say, the cite of star formation is the interstellar gas. The dynamical behavior of the interstellar gas is expected to be significantly different from that of stellar systems because the gas involves its proper fundamental processes.

Inspired by the finding of starburst phenomena in interacting galaxies by Larson and Tinsley (1978), Noguchi and Ishibashi (1986) have numerically investigated the response of interstellar gas and the enhancement of the star formation rate in close encounter models by using the cloud-particle scheme. The cloud-particle scheme has been originally used by Roberts and Hausman (1984) and Hausman and Roberts (1984) in the investigation of the gas response to the rigidly rotating spiral gravitational field of the galaxy. In Noguchi and Ishibashi (1986), the interstellar gas has been treated as a disk system of 8000 cloud particles. Physically each cloud particle can be regarded as a giant molecular cloud. Each cloud particle has a finite radius but no mass. Following cloud processes are included in the numerical code. When two cloud particles collide (i.e., overlap), the radial component of their relative velocity is reduced by 50%. The cloud system thus dissipates excess kinetic energy of random motion by cloud-cloud collisions. It is observationally suggested that cloud-cloud collisions trigger star formation events (Loren 1976; Scoville, Sanders, and Clemens 1986). This process has been incorporated by creating an "OB star" particle (massless) per each collision. It has been assumed that the OB star explodes as a supernova at the end of its lifetime and gives kinetic energy to nearby cloud particles. The consumption of the gas by star formation has not been taken into account, namely, the number of the cloud particles was held constant (to 8000) throughout the simulation.

In the absence of the perturbing galaxy, all the cloud particles are performing nearly circular rotation with small random motions within the disk plane. The rate of the cloud-cloud collisions stays nearly constant with the time so that the energy dissipation due to the cloud-cloud collisions is approximately balanced by the heating due to supernova explosions. The cloud-cloud collision time scale was set to be a typical value in Sb-type spirals. In interaction, a point mass perturber was introduced to disturb the parent galaxy. Therefore, the simulation by Noguchi and Ishibashi (1986) is a gaseous version of the simulation by Toomre and Toomre (1972), although the former has treated only co-planar passages of the perturber.

The result of Noguchi and Ishibashi (1986) is summarized as follows. The gaseous disk develops a tail and a bridge after the perigalactic passage of the perturber as the test-particle disks in Toomre and Toomre (1972) do. As the tail and bridge develop, the cloud-cloud collision rate increases up to about 10 times the pre-encounter value. This is because many cloud particles are gathered into the spiral arm regions and frequently collide with each other. Therefore, most OB stars are formed in the outer disk region where the tidal deformation is the most remarkable. In the late stage of the encounter (later than about 2 disk rotation periods, i.e., about 10^9 years for a typical disk galaxy, from the perigalactic passage), the gas settles into a ring structure due to the energy dissipation.

The most convincing evidence of enhanced star formation in interacting galaxies comes from the observation in the far infrared wavelengths by *IRAS* (Soifer *et al.* 1984).

Young *et al.* (1986) have compared the star formation efficiency defined as the ratio of the far infrared luminosity to the CO luminosity between the samples of interacting/merging and isolated galaxies. They have found that the star formation efficiency in the interacting/merging galaxies is 7-8 times larger than that of the isolated galaxies on the average, in good agreement with the result of Noguchi and Ishibashi (1986). A similar degree of the star formation enhancement has also been found by Sanders *et al.* (1986). Solomon and Sage (1988) have shown by a similar analysis that the star formation efficiency in interacting galaxies is the highest for those galaxies with tidal tails and have suggested that the maximum star formation occurs at the time of tail formation, a result also consistent with the numerical result by Noguchi and Ishibashi (1986).

Although the result by Noguchi and Ishibashi (1986) thus seems to explain the observational results to some extent, several important points remain to be solved as follows. The burst time scale inferred by Larson and Tinsley (1978) seems to be considerably shorter than the model value, a few times 10^8 years. The gas behavior in oblique encounters has not been clarified in the two-dimensional treatment by Noguchi and Ishibashi (1986). However, the most serious discrepancy lies in the spatial distribution of induced star formation, which is discussed in the next sub-section.

2.3 TRIGGERING OF NUCLEAR STARBURSTS AND ACTIVITY

The most remarkable feature of the starbursts in interacting galaxies is that they take place preferentially in the nuclear regions of host galaxies, although there is indication of a large variety (Kennicutt *et al.* 1987, Bushouse 1987). The models in Noguchi and Ishibashi (1986) show little enhancement of star formation in the central region of the disk and seem to be inconsistent with the observed tendency.

Observational studies which suggest the occurrence of starbursts in the nuclear regions include the radio studies by Hummel (1980, 1981) and Condon *et al.* (1982), the infrared studies by Joseph *et al.* (1984), Lonsdale *et al.* (1984), Cutri and McAlary (1985), and Telesco *et al.* (1988), the optical study by Keel *et al.* (1985), and an extensive study by Bushouse (1987) (see Figs. 1 and 2). Nuclear activity such as Seyfert galaxies and quasars is also pointed out its correlation with galaxy-galaxy interactions (Adams 1977; Stockton 1982; Hutchings and Campbell 1983; Kennicutt and Keel 1984; Dahari 1984, 1985a, 1985b, but see Bushouse 1987; Fuentes-Williams and Stocke 1988; MacKenty 1989 for more skeptical view).

It may sound curious that interacting galaxies show the sign of stimulation in their nuclear regions rather than in their outer parts, since the tidal force diminishes in strength with the decreasing distance from the nucleus and vanishes at the nucleus. Some link must exist which connects the tidal force and the central disk part. There have been several attempts to understand the triggering mechanism of nuclear star fomation and activity in interacting galaxies by numerical simulations (Byrd *et al.* 1986; Noguchi 1987, 1988a; Olson and Kwan 1989).

2.3.1 *Models by Noguchi.* Noguchi (1987, 1988a) has stressed the importance of the gas dynamical effects combined with the stellar self-gravity. Based on a series of encounter simulations using galaxy models containing both stars and interstellar gas, he has proposed the following scenario for the triggering of nuclear star formation and activity in interacting galaxies; The tidal force of the perturbing galaxy induces a bar structure in the self-gravitating stellar disk of the perturbed galaxy and this bar in turn induces an infall of the interstellar gas to the nuclear region.

In Noguchi (1988a), the galaxy model to be perturbed consists of a halo and a disk. The halo was treated as a rigid spherical gravitational field which is assumed to remain

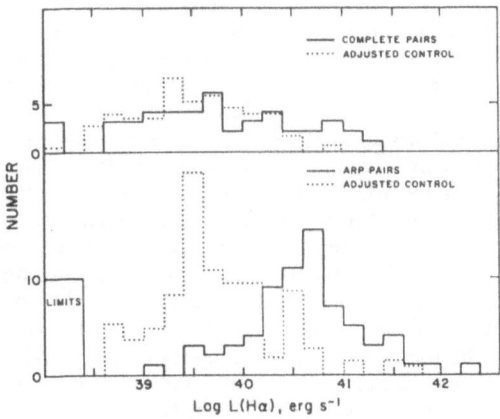

Figure 1. Enhancement of Hα luminosity in the interacting galaxy nuclei (solid histograms) compared with the field galaxy nuclei (dotted histograms) (from Keel *et al*. 1985). The enhancement is in the most part due to active star formation.

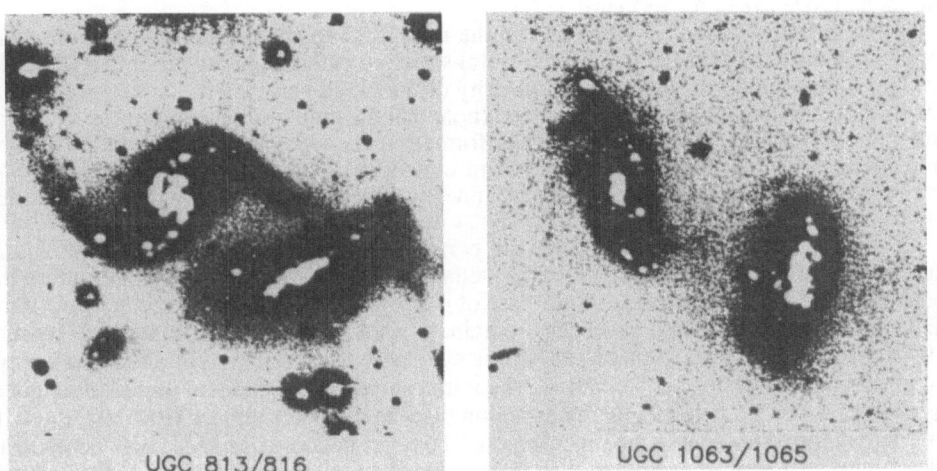

Figure 2. Examples of interacting galaxies showing active star formation near the nucleus (from Bushouse 1987). Hα images (positive contrast) are superposed on the broad-band visible images (negative contrast).

fixed during the interaction. The disk contains stars and the gas. The stellar disk was constructed by 20,000 collisionless particles having the same mass. They move in the halo gravitational field, interacting with each other (and with the perturber in interaction). Therefore, the self-gravity of the stellar disk is taken into account properly. The gravitational field of stars was calculated by the particle-mesh N-body code (Hohl and Hockney 1969; Hockney and Eastwood 1981). The gaseous component was modelled by the cloud-particle scheme as in Noguchi and Ishibashi (1986). The cloud particles are assumed to have no mass. The motion of cloud particles was calculated by using the combined gravitational potential of the stellar disk, the halo, and the perturber. Therefore, the influence of the deformation of the parent disk was taken into account in addition to the direct effect of the tidal force exerted by the perturber. In the absence of the perturbing galaxy, the stellar disk is dynamically stabilized by the halo and maintains nearly axisymmetric shape. Cloud particles are also performing circular rotation within the disk plane. The cloud-cloud collision rates is approximately constant. The perturbing galaxy, assumed to be a point mass, has been introduced in the interaction.

Fig. 3 shows one example (Model A+A1) from Noguchi (1988a). In this model, the stellar disk contains 20% of the total mass of the parent galaxy. The rotation curves and the velocity dispersions for the stellar component are given in Fig. 4. The length is given in units of the initial disk radius of the parent galaxy, and velocities are given in units of the maximum rotational velocity in the disk. The radius of the cloud particles is 8×10^{-3}. The initial velocity dispersion in the cloud system is 0.025. The resulting cloud-cloud collisional time scale is 1.62 in nondimensional units (one rotation period at the outer disk edge is 6.28). The perturber has the same mass as the parent galaxy and moves on a co-planar prograde parabolic orbit with the perigalactic distance twice the disk radius (see Noguchi 1987, 1988a for further details).

In the early phase (T=-0.02, 2.50), the tidal force of the perturber is predominant. Both stars and the gas make two-armed spiral structure extending outward into a tail and a bridge. In the spiral arms, the number density of the cloud particles becomes so large that cloud-cloud collisions are much enhanced (note that the collision rate is roughly square of the number density). Therefore active star formation is expected to occur primarily in the outer region in this phase (e.g., T=2.50) in consistent with the result by Noguchi and Ishibashi (1986). It is also noted that the inner part of the disk is hardly affected in this phase.

About one rotation period after the perigalactic passage (T=6.26), a bar structure begins to evolve in the inner region of the stellar disk. This point is quite different from the behavior of test-particle models. In the case of a test-particle disk, only the outer part of the disk is deformed while the inner part remains almost unchanged (Toomre and Toomre 1972; also see Fig. 3 of Noguchi 1987). It is evident that the self-gravity of the disk plays an important role in this bar formation. The initial amplitude of the tidal perturbation given by the perturber is smaller in the inner region than in the outer region. However, the self-gravity is more dominant in the inner region. The rotation curve due to the disk component alone shown in Fig. 4 indicates that the gravitational field in the inner disk region is virtually governed by the disk component. Therefore the growth of the perturbation is greatly accelerated in the inner region.

After the bar has developed sufficiently, the gas begins to fall to the disk center (T=7.52, 8.78, 11.29, 13.80). This is because the shock induced by the bar removes the kinetic energy and angular momentum from the cloud particles. Gas infall driven by the shock has also been observed in the gas dynamical simulations for imposed bar potentials (e.g., Sanders and Huntley 1976; Sorensen et al. 1976). That the deformation of the stellar disk is essential in this gas infall is demonstrated by Fig. 5, in which the gas response in a

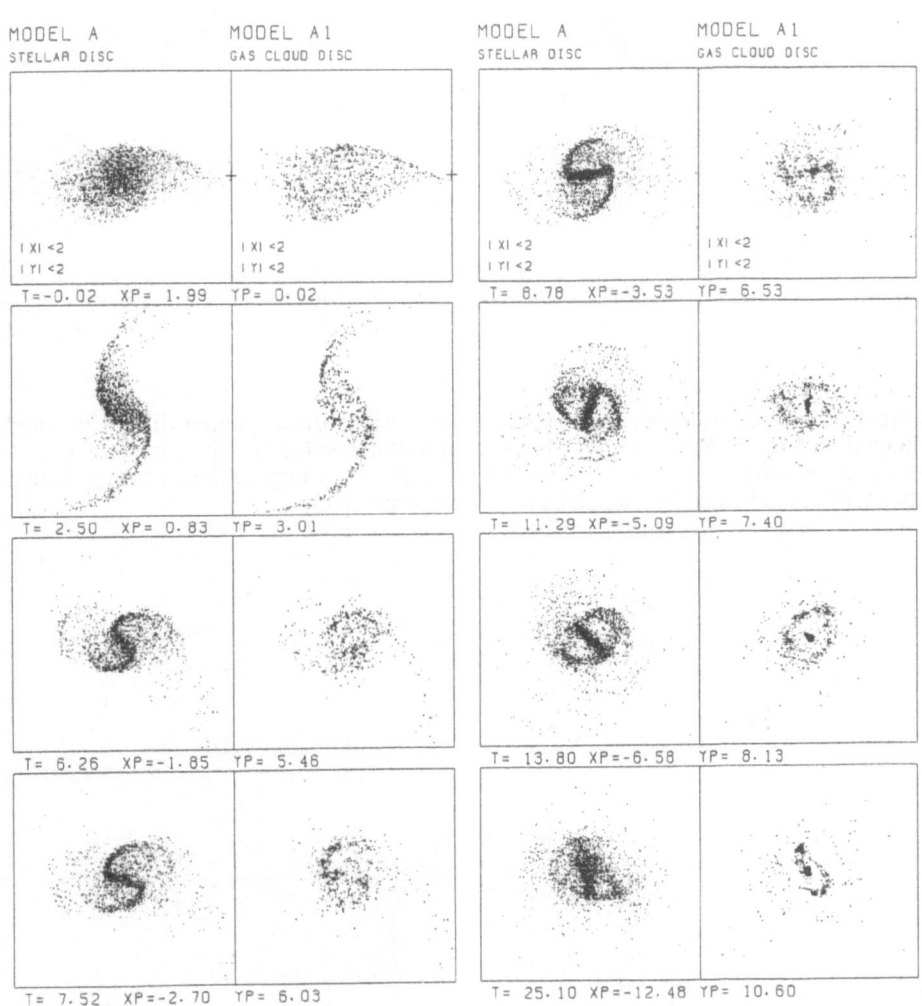

Figure 3. A close encounter model of a disc galaxy containing both stars and gas with an equally massive perturber (from Noguchi 1988a). The time, T, reckoned from the perigalactic passage is given in non-dimensional units. One time unit corresponds to 10^8 years for a galaxy of typical size and mass. XP and YP are X and Y coordinates of the perturber, where X and Y axes are directed to the right and upward, respectively. The disc is rotating counterclockwise.

330

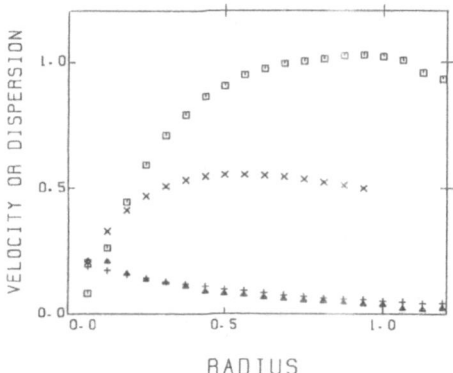

RADIUS

Figure 4. Initial rotation curves and velocity dispersions for the stellar disk of the model indicated in Fig. 3. Squares give the total rotational velocity while crosses give the rotational velocity due to the disc component alone. Tangential and radial velocity dispersions are indicated by triangles and pluses, respectively.

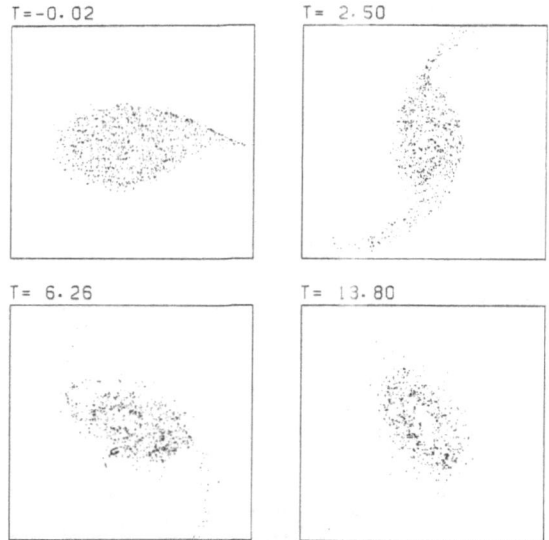

Figure 5. Effect of fixing the stellar disk (from Noguchi 1988a). The galaxy model and the encounter parameters are the same as in Fig. 3, but the stellar disk has been held fixed to nearly axisymmetric initial state during the simulation. This model isolates only the *direct* effect of the tidal force.

model which has a "frozen" stellar distribution is indicated. This model shows that the direct effect of the tidal force alone cannot induce a gas infall.

In Fig. 3, about 20% of the total gas is swallowed within the radius 0.1 (1-2 kpc for a typical galaxy) from the nucleus by T=13 as shown in Fig. 6. Therefore, active star formation is expected to take place in the nuclear region. The maximum infall rate at r=0.1, which occurs around T=10, is $1.1 M_\odot$/yr if we assume that the mass and disk radius of the parent galaxy are $1.8 \times 10^{11} M_\odot$ and 20 kpc, respectively, and the total mass of the gas is $4 \times 10^9 M_\odot$. This infall rate corresponds to the radiation energy of 7×10^{45} erg/sec if the energy conversion efficiency is 0.1. This energy can afford the luminosity of a typical Seyfert galaxy and even that of a quasar. However, caution must be necessary because it is not clear whether the gas accumulated in the central disk region will eventually channeled to the central hypothetical monster itself. On the observational side also, the causal relationship between close encounters and nuclear activity is quite uncertain. This problem will be discussed in section 5. Although the model cannot discriminate between starbursts and nuclear activity, active star formation in the central 1-2 kpc region seems to be a natural outcome in a heavily disturbed gas-rich galaxy.

Finally, it should be noted that the tidal response of a disk galaxy can be very complex in reality. Noguchi (1987, 1988a) has derived the following dependence on model parameters.
(1) The length of the bar induced becomes larger as the mass of the perturber becomes larger.
(2) The pattern speed of the bar (in non-dimensional units) becomes smaller as the disk-to-halo mass ratio becomes smaller.
(3) The length of the bar becomes larger as the outermost radius of the rigid part of the rotation curve becomes larger.
(4) The gas infall is reduced if the inner disk region where the bar is to form is initially devoid of the gas.

These differences derived theoretically may partially explain the diversity of star formation properties and nuclear activity in interacting galaxies observed by Kennicutt et al. (1987) and Bushouse (1987).

2.3.2 *Other Models*. Byrd et al. (1986) have used the particle-mesh N-body code for the polar coordinate system to simulate the tidal response of a self-gravitating disk embedded in a rigid halo. In spite of similar setup of simulations, their results are different from those of Noguchi (1987) in several respects. Their models do not develop so remarkable bars as those in Noguchi (1987). This is presumably because the models by Byrd et al. (1986) have rotation curves strictly flat to the center. Noguchi (1987) suggests that the rigid part of the rotation curve is dynamically favorable to bar formation. Therefore, the absence of bars in Byrd et al. (1986) can be ascribed to the flatness of the rotation law adopted. Byrd et al. (1986) counted the particles which pass through the inner disk boundary and presented the number as the infall rate. The time variation of infall rate given in Byrd et al. (1986) shows irregular variations which bear no clear relationship with the progression of encounter. The mechanism of the claimed infall is not clear. It should also be noted that they did not treat the gas component but a stellar one so that their infall rate cannot be directly related to activity. Their models show continuous expansion of the disk regardless of whether the encounter is prograde or retrograde. Thus the behavior of the models in Byrd et al. (1986) does not seem to permit any simple interpretation.

Olson and Kwan (1989) have carried out simulations which included both stars and gas like in Noguchi (1987, 1988a). Their models have a live halo component instead of a rigid one. Furthermore, their treatment of the gaseous component is more sophisticated than the one in Noguchi (1988a). They considered gas clouds with different masses and took into consideration the collisional coalescence and disruption and the disruption due to the star formation inside massive clouds. They specified parameters for these cloud processes based on the numerical result for the cloud-cloud collision process performed by Lattanzio and Henriksen (1988).

Olson and Kwan (1989) have compared the result for close encounter models with the degree of the observed enhancement of star formation rate in interacting galaxies (e.g., Young *et al.* 1986) and have concluded that the star formation stimulated in interactions should be primarily due to high velocity disruptive collisions rather than coalescence collisions making larger clouds.

They have also found that as the impact parameter of the perturber decreases, the region of induced cloud-cloud collisions becomes more and more concentrated toward the nucleus. Based on this, they claim that bar formation is not necessary for nuclear star formation to be induced. Olson and Kwan (1989) cited the result by Rubin *et al.* (1985) that most disk galaxies have the rotation curves which are flat over 75% of the optical radius unlike Model (A+A1) in Noguchi (1988a), and criticized the argument by Noguchi (1988a) that bar formation is essential ingredient for nuclear starbursts. However, the flatness of rotation curves alone cannot completely rule out the possibility of bar formation. Strictly speaking, the length and the strength of the bar induced depend not only on the shape of the rotation curve but also on the parameters of encounter. Even a disk galaxy having a small portion of rigid rotation can develop a large strong bar if it has been subject to strong tidal force (see Noguchi 1987, Fig. 6b). An extensive statistics on the observational data will be necessary to solve the question whether bars are indispensable or not in stimulating nuclear starbursts.

2.4 TIDALLY-INDUCED BAR STRUCTURE

Another problem closely related with the bar formation discussed above is the origin of barred galaxies. It is well known that a disk with the mass which exceeds the one given by the Ostriker-Peebles (1973) criterion develops a bar spontaneously. This bar instability may be the generating mechanism of barred galaxies. The most direct confirmation of this hypothesis could be obtained by comparing the disk-to-halo mass ratio of barred galaxies with that of normal galaxies: the former should be larger than the latter on the average. Unfortunately it is difficult to measure the total mass of a barred galaxy from kinematical data because considerable non-circular motion exists in the velocity field. van den Bergh (1979) has reported that the fraction of barred galaxies does not differ significantly between different environments such as the field, clusters of galaxies of various richness, and pairs. He argued that the presence or absence of a bar is determined by the intrinsic structure of that galaxy rather than external factors. This result may argue for the bar-instability picture.

The result of Noguchi (1987) suggests that close encounters can also generate barred galaxies from normal galaxies in which the disk mass is sufficiently small and the bar instability did not work. One evidence for this picture may come from a simple statistics that the spiral galaxies contained in *the Atlas of Peculiar Galaxies* (Arp, 1966) show a slight hint of higher incidence of bars compared with the field spiral galaxies (Noguchi 1987). Unfortunately, the statistical significance is very low in this case. The finding of van den Bergh (1979) does not necessarily contradict the interaction picture. Such classification of environments as van den Bergh (1979) did is not necessarily sensitive to the probability of galaxy-galaxy close encounters.

What is the major difference between the spontaneous bars and the tidally-induced bars? It is suggested that the spontaneous bar appearing in a massive disk will have a larger pattern speed than the bar tidally induced in a less massive disk, since the pattern angular

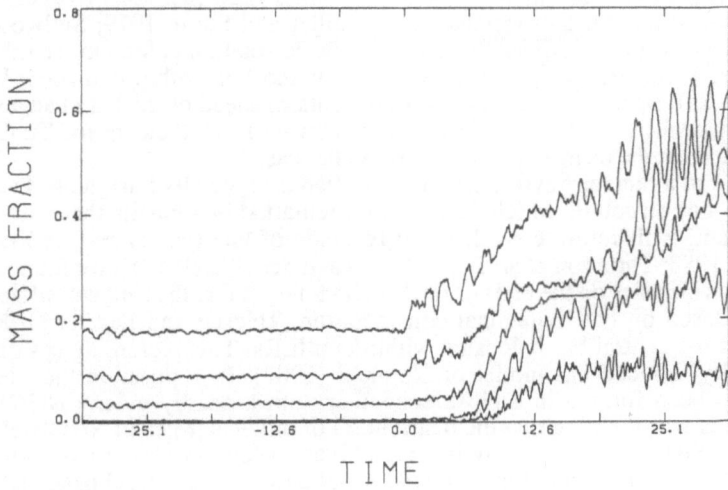

Figure 6. Accumulation of the gas clouds toward the center in the model in Fig. 3. The curves (from bottom to top) indicate the mass fractions (i.e., number fractions) contained within the radii r=0.05, 0.1, 0.2, 0.3, 0.4, respectively.

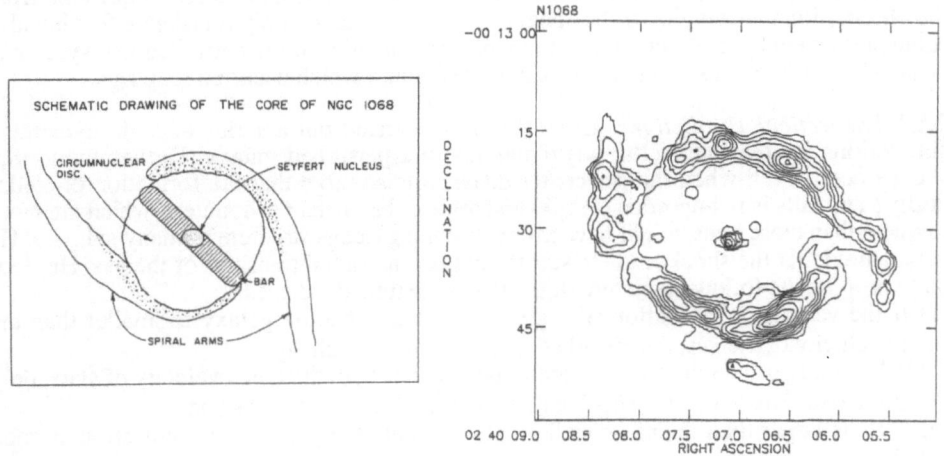

Figure 7. A schematic representation of the inner region of NGC1068 (left, taken from Thronson *et al*. 1989) and the ring of molecular gas (right, from Planesas *et al*. 1989). The length of the bar is about 3 kpc. Compare the molecular gas distribution with the gas cloud disc model at T=13.80 in Fig. 3.

velocity is found to increase as the disk-to-halo mass ratio increases (Noguchi 1987). Numerical simulations for bar-unstable disks (Miller and Smith 1979; Sellwood 1981) show that the spontaneous bar usually ends near the co-rotation point. On the other hand, the bar in Fig. 3, for example, rotates very slowly and lies within two inner Lindblad resonances (ILRs). Because the difference of the pattern speed of the bar changes the gas response dramatically (Combes and Gerin 1985; Schwarz 1984; Sanders and Tubbs 1980), we may infer it from the distribution of the interstellar gas.

The case of a famous Seyfert galaxy NGC1068 is especially intriguing. This galaxy has a central bar structure which is the most remarkably seen in the near infrared wavelengths and is therefore considered to be made of late type stars (see Fig. 7, and Scoville *et al.* 1988; Thronson *et al.* 1989). Myers and Scoville (1987) have found a ring of molecular gas which lies on the ends of the bar and inferred that the ring is confined by the bar forcing. Based on gas dynamical consideration, Telesco and Decher (1988) have suggested that the central bar is located within two ILRs. This picture agrees quite well with the tidally induced bar model of Noguchi (1987). It is possible that the bar in NGC1068 has been formed in a close encounter with a nearby galaxy NGC1055 (or NGC1072). It is worth noting that the distribution of molecular gas in NGC1068 (Myers and Scoville 1987; Planesas, Scoville, and Myers 1989) is akin to the model gas distribution at $T=13.8$ shown in Fig. 3: it consists of a ring and a central peak and the ring lies just on the ends of the stellar (i.e., near infrared) bar. It is plausible that the central peak of molecular gas distribution has resulted from the gas infall induced by the bar.

2.5 END-PRODUCT OF CLOSE ENCOUNTER

What the end-product of a close encounter might look like is an interesting question from the observational viewpoint. If the encounter brings about not only transient effect but also some permanent after-effect on the structure and dynamics of the perturbed galaxy, it may be possible to find out some traces of a close encounter which occurred long ago.

2.5.1 *Theoretical Predictions.* Icke (1985) has carried out a series of hydrodynamical simulations for the interstellar gas dynamics in a perturbed galaxy. He has found that shocks occur even when the pericenter distance is so large that the formation of stellar bridges and tails is unimportant (Fig. 8) and argued that distant encounters, which are more frequent than close ones, might have influenced the gaseous structure of many galaxies. He conjectured that the shock leads to star formation and radial transport of the gas. He went on to propose the following picture of the fate of disturbed galaxies:
(1) If the velocity perturbation, V_p, given by the companion galaxy is smaller than the sound velocity of the gas, the disturbed galaxy remains unchanged.
(2) If V_p is larger than the sound velocity but smaller than the mean velocity of stars, only the gas is redistributed by the shock and the galaxy becomes an S0 galaxy.
(3) If V_p is larger than the mean stellar velocity, both the gas and the stars are perturbed and the galaxy becomes an elliptical.

Although the proposition by Icke (1985) has an attractive feature in its capability of explaining the observed morphological segregation of galaxies, his simulation does not give the rate of star formation and resulting gas depletion quantitatively. Furthermore, it is not clear whether the difference of flatness distribution between spirals, S0s, and ellipticals is well explained by his encounter scenario.

Noguchi (1988b) has investigated a long-term evolution of tidally disturbed galaxy models in Noguchi (1988a). At a few rotation periods after the perigalactic passage, the structural change slows down and the galaxy enters the late phase which typically corresponds to $T=50$ (several times 10^9 years after the perigalactic passage for a disk

335

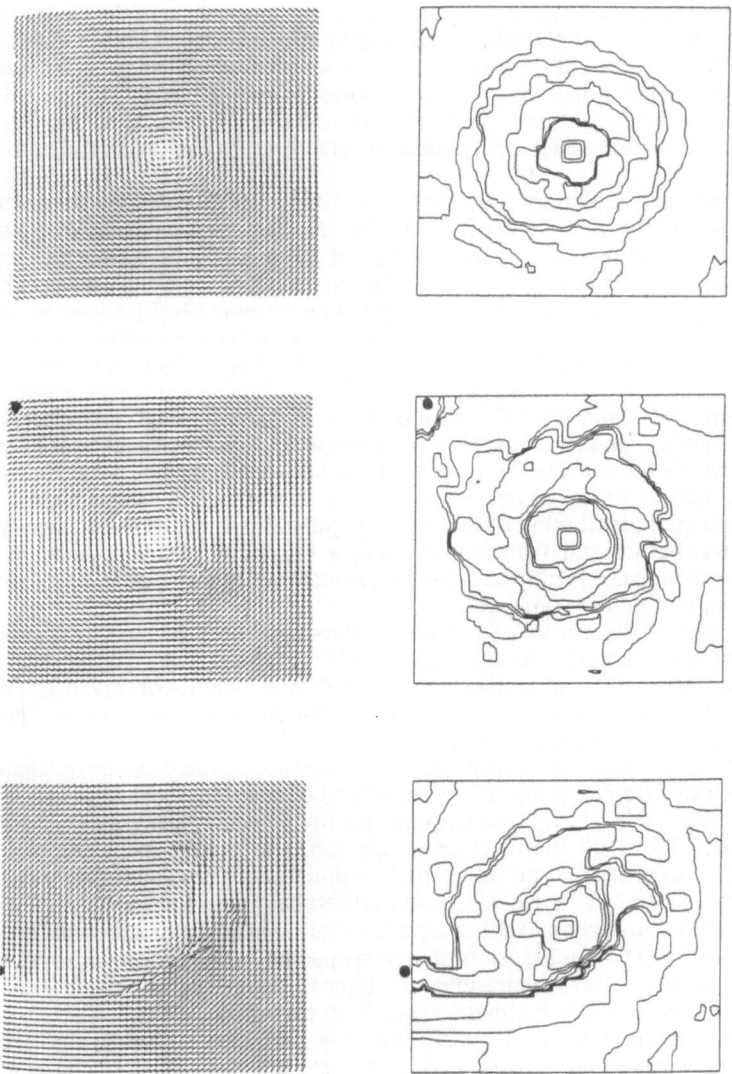

Figure 8. Development of a strong shock in a tidally-perturbed gaseous disc (from Icke 1985). Left and right panels give the velocity field and the density contours, respectively. Filled circle indicates the position of the perturber.

galaxy of typical size and mass). In this phase, the bar dissolves gradually. Fig. 9 shows two snapshots in this phase. The lower panels show the kinematics in the stellar disk. The bar is a dynamically very hot structure: the random motion is comparable with the mean motion in the bar. As the bar dissolves, the heat which was confined to the bar region pervades the whole stellar disk as shown in the lower panels of Fig. 9. This global heating makes the distribution of the stars very smooth because any structures having a short wavelength (such as spiral arms) are dynamically stabilized by the large random motion in the disk.

During the bar dissolution, the stellar disk changes its morphology irregularly with the time (see the stellar disk at $T=30.13$ in Fig. 9). This change shakes up the cloud particles violently and induces a further infall of the gas clouds to the disk center in collaboration with the dissipative cloud-cloud collisions. Note that the most gas has been gathered to the inner disk region at $T=56.50$. Based on the numerical results depicted here, Noguchi (1988b) predicts that the ultimate state of a heavily disturbed disk galaxy is something like an amorphous galaxy with strong nuclear star formation or nuclear activity.

2.5.2 *Observational Evidence*. The terminology of "amorphous" galaxies has been introduced by Sandage and Brucato (1979) in order to establish a new class for some IrrII type galaxies which fall outside the standard morphological sequence from elliptical to spiral galaxies. These galaxies present an amorphous appearance in their background light. Namely, the apparent distribution of old stellar populations is smooth and lacking in any definite spiral structure and therefore resembles elliptical galaxies. In spite of this morphological resemblance to elliptical galaxies, amorphous galaxies often bear evidence of recent star formation such as young clusters.

Several studies suggest that these young stars have been formed by bursts of star formation near the nucleus. Noreau and Kronberg (1987) have found evidence of starbursts in the center of the amorphous galaxy NGC3448 by radio observation. Gallagher and Hunter (1987) have detected active star formation in the nuclei of many amorphous galaxies by broad-band and Hα photometry.

Concerning the origin of amorphous galaxies, galaxy-galaxy close encounters have been often suggested observationally. Gallagher and Hunter (1987) have noted that about half of their sample are located in the environments favorable to galaxy-galaxy interactions (pairs or groups). NGC3448 and its close companion make a pair, for which Noreau and Kronberg (1986) have carried out test-particle simulations and established a dynamical connection between the two galaxies. Cottrell (1978) has proposed that an IrrII galaxy is formed when the old underlying stellar population of a galaxy interacts with the gas and dusts which that galaxy has pulled out from the companion galaxy. Along this line, van den Bergh (1980) speculated that peculiarities of a typical amorphous galaxy, NGC5253, are the results of the close encounter with the gas-rich nearby spiral M83.

In the scenarios of Cottrell (1978) and van den Bergh (1980), the gas in the amorphous galaxy has external origin. In this case, the old stellar component is not necessarily disturbed violently by the companion, from which the gas has been transferred. Therefore, the amorphous galaxy generally had to have a smooth stellar system in prior to the encounter and should have been an elliptical or S0 originally. By contrast, the gas is internal in the picture by Noguchi (1988b). The smooth structure of the stellar component is itself a result of violent tidal disturbance.

A key to discriminate between these two possibilities may come from the observation of the structure and kinematics for the underlying stellar component. If the external gas accretion scenario is correct, we expect that the frequency distribution of flatness and the kinematical properties such as V/σ (V is the maximum rotational velocity and σ is the

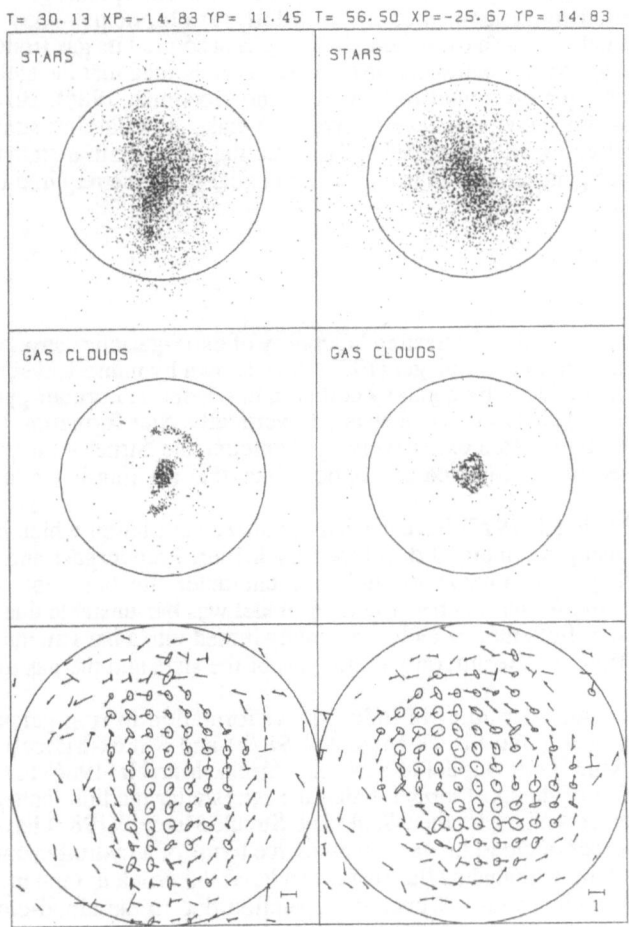

Figure 9. Late phase structure of a tidally-disturbed galaxy with a self-gravitating stellar disc and a gas cloud disc (from Noguchi 1988b). The lower panels indicate the mean velocities (by arrows) and the velocity dispersions (by ellipses) in the stellar disc. The circles indicate the size of the unperturbed disk.

central velocity dispersion) are similar between elliptical/S0 galaxies and amorphous galaxies. Caldwell and Phillips (1989) have found that the amorphous galaxy NGC5253 is primarily supported by random motions, similarly to ellipticals. They guess that the progenitor of this galaxy was a dwarf elliptical and has acquired its gas from outside.

What is expected by the internal scenario? It is easily imagined that the stellar disk will be thickened in an oblique encounter (e.g., Quinn and Goodman 1986). However, it is not clear whether close encounters are so effective as to make a stellar disk similar to elliptical galaxies in its flattening and kinematics. To settle the issue, both a statistics on the star forming amorphous galaxies and an extensive numerical study are required in future.

3. Ring Galaxies

Ring galaxies provide another interesting laboratory of extra-galactic astronomy. Lynds and Toomre (1976) and Theys and Spiegel (1977) have shown by numerical simulations that a ring galaxy can be produced by a galaxy collision in which the intruder galaxy penetrates the disk of the target galaxy at its center nearly vertically. Star formation in ring galaxies has recently been investigated extensively by Appleton and Struck-Marcell (1987a) who have argued, based on the infrared and optical data, that the ring is a cite for active star formation.

Theys and Spiegel (1977) have carried out also a simulation which included gas as well as stars by using the method developed by Miller, Prendergast and Quirk (1970). Unfortunately, they were unable to study the characteristic behavior of the gaseous component in the ring formation, because their model was bar-unstable due to the absence of a halo component. Both the gas and stars rapidly turned into a bar structure and acquired large random motion. As a result, only minor part of the stars and the gas took part in ring formation.

Detailed theoretical studies of induced star formation in ring galaxies have been carried out by Appleton and Struck-Marcell (1987b) and Struck-Marcell and Appleton (1987). They used the "Oort" model of Scalo and Struck-Marcell (1986) for the interstellar gas. The basic idea of the Oort model is that the interstellar medium consists of discrete clouds which interact with each other. Scalo and Strcuk-Marcell (1986) have incorporated fundamental processes of cloud interaction as source terms of a hydrodynamical formalism (i.e., the cloud system is treated as fluid) and rendered the problem numerically tractable. Their formulation includes coalescence or disruption due to the cloud-cloud collisions, collisional energy dissipation, and the cloud disruption and acceleration as a result of star formation events in massive clouds.

Scalo and Struck-Marcell (1986) and Struck-Marcell and Scalo (1987) have found that the only one parameter which essentially determines the qualitative behavior of one-zone Oort models is the ratio of the lifetime, τ_d, of massive star forming clouds to the cloud-cloud collision time scale, τ_c. If $\tau_d/\tau_c \ll 1$, the cloud system is stable. The system relaxes to an equilibrium state within a few cloud collision timescales after it was given a small external disturbance. If $\tau_d/\tau_c > O(1)$, the system exhibits a limit cycle behavior which manifests as repetitive bursts of star formation separated by long quiescent periods. If τ_d/τ_c is sufficiently large, the system shows even a chaotic behavior.

Appleton and Struck-Marcell (1987b) have extended this formalism to a two-dimensional plane and investigated the behavior of the interstellar medium with $\tau_d/\tau_c \ll 1$ (i.e., instantaneous recycling of star forming material) in ring galaxy formation. As expected from the result for one-zone models, their models did not show so large enhancement of the global star formation rate as to explain the observed enhancement of infrared luminosities in actual ring galaxies. Although the local star formation rate can increase by a large amount (up to 20) in the leading edge of the expanding ring, it is offset by a suppression of star formation inside the ring (Fig. 10).

Struck-Marcell and Appleton (1987) have restricted the problem to one dimension (i.e., to exactly on-axis collisions) and investigated the case when τ_d/τ_c is of order unity. In this case, the limit cycle behavior as seen in the one-zone models was triggered by the passage of the intruder galaxy. The star formation rate integrated over the entire target galaxy shows violent oscillations and increases by 2-3 times at its peak value, in consistent with the observed enhancement of infrared luminosity (see Fig. 11). The bursts appear quasi-periodically. Typical time interval between two successive bursts is less than 10^8 years. Such a short period variation of the star formation rate would bring about a large diversity in the appearance of ring galaxies.

The studies by Scalo and Struck-Marcell (1986) and Struck-Marcell and Scalo (1987) seem to indicate that the response of the interstellar gas to the external disturbance is quite sensitive to the gas density. Once the density exceeds a certain critical value, the bursts phenomena seem to be generally inevitable outcome. Bushouse (1987) mentions this burst behavior as a possible cause for the observed diversity in star formation properties of interacting galaxies. Physically, a large value of τ_d/τ_c corresponds to a high density of the interstellar gas and hence to the case of a late type spiral galaxy and/or to a nuclear region. Bushouse (1987) also pointed out that the sensitive dependence on the gas density may explain the preferred star formation in the galactic nuclei.

However, it should be noted here that the models by Olson and Kwan (1989), which used a cloud model quite similar to the Oort model in nature, do not show burst phenomenon such as described by Scalo and Struck-Marcell (1986). Olson and Kwan (1989) have argued that an oversimplified description of the interstellar matter by mean quantities (mass and velocity dispersion) has brought about unrealistic limit cycle behavior in Scalo and Struck-Marcell (1986). Another cause of this discrepancy may lies in the difference of geometry. In a close encounter, each part of the disk responds to the tidal force in a different phase according to the radius and azimuthal angle. Any in-phase behavior is not expected even in a global sense. On the contrary, in an on-axis collision, a ring region with a constant initial radius responds in the same phase.

4. Mergers

4.1 STELLAR DYNAMICS

Galaxies are the systems with many degrees of internal freedom. If the orbital binding energy of two colliding galaxies is sufficiently small, the kinetic energy of their orbital motion can be converted into the kinetic energy of internal motion, and two galaxies finally merge (e.g., Alladin 1965). Galaxy mergers have been considered in relation to the formation of elliptical galaxies (Toomre 1977), cD galaxies (Ostriker and Tremaine 1975), and accretion of satellite galaxies into their parent galaxy (Tremaine 1976).

Toomre and Toomre (1972) and Toomre (1977) have argued that most elliptical galaxies might have been formed by the merger of spiral galaxies, based on the statistics as follows: about 10 on-going mergers evidenced by the presence of tidal tails are recognized among nearby 4000 galaxies. Assuming that the lifetime of tidal tails is about 5×10^8 yr in accordance with the numerical results, the possibility that any galaxy has once experienced

340

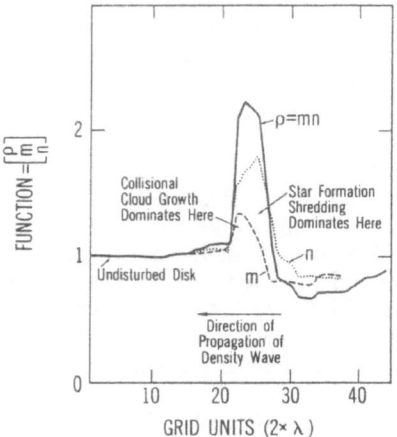

Figure 10. Structure of an expanding ring made by the Oort cloud system (from Appleton and Struck-Marcell 1987b). The mean mass, m, and the mean number density, n, of the clouds are plotted along a slice through the ring. Star formation is stimulated in the leading edge while suppressed behind the ring.

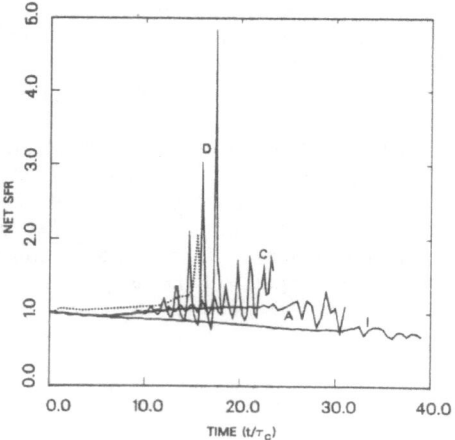

Figure 11. Time variations of the net star formation rate integrated over the whole area of a ring galaxy model (from Struck-Marcell and Appleton 1987). Models A, C, and D have increasing values of τ_d/τ_c in this order.

merger with another galaxy over the past Hubble time becomes up to 20%, considering the time variation of merger probability. This value is close to the fraction of elliptical galaxies in all the galaxies.

This hypothesis has given rise to much controversy regarding the observed properties of elliptical galaxies such as the luminosity-color (or metallicity) relation, the binding energy and scalelength, the galaxian environments, population of globular clusters, metallicity gradient, and kinematical properties (e.g., Ostriker 1980, van den Bergh 1989).

Many numerical studies on mergers have been carried out since late 1970s. Some of these studies (van Albada and van Gorkom 1977, White 1978, 1979; Miller and Smith 1980) have treated the merger of two hot (pressure-supported) systems which correspond to elliptical galaxies. Mergers of two spiral galaxies have proved to be more difficult to simulate numerically because a stable two-component system comprising of a dynamically cold disk and a hot spherical component should be set up prior to merger simulation. Despite such difficulty, several pioneering studies have been carried out so far (Gerhard 1981; Farouki and Shapiro 1982; Negroponte and White 1983). The direct N-body codes (particle-particle) have been usually used in these merger simulations to cover a large range in spatial scale. In merger simulations, both the internal structure of individual galaxies and their relative motion should be treated simultaneously. Although the global dynamics of mergers has been successfully investigated, these simulations were not suited to clarify the detailed change of the internal structure due to small numbers of particles permitted in the direct N-body codes.

The invention of the tree-code (Barnes and Hut 1986) has considerably improved the numerical status. In his merger simulation, Barnes (1988) has used a three-component model which consists of a disk, a bulge, and a dark halo, with each component modelled by a number of self-gravitating particles. Barnes (1988) shows that the merger remnant in his simulation agrees well with the real ellipticals in many respects. In addition to the successful reproduction of $r^{1/4}$-law , two dynamical points which had been considered fatal to the merger hypothesis can be solved. The inclusion of a *live* halo is essential here. According to Barnes (1988), the binding energy and angular momentum of the luminous part which results from the disks and the bulges are largely absorbed by the dark halo components during merger. Therefore, the luminous part in the merger remnant which we can observe optically rotates slowly and has a deep potential well, in a better agreement with the observational data than previous simulations without live halo components. Barnes (1988) has thus shown that the global parameters and properties of ordinary ellipticals can be produced by essentially stellar dynamical mergers of disk galaxies without invoking gas dynamical process such as energy dissipation. Nevertheless, it has not yet been settled whether the most elliptical galaxies are merger remnants.

4.2 GAS DYNAMICS AND INDUCED STAR FORMATION

Mergers as well as close encounters have been recognized to be an efficient trigger of active star formation. In fact, many observational studies mentioned in relation to starbursts in close encounters in section 2 do include merging pairs as a significant fraction of their samples. In the case of mergers also, the observation shows strong concentration of far infrared luminosity to the central region (e.g., Sanders *et al.* 1988). Sanders *et al.* (1988) have suggested the triggering of nuclear activity in advanced mergers from spectroscopic evidence.

Gas dynamical problem in mergers is only poorly understood at present theoretically. Negroponte and White (1983) have found some evidence that the collisions between interstellar gas clouds from two galaxies (which they call inter-galaxy cloud collisions) accelerate the progress of merger. They also found that the gas becomes more strongly concentrated toward the center than the stars after the completion of merger.

Olson and Kwan (1989) have considered a case of merger as well as close encounters in their simulations including both stars and gas. Their simulation shows how the gas behaves differently in the outer and inner regions of a merging system. In the outer part (exterior to 2kpc from the center), the cloud-cloud collision rate increases abruptly at the initial close approach of two galaxies, but decreases rapidly afterwards. This is because the clouds are dispersed into the three-dimensional space above and below the disk plane. On the other hand, the inner region exhibits a continuous enhancement of the collision rate after the initial approach. They have found that the enhancement of the total cloud collision rate over the entire system is smaller in the merger than in the close encounter model and pointed out that this result is consistent with the observation by Solomon and Sage (1988). However, as cautioned by authors, their merger simulation has located the gas clouds only in one of two galaxies and neglects the possible effect of inter-galaxy cloud collisions (Negroponte and White 1983), which are a peculiar process to mergers. Further simulations of mergers involving the gas component would be valuable not only in the deeper understanding of on-going merger candidates such as NGC7252 (Schweizer 1982), but also in settling the controversy on several points such as the globular cluster population and the metallicity gradient in elliptical galaxies.

4.3 ACCRETION OF SMALL SATELLITE GALAXIES

A less dramatic merger with a small satellite galaxy has been investigated by Quinn and Goodman (1986), Hernquist and Quinn (1989), and Hernquist (1989). Quinn and Goodman (1986) have found that as the compact spherical satellite sinks to the inner part of the parent galaxy due to the dynamical friction, the disk component of the parent galaxy is heated and thickens. Bar formation found in Noguchi (1987) has been also observed before the completion of satellite orbital decay (Hernquist and Quinn 1989).

Hernquist (1989) has investigated the gas response in merger with a small satellite galaxy. His simulation, like those of Noguchi (1988a), shows that as the satellite approaches, the star and the gas in the parent galaxy make a bar structure, and the gas finally falls to the disk center. Hernquist (1989) claims that the self-gravity of the gas not considered by Noguchi (1988a) becomes important in the late stage of gas infall. However the importance of the gaseous self-gravity is only circumstantially demonstrated by the ratio of the self-gravitating potential energy to the virial potential energy for the gas, for example. It is not clarified what a role the gaseous self-gravity is playing and how large the effect is. In other words, the possible effect of the self-gravity has not been separated from other effects such as gaseous energy dissipation and the torquing of the stellar bar. Although Hernquist (1989) treated the case of merger unlike Noguchi (1988a), the evolution of the system before the final decay of the satellite is very similar to the one given by Noguchi (1988a). It seems that the essential mechanism for the gas accumulation to the central 1kpc region (occurred in both Noguchi 1988a and Hernquist 1989) is the gaseous energy dissipation combined with the bar forcing.

One motivation for the study by Hernquist (1989) is the observational suggestion that Seyfert galaxies often have amorphous morphology, presumably indicative of recent accretion events, and have a clear excess of faint companions (e.g., Fuentes-Williams and Stocke 1988; MacKenty 1989). However, it should be noted that the amorphous disk morphology may result also from close encounters (Noguchi 1988b), though not excluding the possibility of accretion. Furthermore, Seyfert activity requires the gas fuelling to a far smaller region around the nucleus than reliably described by numerical methods so that the gas infall which appeared in the model cannot be directly related to the onset of Seyfert activity (see the next section).

5. Self-gravity of the Gas

Although the self-gravity of the gaseous component has been sometimes neglected in numerical simulations, there is a growing evidence that the gaseous self-gravity plays an important role in special circumstances.

Molecular gas bars recently observed in the central regions of nearby galaxies (e.g., Ishizuki 1989) suggest that not stellar but gaseous bars are a quite universal phenomenon in the centers of disk galaxies. It is certain that a high mass ratio of the gas to the background stellar component is the cause of bar instability. However, it is still open to question why so much gas has come to reside in the central region. Gaseous bars are not inherent to interacting galaxies but also observed in isolated galaxies such as IC342 and NGC6946. The inward gas flow driven by viscosity (e.g., Fukunaga 1983) may give a partial answer to the gas accumulation in the center of isolated galaxies.

The self-gravity of the gas may provide a key to understanding the fuelling mechanism of active galactic nuclei (AGN). Lin, Pringle, and Rees (1987) have suggested a possibility that the gas infall to the nuclear region of 1-10pc size (from 1kpc) is driven by the spiral structures which appear in a mildly unstable central gas disk. Shlosman, Frank, and Begelman (1989) have proposed a successive bar instability mechanism as follows. If the density enhancement of the central gas disk is sufficiently large, the disk will develop a bar (Ostriker and Peebles 1973). Then this bar will induce gas infall to the inner part of the central disk. Then the inner part becomes bar-unstable and a new bar (much smaller than the first bar) will develop, which induces gas infall again. Bar instability thus can propagate successively toward smaller and smaller radius.

Lin *et al.* (1987) and Shlosman *et al.* (1989) *postulated* density increase in the gas by some mechanism and do not necessarily associate the triggering of AGN with galaxy-galaxy interactions. Numerical studies demonstrate that galaxy close encounters (Noguchi 1988a) or mergers (Hernquist 1989) can give rise to efficient gas increase in the central 1kpc region as postulated by Lin *et al.* (1987) and Shlosman *et al.* (1989). We therefore expect that AGN as well as starbursts may be triggered by galaxy-galaxy interactions.

Observational status is not clear at present. Although Dahari (1984, 1985a, 1985b) claimed overabundance of Seyfert galaxies in galaxies having a sizable companion, his result has been criticized by Fuentes-Williams and Stocke (1988), who have found that Seyferts only appear to show an excess of faint companions, not larger ones. Bushouse (1987) and MacKenty (1989) suggest that close companions are associated more with enhanced star formation than Seyfert activity. It is likewise uncertain whether mergers show higher incidence of AGN compared with isolated galaxies, although Sanders *et al.* (1988) have detected a sign of AGN in many merging systems.

6. Peculiarities in Early-type Galaxies

As indicated by several relevant contributions at this meeting, it is becoming clear that early type galaxies (ellipticals and S0s) constitute more heterogeneous families than previously considered in their interstellar matter and star formation characteristics (e.g., Eskridge and

Pogge 1989; Walsh and Knapp 1989). The galaxy-galaxy interactions may be responsible for some of these peculiarities.

S0 galaxies sometimes show active star formation in their nuclei (Dressel *et al.* 1989; Wrobel 1989). Such S0 galaxies do not fit with the conventional picture that S0 galaxies construct an intermediate class between elliptical and spiral galaxies. Some external origin is suggested to have operated.

One possible explanation may be given by the stripping hypothesis that the gas in the outer part (primarily neutral hydrogen) has been stripped by the ram pressure exerted by intergalactic gas and only the central star forming region has been left. The galaxy-galaxy interactions provide another alternative. Noguchi (1988b) has suggested that a smooth S0-like stellar disk having nuclear gas accumulation (hence star formation) can be produced as an end-product of a violent galaxy encounter. Mergers may also form such a system (Hernquist 1989). Along with a portion of actively star forming amorphous galaxies, these peculiar S0 galaxies may represent aftermath of galaxy-galaxy interactions (it is difficult to draw a definite borderline between these two categories).

Kinematics and structure of the stellar component would place a stringent constraint on any theoretical model. On the observational side, it is highly desired to make a statistics on the flattening of S0 galaxies with active nuclear star formation and a comparison with ordinary spiral galaxies, which may be precursors of those galaxies. At the same time, the possible degree of disk thickening due to galaxy-galaxy interactions (or gas stripping) should be quantitatively assessed by numerical simulations. The relative dominance of random motions to rotations should also be compared between theoretical models and observations.

Counterrotating cores recently found in several ellipticals (Franx and Illingworth 1988; Jedrzejewski and Schechter 1988) and S0 galaxies (Dettmar *et al.* 1989) raise another interesting possibility. Co-existence of two dynamically different subsystems is difficult to understand in the framework of isolated galaxy formation and evolution. As stated by these authors, merger of a small compact galaxy seems to provide a natural interpretation. Capture of the outer envelop of the companion galaxy in a close encounter may provide an alternative.

7. Future Prospect

I have confined myself to peculiarities caused by galaxy-galaxy interactions in this article. However, galaxy-galaxy interactions may not be exceptional events responsible only for a minor category named as "peculiar" galaxies. Not only on the theoretical ground but also on the observational one (e.g., Zepf and Koo 1989), it is considered that the galaxy-galaxy interactions were more frequent in the past. Moreover, galaxy-galaxy interactions are likely to leave their evidence on the internal structure of victim galaxies as discussed here. It will be one of future tasks to assess the importance of galaxy-galaxy interactions in the evolution of "normal" galaxies from a wider viewpoint which takes into account the evolution of galaxian environment, especially, the process of galaxy clustering and the formation of large scale structures. Causal relationship (if any) between starburst phenomena and active galactic nuclei will be another important problem to be attacked, although some preliminary studies have appeared (e.g., Weedman 1983; Norman and Scoville 1988).

I acknowledge the Scientific Organizing Committee of the Conference and Nobeyama Radio Observatory for their financial assistance which enabled my attendance to the meeting.

References

Aarseth, S. J., and Fall, S. D. 1980, *Ap. J.*, **236**, 43.
Adams, T. F. 1977, *Ap. J. Suppl.*, **33**, 19.
Alladin, S. M. 1965, *Ap. J.*, **141**, 768.
Appleton, P. N., and Struck-Marcell, C. 1987a, *Ap. J.*, **312**, 566.
Appleton, P. N., and Struck-Marcell, C. 1987b, *Ap. J.*, **318**, 103.
Arp, H. 1966, *Atlas of Peculiar Galaxies* (Pasadena: California Institute of Technology).
Barnes, J. E. 1988, *Ap. J.*, **331**, 699.
Barnes, J., and Hut, P. 1986, *Nature*, **324**, 446.
Bushouse, H. A. 1987, *Ap. J.*, **320**, 49.
Byrd, G. G., Valtonen, M. J., Sundelius, B., and Valtaoja, L. 1986, *Astr. Ap.*, **166**, 75.
Caldwell, N., and Phillips, M. M. 1989, *Ap. J.*, **338**, 789.
Combes, F. 1978, *Astr. Ap.*, **65**, 47.
Combes, F., anf Gerin, M. 1985, *Astr. Ap.*, **150**, 327.
Condon, J. J., Condon, M. A., Gisler, G., and Puschell, J. J. 1982, *Ap. J.*, **252**, 102.
Cottrell, G. A. 1978, *M.N.R.A.S.*, **184**, 259.
Cutri, R. M., and McAlary, C. W. 1985, *Ap. J.*, **296**, 90.
Dahari, O. 1984, *A.J.*, **89**, 966.
Dahari, O. 1985a, *Ap. J. Suppl.*, **57**, 643.
Dahari, O. 1985b, *A.J.*, **90**, 1772.
Dettmar, R.-J., Dettmar, M. J., and Barteldrees, A. 1989, this symposium.
Dressel, L. L., O'Connell, R. W., Telesco, C. M., and Decher, R. 1989, this symposium.
Eskridge, P. B., and Pogge, R. W. 1989, this symposium.
Farouki, R. T., and Shapiro, S. L. 1982, *Ap. J.*, **259**, 103.
Franx, M., and Illingworth, G. D. 1988, *Ap. J. (Letters)*, **327**, L55.
Fuentes-Williams, T., and Stocke, J. T. 1988
Fukunaga, M. 1983, *Pub. Astr. Soc. Japan*, **35**, 173.
Gallagher, J. S., and Hunter, D. A. 1987, *A.J.*, **94**, 43.
Gerhard, O. 1981, *M.N.R.A.S.*, **197**, 179.
Hausman, M. A., and Roberts, W. W. 1984, *Ap. J.*, **282**, 106.
Hernquist, L. 1989, Nature, 340, 687.
Hernquist, L., and Quinn, P. J. 1989, in *The Epoch of Galaxy Formation*, ed. C. S. Frenk, R. S. Ellis, T. Shanks, A. F. Heavens, and J. A. Peacock (Dordrecht: Kluwer Academic Publ.), p. 427.
Hockney, R. W., and Eastwood, J. W. 1981, *Computer Simulations Using Particles* (New York: McGraw-Hill).
Hohl, F., and Hockney, R. W. 1969, *J. Comp. Phys.*, **4**, 306.
Hummel, E. 1980, *Astr. Ap.*, **89**, L1.
Hummel, E. 1981, *Astr. Ap.*, **96**,111.
Hutchings, J. B., and Campbell, B. 1983, *Nature*, **303**, 584.
Icke, V. 1985, *Astr. Ap.*, **144**, 115.
Ishizuki, S. 1989, this symposium.
Jedrzejewski, R., and Schechter, P. L. 1988, *Ap. J.*, **330**, 687.

346

Joseph, R. D., Meikle, W. P. S., Robertson, N. A., and Wright, G. S. 1984, *M.N.R.A.S.*, **209**, 111.

Keel, W. C., Kennicutt, R. C., Jr., Hummel, E., and van der Hulst, J. M. 1985, *A.J.*, **90**, 708.

Kennicutt, R. C., Jr., and Keel, W. C. 1984, *Ap. J. (Letters)*, **279**, L5.

Kennicutt, R. C., Keel, W. C., van der Hulst, J. M., Hummel, E., Roettiger, K. A. 1987, *A.J.*, **93**, 1011.

Larson, R. B., and Tinsley, B. M. 1978, *Ap. J.*, **219**, 46.

Lattanzio, J. C., and Henriksen, R. N. 1988, *M.N.R.A.S.*, **232**, 565.

Lin, D. N. C., Pringle, J. E., and Rees, M. J. 1988, *Ap. J.*, **328**, 103.

Lonsdale, C. J., Persson, S. E., and Matthews, K. 1984, *Ap. J.*, **287**, 95.

Loren, R. B. 1976, *Ap. J.*, **209**, 466.

Lynds, R., and Toomre, A. 1976, *Ap. J.*, **209**, 382.

MacKenty, J. W. 1989, *Ap. J.*, **343**, 125.

Miller, R. H., Prendergast, K. H., and Quirk, W. J. 1970, *Ap. J.*, **161**, 903.

Miller, R. H., and Smith, B. F. 1979, *Ap. J.*, **227**, 785.

Miller, R. H., and Smith, B. F. 1980, *Ap. J.*, **235**, 421.

Myers, S. T., and Scoville, N. Z. 1987, *Ap. J. (Letters)*, **312**, L39.

Negroponte, J., and White, S. D. M. 1983, *M.N.R.A.S.*, **205**, 1009.

Noguchi, M. 1987, *M.N.R.A.S.*, **228**, 635.

Noguchi, M. 1988a, *Astr. Ap.*, **203**, 259.

Noguchi, M. 1988b, *Astr. Ap.*, **201**, 37.

Noguchi, M., and Ishibashi, S. 1986, *M.N.R.A.S.*, **219**, 305.

Noreau, L., and Kronberg, P. P. 1986, *A.J.*, **92**, 1048.

Noreau, L., and Kronberg, P. P. 1987, *A.J.*, **93**, 1045.

Norman, C., and Scoville, N. 1988, *Ap. J.*, **332**, 124.

Olson, K. M., and Kwan, J. 1989, preprint.

Ostriker, J. P. 1980, *Comments on Ap.*, **8**, 177.

Ostriker, J. P., and Peebles, P. J. E. 1973, *Ap. J.*, **186**, 467.

Ostriker, J. P., and Tremaine, S. D. 1975, *Ap. J. (Letters)*, **202**, L113.

Planesas, P., Scoville, N. Z., and Myers, S. T. 1989, this symposium.

Quinn, P. J., and Goodman, J. 1986, *Ap. J.*, **309**, 472.

Roberts, W. W., and Hausman, M. A. 1984, *Ap. J.*, **277**, 744.

Rubin, V. C., Burstein, D., Ford, W. K., Jr., and Thonnard, N. 1985, *Ap. J.*, **289**, 81.

Sandage, A., and Brucato, R. 1979, *A.J.*, **84**, 472.

Sanders, R. H., and Huntley, J. M. 1976, *Ap. J.*, **209**, 53.

Sanders, R. H., and Tubbs, A. D. 1980, *Ap. J.*, **235**, 803.

Sanders, D. B., Scoville, N. Z., Young, J. S., Soifer, B. T., Schloerb, F. P., Rice, W. L., and Danielson, G. E. 1986, *Ap. J. (Letters)*, **305**, L45.

Sanders, D. B., Soifer, B. T., Elias, J. H., Madore, B. F., Matthews, K., Neugebauer, G., and Scoville, N. Z. 1988, *Ap. J.*, **325**, 74.

Scalo, J. M., and Struck-Marcell, c. 1986, *Ap. J.*, **301**, 77.

Schwarz, M. P. 1984, *M.N.R.A.S.*, **209**, 93.

Schweizer, F. 1982, *Ap. J.*, **252**, 455.

Scoville, N. Z., Matthews, K., Carico, D. P., and Sanders, D. B. 1988, *Ap. J. (Letters)*, **327**, L61.

Scoville, N. Z., Sanders, D. B., and Clemens, D. P. 1986, *Ap. J.*, **310**, L77.

Sellwood, J. A. 1981, *Astr.Ap.*, **99**, 362.

Shlosman, I., Frank, J., and Begelman, M. C. 1989, *Nature*, **338**, 45.

Soifer, B. T. 1984, *Ap. J. (Letters)*, **278**, L71.

Solomon, P. M., and Sage, L. J. 1988, *Ap. J.*, **334**, 613.

347

Sorensen, S.-A., Matsuda, T., and Fujimoto, M. 1976, *Astr. Space Sci.*, **43**, 491.
Stockton, A. 1982, *Ap. J.*, **257**, 33.
Struck-Marcell, C., and Appleton, P. N. 1987, *Ap. J.*, **323**, 480.
Struck-Marcell, C., and Scalo, J. M. 1987, *Ap. J. Suppl.*, **64**, 39.
Telesco, C. M., and Decher, R. 1988, *Ap. J.*, **334**, 573.
Telesco, C. M., Wolstencroft, R. D., and Done, C. 1988, *Ap. J.*, **329**, 174.
Theys, J. C., and Spiegel, E. A. 1977, *Ap. J.*, **212**, 616.
Thronson, H. A., Hereld, M., Majewski, S., Greenhouse, M., Johnson, P., Spillar, E., Woodward, C. E., Harper, D. A., and Rauscher, B. J. 1989, *Ap. J.*, **343**, 158.
Toomre, A. 1977, in *The Evolution of Galaxies and Stellar Populations*, ed. B. M. Tinsley and R. B. Larson (New Haven: Yale Univ. Observatory), p. 401.
Toomre, A., and Toomre, J. 1972, *Ap. J.*, **178**, 623.
Tremaine, S. D. 1976, *Ap. J.*, **203**, 72.
Young, J. S., Kenney, J. D., Tacconi, L., Claussen, M. J., Huang, Y.-L., Tacconi-Garman, L., Xie, S., and Schloerb, F. D. 1986, *Ap. J. (Letters)*, **311**, L17.
van Albada, T. S., and van Gorkom, J. H. 1977, *Astr.Ap.*, **54**, 121.
van den Bergh, S. 1979, *Astron. Nachr.*, **300**, 225.
van den Bergh, S. 1980, *Pub. A. S. P.*, **92**, 122.
van den Bergh, S. 1989, talk at Heidelberg Conference on "Dynamics and Interactions of Galaxies".
Walsh, D., and Knapp, J. 1989, this symposium.
Weedman, D. W. 1983, *Ap. J.*, **266**, 479.
White, S. D. M. 1978, *M.N.R.A.S.*, **184**, 185.
White, S. D. M. 1979, *M.N.R.A.S.*, **189**, 831.
Wrobel, J. M. 1989, this symposium.
Zepf, S. E., and Koo, D. C. 1989, *Ap. J.*, **337**, 34.

MAGNETIC FIELDS IN GALAXIES

Richard Wielebinski

Max-Planck-Institut für Radioastronomie,
Auf dem Hügel 69,
D-5300 BONN 1,   F.R.G.

INTRODUCTION

The search for magnetic fields in the Galaxy was started already at the turn of this century, soon after polarization characteristics of the Felspar crystal were discovered. Additional impetus came from the development of polaroid foil which made this observing technique available even to amateur astronomers. The first substantiated discovery is due to Meyer (1920) who measured the polarization of the Hubble's variable nebula NGC2261. This was an observation of the polarization of a galactic source but it showed that magnetic fields exist and play an important role in the universe. At the same time methods of measuring of the solar magnetic field were being developed. Theoretical arguments for the existence of magnetic fields in galaxies were based on the need of confinement of cosmic particles (e.g. Fermi, 1949). The discovery of a magnetic field in an external galaxy is due to Öhman (1942) who used first a Felspar polarimeter and later a Wollaston prism to observe the polarized emission in Andromeda nebula (M31).

The progress in the measurement of magnetic fields in galaxies using optical polarization methods was slow, since the observations were very difficult. It was the discovery of the radio polarization of the synchrotron emission which added a new and important technique for studying magnetic fields. Also the Zeeman effect in HI clouds (and more recently in OH, H_2O, CCS sources) added new data on magnetic fields in dense molecular clouds in the Galaxy. The progress in the past ten years was basic in giving us an insight into the morphology of the magnetic fields in galaxies.

1. METHODS OF MEASURING MAGNETIC FIELDS

The methods of measurement of the magnetic fields are indirect. Essentially the measurement of the optical, infrared or radio polarization (e.g. Sofue et al., 1986; Beck, 1986; Heiles, 1986) gives us the basic data. In the case of optical polarization it is the alignment of dust grains by magnetic fields that gives an observable effect. At radio frequencies linearly polarized waves are generated by relativistic electrons in magnetic fields.

H. A. Thronson, Jr. and J. M. Shull (eds.), The Interstellar Medium in Galaxies, 349–369.
© 1990 Kluwer Academic Publishers.

A number of effects are responsible for the polarization of optical light. Light scattered by dust grains (Rayleigh scattering) becomes partially polarized with the orientation of the observed polarization perpendicular to a line pointing to the light source. In the case of dust grains aligned in magnetic fields (the Davis-Greenstein effect) we see in the case of the scattered light the polarization perpendicular to the magnetic field while we see the polarization vectors parallel to the field orientation for the directly transmitted light. There are many open questions in the details of the theory of optical polarization generation (e.g. see Purcell, 1979 and Hildebrand, 1988). To study magnetic fields with optical methods a separation of the various effects is necessary.

At radio frequencies the synchrotron emission is emitted with the E vector perpendicular to the orientation of the magnetic field (e.g. Ginzburg and Syrovatskij, 1969). The observed vectors must be corrected for the Faraday rotation which takes place in the galaxy itself, the intergalactic medium and in our Galaxy. To eliminate the Faraday effect we need to have observations at several frequencies. The Faraday effect itself gives us information about the field component parallel to the line of sight. It is also important to consider the effects of different beams for the different frequencies.

The most direct method of measurement of the magnetic fields is the Zeeman effect. The Zeeman effect was observed in molecular clouds as a frequency shift of the opposite circular polarization signals of such molecules as HI, OH, H_2O, CCS etc. Such observations give us information about the magnetic field in the molecular clouds. The magnetic field strength can be inferred from the synchrotron emission intensity (using equipartition arguments).

2. THE MAGNETIC FIELDS IN OUR GALAXY

There are extensive studies of the optical polarization of stars in the Galaxy. The pioneering work of Hiltner (1949) and Hall (1949) was followed by large scale surveys of Behr (1959) and Mathewson and Ford (1970). These data were reanalysed by Ellis and Axon (1978). The general conclusion is that this method gives us at most information about our local neighbourhood. The magnetic field is aligned in general along the galactic plane, but in the direction of l=45°. Beyond a circle of 600 pc the magnetic field is directed towards l=70°. This field configuration is attributed to a local bubble or a single loop of a more general field.

The direct mapping of the polarized radio continuum emission gave us insight into the magnetic fields of galactic objects. The early observations by Mayer et al. (1957) were the first to give information about the magnetic field in the Crab Nebula, a supernova remnant. Galactic radio polarization was discovered by Westerhout et al., 1962) and Wielebinski et al., (1962). The surveys of Berkhuijsen and Brouw (1963), Wielebinski and

Shakeshaft (1964) and Mathewson and Milne (1965) at 408 MHz show the local fields only. The direction of l=140° b=10° is a direction of a unique singularity where we are looking perpendicular to the local magnetic field. Higher frequency surveys (e.g. Spoelstra, 1984; Junkes et al., 1987) show that more distant magnetic fields could be traced at higher radio frequencies.

The studies of Rotation Measures of extragalactic radio sources have given us a some understanding of the large-scale magnetic field of the Galaxy (e.g. Simrad-Normandin et al., 1981). With this method we get information about B|| (field component parallel to line of sight) only. There is a large scale field with numerous 'local' features. The study of RM's should be improved further using a much larger sample of sources. A study of different zones (MacLeod et al., 1988) offers a possibility of understanding some of the details of the magnetic field structure. One of the interesting results from RM studies is that sudden field reversals occur on scales of a few degrees.

Pulsars offer the most direct method of determining of B||. The reason for this is the fact that we can measure both the RM and the Dispersion Measure. From these two pieces of information the value of B|| can be derived. Recent reanalysis of all the available pulsar data by Lyne and Graham Smith (1988) confirmed a magnetic field in the Galaxy of B|| ~ 3μG, directed towards l=90° (i.e. along the local spiral arm). Sudden field reversals (indicated by high positive and negative adjacent rotation measures) are seen in a number of directions.

The measurements of the Zeeman effect have succeeded in HI clouds (e.g. Verschuur, 1979), in OH molecular clouds (e.g. Crutcher et al., 1988) and more recently in H_2O sources (Fiebig & Güsten, 1989). All the Zeeman measurements, maybe the most direct magnetic field determinations, can be made only in a small number of sources. The fields that have been measured are B > 10 μG, with values of ~100μG in some objects. In H_2O maser regions magnetic field values are in the milligauss range.

All the data discussed so far gives us the picture that the magnetic fields in the disc of the Galaxy is **azimuthal**. A recent analysis of Vallee (1988) shows that any deviations of pitch angle of the magnetic field, from the spiral arm are slight, possibly less then 6°. Vallée also deduced a field reversal in the Sagittarius arm. This could support the analysis of Sofue and Fujimoto (1983) who claimed that the magnetic field of the Galaxy is bisymmetric.

The field in the centre of our Galaxy is **in the Z-direction**. The earlier λ 2.8cm observations (Seiradakis et al., 1985) have been substantiated by new λ 9 mm observations (Reich, 1988). The magnetic field in the central nucleus area runs perpendicular to the galactic plane, which may be a part of a more extended poloidal field. This non-thermal emission has also an anomalous (positive), spectral index (Reich et al., 1988).

A model of the magnetic field in the Galaxy is shown in figure 1. The fields in the disc have a uniform component B_u and a turbulent component B_r. Since $B_u||$ (from pulsar rotation measures) is ~ 3µG, we can expect B_u ~ 5µG. Since B_u ~ B_r the total magnetic field in the plane could have the value of B_t~7µG or more.

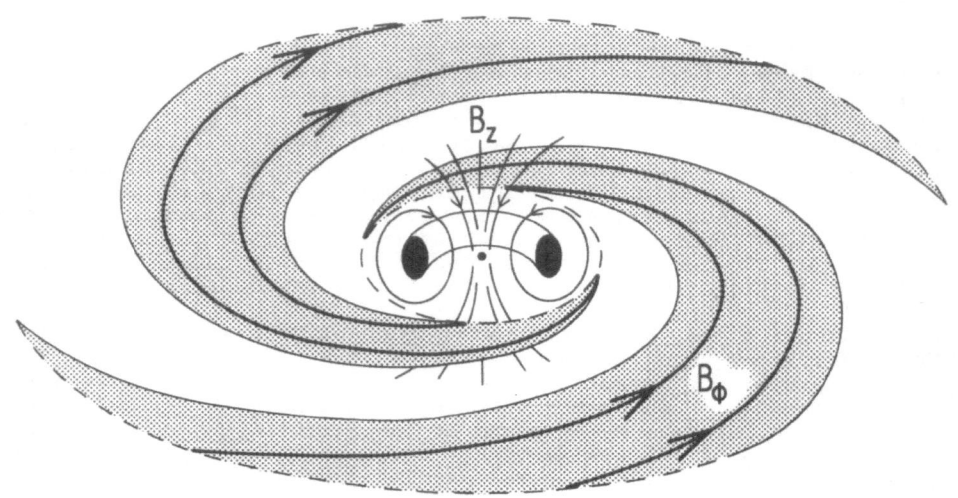

Figure 1. A model of the magnetic field in the Galaxy.

3. THE MAGNETIC FIELDS IN GALAXIES

As mentioned already it was the optical polarization observations that gave us the first information about magnetic fields in external galaxies. Öhman (1942) gave us in addition to the results for M31 the detailed description of all the many problems of the observing technique. The advantages and/or the problems of surface polarimetry compared to the observations of discrete sources (stars, globular clusters) were discussed. The interpetation for the reasons for polarized light had to wait until Davis and Greenstein (1951). The optical polarization studies of galaxies (e.g. Hiltner, 1958; Elvius & Hall, 1965; Appenzeller, 1967; Bingham et al., 1976; Scarrott et al., 1977; Elvius, 1978; Martin & Shawl, 1982; Scarrott et al., 1987) are characterised by ever increasing sensitivity. The photomultiplier has been replaced by a CCD detector. The polarization analyser remained essentially the same;- polarisation foil or a Wollaston prism. Savart plates are also used for studies of stars. A new era of optical polarization observations seems to be at hand in view of the relative availability of medium sized telescopes and sensitive CCD detectors.

The radio observations needed some time to develop sensitive methods to measure polarization in galaxies. The first published result for a galaxy was for M51 by Mathewson et al. (1972) using the then commissioned Westerbork synthesis radio telescope. A follow-up observation of Segalovitz et al. (1976) gave us information about M51 and M81. The Effelsberg 100-m dish has been intensively dedicated to the study of magnetic fields in galaxies since the first results on M31 were published by Beck et al., (1978, 1980). Since that time practically all the large northern galaxies have been mapped in Effelsberg at wavelengths λ λ 11 to 2.8 cm. In the quest of angular resolution the Very Large Array (VLA) has been used, in particular in the D array mode, at lower frequencies. More recently the Parkes radio telescope has been used for polarization mapping of the Magellanic Clouds (Haynes et al., 1986, 1990) and for large southern galaxies (Harnett et al., 1989, 1990). The present data base needs to be expanded both in respect to angular resolution (without loss of sensitivity) and to higher frequencies. Some progress with existing radio telescopes is possible. Given longer integration times for the C and B array mapping at the VLA we should get better information. The 100-m telescope in Effelsberg with a multibeam receiver at λ 9mm wavelength will allow 25" angular resolution practically free of Faraday effects. The commissioning of the Australia Telescope should usher in a new era of studies of southern galaxies.

4. RESULTS FOR SOME INDIVIDUAL GALAXIES

In the following I will describe some of the results for individual galaxies. The order of the galaxies is firstly size, but later some of the types will be described collected in groups. The description of the magnetic field structure will follow the ideas developed from the early observations namely that fieds are either axisymmetic [ASS] or bisymmetric [BSS] spirals. The analysis of the magnetic fields, which was originally developed by Tosa and Fujimoto (1978), involves the study of the Rotation Measure as a function of azimuthal angle θ and is illustrated in figure 2. Further details of such studies can be found in Sofue et al. (1985) and Krause et al. (1989 a,b).

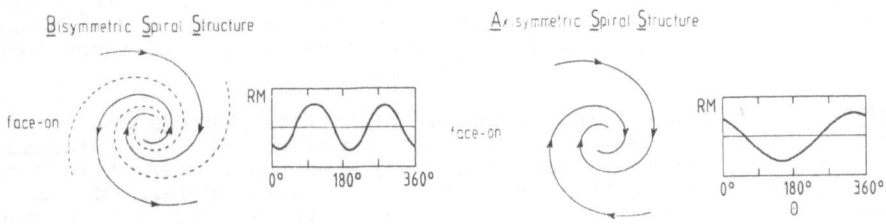

Figure 2.   The basic mode configurations. Rotation Measure as a function of Azimuth θ.

LMC   Optical observations of the Large (and Small) Magellanic Cloud (Schmidt ,1970; Mathewson and Ford, 1970) showed the presence of magnetic fields in both galaxies. The initial interpretation in terms of a 'Pan-Magellanic field' was questioned by Schmidt (1976) who pointed out that the local (foreground) field seems to be also aligned in the LMC-SMC direction. A detailed study of the LMC was recently given by Klein et al. (1989). Radio polarization studies of the LMC (Haynes et al., 1990) show that magnetic fields are seen as a series of filaments originating in 30 Doradus nebula. This in fact agrees with the recent results in HI, CO, FIR, UV etc. These results would suggest that 30 Doradus is the nucleus of the LMC. The filamentary structure of the young components is indeed baffling. These filaments could be spiral arms. This should lead to a reclassification of the LMC to be a 'Spiral' rather then the present classification as 'Irregular'.

SMC   The Small Magellanic Cloud is considered to be the nearest dwarf galaxy. As such the magnetic field structure is of great interest. Optical studies showed some vectors aligned with the 'body' of the SMC, others to be directed towards the LMC (Schmidt 1976). Radio data (Loiseau et al., 1987) shows aligned field in the southern 'body', in agreement with optical data. This indicates a field along the 'body'. However the field is weak, possibly less then 3 µG, as expected in a dwarf galaxy.

M31   This northern spiral has been a subject of extensive study giving possibly the best information of any galaxies to date. The Effelsberg λ λ 11 and 6cm data have now been supplemented with multi-field VLA observations at λ 20cm. The polarized intensity is concentrated to a 'ring' with minima in the direction of the major axis (where Faraday depolarization is expected to be greatest). Due to its inclination the Faraday rotation is strong and can thus be measured with some accuracy. M31 has the prototype 'axisymmetric' spiral field structure (see Beck, 1982 and Beck et al., 1989 and figures 3 and 4). However in detail small wave-like field perturbations are observed.

M33   Multifrequency observations of Buczilowski & Beck, (1987) have now been analysed in some detail. The field in M33 is possibly 'bisymmetric' but this conclusion is only tentative because of problems of sensitivity in this rather low luminosity galaxy. It is difficult in general to determine Faraday rotation in face-on galaxies, in particular when the magnetic field strength is low. A regular field structure is seen in M33 in the northern spiral arm while in the south considerable perturbations are present.

NGC55   This large irregular galaxy is seen edge-on (e.g. Hummel et al., 1986). In the VLA observations no polarized emission was detected. Recent mapping with the Parkes telescope (Harnett et al., 1990) has shown some weak polarized emission in the nuclear area. This would be the second (after the SMC) dwarf galaxy with a confirmed magnetic field.

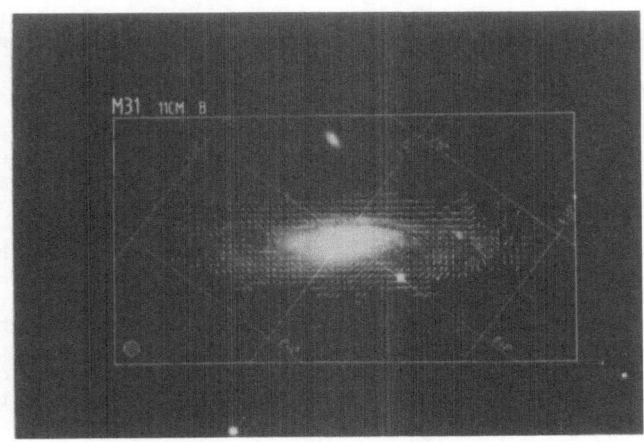

Figure 3. A low resolution map of the magnetic field in M31
(based on Effelsberg data from Beck et al., 1980 with
correction for the Faraday rotation in our Galaxy only)

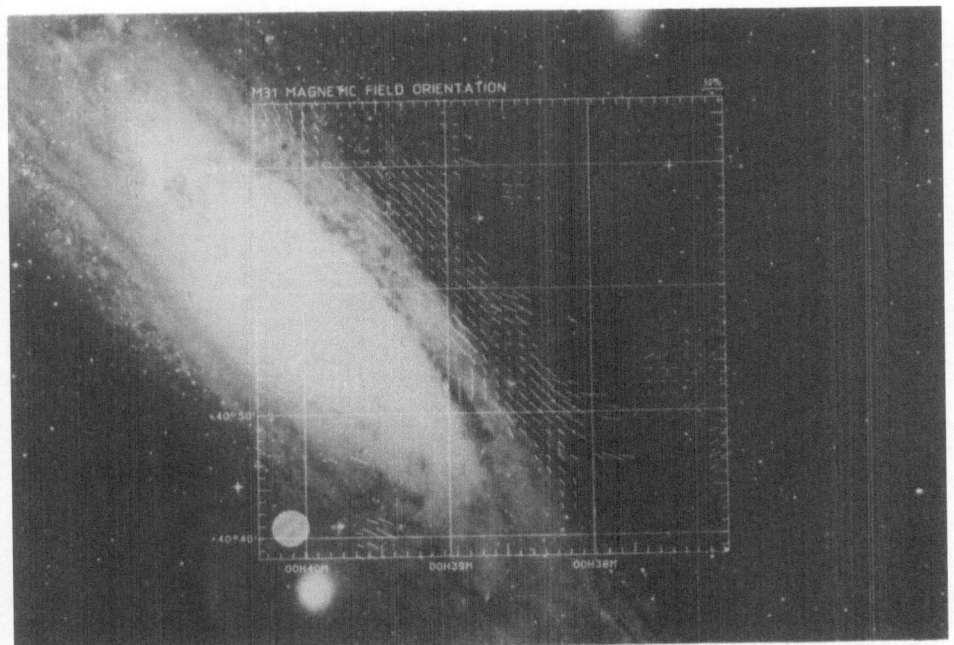

Figure 4. A 'zoom' of a section of M31 with higher
angular resolution (VLA data from Beck et al., 1989).

M101   In M101 the giant HII regions, which have NGC designations themselves, dominate the structure. In spite of this the diffuse nonthermal emission shows two polarization maxima on opposite sides of the nucleus (Gräve et al. 1989). Magnetic fields on a grand scale are present also in this galaxy.

IC342   This galaxy was the subject of detailed studies by Krause et al. (1989a). Both Effelberg and VLA multifrequency data are available. The rotation measure analysis of this galaxy (see figure 5) shows an axisymmetric field. Higher angular resolution observations show that the symmetry on the two opposite sides of the galaxy is quite different. In the South-East a series of very extended filamentary arcs are observed. A polarization maximum in one arc shows zero Faraday rotation with rotation in the same direction on either side. We must be looking into an 'S' like magnetic field filament. In the North-West a very fine filamentary structure is seen with a number of maxima and minima. However the direction of the 'E' vector (i.e. magnetic field) does not change.

M81   This 'grand design' spiral galaxy was subject of multifrequency studies by Krause et al. (1989b). It is the bisymmetric field prototype (see figure 5 and figure 6). However the symmetry is also not perfect. The South-West arm breaks up into two filaments aligned in the direction of the spiral arms. The highest degree of polarization is in the inter-arm region. This is a very significant result pointing to a tangled field in the arms.

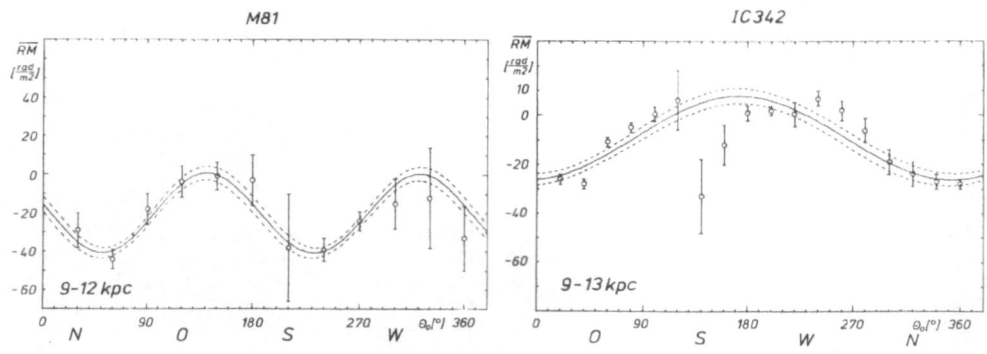

Figure 5. Rotation measure studies for M81 and IC342.
(from Krause et al., 1989a,b)

NGC4258   This galaxy has posed a problem of interpretation in view of its 'anomalous arms' (van der Kruit et al., 1972). The fact that these anomalous arms are highly polarized (van Albada, 1978, Hummel et al., 1989) implies that magnetic fields are involved in the origin of the radio emission in this object. At low angular resolution the two arms show up as maxima of polarisation symmetrically disposed about the nucleus. Rotation measure analysis ba Hummel et al. (1989) implies that these arms are in the plane, or nearly in the plane, of the galaxy.

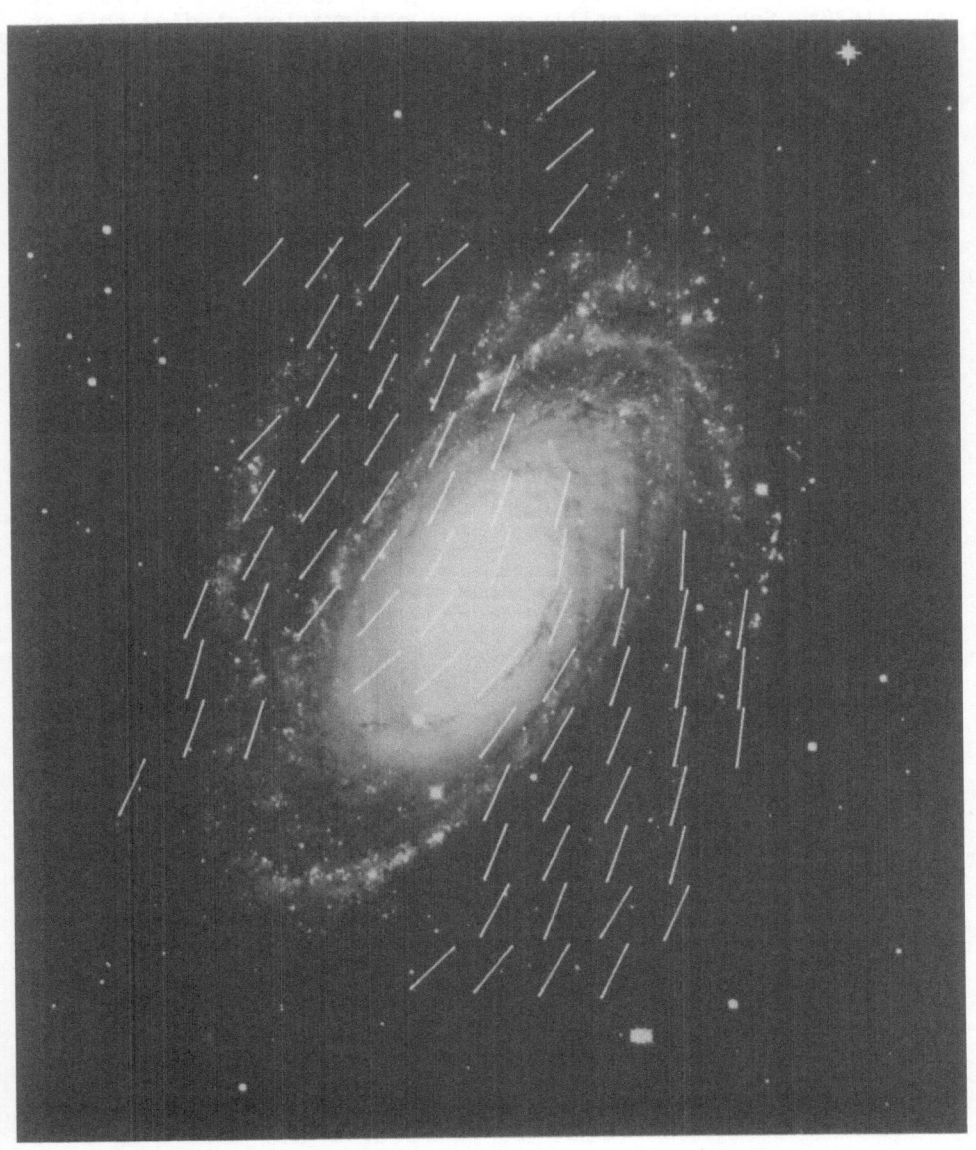

Figure 6. The magnetic field orientation in M81
(Krause et al., 1989b)

NGC6946 This galaxy was one of the earliest to be mapped with polarization information at a high radio frequency (Klein et al., 1982). Subsequent multifrequency observation studies both at the VLA and in Effelsberg (Harnett et al., 1989a) showed that in spite of a regular field structure no decision between axisymmetric of bisymmetric field could be made. The local perturbations in this 'Arp' galaxy make any decision impossible.

M51 There are extensive data for this galaxy both in optical and radio domain. The original Westerbork data (Segalovitz et al., 1976) have been supplemented by Effelsberg and VLA observations. This galaxy was the first to be investigated by Tosa and Fujimoto (1978) for the presence of a bisymmetrical magnetic field. Also excellent optical polarization CCD maps have been made by Scarrott et al. (1987). There is a general agreement between the optical and radio data for most of the galaxy. In the South-West part of the galaxy the optical and radio data disagree (Beck et al., 1988). More recent radio data (Horellou et al., 1990) confirm the BSS magnetic field. The M51 data confirm that the optical and radio polarizations are due to the same magnetic field.

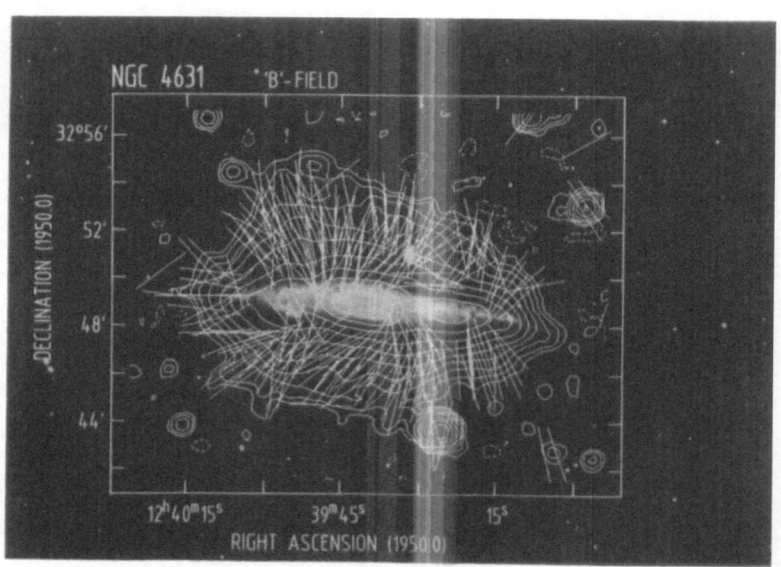

Figure 7. The 'B'-field orientation in NGC4631. (from Hummel et al., 1989; not corrected for Faraday effect)

NGC4631 The question of the structure of the magnetic fields above the plane of a galaxy is of great interest. The closure of magnetic field lines is expected to occur in the halo. The classical edge-on galaxy with a thick disc halo is NGC4631

(Ekers & Sancisi, 1977; Wielebinski & von Kap-herr, 1977). The observations of this galaxy with the VLA (Hummel et al. 1988) showed the existence of halo fields. The orientation of the field is normal to the disc (see figure 7) on the assumption of low Faraday rotation. It seems that Parker instabilities or a galactic wind are pushing the magnetic field above the plane of this galaxy. This observation must however be treated with some caution since NGC4631 has the most extended synchrotron halo. The field strength in the halo may be as much as ~ 2 μG.

NGC891   This nearly perfect edge-on galaxy has been studied in radio continuum (e.g. Allen et al.,1978; Klein et al., 1984). Recent polarization studies by Dahlem, Beck, Hummel, Sukumar, Allen (private communication) indicate a magnetic field away from the plane but not as perpendicular to the plane as in NGC4631. Since data at two frequencies was obtained the confirmation of a low Faraday rotation in the halo (which was so far assumed for NGC4631) has been obtained.

M83   The barred galaxy M83 shows beautiful highly symmetric field structure (Sukumar et al., 1987). More recent VLA observations with higher angular resolution by Sukumar and Allen (1990) indicate that filamentary structure is seen, reminescent of IC342. It is also interesting to note that the aligned magnetic field starts at the optical edge of M83, where the bar structure stops. In the inner parts of this galaxy the field must be quite turbulent.

NGC253, NGC4945   These large southern galaxies are seen nearly edge-on. Radio continuum observation showed in each galaxy two maxima distributed symmetically about the nucleus (Klein et al., 1983, Harnett et al., 1989). The question if this is only a geometrical effect, or a morphological feature are still open.

NGC3628   This edge-on galaxy show some polarization in the halo (M.Krause, private communication). Again the structure is emphasised by two maxima on opposite sides of the nucleus.

M104   The 'Sombrero' galaxy was known for a long time to have a compact (VLBI) source in the nucleus. Optical polarization studies by Scarrott et al. (1987) showed a field along the disc of this galaxy. The disc emission was finally detected by Bajaja et al. (1988) using the high dynamic mode of the VLA. In addition polarized emission perpendicular to the disc was seen in the nuclear area. Recent analysis of the optical polarization by Motsumura & Seki (1989) suggest also a Z-field in the nucleus superimposed on the azimuthal field in the disc. The rotation curve of M104 (e.g. Wagner et al., 1989) suggests a rotating ring surrounding the nucleus.

M82   This mildly active galaxy is the most studied object in all spectral ranges. Although detailed radio continuum observations were published (e.g. Kronberg et al., 1985) none contained polarization data. Optical polarization (Bingham et al., 1976) is dominated by the light scattered by dust, showing a circular vector distribution. Recent CCD observatrons of this object, after subtraction of the scattered light component, showed a Z-field in the nuclear area (Neininger, 1989). The radio continuum spectrum of M82 is very well studied (Klein et al., 1988) indicating that the magnetic field may have mean values of ~ 50μG, the highest of any galaxy. A rotating ring which is seen in all constituents (HI, OH, sub-mm continuum, CO etc.) has been interpreted to be instrumental for the production of this Z-field (Lesch et al., 1989).

Many other galaxies have been observed but not in such detail as those mentioned above. For the general scenario we can conclude that magnetic fields are **azimuthal** in the galaxies except in the nuclear area where fields are **in the Z-direction** . This is in agreement with the model shown for our Galaxy in figure 1.

5. THE MAGNETIC FIELD STRENGTH

The various methods of observing the magnetic fields described at the beginning of this review gave the orientation of the magnetic field only. The only exceptions were the Zeeman effect observations, which are possible in some molecular clouds, and the combination of Pulsar rotation measure and dispersion measure. In general we must envoke the argument of equipartition to determine the magnetic field strength.

The energy spectrum of relativistic electrons is;-

$$N(E)dE = N_0 E^{-v} dE.$$

For equipartition of energy between magnetic field and relativistic electrons we have:-

$$\frac{B^2}{8\pi} = k \int_{E_2}^{E_1} EN(E)dE,$$

where E_1 and E_2 are the lower and upper limits of the energy spectrum (and correspond to the cutoff frequencies v_1 and v_2). The factor k is the ratio of the total energy of cosmic rays to the electron energy. The field strength B (in Gauss) is:-

$$B = 2.3(k \, A \, e \,)^{2/7}$$

(e.g. see Moffat, 1973). In the above equation e is the volume emissivity (in erg sec^{-1} cm^{-3}), A is given by;-

$$A = C \frac{a + 1}{a + 1/2} \frac{v_2^{a+1/2} - v_1^{a+1/2}}{v_2^{a+1} - v_1^{a+1/2}}$$

with $C = 1.057 \times 10^{12}$ cgs units, $a = (v-1)/2$ is the spectral index. The emissivity e is also given by:-

$$e = \frac{4\pi}{1} \int_{v_1}^{v_2} I_v dv$$

where I_v is the intensity and l the length (cm) of the source. The parameters k and v_1(Hz) are taken often as 100 and 10^7 respectively (e.g. Sofue et al., 1986).

Although the fundamental question of the applicability of equipartition is often subject of heated discussions the concept as such has stood many attacks. It is a way an 'economical' concept, where the energy between two reservoirs is minimised.

6. THE INTERPRETATION OF MAGNETISM IN GALAXIES

Two general alternative models for the origin of magnetic fields in galaxies have been proposed;- the primodial model where the fields came through the compression of a relict field or the dynamo scenario where the field is generated through the amplification of a seed field in a galactic rotation. The primodial field model was developed in some detail by Piddington, (1964, 1978, 1981). This seemed at first feasible when the primodial intergalactic field was put at 10^{-8} Gauss or more. A compression by a factor of 100 would bring the fields into the observed range of a few uG. However as a result of more recent observations values for the magnetic field were estimated to be $5 < B_t < 50$ μG. Also more recent estimates of the intergalactic magnetic field suggest that the upper limits are $10^{-9} - 10^{-10}$ G. Also it was pointed out by Parker (1979) that in presence of turbulence the primodial field would be dissipated in 10^7 years. Clearly these two arguments do not allow sufficient room for the primodial field hypothesis.

All the recent theoretical investigations have been directed towards the interpretation of the magnetic fields in galaxies in the context of the turbulent hydromagnetic dynamo theory. Originally the explanation of the magnetism of the Earth, the Sun and Planets determined the development of the dynamo theory. (e.g. Parker, 1955, 1971, 1979; Steenbeck et al., 1966; Krause and Rädler 1980). The application of the dynamo theory to galaxies was investigated by Parker (1971) and Vainshtein and Ruzmaikin (1971). This idea was reinvestigated in some detail in view of the new observational data by Ruzmaikin and Shukurov (1981). The difficulty at first was that the solution of the dynamo equation was for the basic mode only. Since the observations showed magnetic fields with a spiral-like structure this dynamo interpretation was obviously inapplicable. In addition investigations of the Faraday rotation in M51 by Tosa and Fujimoto (1978) suggested that the field was 'bisymmetric', [BSS], i.e. with the field orientation going along a spiral arm into the nucleus and coming out of the nucleus with unchanged field direction. This is in contrast to the axisymmetric [ASS] configuration, where the field lines point in or out of the nucleus.

The basic dynamo equation is given by (e.g. Krause and Rädler, 1980):

$$\frac{dB}{dt} = v \times (u \times B) + kv^2B + v \times (aB)$$

with $\quad v.B = 0$

the first term of the dynamo equation describes the large-scale velocity field (usually given by $u = \Omega \times r$), the second term gives the mean helicity of the turbulence and the third one the turbulent magnetic diffusity.

Parker (1971) has suggested a concept of the 'aw-dynamo', where a mean toroidal magnetic field is generated by the non-uniform (differential) rotation w from an original poloidal magnetic field. The poloidal field is regenerated from the toroidal field by the effects of cyclonic convection (the a-effect). Parker described the dynamo in the 'slab' geometry, i.e. a thin layer of infinite extended electrically conducting gas in cyclonic turbulent motion subject to a large shear.

A spherical dynamo without differential rotation, (which is particularly applicable to the Earth and the Planets), has been investigated by Krause and Steenbeck (1967). This dynamo with constant a leads to an 'a²-dynamo'. A detailed study of the stability of a² dynamos was given by Krause and Meinel (1988).

In fact solutions of the dynamo equation in other geometries are few. An exception, which may be applicable to galaxies, is the solution of the oblate spheroid by Stix (1975). More recently a partial solution for a slender torus was given by Grosser (1988) which may apply to the situation in the nucleus of a galaxy.

Since the observations of magnetic fields in galaxies suggest a dominance of the BSS field structure numerous theoreti-cal papers were published to explain this observational fact. Also the problem of the fields above the plane of the disc (in the halo) were investigated. The solution of the dynamo equation for many modes (Ruzmainin et al., 1985; Baryshnikova et al., 1987; Krasheninnikova et al., 1989) showed that the co-existence of the BSS and ASS modes was possible in the context of the dynamo theory. A special investigation of the dynamo solution leading to the BSS case was given by Sawa and Fujimoto (1986) and Fujimoto and Sawa (1987). The extension to three dimensions was discussed by Sawa and Fujimoto (1987). Further attempts to model three dimensional situations are given by Strachenko and Shukurov (1989). The question of the growth rates of different modes and the stability of the nonlinear dynamo was treated by Brandenburg et al. (1989).

In spite of the great activity in the understanding of the dynamo many questions are open. There is still some lingering hope that compression of the primodial field could be used to explain some of the observed phenomena. The role of reconnections, a perennial discussion point, is still unclear. The applicability of equipartition, so often used to explain observational results, is still not universally accepted. The role of the local magnetic fields (stellar fields, pulsars, supernovae, bi-polar sources, molecular clouds, etc.) in relation to the global magnetic fields is also unclear. The reason for the dominance of the BSS field structure in some galaxies has not been as yet explained. The interplay of theory and observations is doing a lot of good in advancing our understanding but; 'all is not well in the house of magnetic fields'.

7. THE SEED FIELD

Both the primodial field concept and the galactic dynamo theory require some magnetic field to start with. The dynamo has the advantage that it can amplify the seed magnetic field ba a factor of 10^3 or more. In the non-linear dynamo the amplification factor could be even greater. With the accepted values of intergalactic magnetic field of ~10^{-10} Gauss (Ruzmaikin and Sokoloff, 1977) we may have to look for other sources of seed fields for amplification by a dynamo process.

One of the most important processes for the creation of a minute magnetic field is the 'Biermann battery' (Biermann, 1950). When applied to a galaxy the concept requires small turbulent cells which through charge separation would give small currents and hence magnetic fields. The concept of transferring this scenario to galactic scales failed (e.g. Hoyle, 1958) since enormous electromotive force would be required. One way out, which was studied by Hoyle and Ireland (1961), was to postulate helical magnetic fields.

A recent development in this area comes from the observations of CO rings and of poloidal magnetic fields in many mildly active galaxies. CO rings have been seen in the inner parts of M82 (Lo et al., 1987; Nakai et al., 1987; Loiseau et al., 1989), NGC1097 (Gerin et al., 1988), NGC4945, NGC1808, NGC 1068, etc.. Also in these galaxies strong evidence for poloidal magnetic fields (Z-fields) was found either by optical studies or in radio polarization. This scenario was studied by Lesch et al. (1989). By applying the battery effect to give charge separation in the inner CO ring a small seed field can be created. This in turn can be amplified by compression and turbulent stretching. Possibly the poloidal field in the nucleus of a galaxy can in turn be amplified to give the observed azimuthal fields in the spiral arms.

8. THE 'FUTURE' OF MAGNETIC FIELDS

It seems that each decade in astrophysics had its fashion subject. Magnetic fields were 'in' in the 1950's. Then came the gravity fashion with the successes of the density wave theory, mergers etc. The 1970's were the years of the interstellar medium with the nonumental discoveries of molecules, studies of UV absorption lines, the use of the IRAS satellite data etc..

Now in the 1980's it seems we are finding it to be necessary to combine the results on all the various fashions in the hope of understanding the universe. The first IAU symposium on Galactic and Intergalactic Magnetic Fields was held in Heidelberg in June 1989. A simple conclusion can be given after this symposium. We have many astounding pieces of information about the magnetic fields in the Earth, the Sun, the Planets, our Galaxy, and external galaxies. We even know that there is a field in the clusters of galaxies. We have detailed theories that can explain a host of details of various astronomical objects. However we still have a long way to go before we have an overall understanding of the role of magnetic fields in astrophysics.

9. REFERENCES

BOOKS

Parker E.N. 'Cosmical Magnetic Fields' 1979, Claredon Press, Oxford

Krause F. and Rädler K.H. 'Mean Field Magnetohydrodynamics and Dynamo Theory' 1980, Pergamon Press

Beck R. and Gräve R. 'Interstellar Magnetic Fields' 1987, Springer-Verlag

Zeldovich Ya. Ruzmaikin A.A. and Sokoloff D.D. 'Magnetic Fields in Astrophysics' 1987, Gordon & Breach

Asseo E. and Gresillon D. 'Magnetic Fields and Extragalactic Objects' 1987, Cargese Workshop, Edition de Physique

Ruzmaikin A.A. Shukurov A.M. and Sokoloff D.D. 'Magnetic Fields in Galaxies' 1988, Kluwer Academic Publishers

Beck R. Kronberg P.P. and Wielebinski R. 'Galactic and Intergalactic Magnetic Fields', Proceedings I.A.U. Symposium No.140, Kluwer Academic Publishers, (in press)

REVIEW ARTICLES

Gardner F.F. and Whiteoak J.B. 1966, Ann. Rev. Astron. Astrophys., **4**, 245

Heiles C. 1976, Ann. Rev. Astron. Astrophys., **14**, 1

Verschuur G.L. 1979, Fund. Cosmic Phys., **5**, 113

Blandford R.D., 1983, Astron.J., **88**, 245

Sofue Y., Fujimoto M. and Wielebinski R. 1986, Ann. Rev. Astron. Astrophys., **24**, 459

Beck R. 1986, IEEE Trans. Plasma Science, **PS-16**, 740

Asseo E. and Sol H. 1987, Physics Reports, **148**, 307

Heiles C. 1987, in "Interstellar Processes" eds. D.J.Hollenbach and H.A.Thronson Jr., p171, D.Reidel Publ. Comp.

Zweibel E.G. 1987, in "Interstellar Processes" eds. D.J.Hollenbach and H.A.Thronson Jr., p195, D.Reidel Publ. Comp.

Ruzmaikin A.A., Sokoloff D.D. and Shukurov A.M. 1988, Nature, **336**, 341

GENERAL REFERENCES

van Albada G.D. 1978, Ph.D. Thesis, Leiden University

Allen R.J., Baldwin J.E., Sancisi R. 1978, Astron. Astrophys., **62**, 397

Appenzeller I. 1967, Publ. Astron. Soc. Pacific, **79**, 600

Bajaja E., Dettmar R-J., Hummel E., Wielebinski R. 1988, Astron. Astrophys., **202**, 35

Baryshnikova Y., Ruzmaikin A.A., Sokoloff D.D., Shukurov A. 1987, Astron. Astrophys., **177**, 27

Beck R. 1982, Astron Astrophys., **106**, 121

Beck R., Berkhuijsen E.M., Wielebinski R. 1978, Astron. Astrophys., **68**, L27

Beck R., Berkhuijsen E.M., Wielebinski R. 1980, Nature, **283**, 272

Beck R., Klein U., Wielebinski R. 1987, Astron. Astrophys., **186**, 95

Beck R., Loiseau N., Hummel E., Berkhuijsen E.M., Gräve R., Wielebinski R. 1989, Astron. Astrophys., (in press)

Behr A. 1959, Veröff. U. Sternw. Göttingen, No.126

Berkhuijsen E.M., Brouw W.N., 1963, Bull. Astron. Inst. Netherland, **17**, 185

Biermann L 1950, Z.Naturf., **5a**, 65

Bingham R.G., McMullan D., Pallister ,W.S., White C., Axon D.J., Scarrott S.M. 1976, Nature, **256**, 463

Brandenburg A., Krause F., Meinel R., Moss D., Tuominen I. 1989, Astron. Astrophys. **213**, 411

Buczilowski U., Beck R. 1987, Astron. Astrophys. Suppl., **68**, 171

Crutcher R.M., Kazes H., Troland T.H. 1987, Astron. Astrophys., **181**, 119

Davis L., Greenstein J.L. 1951, Astrophys. J., **114**, 206

Ekers R., Sancisi R. 1977, Astron. Astrophys., **54**, 973

Ellis R.S., Axon D.J. 1978, Astrophys. Space Sc., **54**, 425

Elvius A. 1978, Astron. Astrophys., **65**, 233

Elvius A., Hall J.S. 1965, Astron. J., **70**, 138

Fermi E. 1949, Phys. Rev., **171**, 1169

Fiebig D., Güsten R. 1989, Astron. Astrophys., **214**, 333

Fujimoto M., Sawa T. 1987, Publ. Astron. Soc. Japan, **39**, 375

Gerin M., Nakai N., Combes F. 1988, Astron. Astrophys., **201**, 47

Ginzburg V.L., Syrovatskij S.I. 1969, Ann. Rev. Astron. Astrophys., **7**, 375

Gräve R., Klein U., Wielebinski R. 1989, (in preperation)

Grosser H. 1988, Astron. Astrophys., **199**, 235

Hall J.S. 1949, Science, **109**, 166

Harnett J.I., Beck R., Buczilowski U. 1989a, Astron. Astrophys., **208**, 32

Harnett J.I., Haynes R.F., Klein U., Wielebinski R. 1989b, Astron. Astrophys., **216**, 39

Harnett J.I., Haynes R.F., Wielebinski R., Klein U. 1990, Proc. Astron. Soc. Australia, (in press)

Haynes R.F., Klein U., Wielebinski R., Murray J.D. 1986, Astron. Astrophys., **159**, 22

Haynes R.F. et al., 1990, Proc IAU Symp. No.140 (in press)

Hildebrand R.H. 1988, Astro. Lett. and Communicat., **26**, 265

Hiltner W.A. 1949, Science, **109**, 471

Hiltner W.A. 1958, Astrophys. J., **128**, 9

Horellou C., Beck R., Klein U., Krause M. 1990, Proc. IAU Symp. No.140, (in press)

Hoyle F. 1958, in "La Structure et l'Evolution d'Univers", XI Solvey Conference, p53

Hoyle F., Ireland J.G. 1961, Mon. Not. Roy. astr. Soc., **122**, 35

Hummel E., Dettmar R-J., Wielebinski R. 1986, Astron. Astrophys., **166**, 97

Hummel E., Lesch H., Wielebinski R., Schlickeiser R. 1988, Astron. Astrophys., **197**, L29

Hummel E., Krause M., Lesch H. 1989, Astron. Astrophys., **211**, 266

Junkes N., Fürst E., Reich W. 1987, Astron. Astrophys. Suppl., **69**, 451

Klein U., Beck R., Buczilowski U., Wielebinski R. 1982, Astron. Astrophys., **108**, 176

Klein U., Urbanik M., Beck R., Wielebinski R. 1983, Astron. Astrophys., **127**, 177

Klein U., Wielebinski R., Beck R. 1984, Astron. Astrophys., **133**, 19

Klein U., Wielebinski R., Morsi H.W. 1988, Astron. Astrophys., **190**, 41

Klein U., Wielebinski R., Haynes R.F., Malin D.F. 1989, Astron, Astrophys., **211**, 280

Krasheninnikova Y., Ruzmaikin A.A., Sokoloff D.D., Shukurov A. 1989, Astron. Astrophys., **213**, 19

Krause F., Steenbeck M. 1967, Z.Naturf. **22a**, 671

Krause F., Meinel R. 1988, Geophys. Asrophys. Fluid Dyn., **43**, 95

Krause M., Hummel E., Beck R. 1989b, Astron. Astrophys., **217**, 4

Krause M., Beck R., Hummel E. 1989a, Astron. Astrophys., **217**, 17

Kronberg P.P., Biermann P., Schwab F.R. 1985, Astrophys. J., **291**, 693

van der Kruit P.C., Oort J.H., Mathewson D.S. 1972, Astron. Astrophys., **21**, 169

Lesch H., Crusius A., Schlickeiser R., Wielebinski R. 1989, Astron.Astrophys., **217**, 99

Lo K.Y., Cheung K.M., Masson C.R., Phillips T.G., Scott S.L., Woody D.P. 1987, Astron. J., **312**, 574

Loiseau N., Klein U., Greybe A., Wielebinski R., Haynes R .F. 1987, Astron. Astrophys., **178**, 62

Loiseau N., Nakai N., Sofue Y., Wielebinski R., Reuter H-P., Klein U. 1989, Astron. Astrophys., (in press)

Lyne A.,   Smith Graham F.   1989,   Mon. Not. Roy. astr. Soc., **237**, 533

MacLeod J.M.,   Vallee J.P.,   Broten N.W. 1988, Astron. Astrophys. Suppl., **74**, 97

Martin P.G., Shawl S.J. 1982, Astrophys. J., **253**, 86

Mathewson D.S., Milne D.K. 1965, Austr. J. Phys., **18**, 635

Mathewson D.S., Ford V.L. 1970, Astron. J., **75**, 778

Mathewson D.S., Ford V.L. 1970, Mem. Roy. astr. Soc., **74**, 139

Mathewson D.S.,   van der Kruit P.C.,   Brouw W.N.   1972,   Astron. Astrophys., **17**, 468

Matsumuna M., Seki M. 1989, Astron. Astrophys., **209**, 8

Mayer C.H.,   McCullough T.P.,   Sloanaker R.M. 1957, Astrophys. J., **126**, 468

Meyer   1920, Lick Obs. Bull., **10**, 68

Moffat A.T.   1973,   in   "Galaxies and the Universe", eds. Sandage et al., **9**, Chapter 7, Chicago Uni. Press

Nakai N., Hayashi M., Handa T., Sofue Y., Hasagawa T. 1987, Publ. Astron. Soc. Japan, **39**, 680

Neininger N. 1989, Dipl. Thesis, Bonn University

Öhman Y. 1942, Stockholm Obs. Bull., **10**, 68

Parker E.N. 1955, Astrophys. J., **122**, 293

Parker E.N. 1971, Astrophys. J., **163**, 255

Piddington J.H. 1964, Mon. Not. Roy. astr. Soc., **128**, 345

Piddington J.H. 1978, Astrophys. Space Sci., **59**, 237

Piddington J.H. 1981, Astrophys. Space Sci., **80**, 457

Purcell E.M. 1978, Astrophys. J., **231**, 404

Reich W. 1989, Proc IAU Symp. No.136 (in press)

Reich W., Sofue Y., Wielebinski R., Seiradakis J.H. 1988, Astron. Astrophys., **191**, 303

Ruzmaikin A.A., Sokoloff D.D. 1977, Astron. Astrophys., **58**, 147

Ruzmaikin A.A., Shukurov A.M. 1981, Sov. Astron., **25**, 553

Ruzmaikin A.A.,   Sokoloff D.D.,   Shukukov A.M. 1985, Astron. Astrophys., **148**, 335

Sawa T., Fujimoto M. 1986, Publ. Astron. Soc. Japan, **38**, 132

Sawa T., Fujimoto M. 1988, in Proc. International Workshop, Cargese, eds. E.Asseo and H.Sol

Scarrott S.M., White C., Pallister W.S., Solinger A.B. 1977, Nature, **265**, 32

Scarrott S.M., Ward-Thompson D., Warren-Smith R.F. 1987, Mon. Not. Roy. astron. Soc., **224**, 299

Schmidt Th. 1970, Astron. Astrophys., **6**, 294

Schmidt Th. 1976, Astron. Astrophys. Suppl., **24**, 357

Segalovitz A., Shane W.W., de Bruyn A. 1979, Nature, **264**, 272

Simard-Normandin M., Kronberg P.P., Button S. 1981, Astrophys. J. Suppl., **45**, 97

Sofue Y., Fujimoto M. 1983, Astrophys. J., **265**, 722

Sofue Y., Klein U., Beck R., Wielebinski R. 1985, Astron. Astrophys., **144**, 257

Spoelstra T.A.-Th. 1984, Astron. Astrophys., **135**, 238

Steenbeck M., Krause F., Rädler K.H. 1966, Z.Naturf., **21a**, 369

Stix M. 1975, Astron. Astrophys., **42**, 85

Strachenko S.V., Shukurov A.M. 1989, Asron. Astrophys., **214**, 47

Sukumar S., Klein U., Gräve R. 1987, Astron. Astrophys., **184**, 71

Sukumar S., Allen R.J. 1990, Proc. IAU Symp. No.140, (in press)

Tosa M., Fujimoto M. 1978, Publ. Astron. Soc. Japan, **30**, 315

Vainstein S.I., Ruzmaikin A.A. 1971, Astron. J. (UdSSR), **48**, 902

Vallee J.P. 1988, Astron. J., **95**, 750

Wagner S.J., Dettmar R-J., Bender R. 1989, Astron. Astrophys., **215**, 243

Westerhout G., Seeger Ch.L., Brouw W.N., Tinbergen J. 1962, Bull. Astron. Inst. Neth., **16**, 187

Wielebinski R., Shakeshaft J.S., Pauliny-Toth I.I.K. 1962, Observatory, **82**, 158

Wielebinski R., Shakeshaft J.S. 1964, Mon. Not. Roy. astron. Soc., **128**, 19

Wielebinski R., von Kap-herr A. 1977, Astron. Astrophys., **59**, L17

The Early ISM and Galaxy Formation

Simon D.M. White
Steward Observatory,
University of Arizona.

ABSTRACT. I review current ideas about galaxy formation, concentrating on when and how it occurred, and what it might look like. As theoretical ideas have been refined, it has become clear that galaxy formation may be both less spectacular and more recent than originally predicted. Nevertheless, all theoretical predictions for the appearance of young galaxies should be treated with caution. There is enormous latitude in predictions of their redshifts, their morphology, their star formation rates, the relevant stellar initial mass functions, and the extent to which their spectra are affected by radiative reprocessing. One might hope to clarify some of these questions by appealing to observations of analogous processes in nearby objects, but the situation is still murky. Current observations of distant objects do seem to have detected the formation of a major component of galaxies. Galaxy formation has become an observational and rapidly developing science. I suggest how these data may be interpreted within the Cold Dark Matter theory for the origin of structure.

1. What is galaxy formation?

As we go back in time we eventually get to an epoch when there is no ISM in galaxies, not because of the lack of the M, but because of the lack of the S. We would probably say that galaxies didn't exist at such epochs, and that their *formation* occurred later when conversion of gas into stars really got under way. On the other hand some ellipticals are being produced today by the merging of two or more spirals, and it would seem reasonable to identify the present as their time of formation, even though almost all their stars are old. Since the assembly of a galaxy and the formation of its constituent stars may occur at different epochs and may each be spread out over cosmologically significant timescales, galaxy formation is better thought of as an ongoing process than as a specific event.

Different galaxy formation pictures can be classified by ranking them according to various characteristic timescales. Thus the old and highly schematic models of Partridge and Peebles (1967) envisaged a nearly homogeneous protogalactic perturbation which suddenly turned into stars when it reached its maximum size (and thus *minimum* density) and then fell back on itself to form a galaxy. All stars were formed in a short period of time, $\Delta t_\star$, when the age of the universe, $t_\star$, was about equal to a typical orbital time in

H. A. Thronson, Jr. and J. M. Shull (eds.), The Interstellar Medium in Galaxies, 371–386.
© 1990 *Kluwer Academic Publishers.*

observed galaxies, t_{orb}. The cooling time of the protogalactic material, t_{cool}, was short, so that pressure effects were unable to inhibit star formation, and the galaxy came to its final equilibrium somewhat later, at time, t_{equ}. Hence $t_{cool} < \Delta t_\star < t_\star \sim t_{orb} \sim t_{equ}$. These models have very high luminosities because of the (arbitrary) assumption that $\Delta t_\star$ is short implies large star formation rates. On the other hand, $t_\star \sim t_{orb}$ requires formation at such high redshifts ($z \sim 20\text{--}50$) that the luminous phase will only be detectable in the infrared. This kind of ranking of timescales is implicit in all "dissipationless" models for elliptical galaxy formation (e.g. van Albada 1982; McGlynn 1984; Aguilar and Merritt 1989). Such models are incomplete because they ignore the obvious question of how the stars formed. This would seem to require substantial inhomogeneities within the protogalaxy before its collapse.

Much more realistic protogalaxy models were made by Meier (1976) based on simulations by Larson (1974a). Turbulent pressure and radiative processes slowed down the collapse in these models and allowed it to continue to scales much smaller than the initial protogalaxy. This results in star formation and equilibration timescales which are longer than the dynamical time of the final galaxy; $t_{orb} < t_{cool} \leq \Delta t_\star \leq t_\star \sim t_{equ}$. Thus when scaled to real galaxies, these models predict lower star formation rates and luminosities, but at more accessible redshifts. Their main limitations are the lack of any *a priori* justification for the assumed initial conditions, or of any treatment of the dark matter component. In addition their modeling of hydrodynamic and star formation processes is purely phenomenological. One result of this is that the observable properties of the models are determined by arbitrary choices of initial and modeling parameters.

The first attempts to treat realistic levels of inhomogeneity, to account consistently for the gravitational effects of dark matter, and to predict the observable properties of such young galaxies, came with the models of Carlberg, Lake and Norman (1976), Baron and White (1987), Lake and Carlberg (1988), and Katz (1989). These authors all found star formation to extend over periods of order the protogalactic collapse time. The models which were able to treat substantial precollapse inhomogeneities also found star formation to start before nominal turnround of the protogalaxy as gas cooled and condensed within the small subunits which merged into the final system. Katz was able to show that gas from relatively diffuse substructures would settle into a disk soon after the main collapse, and would continue to form stars in a flattened system. Apart from the possibility that a significant number of stars might form at very early times, the timescales in these models rank in the same way as in Larson's models; $t_{orb} < t_{cool} \leq \Delta t_\star \leq t_\star \sim t_{equ}$.

In simulations of this kind Katz (1989) obtained ellipsoidal systems from collapses with well defined and concentrated subunits which were able to lose their orbital angular momentum and form large numbers of stars before the final consolidation of the system. This is really an example of elliptical galaxies forming from mergers of preexisting stellar systems, a hypothesis that has a long and controversial history, particularly when posed in the extreme form that all ellipticals had progenitors resembling present-day spiral galaxies (Toomre and Toomre 1972; Toomre 1977; Efstathiou and Jones 1980; Ostriker 1980; Tremaine 1981; White 1982,1983; Schweizer 1983,1990; Barnes 1988). No merger advocate probably ever believed this extreme view, and Katz's simulations illustrate the opposite extreme where the merging is scarcely differentiated from the first collapse of structures. Nevertheless, it is clear that *some* ellipticals are being formed today by mergers of spirals. For these objects the equilibration timescale greatly exceeds our other characteristic times; $t_{orb} < t_{cool} \leq \Delta t_\star \leq t_\star \ll t_{equ}$.

Yet another possible mode of galaxy formation is illustrated by cooling flows (Fabian, Nulsen and Canizares 1982,1984; Sarazin 1987; Fabian, Arnaud and Thomas 1987). In these systems a quasistatic atmosphere of gas is apparently radiating its binding energy and settling to the center of the cluster in which it resides. The cooling radiation is observed directly and the inferred inflow rates can be hundreds of solar masses per year. Detailed analysis suggests that mass is dropping out of the flow over a wide range of radii, and presumably turning into stars. However, although some evidence for star formation is seen, it is clear that massive stars must be considerably underrepresented relative to other known regions of star formation. With this worrying caveat it is easy to make models in which the cooling flow is responsible for forming the observed central galaxy. In this kind of model, the gas cooling time and the star formation times are equal to the present age of the universe even though the dynamical time of the galaxy is much shorter; $t_{orb} \ll t_{equ} < t_{cool} \sim \Delta t_\star \sim t_\star$.

These examples show that different definitions of galaxy "formation", formation of the first stars, or of most of the stars, or assembly and equilibration of the observed structure, can lead to different estimates of the relevant epoch. Thus in discussions of galaxy formation and its observable consequences, it is important to be clear what is meant. Throughout this article I shall define the epoch of most rapid star formation as the epoch of galaxy formation. This is the time when a galaxy is likely to be most luminous, although not necessarily most easily visible; dust may reprocess much of the light from young stars into the far infrared. Note that the emission from such a young galaxy may come from several clumps spread over a region significantly larger than final galaxy. This possibility only applies to spheroidal components, since it seems almost certain that thin disks must be assembled *before* they are turned into stars. Note also that this definition may be difficult to interpret if star formation occurs in bursts. Further, it implies that many late-type galaxies are forming today (see, for example, Kennicutt 1983).

2. What do young galaxies look like?

In most large galaxies gas is a small fraction of the total mass, and star formation rates are typically a few solar masses per year. However, at the time of galaxy formation much more of the mass was gaseous, and star formation was probably both more violent and more rapid. The ISM activity associated with galaxy formation may thus have been much stronger than we are used to seeing in nearby objects. Because conditions at this epoch are uncertain and, perhaps, quite unlike those in well studied regions of star formation, it is difficult to predict the appearance of young galaxies. The most obvious procedure would be to argue by analogy, using nearby systems where the physical conditions seem most clearly akin to young galaxies. However, this depends on having a clear idea of *how* galaxies form, and, as we have seen, there are a number of viable possibilities.

Currently, one of the most popular models for the formation of structure is the Cold Dark Matter model (Blumenthal *et al.* 1984; Davis *et al.* 1985). In this theory, as in other hierarchical clustering models, the first collapse of a protogalaxy is expected to be a highly inhomogeneous affair, and substantial bursts of star formation may occur during the collision and amalgamation of sublumps. The most obvious nearby place to look for analogous physical conditions is in collisions and mergers of gas rich galaxies. Such mergers are indeed associated with some of the most active regions of star formation known, namely high luminosity IRAS galaxies (Soifer *et al.* 1986; Sanders *et al.* 1988). In these systems most of the light from young stars is reprocessed by dust and comes out in the far infrared. Clearly if this were also true in young galaxies, they would be very hard to see at high redshift. However, since the IRAS sample was selected for high far IR flux, it will be biased in favor of dusty objects. Interacting systems selected purely on the basis of optical morphology, show considerable scatter in their optical colors (Larson and Tinsley 1978). In addition, while many show evidence for enhanced star formation, others show none, and in the extensive survey of Bushouse, Lamb and Werner (1988) all galaxies with a substantial IR excess also show enhanced H-α emission.

An even more pessimistic paradigm for forming galaxies is offered by cooling flows. These are alleged to be producing the most luminous elliptical galaxies known, and yet the evidence for star formation is subtle enough for its interpretation to be highly controversial. We would certainly be unable to recognise such a "young" galaxy at high redshift, since we cannot agree on its status when it is in our own backyard. Fortunately one can argue that even if current star formation in these objects has an IMF which is highly deficient in massive stars, this cannot always have been the case. This follows from the observation that bright spheroidal systems are at least as metal-rich as the Sun (*e.g.* Terlevich *et al.* 1981). The number of massive stars required to produce these metals is similar to that predicted by a standard solar neighborhood IMF. A substantial blue population must therefore have been present at some earlier stage of the evolution of these systems; star formation in all spheroidal systems cannot always have been as inconspicuous as it now seems to be in cooling flows.

Observed optical emission from galaxies at high redshift comes from the UV continuum of young stars, together with some line emission from surrounding HII regions. Lyman α is expected to be the dominant emission line, but because of resonant scattering in surrounding neutral gas, its effective path length for escape from the galaxy can be very long. As a result it is very susceptible to suppression by absorption onto dust grains. It is often stated that significant L-α equivalent widths and UV continuum luminosities are only to be expected in systems with metallicities less than 1% of solar (e.g. Koo 1986). However, IUE observations of nearby extragalactic HII regions suggest a rather less pessimistic conclusion. Hartmann *et al.* (1988) analyse all previous data in addition to their own, and find that although the equivalent width of L-α is strongly anticorrelated with metallicity, it reaches detectable strengths once the metal content drops to about a tenth solar. The effect of dust on the UV continuum is probably much weaker. It is difficult to know how to extend these results to forming galaxies. The amounts of UV and L-α which escape undoubtedly depend critically on how the star-forming regions and the remaining gas and dust are mixed; these may be very inhomogeneous objects. Koo (1986) concluded that to avoid conflict with the null results of primeval galaxy searches, either L-α suppression must be efficient, or galaxy formation must occur at $z > 6$. However, Baron and White (1987) gave an explicit counterexample which evades these limits even though galaxy formation occurs at $z \sim 2$, and the effects of dust are assumed negligible. Koo's high estimates of protogalactic luminosity were a result of assuming rapid formation, $\Delta t_* \ll t_*$.

A more optimistic assessment of the visibility of young galaxies is also suggested by the properties of known galaxies at high redshift. The recent review of Spinrad (1989) discusses 18 objects with redshifts exceeding 1.6, the most distant at $z \sim 3.5$. Almost all these systems are radio galaxies, and they all show strong L-alpha emission. In fact this emission is often so strong, so broad, and so spatially extended that it is difficult to interpret in terms of star-forming regions within the main body of a galaxy. In many cases, a close association with the central source and its associated jets and radio lobes seems to be indicated (*e.g.* McCarthy and van Breugel 1989). The interpretation of the continuum radiation in these sources is also controversial. In some cases, notably 3C326.1 (McCarthy *et al.* 1987), the continuum is very weak, and it may be reasonable to consider the system as truly in its initial formation stages. In most cases, although there is a substantial blue continuum which almost certainly comes from a large population of unobscured young stars, the spectrum rises substantially into the near infrared. K-band images show significantly less irregularity than CCD frames in the optical, and both the shape of the spectrum and the uniformity of the absolute magnitudes at K seem to indicate the presence of a substantial "old" (*i.e.* more than 1 to 2 Gyrs) stellar population (Lilly 1989; note that the K-band flux of 0902+34 at $z = 3.4$ has recently been revised downward, thus weakening Lilly's case (Eisenhardt, private communication)). In contrast the "field" galaxy with a tentative redshift of 3.4 found by Cowie and Lilly (1989) has no associated radio emission and no continuum rise in the infrared. If confirmed it will be a clear example of a large galaxy in its major period of star formation and with very little apparent obscuration by dust. The very large number of faint blue galaxies found by Tyson (1988) suggest that this object may not be untypical (see below).

If obscuration does indeed have only minor effects, the spectra of young galaxies can be predicted using evolutionary synthesis programs. Figure 1 shows how the spectrum of an ongoing burst of star formation changes according to an updated version of Bruzual's (1983) code. This diagram does not include the expected line emission from HII regions surrounding the young stars. The continuum is relatively flat between 1000 and 3800 Åfor ages between 0.1 and 2 Gyrs. The assumed IMF affects the overall continuum level quite strongly but has rather less effect on its slope. With this kind of model one can use simulations to predict the appearance of young galaxies in detail. Based on a rather crude simulation of protogalactic collapse and star formation, Baron and White (1987) concluded that most modern theories predict predict young galaxies to be fainter, more irregular and at lower redshift than the old models of Partridge and Peebles (1967) or Meier (1976). More recently, Katz (1989) has carried out more sophisticated simulations of dissipative collapse and star formation. For the first time these led without fudging to objects with the size and structure of both elliptical and spiral galaxies. Figure 2 shows the observed I-band surface brightness predicted for one of his "spiral" protogalaxies seen at $z = 2.3$, just as it is collapsing into a single unit. Its star formation rate at this time is 60 $M_\odot$/yr, its I apparent magnitude is 23.5, and its peak surface brightness is about 22 (with 0.2 arcsec resolution). The irregular morphology and apparent magnitude of the system are very similar to the Baron and White model. However, it is smaller and so has higher surface brightness. To the extent that it produces a better looking final galaxy, it is probably more realistic.

3. When was galaxy formation?

The early galaxy formation models assumed stars to form over a relatively short timescale. As a result protogalaxies were predicted to be very bright, and when searches for them

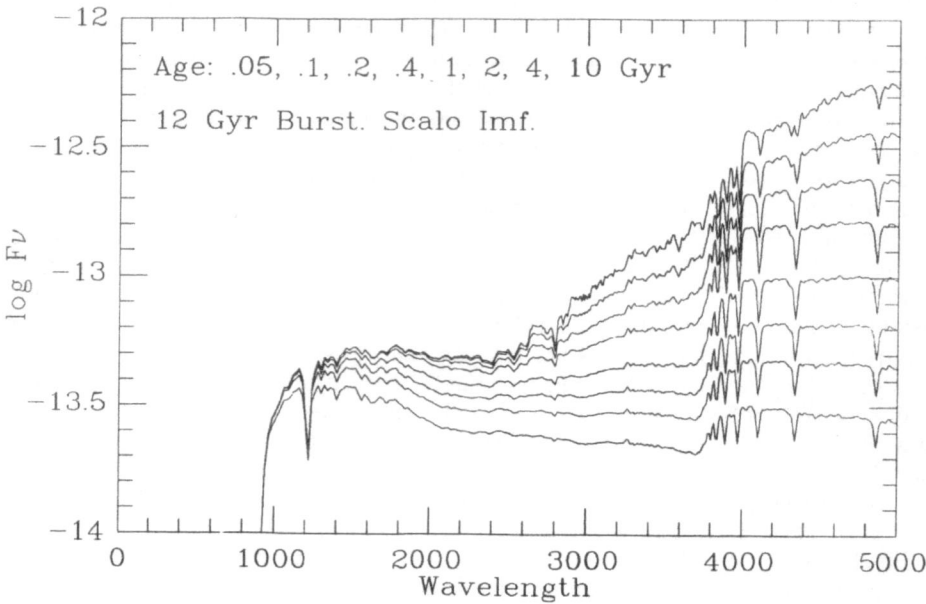

Fig. 1 Synthetic spectra for a region with constant star formation at the ages indicated. The initial mass function advocated by Scalo (1986) was assumed in making these predictions. The IMF dependence of the results can be gauged by comparison with the similar diagram for a Salpeter IMF presented by White (1989). These spectra were provided by Gustavo Bruzual from the most recent version of his synthesis program.

yielded null results the seekers generally concluded that the main epoch of star formation must be at $z > 6$ where its light could be hidden behind the near IR sky background (*e.g.* Koo 1986). However, reviewing more recent theories, Baron and White (1987) concluded that most would predict substantial galaxy formation activity at lower redshifts and at luminosities which did not conflict with observational limits. Thus it is interesting to ask whether there is positive evidence for galaxy formation at observationally accessible epochs. There is ample evidence that some aspects of the galaxy population are changing on rather short timescales. For example the effect originally discovered by Butcher and Oemler (1978), that the galaxy population in clusters at $z = 0.2 - 0.5$ differs substantially from that in nearby clusters, has been amply confirmed by more recent data (Gunn 1989).

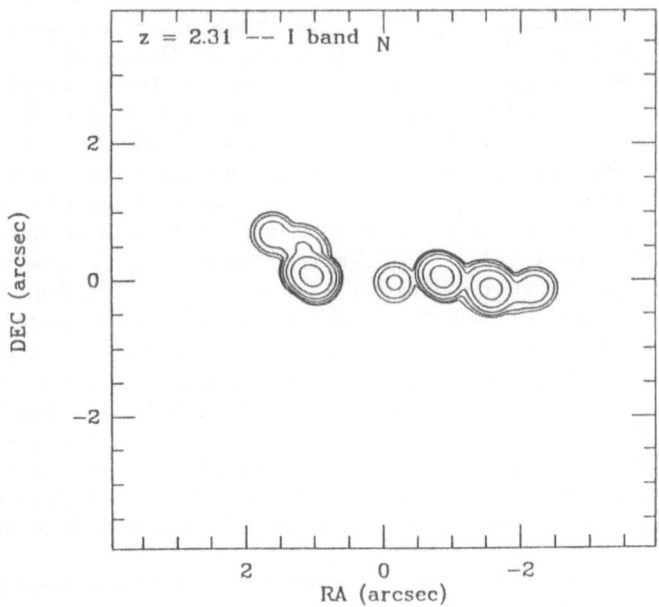

Fig. 2 A protospiral containing $7 \times 10^{10}\,M_\odot$ of gas an ten times as much dissipationless dark matter is seen at the instant of first collapse during its first major episode of star formation (from Katz 1989).

Surveys of field galaxies show that bursts of star formation were apparently much more common at $z \sim 0.3$ than nearby, although this behavior appears to be limited to lower luminosity galaxies (Colless, Ellis and Taylor 1989). The abundance of QSO's increases by almost an order of magnitude from $z = 0$ to $z = 0.5$ (Green 1989). All these effects demonstrate that quite a lot has changed in the last 1/3 of a Hubble time, and they suggest that in the recent past galaxies had significantly greater amounts of gas, higher star formation rates, and perhaps more frequent interactions. In a similar vein, the steep decline in QSO abundance from $z = 2$ to $z = 4$ (Green 1989) is most simply attributed to the fact that this epoch corresponds to the collapse of the massive galaxies which host such objects (Efstathiou and Rees 1988).

In discussions of galaxy formation there is a tendency to put considerable emphasis on elliptical galaxies and spiral bulges, and to pay less attention to disks, even though disks contain roughly half the stars in the universe (Schechter and Dressler 1987). This is because standard evolutionary models (*e.g.* Tinsley 1980; Bruzual 1983) assign spheroids very high luminosities in their initial star formation phase, which make them easily visible to high redshift. As noted above these luminosities are not *required* by theory or data. For at least one galactic disk, our own, we have quite detailed information on the star formation history. It appears that about half the stars in the solar neighborhood have formed in the last 5 Gyrs, and that the average star formation rate may never have been much higher than it is now (Carlberg *et al.* 1985). The current star formation rates inferred in many other late-type galaxies are consistent with a similar history (Kennicutt 1983). Thus with my adopted definition, the formation epoch for disks, the time of maximum star formation activity, probably varies from $z = 0$ up to a redshift of at least 2. Theoretical arguments based on the angular momenta of disks suggest they they cannot have been assembled much before a redshift of 2 or 3 (Fall and Efstathiou 1980; Baron and White 1987). Figure 3 shows the star formation rates in two of Katz's (1989) models, one which makes a spiral and another which makes a similar mass elliptical. In neither system does the star formation rate reach very extreme values, and the peak rate actually occurs in the spiral, rather than the elliptical, during a burst which immediately follows the formation of a flat disk.

Data from quite a different source suggest that disks may have been assembled by $z = 2$, but that they had hardly begun to make stars. Beyond $z = 2$ the line of sight to about one QSO in five interects a neutral hydrogen cloud with a column density exceeding $2\ 10^{20}$ neutral atoms per cm^2 (Wolfe 1989). For such a high column density the width of the absorption line is determined by the damping wings of L-α rather than by doppler broadening, giving a direct estimate of the column density. Absorption lines measurements at 21 cm show these systems to have small velocity dispersions, similar to galactic disks, and in one case they demonstrate the absorbing cloud to be at least several kiloparsecs across. The total amount of neutral hydrogen observed in these systems (per Mpc^3) exceeds that seen in the local universe by an order of magnitude and is comparable to the amount of material observed locally in stars. This gas could therefore provide enough raw material to make all galaxies, or at least all galactic disks. Its relatively unprocessed state is confirmed by its low metallicity (*e.g.* Pettini, Boksenberg and Hunstead 1989), its low but measurable dust to gas ratio (Fall, Pei and McMahon 1989), and its very low molecular content (Black, Chaffee and Fdlz 1987). Even if this gas has not turned into stars since a redshift of two, it must have somehow been reheated and dispersed; this would seem to require much more energetic star formation than is currently seen in galaxies.

It is also possible that we have already detected light from much of the star formation that ever took place in galaxies. Galactic photometry is now possible to magnitudes fainter than 26. Counts to these limits find more than 10^5 galaxies per square degree, and fainter than about 23 most objects have rather blue U-B and B-R colors (Tyson 1988; Majewski 1989; but *cf* Cowie 1989). Assuming, as seems likely from current spectroscopy to B=22.5, that these blue galaxies are at redshifts beyond $z = 0.5$, the observed light at U and B must come almost entirely from young stars. As pointed out by Cowie (1988, 1989), a fortunate cancellation of factors then implies that the contribution of this population to the mean extragalactic background light is a direct measure of the total number of massive stars formed per Mpc^3, or equivalently of the total mass of metals produced per Mpc^3.

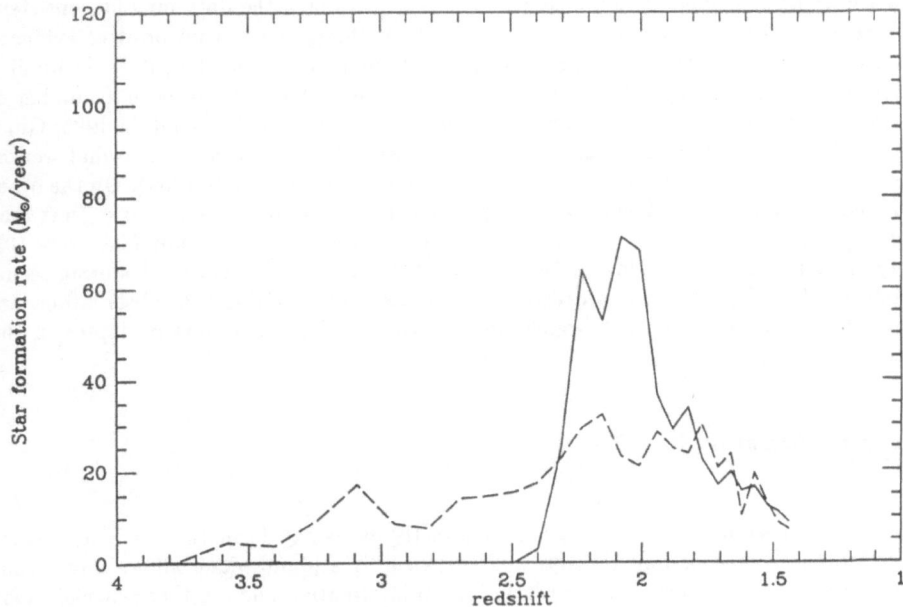

Fig. 3 These simulations by Katz (1989) followed two protogalaxies differing only in the amplitude of the small-scale structure imposed on their initia conditions. The model with little small-scale structure (solid-line) formed few stars until it collapsed as an almost coherent unit. At this stage it was very similar to the model of Figure 2. The second and larger burst of star formation occurred as gas settled into a thin disk. Later star formation was mainly confined to this disk. The more inhomogeneous model (dashed line) formed many stars in the subclumps present at early times. These merged to give an ellipsoidal system which was more compact and more slowly rotating than the "spiral." Note that these models predict the spiral to get brighter than the elliptical contrary to most standard evolutionary models.

This result is independent of the actual redshift of the observed galaxies. For a standard IMF the mass of stars formed in observed events is a substantial fraction of the entire stellar mass of galaxies, although the uncertainties probably allow values ranging from 10% to 100%. Note that the galaxies observed at U are almost certainly closer than $z = 3$, since the Lyman break must be shortward of 3700 Å. In addition, note that star formation events can be seen only if they are not obscured by dust. Apparently, many or most star formation regions remained unobscured throughout much of galaxy formation.

From all these lines of evidence it seems clear that many aspects of the galaxy population have been shaped at relatively recent times. Although the data may be consistent with the formation of most stars since a redshift of three, there is no positive evidence that we have detected the formation of spheroidal components. Rather, it could be disk formation only that occurs at low redshift. Indeed the stellar population in a number of elliptical galaxies at $z = 0.6$ - 0.8 is clearly at least 2 or 3 Gyrs old (Hamilton 1985; Gunn 1989), and a number of radio galaxies at much higher redshift also show somewhat weaker evidence for substantial populations of similar age (Spinrad 1989; Lilly 1989). On the other hand, the relationship between galaxy morphology and environment seems to suggest that no galaxy population has its characteristics defined at very high redshift (*i.e.* $z > 10$); galaxy clustering has such a low amplitude that location with respect to clustering seems unlikely to affect high redshift processes. In any case, it is exciting that observations are now opening a window through which we can study at least one major chapter in the genesis of galaxies.

4. Galaxy formation in CDM

As more data become available which bear directly on galaxy formation, it is important to develop the corresponding theoretical framework to a point which allows direct confrontation with observation. Until recently lack of information allowed theorists enormous freedom. As a result a very wide range of galaxy formation pictures were proposed, often with, at best, a sketchy evaluation of their consequences. On scales larger than individual galaxies the theoretical situation is simpler, both because structures are only mildly nonlinear, and because gravity dominates much of the evolution. Three-dimensional surveys of the galaxy distribution and large catalogs of direct distance measurements have greatly improved our picture of large-scale structure; advances in computer power and numerical techniques have similarly improved our ability to model its evolution (see the many excellent contributions in Rubin and Coyne 1989). As part of this development a "standard" picture for the formation of large scale structure has emerged – the Cold Dark Matter (CDM) model (Blumenthal *et al.* 1984; Davis *et al.* 1985; White *et al.* 1987). This model reached preeminence because it is well specified, is physically simple, has relatively few parameters, and can be *demonstrated* to agree with many aspects of the galaxy distribution. This does not, of course, mean that it is correct; its fundamental assumptions that the Universe is closed and is dominated by some entirely new form of matter are inherently implausible. However, it is falsifiable, and eminently attackable. Its survival for five years suggests that it may contain some elements of the truth. It is clearly of interest to ask whether the data discussed above are consistent with this theory.

Galaxy formation in the CDM theory is a specific case of the general picture outlined by White and Rees (1978). The gravitationally dominant dark matter background clusters hierarchically into larger and larger aggregates. These are identified first as galaxy halos, then as galaxy groups. The gas clusters with the dark matter, but, as successive generations of halos form, it is able to dissipate and sink to the center. There it either becomes self-gravitating and fragments into stars, or it reaches centrifugal equilibrium in a disk where it can also make stars (Fall 1979; Gunn 1981; Faber 1981). The CDM theory specifies the detailed statistical properties of the initial conditions from which this evolution is supposed to occur. In principle, it is possible to simulate later nonlinear evolution to check the theory against the present universe. So far, however, most work has concentrated on the evolution of the dark matter component; this is easier to model, but is also, unfortunately, invisible. It is clear that the theory produces dark halos with the abundances and physical properties inferred for real systems ranging from individual galaxy halos to large clusters (Frenk *et al.* 1985,1988; White *et al.* 1987a,b). Large galaxies are clearly predicted to form very recently ($z \sim 1$ - 2) (Frenk *et al.* 1985; Carlberg and Couchman 1989).

Simulation of the evolution of the visible part of the universe is only just beginning. Carlberg and Couchman (1989) carried out cosmological simulations of moderately large volumes including a dissipative "gaseous" component. These models were marginally able to resolve individual galaxies; they gave very interesting results about the galaxy distribution and some information about when galaxies form, but they were too coarse to follow the formation of individual objects. Katz's (1989) work is complementary to this. His simulations of the collapse of individual galaxies use substantially improved techniques for treating hydrodynamics, dissipation, star formation, feedback, and stellar evolution. As noted above, his results prove for the first time that it is indeed possible to form thin rotationally supported disk systems from the collapse and dissipation of "generic" inhomogeneous and slowly rotating mixtures of dark matter and gas. The standard assumption of earlier work is thus verified, together with a lot of new information on the details of the mechanism. Katz has not yet applied these techniques to initial protogalactic perturbations with the statistical properties predicted by CDM, but will do so shortly. Unfortunately, it will not soon be feasible to combine the programs of Katz and of Carlberg and Couchman, and so to follow galaxy formation in detail within a representative volume of space. Such a synthesis is currently best made by using analytical methods tailored to fit simulation results where these are available.

Carlos Frenk and I are currently involved in a project to carry out such a synthesis. Preliminary results are reported in White (1989), and it does not seem worth repeating them in detail here. Instead, I will limit myself to describing the physical processes which must be included in such a synthesis, the extent to which the models can be checked against numerical experiment, the steps which are needed to obtain predictions that can be compared directly with observation, and the general picture of galaxy formation which seems to be emerging.

The first step is to find a description of the nonlinear clustering of the dark matter in order to specify the abundance, structure and characteristic parameters of the dark halos present at any given time. Most work on this has used the simple theoretical framework proposed by Press and Schechter (1974) to specify halo abundances. Useful formulae for implementing this machinery in the CDM case are given by Cole and Kaiser (1988) and Narayan and White (1988). The internal structure of the halos is usually taken to be that of a singular isothermal sphere, thus adopting the radial structure (and the lack of internal substructure) found in numerical experiments (Frenk *et al.* 1985,1988; Quinn, Salmon and Zurek 1986) but ignoring the substantial ellipticities of the simulated halos. It is important to note that while the resulting predictions for halo abundance have been checked against numerical work for some ranges of parameters (Narayan and White 1988; Efstathiou and Rees 1988; Carlberg and Couchman 1989) their general applicability is still quite uncertain. In addition, this theory does not describe the dynamical evolution of halos, and so gives little clue to the origin of galactic morphological differences and their relation to environment.

The gas within these dark halos can radiate, sink towards the center, and begin to form stars. A simple estimate of the mass of gas that cools in any given halo is given by the mass within the radius at which the cooling time equals the halo's age. However, this leads to a contradiction when the effect is integrated over the history of the clustering hierarchy - all the gas available cools off during the early stages of evolution, leaving no material to make large galaxies or the intergalactic medium in clusters. The most obvious way to avoid this problem is to invoke substantial feedback from star formation which expels most of the gas from small objects (Larson 1974b; White and Rees 1978; Dekel and Silk 1986). Such feedback is certainly expected, given our knowledge of the local interstellar medium, but the exact manner in which it interacts with the cooling gaseous halo surrounding a forming galaxy is obviously highly uncertain. With a plausible model it is possible to get the right total number of stars, but then two complementary problems arise. The first is that the theory still predicts large numbers of stars to form in small units at early times. If these objects survive, they give rise to a galaxy luminosity function containing too many faint systems. Although this may perhaps be cured by merging to form larger objects, excessive merging is a serious danger, since it may lead to the disruption of spiral disks by infalling satellites (Peebles 1988) or to the formation of single supergalaxies, rather than of virialised clusters of distinct galaxies (White and Rees 1978). The simulations of Carlberg and Couchman (1989) are encouraging in this respect, because they show that "galaxies" are indeed more resistant to merging than their halos, apparently confirming the original speculation of White and Rees. Thus with "enough" merging, but not "too much", CDM may give the correct galaxy luminosity function. It will be very difficult for simulations to give a credible evaluation of this possibility.

There are some obvious observational tests of the framework so far. Many galaxies should still be surrounded by accreting gaseous halos, and their star formation activity should heat this gas out to the radius of order 100 kpc where its cooling time equals the Hubble time. On the isothermal sphere model the gas temperature is related to the circular velocity of the halo through,

$$T = 0.12(V_c/200\text{km/s})^2\text{keV}.$$

The emission around all but the largest spirals should be too cool for detection by the Einstein Observatory, but many bright ellipticals should have been visible. The observational situation is reviewed by Fabbiano (1989). Although no gaseous halo has been detected around a spiral, the low predicted temperatures avoid any clear conflict with the straightforward theoretical prediction. Many early-type galaxies do have detected halos, but, except for central cluster galaxies like M 87, the amount of gas involved is significantly less than the expected $\sim 10\%$ of the CDM mass. To accomodate this one must argue that the original gaseous halos of these ellipticals (and perhaps also their CDM halos) were lost through interactions with the environment. This is quite possible but seems contrived. Some positive evidence that star formation activity does heat a large surrounding halo comes from [Mg II] absorption lines in QSO spectra. These systems are typically found to originate at distances of order 50 kpc from bright, actively star-forming galaxies (Bergeron 1988).

Late galaxy formation in CDM models is illustrated both by the simulations of Carlberg and Couchman (1989) and the analytic models of White (1989); most of the observed stellar population in galaxies is predicted to form after $z = 3$, and so may be observable in the optical. To get a definite prediction it is necessary to make assumptions about the

IMF and about the effects of dust. Adopting a standard Solar neighborhood IMF and assuming obscuration to be negligible, White's (1989) model was able to match Tyson's (1988) counts of faint galaxies surprisingly well; the many faint blue objects counted ($\sim 10^5$ gal./squ.degree at B=26) were identified as star-forming regions at redshifts of 1 to 3. This success is interesting since the model parameters were chosen to fit galaxy clustering and the current luminosity density of the Universe, rather than the abundance of galaxies. The model neatly avoids Koo's (1989) observation that, given the density of galaxies seen locally, there is not enough volume in an Einstein-de Sitter universe to fit the number of objects counted by Tyson. In a CDM universe galaxy number is not conserved, and there were more galaxies at $z = 2$ than there are today. As noted above, the merging which has occurred at relatively recent epochs is difficult to model; further work is clearly needed, as is more detailed work on the evolution of the stellar populations (as in Tinsley 1980, Bruzual 1983, and Rocca-Volmerange and Guiderdoni 1987). Only then will it be possible to compare the model in detail with nearby galaxies.

Let me conclude by summarising my current understanding of how real galaxies could have formed in a CDM universe. Their characteristic stellar mass reflects the requirement that gas could cool in the time available from the densities and temperatures which arose during hierarchical clustering. At early times all the gas in small halos cooled as they collapsed; however, only a small fraction turned into stars because of the energy produced by supernovae. Large halos formed relatively late, and only a fraction of their gas cooled, much of it well after collapse. Even in these objects, feedback may have been significant. Galactic angular momentum was created by tidal fields acting on protogalactic perturbations. Observed specific angular momenta can be much smaller than the mean acquired by protogalactic perturbations because nonlinear effects during collapse concentrated angular momentum in the outer regions of halos. When stars formed before an object came to equilibrium, an ellipsoidal system was produced. This process may have resembled an inhomogeneous, quasi-dissipationless collapse, or the merging of preformed galaxies. When gas cooled after equilibration of the halo, it settled into a disk before turning into stars. Often both processes occurred in the same halo. Gas in the center of protogalactic subunits formed stars which lost energy and angular momentum to the dark matter as the subunits merged; thus a bulge was produced. Gas from less dense regions lost less angular momentum, and cooled later to form a disk at larger radii. In protocluster regions massive galaxies formed earlier than in protovoids, but then had to survive in a dynamically active environment. As a result, the violent ellipsoid formation phase produced higher mass objects, but the later disk formation phase was truncated. Many of the spheroids which did manage to form disks had their continuing supply of gas removed; they became S0 galaxies. By contrast, in low density regions massive halos formed later and often still remain undisturbed. Hence spheroids are smaller in such regions and almost always have disks.

Numerical simulations have demonstrated that most of the processes in this story work at least qualitatively. The advantage of the CDM model is that it is specific enough for there to be some hope of testing that they also work quantitatively. The most vulnerable point for the model seems to be its prediction that much of galaxy formation occurs at redshifts which are now observable. These predictions need to be refined, but they reinforce a major conclusion that this author draws from the data discussed above; galaxy formation has finally become a subject for detailed observational study, rather than for purely theoretical speculation.

384

I thank Neal Katz for permission to show some of his results in advance of publication. This work was supported by NASA Astrophysical Theory grant NAGW-763

5. References

Aguilar, L., and Merritt, D. 1989, *Ap. J.*, in press.

Barnes, J. 1988, *Ap. J.*, **331**, 699.

Baron, E., and White, S.D.M. 1987, *Ap. J.*, **322**, 585.

Bergeron, J. 1988, in *High Redshift and Primeval Galaxies*, eds. J. Bergeron, D. Kunth, B. Rocca-Volmerange, and J. Tran Thanh Van, p. 59.

Black, J. H., Chaffee, F. H., and Foltz, C. B. 1987, *Ap. J.*, **312**, 50.

Blumenthal, G. R., Faber, S. M., Primack, J. R., and Rees, M. J. 1984, *Nature*, **311**, 527.

Bruzual A., G. 1983, *Ap. J. (Letters)*, **273**, 105.

Bushouse, H. A., Lamb, S. A., and Werner, M. W. 1988, *Ap. J.*, **335**, 74.

Butcher, H., and Oemler, A. 1978, *Ap. J.*, **219**, 18.

Carlberg, R. G., and Couchman, H. 1986, *Ap. J. (Letters)*, **300**, L1.

Carlberg, R. G., and Couchman, H. 1989, *Ap. J.*, **340**, 47.

Carlberg, R. G., Dawson, P. C., Hsu, T., and VandenBerg, D. A. 1985, *Ap. J.*, **294**, 674.

Cole, S., and Kaiser, N. 1988, *M.N.R.A.S.*, **233**, 637.

Colless, M., Ellis, R. S., and Taylor, K. 1989, in *The Epoch of Galaxy Formation*, eds. C. S. Frenk, R. S. Ellis, T. Shanks, A. F. Heavens, and J. A. Peacock (Kluwer), p. 39.

Cowie, L. L. 1988, in *The Post-Recombination Universe*, eds. N. Kaiser and A. Lasenby (Kluwer), p. 1.

Cowie, L. L. 1989, in *The Epoch of Galaxy Formation*, eds. C. S. Frenk, R. S. Ellis, T. Shanks, A. F. Heavens, and J. A. Peacock (Kluwer), p. 39.

Cowie, L. L., and Lilly, S. J. 1989, *Ap. J. (Letters)*, **336**, L41.

Davis, M., Efstathiou, G., Frenk, C. S., and White, S.D.M. 1985, *Ap. J.*, **292**, 371.

Dekel, A., and Silk, J. 1986, *Ap. J.*, **303**, 39.

Efstathiou, G., and Jones, B.J.T. 1980, *Comments Ap.*, **8**, 165.

Efstathiou, G., and Rees, M. J. 1988, *M.N.R.A.S.*, **230**, 5P.

Fabbiano, G. 1989, *Ann. Rev. Astr. Ap.*, **27**, in press.

Faber, S. M. 1981, *Astrophysical Cosmology, Proceedings of the Vatican Study Week on Cosmology and Fundamental Physics*, eds. M. S. Longair, G. V. Coyne, and H. A. Brück (Citta del Vaticano: Pontifica Academia Scientiarum).

Fabian, A. C., Arnaud, K. A., and Thomas, P. A. 1987, in *Dark Matter in the Universe*, eds. J. Kormendy and G. Knapp (Dordrecht: Reidel), p. 201.

Fabian, A. C., Nulsen, P.E.J., and Canizares, C. R. 1982, *M.N.R.A.S.*, **201**, 933.

Fabian, A. C., Nulsen, P.E.J., and Canizares, C. R. 1984, *Nature*, **310**, 733.

Fall, S. M. 1979, *Nature*, **281**, 200.

Fall, S. M., and Efstathiou, G. 1980, *M.N.R.A.S.*, **193**, 189.

Fall, S. M., Pei, Y., and McMahon, R. G. 1989, *Ap. J. (Letters)*, **341**, L5.

Frenk, C. S., White, S.D.M., Davis, M., and Efstathiou, G. 1988, *Ap. J.*, **327**, 507.

Frenk, C. S., White, S.D.M., Efstathiou, G., and Davis, M. 1985, *Nature*, **317**, 595.

Green, R. F. 1989, in *The Epoch of Galaxy Formation*, eds. C. S. Frenk, R. S. Ellis, T. Shanks, A. F. Heavens, and J. A. Peacock (Kluwer), p. 121.

Gunn, J. E. 1981, *Astrophysical Cosmology, Proceedings of the Vatican Study Week on Cosmology and Fundamental Physics*, eds. M. S. Longair, G. V. Coyne, and H. A. Brück (Citta del Vaticano: Pontifica Academia Scientiarum).

Gunn, J. E. 1989, in *The Epoch of Galaxy Formation*, eds. C. S. Frenk, R. S. Ellis, T. Shanks, A. F. Heavens, and J. A. Peacock (Kluwer), p. 167.

Hamilton, D. 1985, *Ap. J.*, **297**, 371.

Hartmann, L. W., Huchra, J. P., Geller, M. J., O'Brien, P., and Wilson, R. 1988, *Ap. J.*, **326**, 101.

Katz, N. 1989, Ph.D. thesis, Princeton University.

Kennicutt, R. C. 1983, *Ap. J.*, **272**, 54.

Koo, D. 1986, in *Spectral Evolution of Galaxies*, eds. C. Chiosi and A. Renzini (Dordrecht: Reidel), p. 419.

Koo, D. C. 1989, in *The Epoch Of Galaxy Formation*, eds. C. S. Frenk, R. S. Ellis, T. Shanks, A. F. Heavens, and J. A. Peacock (Kluwer), p. 85.

Lake, G., and Carlberg, R. G. 1988a, *A. J.*, **96**, 1581.

Larson, R. B. 1974a, *M.N.R.A.S.*, **166**, 585.

Larson, R. B. 1974b, *M.N.R.A.S.*, **169**, 229.

Larson, R. B., and Tinsley, B. M. 1978, *Ap. J.*, **219**, 46.

Lilly, S. J. 1989, in *The Epoch of Galaxy Formation*, eds. C. S. Frenk, R. S. Ellis, T. Shanks, A. F. Heavens, and J. A. Peacock (Kluwer), p. 71.

Majewski, S. R. 1989, in *The Epoch of Galaxy Formation*, eds. C. S. Frenk, R. S. Ellis, T. Shanks, A. F. Heavens, and J. A. Peacock (Kluwer), p. 101.

McCarthy, P. J., Spinrad, H., Djorgovski, S., Strauss, M. A., van Breugel, W., and Liebert, J. 1987, *Ap. J. (Letters)*, **319**, L39.

McCarthy, P. J., and van Breugel, W. 1989, in *The Epoch of Galaxy Formation*, eds. C. S. Frenk, R. S. Ellis, T. Shanks, A. F. Heavens, and J. A. Peacock (Kluwer), p. 63.

McGlynn, T. 1984, *Ap. J.*, **281**, 13.

Meier, D. 1976, *Ap. J.*, **207**, 343.

Narayan, R., and White, S.D.M. 1988, *M.N.R.A.S.*, **231**, 97P.

Ostriker, J. P. 1980, *Comments Ap.*, **8**, 177.

Partridge, R. B., and Peebles, P.J.E. 1967, *Ap. J.*, **187**, 425.

Peebles, P.J.E. 1988, in *Large Scale Structures of the Universe*, eds. J. Audouze, M. C. Pelletan, and Szalay, A. S. (Kluwer), p. 495.

Pettini, M., Boksenberg, A., and Hunstead, R. W. 1989, in *The Epoch of Galaxy Formation*, eds. C. S. Frenk, R. S. Ellis, T. Shanks, A. F. Heavens, and J. A. Peacock (Kluwer), p. 107.

Press, W. H., and Schechter, P. L. 1974, *Ap. J.*, **187**, 425.

Quinn, P. J., Salmon, J. K., and Zurek, W. H. 1986, *Nature*, **322**, 329.

Rocca-Volmerange, B., and Guiderdoni, B. 1987, *Astr. Ap.*, **186**, 1.

Rubin, V. C., and Coyne, G. V. 1989, editors of *Large-Scale Motions in the Universe*, Princeton University Press.

Sanders, D. B., Soifer, B. T., Elias, J. H., Madore, B. F., Matthews, K., Neugebauer, G., and Scoville, N. Z. 1988, *Ap. J.*, **325**, 74.

Sarazin, C. L. 1987, in *Dark Matter in the Universe*, eds. J. Kormendy and G. Knapp (Dordrecht: Reidel), p. 183.

Scalo, J. M. 1986, *Fund. of Cosmic Phys.*, 11, 1.

Schweizer, F. 1983, in *Internal Kinematics and Dynamics of Galaxies*, eds. E. Athanassoula (Dordrecht: Reidel), p. 319.

Schweizer, F. 1990, in *Dynamics and Interactions of Galaxies*, ed. R. Wielen (Springer), in press.

Soifer, B. T., Sanders, D. B., Neugebauer, G., Danielson, G. E., Lonsdale, C. J., Madore, B. F., and Persson, S. E. 1986, *Ap. J. (Letters)*, **303**, L41.

Spinrad, H. 1989, in *The Epoch of Galaxy Formation*, eds. C. S. Frenk, R. S. Ellis, T. Shanks, A. F. Heavens, and J. A. Peacock (Kluwer), p. 39.

Terlevich, G., Davies, R. L., Faber, S. M., and Burstein, D. 1981, *M.N.R.A.S.*, **196**, 381.

Tinsley, B. M. 1980, *Ap. J.*, **241**, 41.

Toomre, A. 1977, in *The Evolution of Galaxies and Stellar Populations*, ed. B. M. Tinsley and R. B. Larson (New Haven: Yale Univ. Obs.), p. 401.

Toomre, A., and Toomre, J. 1972, *Ap. J.*, **179**, 623.

Tremaine, S. D. 1981, in *Structure and Evolution of Normal Galaxies*, eds. S. M. Fall and D. Lynden-Bell, p. 67.

Tyson, J. A. 1988, *A. J.*, **96**, 1.

van Albada, T. S. 1982, *M.N.R.A.S.*, **201**, 939.

White, S.D.M. 1982, in *The Morphology and Dynamics of Galaxies*, ed. L. Martinet and M. Mayor (Sauverny: Geneva Obs.), p. 289.

White, S.D.M. 1983, in *Internal Kinematics and Dynamics of Galaxies*, ed. E. Athanassoula (Dordrecht: Reidel), p. 337.

White, S.D.M. 1989, in *The Epoch of Galaxy Formation*, eds. C. S. Frenk, R. S. Ellis, T. Shanks, A. F. Heavens, and J. A. Peacock (Kluwer), p. 15.

White, S.D.M., Davis, M., Efstathiou, G., and Frenk, C. S. 1987a, *Nature*, **330**, 451.

White, S.D.M., Frenk, C. S., Davis, M., and Efstathiou, G. 1987b, *Ap. J.*, **313**, 505.

White, S.D.M., and Rees, M. J. 1978, *M.N.R.A.S.*, **183**, 341.

Wolfe, A. M. 1989, in *The Epoch of Galaxy Formation*, eds. C. S. Frenk, R. S. Ellis, T. Shanks, A. F. Heavens, and J. A. Peacock (Kluwer), p. 101.

Gaseous Halos and Disks of Galaxies at Large Redshifts

A. M. Wolfe

University of California, San Diego
Department of Physics

ABSTRACT. QSO absorption systems selected for MgII and damped Lyα lines are likely to be associated with high redshift galaxies. CCD imaging of fields surrounding QSOs with MgII absorption reveal that a luminous galaxy at the MgII redshift is present in almost every case. The impact parameter of the QSO further suggests that galaxies at $z \sim 1$ are surrounded by huge halos of gas. The damped Lyα systems are cold and quiescent disk-like configurations detected at $z \sim 2.5$. The mass per unit comoving volume is similar to that contributed by stellar disks of spiral galaxies. This suggests that the damped systems are the progenitors of galactic disks. The damped systems apparently contain μGauss magnetic fields, which is difficult to understand in the context of a galactic dynamo. There are recent hints that the metallicity of these initial disks may be ~ 10 % solar.

1. Introduction

The aim of this paper is to discuss recent developments concerning the properties of diffuse gas at large redshifts. There is a growing body of evidence which suggests that most of this gas, at least that which is metal enriched, resides in the interstellar media of high-redshift galaxies. In fact many properties of the high-z gas bring to mind the disk-halo configuration of our own Galaxy and of most nearby spirals.

The relevant gas has been detected in the absorption spectra of QSOs. Statistical analysis of the data shows that most of the gas is in redshift systems which are unassociated with the background QSO, and which are randomly distributed along the line of sight (Sargent *et al.* 1980). The complexity of the spectra further indicates that the absorption systems are not drawn from a unique parent populations of clouds. Rather there are at least 4 distinct cloud populations contributing to any line of sight (cf. Lanzetta 1988a). Figure 1 shows a QSO spectrum with absorption lines arising from these populations:

(a) Lyα forest systems: These unidentified redshift systems are selected from the high density of absorption lines blueward of Lyα emission. The lines are undoubtedly Lyα, and the absence of heavy elements may indicate a primordial chemical composition (cf. Sargent *et al.* 1980; but see Tytler 1988). The Lyα forest systems comprise the most numerous cloud population known, with the number of clouds per unit redshift interval $dN/dz = 55$

H. A. Thronson, Jr. and J. M. Shull (eds.), The Interstellar Medium in Galaxies, 387–403.
© 1990 *Kluwer Academic Publishers.*

388

for restframe equivalent widths $W(Ly\alpha) \geq 0.3$ Å and average redshift $\langle z \rangle = 2.4$. The HI column densities $N(HI)$ typically exceed 10^{13} cm^{-2}.

(b) Metal-line systems: These redshift systems are selected from the appearance of 2 prominent resonance doublets: $CIV\lambda\lambda1548, 1551$ and $MgII\lambda\lambda2796, 2803$. The doublets are easiest to detect when they occur redward of $Ly\alpha$ emission where confusion noise from $Ly\alpha$ forest lines is absent. The metal-line systems are far less numerous than the $Ly\alpha$ forest clouds, with $dN/dz \approx 1$ for $W(CIV$ or $MgII) \geq 0.3$ Å at $\langle z \rangle \approx 2$ (Lanzetta, Turnshek, and Wolfe 1987; Sargent, Boksenberg, and Steidel 1988; Sargent, Steidel, and Boksenberg 1988). The ionic column densities typically exceed 10^{13} cm^{-2}.

(c) Lyman limit systems: The Lyman limit systems are found through the appearance of a Lyman-limit discontinuity in the QSO spectrum (Tytler 1982). Thus in common with the $Ly\alpha$ forest clouds, the Lyman limit systems are selected on the basis of HI rather than metal content. However, their incidence along the line of sight, $dN/dz \approx 1.5$ for $N(III) \geq 2\times10^{17}$ cm^{-2} and $z = [0.3, 3.0]$ (Lanzetta 1988b; Sargent, Steidel, and Boksenberg 1989) suggests that they are more closely related to the metal-line systems. In fact arguments based on ionization equilibria (Wolfe 1985) and global statistics (Lanzetta 1988b) support the idea that the detected MgII systems are a subclass of the Lyman-limit population.

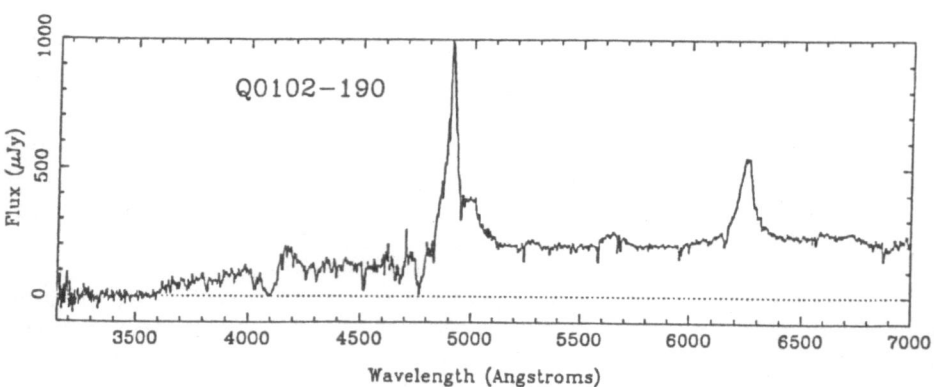

Figure 1. Low resolution ($\Delta\lambda = 5$ Å) spectrum of Q0102-190 (Sargent, Steidel, and Boksenberg 1989) illustrating absorption lines arising in cloud populations 1 → 4. The emission features at 4900 and 6210 Å are $Ly\alpha$ and CIV. Spectral noise blueward of $Ly\alpha$ emission is due to unresolved $Ly\alpha$ forest lines. Decline of continuum blueward of 3650 Å is due to Lyman-limit absorption at $z = 2.94$. MgII doublet absorption at $z = 1.0262$ is present at 5665, 5680 Å . CIV doublet absorption at $z = 2.84$ occurs at 5950, 5951 Å . The strong absorption trough at 4100 Å is due to damped $Ly\alpha$ at $z = 2.37$.

(d) Damped Lyα systems: The selection criterion for these systems is also based on HI content ; namely, N(HI) must be large enough for the Lyα absorption line to be broadened by radiation damping (Wolfe *et al.* 1986). However, in every case studied so far, metal lines with ionization state similar to the MgII systems are found at the damped Lyα redshift. The damped systems occur infrequently, with dN/dz $\approx$ 0.3 for N(HI) $\geq$ 2×10^{20} cm^{-2} at $\langle z \rangle$ = 2.3 (Turnshek *et al.* 1989).

The focus of this paper is on cloud populations associated with one or more components of observable galaxies. In particular the MgII systems will be examined because of direct evidence linking them to galactic halos (Bergeron 1988). The damped Lyα systems will be discussed because of evidence suggesting that they are the progenitors of galactic disks, and that they are physically connected to galactic halos (Wolfe 1988). The Lyα forest clouds will be ignored, since they are probably drawn from an intergalactic population unassociated with galaxies (Sargent *et al.* 1980). The CIV systems and the Lyman-limit systems will not be emphasized because the evidence relating them to galactic halos is not as strong as for MgII.

2. MgII Systems as Gaseous Galatic Halos

The nature of the MgII systems was revealed through two types of studies. In the first type, spectra were obtained for a large sample of QSOs, and a significant number of MgII absorbers were discovered. In some cases the background QSO was bright, and high resolution spectra were acquired to deduce physical conditions within the absorber. In the second type of investigation, specific absorption systems were selected in order to search for emission from the related galaxy.

2.1 PROPERTIES OF MGII SYSTEMS DEDUCED FROM QSO SPECTROSCOPY

Recently a number of groups surveyed the redshift interval z = [0.2, 2.0] for MgII doublet absorption (Tytler *et al.* 1987; Bergeron and Boisse 1984; Lanzetta *et al.* 1987; Sargent, Steidel, and Boksenberg 1988). The surveys produced several important statistics. For example dN/dz $\approx$ 1 at z $\approx$ 1 for lines with W(MgII) $\geq$ 0.3 Å . The implication is that the cross-sectional radius R $\geq$ few$\times$R$_{H_o}$, where R$_{H_o}$ is the Holmberg radius, if the MgII systems are spherical halos with the comoving density of galaxies. Furthermore the functional form of dN/dz implies that the halos evolve in time, since dN/dz $\propto$ $(1+z)^\gamma$ where γ ranges between 1.5 (Sargent, Steidel, and Boksenberg 1988) and 2.4 (Lanzetta, Turnshek, and Wolfe 1987) and no evolution is signified by γ = [0.5, 1.0]. This result must be treated with caution, however, because the inferred γ's differ from unity at less than the 2-σ level of significance.

The spectroscopic studies are also useful for restricting physical properties of the absorbing gas. First, high-resolution observations of a few objects show that MgII lines which are unresolved at low resolution break up into number of distinct and narrow velocity components with fwhm, Δv $\approx$ 25 km s^{-1} (Bergeron *et al.* 1988). As a result the total equivalent width is proportional to the number of components. Secondly Lanzetta *et al.* (1987) compared the strength of absorption by low-ions and high-ions by measuring the equivalent widths of the CIV doublet and CIIλ1334 in $\approx$ 30 MgII systems. A previous comparison between the CIV doublet and CII showed that W(CIV) > W(CII) in redshift systems *selected for CIV absorption* (Wolfe 1983). This argued against a galactic halo origin, because W(CII) > W(CIV) for high latitude gas in the Galaxy (Savage 1988). When the same comparision was carried out for the MgII systems, Lanzetta *et al.* found that in general W(CII)> W(CIV): in some instances CIV was so weak that it would have been missed in most CIV surveys. As a result gas in MgII selected systems more closely resembles gas in the halo of the Galaxy. However, a Kolmogorov-Smirnov test shows that absorption lines from the Galactic halo and the MgII systems are unlikely to be drawn from the same parent distribution (Lanzetta 1988a). The discrepancy is due to strong CII lines (say W(CII) $\geq$ 1 Å) which are present in the MgII systems, but absent in the Galaxy. The difference need not rule out a halo origin for the MgII systems, since the velocity interval sampled by the Galactic line of sight to the LMC may be smaller than that sampled by paths through halos at large redshifts.

Unfortunately a crucial property of the MgII systems, the metal abundance, is highly uncertain. One problem is that the optical depth of MgII gas at the Lyman limit is not very large (Bechtold *et al.* 1984), and as a result neither H^0 nor Mg^+ are the dominant ionization stages of H and Mg, in contrast with Galactic HI regions (Bergeron and Stasinka 1986). Consequently photoionization models, which make use of assumptions about the spectrum of the background ionizing radiation field etc., must be used to predict the ratios, H/H^0 and Mg/Mg^+ which are then combined with the observables, N(MgII) and N(III), to find Mg/H. However, neither column density is accurately known, because both transitions are highly saturated.

2.2 DETECTION OF GALAXIES ASSOCIATED WITH MGII SYSTEMS

To find galaxies associated with MgII systems, candidates are selected from CCD frames of fields surrounding the background QSOs. Because the MgII statistics indicate that QSO sightlines should typically be displaced by 3 — 10 arcsec from the galaxy centers (Bergeron 1988), objects near the QSO are initially chosen as candidates. A spectrum of the candidate is then acquired to search for evidence of the MgII redshift. So far spectra have been accumulated for 13 candidates (Bergeron 1988). In 10 cases emission and/or absorption features are found with redshifts in the interval z = [0.16, 0.79]. The redshifts agree well with the MgII redshift (to within $\approx$ 100 km s^{-1}). Figure 2 shows the CCD image for the field surrounding 4C 55.27, a QSO with MgII absorption at z=0.373 (Miller, Goodrich, and Stephens 1987). Object A, has been identified with a galaxy at the same redshift from the spectrum in Fig. 3. Although the galaxy is too distant to identify, the shape of the spectrum and the strengths of Mg H, Mg Ib, and Hβ absorption are typical of luminous ellipticals (Miller, Goodrich, and Stephens 1987).

The properties of the detected MgII galaxies are somewhat different than predicted. For example the galaxies are more luminous than expected (Bergeron 1988). The average absolute red magnitude, $\langle M(r) \rangle$ = -21.1, for H_0 = 50 km s^{-1} Mpc^{-1}. At $\langle z \rangle$ = 0.4, the average redshift of the galaxies, this corresponds to an intrinsic magnitude in the B band. As a result the majority of the MgII galaxies are more luminous than L. even though galaxies as faint as the LMC could have been detected. On the other hand, standard galaxy statistics predict that most of the absorbing galaxies should have luminosities L $\geq$ 0.3×L$_*$. The implication is that there is a cutoff to the luminosity function of MgII galaxies at the faint end.

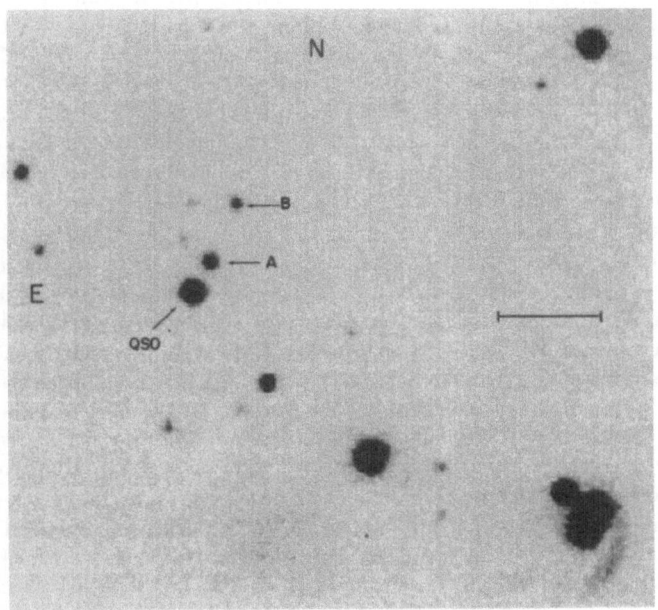

Figure 2. Red CCD image of field surrounding 4C 55.27. Object A, the MgII galaxy, is 5 arcsec from the QSO. Length of horizontal line is 15 arcsec (Miller, Goodrich, and Stephens 1987).

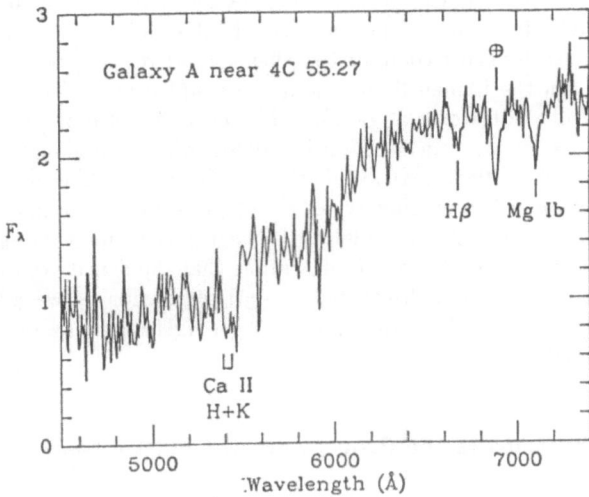

Figure 3. Spectrum of object A. Labeled absorption features identify this as a galaxy at $z=0.373$ (Miller, Goodrich, and Stephens 1987).

The spectra of the MgII galaxies are different from the spectra exhibited by nearby galaxies. In every case there is a strong blue continuum down to the U band. This is in contrast with normal galaxies which are generally faint at short wavelengths. Moreover, while 50 % of the galaxies show stellar absorption lines, in 80 % of the objects [OII] $\lambda 3727$ emission is present. While this differs from nearby galaxies, which exhibit stellar absorption more often than [OII] emission, recent observations of field galaxies at z $\approx$ 0.5 show that they too exhibit a high incidence of emission lines (Dressler 1989). Field galaxies are relevant here, because there is no evidence for clustering in the MgII galaxy sample.

The average impact parameter separating the QSO line-of-sight from the galaxy center, $\langle b \rangle = (2.3 \pm 0.8) \times R_{Ho}$, where $\sigma_b = 0.8 \times R_{Ho}$ is computed from the scatter of impact parameters reported by Bergeron (1988). By comparison, the impact parameter predicted at $\langle z \rangle = 0.4$, and for W(MgII) $\geq$ 4 Å , the smallest MgII equivalent width in the corresponding redshift systems, is $3.5 \times R_{Ho}$, where the W(MgII) $\propto$ b^{-2} correlation discussed below was used to extrapolate dN/dz to W(MgII) = 0.4 Å . Thus the observed impact parameter is within $\sim$ 1-σ of the galaxy cross-sectional radius predicted by combining the Schecter luminosity function of galaxies and the Holmberg relation between radius and luminosity, $R_{Ho} \propto L^{0.4}$, with dN/dz observed for MgII absorption systems (Bergeron 1988). Agreement this good is remarkable in such a speculative subject. The implication is that the majority of galaxies within the redshift interval of the survey are embedded in huge envelopes of gas. Furthermore the projected 2 dimensional covering factor, f, of the gas must be near unity, or else the MgII cross-sectional radius, $R \gg 3.5 \times R_{Ho}$, which is difficult to understand. The presence of multiple cloud components along any line of sight is additional evidence for f $\approx$ 1, since only single clouds would be detected if the covering factor were low.

The large covering factor means that that the cloud collision time, $t_{coll} \approx R/(f\sigma_{MgII})$, cannot exceed the cloud crossing time, $t_{cross} = R/\sigma_{MgII}$, where σ_{MgII} is the velocity dispersion of MgII clouds within a single redshift system (Lanzetta 1989). But since $t_{cross} \approx 10^9$ yr, the collision time is less than the time interval over which MgII systems have been observed; i.e., 10^{10} yr. The collision time is the characteristic time for dissipation of the kinetic energy of the MgII cloud system. Since the free fall time is $\sim t_{cross}$, the MgII halos will collapse on time-scales short compared to their apparent age. On the other hand the absence of evolution in the Lyman limit population coupled with the evidence for evolution in the MgII clouds indicates that gaseous halos do not evolve in size, but rather that individual clouds are decreasing in MgII optical depth while remaining optically thick at the Lyman limit (Lanzetta 1988b). To avoid collapse, the mass or kinetic energy of a single MgII halo must be replenished on time scales $\leq 10^9$ yr. Perhaps the mass influx is in the form of a cooling flow similar to those inferred for nearby rich clusters of galaxies (Fabian 1989). In any case Lanzetta and Bowen (1989) argue that the density of the MgII gas falls off with radius as $\rho \propto r^{-3}$. They discovered an anti-correlation between W(MgII) and b of the form W(MgII) $\propto$ b^{-2}, where they interpret W(MgII) as being proportional to the surface density of MgII clouds.

3. Damped Lyα Systems as the Progenitors of Galactic Disks

Damped Lyα systems are a population of high-redshift gas layers which are the likely progenitors of the disks of spiral galaxies; i.e, disks of *stars* as well as gas. Although the damped systems differ physically from the MgII absorbers, the two coexist in many instances.

3.1 WHAT IS A DAMPED Lyα SYSTEM ?

The operational definition of a damped Lyα system is an HI layer that produces a Lyα absorption line broadened by radiation damping. In principle an III layer with velocity dispersion, σ_{HI}, is optically thick in the damping wings of the Lyα profile function if N(HI) $\geq 10^{18}$cm$^{-2}(\sigma_{HI}/10$ km s$^{-1})^2$. In practice it is difficult to distinguish damped profiles from profiles in which the principal broadening mechanism is the Doppler effect, e.g., profiles with narrow velocity components distributed across large velocity intervals, unless the damping wings are detected over many resolution elements. For spectra acquired with $\Delta\lambda \geq 1$ Å and signal-to-noise ratios exceeding 10, N(HI) in the damped system must exceed $\sim 10^{20}$ cm^{-2}. The HI threshold used in existing surveys for damped Lyα systems is N(HI) $> 2\times10^{20}$ cm^{-2} which coincides with the HI isophote out to which most spiral galaxies have been investigated (cf. Bosma 1981; Wolfe et $al.$ 1986; Sargent, Steidel, and Boksenberg 1989a).

The spectroscopic signatures of a damped Lyα system are (i) a broad Lyα absorption trough ($\Delta v_{HI} \geq 1800$ km s^{-1}) with a black core and divergent wings characteristic of radiation damping, (ii) narrow metal lines ($\Delta v_{metals} \leq 100$ km s^{-1}) with velocity centroids v_{metals} that differ from v_{HI}, the velocity centroid of the Lyα line, by $|v_{HI}-v_{metals}| < \Delta v_{metals}$ (Turnshek et $al.$ 1989), (iii) curve of growth consistency of the higher order Lyman lines, and (iv) if the background QSO is a bright radio source, the appearance of 21 cm absorption at the Lyα redshift.

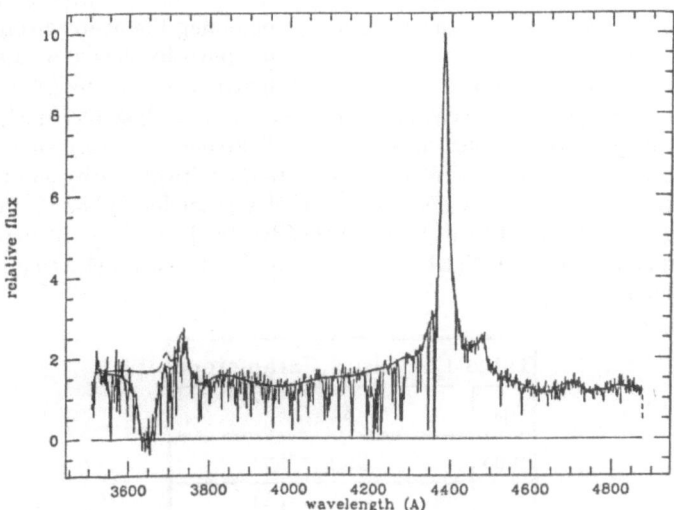

Figure 4. Intermediate resolution ($\Delta\lambda = 1$ Å) spectrum of GC 1215+33 acquired at MMT. Damped Lyα absorption line at z=2.006 is strong feature at 3650 Å . Narrow feature near 4587 Å is SiIIλ 1526 at the same redshift (Wolfe, Turnshek, and Lanzetta 1990).

Identification criteria (i) and (ii) are the ones used most often to confirm damped Lyα redshift systems. The significance of (i) is obvious. Criterion (ii) was adopted to filter out multi-component systems in which $\Delta v_{HI} \propto \Delta v_{metals}$ (cf. Bechtold, Green, and York 1987); i.e., absorption systems lacking components with N(HI) $\geq 2 \times 10^{20}$ cm^{-2}. The problem with criterion (iii) is that the higher order Lyman lines are weak and thus difficult to extract from the Lyα forest. Criterion (iv), the detection of 21 cm absorption, is a sufficient but not necessary condition for the existence of a damped Lyα system: there are many reasons why a damped system towards a radio-bright QSO need not create a detectable 21 cm absorption line (cf. Wolfe 1980). Criteria (i) and (ii) are illustrated by the damped Lyα system in fig. (4).

3.2 GLOBAL PROPERTIES OF DAMPED Lyα SYSTEMS

The most suprising fact to emerge from the various surveys is that absorption systems with damped Lyα are *not* uncommon. That is, absorbers with N(III) $\geq 2 \times 10^{20}$ cm^{-2} are detected 5 to 8 times more frequently along the line of sight than expected for galactic disks. More specfically, the statistic derived from the 16 damped systems detected along the redshift path, $z_{path} = 55$, of the Lick survey is dN/dz = 0.29±0.07 within an average redshift interval z = [1.8, 2.6] for N(III) $\geq 2 \times 10^{20}$ cm^{-2} (Wolfe *et al.* 1986; Turnshek *et al.* 1989). By contrast the number of disks expected down to the same HI contour is no more than 3. A straightfoward, though not unique, interpretation of the data is that the comoving density of the damped systems equals that of spiral galaxies, and as result their radii, R$> 3.5 \times$R$_{Ho}$ (cf. Wolfe 1988). In other words the damped systems may be giant III disks. The incidence of damped systems along the line of sight is also compatible with an inflated population of gas rich dwarf galaxies (Tyson 1988).

The mass content of the damped systems can be estimated without assumptions about the size or the geometry of the absorbers. Combining the average column density $\langle N(III) \rangle = 10^{21}$ cm^{-2} with f, the fraction of the sky occupied by the damped systems, one finds that the mass density averaged along the past lightcone is given by $\langle \rho \rangle = f \langle \Sigma \rangle / \langle c \Delta t \rangle$. Here $\langle \Sigma \rangle$ is the mass surface density, including He, corresponding to $\langle N(III) \rangle$, and $\langle \Delta t \rangle$ is the average time interval sampled along the past lightcone. The corresponding density parameter, Ω_{damp}, is deduced by normalizing $\langle \rho \rangle$ with the critical density and transforming the result to the present epoch by adopting a suitable cosmology. The results, shown in table 1, are slightly different than in Wolfe (1988) due to the subsequent detection of an additional damped system with N(HI) $\geq 2 \times 10^{20}$ cm^{-2} (cf. Turnshek *et al.* 1989). Note that h=H$_0$/100 km s^{-1} Mpc^{-1}.

Table 1. Density Parameters	
q_0	Ω_{damp}
0.05	1.6×10^{-3}h^{-1}
0.5	2.7×10^{-3}h^{-1}

Ω_{damp} is comparable to $\Omega_{lumdisk}$, the density parameter due to luminous matter, i.e., stars, in the disks of spiral galaxies, where $\Omega_{lumdisk} = 2 \times 10^{-3}$h^{-1} (cf. Wolfe 1988). The implication is that both the damped Lyα systems and the stars residing in current galactic disks are made up of the same baryons. This suggests that the damped Lyα systems consist of protogalactic matter which evolves into the disks of spiral galaxies through star formation and other processes. A VLBI study of the 21 cm absorption line in one damped systems reveals that in this case the protogalactic matter is in the form of a giant HI disk (see §3.3).

Another global statistic of importance is the redshift distribution. If the damped systems evolve into galactic disks, their cross-sections must decrease with time. As a result dN/dz should increase more rapidly with redshift than predicted for invariant disks that are embedded in a Friedman cosmological model. Furthermore dN/dz should decrease with z above some redshift, if there is a well defined epoch of disk formation. The present sample of damped systems is too small to detect either evolutionary trend. Because the size of this sample should triple with the completion of a new survey for damped systems, the detection of evolution in the damped population may soon be within reach.

3.3 HYDROGEN CONTENT OF DAMPED Lyα SYSTEMS

To deduce physical conditions in the absorbing gas, accurate spectra have been obtained for most of the known damped Lyα systems. N(HI) has been inferred by fitting Voigt damping profiles to Lyα lines similar to the one shown in fig. 4. The column densities range from the threshold, N(HI) $= 2 \times 10^{20}$ cm^{-2} , to 5×10^{21} cm^{-2} with a mean, $\langle$N(HI)$\rangle$ $= 10^{21}$ cm^{-2} . The Lyman limit optical depth of this neutral gas, $\tau_{LL} \sim 10^4$. The large τ_{LL} means that the ionization state of each element will be completely dominated by low ions such as H^0, Fe$^+$, C$^+$, Si$^+$, etc. As a result the kinetic temperature of the gas, $T_k \leq 10^3$ K. By contrast, in the majority of the CIV and MgII systems $\tau_{LL} << 10^4$. Thus in most metal-line systems, hydrogen is mainly ionized, the ionization state of the metals is highly uncertain, and $T_k \approx 10^4$ K.

Damped Lyα systems with 21 cm absorption can be probed with highly sensitive techniques not available at optical wavelengths (cf. Briggs 1988). For example radio autocorrelation spectrometers afford a spectral resolution of $\Delta v \geq 1$ km s^{-1}. Spectra acquired at this resolution give a detailed picture of the velocity fields, because 21 cm lines are rarely, if ever, saturated. So far 21 cm absorption has been detected in 3 damped Lyα systems. The velocity profiles have been fitted with gaussians characterized by $\sigma \sim 10$ km s^{-1}. The HI in one system is distributed into 2 components separated by 15 km s^{-1}, where the dispersion of each component is 5 km s^{-1} (Wolfe et al. 1985). Therefore the HI component of the damped Lyα systems is a quiescent gas. It is also a cold gas. Comparison between N(HI) deduced from the damped profiles and τ_{21}, the 21 cm optical depth, results in limits on the temperature of the gas, since $\tau_{21} \propto$ N(HI)/T_s, where T_s, the spin temperature, is coupled to the kinetic temperature, T_k by particle collisions. The best limit determined so far is $T_s \leq 600$ K for the z=2.04 damped system towards PKS 0458-02. The same system also appears to be a large disk-like structure (Briggs et al. 1989). VLBI observations near 467 MHz, the frequency of the redshifted 21 cm line, show that the background radio source consists of a core and a jet which extends for 2 arcsec. VLBI observations in the line reveal that the line depth and velocity structure of the fringe visibility spectrum is identical to the Arecibo single-dish spectrum. As a result the absorbing HI must extend across most of the extended jet; i.e., across 8h^{-1} kpc. However, the extent of the gas along the line of sight is $<< 8h^{-1}$ kpc. The strong self-gravity due to its $\Sigma = 60\,M_O\,\mathrm{pc}^{-2}$ surface density and the low velocity dispersion imply a hydrostatic scale height less than 500 pc for the absorbing HI. As a result the absorber is a disk-like structure with a radius large compared to the radii of dwarf galaxies.

The properties of the HI in the damped systems bring to mind the cold and quiescent HI disk of our Galaxy. Because H_2 molecules have been detected along Galactic sightlines with $N(HI) \geq 2 \times 10^{20}\,\mathrm{cm}^{-2}$, it is reasonable to extend the Galactic analogy and search for H_2 in the damped systems. Sensitive searches for Lyman and Werner band absorption in the Lyα forest region of the spectra of 2 QSOs with damped systems at z=2.3 and z=2.8 have yielded null results (Black, Chaffee, and Foltz 1987; Lanzetta, Wolfe, and Turnshek 1989): The detection of H_2 absorption in a third damped system (Foltz $et\ al.$ 1987) will not be discussed here, since the proximity of the absorption and emission redshifts suggests that the physical state of the absorber may be influenced by the QSO. The absence of molecular transitions leads to conservative upper limits of $2 \times \mathrm{n}(H_2)/\mathrm{n}(HI) < 10^{-5}$ to the molecular content of the gas in the 2 damped systems. The absence of molecules indicates that the physical state of the damped systems differs markedly from the HI disk of the Galaxy. Galactic sightlines through similar HI columns, $N(HI) \approx 10^{21}\,\mathrm{cm}^{-2}$, encounter molecular fractions $2 \times \mathrm{n}(H_2)/\mathrm{n}(HI) > 10^{-1}$ (Scoville and Sanders 1987). The difference is probably related to the low dust content of the damped systems (see §3.5).

3.4 METAL LINE ABSORPTION IN DAMPED Lyα SYSTEMS

Heavy element absorption is a generic feature of the damped Lyα systems. Low-ion transitions such as CIIλ1335, SiIIλ1206, FeIIλ1608, etc. are invariably found at the redshift predicted by the damped Lyα trough. In fact the coincidence of narrow heavy-element lines at the redshift centroid of the broad Lyα trough is a principal signature of a damped Lyα redshift system. A significant fraction of the damped systems also exhibit high-ion transitions such as CIV$\lambda\lambda$1550 and SiIV$\lambda\lambda$1390 (cf. Turnshek $et\ al.$ 1989). Thus the metal lines appear to form in mixed ionization sytems; i.e., in gas similar to the clouds detected in MgII surveys. Quantitative comparisons between the two populations confirm this impression (Lanzetta 1988a). Figure (5) shows the frequency distributions of FeIIλ2382 equivalent widths for samples of damped Lyα systems and MgII systems. The curves are maximum likelihood fits of the function n(W)=exp(-W(FeII)/W_*)/W_* to each distribution. The results are $W_* = 0.66 \pm 0.13$ Å for the MgII sample and $W_* = 0.96 \pm 0.3$ Å for the damped sample. In other words there is no evidence for a significant difference between W_* characterizing each sample (Lanzetta 1988a). The ionization state of each sample provides an additional statistical test. CIV and CII equivalent widths were measured for redshift systems in each sample and plotted against each other. A two-dimensional version of the Kolmogorov-Smirnov test shows that the two distributions are statistically compatible (Lanzetta 1988).

The above discussion demonstrates that the metal properties of the MgII and damped Lyα systems are quite similar despite the vast difference in HI content. This apparent paradox is a consequence of the multi-component nature of the gas which gives rise to the metal lines. High resolution observations of the two populations show that in both cases the metal lines are caused by narrow velocity components distributed across a velocity interval, Δv_{tot}. The stronger lines have many components distributed over large Δv_{tot}, while the weaker lines have fewer components distributed across small Δv_{tot}. The HI content of each velocity component in a MgII system is characterized by $\tau_{LL} \sim 1$. In that event Lyα, like the stronger metal lines, will be saturated, and the equivalent widths of both transitions will be proportional to the number of components. As a result $W(Ly\alpha) \propto W(MgII)$. On the other hand, 21 cm observations show that the bulk of the HI in the damped Lyα systems is confined to a velocity interval $\Delta v_{HI} \ll \Delta v_{metals}$. Therefore what distinguishes a damped system from a MgII systems is the added presence of one or two opaque components in which $\tau_{LL} \sim 10^4$. Because $\Delta v_{HI} \ll \Delta v_{tot}$, the opaque components have no effect on the total equivalent width of the metal lines. However, absorption in the opaque components dominates $W(Ly\alpha)$, because the line width caused by radiation damping in a single opaque component exceeds the velocity interval absorbed by all the components.

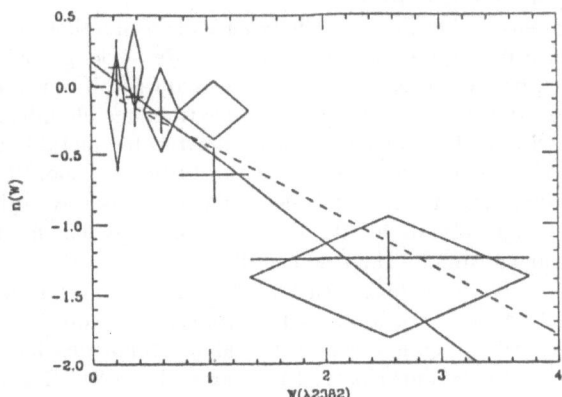

Figure 5. Equivalent width distributions of FeIIλ2382 rest-frame equivalent widths. Crosses correspond to MgII redshift systems and polygons to damped Lyα systems. Maximum likelihood fit of exponentials to MgII and damped points are given by solid line and dashed line (Lanzetta 1988a).

The similarities between the MgII and damped Lyα systems suggest that HI disks which create the damped Lyα lines are embedded in turbulent gaseous halos which determine the strengths of low-ion metal lines. Lanzetta et $al.$ (1987) suggested a model in which the weak MgII lines form in the outer halo where the line of sight encounters a small number of clouds, while the strong MgII lines arise in the inner halo where large numbers of clouds are encountered. One consequence of this model is that the impact parameter through the HI disk should increase with decreasing $W(MgII)$. Because $N(HI)$ should also be smaller at large radii, $N(HI)$ should be correlated with $W(MgII)$. However, no such correlation is observed in the damped sample. To the contrary, some of the highest $N(HI)$ in the sample are associated with some of the lowest $W(MgII)$, $W(CII)$, $W(SiII)$, etc (Turnshek et $al.$ 1989). This suggests that in damped Lyα systems with weak low-ion lines, both Lyα and the low-ion lines form in the disk: the low-ion lines are weak, because they are saturated transitions arising in gas with low velocity dispersion. In principle the halo could be optically thin to low-ion transitions because it is highly ionized. That explanation is ruled out by the absence of CIV or SiIV absorption in damped systems with weak low-ion lines (cf. Turnshek et $al.$ 1989). A more likely explanation is that gaseous halos do not envelop the HI disks which give rise to damped systems with weak metal lines.

3.5 METAL ABUNDANCES OF DAMPED Lyα SYSTEMS

The metal abundances of the damped Lyα systems are fundamental in understanding the chemical evolution of galaxies. Abundance studies of stars in the disk of the Galaxy indicate an age-metallicity relationship that requires an initial disk metallicity, [Fe/H] > -1 (Twarog 1980; Carlberg *et al.* 1987). Since damped Lyα systems are likely to be the progenitors of galactic disks and because they have been detected out to z = 3.5 (cf. Sargent, Steidel, and Boksenberg 1989), the metallicities of the highest redshift damped systems are potential indicators of the initial metallicity of galactic disks. Furthermore, studies of metal abundances over a wide range of damped Lyα redshifts could yield the history of disk chemical evolution directly.

The damped systems are ideal sites for abundance determinations because, in contrast to most metal-line absorption systems, the ionization state of the gas is precisely known. On the other hand the presence of turbulent gas associated with the quiescent disk prevents one from using simple curve of growth techniques to extract the abundances of prominent elements. We have seen how the equivalent widths of saturated metal line transitions will be dominated by the low column density gas in the turbulent halo. Thus it makes no sense to compare column densities of metals determined in this way with the HI column densities arising in the high column-density disk. Investigations that fail to recognize this dichotomy inevitably underestimate the true metallicity of the damped systems.

Three techniques have been suggested to get around this problem. The first is to work with unsaturated lines created by weak transitions: the equivalent widths of metal lines arising in ion X_i will be dominated by the high-column density disk component, because $W(X_i) \propto N(X_i)$ on the linear part of the curve of growth. The second idea is to carry out high resolution echelle observations ($\Delta\lambda \approx 0.1$ Å) which can distinguish between the various velocity components in the metal lines, and then compare them with 21 cm absorption profiles to determine which of the velocity components is the quiescent disk. The third idea is to concentrate on damped Lyα systems with weak metal lines, because the absorption likely arises in pure disk systems.

The first technique was used to obtain reliable column densities of Zn, Cr, and Ni in 3 damped Lyα systems towards 2 bright QSOs. Meyer and York (1987) observed PKS 0528-250 and detected Ni$^+$ at z=2.81 and Cr$^+$ at z=2.41. Pettini and Hunstead (1989) detected Zn$^+$ and Cr$^+$ at z=2.31 towards PHL 957. Each of these ions dominates the ionization state of its species in the highly opaque HI region where damped Lyα absorption occurs. The detection of Zn is particularly valuable, because it is undepleted in the ISM of the Galaxy and probably reflects the true metallicity of the gas. For this reason and because the z=2.81 system in PKS 0528-250 may be associated with the background QSO, the following discussion will focus on PHL 957.

Pettini and Hunstead (1989) acquired a spectrum of PHL 957 with intermediate resolution ($\Delta\lambda = 1$ Å) at Palomar. Although all absorption features were unresolved, the high signal-to-noise ratio of ≈ 90 per resolution element was adequate for detecting weak ($W \approx 0.04$ Å) ZnII and CrII lines. The observed ratio of CrIIλ2055.596 to CrIIλ2065.501 indicates that the CrII lines are unsaturated. While the blending of ZnIIλ2062.03 with CrIIλ2061.58 prohibits a direct measure of the ZnII saturation, the authors show that the velocity dispersion of the gas must be unacceptably low for the ZnII transitions to be optically thick. Pettini and Hunstead then derived the Zn and Cr abundances for the z=2.31 system by comparing the unsaturated Zn$^+$ and Cr$^+$ column densities with the

N(HI)=2.5×10^{21} cm^{-2}derived from the damped Lyα line. The results are shown in table 2. If one assumes that Zn/H is a genuine measure of iron peak metallicity, then the Zn abundance may be factors of 10 to 20 below solar. The table also shows that the Cr abundance is depleted by a further factor of 4 relative to Zn. Because Cr is depleted by a factor of 100 relative to Zn in the ISM, the implication is that (D/G), the dust-to-gas ratio in the damped system, is far below that in the ISM. This is consistent with recent statistical tests, which show that $(D/G) \leq 0.1\times(D/G)_{Galaxy}$ in the damped population (Fall, Pei, and McMahon 1989).

Table 2. Metal Abundances at z=2.3

Element X	X/H	$(X/H)_\odot$	Delpletion
Zn	$(1.6 \pm 0.3)\times10^{-9}$	38×10^{-9}	24
Cr	$(5.9 \pm 0.9)\times10^{-9}$	580×10^{-9}	100

The PHL 957 observations provide the first reliable hint that metal abundances in protogalactic disks are ~ 10 % solar. To confirm this result it is necessary to obtain column densities of more abundant metals such as C, O, Fe, and Si. Since the absorption lines created by these elements will be optically thick in the low-column density halo, the halo component contributes signficantly to the line profile. As a result curve of growth analyses of unresolved profiles will be unreliable for abundance studies of the disk. Bechtold and Wolfe (1989) recently obtained high-resolution echelle spectra of PKS 1157+014 in order to distinguish the halo from the disk components in the damped Lyα system at z=1.94. The redshift of the disk component is determined precisely by the 21 cm absorption line detected at the damped Lyα redshift (Wolfe, Briggs, and Jauncey 1981). Analysis of the spectra is currently underway. Turnshek et $al.$ (1989) used unresolved line profiles to determine abundances in 2 damped Lyα systems (towards Q1337+113 and Q1151+068) with velocity dispersions $\sigma \approx 10$ km s^{-1}; i.e., damped systems without turbulent halo gas. Curve of growth studies of both systems yield metal abundances between 10^{-3} to 1 times solar. Because the high abundances were determined from the single transitions observed in C$^+$ and O^0, they need to be confirmed by weaker transitions of the same elements that may provide additional constraints on the curve of growth.

3.5 MAGNETIC FIELDS IN PROTOGALACTIC DISKS

Surveys for Faraday rotation in the polarized radio flux of radio-bright QSOs (Welter, Perry, and Kronberg 1984) give strong evidence that microGauss magnetic fields are a generic property of the damped Lyα systems, but are rarely present in typical metal line systems (Wolfe 1988). The evidence is statistical. Out of the 116 QSOs observed, Faraday rotation was detected in 23 (at the 3-σ level of statistical significance). Because each of the 116 QSO sightlines are likely to encounter 1 or more metal systems, the probability for Faraday rotation is less than 20 % in a typical metal line system. Thus it was suprising to find Faraday rotation in all 5 QSOs in the sample known to exhibit damped Lyα absorption (Wolfe 1988).

The argument can be put more precisely. Assume as a null hypothesis, that the

Faraday properties of damped systems and typical metal-line systems are the same. To check this hypothesis we require the probability per line of sight for detecting a Faraday event. The first step in such a computation is to construct the sampling function g(RRM). The sampling function is the number of sightlines along which the residual rotation measure, RRM, i.e., the rotation measure arising in magneto-ionic media outside the Galaxy (cf. Welter, Perry, and Kronberg 1984), can be detected at greater than 3-σ significance. The probability per line of sight, p_F, for a Faraday detection is the statistically weighted sum of $1/g(RRM_i)$ where RRM_i is the RRM of the ith QSO with statistically significant Faraday rotation. The result is that $p_F = 0.179 \pm 0.005$. Since the probability for detecting Faraday rotation in 5 QSOs, $P_5 = (p_F)^5$, $P_5 = 1.85 \times 10^{-4}$; i.e., the null hypothesis can be confidently rejected.

As a result sightlines towards QSOs with damped Lyα absorption have a higher probability of encountering magneto-ionic media than any other sightline. The implication is that damped Lyα systems typically contain **B** fields which are coupled to the HI by ions and electrons in the partially ionized gas. The magnitude of $\langle B_\parallel \rangle$, the average line-of-sight component of the magnetic field, can be estimated by noting that $RRM \propto N_e \times \langle B_\parallel \rangle$ where N_e is the electron column density. Although N_e is unknown, it is safe to assume that $n_e/n(H)$, the ionized fraction of the gas, is unlikely to exceed 0.1 when $\tau_{LL} > 10^4$. Consequently $\langle B_\parallel \rangle$ exceeds $\sim 10^{-6}$ Gauss (Wolfe 1988).

The detection of μGauss magnetic fields in damped Lyα systems with z $\approx$ 2 is difficult to understand. The dynamo theory for amplification of the Galactic magnetic field predicts that primordial fields with $|\mathbf{B}| \sim 10^{-20}$ Gauss are amplified by differential winding of the rotating disk, and that the e folding time for amplification is one rotation period (Parker 1979). The problem is that the age of a disk at z $\approx$ 2 is not very large in units of rotation period, P, especially if the disk radius R $> 3.5 \times R_{H_o}$ whereupon P $\sim 10^9$ yr. Since Faraday rotation has been detected towards QSOs exhibiting damped Lyα absorption out to z=2.04 (Wolfe 1988), dynamo amplification of primordial fields is unlikely to generate the observed rotation measures. The dilemma will become even more severe if Faraday rotation is detected in a damped Lyα system with z $\approx$ 3 that was recently discovered towards a radio-bright QSO in a new survey for damped Lyα absorption systems.

The Faraday statistics raise a further question; namely, what is the nature of the 18 rotation measures detected towards QSOs without damped Lyα absorption ? The above analysis shows that a magneto-ionic medium that gives rise to detectable rotation measures is not a generic property of galactic halos: otherwise Faraday rotation would have been detected more often than observed. Three explanations come to mind. First, it is possible that some rotation measures arise in low-z damped Lyα systems. At z $<$ 1.8, Lyα is unobservable, because it is shifted blueward of the atmospheric cutoff. The low-z damped systems could be detected in 21 cm absorption, but a systematic search for radio absorption lines in the sample QSOs has not been attempted. Second, other rotation measures could arise in the outer regions of high redshift disks where N(HI) is less than the 2×10^{20} cm^{-2} threshold required for entry into the damped Lyα population. The RRM could remain large despite the drop in HI column density, because the fractional ionization of the gas is expected to increase as its density decreases in the face of ionizing background radiation fields. Third, the remaining rotation measures could be due to the integrated effect of multiple halos along the line of sight. Suppose the rotation measure of a single halo is finite, but below the detection threshold. The magnitude of dN/dz for MgII absorbers assures that an occasional QSO line of sight will intercept a large number of halos. As a result the net RRM along the line of sight could exceed the threshold for detection. In this case the magnitude of RRM would be correlated with the number of MgII absorbers.

4. Conclusions

This paper focused on specific populations of QSO absorption systems which are sensitive probes of high-redshift galaxies: the MgII systems and the damped Lyα systems. The MgII absorption systems are key objects at $z \leq 1$ (cf. Bergeron 1988). Their association with galaxies with luminosities $L \approx L_*$ demonstrates that a luminous galaxy is a necessary condition for MgII absorption. The large impact parameters and the high incidence of detection further suggests that most ordinary galaxies at $z \leq 1$ are luminous objects which are surrounded by huge halos of gas, as predicted by Bahcall and Spitzer (1969). An alternative idea, that dwarf galaxies are the absorption site (cf. York *et al.* 1986), seems implausible. The large MgII covering factor requires a minimum of $\sim (R/R_{dwarf})^2$, i.e., 10^3, dwarfs per L_* galaxy, whereas Bergeron's survey fails to find any satellite galaxies down to LMC luminosities. The large covering factor further indicates that the MgII cloud system is collision dominated, and as a result will collapse on time scales of $\sim 10^9$ yr unless the gas is replenished with energy and/or mass. The evidence for recent star formation in the MgII systems may be an indicator of energy deposition in the halo via supernova shocks, etc. While the association of MgII systems with galaxies is established, the nature of CIV selected systems with W(CIV) > W(CII) is unclear. Future observations with HST will show whether or not low-redshift CIV systems also reside in galactic halos.

Damped Lyα systems have been studied extensively at redshifts between 1.8 and 3, although individual objects may have been detected with redshifts higher than 4 and as low as 0.4 (the latter through the detection of 21 cm absorption). Physical conditions within the damped systems imply that they are cold HI layers that closely resemble the gaseous disk of the Galaxy. In one case a VLBI study of the associated 21 cm absorption line indicates that the damped system is a disk-like configuration larger than an ordinary galaxy. The implication is that the damped systems are unlikely to be gas-rich dwarfs. The metal content of the damped systems also brings to mind galactic disks, because metals in a low ionization state are detected in every case. There is further evidence that some of the metals are located in associated MgII-type halos. There are recent indications that the metallicity of the the damped systems is about 10 % solar, an important factor for models of chemical evolution. A further similarity is in magnetic field strength which appears to approach the Galactic value, although that is difficult to understand at high redshifts in the context of the galactic dynamo model. The damped systems differ from local disks in that they are devoid of molecules and contain little, if any, dust. Moreover the gas content of the damped systems exceeds that in nearby spirals. In fact the mass per unit comoving volume of the damped systems is similar to that in the stellar disks of ordinary galaxies, suggesting that most of the gas has been converted into stars over the past Hubble time. A crucial piece of evidence that is missing in this picture is any direct evidence for the stellar content of the damped galaxies. Surveys for emission from galaxies associated with damped systems are currently in progress.

I wish to thank Ken Lanzetta for valuable comments. This research was partially supported by NSF grant AST8715070.

5. References

Bechtold, J., Green, R.F., Weymann, R.J., Schmidt, M., Estabrook,F.B., Sherman, R.D., Wahlquist, H.D., and Heckman, T.M. 1984, *Ap.J.*, **281**, 76.

Bechtold, J., and Wolfe, A. M. 1989, in preparation.

Bechtold, J., Green, R. F., and York, D. G. 1987, *Ap.J.*, **312**, 50.

Bergeron, J. 1988, in *IAU Symposium No. 130, Evolution of Large Scale Structure in the Universe*, ed. J. Audouze and A. Szaly (Dordrecht: Reidel), p. 343.

Bergeron, J., and Boisse, P. 1984, *Astr.Ap.*, **133**, 134.

Bergeron, J., Boulade, O., Kunth, D., Boksenberg, A., and Vigroux, L. 1988, *Astr.Ap.*, **191**, 1.

Bergeron, J., and Stasinska, G. 1986, *Astr.Ap.*, **169**, 1.

Black, J. H, Chaffee, F. H., and Foltz, C. B. 1987, *Ap.J.*, **312**, 50.

Bosma, A. 1981, *A.J.*, **86**, 1825.

Briggs, F. H. 1988, in *QSO Absorption Lines: Probing the Universe*, ed. J. C. Blades, D. A. Turnshek, and C. A. Norman (Cambridge: University Press), p. 275.

Briggs, F. H., Wolfe, A. M., Liszt, H. S., Davis, M. M., and Turner, K. C. 1989, *Ap.J.*, **341**, 650.

Carlberg, R. G., Dawson, P. C., Hsu, T., and VandenBerg, D. A. 1985, *Ap.J.*, **294**, 674.

Dressler, A. 1989, private communication.

Fabian, A. 1989, this volume.

Fall, S. M, Pei, Y. C., and McMahon, R. 1989, *Ap.J.Lett.*, **341**, L5.

Foltz, C. B., Black, J. H., and Chaffe, F. H. 1988, *Ap.J.*, **324**, 267.

Lanzetta, K. M. 1988a, unpublished Ph.D. thesis, University of Pittsburgh.

Lanzetta, K. M, 1988b, *Ap.J*, **322**, 96.

Lanzetta, K. M. 1989, private communication.

Lanzetta, K. M., and Bowen, D. 1989, private communication.

Lanzetta, K. M., Turnshek, D. A., and Wolfe, A. M. 1987, *Ap.J*, **322**, 739.

Lanzetta, K. M., Wolfe, A. M., and Turnshek, D. A. 1989, *Ap. J.*, **344**, 277.

Meyer, D., and York, D. G. 1987, *Ap.J.Lett.*, **319**, L45.

Miller, J. S., Goodrich, R. W., and Stephens, S. A. 1987, *A.J.*, **94**, 633.

Parker, E. N. 1979, in *Cosmical Magnetic Fields* (Oxford: Clarendon Press), p. 616.

Pettini, M., and Hunstead, R. W. 1989, accepted for publication in the Ap.J.

Sargent, W. L. W., Young, P. J., Boksenberg, A., and Tytler, D. 1980, *Ap.J.Suppl.*, **42**, 42.

Sargent, W. L. W., Boksenberg, A., and Steidel, C. C. 1988, *Ap.J. Suppl.*, **68**, 539.

Sargent, W. L. W., Steidel, C. C., and Boksenberg, A. 1988 *Ap.J.*, **334**, 22.

Sargent, W. L. W., Steidel, C. C., and Boksenberg, A. 1989 *Ap.J. Suppl.*, **69**, 703.

Savage, B. D. 1988, in *QSO Absorption Lines: Probing the Universe*, ed. J. C. Blades, D. A. Turnshek, and C. A. Norman (Cambridge: University Press), p. 195.

Scoville, N. Z., and Sanders, D. B. 1987, in *Interstellar Processes*, ed. D. J. Hollenbach and H. A. Thronson (Dordrecht: Reidel), p. 21.

Turnshek, D. A., Wolfe, A. M., Lanzetta, K. M., Briggs, F. H., Cohen, R. D., Foltz, C. B., Smith, H. E., and Wilkes, B.J. 1989, *Ap.J.*, Sept. 15 issue.

Twarog, B. A. 1980, *Ap.J.*, **242**, 242.

Tyson, N. D. 1988, *Ap.J.Lett*, **329**, L57.

Tytler, D. 1982, *Nature*, **298**, 427.

Tytler, D. 1988, in *QSO Absorption Lines: Probing the Universe*, ed. J. C. Blades, D. A. Turnshek, and C. A. Norman (Cambridge: University Press), p. 179.

Tytler, D., Boksenberg, A., Sargent, W. L. W., Young, P. J., and Kunth, D. 1987, *Ap.J.Suppl*, **64**, 667.

Welter, G. L., Perry, J. J., and Kronberg, P.P. 1984, *Ap.J.*, **279**, 19.

Wolfe, A. M. 1980, *Phys.Scripta*, **21**, 744.

Wolfe, A. M. 1983, *Ap.J.Lett.*, **268**, L1.

Wolfe, A. M. 1985, in *Gaseous Halos of Galaxies*, eds. J. N. Bregman and F. J. Lockman (NRAO: Charlottesville), p. 259.

Wolfe, A. M. 1988, in *QSO Absorption Lines: Probing the Universe*, eds. J. C. Blades, D. A. Turnshek, and C. A. Norman (Cambridge: University Press), p. 296.

Wolfe, A. M., Briggs, F. H., and Jauncey, D. L. 1981, *Ap.J.*, **248**, 460.

Wolfe, A. M., Briggs, F. H., Turnshek, D. A., Davis, M. M., Smith, H. E., and Cohen, R. D.1985, *Ap.J.Lett.*, **294**, L67.

Wolfe, A. M., Turnshek, D. A., Smith, H. E., and Cohen, R. D. 1986, *Ap.J.Suppl*, **61**, 249.

Wolfe, A. M., Turnshek, D. A., and Lanzetta, K. M. 1990, in preparation.

York, D. G., Dopita, M., Green, R. F., and Bechtold, J. 1986, *Ap.J.*, **311**, 610.

Large Scale Star Formation and the Interstellar Medium

Robert C. Kennicutt, Jr.
Steward Observatory
University of Arizona
Tucson, AZ 85721

ABSTRACT. The development of empirical techniques for measuring star formation rates in galaxies has made it possible to characterize the basic star formation properties of the Hubble sequence, and to explore the physical process which regulate the star formation on large scales. I review the observational techniques which are currently used to measure star formation rates, and what these observations tell us about the global star formation properties of galaxies. I then discuss observational efforts to understand the dependence of the star formation rate on the physical properties of the interstellar medium.

1. Introduction

In the last decade the study of star formation in external galaxies has blossomed from an area pursued by a handful of optical observers to a major subfield encompassing virtually every wavelength regime. This widespead interest can be traced to the development of new observational techniques for measuring star formation rates (SFRs) in galaxies, and to relevance of such data for a broad spectrum of astrophysical problems, ranging from the study of star formation and the interstellar medium in our own Galaxy to the evolution of galaxies and quasars.

This paper will concentrate on observations of SFRs in galaxies, and the relationships between the global SFR and the properties of the interstellar medium. The review by Michael Dopita in this volume will emphasize the theoretical aspects of these problems. I will begin here with a progress report on the observational techniques used to measure global SFRs in galaxies, and a discussion of the reliability of these data. I will then briefly summarize what these observations tell us about the large scale star formation properties of galaxies along the Hubble sequence. Finally I will review in some detail what is known about the relationships between the SFR and the physical properties of the interstellar gas. Throughout I will emphasize the properties of normal galaxies, and only briefly touch on the active field of starbursts, since the latter is covered elsewhere in this book.

Several conferences on star formation in galaxies have been held recently, and the published proceedings of these meetings offer an excellent introduction to the field. I particularly recommend *Star Formation in Galaxies* (Lonsdale Persson 1987), *Galactic*

H. A. Thronson, Jr. and J. M. Shull (eds.), The Interstellar Medium in Galaxies, 405–435.

and Extragalactic Star Formation (Pudritz and Fich 1988), and *Starbursts and Galaxy Evolution* (Thuan, Montmerle, and Van 1987).

2. Observational Techniques

Ten years ago the only method for estimating the total SFR in a galaxy was from modelling of its integrated colors, but since then several more direct techniques have been developed. Below I describe each of the methods (roughly in order of their current usage), and evaluate their relative advantages and limitations. I will then discuss the efforts that have been made to test and compare the various SFR scales, and will conclude with an assessment of the overall reliability of current SFRs.

2.1 LYMAN CONTINUUM PHOTON COUNTING

This is currently the most widely applied technique, especially for applications where spatially resolved data are required. For an ionization bounded H II region the number of photons emitted in a recombination line or in the nebular continuum is proportional to the number of Lyman continuum ionizing photons emitted by the exciting stars, and since the aggregate photoionization of giant H II regions and galaxies usually requires several hundred to thousand OB stars, the total nebular luminosity can be used to directly infer the massive SFR. The most commonly used tracer is the Hα emission line, but observations of the Brackett and Paschen lines or the thermal radio continuum are also used.

2.1.1 *H-alpha.* Atlases of Hα photographs of galaxies by Hodge, Courtes, and others provided the principal source of data on star formation in galaxies for two decades, and the development of low-noise CCD detectors has made Hα imagery a routine and widely applied technique for mapping the massive star formation in galaxies. Photometric surveys of integrated Hα fluxes of galaxies provided the first comprehensive set of data on their total SFRs (e.g., Kennicutt and Kent 1983, Kennicutt 1983, Gallagher, Hunter, and Tutukov 1984).

The main advantages of Hα as a star formation tracer are its sensitivity— several percent of the ionizing luminosity of a galaxy is re-emitted in this single line— and the relative ease with which observations can be obtained. Nearby galaxies can easily be mapped at arcsecond resolution with 1m-class telescopes, and detection of the line in distant galaxies is mainly limited by the redshifting of the line out of the optical window, rather than by sensitivity considerations. The primary limitation of the technique is extinction, which can average from a few tenths of a magnitude in irregular galaxies such as the Magellanic Clouds (Caplan and Deharveng 1986), to 1–3 magnitudes in normal spirals (Israel and Kennicutt 1980, Kennicutt 1983, Kaufman *et al.* 1987, van der Hulst *et al.* 1988), and much more in highly obscured infrared-luminous galaxies. When interpreting SFRs based on Hα measurements it is important to take this extinction into account.

The Hα flux provides a direct estimate of the total mass of ionizing stars ($M >$ $10 - 15 M_{\odot}$), a population which comprises approximately 10–20% of the total stellar mass for a conventional initial mass function (e.g., Scalo 1986), and consequently the bulk of the total SFR must be extrapolated or measured using other techniques. This renders the SFRs derived from Hα or the other Lyman photon tracers very sensitive to both the slope and the upper mass limit of the initial mass function (IMF). Uncertainty in the form and variation of the IMF is easily the largest source of systematic error in SFRs derived from this technique, or any other currently available technique for that matter. This problem will be addressed explicitly in ¶2.6.

2.1.2 *Thermal Radio Continuum.* Measurements of the integrated thermal radio continuum emission of galaxies have been made by several workers (e.g., Gioia, Gergorini, and Klein 1982; Israel and van der Hulst 1983, Berkehuijsen 1983, Klein 1986, Klein and Gräve 1986, Duric, Bourneuf, and Gregory 1988), and the same method has been applied to individual H II regions by a several authors. The intensity of the thermal radio emission scales directly with the Balmer line flux, with only a weak temperature dependence, so accurate radio measurements can provide the same information as $H\alpha$, without the extinction problems. In fact comparisons of thermal radio and Balmer fluxes of H II regions and galaxies provide the most accurate means of calibrating the $H\alpha$ extinction corrections.

The main drawback of this technique is the contamination of the free-free emission by nonthermal synchrotron emission, which is usually the dominant component at centimeter wavelengths. Extracting the thermal component requires precise flux measurements at several frequencies, and the typical uncertainties in the net fluxes and SFRs range from of order 10% in irregulars to 30–100% in spirals, where the nonthermal contamination is most severe (Israel and van der Hulst 1983). Nevertheless these data currently provide the largest body of extinction-free data on SFRs in galaxies, especially for irregular galaxies.

2.1.3 *Near-Infrared Recombination Lines.* The extinction problem can also be avoided by observing the near-infrared hydrogen recombination lines, and measurements of the Brackett and Paschen lines have proven to be especially valuable for studying the highly obscured star forming regions in many starburst galaxies (Telesco 1988). The weakness of the lines, typically 1–10% of $H\alpha$, has limited their application thus far to unusually high surface brightness regions, such as nearby giant H II regions (e.g., Joy and Lester 1988, Skillman and Israel 1988) and starburst regions (e.g., DePoy 1987, Turner, Ho, and Beck 1987; Kawara, Nishida, and Phillips 1989), but the rapidly improving performance of large-format IR detector arrays should change this situation over the next few years. Global measurements of the integrated fluxes of galaxies in these lines would be especially valuable for deriving extinction-corrected SFRs, and such a program is being undertaken by T. Greene and E. Young at the University of Arizona.

2.2 ULTRAVIOLET CONTINUUM FLUXES

Similar information on the massive SFR can be obtained by directly observing the ultraviolet continnum longward of the Lyman break. Integrated fluxes of nearby galaxies in the 1500-2500 Å region are available from the OAO-2 and ANS satellites (Code and Welch 1982, Wesselius *et al.* 1982), and subsequent surveys in the 1500-3200 Å range have been conducted by groups at Marseille using balloon-borne telescopes (Donas *et al.* 1987), and at Goddard using sounding rocket flights (Smith and Cornett 1982, Stecher *et al.* 1982, Bohlin *et al.* 1983; Hill, Bohlin, and Stecher 1984; Smith, Cornett, and Hill 1987). The IUE satellite has also been used to study the SFRs in selected galaxies (e.g., Huchra *et al.* 1983, Lamb *et al.* 1985, Rocca-Volmerange and Guiderdoni 1987; Fanelli, O'Connell, and Thuan 1988). The anticipated launches of the Hubble Space Telescope and the Ultraviolet

Imaging Telescope on the ASTRO platform should add considerably to these data, and make available for the first time spatially resolved maps of the UV emission in a diverse sample of galaxies.

Compared to the Lyman photon counting techniques this method has several advantages; all of the massive star formation is measured, not just the fraction occurring in H II regions, and the flux at these longer wavelengths traces a somewhat broader range of stellar masses. It is also the technique which is most easily extended to high redshifts. The principle limitation of the technique is the severe extinction at UV wavelengths. Simple single-screen reddening and extinction models, which sometimes may be adequate at optical wavelengths, cannot be trusted in the UV, mainly because the extinction across the disks is highly variable, and the integrated UV fluxes will be strongly weighted to the least obscurred patches. At present it is necessary to model the extinction in order to derive reliable SFRs (e.g., Mochkovitch and Rocca-Volmerange 1984, Donas *et al.* 1987; Buat, Donas, and Deharveng 1987), but when spatially resolved UV data become available it may become possible to measure these corrections directly. When the extinction is taken into account, the UV fluxes yield SFRs which are in good agreement with those obtained from Hα measurements (see ¶2.6).

2.3 FAR INFRARED CONTINUUM

Considerable effort is currently being directed at calibrating the far infrared (FIR) thermal dust emission in galaxies (20–400 μm) as a quantitative SFR tracer. The utility of the FIR emission for this purpose was demonstrated in spectacular fashion by the early results from the IRAS survey, which revealed that the FIR component often dominates the bolometric luminosity of galaxies, especially in the most luminous starbursting systems. There now exists an extensive literature on the FIR properties of galaxies, which I can only briefly cover here, but several comprehensive reviews of the subject are available, including Soifer, Houck, and Neugebauer (1987), Telesco (1988), and Rowan-Robinson's paper in this volume.

In principle the FIR emission offers several distinct advantages over the nebular and UV star formation tracers discussed above. The absorption cross section of the dust grains is high over a broad range in wavelengths through the UV and visible, and hence the dust emission traces a much larger fraction of the stellar mass spectrum. The FIR is clearly the wavelength of choice in galaxies where most of the radiation from young stars is absorbed by dust, especially if the optical depth of the dust is sufficiently high that the grains compete effectively with the gas for ionizing photons. In such cases even the IR emission lines and free-free radio luminosities will be affected by the extinction, and the FIR fluxes offer the only reliable means of measuring the total SFR. This situation often characterizes compact H II regions in our Galaxy, and it probably applies in many starburst galaxies, as well as in the nuclear environments of many normal spirals.

Unfortunately the extraction of quantitative SFRs from broadband FIR fluxes has proven to be far more problematic than was initially anticipated. It is now known that the integrated FIR emission of galaxies arises from a combination of hot dust locally heated by young stars, and a cooler "cirrus" component heated by the general interstellar radiation field. These models show that in normal galaxies the dust is heated by stars with a large range of masses and ages, and that the fraction of FIR emission which is directly tied to recent star formation can vary from nearly 100% in starbursting regions down to a few percent or less in early-type galaxies (e.g., Lonsdale and Helou 1987, Walterbos and Schwering 1987, Buat and Deharveng 1988, Salzer and MacAlpine 1988; Bothun, Lonsdale, and Rice 1989, Thronson *et al.* 1989*a*). The realization that the total FIR luminosities of galaxies are *not* necessarily proportional to their SFRs has been reflected only gradually in the literature, and one should be wary of earlier papers in which the FIR data were overinterpreted.

If the heating of the dust can be modelled with sufficient accuracy it should be possible to extract the FIR component due to recent star formation, and use it to measure the SFR. The studies cited above indicate that this can be done in cases where the dust heating is dominated by young stars, such as starburst galaxies, but in most normal galaxies, where young stars may be responsible for only a small part of the heating, the derived SFRs are very sensitive to errors in the IRAS fluxes, and to the assumptions built into the radiative transfer models. Two approaches are currently being taken to address this problem. One is to combine the IRAS data with independent measurements of the star formation and gas distributions, in an effort to empirically constrain the FIR emission models. Factors which need to be taken into account include the variations in the stellar populations which contribute to the radiation field (both the age mix and the spatial distribution), and the mix of dust between the diffuse clouds and dense, cold molecular regions. Another approach is to suppplement the IRAS 12–100 μm data with submillimeter observations of the cold dust (Eales, Wynn-Williams, and Duncan 1989, Thronson *et al.* 1989*b*), or higher resolution observations in the 20–300 μm region (Thronson *et al.* 1988, 1989*a*), in order to impose additional constraints on the models. I refer the reader to the reviews listed above for more details.

The extraction of reliable SFRs from the IRAS data has proven to be challenging problem, but the potential payoff from such an effort is enormous. Several thousand galaxies were detected in this all-sky survey, and the database is free of many of the selection biases which influence the existing optical and UV surveys.

2.4 RESOLVED STARS

The most direct possible approach to measuring the SFR in a galaxy would be to resolve the individual stars, and use stellar photometry and spectroscopy to derive the SFRs and IMFs directly. Photometric surveys, most notably by Freedman (1985), have provided detailed data on the luminosity functions and spatial distributions of massive stars in the nearest galaxies, but as discussed by Massey (1985), spectral types for complete samples are required if accurate absolute SFRs are to be determined.

Such a survey is currently being undertaken for the Magellanic Clouds by Massey, Garmany, Parker, and DeGioia-Eastwood (Massey *et al.* 1989*a*, *b*). This group has supplemented existing photometric and spectroscopic surveys of the field OB stars in the LMC and SMC with new UBV photometry for most of the stellar associations and young clusters in the galaxies, including those located inside H II regions, where previous surveys were very incomplete. Spectra are then obtained for candidate stars revealed by the color magnitude diagrams, resulting in a complete mass-limited sample of stars in those regions. This enormous undertaking will provide the most detailed data available on the star formation properties and IMF in a galaxy, and will provide a fundamental calibration and zeropoint for the other techniques discussed here.

2.5 BROADBAND VISIBLE COLORS AND MAGNITUDES

For many galaxies the only data available are broadband (usually UBV) photometry, and a question frequently asked is whether these data can be used to estimate a quick-and-dirty SFR. Models of the integrated colors of galaxies (e.g., Roberts 1963, Searle, Sargent, and Bagnuolo 1973, Huchra 1977, Larson and Tinsley 1978, Tinsley and Danly 1980, Rocca-Volmerange, Lequeux, and Marcherat-Joubert 1981, Mochkovitch and Rocca-Volmerange 1984) provided much of our current understanding of the evolutionary properties of disks, as well as the first quantitative estimates of global SFRs in galaxies. Although the errors in the derived SFRs are much higher than those derived from Hα or UV fluxes, the mean rates are very similar (Kennicutt 1983).

Larson and Tinsley (1978) and Tinsley and Danly (1980) have published models giving the SFR per unit disk luminosity as a function of the disk color, and these can be used to estimate approximate SFRs for galaxies with published broadband photometry. The SFR-color relations are very steep— a change of a tenth of a magnitude in B–V color may correspond to a difference of as much as 1–2 orders of magnitude in the SFR— so the SFRs derived in this way should only be regarded as rough estimates. For the same reasons the SFRs are subject to large systematic errors due to reddening, bulge contamination, nuclear emission, etc., and consequently this method should be avoided when quantitative SFRs are needed. When combined with the other star formation tracers described above, however, these data can provide very powerful constraints on the average IMF in galaxies, as discussed below.

It recently has become fashionable to use the integrated blue luminosity of a galaxy alone as an measure of the SFR. This alarmingly frequent practice is *horribly* inaccurate, and it usually leads to misleading conclusions regarding the star formation properties of the galaxies in question. The SFR per unit blue luminosity in spiral and irregular galaxies ranges over 3 orders of magnitude (¶2), and even among galaxies of the same Hubble type the ratio varies by factors of 10–100. Consequently the blue luminosity of a galaxy provides little meaningful constraint on its SFR.

2.6 ACCURACY OF GLOBAL STAR FORMATION RATES

2.6.1 *Comparisons of Different Techniques.* The availability of several independent methods for measuring SFRs in galaxies is now making it possible to obtain external checks on the accuracy of the SFR scales. Most of the comparisons to date show gratifying consistency, at least for the massive SFRs.

Buat, Donas, and Deharveng (1987) have compared SFRs derived from measurements of the UV continuum at 2000 Å with the Hα-derived SFRs from Kennicutt (1983), and find good agreement when a consistent IMF and extinction corrections are applied. These are also consistent with SFRs derived from modelling of the optical and near-IR colors of galaxies, implying that the integrated energy distributions of disk galaxies from the Lyman continuum to the near infrared can be understood with a common set of SFRs and an IMF with approximately a Salpeter slope between 1–100 $M_\odot$ (Kennicutt 1983).

Massey *et al.* (1989*a, b*) have directly tested the Hα-derived massive SFRs in two giant H II regions in the LMC and SMC, by comparing the ionizing fluxes predicted from their Hα fluxes with their actual stellar contents. In one case the predicted and observed ionizing fluxes agree to within the errors, while in the other the Hα flux underestimates the actual OB stellar content by 40%. A larger sample is needed to evaluate the overall reliability of the Lyman photon based SFR scale, but the verification of the SFRs within their quoted errors so far is gratifying nevertheless.

Several comparisons have been made of SFRs derived from Hα and FIR emission in different types of galaxies. Hunter *et al.* (1986), Thronson *et al.* (1987, 1988), and Belfort, Mochkovitch, and Dennefeld (1987) combined IRAS and/or KAO observations of blue, active star forming galaxies with single component dust models to infer their SFRs, and found that these are generally consistent with those derived from extinction-corrected Hα fluxes. This suggests that dusty, optically-hidden star forming regions are not a significant contributor to the total SFR in these galaxies, and it confirms the utility of the FIR flux as a quantitative star formation tracer in cases where the UV radiation field is dominated by young stars. Lonsdale and Helou (1987) reached similar conclusions using a more sophisticated two-component FIR model, but they also found that in normal spirals the FIR fluxes are much higher than expected from the Hα emission. They offer several possible explanations for the excess infrared emission, including high extinction of the Hα emission, trapping of ionizing photons within the H II regions, or the presence of a luminous FIR component which is not taken into account in their model.

2.6.2 *The Initial Mass Function.* An obvious weakness in all of these comparisons is that they only test the reliability of the SFRs derived for the massive end of the stellar population, at best for stars above 5–10 $M_\odot$. Unfortunately most of the techniques available for measuring the SFR, including the nebular emission lines and the UV and FIR continua, are heavily weighted to the most massive stars. Up to 80–90% of the total stellar mass is not reflected in these tracers, and hence determining a total SFR from these data requires an uncomfortably large extrapolation of the IMF.

Fortunately the visible continuum fluxes and colors of galaxies provide constraints on the star formation and SFR at lower masses, and greatly reduce these systematic errors. In most disks the optical continuum is dominated by intermediate mass main sequence stars ($\sim$5–10 $M_\odot$) and disk giants ($\sim$0.7–3 $M_\odot$), and combining the continuum fluxes with the Hα or UV fluxes allows one to constrain the IMF slope between approximately 1–30 $M_\odot$, and derive the total SFR down to approximately 1 $M_\odot$ (Kennicutt 1983). This represents approximately 60% of the total star formation for a Scalo (1986) IMF, and hence reduces the error in the total SFR from factors of several to ± 50% or less.

This conclusion only applies to the disk-averaged properties of normal spirals, however. The modelling procedure described above cannot be applied to individual star forming regions (their continua are dominated by massive stars), and consequently very little is known about the systematic changes in the IMF as functions of metallicity, galaxy type, or star formation environment. There is indirect evidence for a systematic change in the IMF with abundance (Shields and Tinsley 1976, Vílchez and Pagel 1988), IMFs enriched in massive stars in starburst galaxies (e.g., Rieke *et al.* 1980; Gehrz, Sramek, and Weedman 1983; Augarde and Lequeux 1985, Kennicutt *et al.* 1987), and IMFs depleted in massive stars in cooling flows (see Sarazin's review), but these results are highly model dependent and are not yet conclusive. Several observational programs directed at constraining the variation of the IMF are under way, including direct measurements of the stellar luminosity and mass functions in star clusters in the Galaxy and nearby galaxies (e.g., Mateo 1988, Massey 1989*a, b*), and measurements of the integrated spectra of star forming regions (e.g., Villiafond 1988; Kennicutt *et al.* , in preparation).

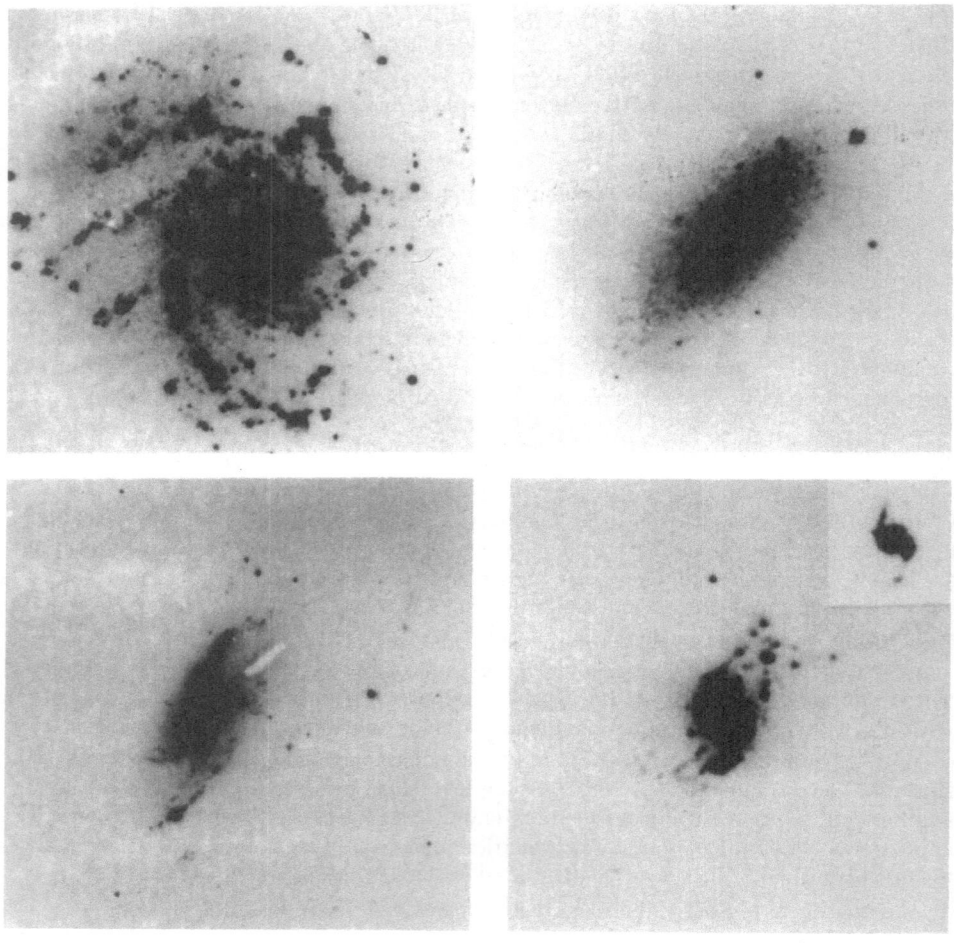

Fig. 1. Hα photographs of 4 nearby spiral galaxies, illustrating the diversity of star formation properties. The panels are reproduced to the same linear scale. Top left: NGC 628 (Sc); Top right: NGC 2841 (Sb); Lower left: NGC 4258 (Sb); Lower right: NGC 3310 (Sbc).

3. Results: Global Star Formation Properties of Normal Galaxies

3.1 INTEGRATED STAR FORMATION RATES

Total SFRs have been measured for a few hundred galaxies using one or more of the techniques described in the previous section, and these have been used to define the basic star formation properties of the Hubble sequence. The most extensive compilations are based on $H\alpha$ measurements, including large surveys of spirals by Kennicutt (1983) and Kennicutt and Kent (1983), and irregulars by Hunter, Gallagher, and Rautenkranz (1982), Gallagher, Hunter, and Tutukov (1984), Hunter and Gallagher (1985), and Gallagher and Hunter (1989).

The availability of CCD detectors for $H\alpha$ observations and the IRAS FIR data has made it possible to extend these surveys to galaxies covering the entire range of Hubble types and luminosities. Pogge and Eskridge (1987) have mapped the $H\alpha$ emission in a sample of S0 galaxies, and analyses of the IRAS emission of lenticulars also show strong hints of star formation in many cases (Dressel 1988, Bally and Thronson 1989; Dressel, O'Connell, and Telesco 1989, Thronson et al. 1989c, d). Several deep $H\alpha$ surveys of these galaxies are currently under way. It is also possible now to study the star formation properties of the faintest dwarf galaxies (e.g., Hunter and Gallagher 1985, Klein and Gräve 1986, Thronson and Telesco 1986; Hodge, Lee, and Kennicutt 1989a, b; Salzer, MacAlpine, and Boroson 1989). These galaxies offer the opportunity to study the star formation processes in interstellar environments which are very different from those in spirals.

As an illustration of the diversity of SFRs in galaxies, Figure 1 shows $H\alpha$ photographs of 4 spirals in the survey of Kennicutt (1988). The panels have been reproduced to the same linear scale, so that the properties of the H II region populations can be directly compared. The Hubble types of these galaxies range from early Sb (NGC 2841) to Sc (NGC 628). The lower right panel illustrates an example of a nearby starburst galaxy, NGC 3310.

A useful parameter for characterizing the diversity in star formation properties is the total SFR, integrated over all stellar masses, and and normalized to a fixed galaxy luminosity, for the purposes of this discussion $M_B = -21$. Expressed in these terms the total SFRs range from about 0.1–1 $M_\odot yr^{-1}$ in S0/a–Sa galaxies to 1–100 $M_\odot yr^{-1}$ in Sc–Irr galaxies (Kennicutt 1983, Caldwell et al. 1990). The mean rates are strongly correlated with Hubble type, but vary by 1–2 orders of magnitude within each type. Both of these trends are illustrated in Figure 2, which shows the distribution of integrated $H\alpha$ equivalent widths in the original survey of Kennicutt and Kent (1983). The equivalent width, defined as the flux of the emission line divided by the continuum intensity at 6563 Å, provides a convenient luminosity-normalized index of the relative SFR in the galaxies. SFRs derived using the other tracers show a similar range and diversity in properties.

Spatially resolved $H\alpha$ observations show that this thousandfold increase in integrated SFR per unit area or disk luminosity along the Hubble sequence is due to a combination of an increase in the number of star forming regions per unit area, and an increase in the characteristic masses and luminosities of the individual H II regions (Kennicutt, Edgar, and Hodge 1989). Both of these trends are evident in Figure 1, and they are shown quantitatively in Figure 3. It is assumed that the changes in H II region properties reflect proportional changes in the total SFRs in the star forming regions, but it is possible that other effects, such as changes in extinction or the IMF, may also be important.

414

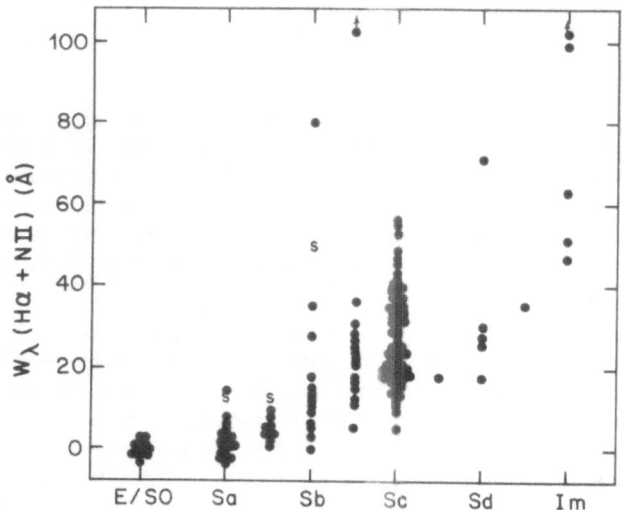

Fig. 2. Distribution of integrated Hα emission line equivalent widths of galaxies in the survey of Kennicutt and Kent (1983).

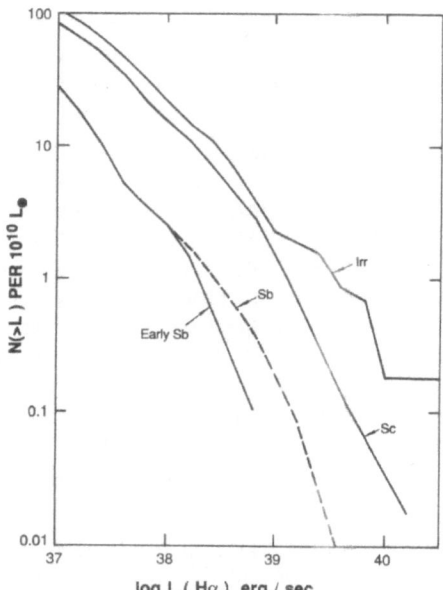

Fig. 3. Comparison of integral luminosity functions of H II regions, grouped by Hubble type, from Kennicutt *et al.* (1989).

In addition to the trends in luminosity and mass of the star forming regions, there are significant differences in the structure of the star forming regions in different environments (e.g., Kennicutt 1984, Hodge *et al.* 1989b). If these large changes in H II region sizes and morphologies are at all representative of the conditions in the atomic and molecular components, they would imply that the existence of major changes in the structure of the ISM with galaxy type and luminosity.

3.2 FACTORS INFLUENCING THE STAR FORMATION RATE

What physical processes are responsible for the large dispersion in global SFRs in galaxies? Several factors appear to contribute. The amount of interstellar gas is one of the primary parameters, of course, and the role of the ISM in regulating the SFR will be discussed in detail in the next section. Other galaxy properties which have been investigated include environmental effects, such as close companions and interactions with an intercluster medium, and the presence of bars and spiral arms.

Of these the most unambiguous influence is tidal interactions. Several studies have convincingly demonstrated that interactions can induce large bursts of star formation (e.g., Larson and Tinsley 1978; Lonsdale, Persson, and Mathews 1984, Bushouse 1987, Kennicutt *et al.* 1987; Bushouse, Lamb, and Werner 1988), especially in the nuclear regions (e.g., Joseph *et al.* 1984, Cutri and McAlary 1985, Keel *et al.* 1985, Bushouse 1987). The range in responses to the interactions is large— some galaxies do not appear to be affected at all— but in extreme cases bursts of more than 100 times the previous SFR are observed. The cluster environment also appears to influence the global star formation properties of galaxies in at least some cases, and this topic is addressed in detail by Kenney elswhere in this conference.

The role of spiral structure and bars in the large scale star formation is more controversial. Elmegreen and Elmegreen (1986), McCall and Schmidt (1986), and Elmegreen (1988) have argued that the total SFRs in well developed "grand-design" spiral galaxies are essentially the same as in galaxies with patchy spiral patterns. Romanishin (1985) found a statistically significant SFR in galaxies with strong spiral patterns, but the enhancement was only ~ 30%. There are several possible interpretations of this result. It may be that spiral density waves merely reorganize star formation that would otherwise have taken place elsewhere in the disk, as favored by Elmegreen (1988). However there is accumulating evidence for direct density wave triggering of star formation in nearby spirals such as M51 (Vogel, Kulkarni, and Scoville 1988, Tilanus and Allen 1989, Lord and Young 1989). Kennicutt (1989) has argued that the importance of density wave triggering may depend critically on the stability properties of the ambient ISM (see ¶4.5), with the arms playing only a minor role in gas-rich galaxies, but serving as the primary star formation triggering mechanism in many early-type, gas-poor galaxies.

The role of bars in triggering global star formation is even more ambiguous. Barred galaxies appear to exhibit excess 10–25 μm FIR emission relative to non-barred galaxies (Hawarden *et al.* 1986, Devereux 1987), and this has been interpreted by some as evidence for enhanced global star formation. Devereux (1987) has emphasized, however, that much of the IR enhancement occurs in the nuclear regions of early-type spirals (late-type barred galaxies show no significant enhancements), and he attributes the excess emission to a combination of nonthermal nuclear activity and star formation. Pompea and Rieke (1989) made near-IR maps of a sample of nearby IR-luminous galaxies, and found bars in only a fraction of the cases. It seems likely that bars do lead to enhancements of star formation in some galaxies, but the nature and extent of the effect is unclear.

3.3 EVOLUTIONARY PROPERTIES: STAR FORMATION HISTORIES

The mean SFRs derived from these studies can be compared with the total masses of stars and gas, in order to derive the characteristic time scales for the past and future star formation. For example the current SFR in a galaxy can be compared with its past SFR, averaged over the age of the disk:

$$b \equiv \frac{R}{< R >_{past}} = \frac{R \cdot T}{M_{disk}}$$

where M_{disk} is the total stellar mass of the disk, and T is the age of the disk. A characteristic time scale for the star formation in the disk is:

$$\tau_{disk} = b \cdot T$$

In normal spirals these time scales range from roughly 1 Gyr in Sa-Sab galaxies to 10 Gyr or more in some Sc-Irr galaxies, with the birthrate parameter b ranging from order 0.1 in early type spirals to an average near unity in Sc galaxies (Kennicutt 1983, 1986; Gallagher, Hunter, and Tutukov 1984). In the Galaxy, where the birthrate history can be derived directly from the stellar population in the solar neighborhood, a variety of studies show that $b \simeq 0.3 - 1$ (Twarog 1980, Scalo 1986, Barry 1988, Noh and Scalo 1989). Thus the disks of most spiral galaxies, including the Milky Way, have formed stars at a relatively constant rate over their lifetimes. Indeed in many late-type spirals and irregular galaxies the current rate is higher than the average past rate ($b > 1$).

These conclusions apply to the steady-state star formation histories of galaxies, averaged over time scales of 10–100 Myr. Variations in the SFR on shorter time scales are clearly present in some galaxies, and these account for part of the dispersion in SFRs within a given Hubble type (Fig. 2). There is considerable disagreement, however, about the importance of star formation bursts in the general evolution of galaxies. Irregular galaxies, for example, exhibit a large range in photometric properties, and Bagnuolo (1976) argued on the basis of detailed evolutionary models that a significant fraction of the stars in these galaxies form in bursts, while Gallagher et al. (1984) and Hunter and Gallagher (1985), using similar data, concluded that bursts are relatively unimportant. The disagreement can be traced in part to differences in the models used to interpret the data; the latter studies, for example, were heavily based on long-term SFRs derived from blue luminosities (¶2.5), without accounting for luminosity evolution during the bursts, and this will tend to underestimate any real variation in the SFR. The most direct evidence for the importance of bursts comes from studies of the age distributions of stars and clusters in the Magellanic Clouds (e.g., Hodge 1973, Frogel 1984), but the role of bursts in irregular galaxies as a class has yet to be defined unambiguously.

It has generally been assumed that spiral galaxies are much more homogeneous in their photometric properties than irregulars, and hence that starbursts were a less important contributor to their evolution. The discovery of a significant population of low surface brightness spirals, however, has demonstrated that spirals possess extremes in stellar populations which rival those in irregulars (e.g., Romanishin, Strom, and Strom 1983, Bothun *et al.* 1987, van der Hulst *et al.* 1987, Schombert and Bothun 1988). This circumstantial evidence, combined with suggestions of bursting behavior in the age distributions of stars in the solar neighborhood (Scalo 1986, Barry 1988) led Bothun (1987, 1990) and Scalo (1988) to suggest that bursts may be responsible for a significant fraction of the star formation in spirals as well. If correct this would imply that existing surveys of SFRs in galaxies, which are all based on optical catalogs, would be systematically biased toward the most active (bursting) systems. Tinsley and Danly (1980) and Kennicutt (1983) have argued against any strong bias among the existing surveys of spirals. It is clear that starbursts contribute at some level to the global star formation in spirals— tidal interactions, for instance, account for a few percent of all star formation in disks (Larson and Tinsley 1978, Kennicutt *et al.* 1987)— but the role of bursts in normal, isolated spirals is less clear. If the low surface brightness spirals others turn out to comprise a significant fraction of the total space density of galaxies, then the role of bursts will have to be reassessed.

One can derive an estimate for the future star formation lifetimes of galaxies, by comparing the current SFRs with the total masses of interstellar gas:

$$\tau_{gas} = \frac{M_{gas}}{R} \cdot r$$

where M_{gas} includes both atomic and molecular gas, and r is a correction factor to account for recyling of the gas through multiple generations of stars. For disk galaxies τ_{gas} ranges from less than 1 Gyr in many starbursting systems to over 10 Gyr in low surface brightness and gas rich red spirals, with a median (for Shapley-Ames or RC2 selected objects) of 4-5 Gyr (Larson, Tinsley, and Caldwell 1980, Kennicutt 1983). Similar calculations applied to dwarf irregular galaxies generally yield longer consumption times (Sandage 1986).

The times for spirals are considerably shorter than derived previously (Roberts 1963), and suggest the "burnout" of a significant fraction of the current-day spirals in a fraction of a Hubble time, unless the gas supply is replenished from outside. The effective consumption times are even shorter when one takes into account that much of the gas lies outside of the star forming disk, and that star formation will probably be abruptly curtailed long before all of the gas is consumed. The interpretation of these time scales remains controversial, however.

4. Dependence of Star Formation Rates on ISM Properties

4.1 NATURE OF THE PROBLEM

One of the most fundamental problems in this field is the relationship between the global star formation and the physical properties of the interstellar gas. Observations of the correlations between the SFR and the ISM properties can provide data which are relevant to a number of astrophysical problems, including the triggering and regulation of star formation, and the star formation efficiency in different galactic environments. The relationships are of particular importance for understanding galactic evolution, because parametrizations between the SFR and the density or other properties of the ISM form the basis of virtually all contemporary galactic evolution models.

The advent of arcsecond-resolution aperture synthesis observations in H I and CO allows us to study the SFR-ISM relations over an enormous range of scales, from $1 - 10^5$

pc. Most of the currently available data on the SFRs and gas contents of galaxies are integrated measurements. Unfortunately these data are the least useful for understanding the star formation process itself, because the data are averaged over entire disks, but they are valuable for testing some aspects of galactic evolution models. At the other extreme, high resolution measurements ($\ll$ 100 pc), which resolve the cloud structure of the ISM, can provide detailed information on local star formation processes, the star formation efficiency, and the structure of individual regions, but they are less useful for characterizing the global processes which appear to regulate the long term evolution of disks. For the latter the most useful observations may be galaxy-wide observations with intermediate resolution, of order .1–1 kpc. Measurements on the kiloparsec scale are sensitive to the collective processes such as gravitational stability or feedback by the ISM, and on scales of order 100 pc one can study more localized processes such as supernova shocks, spiral arm perturbations, or cloud-cloud collisions.

I mention this partly to illustrate the rich variety of physical processes which can now be studied by high-resolution observations of external galaxies, but also to emphasize the need in such studies to match the observations to the physical regime of interest. As will be seen later, many of the processes which regulate the star formation on intermediate scales are highly nonlinear, and consequently the SFR-ISM correlations averaged over global scales may look very different from the local relations. For example, if one wishes to compare the star formation efficiencies in molecular clouds in different galaxies, it is imperative to measure the efficiencies on the cloud scale, not summed over an entire interstellar medium. By the same token if one wishes to test the large-scale parametrizations of the SFR which are relevant to galactic evolution models, these are best measured on galactic scales, rather than on a cloud-by-cloud basis.

4.2 THE SCHMIDT LAW

Most current models for large-scale star formation in galaxies are based on the parametrization suggested by Schmidt (1959), in which the SFR is assumed to scale as a simple power of the gas density:

$$R = a\rho_g^n$$

In external galaxies the primary observable parameters are the projected surface densities of young stars and gas, and for most of this discussion we will consider a modified form of the Schmidt law:

$$\Sigma_R = A\Sigma_{gas}^N$$

where the Σ's denote surface densities, and an upper case N is used to distinguish the index from the volume density index n. Note that for nearly linear laws ($n \sim 1$), $N \sim n$.

Since it was proposed 30 years ago the Schmidt law has been used almost exclusively to describe the large scale star formation and evolution of galaxies, and there have been dozens of studies directed at quantifying the relation and measuring the value of N, based on correlations of SFRs with column densities or total masses of gas in nearby galaxies. These observations, however, have not clarified the nature of the star formation law. Figure 4 shows a histogram of the empirically determined values of N which I was able to compile from the literature (see Berkhuijsen 1977 and Freedman 1984 for references to early papers). The best fitting indices show a very large dispersion $0 < N < 4$, with a broad peak around Schmidt's (1959) value $n = 2$.

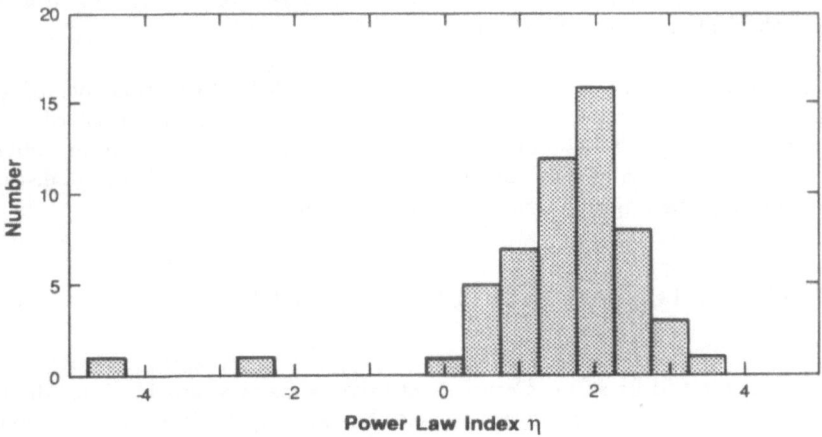

Fig. 4. Histogram of empirically determined values for the power law index N in the Schmidt law.

Some of the scatter in Fig. 4 is due to observational selection effects, as discussed by Madore (1977) and Freedman (1984). Freedman, for instance, found that the best fitting value of N derived from *the same data* increased from 1.1 to 2.3 when the resolution of the data was degraded from 250 to 600 pc. This trend may be partly physical, reflecting a transition from correlating the star formation properties of individual cloud complexes, where a constant star formation efficiency would produce a nearly linear Schmidt law, to measuring global SFR-density correlations on the largest scales. Another selection effect arises when the SFR is correlated only with the atomic or molecular densities, as was the case for most of the previous studies. Even when these problems are taken into account, however, there is still a large dispersion in the derived values of N, even within the same galaxy (e.g., Madore, van den Bergh, and Rogstad 1974, Freedman 1984), and this suggests that either the form of the Schmidt law varies systematically within and between galaxies, or that a monotonic power law approximation is inadequate.

There are several other general star formation properties of galaxies which are difficult to reconcile with a universal Schmidt law. For example, the similar radial profiles of the SFR and CO emission in many galaxies, as well as the relatively constant birthrate histories of disks, are both best understood by a relatively weak SFR-density dependence ($n \simeq 1$), whereas the rapid increases of the SFR in spiral arms and interacting galaxies require a very steep law ($n \gg 2$). Other observations, such as the virtual absence of star formation in many gas-rich S0 and Sa galaxies, are inconsistent with any form of the Schmidt law (e.g., Schommer and Bothun 1983, van der Hulst *et al.* 1987).

Until recently, attempts to better define the empirical star formation law and resolve these inconsistencies were severely handicapped by the lack of a homogeneous data set on the SFRs and total gas contents (atomic *and* molecular) for a diverse sample of galaxies. The recent publication of several large surveys of H I, CO, and Hα emission in nearby galaxies has made it possible to reassess the SFR-density relation, as well as to isolate the underlying physical processes which regulate the SFR.

The observational studies can be conveniently divided into comparisons of global SFRs and gas contents, integrated over entire galaxies, and spatially resolved studies, either on a point by point basis or averaged as a function of radius within individual objects. I will discuss each of these areas separately, and conclude with a brief discussion of the possible physical origins of the observed star formation law.

4.3 GLOBAL STAR FORMATION VS ISM CORRELATIONS

Comparisons of total SFRs and gas contents of galaxies may not provide much direct information on the local star formation law, but they are important for galaxy evolution modelling, because they test whether simple 'single-zone' models can realistically characterize the average star formation properties of disks.

Until a few years ago the only global measurements of gas contents of galaxies were from 21 cm H I surveys, and comparisons of the integrated SFRs and total H I contents showed only a weak correlation (e.g., Lequeux 1979, Kennicutt and Kent 1983). The availability of total CO fluxes for large samples of galaxies has made it possible to correlate the total SFR with the total gas content, as well as with the individual atomic and molecular components.

Several studies have correlated the total SFR, as indicated by integrated $H\alpha$, FIR, or blue luminosities, with the total mass of molecular gas, as inferred from the luminosity of the CO $(1 \rightarrow 0)$ line (e.g., Young and Scoville 1982b, Sanders and Mirabel 1985, Young et al. 1986, Tacconi and Young 1987, Kenney and Young 1989, Solomon and Sage 1988). Figure 5a shows an typical FIR vs CO comparison, with data taken from Tacconi and Young (1987) and Solomon and Sage (1988). The total FIR and CO luminosities invariably show a tight, nearly linear correlation (the least-squares fit has slope 0.92). When the same comparison is made using total H I luminosities, a linear correlation is also apparent, but with a much larger scatter. This scatter is reduced, however, when the comparison is restricted to the mass of H I in the inner star forming disk (Kenney and Young 1988).

The strong correlation illustrated in Fig. 5a has been cited in the above studies as evidence for a global Schmidt law with index $n = 1$, and a constant 'star formation efficiency' in galaxies. However several workers have pointed out that the strong linear correlations in these mass-dependent quantities are artifacts of simple scaling, the fact that "bigger galaxies have more of everything" (Stark et al. 1986, Kennicutt 1989), rather than physical relationships. As an illustration, Fig. 5b shows the same FIR and CO data as in Fig. 5a, but this time I have added calculated (back-of-the-envelope) luminosities for a few other objects, in ascending order a cigar, a Jeep, a forest fire, Venus, and all of the galaxies in the observable universe. Note the tight linear correlation (slope = 0.97), extending over 50 orders of magnitude in luminosity. Clearly such correlations tell us little more than that some galaxies (or fires) are bigger than others.

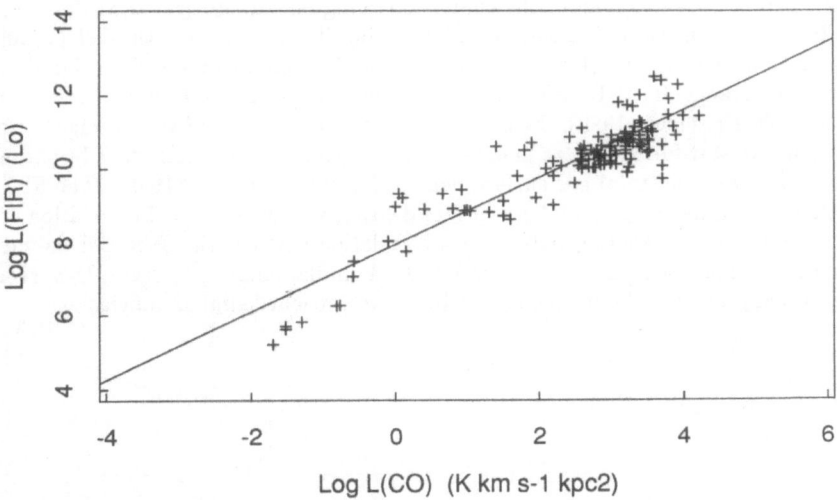

Fig. 5. (a) Correlation between total FIR luminosity and CO luminosity for galaxies, with a least-squares fit. Data from Tacconi and Young (1987) and Solomon and Sage (1988). Upper limits included.

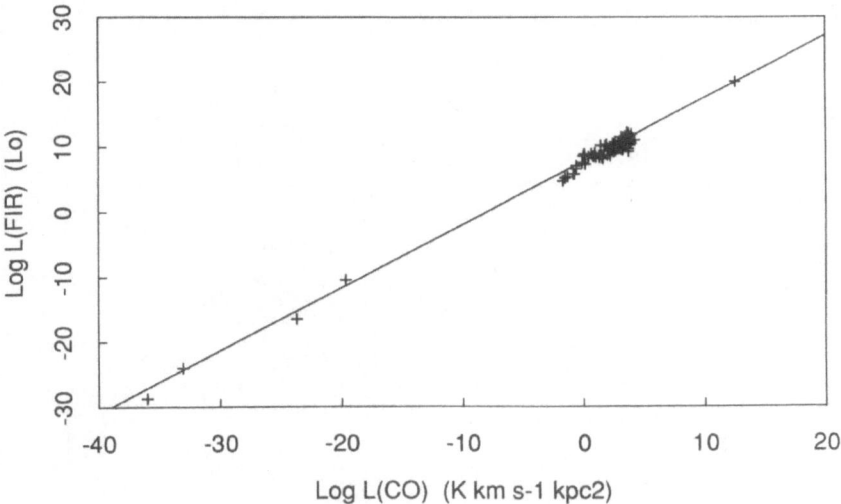

Fig. 5. (b) Same data as above, but also including calculated luminosities for (increasing order) a burning cigar, a Jeep Cherokee which has been sitting in a closed garage for an hour with the engine running, the 1988 Yellowstone Park forest fire, Venus, and the observable universe (assumed to be 10^{11} average galaxies). The line is a least-squares fit.

When care is taken to remove the effects of scaling on the integrated SFRs and gas contents, by normalizing the quantities to galaxy luminosity or the area of the disk (yielding average surface densities), the character of the correlations change markedly (Guiderdoni and Rocca-Volmerange 1985; Donas *et al.* 1987, Kennicutt *et al.* 1987, Buat, Deharveng, and Donas 1989, Kennicutt 1989). Figure 6 shows the dependence of the average surface density of massive star formation, as indicated by the mean Hα or FIR surface brightness, on the mean H I, H$_2$, and total gas surface densities, from Kennicutt (1989). The SFR is best correlated with the average total gas surface density, and shows a good correlation with mean HI surface density. There is only a weak correlation between the SFR and the mean surface density of molecular gas (inferred from CO). A similar analysis by Buat, Deharveng, and Donas (1989), using UV data instead of Hα fluxes, reached similar conclusions.

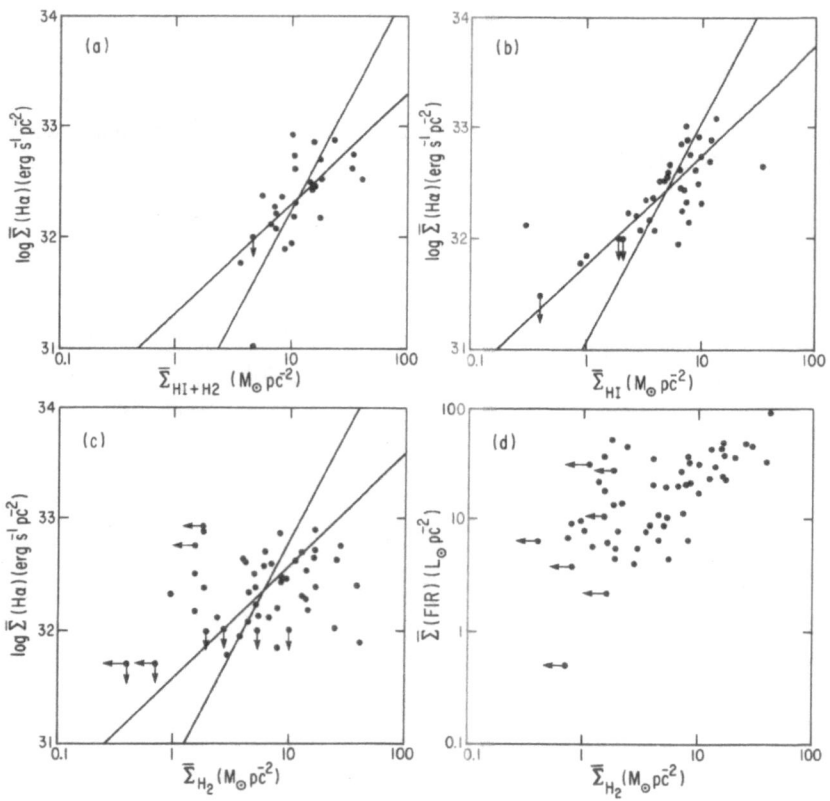

Fig. 6. Correlation of mean SFR per unit disk area, as indicated by Hα or far infrared surface brightness, on the mean atomic, molecular, and total gas surface densities.

The correlation between the average SFRs and the total gas densities is gratifying, because it shows there is at least some empirical basis for single-zone galactic evolution models. The physical interpretation of the other correlations is unclear. For example, the relatively good Hα vs HI correlation could be seen as evidence for a causal relation between atomic gas density and the SFR, akin to a Schmidt law, but it just as possible that the amount of HI in a galaxy is controlled by the SFR, through photodissociation of molecular clouds by hot stars (Shaya and Federman 1987, Tilanus and Allen 1989). It is bad enough that the physical interpretation of these global relations is unclear, but even the causal direction is ambiguous!

The most puzzling result in Fig. 6 is the poor correlation between the SFR and the H_2 surface density, especially in view of the excellent correlation between the radial profiles of CO and Hα in many galaxies (next section). This is not entirely a new result, however. Kenney and Young (1989), for instance, noted the existence of a significant population of active star forming but CO-poor galaxies, and it is these objects which are responsible for much of the scatter in Fig. 6. The fact that the FIR and Hα show a similar lack of correlation with CO (they are much more strongly correlated with each other) demonstrates that the dispersion is not due to extinction or some other spurious effect in the SFR tracers. There are many possible physical explanations for this result, including variations in the CO/H_2 conversion factor or the IMF, but I suspect that much of the scatter may simply reflect nonlinearities in the large-scale star formation law, which smear out any physical SFR-density correlation when these quantities are averaged over entire disks.

The main intent of this discussion was not to offer the final word on the nature of the global SFR-ISM correlations in galaxies, but rather to emphasize the care which must be taken in drawing causal inferences from correlations between the integrated SFR and ISM properties of galaxies. Several coordinated Hα, FIR, HI, and CO surveys are currently under way, and these data should help to better define both the empirical correlations themselves and their interpretation. Especially needed are more spatially resolved HI and CO observations for galaxies in common— the sample of objects with measured mean total gas densities is very small— and observations of galaxies spanning a larger range of SFRs and gas densities.

4.4 SPATIALLY RESOLVED OBSERVATIONS

More direct information on the form of the star formation law can be obtained from comparisons of spatially resolved SFRs and gas densities within individual galaxies. Numerous studies correlating the local density of stars or HII regions with the column densities of HI were published during the 1960's and 70's (cf. Fig. 4), and this work has been discussed extensively elsewhere (e.g., Hamajima and Tosa 1975, Madore 1977, Talbot 1980, Freedman 1984). Here I will concentrate on recent attempts to use intermediate resolution (order .1–1 kpc) data to characterize the large-scale star formation law.

Several workers have investigated the relationship between the radial distributions of molecular and/or atomic gas with various star formation tracers. In most spiral galaxies the SFR is well represented by an exponential disk (Hodge and Kennicutt 1983, Freedman 1984, Kennicutt 1989), sometimes with a central depression, and in many spirals this distribution coincides closely with the radial distribution of CO emission (e.g., Young and Scoville 1982a, Scoville and Young 1983, DeGioia-Eastwood et al. 1984, Tacconi and Young 1986, Kenney 1987, Kennicutt 1989). The HI in these galaxies, on the other hand, shows

424

are relatively flat radial distribution, and in most cases is virtually uncorrelated with the SFR distribution. Note that this is the opposite of what was observed in the mean surface densities (Fig. 6). It is now known that there are many galaxies, especially early-type spirals and low-luminosity systems, where the CO emission is confined to a ring or a small inner disk, respectively (e.g., Young 1987), and in those systems the CO profiles are often uncorrelated with the SFR on large scales (Kennicutt 1989).

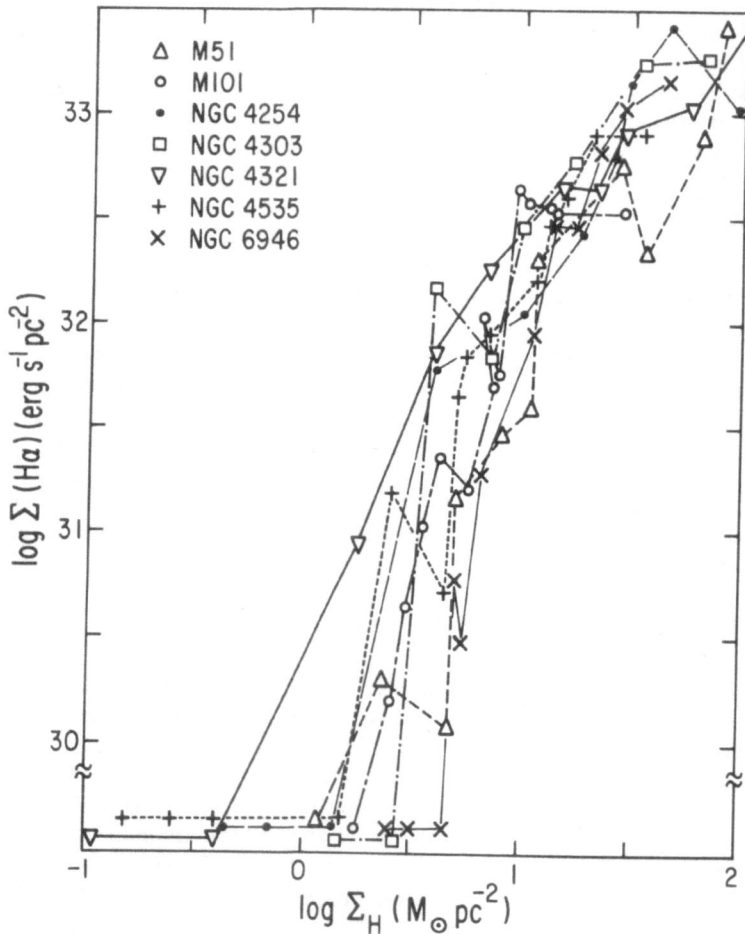

Fig. 7. Correlation of Hα surface brightness with total gas surface density in 7 Sc galaxies. Points are azimuthal averages at constant galactocentric radius, so factors other than gas density may be influencing the SFR.

Probably the most significant recent development has been the recognition of the importance of star formation thresholds. Several workers have combined H I maps of nearby galaxies with data on the Hα emission or resolved stars, and found that star formation tends to occur only in regions above a minimun H I column density, usually in the range $10^{20} - 10^{21}$ H cm^{-2} (Davies, Elliott, and Meaburn 1976, Hunter and Gallagher 1986, Guiderdoni 1987, Skillman 1987, van der Hulst *et al.* 1987). Along similar lines Nakano *et al.* (1987) and Ohta, Sasaki, and Saito (1989) found that molecular cloud formation in M31 and IC10 requires a threshold H I density of order 10^{21} H cm^{-2}. It would be interesting to test whether the thresholds for the formation of molecular clouds and stars are identical, but unfortunately the existing CO data for most galaxies do not extend to sufficiently faint limits to answer this question.

Kennicutt (1989) extended this analysis by compiling radial profiles of Hα, HI, and CO emission for approximately 20 field spirals with active star forming disks. All showed radial cutoffs in the SFR well within the gas disks, but with a large range of threshold densities, $3 - 50 \times 10^{20}$ H cm^{-2}, or $2 - 40 M_\odot$pc^{-2}. The densities quoted are total (atomic plus molecular) gas densities; most of the thresholds occurred at radii where the gas is predominantly atomic, but in at least two cases the thresholds occurred well within the optical disk, where the much of the gas is molecular.

The effect of the star formation thresholds on the SFR vs gas density law is illustrated in Figure 7, which shows such curves for 7 Sc galaxies in the Kennicutt (1989) study. At high gas densities, well above the threshold, the SFR-density relation is well represented by a Schmidt law with a mean index $N \simeq 1.3$, and nearly constant zeropoint (the upper parts of the curves in Fig. 7 nearly coincide). At low densities, well below the threshold, star formation is almost completely suppressed, while near the threshold the slope of the SFR-density relation is much steeper than a normal Schmidt law. Fig. 7 only shows late-type galaxies with active star forming disks; when the same data are compared for early type, less active star forming disks, the upper, linear part of the star formation law is absent, but the other two regimes are present (Kennicutt 1989).

This existence of thresholds which divide the response of the SFR on gas density into (relatively) linear and nonlinear regimes may help to resolve many of the inconsistencies with the Schmidt law which were discussed earlier. The existence of the thresholds is now firmly established observationally, and the variation in densities from one galaxy to another is clearly real. This may provide clues to the origin of the thresholds and eventually to the overall form of the star formation law.

4.5 THEORETICAL BASIS OF THE STAR FORMATION LAW

There is substantial evidence associating the star formation thresholds in galaxies with large scale gravitational instabilities in the gas disks, as originally suggested by Quirk (1972), and subsequently discussed by a number of workers, including Elmegreen (1979, 1989), Fall and Efstathiou (1980), Cowie (1981), Balbus and Cowie (1985), Lacey and Fall (1985), and Larson (1988). Guiderdoni (1987) noted that the star formation thresholds occur at column densities which are of the same order as those expected from gravitational stability considerations (Goldreich and Lynden-Bell 1965), and Zasov and Simikov (1988), Kennicutt (1989), and Oey and Kennicutt (1989) have modelled the stability of the gas disks in nearby galaxies, and compared the derived stability properties with the observed distributions of gas and star formation.

The condition for gravitational stability of a gas disk can be expressed as a critical surface density using the Toomre (1964) criterion:

$$\Sigma_c = \alpha \frac{\kappa c}{\pi G}$$

where κ is the epicyclic frequency, c is the velocity dispersion of the gas disk, and α is a constant of order unity which takes into account deviations from the pure, thin disk approximation in the Toomre calculation. The most important such deviation is the presence of two-fluid instabilities between the stellar and gaseous disks, but the calculations of Jog and Solomon (1984) indicate that deviations from the Toomre condition should be small in most realistic cases.

The stability properties of a disk can be calculated directly from its rotation curve, if the velocity dispersion of the gas is known. Observations of H I and CO disks in external galaxies (Lewis 1975, van der Kruit and Shostak 1984, Murray and Dickey 1989) and our own Galaxy (e.g., Clemens 1985) show that the disks are nearly isothermal, with dispersions in the range $4 - 10$ km s^{-1}. Zasov and Simikov (1988), Kennicutt (1989), and Oey and Kennicutt (1989) combined the Toomre criterion above (assuming a constant velocity dispersion) with published rotation curves for nearby galaxies, in order to derive the radial dependence of the critical densities, and compared these with the actual gas distributions. Using the stability criterion for an isothermal Toomre disk is clearly oversimplistic, but this approach has the advantage of minimizing the number of assumptions and free parameters in the model, and serves to test whether the gravitational stability criterion is at least qualitatively applicable.

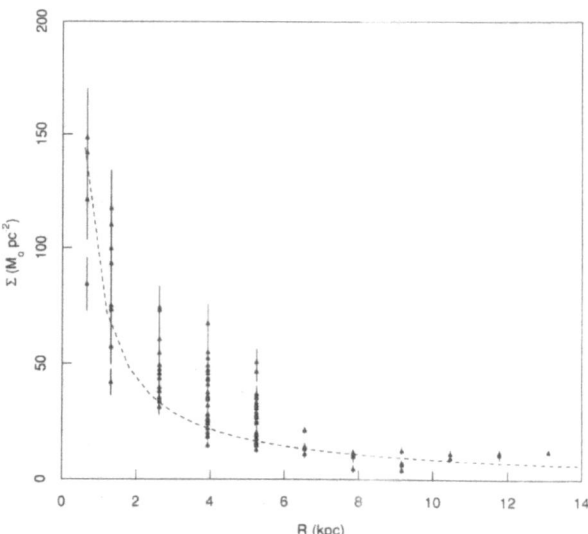

Fig. 8. Observed gas surface densities in NGC 6946, plotted as a function of galactocentric radius. The curve shows the calculated radial behavior of the gravitational threshold density, using an isothermal Toomre model.

As an example, Figure 8 shows the radial dependence of the stability density Σ_c in the nearby spiral NGC 6946, calculated by Oey and Kennicutt (1989) using the Toomre criterion above with $c = 6$ km s^{-1} and $\alpha = 0.7$ (see below). The points show the actual measured total gas densities, as measured by Tacconi-Garman (1988). This comparison reveals two significant trends. First, the gas distribution closely follows the stability limit; the absolute surface density of gas drops by nearly two orders of magnitude, but the ratio of the gas density to the critical density for stability does not change my more than factors of a few. There is also a tendency for the density to drop below the threshold at large radii, near the location of the observed thresholds.

These same trends are shown for a large sample of active star forming spirals in Figure 9. Here what is plotted is the ratio of the gas density to the Toomre density ($c = 6$ km s^{-1}, $\alpha = 1$), what is often referred to as the stability parameter "Q" in stellar dynamics, as a function of radius (Kennicutt 1989). In this case the radii are normalized to the observed edges of the H II region disks, *i.e.*, the observed threshold radii. Again the gas disks closely follow the stability limit— the surface density is rarely more than a factor of two above or below the value required for stability— and there is a single value for Q, near 0.7, which lies near the observed star formation thresholds in all of the galaxies. A similar plot can be made for galaxies which are characterized by low disk SFRs, and this is illustrated in Fig. 10. In this case the gas disks lie near, but usually below the predicted star formation thresholds, so the simple model is at least qualitatively consistent with the gross star formation properties of the disks.

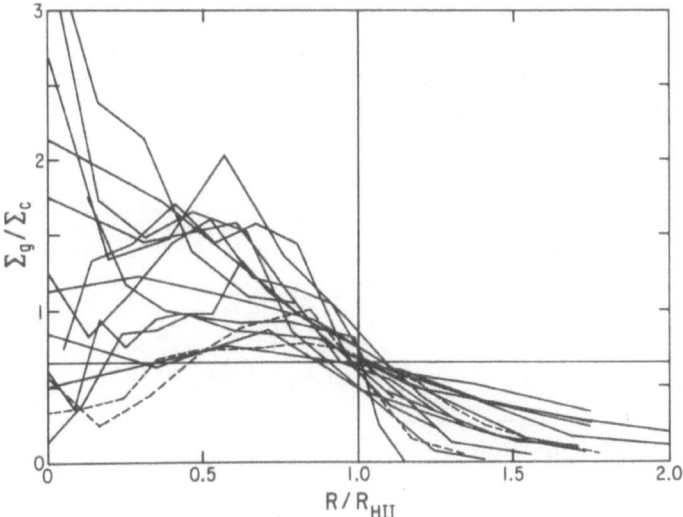

Fig. 9. Radial dependence of the ratio of gas surface density to the critical density for gravitational stability (Toomre criterion), for galaxies with high disk SFRs. The radial coordinate is normalized to the observed threshold radius. The horizontal line shows the value of Q (0.7) which best fits the threshold radii.

428

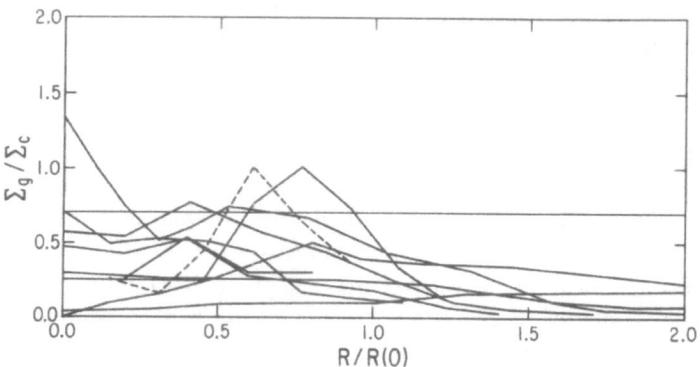

Fig. 10. Same plot of Q_{gas} vs radius as in Fig. 9, but for galaxies (most Sa-Sb) with low disk SFRs, or with star formation confined to spiral arms. The horizontal line shows the value of Q which best fits the star formation thresholds in the active star forming disks. Radii are normalized to the isophotal radius.

This simple gravitational model reproduces the observed threshold densities and radii with surprising accuracy, as illustrated in Figure 11 (from Kennicutt 1989). If this interpretation is correct, then the principal mechanism which determines whether global star formation takes place in a disk is large-scale gravitational stability, as suggested by Quirk (1972), Fall and Efstathiou (1980), and Lacey and Fall (1985).

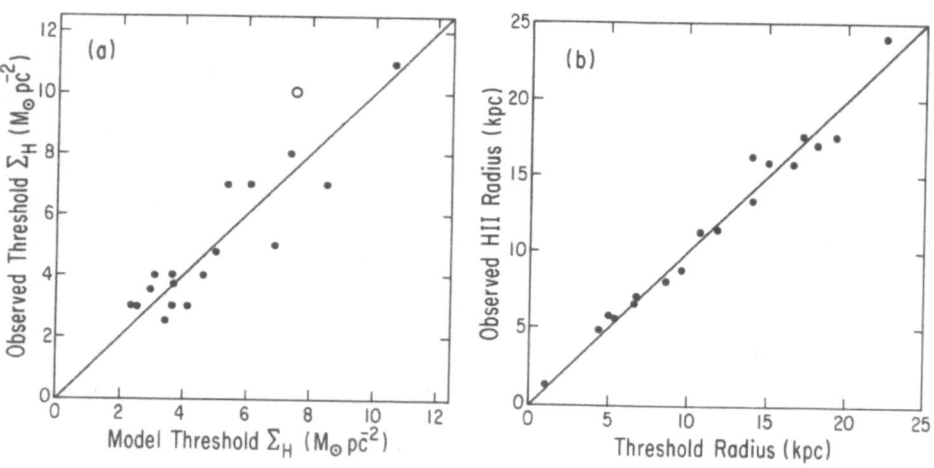

Fig. 11. Comparison of the observed threshold densities (left) and radii (right) with those predicted by the gravitational stability model discussed in the text.

Other physical mechanisms are needed to understand the form of the star formation law above the threshold. Unfortunately the roughly power-law shape of the law at high gas densities is consistent with any number of physical processes, including a cloud-cloud collision model (e.g., Scoville and Good 1987), a self-regulating feedback model (e.g., Talbot and Arnett 1975, Franco and Cox 1983, Dopita 1985, Silk 1987), or a variety of other mechanisms (Larson 1988). I refer the reader to Dopita's review in this volume for a more detailed discussion of this subject.

The implications of these results for the interpretation of the star formation law are summarized schematically in Figure 12. In the Schmidt law scenario the SFR at any point in a galaxy is determined only by the gas density, according to a unique power-law relation, but if the Quirk picture is correct the SFR depends on at least two parameters, the local density and a dynamical parameter, the ratio (Q) of the density to the threshold value. For a flat rotation curve the threshold density goes roughly as $1/r$, so as one proceeds from the outer edge of galaxy inward, the SFR vs density relation will shift to the right in Fig. 12.

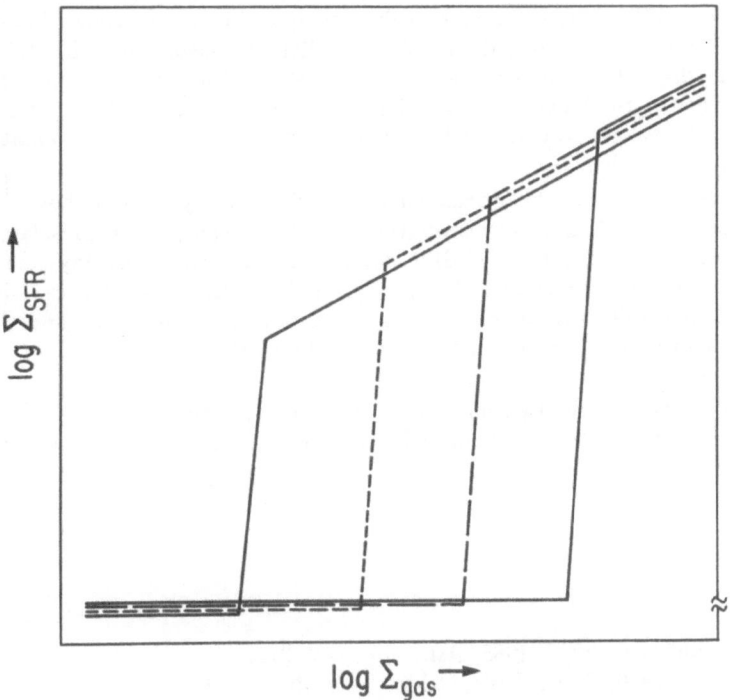

Fig. 12. Schematic sketch of the behavior of the SFR vs gas density law suggested by the stability model. Each curve represents the star formation law at a particular galactocentric radius. In most galaxies the curves will shift to the right with decreasing radius.

In this picture the sensitivity of the SFR to the gas density at a given location will depend on the stability parameter Q. For $Q > 1$ the SFR will increase as a mild power (1–2) of the density, while in regions where $Q < 1$ (0.7 for the velocity dispersion we assumed) the SFR will exhibit a strongly nonlinear increase with density. The SFR will be strongly suppressed where $Q \ll 1$. This latter regime will be important in the outer H I disks of most spiral galaxies, but in the inner disks the gas closely tracks the stability limit (Figs. 8-9); hence the SFR should lie near the transition between the high density power-law regime and the threshold regime throughout the visible disks. The consequences of the thresholds may be particularly interesting in the *innermost* regions, because there the thesholds densities tend to be very high (cf. Fig. 8), and in such regions the gas could accumulate to relatively high densities without forming stars. Once triggered the star formation would be fueled by a large, dense supply of gas.

If these results are correct, they may partly explain why previous observations failed to uncover a universal star formation law; the law itself changes in different interstellar environments (though in a readily predictable manner). The results may also render moot the question of whether star formation is a global or locally-determined process. In the Quirk picture, if the gas density lies above the stability limit, global star formation will occur independently of local conditions, but on the other hand where the gas density is slightly below the stability limit (e.g., the galaxies in Fig. 8b), global star formation will be suppressed, and local perturbations, such as compression by supernova shocks or spiral arms, will be necessary to drive the density of individual regions over the stability limit. Consequently there should be situations where local conditions are unimportant (e.g., in gas-rich galaxies or the inner molecular ring in our Galaxy), and others (e.g., early-type spirals, the solar neighborhood?), where local perturbations play the dominant role in triggering star formation.

The main purpose of this discussion was to illustrate by example how quantitative observations of the SFRs and gas distributions in galaxies can be used to isolate the physical mechanisms which regulate the SFR. We are still far away from any comprehensive understanding of the large scale SFRs in galaxies, but over the next decade it may become possible to supplant the empirical parametrizations which presently underlie galactic evolution models with a bona-fide astrophysically based theory.

Acknowledgements. I gratefully acknowledge the support of the National Science Foundation, through Grant AST86-13257. Special thanks to John Black for checking the calculations which went into Figure 5.

5. References

Augarde, R., and Lequeux, J. 1985, *Astr. Ap.*, **147**, 273.
Bagnuolo, W. G. 1976, Ph.D. thesis, California Institute of Technology.
Balbus, S. A., and Cowie, L. L. 1985, *Ap. J.*, **297**, 61.
Bally, J., and Thronson, H. A. 1989, *A. J.*, **97**, 69.
Barry, D. C. 1988, *Ap. J.*, **334**, 446.
Belfort, P., Mochkovitch, R., and Dennefeld, M. 1987, *Astr. Ap.*, **176**, 1.
Berkhuijsen, E. M. 1977, *Astr. Ap.*, **57**, 9.

Berkhuijsen, E. M. 1983, *Astr. Ap.*, **127**, 395.

Bohlin, R. C., Cornett, R. H., Hill, J. K., Smith, A. M., and Stecher, T. P. 1983, *Ap. J. (Letters)*, **274**, L53.

Bothun, G. 1987, in *Nearly Normal Galaxies*, ed. S. M. Faber, (New York: Springer-Verlag), p. 184.

Bothun, G. D. 1990 , in *The Evolution of the Universe of Galaxies: The Edwin Hubble Centennial Symposium*, ed. R. G. Kron, (Provo: A.S.P. Conference Series), in press.

Bothun, G. D., Impey, C. D., Malin, D. F., and Mould, J. R. 1987, *A. J.*, **94**, 23.

Bothun, G. D., Lonsdale, C. J., and Rice, W. 1989, *Ap. J.*, **341**, 129.

Buat, V., and Deharveng, J. M. 1988, *Astr. Ap.*, **195**, 60.

Buat, V., Deharveng, J. M., and Donas, J. 1989, *Astr. Ap.*, in press.

Buat, V., Donas, J., and Deharveng, J. M. 1987, *Astr. Ap.*, **185**, 33.

Bushouse, H. A. 1987, *Ap. J.*, **320**, 49.

Bushouse, H. A., Lamb, S. A., and Werner, M. W. 1988, *Ap. J.*, **335**, 74.

Caldwell, C. N., Kennicutt, R. C., Schommer, R. A., and Phillips, A. 1990, in preparation.

Caplan, J., and Deharveng, L. 1986, *Astr. Ap.*, **155**, 297.

Code, A. D., and Welch, G. A. 1982, *Ap. J.*, **256**, 1.

Cowie, L. L. 1981, *Ap. J.*, **245**, 66.

Cutri, R. M., and McAlary, C. W. 1985, *Ap. J.*, **296**, 90.

Davies, R. D., Elliott, K. H., and Meaburn, J. 1976, *Mem. R.A.S.*, **81**, 89.

DeGioia-Eastwood, K., Grasdalen, G. L., Strom, S. E., and Strom, K. M. 1984, *Ap. J.*, **278**, 564.

DePoy, D. L. 1987, Ph.D. thesis, University of Hawaii.

Devereux, N. 1987, *Ap. J.*, **323**, 91.

Donas, J., Deharveng, J. M., Laget, M., Milliard, B., and Huguenin, D. 1987, *Astr. Ap.*, **180**, 12.

Dopita, M. A. 1985, *Ap. J. (Letters)*, **295**, L5.

Dressel, L. L. 1988, *Ap. J. (Letters)*, **329**, L69.

Dressel, L. L., O'Connell, R. W., and Telesco, C. M. 1989, in *The Interstellar Medium in External Galaxies*, ed. D. Hollenbach and H. Thronson, NASA Conference Publications, in press.

Duric, N., Bourneuf, E., and Gregory, P. C. 1988, *A. J.*, **96**, 81.

Eales, S. A., Wynn-Williams, C. G., and Duncan, W. D. 1989, *Ap. J.*, **339**, 859.

Elmegreen, B. G. 1979, *Ap. J.*, **231**, 372.

Elmegreen, B. G. 1988, in *Comets to Cosmology*, (New York: Springer-Verlag Lecture Notes in Physics), **297**, 186.

Elmegreen, B. G. 1989, *Ap. J.*, **344**, in press.

Elmegreen, B. G., and Elmegreen, D. M. 1986, *Ap. J.*, **311**, 554.

Fall, S. M., and Efstathiou, G. 1980, *M.N.R.A.S.*, **193**, 189.

Fanelli, M. N., O'Connell, R. W., and Thuan, T. X. 1988, *Ap. J.*, **334**, 665.

Franco, J., and Cox, D. P. 1983, *Ap. J.*, **273**, 243.

Freedman, W. L. 1984, Ph.D. thesis, University of Toronto.

Freedman, W. L. 1985, *Ap. J.*, **299**, 74.

Frogel, J. A. 1984, *Pub. A.S.P.*, **96**, 856.

Gallagher, J. S., Hunter, D. A., and Tutukov, A. V. 1984, *Ap. J.*, **284**, 544.

432

Gallagher, J. S., and Hunter, D. A. 1989, *A. J.*, in press.
Gehrz, R. D., Sramek, R. A., and Weedman, D. W. 1983, *Ap. J.*, **267**, 551.
Gioia, I. M., Gegorini, L, and Klein, U. 1982, *Astr. Ap.*, **116**, 164.
Goldreich, P., and Lynden-Bell, D. 1965, *M.N.R.A.S.*, **130**, 97.
Guiderdoni, B. 1987, *Astr. Ap.*, **172**, 27.
Guiderdoni, B., and Rocca-Volmerange, B. 1985, *Astr. Ap.*, **151**, 108.
Hamajima, K., and Tosa, M. 1975, *Pub. A.S. Japan*, **27**, 561.
Hawarden, T. G., Mountain, C. M., Leget, S. K., and Puxley, P. J. 1986, *M.N.R.A.S.*, **221**, 41.
Hill, J. K., Bohlin, R. C., and Stecher, T. P. 1984, *Ap. J.*, **277**, 542.
Hodge, P. W. 1973, *A. J.*, **78**, 807.
Hodge, P. W., and Kennicutt, R. C. 1983, *Ap. J.*, **267**, 563.
Hodge, P., Lee, M. G., and Kennicutt, R. C. 1989*a*, *Pub. A.S.P.*, **101**, 32.
Hodge, P., Lee, M. G., and Kennicutt, R. C. 1989*b*, *Pub. A.S.P.*, **101**, 640.
Huchra, J. P. 1977, *Ap. J.*, **217**, 928.
Huchra, J. P, Geller, M. J., Gallagher, J., Hunter, D., Hartmann, L., Fabbiano, G., and Aaronson, M. 1983, *Ap. J.*, **274**, 125.
Hunter, D. A., Gallagher, J. S., and Rautenkranz, D. 1982, *Ap. J. Suppl.*, **49**, 53.
Hunter, D. A., and Gallagher, J. S. 1985, *Ap. J. Suppl.*, **58**, 533.
Hunter, D. A., and Gallagher, J. S. 1986, *Pub. A.S.P.*, **98**, 5.
Hunter, D. A., Gillett, F. C., Gallagher, J. S., and Low, F. J. 1986, *Ap. J.*, **303**, 171.
Israel, F. P., and Kennicutt, R. C. 1980, *Ap. Letters*, **21**, 1.
Israel, F. P., and van der Hulst, J. M. 1983, *A. J.*, **88**, 1736.
Jog, C. J., and Solomon, P. M. 1984, *Ap. J.*, **276**, 127.
Joseph, R. D., Meikle, W. P. S., Robertson, N. A., and Wright, G. S. 1984, *M.N.R.A.S.*, **209**, 111.
Joy, M., and Lester, D. F. 1988, *Ap. J.*, **331**, 145.
Kaufman, M., Bash, F. N., Kennicutt, R. C., and Hodge, P. W. 1987, *Ap. J.*, **319**, 61.
Kawara, K., Nishida, M., and Phillips, M. N. 1989, *Ap. J.*, **337**, 230.
Keel, W. C., Kennicutt, R. C., Hummel, E., and van der Hulst, J. M. 1985, *A. J.*, **90**, 708.
Kenney, J. D. 1987, Ph.D. thesis, University of Massachusetts.
Kenney, J. D., and Young, J. S. 1988, *Ap. J.*, **326**, 588.
Kennicutt, R. C. 1983, *Ap. J.*, **272**, 54.
Kennicutt, R. C. 1984, *Ap. J.*, **287**, 116.
Kennicutt, R. C. 1986, in *Stellar Populations*, ed. C. Norman, A. Renzini, and M. Tosi, (Cambridge: Cambridge University Press), p. 125.
Kennicutt, R. C. 1988, *Ap. J.*, **334**, 144.
Kennicutt, R. C. 1989, *Ap. J.*, **344**, in press.
Kennicutt, R. C., and Kent, S. M. 1983, *A. J.*, **88**, 1094.
Kennicutt, R. C., Edgar, B. K., and Hodge, P. W. 1989, *Ap. J.*, **337**, 761.
Kennicutt, R. C., Keel, W. C., van der Hulst, J. M., Hummel, E., and Roettiger, K. A. 1987, *A. J.*, **93**, 1011.
Klein, U. 1986, *Astr. Ap.*, **168**, 65.
Klein, U., and Gräve, R. 1986, *Astr. Ap.*, **161**, 155.

Lacey, C. G., and Fall, S. M. 1985, *Ap. J.*, **290**, 154.

Lamb, S. A., Gallagher, J. S., Hjellming, M. S., and Hunter, D. A. 1985, *Ap. J.*, **291**, 63.

Larson, R. B. 1988, in *Galactic and Extragalactic Star Formation*, ed. R. E. Pudritz and M. Fich (Dordrecht: Kluwer), p. 435.

Larson, R. B., and Tinsley, B. M. 1978, *Ap. J.*, **219**, 46.

Larson, R. B., Tinsley, B. M., and Caldwell, C. N. 1980, *Ap. J.*, **237**, 692.

Lequeux, J. 1979, *Revista. Mex. Astron. Ap.*, **4**, 325.

Lewis, B. M. 1975, *Astr. Ap.*, **44**, 147.

Lonsdale-Persson, C. J. 1987, *Star Formation in Galaxies*, NASA Conference Publication 2466.

Lonsdale, C. J., and Helou, G. 1987, *Ap. J.*, **314**, 513.

Lonsdale, C. J., Persson, S. E., and Mathews, K. 1984, *Ap. J.*, **287**, 95.

Lord, S. D., and Young, J. S. 1989, *Ap. J.*, in press.

Madore, B. F. 1977, *M.N.R.A.S.*, **178**, 1.

Madore, B. F., van den Bergh, S., and Rogstad, D. H. 1974, *Ap. J.*, **191**, 317.

Massey, P. 1985, *Pub. A.S.P.*, **97**, 5.

Massey, P., Garmany, C. D., Silkey, M., and DeGioia-Eastwood, K. 1989, *A. J.*, **97**, 107.

Massey, P., Parker, J. W., and Garmany, C. D. 1989, *A. J.*, in press.

Mateo, M. 1988, *Ap. J.*, **331**, 261.

McCall, M. L., and Schmidt, F. H. 1986, *Ap. J.*, **311**, 548.

Mochkovitch, R., and Rocca-Volmerange, B. 1984, *Astr. Ap.*, **137**, 298.

Nakano, M., Ichikawa, T., Tanaka, Y. D., Naki, N., and Sofue, Y. 1987, *Pub. A.S. Japan*, **39**, 57.

Noh, H.-R., and Scalo, J. 1989, *Ap. J.*, in press.

Oey, M. S., and Kennicutt, R. C. 1989, in *The Interstellar Medium in External Galaxies*, ed. D. Hollenbach and H. A. Thronson, NASA Conference Publication Series, in press.

Ohta, K., Sasaki, M., and Saito, M. 1989, *Pub. A.S. Japan*, **40**, 653.

Pogge, R. W., and Eskridge, P. B. 1987, *A. J.*, **93**, 291.

Pompea, S. M., and Rieke, G. H. 1989, *Ap. J.*, in press.

Pudritz, R. E., and Fich, M. 1988, *Galactic and Extragalactic Star Formation*, (Dordrecht: Kluwer).

Quirk, W. J. 1972, *Ap. J. (Letters)*, **176**, L9.

Rieke, G. H., Lebofsky, M. J., Thompson, R. I., Low, F. J., and Tokunaga, A. T. 1980, *Ap. J.*, **238**, 24.

Roberts, M. S. 1963, *Ann. Rev. Astr. Ap.*, **1**, 149.

Rocca-Volmerange, B., and Guiderdoni, B. 1987, *Astr. Ap.*, **175**, 15.

Rocca-Volmerange, B., Lequeux, J., and Marcherat-Joubert, M. 1981, *Astr. Ap.*, **104**, 177.

Romanishin, W. 1985, *Ap. J.*, **289**, 570.

Romanishin, W., Strom, K. M., and Strom, S. E. 1983, *Ap. J. Suppl.*, **53**, 105.

Salzer, J. J., and MacAlpine, G. M. 1988, *A. J.*, **96**, 1192.

Salzer, J. J., MacAlpine, G. M., and Boroson, T. A. 1989, *Ap. J. Suppl.*, **70**, 479.

Sandage, A. 1986, *Astr. Ap.*, **161**, 89.

Sanders, D. B., and Mirabel, I. F. 1985, *Ap. J. (Letters)*, **298**, L31.

434

Schombert, J. M., and Bothun, G. D. 1988, *A. J.*, **95**, 1389.
Schommer, R. A., and Bothun, G. D. 1983, *A. J.*, **88**, 577.
Searle, L., Sargent, W. L. W., and Bagnuolo, W. G. 1973, *Ap. J.*, **179**, 427.
Scalo, J. M. 1986, *Fund. Cos. Phys.*, **11**, 1.
Scalo, J. M. 1988, in *Evolution of Galaxies*, ed. J. Palous, (Prague: Astronomical Institute), p. 101.
Schmidt, M. 1959, *Ap. J.*, **129**, 243.
Scoville, N. Z., and Good, J. C. 1987, in *Star Formation in Galaxies*, ed. C. J. Lonsdale Persson, NASA Conference Publication 2466, p. 3.
Scoville, N. Z., and Young, J. S. 1983, *Ap. J.*, **265**, 148.
Shaya, E. J., and Federman, S. R. 1987, *Ap. J.*, **319**, 76.
Shields, G. A., and Tinsley, B. M. 1976, *Ap. J.*, **203**, 66.
Silk, J. 1987, in *Star Forming Regions*, ed. M. Peimbert and J. Jugaku (Dordrecht: Reidel), p. 663.
Skillman, E. D. 1987, in *Star Formation in Galaxies*, ed. C. J. Lonsdale Persson, NASA Conference Publication 2466, p. 263.
Skillman, E. D., and Israel, F. P. 1988, *Astr. Ap.*, **203**, 226.
Smith, A. M., and Cornett, R. H. 1982, *Ap. J.*, **261**, 1.
Smith, A. M., Cornett, R. H., and Hill, R. S. 1987, *Ap. J.*, **320**, 609.
Soifer, B. T., Houck, J. R., and Neugebauer, G. 1987, *Ann. Rev. Astr. Ap.*, **25**, 187.
Solomon, P. M., and Sage, L. J. 1988, *Ap. J.*, **334**, 613.
Stark, A. A., Knapp, G. R., Bally, J., Wilson, R. W., Penzias, A. A., and Rowe, H. E. 1986, *Ap. J.*, **310**, 660.
Stecher, T. P., Bohlin, R. C., Hill, J. K., and Jura, M. A. 1982, *Ap. J. (Letters)*, **255**, L99.
Tacconi-Garman, L. J. 1988, Ph.D. thesis, University of Massachusetts.
Tacconi, L. J., and Young, J. S. 1986, *Ap. J.*, **308**, 600.
Tacconi, L. J., and Young, J. S. 1987, *Ap. J.*, **322**, 681.
Talbot, R. J. 1980, *Ap. J.*, **235**, 821.
Talbot, R. J., and Arnett, W. D. 1975, *Ap. J.*, **197**, 551.
Telesco, C. M. 1988, *Ann. Rev. Astr. Ap.*, **26**, 343.
Thronson, H. A., Hunter, D. A., Telesco, C. M., Harper, D. A., and Decher, R. 1987, *Ap. J.*, **317**, 180.
Thronson, H. A., Hunter, D. A., Telesco, C. M., Greenhouse, M., and Harper, D. A. 1988, *Ap. J.*, **334**, 605.
Thronson, H. A., Hunter, D. A., Casey, S., Latter, W. B., and Harper, D. A. 1989*a*, *Ap. J.*, **339**, 803.
Thronson, H. A., Hunter, D. A., Casey, S., and Harper, D. A. 1989*b*, *Ap. J.*, in press.
Thronson, H. A., Bally, J., and Hacking, P. 1989*c*, *A. J.*, **97**, 363.
Thronson, H. A., Tacconi, L., Kenney, J., Greenhouse, M., Margulis, M., Tacconi-Garman, L., and Young, J. 1989*d*, *Ap. J.*, in press.
Thronson, H. A., and Telesco, C. M. 1986, *Ap. J.*, **311**, 98.
Thuan, T. X., Montmerle, T., and Van, J. T. T. 1987, *Starbursts and Galaxy Evolution*, (Paris: Editions Frontieres).
Tilanus, R. P. J., and Allen, R. J. 1989, *Ap. J. (Letters)*, **339**, L57.

435

Tinsley, B. M., and Danly, L. 1980, *Ap. J.*, **242**, 435.

Toomre, A. 1964, *Ap. J.*, **139**, 1217.

Turner, J. L., Ho, P. T. P., and Beck, S. C. 1987, *Ap. J.*, **313**, 644.

Twarog, B. A. 1980, *Ap. J.*, **242**, 242.

van der Hulst, J. M., Kennicutt, R. C., Crane, P. C., and Rots, A. H. 1988, *Astr. Ap.*, **195**, 38.

van der Hulst, J. M., Skillman, E. D., Kennicutt, R. C., and Bothun, G. D. 1987, *Astr. Ap.*, **177**, 63.

Viallefond, F. 1988, in *Galactic and Extragalactic Star Formation*, ed. R. E. Pudritz and M. Fich, (Dordrecht: Kluwer), p. 439.

Vílchez, J. M., and Pagel, B. E. J. 1988, *M.N.R.A.S.*, **231**, 257.

Vogel, S. N., Kulkarni, S. R., and Scoville, N. Z. 1988, *Nature*, **334**, 402.

Walterbos, R. A. M., and Schwering, P. B. W. 1987, *Astr. Ap.*, **180**, 27.

Wesselius, P. R., van Duinen, R. J., de Jonge, A. R. W., AAdlers, J. W. G., Luinge, W., and Wildeman, K. J. 1982, *Astr. Ap.*, **49**, 427.

Young, J. S. 1987, in *Star Formation in Galaxies*, ed. C. J. Lonsdale Persson, NASA Conference Publication 2466, p. 197.

Young, J. S., Schloerb, F. P., Kenney, J. D., and Lord, S. D. 1986, *Ap. J.*, **304**, 443.

Young, J. S., and Scoville, N. 1982*a*, *Ap. J.*, **258**, 467.

Young, J. S., and Scoville, N. 1982*b*, *Ap. J. (Letters)*, **260**, L11.

Zasov, A., and Simikov, S. 1988, *Astrophysica*, **29**, 190.

The Star - Gas Cycle In Galaxies

Michael A. Dopita,
Australian National University

ABSTRACT. The star-gas cycle is central to an understanding of the evolution of disk galaxies. In this paper, we investigate three key aspects. First, the effects of mass loss and initial abundance on the chemical yield of, and the evolution of, the high-mass stars. Second, we investigate star formation, and demonstrate that this is likely to be a bimodal process. The stars with masses $< 1M_\odot$ are formed in dense molecular clouds and determine the stochastic equilibrium of these clouds. On the other hand, the stars with masses $> 1M_\odot$ are formed predominantly in cloud-cloud collisions which destroy the parent clouds. Massive stars regulate the phase structure of the ISM in the galactic disks. The final sections investigate the chemical evolution of both our solar neighbourhood, and of the Magellanic Clouds. The hazy concept of "metallicity" is inappropriate to an understanding of chemical evolution. However, the metallicity - metallicity relationships may be used to put useful constraints on past populations of stars in these systems, and to lay the basis for a clearer understanding of the chemical evolution of other disk systems.

1. Introduction

The physics of the star-gas cycle is central to an understanding of both the physical and chemical evolution of galaxies. However, this is an area in which many of the various key topics in modern astrophysics meet. To illustrate this, let us list a few of the problems that must be solved in order to provide an adequate description of the star-gas cycle.

What are the mechanisms of star formation? Is this a unique process, or is star formation bimodal? The structure of the interstellar medium (ISM) itself is profoundly modified by the mass loss and subsequent explosion of massive stars. How then is this influenced by the state of chemical evolution of the medium, and, indeed, how are the chemical yields from the massive stars in turn influenced by this evolution? How important is the past rate of star formation in determining the current rate of chemical evolution by the delayed feedback of C,N and He-enriched gas from low-mass stars? The chemical evolution of the medium is also influencd by infall of pristine matter, and by radial flows in galaxies. How important are these, and what observational diagnostics may be devised? Are these radial flows important in influencing the local rate of star formation in galaxies, through, for example, local shocks? If our current understanding of these topics is less than perfect, this reflects more upon the imperfections in the

H. A. Thronson, Jr. and J. M. Shull (eds.), The Interstellar Medium in Galaxies, 437–472.
© 1990 *Kluwer Academic Publishers.*

438

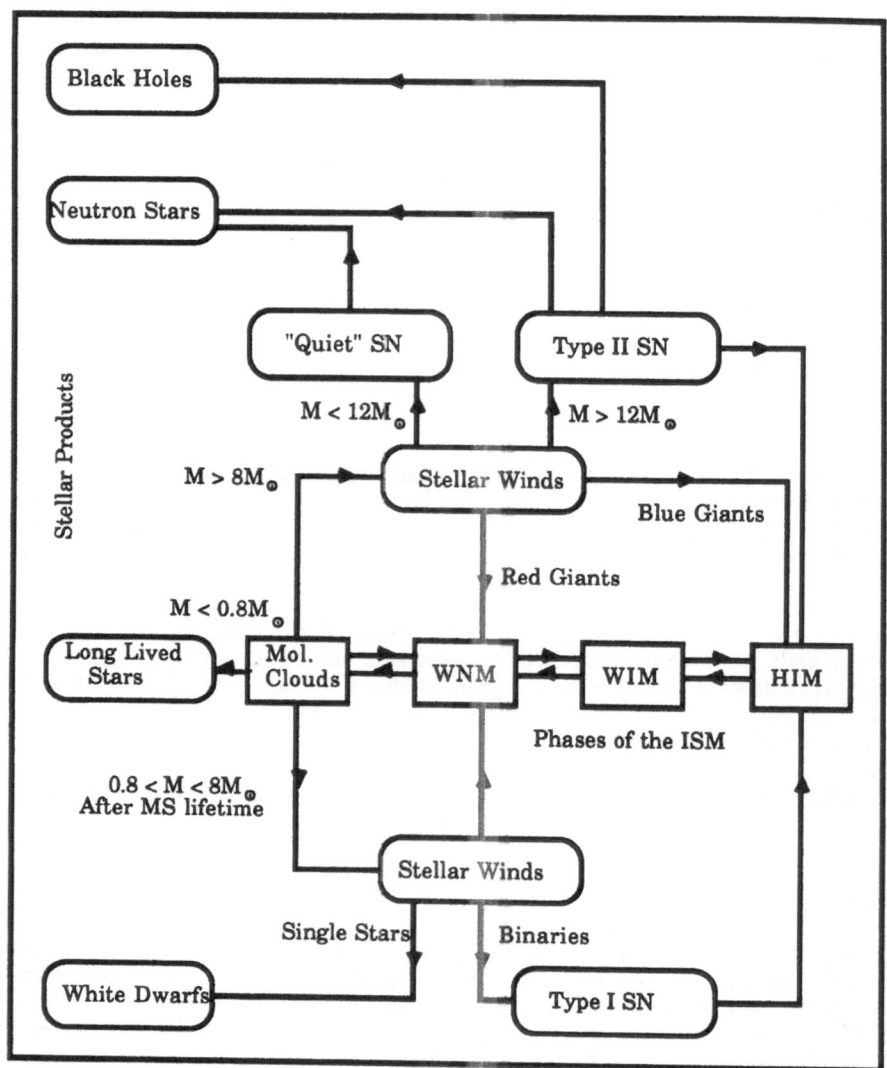

Figure 1: The Star - Gas Bicycle.

synthesis that has been so far achieved, rather than upon the absence of any key element in the picture.

In practice, given the quite different evolutionary histories and stellar remnants of the high-mass and the lower mass stars, one should strictly refer not to a star-gas cycle, but to a star-gas bi-cycle. This is illustrated in Figure 1. The various components of this figure are discussed in detail below.

In this Review, I will concentrate on specific topics relevant to mass exchange, the energy balance, star formation processes and the chemical evolution of the interstellar medium. Where possible, I will give simple parametric representations of the basic physics so as to be free to emphasise the key variables in this very complex exchange.

2. Mass Loss and Energy Input from Stars

2.1. MASSIVE STARS

Mass loss is a universal feature in all high-luminosity stars, be they on the main sequence, helium-burning giants, supergiants or Wolf-Rayet stars (see reviews by Barlow,1982; Chiosi and Maeder,1986; Conti, 1981; Dupree, 1986). Many formulations for mass-loss "laws" have been proposed which are neatly summarized in the review of Chiosi and Maeder (1986). However, it has become increasingly clear that, except for possibly the case of the red supergiants, the winds are radiatively-driven according to the theory developed by Abbott (1978, 1982) and refined by Kudritski, Pauldrach and Puls (1987). The modified radiatively-driven wind theory gives, ignoring the very weak effective temperature dependence, the following expressions for the mass-loss rate ($M_\odot$ yr^{-1}) and terminal velocity (km.s^{-1}):

$$log(\dot{M}) = -14.85 + 1.98\ log\ (L/L_\odot) - 1.03\ log(M_{eff}/M_\odot) + 0.465\ log\ (Z/Z_\odot) \qquad (2.1)$$

and
$$log(V_\infty) = 3.28 + 0.178\ log\ (M/M_\odot) + 0.138\ log\ (Z/Z_\odot) \qquad (2.2a)$$

where $M_{eff} = M(1-\Gamma)$ and $\Gamma = \sigma_e L / 4\pi GMc$. The expression for V_∞ applies for main sequence stars. For the Blue Supergiants, the terminal velocity can be inferred from the mass-loss formula of Wilson and Dopita (1985), which showed that the momentum in the wind is directly coupled to the momentum in the radiation field, as indeed it should be if radiatively-driven wind theory is correct. In this case we get, including the Kudritski, Pauldrach and Puls (1987) abundance dependence:

$$log(V_\infty) = -7.98 + log\ (L/L_\odot) - log(\dot{M}) + 0.138\ log\ (Z/Z_\odot) \qquad (2.2b)$$

The initial zero-age luminosity and the lifetime on the main sequence can be represented by (Chiosi and Maeder 1986; Maeder 1983,1987; Maeder and Meynet 1987):

$$log\ (L/L_\odot) = -0.148 + 4.62\ log\ (M/M_\odot) - 0.747\ \{\ log\ (M/M_\odot)\ \}^2 \qquad (2.3)$$

$$\tau_H = 4.53\ (M/40M_\odot)^{-0.97}\ Myr \qquad\qquad M \leq 40\ M_\odot \qquad (2.4)$$

$$\tau_H = 4.53\ (M/40M_\odot)^{-0.43}\ Myr \qquad\qquad M > 40\ M_\odot$$

and the Helium-burning luminosity and lifetime by:

$$log\ (L/L_\odot) = -0.678 + 5.98\ log\ (M/M_\odot) - 1.245\ \{\ log\ (M/M_\odot)\ \}^2 \qquad (2.5)$$

$$\tau_{He} = 0.856\ (M/40M_\odot)^{-0.97}\ Myr \qquad\qquad M \leq 44\ M_\odot \qquad (2.6)$$

$$\tau_{He} = 0.78\ Myr \qquad\qquad M > 44\ M_\odot$$

provided that mass-loss does not terminate either the hydrogen-burning or helium-burning phases. Expressions (2.1) through (2.6) are sufficient to compute the mechanical luminosity input by stellar winds up until the time that the star becomes a Red Supergiant, or a Wolf-Rayet star. In these cases the mass-loss rate can be empirically fitted by the relation:

$$log(\dot{M}) = -11.5 + 1.25 \log (L/L_{\odot}) \tag{2.7}$$

which seems to fit equally well to *both* the Red Supergiants and the Wolf-Rayet stars in the the compilation of de Jager *et al.* (1985). Our various mass-loss formulations are shown as a function of luminosity in Figure (2).

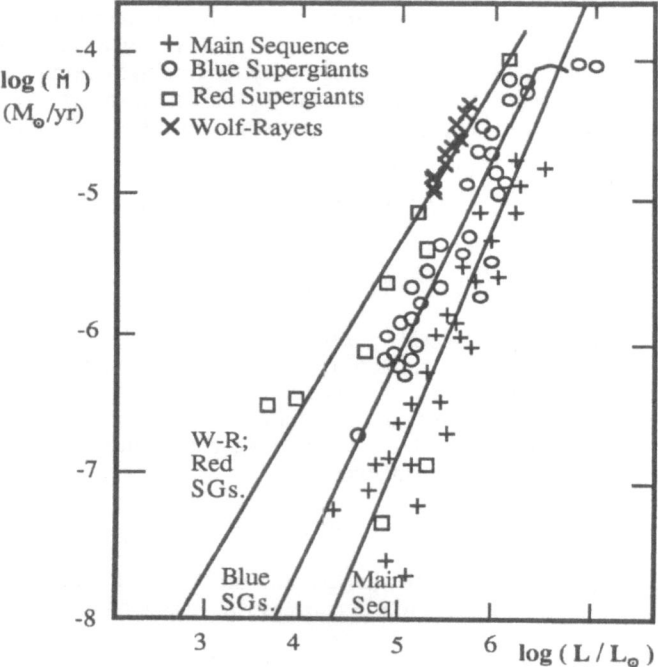

Figure 2: Mass-loss as a function of luminosity for the various types of intrinsically luminous stars from the compilation de Jager *et al.* (1985). Our various mass-loss formulae are labelled.

The very high mass loss rate implied by (2.7) results in a large fractional mass-loss from the central star, despite the relative brevity of these late stages of evolution. During the Wolf-Rayet stage of evolution, the star moves down a relatively narrow mass-luminosity band given by (Maeder, 1983; Maeder and Meynet, 1987):

$$log (L/L_{\odot}) = 4.02 + 1.34 \log (M/M_{\odot}) \tag{2.8}$$

The terminal velocity in the wind is of order 2800km.s^{-1} during this phase. In fact a slightly better representation which takes account of the variation with mass, might be:

$$log(V_\infty) = 4.3 - 0.18 \, log \, (L/L_\odot) \tag{2.9}$$

In Figure (3) the evolution of mechanical luminosity for various mass stars of initially solar composition are shown. This neatly illustrates the various fates of the stars of different initial mass. The 15M$_\odot$ star evolves directly to the Red Supergiant phase, where its input of mechanical energy falls abruptly. The 25M$_\odot$ star survives for some time as a helium-burning Blue Supergiant, before crossing to the red in the H-R Diagram. The 40M$_\odot$ star returns to the blue as a WN-type Wolf-Rayet star, and the 100M$_\odot$ star evolves directly to the Wolf-Rayet phase, becoming a WN-type star before losing enough mass to be seen as a WC-type, and, possibly, a WO-type at the very end of its lifetime.

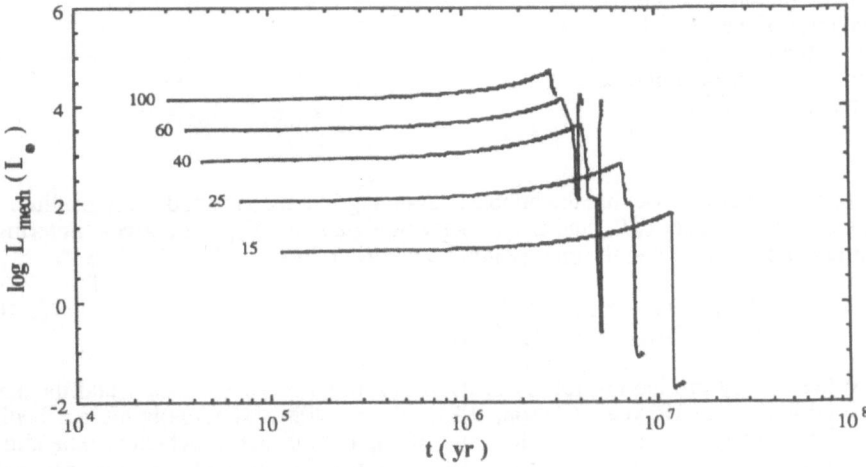

Figure 3: The variation of mechanical luminosity as a function of age for massive stars.

The energy input increases greatly with increasing mass. As a result, massive stars dominate in terms of their energy input to the ISM prior to the supernova event, despite the relative scarcity of high-mass stars and their shorter lifetimes. In Figure 4, we show the relative energy input to the ISM by stars of different mass, assuming that the Initial Mass Function (IMF) for massive stars is a power law with slope $\alpha = -1.8$. The energy input is dominated by stars with masses above 50M$_\odot$, which lose matter at a mean velocity of about 2000km.s^{-1}.

For a star with initial mass of 40M$_\odot$, these figures imply that an energy roughly equivalent to the kinetic energy imparted by the subsequent supernova explosion will be deposited into the ISM. Indeed, the effect on the ISM may well be considerably greater than that produced in the final supernova explosion, since stellar winds couple their momentum to the ISM better than do supernovae. The radius and velocity of expansion of

442

a mass-loss bubble are related in a simple way with the energy input from the central star, and the mass distribution about it (e.g. Dyson, 1981; Kwok and Volk, 1985; Volk and Kwok, 1985).

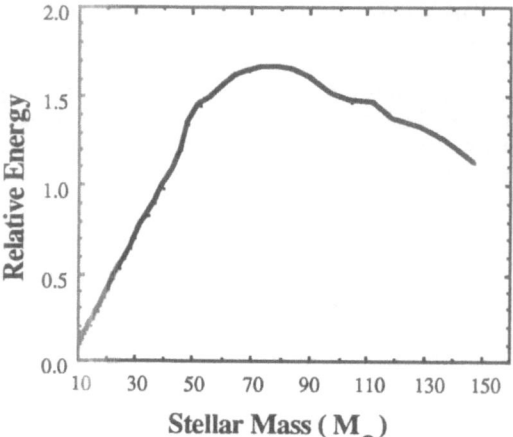

Figure 4: The relative energy input into the ISM by stellar winds during the stellar lifetime as a function of initial mass of the star. A high-mass slope of the IMF of -1.8 has been assumed. Note the dominance of massive stars in this.

In the particular case that the bubble is evolving into a constant density medium, the radius of the outer shock, R , and the velocity of expansion, V_{exp} , are given in terms of the ambient density, n, and the energy input rate, dE/dt by;

$$R = 1.7 \ (\ dE/dt_{36} \ / \ n \ )^{1/5} \ t_4^{3/5} \quad \text{pc.} \qquad (2.10)$$
$$V_{exp} = 100 \ (\ dE/dt_{36} \ / \ n \ )^{1/5} \ t_4^{-2/5} \quad \text{km.s}^{-1}$$

where the energy production rate is measured in units of 10^{36} ergs.s^{-1}, and the age is measured in units of 10^4 years (Dyson, 1981). Thus, a $40 M_\odot$ star, evolving in a medium of ambient density of 1 cm^{-3}, may blow a cavity up to 50pc across which is expanding at about 20km.s^{-1}. This implies that the momentum coupling factor of the interstellar medium with the wind and therefore with the momentum carried by the radiation field of the original star is of order unity.

2.2. METALLICITY DEPENDENT EFFECTS.

The equations given above imply that initial metallicity affects the specific energy delivered to the ISM, the period of time that a star spends in the various evolutionary stages, and the chemical yields in both the stellar winds and in the ultimate explosion. The particular case of Supernova 1987A has concentrated attention on this problem, since this occurred in a B3 supergiant progenitor rather than in a red supergiant, as might have been expected. It is clear that the relatively compact nature of the precursor is related in some intimate fashion with the fact that it occured in the LMC, with a metallicity of [Fe/H] ~ -0.3. A concise review of the various important factors is given by de Loore and Doom (1988).

In order to develop a broad understanding of the effects of metallicity, the results of Maeder (1987) and Maeder and Meynet (1987) were cast in the form of parametric equations describing the composition and extent of the various nucleosynthetic zones within the star as a function of mass and of time, and then the outer layers were stripped according to the equations given above. Figure 5 shows a typical result, for the variation of the chemical yields in the stellar winds as a function of initial mass for stars of solar metallicity. Note that, for this metallicity, this scenario suggests that single stars can make the transition to N enhanced supergiants ($M > 22M_\odot$), they become late type WN stars for $M > 30M_\odot$, early type WN stars for $M > 40M_\odot$, may become WC stars if $M > 44M_\odot$, and finally, may even become WO stars provided $M > 52M_\odot$. However, these limits refer only to the state of the pre-supernova star, and so the effective observational limits may be somewhat higher. These limits are in fair agreement with what is known about the morphology of the upper part of the H-R Diagram (Humphreys, 1982, 1984; Humphreys and McElroy, 1984). The relationship between atmospheric chemical composition, and the position of the star on the H-R diagram is reviewed in some detail by Maeder (1988).

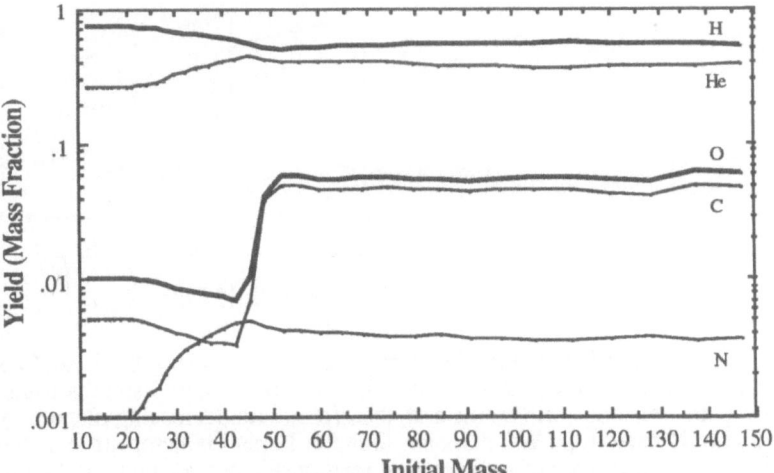

Figure 5: The variation of the lifetime-averaged chemical yields of various elements in the winds of stars with initially solar metallicity as a function of their initial mass. Note that the products of hydrogen burning appear at the surface for stars more massive than $21M_\odot$, and products of helium burning appear in the atmospheres of stars more massive than $42M_\odot$.

The expected effects of the initial metallicity on the final state of the pre-supernova star are shown as a function of mass in Figure 6. This diagram cannot be taken very seriously, since the results obtained depend critically on the spatial extent of the various convective zones within the star, which theoretically is determined by the convective criteria adopted in the models. Nevertheless, various qualitative results may be inferred.

Note that single WC stars cannot occur in systems with metallicity less than [Z] ~ -0.8. At Magellanic Cloud abundances (SMC; [Z] ~ -0.55, LMC; [Z] ~ -0.28; Russell,

444

Bessell and Dopita, 1988), the ratio of WC to WN stars is expected to be large. For metallicities higher than solar, Figure 6 would imply that the ratio of WC to WN stars would become larger than, or about equal to unity.

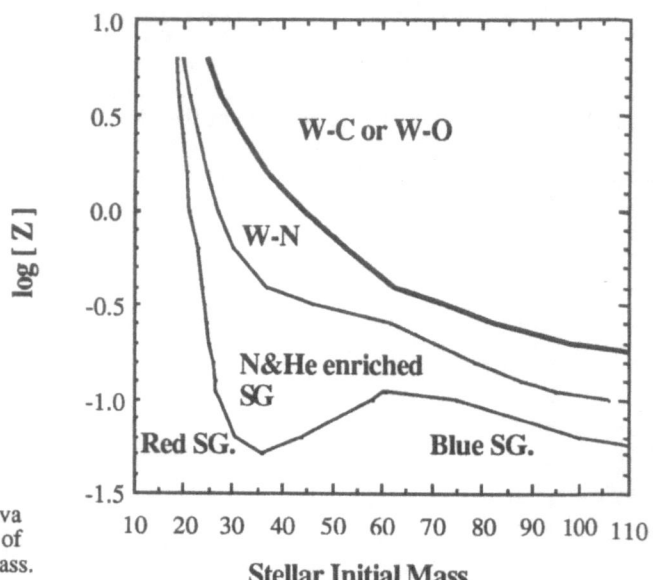

Figure 6: The pre-supernova state of stars as a function of initial metallicity and of mass.

All this is in substantial agreement with observation. In the Galaxy (Hidayat, Supelli and van der Hucht, 1982; Hidayat *et al.* 1988), the ratio of WN/WC stars is found to rapidly increase with Galactocentric radius and, therefore, with decreasing metallicity. It is interesting that only early-type WC stars are found at larger Galactocentric distance. Since, even for W-R stars we expect some kind of relationship between terminal velocity and escape velocity, this implies that only the most massive of stars, with the largest core masses, can become WC stars at low metallicity. The scenario for the production of WR stars is therefore that put forward by Schild and Maeder (1984).

In the LMC, the ratio of of WN/WC stars is ~3, and in the SMC rises as high as ~7 (see van der Hucht 1981 and references therein). A similar gradient in the WN/WC ratio has been found in the case of M31 by Moffat and Shara (1987). With these observations, an approximate relationship between WN/WC ratio and metallicity can be inferred and this is given in Table 1.

log $(Z/Z_\odot)$	+0.3	+0.15	+0.0	-0.15	-0.3	-0.6
N (WN) / N(WC)	0.25	0.35	0.89	2.0	4.5	7.5

Table 1: The observational ratio of WN/WC stars as a function of metallicity.

Figure 6 implies that only a small fraction of massive stars will explode as SN with WR star progenitors. For an IMF of slope -1.8, and assuming that stars more massive than $12.5M_\odot$ give Type II events, than about 20% of SN events will be of this type at solar metallicity, falling to only about 5% at SMC metallicity. However, at higher metallicities than the sun, a considerably larger fraction of stars end their lives as WR stars. Furthermore, the strong mass dependence of mass-loss tends to drive the cores towards a common mass at the end of the stellar lifetime, of the order of 6 - 12 $M_\odot$. These figures agree well with those predicted by Langer (1989), for the same basic physical reason. In the subsequent supernova explosion, the compact nature of the precursor ensures that the lightcurve is essentially powered by the radioactive ^{56}Co, which, by analogy with SN 1987A would be of order of $0.1M_\odot$. Thus, these stars would be expected to give a Type Ib or Type Ic display, depending on whether the outer layers are still He rich, or else have been stripped. It is not yet clear whether the features of the light curve can be well fitted by such a large core mass (Ensman, 1989), but these may still be uncertain by a factor of two.

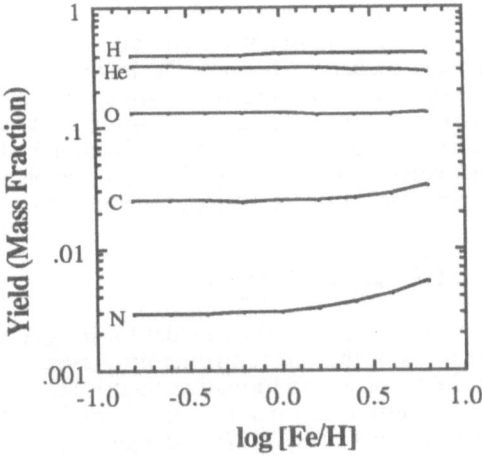

Figure 7: The variation of the Chemical Yields (Wind+SN) for massive stars as a function of metallicity. Mass-loss during the stellar lifetime only has a appreciable effect on yields for metallicity greater than solar.

Despite the importance of mass loss in the final phases of evolution of massive stars, the chemical yields of the lighter elements are not very much affected by this mass loss. This is because the bulk of nucleosynthesis occurs during the much longer hydrogen-burning phases, or in the helium-burning phases before the WC phase. The yields for H, He, and the CNO group of elements are shown in figure 7 as a function of metallicity, for an IMF of slope -1.8, neglecting the effects of explosive nucleosynthesis. Note that variations of yield may be neglected below solar metallicity. The effect of mass loss is to make it appear that the population enriching the ISM has in fact a steeper IMF, since, in the presence of mass-loss, high mass stars evolve into stars with smaller core mass.

2.3 THE EFFECT OF LOW-MASS STARS.

The main source of energy input to the interstellar medium of galaxies from the low-mass component is through the Type I supernova explosions. Whatever the physics of thies events, be they carbon deflagration or detonation binaries, or Type I1/2 explosions of isolated intermediate mass stars (Nomoto, Thielemann and Yokoi, 1984), it is clear that they represent and older population of stars. As far as the interstellar medium of disk galaxies are concerned, this means that stellar orbital diffusion will have had time to act (see Section 4.2, below). This will ensure that the scale height for the Type I supernova explosions is much greater 'than the scale height for the disk gas and that the Type I supernovae are more important in heating and maintaining the Galactic Halo (see Section 4.4), than in heating the disk ISM. As the galaxy evolves, the specific energy per unit mass of gas in the ISM would continually increase, unless the times scale for production of the Type I supernovae is shorter than the timescale for gas depletion in the galaxy. It is therefore important to determine these relative timescales to discover if late-type galaxies can produce galactic winds.

As far as the chemical evolution of the disk is concerned, the low-mass stars play an important rôle in determining the abundances of Carbon and of Nitrogen through the three-phase dredge-up processes which occur during the asymptotic giant branch (AGB) phase of evolution (see reviews by Iben and Renzini, 1983 and Iben, 1987), and the subsequent ejection of these elements as winds and planetary nebulae.

Perhaps the most notable effect of the low mass star is a result of their comparatively long main-sequence lifetimes. The low mass stars therefore serve to "lock up" a sample of the interstellar medium out of which they were formed, and store this in an (almost) pristine state before releasing it during the Giant and AGB phases of evolution. This material returns to a medium which is strongly depleted in gas, and which chemically, is more evolved. Thus, the rate of chemical evolution of the interstellar medium is slowed, and more gas is made available at late times to continue star formation.

3. Star Formation in Galaxies

There can be no successful model for the evolution of disk galaxies without a correct description of the star formation processes. As Silk (1985), pointed out, the essential ingredients of a star formation theory are the initial mass function (IMF), the star formation efficiency and the rate of star formation. Most models of galactic evolution (Audouze and Tinsley 1977; Vader and de Jong 1981) have tended to assume a constant IMF and to reduce the star formation problem to a simple "prescription" of the rate as a simple power law of the local HI gas density (Schmidt 1959), or of HI surface density (Sanduleak 1969; Hamajima and Tosa 1975). However, it has become apparent that the molecular component of gas can exceed the atomic contribution by a considerable margin (e.g. Young and Scoville 1982; Scoville and Young 1983, Sanders, Solomon and Scoville 1984), and a better correlation is claimed between the total gas surface density and star formation rates (Talbot 1980; DeGioia-Eastwood et al. 1984, Kennicutt 1989) and this has been used in the more recent models of galactic chemical evolution.

The rôle of the IMF has been receiving increasing attention in recent papers. In our own Galaxy, star formation may well have a bimodal character, with high mass stars being preferentially formed in the vicinity of the spiral arms but low-mass stars being formed throughout the disk (Güsten and Mezger 1982). If the CO-emitting molecular clouds map star-formation regions, then their distribution in the Galaxy appears to offer convincing support of the bimodal hypothesis (Scoville and Good 1987). The CO clouds are clearly divided into two populations which reflect their kinetic temperatures. The warm molecular clouds are clustered, are associated with HII regions and form a spiral arm population. The cold core clouds are distributed throughout the disk. Scoville Sanders and Clemens (1986) argue that, since the star formation efficiency for massive stars appears to decrease as the mass of the parent cloud increases, the formation of these stars must be triggered by an external cause, such as cloud-cloud collisions, rather than internally as in the sequential star formation models.

The apparent segregation of the high-mass and low-mass modes of star formation becomes even more pronounced in starburst regions. Here, several analyses suggest that, in these regions, only the high mass stars are being formed and that the low mass cutoff in the IMF is of order 3 solar masses (Rieke *et al.* 1980,1985; Olaffsson, Bergrall and Ekman 1984; Augarde and Lequeux 1985).

Larson (1986) has shown that, in a model of the solar neighbourhood with a locally decreasing global rate of star formation, the IMF must be a double peaked bimodal. In such a model, the high mass mode is more important over the lifetime of the disk, and stellar remnants can account for the unseen mass in the solar neighbourhood. The division between the "high" and "low" mass sections of the IMF occurs at about one solar mass, so that, from the point of view of galactic chemical evolution, only the high mass mode of star formation is important.

Although, in our own Galaxy, the two modes of star formation appear to be spatially distinct, with the high-mass stars preferentially formed near spiral arms, it is not necessary, or even desirable, to associate this with a density wave trigger. Elmegreen (1986) has shown that galaxies of the same Hubble types with and without a density wave have effectively identical star formation rates. He argues that the rôle of the density wave is one of spatial ordering rather than one of triggering.

In an earlier paper (Dopita 1985), a model for a high-mass mode of star formation was presented. This relied on the idea that high-mass stars control the phase properties and the velocity dispersion in the vertical, axial or w plane, and that this produces a self-regulating feedback to the star formation process. A rather different self-regulating process has been offered for individual molecular clouds by Franco and Cox (1983), who developed on ideas by Norman and Silk (1980). In this, the momentum input by winds from low mass young stars supplies the turbulent support for the cloud. It is possible that these two feedback processes work simultaneously to control the high-mass and low-mass modes of star formation respectively, as discussed below.

3.1. THE LOW-MASS MODE OF STAR FORMATION

It has long been recognised that the lifetime of molecular clouds is at least an order of magnitude longer than their free-fall timescales (Kwan 1979; Blitz and Shu 1980) and that the typical turbulent velocities are highly supersonic. It is clear that an energy source is required to give the required turbulent support. Norman and Silk (1980) and Franco (1983) suggested that the winds from young stellar objects might provide this energetic input. Franco and Cox (1983) were able to derive a stellar birthrate on the assumption that this turbulent input also serves to regulate the rate of low-mass star formation within molecular clouds. Essentially, the structure of a such a cloud at any instant can be regarded as a set of interlocking shells of compressed gas, orbiting each other under their mutual gravitational attraction. Amongst these, just a sufficient number are in a state of collapse under their self-gravity to provide enough new stars for the turbulent support. Thus, in the absence of any external perturbation, the cloud is converted into stars at a nearly constant rate per unit mass.

The details of this model depend on complex physics of the growth of gravitational instabilities on the one hand, and the transport of turbulence through the various spatial scales, on the other. We do not propose to review these in detail, but instead, refer the reader to the the review by Falgarone and Puget (1988). However, direct observational evidence for our simple physical model is forthcoming from two separate directions. Firstly, Fukui et al. (1986) have shown that in one cloud, the Orion Southern molecular cloud, the energy input from the CO outflow sources found in an unbiased survey is sufficient to balance the cloud against turbulence dissipation, provided that the timescale over which this operates is an order of magnitude greater than the free-fall timescale. A very similar result was obtained by Lada (1988) for the Mon OB1 giant molecular cloud. Second, the observed mass / radius or velocity dispersion / radius relations also support this picture. Larson (1981) found from observation that $M \propto R^2$ and that the velocity dispersion, $\Delta v \propto R^{1/2}$. Chièze (1987) has shown that this is exactly what would be expected if the interstellar clouds are close to gravitational instability in a constant pressure environment, and suggests that the sub-condensations may form a gravitational N-body system in a quasi-static virialised condition, which will leave the scaling relationships unchanged for the individual fragments. In this case the following relationships apply;

$$M/M_\odot = 142 \ (P \ / \ 3800 \ K \ cm^{-3})^{1/2} \ (R \ / \ pc \ )^2 \qquad (3.1)$$

$$(\Delta v \ / \ km.s^2 \ ) = \ 0.68 \ (P \ / \ 3800 \ K \ cm^{-3})^{1/4} \ (R \ / \ pc \ )^{1/2} \qquad (3.2)$$

These relationships conform closely to the observational material (e.g. Dame et al. 1986), which gives some confidence in the hypothesis.

In order to maintain the turbulence in the cloud, the dissipation rate in shocks must match the energy input rate through bipolar flows etc. Thus, the star formation rate per unit volume in a cloud of density ρ_g is given by a relation of the form:

$$d\rho_* \ / \ dt = \psi(Z) \ \rho_g \ \Delta v^2 \qquad (3.3)$$

where ψ is a complex function which depends on the cooling rate in the cloud, and is therefore a function of chemical composition. Some of the factors which affect this are discussed in section 3.3, below. Using the scaling relationships (3.1) and (3.2) gives the specific star formation rate (rate per unit mass):

$$(1/M) \ (dm_* \ / \ dt) = const. \ \psi(Z) \ P \ M^{1/2} \qquad (3.4)$$

This demonstrates that the relative astration efficiency in the low mass mode depends on the chemical composition and size distribution of the molecular clouds, and upon the stochastic pressure set up by the high mass star formation activity.

3.2. THE FORMATION OF HIGH-MASS STARS.

The energetic proccesses (winds, ionising radiation and supernova explosions) associated with the young, massive stars exercise the fundamental control of the phase properties and pressure of the interstellar medium (ISM). This is the basis of the multi-phase models (Field Goldsmith and Habing 1969; Cox and Smith 1974; McKee and Ostriker 1977; Cox 1979,1980), although these differ in emphasis, and in the details of how such a multi-phase medium is set up and maintained. Dopita (1985a) argued that these energetic proccesses also stir up the ISM to maintain the vertical w-velocity dispersion in disk galaxies. By assuming an equipartion between turbulent and thermal pressures in the disk medium he was able to develop a model for the rate of formation of massive stars. Here we abandon this as an arbitary assumption, and propose a model in which star formation is induced in a shock compressed layer by cloud-cloud collisions. This type of model is more satisfactory from a philosophical viewpoint in that it invokes a known physical process for star formation, namely, the development of large scale gravitational instabilities in the shocked layer (Mouschovias, Shu and Woodward 1974; Elmegreen, 1979,1982; Cowie, 1981; Balbus and Cowie, 1985).

Cloud-cloud collisions reduce the momentum, and therefore, the velocity dispersion of the gas in the vertical (w-plane). Thus, energetic processes associated with the high-mass mode of star formation must, in the steady state, feed as much momentum into the gas of the ISM as is being lost in cloud-cloud collisions. Since we have shown in Section 2 (above), that the momentum carried in the radiation field of a massive star is transferred to the ISM with a high efficiency factor, we would expect therefore that the momentum input is directly coupled with the birthrate. If $d\sigma*/dt$ is the surface rate of star formation, and σ_g is the surface density of gas, then the assumption that the star formation rate is proportional to the the cloud - cloud collision timescale τ_{cc}, gives:

$$(d\sigma*/dt) = \beta \, \sigma_g \, / \, \tau_{cc} \tag{3.5}$$

where the constant of proportionality β is composed of both a "spontaneous" term and a "stimulated" term which accounts for the fact that a burst of star formation may induce a local overpressure leading to cloud crushing and induced star formation in its vicinity. These processes represent the basis for the model of stochastic self-propogating star formation (Gerola and Seiden 1978; Seiden and Gerola 1979; Feitzinger et al. 1981) which has enjoyed some success in reproducing the structural features of both spiral and irregular disk galaxies. Here we assume that the coefficient of stimulated star formation is linearly related to the spontaneous term, so that β is not too sensitive to the galaxian environment. This is a rather weak assumption, but can be somewhat justified by the argument that stimulated star formation is ultimately controlled when breakout of the gas layer occurs, allowing the local overpressure to drain into the halo. This sets the limit on the spread of stimulated star formation events.

The intrinsic assumption in equation (3.5) is that a multi-phase medium has been set up in the first place. In practice, the gaseous disk itself can only become unstable to its self-gravity above a critical surface density. That this would lead to an outer observable edge for spiral galaxies was first pointed out by Fall and Efstathiou (1980). Kennicutt (1989 and this conference) has shown observationally that high-mass star formation shows a threshold effect in the outer portions of disk galaxies. This can be understood as resulting from the onset of large-scale gravitational instability above a critical density threshold, $\sigma_g(crit)$, given by the Toomre criterion (Toomre, 1964; Quirk, 1972; Cowie, 1981);

$$\sigma_g(crit) = \alpha \, \kappa \, <v_g> \, / \, 3.36 \, G \tag{3.6}$$

where κ is the epicyclic frequency, $<v_g>$ is the velocity dispersion in the gaseous layer, and α is a dimensionless constant near unity. An equation such as (3.6) can explain the very steep outer boundaries observed in the star-forming activity and the luminosity profiles of disk galaxies.

Only those cloud-cloud collisions that are appreciably supersonic with respect to the cloud gas will result in high-mass star formation. Lower velocity collisions will induce turbulent merging of the clouds. Massive clouds will continue to increase their mass by gravitational accretion of smaller clouds and of "cirrus" until high mass stars form spontaneously within them, or else two massive clouds collide. Thus, the size spectrum of the molecular clouds is determined by a gravitational cascade fom smaller to larger clouds. This model was first proposed many years ago by Oort (1954), and Penston et al. (1969) showed that it results in a power-law number / mass distribution.

Since the cloud-cloud collisions are radiative, the physical parameter which is conserved in the collision is the net momentum. In the steady state disk, therefore, the modulus of the sum of the momentum vectors of the individual gas clouds is maintained at a constant value. Thus, the rate of injection of momentum to the gas by radiation, mass loss and the supernova explosions of the young stellar population must just match the loss of momentum in cloud-cloud collisions;

$$\gamma v_{ej} (d\sigma*/dt) = \sigma_g v_g / \tau_{cc} \tag{3.7}$$

where v_g is the vertical w-velocity dispersion of the gaseous layer v_{ej} is a characteristic velocity of ejection of matter and γ is a coupling constant. To the extent that the IMF and the energy yield from the high mass stellar population does not depend on metallicity, γ and v_{ej} will be independent of galaxian environment.

Equations (3.5) and (3.7) imply that the vertical velocity dispersion of the gaseous disk will be maintained at a constant value, a result which appears to be valid for all disk galaxies measured so far (see section 4 , below).

Star formation pressurises the ISM in the plane, and, because the hot gas from supernova explosions can bubble up to form a hot halo to the galaxy, the scale height for pressure variation is long compared with the matter scale height. Thus, in regions where the hot coronal gas generated by supernova explosions is incapable of driving a galactic wind the stochastic pressure, P, is simply proportional to the surface rate of massive star formation divided by the mass scale height;

$$P = \alpha (d\sigma*/dt) / z_* \tag{3.8}$$

On the van der Kruit and Searle (1981a,b,1982) disk model the solution to the Poisson equation gives a z-density distribution of matter $\rho(z) = \rho(0). \text{sech}^2(z / z_*)$. This function tends to an exponential at $z > z_*$. The total matter surface density σ_t is related to the midplane velocity dispersion of the stars, v_*, by ;

$$v_*^2 = \pi G \, \sigma_t \, z_* \tag{3.9}$$

The gas velocity dispersion is considerably less than v_* generally, therefore the gaseous scale height is also less. The relation between the velocity dispersion of the gas and its scale height z_g is given by;

$$v_g^2 = \pi G \, z_g \, [\sigma_g + (\sigma_t - \sigma_g)z_g / z_*]$$

at early times, when the disk is entirely gaseous, and at late times when the self-gravity of the gas is small compared with the self gravity of the stars in the gas layer, this equation is well represented by the approximation:

$$v_g^2 = \pi G \; \sigma_t \, z_g^2 \, / \, z* \qquad\qquad (3.10)$$

In the case that the disk matter distribution is exponential, and the gas can be regarded as a subpopulation in the same potential, but with different scale height, the mid-plane velocity dispersion is given (exactly) by:

$$v_g^2 = 2\pi G \; \sigma_t \, z_g^2 \, / \, (z* + z_g) \qquad\qquad (3.10a)$$

where the scale heights are are now the scale heights of the exponential, half that of the previous definition. The one-dimensional velocity dispersion of the Giant Molecular Clouds (GMC's) in our solar neighbourhood has been measured by Stark and Brand (1989) at 8 km.s^{-1} (RMS). This is the same as the vertical velocity dispersion of the LMC GMC's (Cohen *et al.* 1988), and similar to the HI vertical velocity dispersions measured in other galaxies, as would be expected if the velocity ellipsoid is symmetric.

To solve the above equations in terms of the total star formation rate we require an expression for the cloud-cloud collision timescale, τ_{cc}. If the area filling factor of molecular clouds in the gaseous disk is f, then the cloud-cloud collision timescale is given by:

$$\tau_{cc} = 2 \, z_g \, / \, v_g f \qquad\qquad (3.11)$$

However, we know from the previous section that the clouds act as if they are in marginal equilibrium against collapse with the external pressure, and that they obey the scaling relationships, (3.1) and (3.2) which imply a constant effective surface density for all clouds at a given pressure. Thus the filling factor is given by:

$$f = f_o \, (P \, / \, P_o)^{-1/2} \, (\, \sigma_g \, / \, \sigma_o \,) \qquad\qquad (3.12)$$

where P_o and σ_o are a reference pressure and surface density, respectively. Solving for the star formation rate in equations (3.5, 3.8, 3.10a, 3.11 and 3.12) yields:

$$d\sigma*/dt = (\beta/2 f_o \, \sigma_o)^{2/3} \, (2\pi G \, P_o \, / \, \alpha)^{1/3} \, (z* \, / \, (z* + z_g))^{1/3} \, \sigma_t^{1/3} \, \sigma_g^{\,4/3} \quad (3.13)$$

As might have been expected on purely phemenological grounds, the star formation rate depends primarily on the local surface density of gas. However, there is also a dependence on total surface density, and upon the relative scale heights of the gas and of the stars. Thus, as stellar diffusion changes these, there will be a corresponding, but weak, evolution in the specific star-formation activity. High mass star-formation according to (3.13) will follow a Schmidt-Type law with index N= 4/3. This is in good agreement with Kennicutt's (1989) data for disk galaxies, which show N = 1.3 ±0.3.

3.3. HOW MAY THE CURRENT RATE OF STAR FORMATION BE MEASURED?

There is little problem in measuring the rates of star formation for stars massive enough to substantially ionise the surrounding ISM. Provided that the H II regions formed absorb all the UV photons produced by these stars, then the number of recombinations per unit time matches the ionisation rate, which in turn measures the available surface area of ionising stars. This is a function of the mean lifetime of massive stars, the slope of the IMF, but, most importantly, is directly proportional to the birthrate of massive stars. Thus a measure of the flux in radio continuum emission, or in a recombination line will give a good estimate of the birthrate of massive stars. This is the basis of the technique used by Kennicut and Kent (1983) to determine global rates of star formation in external galaxies from the Hα line emission, or by Viallefond and Goss (1986) who used the thermal radio continuum. The non-thermal radio continuuum can also be used to estimate star formation rates, provided that this arises from the cosmic ray electrons produced in supernova explosions (Wunderlich, Klein and Wielebinski, 1987).

A slightly more indirect way of measuring star-formation rates in the upper main sequence is to use the reprocessing of the UV light from young stars by dust. The volume emissivity in the far-IR measures the rate of dust heating and is proportional to the product of the intensity of the local radiation field and the density of the dust. Thus, for a given dust content, the far-IR emission is directly proportional to the space density of young stars. In regions of high specific star formation, the dust temperature should be strongly correlated with excess far-IR emission, as indeed is observed (Young *et al.* 1986; Mirabel and Sanders, 1988). Problems will arise in this method as a result of changes in the geometrical relationship of the dust with repect to the star-forming region, or as a result of changing metallicity, which changes both the quantity and the chemical makeup of the dust. In particular, we would expect this method to give higher rates of star formation in young, dust-embedded starburst galaxies, and lower rates in low metallicity blue compact galaxies.

What does the CO flux measure? The molecular clouds in galaxies dissipate their internal turbulence by internal shocks, which radiate predominantly in CO lines at solar abundances and at the densities which are typical for these clouds. Thus, the CO emission is an indicator of the rate of energy dissipation within molecular clouds rather than a direct indicator of the total mass of the cloud. However, if molecular clouds are in virial equilibrium, and if low-mass star formation drives turbulence,there should be a close correlation between the virial mass of molecular clouds and their CO luminosity. Such is indeed observed to be the case by Solomon *et al.* (1987). It therefore seems plausible that the CO emission is a measure of the rate of low-mass star formation within molecular clouds.

Usually, the CO luminosity is related to the molecular mass in hydrogen. The argument is based on the definition of the CO integrated luminosity and on arguments of virial equilibrium (*e.g.* Dickman, Snell and Schloerb, 1986: Young *et al.* 1986). The CO luminosity, L_{CO}, can be written in terms of a function of the relative abundance of the CO molecule, $Z(CO)$, the radius of the molecular cloud, R, the surface density of the cloud, σ_g, the linewidth, Δv, and as a function of the gas temperature, T_g;

$$L_{CO} \propto \pi R^2 \, \Delta v \, T_g \tag{3.14a}$$
$$\text{or} \quad L_{CO} \propto \pi R^2 \, \sigma_g \, F\{Z(CO)\} \, f(T_g) \tag{3.14b}$$

where the (3.14a) refers to a cloud which is optically thick in ^{12}CO, and (3.14b) is applicable to the optically thin case. Young (1988 and this conference), would argue that for optically thick clouds, these latter factors tend to cancel out, and therefore L_{CO} directly measures $M(H_2)$ directly. This view is also supported by Knapp (this conference). However, it may be more physically instructive to accept the Chièze (1987) proposition that the interstellar clouds are close to gravitational instability in a constant pressure environment, which gives the observed scaling relationships between mass, velocity dispersion and radius $M \propto R^2$; $\Delta v \propto R^{1/2}$ (c.f. equations 3.1 and 3.2). From equation (3.14a) we expect that for the local molecular clouds in our Galaxy, for which the abundance effects dissapear, $L_{CO} \propto T_g M^{5/4}$, which agrees to high precision with the observational result (Solomon *et al.* 1987). This is demonstrated in Figure 8, below.

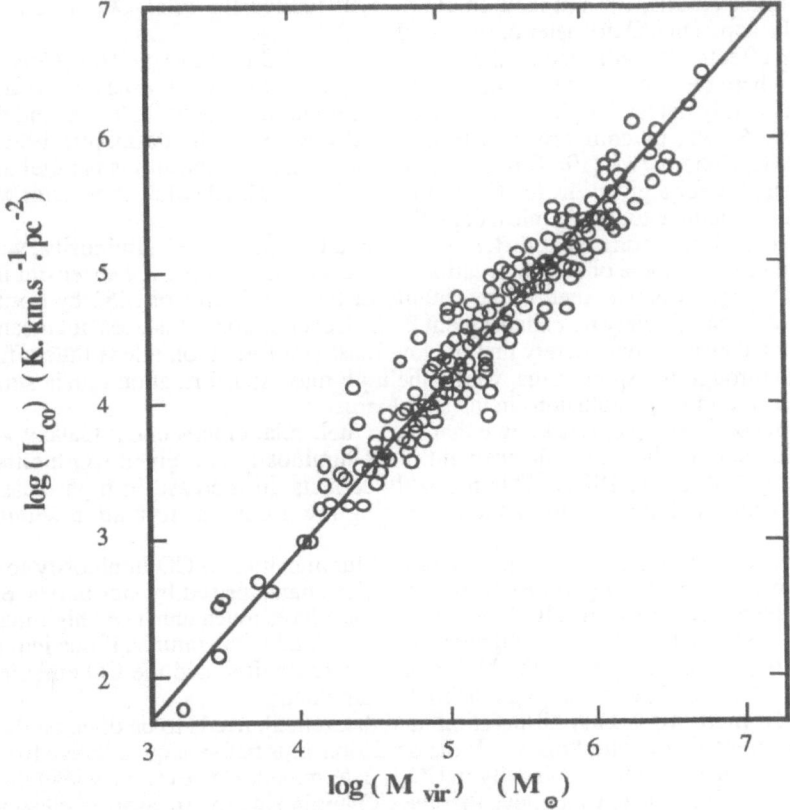

Figure 8: The virial mass / CO luminosity relationship for molecular clouds in the galaxy as observed by Solomon *et al.* (1987) compared with the theoretical relationship derived in the text (L_{co}= const. $M^{5/4}$).

For optically thick clouds, there is therefore little doubt that the CO line locally measures the energy input from low-mass star formation. To the extent that the molecular clouds are optically thick, and that the size spectrum of molecular clouds is invariant with galaxian environment, the line may *also* be used to measure the molecular gas content in metal-rich galaxies.

Turning now to the question of clouds with intrinsically low CO abundance. Here the optically thin case is much more likely to apply. At these low abundances, the luminosity will still scale with the mass of the cloud, but the absolute value of the luminosity is reduced by a factor $f\{Z(CO)\}\ T_g^{-1} f(T_g)$. The factor $f\{Z(CO)\}$ is not expected to be linear with abundance, but rather, to be a steeper function. Maloney and Black (1988) have shown that the increased transparency of the clouds in the UV will ensure that the CO abundance in the cloud drops rapidly for abundances below solar, but that the number density of molecular hydrogen is much less affected. Indeed, even for optically thick clouds, the reduced shielding will reduce the total CO emission by reducing the apparent CO diameter of the cloud.

These effects are confirmed by the observations of CO in the Magellanic Clouds. In the SMC, where the C abundance in depleted by a factor of ten with respect to solar, CO emission is barely detectable. In the LMC, the C depletion is about 0.55 dex. and the O depletion 0.35 dex., by comparison with the local ISM (Russell and Dopita 1989a,b). For the LMC, Cohen *et al.* (1988) find that the L_{CO} vs. Δv relationship is parallel to the Galactic one, but for a given Δv, the CO luminosity is some 0.7dex lower, a factor which is similar in magnitude to the chemical depletion.

Armed with this somewhat different interpretation of the CO luminosity, we can make more sense of some of the observations. For example, a strong enhancement in the Hα surface brightness was seen in the vicinity of the spiral arms of M51 by Lord and Young (1987), but virtually no enhancement in the CO emission. This does not mean that there is not a density wave, merely that the low-mass star formation rate is little affected by passage through the spiral arms, whilst the high-mass star formation rate is strongly enhanced by cloud-cloud collisions in the spiral arms.

As a second example, it is known that those molecular clouds in our Galaxy which contain HII regions show an enhancement in IR-luminosity at a given CO luminosity (Solomon and Mooney, 1988). This probably reflects an increase in high-mass star formation which is independent of the continuing low-mass star formation within the cloud.

Third, consider the relationship between IR luminosity and CO luminosity for 150 galaxies observed by Young(1988). Those galaxies characterised by star bursts, either internally generated, or as a result of interaction, must have much enhanced high-mass to low-mass star formation ratios and higher molecular cloud temperatures, if one interprets the IR heating as resulting from the high mass star formation, and the CO emission as measuring the dissipation of energy in the molecular clouds.

Clearly, more work, both observationl and theoretical, needs to be done on the CO luminosity - metallicity - virial mass - HI mass relationships before a quantitative function connecting H$_2$ mass and the luminosity in CO is determined. However, provided that we can measure O abundance, we believe that the CO luminosity, by measuring the amount of turbulent dissipation in molecular clouds, is nonetheless a very useful indicator of low-mass star formation rates in external galaxies.

4. The Structural Evolution of Galactic Disks.

4.1 THE EVOLUTION OF THE GASEOUS DISK

Equations (3.1) and (3.2), which depend on the assumptions that high mass star formation is initiated by cloud-cloud collisions, and that high mass star formation also maintains the vertical or axial velocity dispersion in the gas layer, together imply an immediate observational consequence, namely ;

$$v_g = \beta \gamma v_{ej} \qquad (4.1)$$

that is to say, the vertical velocity dispersion of the gas in *all galaxies* will be the same, and independent of the radial coordinate. This is an observed property of all disk galaxies, covering a wide variety of morphological type, which have so far been observed (van der Kruit and Shostak 1984 and references therein, Meatheringham *et al.* 1987, Stark and Brand 1989). The HI velocity dispersion is of order 6-10 km.s^{-1}, and shows little variation between the arm and interarm regions. However, it is seen to locally increase in regions of active star formation, such as the 30 Doradus region in the LMC. This is entirely consistent with our hypothesis.

The total surface density, $\sigma_T(r)$, in the disk is observed to decline radially outwards according to an exponential law with scale length R_o;

$$\sigma_T(r) = \sigma(0) \, exp[\, -r / R_o] \qquad (4.2)$$

(see, for example, van der Kruit and Searle, 1981a,b). Thus, equation (3.5) with a constant velocity dispersion in the gas layer implies that the vertical scale height in the gas varies as;

$$z_g = z_g(0) \, exp[-r / 2R_o] \qquad (4.3)$$

Such a variation is in fair agreement with the observations of the thickness of the HI layer in our Galaxy, assuming a disk scale length of order 4 kpc (Downes and Güsten, 1982).

Turning now to the question of the radial distribution of gas in the disk. It has been known for many years that the radial distribution of the luminous part of a galaxy can be well represented by an exponential function (Freeman 1971). It is generally thought that this is a comes about as a natural consequence of the collapse, because an exponential disk has almost the same distribution of specific angular momentum as a self-consistent rotating spheroid of gas. As far as the gaseous part is concerned, the current distribution has been altered by star formation, infall, and by the effects of stellar orbital diffusion . These topics are discussed in detail elsewhere in this review. However, radial flows and mixing of material may also also a natural consequence of viscosity in the gaseous part of a galactic disk. Recent work by Lin and Pringle (1987), and by Yoshii and Sommer-Larsen (1988, 1989a,b), has shown that, even if the galaxy collapses to a disk with a non-exponential radial profile, then the exponential profile of the stellar disk can be accounted for if the viscous timescale is of the same order of size as the star-formation timescale. This model could account for the extended outer HI envelopes seen in many galaxies (Bosma, 1978,81; Wevers, 1984).

Powerful radial motions in the gas will be driven by bar-like deformations in the potential. This is the probable mechanism which cleans out the gas from the inner portions of barred spirals, and which feeds nuclear Seyfert and starburst activity in these galaxies.

4.2 STELLAR ORBITAL DIFFUSION

A classic problem in stellar dynamics is that the velocity dispersion and scale height of stars appears to increase with age (Wielen 1977; Twarog 1980; Wielen and Fuchs 1983). One possibility is that the observed velocity dispersions reflect the velocity dispersion at birth (Larson 1976; Gilmore and Wyse 1986). However, the vertical velocity dispersion of the gaseous thin disk is approximately constant for all galaxies. Since this result is time, position and environment independent, it is not possible to sustain the argument that velocity dispersions have remained unchanged with time. A likely explanation for the scale height / velocity dispersion / age relationships observed in our Solar neighbourhood is that the velocity dispersion is increased with time by stellar orbital diffusion caused by gravitational scattering by giant molecular clouds. Such an evolution was originally proposed by Spitzer and Schwarzschild (1951, 1953), and the idea has been further developed by Wielen (1977), Vader and de Jong (1981) Lacey (1984) and Villumsen (1985). According to this theory, stars are born with some initial velocity dispersion $V_*(0)$, which we can take to be the same as the dispersion in the gaseous thin disk. At time t, they will have aquired a velocity dispersion $V_*(t)$ given by:

$$V_* (t) = V_* (0) [1 + t / t_{diff}]^{1/3} \qquad (4.4)$$

this equation remains valid only for so long as the scattering clouds and the stars can be considered to remain in the same layer. The breakdown of this assumption will lead to a change in the exponent. Weilen (1977) finds empirically that an exponent of 1/2 gives a good fit to the observations, whereas both Lacey (1984) and Villumsen (1985) find that a rather lower exponent in the range 0.25-0.35 is a better fit to the theoretical models. The diffusion or scattering timescale depends on the characteristic mass, M_c, of the scattering centers (molecular clouds or spiral arm density perturbations) and the local density. The Spitzer - Schwarzschild formulation gives;

$$t_{diff} = 4 V_* (0)^3 / 3\pi^{3/2} G^2 M_c \rho \ln[a] \qquad (4.5)$$

where a is an impact parameter. Locally, t_{diff} is of order 5×10^7 years. For the purpose of generality, it is convenient to cast the diffusion time in terms of the diffusion timescale of a disk initially entirely gaseous, $t_{diff}(0)$, which can be related to the initial diffusion timescale at a reference point in the galaxy, t_r;

$$t_{diff}(0) = \varphi V_g (0)^3 / \rho_g(0) = t_r [\sigma_r / \sigma_T]^2 \qquad (4.6)$$

To a first approximation, we can assume that the conversion of matter into stars and remnants proceedes exponentially with time, and that the gas collapses to a thin disk on a timescale which is short in comparison to the current age of the galaxy. On this basis, equations (4.4) to (4.6) can be used to compute the time evolution of the mean stellar velocity dispersion V_*. From many such runs with different gas depletion and infall timescales, an adequate analytic fit to V_* is found to be;

$$V_* = V_g [1 + (t / t_r)(\sigma_T / \sigma_r)^2]^m \qquad (4.7)$$

where the exponent, m varies between 0.25 and 0.31, in agreement with Lacey(1984) and Willumsen(1985). With the particular value m=0.25, a very interesting result is found. From equation (3.4), (3.5) and (4.7) it follows that, at any radial position in the galaxy, where the surface density is $\sigma_T(r)$, the scale height of the stars, $z_*(r,t)$, is given by:

$$z_* (r,t) = (V_g^2 / \pi G \sigma_T(r,t)) [1 + (t / t_r) (\sigma_T(r,t) / \sigma_r)^2]^{1/2} \qquad (4.8)$$

Since $t_r \ll t$ at the current time, this simplifies to:

$$z_* (r,t) = z_* (t) = (V_g^2 / \pi G \sigma_r) (t / t_r)^{1/2} \qquad (4.9)$$

Thus we have the result that, in a galaxy in which the disk formed at a particular epoch, the current scale height of the stars depends only on the age of the disk, and is independent of the radial coordinate in the galaxy. The variation in the exponent m from the value of 1/4 is so weak that this result is essentially independent of the actual value of m (see also Lacey and Fall 1983). This result is exactly what is required to explain the results of van der Kruit and Searle (1981a,b, 1982), who found that the observed light distribution in edge-on galaxies could be best fitted by a model in which the stellar scale height is constant with radius.

The effect of the stellar diffusion on the gaseous component cannot be neglected. At late times, there is an asymmetric drift of the stars relative to the gas, since the stellar component has now been perturbed into elliptical orbits. The effect of encounters will be to produce a dynamical friction on the gas clouds, which will have the result of shedding some of their angular momentum about the galaxy. Thus, a net inwards radial flow of gas is induced. This process will feed gas in towards the centre of the galaxy, even in the absence of a bar-like instability such as mentioned in the previous section.

4.3 INFALL

Infall refers to an extended period of mass accretion and has been considered by a variety of authors (Larson 1972; Searle 1972; Lynden-Bell 1975; Chiosi 1980; Tinsley 1980; Vader and de Jong 1981; Lacey and Fall 1983; Lacey 1985). The major motivation for its introduction is to attempt to account for the metallicity distribution amongst the GK dwarfs in the solar neighbourhood, frequently referred to as the "G dwarf problem " (Schmidt 1959; Bond 1970; Pagel and Patchett 1975). The infall is generally assumed to be an exponentially decreasing function of time, although Chiosi (1980) allows, in addition, a low level of continuous inflow. In order to come close to fitting the GK metallicity distribution with this process alone, long infall timescales are required, of order 3-6 Gyr. Such long timescales imply that matter falling into the disk cannot be a leftover of some initial disk configuration, but would have to be derived from very large initial radii. The free-fall timescale for a galaxy is $t_{ff} = 1.65 \ (R_{100}^3 / M_{11})^{1/2}$ Gyr, where the radius R is measured in units of 100kpc and the mass M is in units of 10^{11} solar masses.Since the mass of our Galaxy is about 4×10^{11} solar masses, the protogalaxy would have had to have a diameter of order 300 - 600kpc in this model. Such a value is not absolutely excluded, but it does seem uncomfortably large.

4.4 GAS PHASES AT LARGE SCALE HEIGHTS

For the Galaxy, it has become clear that the hot corona of shock-heated gas, first suggested by Spitzer (1956), does indeed exist. The presence of this gas is evident in the soft X-ray observations (Tanaka and Bleeker 1977), and in observations of OVI absorption (Jenkins 1978). As it cools, it gives rise to absorption in highly ionised species such as N V, C IV and Si IV which can be observed with the IUE Satellite (Savage and de Boer 1978,82; York et al. 1982; Pettini et al. 1982; de Boer and Savage 1983). From this work, it is evident that this gas has a (local) scale height for cooling of about 3-4 kpc, and is denser and more confined to the disk towards the inner parts of the Galaxy. These observations represent the experimental justification for eqn.(2.3); that the coronal gas is supplied and heated by supernova explosions in the disk and is responsible for pressurising the disk cloud and intercloud matrix as in the theories of the interstellar medium by McKee and Ostriker (1977) or by Cox (1981). In the Galaxy, the rate of mass exchange between the halo and the disk is of the same order as the star formation rate.

The energy input from Type I supernovae may be particularly important in determining the phase structure of the ISM for heights of order 300 pc and greater above the galactic midplane. In this conference, Don Cox (this conference) has emphasied the fact that there exists a warm, neutral medium (WNM) with a density of order 0.2 cm^{-3} and a scale height of about 300 pc (the so-called Lockman Layer), and that a warm, ionised medium (WIM) with a typical density of order 0.025 cm^{-3} and a scale height of 1500 pc (the so-called Reynolds Layer) must also exist if the diffuse Hα observations are to be explained. The Type I supernovae are thought to result from relatively low mass binary systems, and the characteristic evolutionary lifetime is of order 1Gyr, or greater (see below). In this times, stellar diffusion will have increased the w-velocity dispersion to about 25km.s^{-1}, compared with 8km.s^{-1} for the molecular cloud layer. The scale height of the population at the time of explosion is therefore about 1kpc, and the Type I events is therefore very important in heating and supporting the WNM and the WIM as well as resupplying the hot coronal gas. The collisionless shock emission associated with the Type I supernovae furnishes direct evidence that these occur in the WNM or WIM. The Hα data give $< n_{HI} > \sim 0.06$ cm^{-3} (Tuohy *et al.* 1982); whereas the X-Ray data suggest $< n_{HI} + n_{HII} > \sim 0.3$ cm^{-3} (Long, Dopita and Tuohy 1982). These figures are consistent with those given above, but suggesti that the WNM and WIM may not exist as separate phases. If this is the case, then the partial ionisation may be the result of photoionisation by diffuse X-rays (Dopita 1985b) rather than from photoionisation by hot stars in the plane as suggested by Cox (this conference).

If the gravitational binding energy of the hot coronal gas is smaller than its thermal energy less the radiative losses as it streams out into the halo, then a galactic wind, rather than a steady state "galactic fountain" will result. The conditions under which this will occur were discussed in an elegant paper by Chevalier and Oegerle (1979). In our local solar neighbourhood, and in the inner regions of the Galaxy, this condition does not appear to be met, but in the outer parts of the Galaxy, in star-burst systems, or in low-mass systems, such a wind may be driven.

5. The Chemical Evolution of Galactic Disks.

In the past there has been a tendency to separate disk evolution into three largely independent areas of study; star formation, disk kinematics and the chemical evolution problem (for exceptions to this, see Vader and de Jong, 1981, Lacey and Fall 1983 and Gilmore and Wyse 1986). Insofar as chemical evolution models is concerned, there has been a tendency to follow, in general, the prescriptions set out, for example, by Tinsley (1980). In these "classical" models, the star formation problem is reduced to a rate cast in terms of the local HI and / or molecular gas density (Schmidt 1959; Guibert, Lequeux and Viallefond 1978), or surface density (Sanduleak 1969; Hamajima and Tosa 1975). The chemical evolution is treated on a "closed box" assumption, i.e. no radial exchange of matter is permitted. The additional assumption of instantaneous mixing and recycling is frequently used in analytic models (Pagel and Patchett 1975; Audouze and Tinsley 1977; Rana and Wilkinson 1986; Edmunds 1987), although considerations of stellar lifetimes of low-mass stars in comparison with disk lifetimes clearly invalidate this assumption.

The simple model resulting from these approximations is incompatible with the observational restraints derived from the solar neighbourhood, or from what is known about our Galaxy as a whole (Miller and Scalo 1979; Tinsley 1980; Twarog 1980; Tosi 1982). The fit can be improved in a variety of ways, which fall into two main classes; those which introduce bulk gas flows, and those that change the initial mass function or star formation rate. In the first class can be found the models of Lynden Bell (1975);

Choisi (1980) and Choisi and Matteucci (1981) which allow for infall over an extended period of time. Gilmore and Wyse (1986) allow for a two-zone disk structure, with a thick disk forming before a thin disk, although it is very difficult to see how a thick disk could be supported for the period of time required. Mayor and Vigroux (1981) and Lacey and Fall (1985) have investigated the effect of radial gas flows, which gives in effect, a multizone model with exchange.

Radial flows and mixing of material can occur as a natural consequence of viscosity in the gaseous part of a galactic disk, as a consequence of stellar diffusion and/or bar instabilities, or else through the redistribution of angular momentum and through thermal instabilities in a hot halo. It is clear that the Pandora's "closed box" has been opened, and future models of galactic chemical evolution will have to be run at much higher levels of sophistication in the input physics.

Since the effective chemical yield depends critically on the shape and time-evolution (if any) of the IMF and on the local efficiency of star-formation, it is clear that a good model for the chemical evolution of galaxies requires an adequate description of the physics of the star-formation process. The effect of changing the efficiency of star formation has been discussed by Tozi and Diaz (1984) who assume infall and a star formation rate which decreases with time, but which is coupled to gas content only through the initial parameters. It has been argued(Terlevich and Melnick 1983 ; cited by Matteuchi and Tornambé, 1984) that the slope of the initial mass function may be very sensitive to the metallicity. However, if such a variation exists, it must be very much weaker originally proposed. The possibility that the efficiency of star formation may change with position in a galaxy has been fairly extensively investigated. Wyse (1986) and Wyse and Silk (1989) suggest that it is influenced by the local angular frequency of the gas. In the first paper Wyse assumes that the molecular gas content is coupled to the atomic gas through the local angular frequency, in an attempt to explain the very different radial distributons in HI and CO emissions, and in the rates of star formation for young stars, and in the second of these papers, a simple modification to a local Schmidt law was made by introducing a multiplier factor which depended on the local angular frequency. Recently, Kennicutt (1989), has made an objective study of the Schmidt-type star formation laws using his new Hα data in conjunction with H I and CO survey results of many disk galaxies. A power law with index slightly larger than unity, about 1.3±0.3, is indicated

The hypothesis of bimodal star formation has in recent years recieved considerable observational support and has been the subject of evolutionary models (Güsten and Mezger, 1983; Larson 1986; Wyse and Silk 1987). The major problem here is to avoid ad-hoc assumptions about the relative star formation efficiencies in the high and low mass modes.

The observational material that has been accumulated in our solar neighbourhood over the years is sufficient to place very severe restraints on any model of galactic evolution (see the review by Wheeler and Sneden 1989). The end point for models is determined by the measured age, the metallicity, the present-day local gas and stellar content and scale height, estimates of the mass fraction in stellar remnants, the rate of star formation and the gas depletion timescale. The past history of the local disk can be inferred by age/metallicity, metallicity/height, stellar dynamics/age and element abundance ratio/metallicity relationships, or by the metallicity distributions of long-lived stars. Since there is absolutely no reason to suppose that our solar neighbourhood is in any way peculiar in its properties, it is therefore reasonable to hope that a galactic evolution model which can successfully account for all the locally observed relationships, should also be capable of describing the absolute values and the radial variation of the observable parameters in this and other galaxies.

To this end, a chemical evolution code was developed along the general lines described by Matteucci and Tornambé (1985) or Matteucci and Greggio (1986). This allowed for infall, for the delay in the timescale for return of gas from envelope ejection from the lower-mass stars, and for the variation of yield with metallicity discussed above. Radial flows are not taken into account. The star formation is assumed bimodal in the sense of Larson (1986). That is to say, the high-mass mode of star- formation applies essentially above one solar mass, and determines all the chemical yields; whereas the low-mass mode of star formation applies below one solar mass and serves essentially to lock up gas in very long-lived stars. The ratio of the star formation rated for the two modes is maintained as a free parameter. The shape of the IMF for the low-mass component is immaterial, since it has no effect on the chemical evolution. The form adopted for the high-mass IMF is:

$$dN(m) \ /d(log(m))= const \ [1-exp\{-(m-m_{low}/2)/2m_{low}\}] \ m^{-\alpha} \qquad (5.1)$$

The lower mass cutoff, m_{low}, is chosen to be one solar mass, so that the term in square brackets serves to give a peak near $1M_{\odot}$, and to truncate the IMF below $0.5M_{\odot}$. At high masses, the IMF becomes a simple power law. In the Solar neighbourhood, the slope of this is fairly well determined by by observation to lie in the range $1.5 \le \alpha \le 2.0$ (Burki 1977; Scalo 1986; Larson 1986 and references therein). In the Magellanic Clouds, the slope appears to be somewhat steeper (Humphreys and McElroy, 1984), possibly as high as 2.5 in the case of the SMC.

We have modelled the abundance variations of He, C, N, O and Fe. This list of elements encompasses the three main nucleosynthetic sources of the elements; C and N are produced in the low to intermediate mass range, O is produced in massive stars, and Fe by Type I supernovae, the precursors of which are discussed below. Included in the iron peak elements would be Cr, Ni and Zn, whereas the α- elements, Ne, Mg, S, Si, Ca and Ti will enjoy a similar enrichment history to O.

The metallicity-metallicity diagrams presented by Wheeler and Sneden (1989), put very strong constraints on the parameters which apply in the solar neighbourhood, and we use these in the attempt to carry through this exercise.

5.1. OXYGEN

Oxygen has perhaps the simplest enrichment history of the elements we are considering, since it is produced only in massive stars with masses greater than, say, $12M_{\odot}$. In this sense it it "the right stuff" (Wheeler and Sneden, 1989) for probing the past history of high-mass star formation. The bulk yield for the massive stars is shown in figure 6 for an IMF of slope 1.8. In our Galaxy, the total yield (when star formation of all possible masses is taken into account), is only ~ 0.015, and for the Magellanic Clouds is as low as 0.005. These figures restrain the slope of the IMF and the ratio of high-mass to low-mass star formation modes.

5.2 IRON

Iron is produced at very low abundance in Type II supernova explosions. However the majority of this element comes from Type I supernova explosions. These may either be of the so-called Type I1/2, the carbon deflagration of the degenerate core of a single star of mass $\sim 8\text{-}12M_{\odot}$, or else the detonation or deflagration of a white dwarf component of binary star, once its mass exceeds the Chandrasekar Mass as a result of accretion Nomoto, 1982a,b; Nomoto, Thielemann and Yokoi 1984).

The low-abundance asymptotic value of the [O/Fe] ratio can put very strong constraints on the production of iron in normal Type II or in Type I1/2 events. First, consider the Type I1/2 events. Each of these will produce ~ $0.5 M_\odot$ of Fe (Nomoto, Thielemann and Yokoi 1984). This iron is produced in a timescale comparable to the evolution timescale of the more massive stars, and so will have a profound effect on the asymptotic [O/Fe] ratio. The range of stellar masses that might possibly become Type I1/2 supernovae has been found to be increasingly broad as the metallicity is decreased (Tornambé, 1984). If the mass range was 8-12 $M_\odot$, then the asymptotic [O/Fe] ratio would be -0.58, and if only 9-11 $M_\odot$, then it would be -0.34 as a result of the Type I1/2 events alone. This compares with the observed value of ~ +0.5. We therefore conclude that Type I1/2 events do not occur to any appreciable extent, since otherwise the galaxy would be drowning in excess iron! This conclusion is drawn in sharp distiction to that of Matteucci and Tornambé (1985), for reasons that are not at all evident to us.

Assuming that Type I1/2 events, if they exist, are unimportant, then the asymptotic value of the [O/Fe] is porbably telling us more about the relative production of Fe in massive stars. The explosion of SN 1987A allows us to estimate the value of this asymptotic ratio. From the photometry of the supernova during the nebular phase, when the light curve is powered by the radioactive decay of ^{56}Ni, Dopita et al. (1988) estimate that 0.085±0.029 $M_\odot$ of radioactive nickel was produced in the original explosion. Theoretical models of the progenitor and the subsequent explosion show that 1.7±0.3 $M_\odot$ of oxygen were produced. This corresponds to an [O/Fe] = 0.65±0.25. This is very similar to the asymptotic value observed for extreme Pop II halo stars in the Galaxy; [O/Fe] = 0.5±0.2 (Sneden, 1985; Wheeler and Sneden 1989). Therefore, to the extent that one believes that SN 1987A represented a typical Type II event, the implication is that Type II supernova events dominated the nucleosynthesis in the collapse phase of our galaxy. The [O/Fe] value falls towards its solar value only as the iron produced in the Type I carbon deflagration events became important.

The actual evolution of the [O/Fe] ratio with metallicity is driven by the slope of the IMF and by the relevant timescales; the timescale for infall of halo gas, and the deflagration timescale. The deflagration timescale is the characteristic period between the formation of a white dwarf and its subsequent deflagration. This depends critically on the nature of the precursor binary system (Iben and Tutukov, 1984; Matteucci and Greggio, 1986). If the companion star is a low-mass star, and the white dwarf had a more massive precursor, which put it near the Chandrasekar mass, then it is possible that mass-accretion during the giant phase of the companion would push it over the limit to produce a carbon deflagration supernova. However, a more likely outcome is the formation of a common envelope, which is expelled as the two stars spiral together. When the envelope is lost, we have remaining a doubly-degenerate system which will eventually spiral together due to gravitational wave radiation until matter transfer can start again, resulting in a Type I event when enough has been accreted. The timescale for gravitational wave losses depends as the fourth power of the semi-major axis of the system, and so may be highly variable from system to system. This is the probable explanation of why Type I supernovae can occur so frequently in elliptical galaxies.

Given these physical complexities, the deflagration time must be treated as a free variable. The Type I precursors are assumed to be a exponentially decaying fraction (about 1-4%) of all the white dwarfs which are formed. The actual percentage required is influenced by the infall timescale and by the deflagration timescale itself, since [O/Fe] ~0.0 at [Fe/H]~0.0. In figure 9 is shown the comparison between the observed [α/Fe]

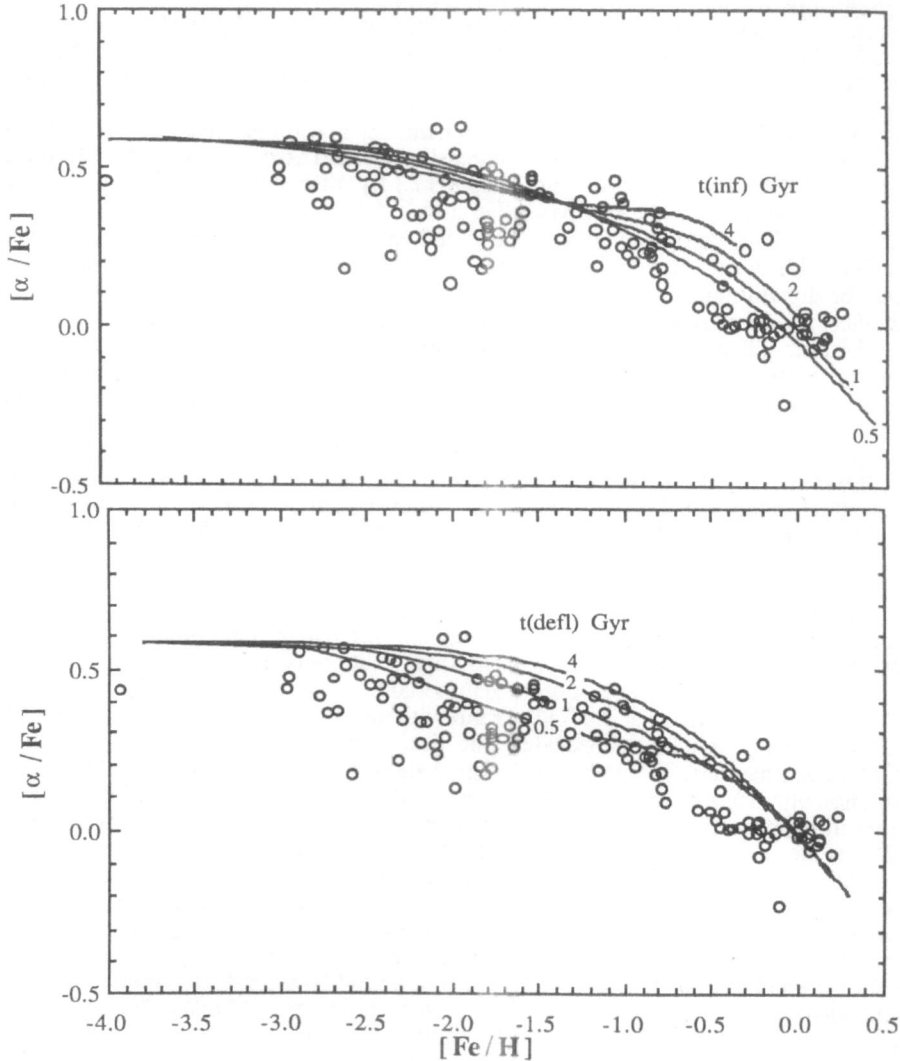

Figure 9: The [α/Fe] vs. [Fe/H] relationship for old disk stars in the solar neighbourhood (Wheeler and Sneden 1989) compared with models with different infall and deflagration timescales.

ratio for old disk stars from the compilation by Wheeler and Sneden (1989), and our models with varying deflagration and infall times are shown as solid lines. The implicit assumption here is that $[\alpha/O] = 0.0$ at all metallicities. The observations do not constrain either of these timescales very tightly, but it is clear that $0.5 \lesssim t_{defl} \lesssim 4$ Gyr, and $t_{infall} <$ 4 Gyr gives a satisfactory fit.

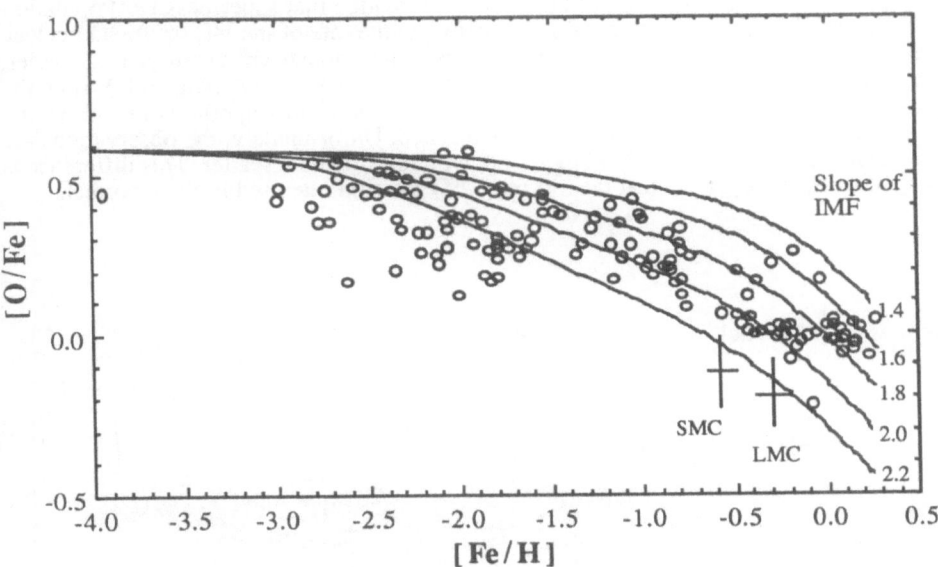

Figure 10: The $[\alpha/Fe]$ vs. $[Fe/H]$ relationship for old disk stars in the solar neighbourhood (Wheeler and Sneden 1989), and for the current values $[O/Fe]$ in the interstellar medium of the Magellanic Clouds (Russell and Dopita, 1989) compared with models with slopes of the IMF. We assume that the α elements are enriched the same as O, and the infall and deflagration timescales are both set equal to 1Gyr.

In the Magellanic Clouds, Russell and Dopita (1989a,b) find that $[O/Fe]$ is somewhat less than unity, whereas $[Fe/H]$ is also less than unity. Such a result can be explained simply as a result of relative youth, longer infall timescale and somewhat steeper IMF. The strong effect of the IMF on the $[O/Fe]$ vs. $[Fe/H]$ relationship is illustrated in figure 10. A slope of 2.3 would be sufficient of itself to explain the LMC and SMC results. These results are in accord with the direct observational meterial (Stryker and Butcher 1981; Hawkins and Brück, 1982,84; Frogel and Blanco 1984; Rocca-Volmerange *et al.* 1981; Humphreys and McElroy 1984).

5.3. CARBON.

The constancy of the [C/Fe] ratio over a very wide range of [Fe/H] is an extraordinary outcome of the metallicity-metallicity relationships in the solar neighbourhood (Wheeler and Sneden 1989). The implication of this, since Fe is produced in stars of both high and in stars of low mass after a long latency period, must be that carbon shares both a short-lived and a long-lived source of enrichment.

In the case of the high-mass stars, the asymptotic values of the [C/Fe] or [C/O] ratios can once again be used as a test of the chemical evolution in massive Population II stars, since the products of this evolution are recycled promptly back to the ISM. The [C/O] ratio, in particular, is important because, provided that supermassive (Population III) objects are not formed, this ratio is sensitive to the rate of the $^{12}C\ (\alpha,\gamma)\ ^{16}O$ nuclear reaction, which has been the subject of considerable debate in recent years (Fowler, Caughlan and Zimmerman, 1975; Fowler, 1985; Thielemann and Arnett, 1985). With the prescription based on the Maeder and Meynet models of pre-supernova evolution, we find an asymptotic value of [C/O] = -0.4 is reached. Unfortunately, the observations are less clear, but a value of ~-0.2 is suggested, with considerable scatter. This difference is insignificant, and suggests that the $^{12}C\ (\alpha,\gamma)\ ^{16}O$ reaction rate used is about correct.

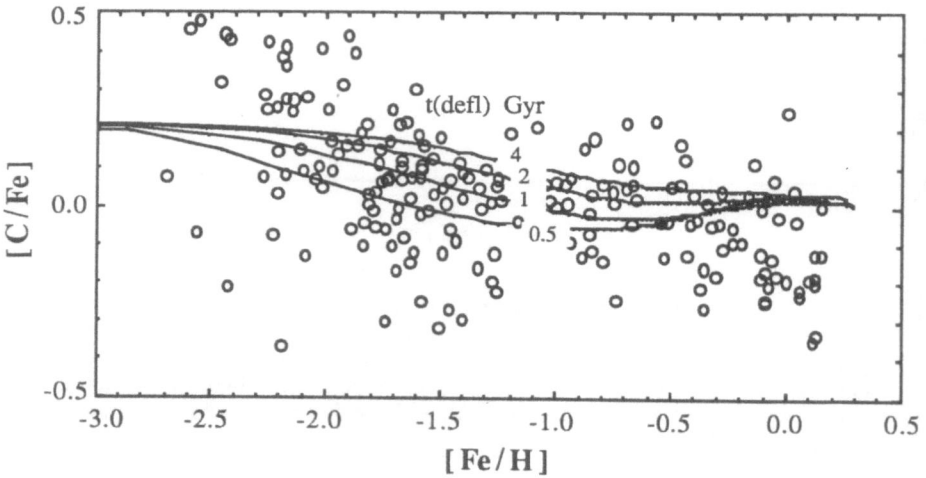

Figure 11: The variation of [C/Fe] as a function of [Fe/H] for various different deflagration timescales, plotted with the observed points for old disk stars.

The stars which are the main source of C at higher abundance would have to have fairly low mass, on average. From Iben and Tutukov (1985), the main-sequence lifetimes of stars of mass m solar masses is given by;

$$\tau_{MS} = 2.64.m^{-2.16} \ Gyr \qquad\qquad m > 2.3M_\odot \qquad\qquad (5.1a)$$
$$\tau_{MS} = 11.0.m^{-3.50} \ Gyr \qquad\qquad m \le 2.3M_\odot \qquad\qquad (5.1b)$$

Since the deflagration timescale is 1-4 Gyr, approximately, the mean mass of the stars contributing to the cabon enrichment has to be 1.3 - 2.0 $M_\odot$. Thus we have the not too surprising result that carbon stars are a major source of carbon enrichment!

In carbon stars, the material which is later ejected into the ISM in the form of a planetary nebula, is first dredged up into the hydrogen-rich layers in the period of expansion and cooling following the helium thermal pulses, which serves to temporarily extinguish hydrogen burning, and lower the base of the convective envelope. The theoretical details of this depend on parameters such as the core mass, the metallicity, and on the ratio of the mixing length to the scale height, amongst other variables (Iben and Truran, 1978; Iben, 1981,1987; Wood 1981; Renzini and Voli 1981; Iben and Renzini 1983). In general, for a given core mass dredge-up is favoured at low metallicity, a larger mixing length parameter, or for larger envelope mass. Again, we are forced to a simple formulation in which the envelope C/O ratio is assumed constant up to the upper mass limit. The value of the ratio is that which gives [C/O] = 0.0 at [O/H] = 0.0. In figure 11 we show the effect of the deflagration time on the [C/Fe] ratio, plotted against the observed values (Wheeler and Sneden, 1989).

5.4. NITROGEN

Both carbon and nitrogen are produced largely in intermediate-mass stars during three convective dredge up episodes (Iben, I.Jr. 1975; Iben and Truran 1978; Renzini and Voli 1981). During the first of these, ^{14}N is enhanced at the expense of ^{12}C. In the second phase is initiated following the ignition of the He and occurs in stars more massive than 3-5$M_\odot$. The result is also to dredge up nitrogen and helium. The third phase occurs during AGB evolution, and is a result of the helium shell flashes, which turn off the hydrogen burning, and allow the convective zone to penetrate all the way into the intershell region where incomplete helium burning has occurred. Correspondingly, each shell flash allows significant amounts of ^4He and ^{12}C to reach the surface.

Nitrogen presents something of a problem. All the data on HII regions in other galaxies show that, in general, [N/O] is positively correlated with [O/H], at least for [O/H] > -0.8 (Pagel 1985). Russell and Dopita (1989) found that the data for SNR in the Magellanic Clouds yield much higher [N/O] ratios for a given [O/H] ratio. It is possible that this is a modelling problem, but if so, the cause remains obscure. It is more probable that we are seeing a genuine abundance difference between these objects. This would be the result of dredge up of partially hydrogen-burnt material into the pre-supernova envelope of the precursor massive star. Such a process is known to occur in many hot stars (Walborn, 1988). If this material was subsequently ejected in a stellar wind, this material could come to contaminate the surrounding ISM which is later shocked by the supernova blast wave. Evidence to support this model comes from several sources. Nitrogen-rich knots are observed in several supernova remnants, most notably in Pupp A. Also, the precursor of SN 1987A had a nitrogen-rich envelope (Walborn, 1988) and the subsequent UV pulse illuminated and photoionised a highly nitrogen-enriched ejecta about the star (Kirshner, 1988).

If direct dredge-up of partially hydrogen-burnt material can occur in massive stars, then this would represent a primary source of nitrogen. Similarly, the hot bottom burning in intermediate stars is a primary process. However, the slope of the [N/O] vs. [O/H] ratio implies that nitrogen is produced mostly like a secondary element, at least down to

SMC abundances. We are thus presented with an inconsistency which is exemplified by the comparison of our "best" model for the enrichment of the solar nebula, and the data from HII regions from Pagel (1985) and Garnett (1989), shown in Figure 12. It should be recognised that Figure 12 is not strictly comparable with the previous metallicity - metallicity plots, since the data here refer to the final chemical states of many galaxies, rather than the chemical evolution of a single galaxy. Nevertheless, for the low metallicity HII Dwarf galaxies, we would expect that the asymptotic ratio of [N/O] would tend towards the element yield ratio for a young stellar population.

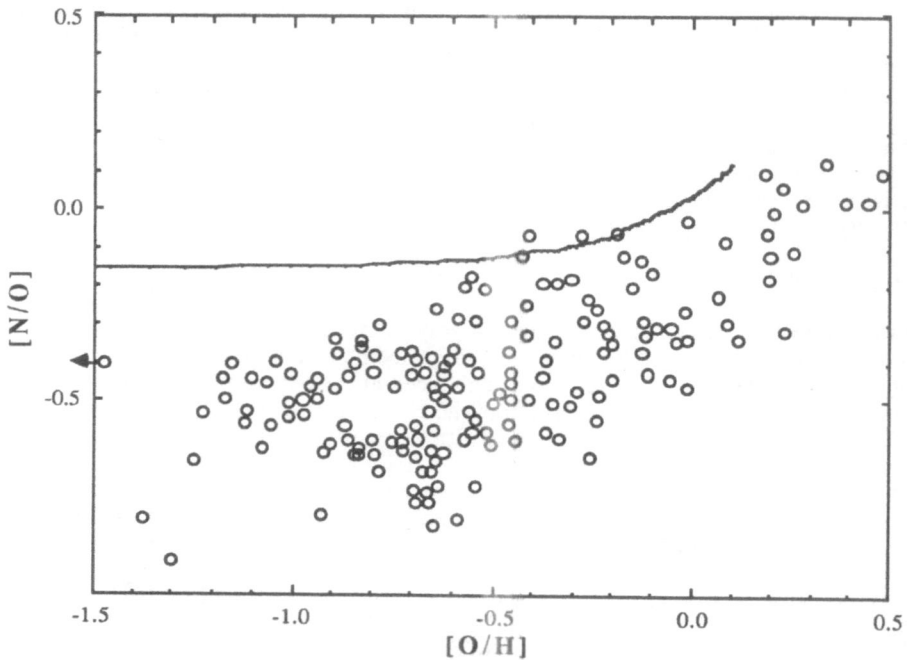

Figure 12: The variation of [N/O] as a function of [O/H] for our standard model, plotted with the observed points for HII regions in external galaxies from Pagel (1975), with low abundance points added from Garnett (1989). From this, it may be concluded that the yield of nitrogen by hot-bottom burning must be lower at lower abundance.

Interestingly enough, the estimated yield of nitrogen from the massive stars alone would lead to an asymptotic log(N/O) ratio of about -1.6 or [N/O] about -0.7 (*c.f.* Figure 6). This would be quite consistent with the HII region data. Therefore, assuming that the evolution theory for massive stars is about correct, then the observational material implies that the dredge up of ^{14}N into the atmospheres of the heavier intermediate-mass stars is relatively inefficient at low metallicities.

6. Conclusions

The star-gas cycle (or bicycle) is the interface between stellar and interstellar astrophysics. I hope that I have shown some of the ways in which this interface can be used as a test of, and for the refinemement of, theories in both of these disciplines. In particular, it is evident that evolutionary scenarios for the massive stars can be tested by observations of the statistics of supernovae of classes Type II, Type Ib and Type Ic; by the ratio of WC / WN / Red Supergiants / Blue Supergiants, and the variation of these with metallicity; and by the asymptotic ratios of elemental abundances in low metallicity galaxies and in galactic halos. An improved understanding of both the formation and evolution of the massive stars would then lead in turn to an improved description of the energy input to, and therefore to our understanding of the phase structure of, the interstellar medium.

It seems clear that the models that have been used to investigate the chemical evolution of galaxies are, for the most part, inadequate.We need an accurate theoretical model for the stellar diffusion processes, and for the effective viscous dissipation in the gaseous component. Inclusion of these processes will lead to the opening of the "closed box" model, and to much more realistic models for the evolution of galactic disks.

As far as the low mass stars are concerned, it is clear that major theoretical uncertainties remain in the treatment of dredge-up and its variation with metallicity, and in the evolution of lower mass stars towards Type I supernova events. Until these are improved, it will be difficult to model the chemical evolution of C, N, Fe and of the s-process elements in galactic disks. We may hope that improved statistics on Type I supernovae, studies of the luminosity distribution of of C-stars, and chemical abundance analyses of planetary nebulae using UV data obtained with HST will go a long way towards providing new observational constraints for this theory.

Acknowledgements
Much of this paper was written during my study leave in Baltimore. I would like to thank the Johns Hopkins University for their award of a Visiting Fellowship, and for the hospitality shown by the staff there and at the Space Telescope Science Institute during my visit.

References

Abbott, D.C., 1978, *Astrophys. J.* , **225**, 893.
Abbott, D.C., 1982 in *"Wolf-Rayet Stars: Observations, Physics, Evolution"*, eds. C.de Loore, A. J. Willis, (Dordrecht:Reidel), p185.
Audouze, J., and Tinsley, B.M. 1977, *Ann. Rev. Astron. Astrophys.*, **14**, 43.
Augarde,R. and Lequeux,J. 1985, *Astron. Astrophys.* ***
Balbus, S.A. and Cowie,L.L. 1985, *Astrophys. J.*, **277**,550.
Barlow, M.J., 1982, in *"Wolf-Rayet Stars: Observations, Physics, Evolution"*, eds. C.de Loore, A. J. Willis, (Dordrecht:Reidel), p149.
Blitz, L. and Shu, F.H. 1980, *Astrophys. J.*, **238**, 148.
de Boer, K.S., and Savage, B.D. 1983, *Astrophys. J.*, **265**, 210.
Bond, H.E. 1970, *Astrophys. J. Suppl. Ser.*, **22**, 117.
Bosma, A. 1978, Ph.D. Thesis, University of Groningen.
_____. 1981, *A.J.*, **86**, 1825.
Burki, G. 1977, *Astron. Astrophys.*, **57**, 135.
Chevalier, R.A. and Oegerle, W.R. 1979, *Astrophys. J.*, **227**, 398.

Chièze, J-P. 1987, *Astron. Astrophys.*, **171**, 225.

Chiosi, C. 1980, *Astrophys. J.*, **83**,206

Chiosi, C. and Maeder, A., 1986, *Ann. Rev. Astron. Astrophys.*, **24**, 329.

Chiosi,C. and Matteucci, F.M. 1982, *Astron. Astrophys.*, **105**, 140.

Cohen, R.S., Dame, T.M., Garay, G., Montani, J., Rubio, M., and Thaddeus, P. 1988, *Astrophys. J. (Lett.)*, **331**, L95.

Conti, P.S., 1981, in *"Effects of Mass Loss on Stellar Evolution"*, ed. C. Chiosi, R. Stalio, Dordrecht:Reidel, p

Cowie, L.L. 1981, *Astrophys. J.*, **245**, 66.

Cox, D.P. 1979, *Astrophys. J.*, **234**, 863.

_____. 1980, *Astrophys. J.*, **245**, 534.

Cox, D.P. and Smith, B.W. 1974, *Astrophys. J. (Lett.)*, **189**, L105.

Dame, T.M., Elmegreen, B.G., Cohen, R.S., and Thaddeus, P. 1986, *Astrophys. J.*, **305**, 892.

DeGioia-Eastwood, K., Grasdalen, G.L., Strom,S.E. and Strom, K.M. 1984, *Astrophys. J.*, **278**, 564

de Jager, C, Nieuwenhuijzen, H., and van der Hucht, K.A. 1987, *Astron. Astrophys. Suppl. Ser.*

Dickman, R., Snell, R., and Schloerb, P. 1986, *Astrophys. J.*, **309**, 326.

Dopita, M.A. 1985a, *Astrophys. J. (Lett.)*, **295**, L5.

Dopta, M.A. 1985b, in *"Birth and Evolution of Massive Stars and Stellar Groups"*, eds. W. Boland and H. van Woerden, Reidel:Dordrecht, p269.

Dopita, M.A., Dawe, J.A., Achilleos, N., Brissenenden, R.J.V., Flynn, C., Meatheringham, S.J., Rawlings, S., Tuohy, I.J., McNaught, R.D., Coates, D.W., Hancy, S., Thompson, K., and Shobbrook, R.R., 1988, *Astron. J.*, **95**, 1717.

Downes, D. and Güsten, R. 1982, *Astron. Gesellschaft*, **57**, 207. *Astrophys. J. (Lett.)* (in press).

Dupree, A. K., 1986, *Ann. Rev. Astron. Astrophys.*, **24**, 377.

Dyson, J.E., 1981 in *"Exploring the Universe"* ed. F.D. Kahn, (Dordrecht:Reidel), p 125.

Edmunds, M.G. 1987, (private communication)

Elmegreen, B.G. 1979, *Astrophys. J.*, **231**,372.

_____. 1982, *Astrophys. J.*, **253**,655.

_____. 1986, IAU Symposium #115, *"Star Forming Regions"*, Tokyo, Japan 11-15 Nov., eds M.Peimbert and J.Jugaku, Reidel:Dordrecht.

Elmegreen, B.G. and Elmegreen, D.M. 1978, *Astrophys. J.*, **220**, 1051.

Ensman, L. 1989 *Astrophys. J.*, (in press).

Falgarone, E. and Puget, J.L. 1985, *Astron. Astrophys.*, **142**, 157.

Falgarone, E. and Puget, J.L. 1988, in *"Galactic and Extragalactic Star Formation"*, eds. R.E. Pudritz and M. Fich, (Kluwer:Dordrecht), p 195.

Fall, S.M., and Efstathiou, G. 1980 *Mon. Not. Roy. Ast. Soc*, **193**, 189.

Fukui,Y., Sugitani, K., Takabe, H., Iwata, T., Mizuno, A., Ogawa, H. and Kawataba, K. 1986, *Astrophys. J. (Lett.)*, **311**, 85.

Field, G.B., Goldsmith, D.W., and Habing, H.J. 1969, *Astrophys. J. (Lett.)*, **155**, L149.

Feitzinger, J.V., Glassgold, A.E., Gerola, H. and Sieden, P.E. 1981, *Astron. Astrophys.*, **98**, 371.

Fowler, W.A. 1985, in *"Nucleosynthesis: Challenges and New Developments"* , eds. W.D. Arnett and J.W. Truran, (U. of Chicago:Chicago), p9.

Fowler, G.M., Caughlan, G.R. and Zimmerman, B.A. 1975,
    Ann. Rev. Astron. & Astrophys., **13**, 69.
Franco, J. 1983, *Astrophys. J.*, **264**, 508.
Franco, J., and Cox, D.P. 1983, *Astrophys. J.*, **273**, 243.
Freeman, K.C. 1971, *Astrophys. J.*, **160**, 811.
Frogel, J.A., and Blanco, V.M. 1983, *Astrophys. J. (Lett.)*, **274**, L57.
Garnett, D.G. 1989, Thesis, University of Texas at Austin.
Gerola, H. and Seiden, P.E., 1978, *Astrophys. J.*, **223**,129
Gilmore, G. and Wyse, R.F.G. 1986, *Nature,* **322**, 806.
Güsten, R., and Mezger, P.G. 1982, *Vistas in Astronomy,* **26**, 159.
Guibert, J., Lequeux, J. and Viallefond, F. 1978, *Astron. Astrophys.*, **68**, 1.
Hamajima,K., and Tosa, M. 1975,*Pub. Astron. Soc. Japan*, **27**,561.
Hawkins, M.R.S., and Brück, M.T. 1982, *Mon. Not. Roy. Ast. Soc.*, **198**, 935.
_____.1984, in *"Structure and Evolution of the Magellanic
    Clouds"*, eds. S. van den Bergh and K.S. de Boer, (Dordrecht:Reidel), p101.
Hidayat, B. Supelli. K., and van der Hucht, K.A., 1983 in *"Wolf-Rayet Stars:
    Observations, Physics, and Evolution"*, eds. C.W.H. de Loore and A.J. Willis,
    (Dordrecht:Reidel), p 27.
Hidayat, B., Admiranto, A.G., Supelli, K.R., and van der Hucht, K.A., 1988, in
    "Atmospheric Diagnostics of Stellar Evolution", ed. K. Nomoto,
    (Springer-Verlag:Berlin), p 152.
van der Hucht, K.A., 1981, in the ESO Workshop on *"The Most Masive Stars"*,,
    eds. S. D'Odorico, D. Baade, K. Kjär, (ESO:Garching), p 157.
Humphreys, R.M. 1982, in *"The Most Massive Stars"*,
    ed S. D'Odorico, D. Baade, K. Jjär, (ESO:Garching), p 5.
Humphreys, R.M. 1984, in *"Observational Tests of Stellar Evolution Theory"*,
    ed. A. Maeder, A. Renzini, (Dordrecht:Reidel), p 279.
Humphreys, R.M. , and McElroy, D.B. 1984, *Astrophys. J.*, **284**, 565.
Iben, I.Jr. 1975 *Astrophys. J.*, **196**, 525.
_____. 1987 in *"Late Stages of Stellar Evolution"*, eds. S. Kwok and S.R. Pottasch,
    (Reidel:Dordrecht), p175.
_____. 1981, in*"Physical Processes in Red Giants"*, ed I.Iben Jr. and A. Renzini
    (Dordrecht:Reidel), p3.
Iben, I. Jr. and Renzini, A. 1983, *Ann. Rev. Astron. Astrophys.*, **21**, 271.
Iben, I. Jr., and Truran, J.W., 1978, *Astrophys. J.*, **220**, 980.
Iben, I., Jr. and Tutukov, A.V. 1985, *Astrophys. J. Suppl. Ser.,* **58**, 661.
Jenkins, E. 1978, *Astrophys. J.*, **220**, 107.
Kennicutt, R. 1989, *Astrophys. J.*, (in press).
Kennicutt, R. and Kent, S. 1983, *Astron. J.*, **88**, 1094.
Kirshner, R.P. 1988 in *"Supernova 1987A in the Large Magellanic Cloud"*,
    eds. M. Kafatos and A. Michalitsianos, (Cambridge UP:Cambridge), p 87.
van der Kruit, P.C. and Shostak, G.S. 1984, *Astron. Astrophys.,* **134**, 258.
van der Kruit, P.C. and Searle, L. 1981a, *Astron. Astrophys.,* **95**, 105.
_____. 1981b, *Astron. Astrophys.,* **95**, 116.
van der Kruit, P.C. and Searle, L. . 1982, *Astron. Astrophys.,* **110**, 61.
Kudritski, R.P., Pauldrach, A., and Puls, J., 1987, *Astron. Astrophys.*, **173**, 293.
Kwan, J. 1979, *Ap.J.,* **229**, 567.
Kwok, S. and Volk, K., 1985, *Astrophys. J.*, **299**, 191.
Lacey, C.G. 1984, Ph.D. thesis, University of Cambridge.

Lacey, C.G. and Fall, S.M. 1983, *M.N.R.A.S.*, **204**, 791.
_____. 1985, *Astrophys. J.*, **290**, 154.
Lada, C.J. 1988, in in *"Galactic and Extragalactic Star Formation"*, eds. R.E. Pudritz and M. Fich, (Kluwer:Dordrecht), p 5.
Langer, N. 1989 *Astron. Astrophys.*, (in press).
Larson, R.B. 1972, *Nature Phys. Sci.* **236**, 7.
_____. 1976, *M.N.R.A.S.*, **176**, 31.
_____. 1981, *M.N.R.A.S.*, **194**, 809.
_____. 1986, *M.N.R.A.S.*, **218**, 409.
Lin, D.N.C. and Pringle, J.E. 1987,*Astrophys. J.*, **320**, L87.
de Loore, C., and Doom, C. 1988, in *"Atmospheric Diagnostics of Stellar Evolution"*, ed. K. Nomoto, (Springer-Verlag:Berlin), p 246.
Long, K.S., Dopita, M.A., and Tuohy, I.R., 1982, *Asrtrophys. J.*, **260**, 202.
Lynden-Bell, D. 1975, *"Vistas in Astronomy,* **19**, 299.
McKee, C.F. and Ostriker, J.P. 1977, *Astrophys. J.*, **218**, 148.
Maeder, A. 1983, *Astron. Astrophys.*, **120**, 149.
Maeder, A. 1987, *Astron. Astrophys.*, **173**, 247.
Maeder, A. 1988, in*"Atmospheric Diagnostics of Stellar Evolution"*, ed. K. Nomoto, (Springer-Verlag:Berlin), p 79.
Maeder, A. and Meynet, G., 1987, *Astron. Astrophys.*, **182**, 243.
Maloney, P., and Black, J.H., 1988, *Astrophys. J.*, **325**, 389.
Matteucci, F. and Greggio, L. 1986, *Astron. Astrophys.*, **154**, 279.
Matteucci, F. and Tornambé, A. 1985, *Astron. Astrophys.*, **142**, 13.
Mayor M. and Vigroux, L. 1981,*Astron. Astrophys.*, **98**, 1.
Meatheringham, S., Dopita, M.A., Ford, H.C. and Webster, B.L. 1987, *Astrophys. J.*, **327**, 639.
Miller, G.E. and Scalo, J.M. 1979, *Astrophys. J. Suppl.*, **41**, 513.
Mirabel, I.F. and Sanders, D.P. 1988, in *"Galactic and Extragalactic Star Formation"*, eds. R.E. Pudritz and M. Fich, (Kluwer:Dordrecht), p 551.
Moffat, A.F.J. and Shara, M.M., 1987, *Astrophys. J.*, **320**, 266.
Mouschovias, T. Ch., Shu, F.H. and Woodward,P. 1974, *Astr.Astrophys.*, **33**, 73.
Nomoto, K. 1982a, *Astrophys. J.*, **253**, 798.
_____ . 1982b, *Astrophys. J.*, **257**, 780.
Nomoto, K., Thielemann, F-K., and Yokoi, K. 1984, *Astrophys. J.*, **286**, 644.
Norman, C. and Silk, J. 1980, *Astrophys. J.*, **238**, 158.
Olaffsson, K., Bergrall, N. and Ekman, A., 1984, *Astron. Astrophys.*, **137**, 327.
Oort, J.H., 1954, *Bull. Astr. Instts. Neth.*, **12**, 177.
Pagel , B.E.J., 1985, in *"Production and Distribution of C,N, O Elements"*, eds I.J. Danziger, F. Matteucci, and K. Kjär, ESO:Garching p155.
Pagel, B.E.J. and Patchett, B.E. 1975, *M.N.R.A.S.*, **172**, 13.
Penston, M.V., Munday, V.A., Stickland, D.J., and Penston, M.J., 1969, *Mon. Not. Roy, Ast. Soc.*, **142**, 355.
Pettini, M. *et al.* 1982, *Mon. Not. Roy. Ast. Soc.*, **199**, 409.
Quirk, W.J. 1972, *Astrophys. J.*, **176**, L9.
Rana, N.C., and Wilkinson, D.A. 1986, *M.N.R.A.S.*, **218**, 497.
Reike, G.H., Catri, R.M., Black, J.H., Kailey, W.F., Mc.Alary, C.W., Lebofsky, M.J. and Elston, R. 1980, *Astrophys. J.*, **290**, 116.
Renzini, A. and Voli, M. 1981, *Astron. Astrophys.*, **94**, 175.
Rieke,G.H., Lebofsky,M.J., Thomson, R.I., Low,F.J. and Tokunaga,A.T. 1985, *Astrophys. J.*, **238**, 24.

Rocca-Volmerange, B., Lequeux, J. and Maucherat-Joubert, M. 1981,
    Astron. Astrophys., **104**, 177.
Russell, S.J., Bessell, M.S., and Dopita, M.A., 1989, *Astrophys. J.,* (in press).
Russell, S.J. and Dopita, M.A., 1989a,b, *Astrophys. J.,* (in press).
Sanders, D.B., Solomon, P.M., and Scoville, N.Z. 1984, *Astrophys. J.,* **276**,182.
Sanduleak,N. 1969, *Astron. J.,* **74**,47.
Savage, D.B., and de Boer, K.S. 1979, *Astrophys. J. (Lett.),* **230**, 460.
_____. 1981, *Astrophys. J.,* **243**, 460.
Scalo, J.M. 1986, *Fund. Cosmic. Phys.,* **11**, 1.
Schild, H., and Maeder, A., 1984, *Astron. Astrophys.,* **136**, 237.
Schmidt, M. 1959, *Astrophys. J.,* **129**,243.
Scoville, N.Z. and Good, J.C. 1987, *Astrophys. J.* ***
Scoville, N.Z., Sanders, D.B. and Clemens, D.P. 1986,
    Astrophys. J. (Lett.), **310**, L77.
Scoville, N.Z., and Young, J.S. 1983, *Astrophys. J.,* **265**,148.
Searle,L. 1972 in IAU Coll. #17, *"L'Age des Etoiles"*,
    eds. G. Gayrel de Strobel and A.M. Delplace, p. LII
Seiden, P.E. and Gerola, H. 1979, *Astrophys. J.,* **233**, 56.
Silk,J. 1985, IAU Symposium #115, *"Star Forming Regions"*,
    eds. M.Peimbert and J.Jugaku, (Reidel:Dordrecht), p663.
Solomon, P.M., Rivolo, A.R., Barrett, J.W., and Yahil, A.1987,
    Astrophys. J., **319**, 730.
Solomon, P.M., and Mooney, T.J. 1988, in *"Galactic and Extragalactic Star Formation"*,
    eds. R.E. Pudritz and M. Fich, (Kluwer:Dordrecht), p 589.
Sommer-Larsen, J., and Yoshii, Y. 1989, *Mon. Not. Roy. Ast. Soc.,* **238**, 133.
Spitzer, L., 1956, *Astrophys. J.,* **124**, 20.
Spitzer,L.and Schwartzschild,M. 1951, *Astrophys. J.,* **114**,106.
_____. 1953, *Astrophys. J.,* **118**, 106.
Stark, A.A. and Brand, J. 1989, *Astrophys. J.,* **339**, 763.
Stryker, L.L. and Butcher, H.R. 1981, in *"Astrophysical Parameters for Globular
    Clusters"* , eds. A.G.D. Philip and D.S. Hayes, (Schenectady:L.Davis), p201.
Talbot,R.F. 1980, *Astrophys. J.,* **235**, 821.
Tanaka, Y., and Bleeker, J.A.M. 1977, *Space Science Rev.,* **20**, 815.
Thielemann, F-K., and Arnett, W.D., 1985, in *"Nucleosynthesis: Challenges and New
    Developments"* , eds. W.D. Arnett and J.W. Truran,(U.Chicago:Chicago), p151.
Thielemann, F-K., Nomoto, K. and Yokoi, K. 1986, *Astr. Astrophys.,* **158**, 17
Tinsley, B.M. 1980, *Fund. Cosmic Phys.,* **5**, 287.
Toomre, A. 1964, *Ap. J.,* **139**, 1217.
Tosi, M. and Diaz, A. 1985, *M.N.R.A.S.,* **217**, 571.
Tornambé, A. 1984, *Mon. Not.R.oy. Ast. Soc.,* **206**, 867.
Tuohy, I.R., Dopita, M.A., Mathewson, D.S., Long, K.S., and Helfand, D.J., 1982
    Astrophys. J., **261**, 473.
Twarog, B.A. 1980, *Astrophys. J.,* **242**, 242.
Vader,J.P., and de Jong, T. 1981, *Astron. Astrophys.,* **100**, 124.
Viallefond, F. and Goss, W.M. 1986, *Astron. Astrophys,* **154**, 357.
Villumsen, J.V. 1985, *Astrophys. J.,* **290**, 75.
Volk, K., and Kwok, S., 1985, *Astron. Astrophys.,* **153**, 79.
Walborn, N.R. 1988, in *"Atmospheric Diagnostics of Stellar Evolution"*,
    ed. K. Nomoto Spinger-Verlag:Berlin, p70.
Wevers, B.M.H.R. 1984, Thesis, University of Groningen

472

Wielen, R. 1977, *Astron. Astrophys.,* **60**, 263.
Weilen, R. and Fuchs, B. 1983, in *"The Milky Way: Structure, Kinematics and Dynamics,* ed. H. van Woerden (Dordrecht:Reidel).
Wood, P.R. 1981, in *"Physical Processes in Red Giants",* eds. I.Iben, Jr., and A. Renzini, (Dordrecht:Reidel), p135.
Wunderlich, E., Klein, U. and Wielebinski, R. 1987, *Astron. Astrophys. Suppl.,* **69**, 487.
Wyse, R.F.G. 1986, *Astrophys. J. (Lett.),* **311**, L41.
Wyse, R.F.G. and Silk, J. 1989, *Astrophys. J.,* **339**, 700.
York, D.G., Blades, J.C., Cowie, L.L., Morton, D.C., Songaila, A., and Wu, C.-C. 1982 *Astrophys. J.,* **255**, 467.
Yoshii, Y. and Sommer-Larsen, J. 1989a, *M.N.R.A.S.,* **236**, 779.
_____. 1989b, *M.N.R.A.S.,* (in press).
_____. 1989c, *M.N.R.A.S.,* (in press).
Young, J.S., and Scoville, N.Z. 1982, *Astrophys. J.,* **258**, 467.
Young, J.S., Schloerb, F.P., Kenney, J., and Lord, S. 1986, *Astrophys. J.,* **304**, 443.
Young, J.S. 1988, in *"Galactic and Extragalactic Star Formation",* eds. R.E. Pudritz and M. Fich, (Kluwer:Dordrecht), p 551.

Measuring Atomic Hydrogen Masses Using the 21-cm Line

John M. Dickey
University of Minnesota

ABSTRACT. Emission in the 21-cm line of neutral hydrogen is very simple to use to estimate masses and column densities of the neutral atomic gas phase in the Milky Way and other galaxies. The only serious uncertainty is in the optical depth (self-absorption) correction to apply. Absorption surveys of the galaxy suggest that this effect is not very serious, but data at very low latitudes is sparse, and the abundance of very high optical depth regions is not well known. In other galaxies statistics of 21-cm line flux vs. inclination show that it is only for inclinations less than about 10^{0} to the line of sight that self-absorption becomes severe. For high inclinations a simple correction factor is sufficient for converting observed 21-cm line flux to H I mass.

1. Basic Physics

The 21-cm line of neutral hydrogen (H I) makes a good tracer of the mass of neutral atomic gas in the interstellar medium of galaxies. This is in part because it comes from the dominant atom of this phase, and in part because it is conveniently placed at a frequency (1420.405752 MHz) which is easy to observe from large ground based telescopes. But what makes this line seem almost "a gift from above" is the simplicity of the physics of its excitation and radiative transfer. The spontaneous deexcitation rate is so low that collisions generally keep the level populations in equilibrium with the kinetic temperature, so that measuring the **excitation**

473

H. A. Thronson, Jr. and J. M. Shull (eds.), The Interstellar Medium in Galaxies, 473–481.
© *1990 Kluwer Academic Publishers.*

temperature (often called the "spin temperature" because it is a spin flip transition) gives a good estimate of the **kinetic** temperature.

The energy separation of the levels is so tiny, $6 \cdot 10^{-6}$ ev = 0.068 K·k which is always much less than the spin temperature, that to first order there is no dependance on temperature in the level population ratio, so that the emission coefficient is proportional to the density (n) alone :

$$j_v = n \cdot 1.6 \cdot 10^{-33} \text{ erg sec}^{-1} \text{ Hz}^{-1} \text{sterrad}^{-1} \tag{1}$$

This makes the brightness, which in the optically thin limit is just the column integral of the emission coefficient, proportional to the column density of gas only. Expressed in terms of the integral of the brightness temperature T_B over the velocity width of the line this becomes :

$$N = 1.83 \cdot 10^{18} \text{ cm}^{-2} \cdot \frac{\int T_B (v) \, dv}{K \text{ km s}^{-1}} \tag{2}$$

as long as the optical depth τ is low. How much correction to make for finite optical depth ("self-absorption") is one of the interesting questions about the 21-cm line which has been discussed for a long time.

2. Galactic Surveys

In the early 21-cm emission surveys of the Milky Way it was noticed that the highest brightness temperature ever seen was about 125 K, which led to the incorrect assumption that this was the spin temperature of the H I gas everywhere. This provided an optical depth estimate for the 21-cm line everywhere **else**, and so allowed a rough column density correction factor for self-absorption to be estimated. This came out to be typically a factor of 1.4 (Kerr and Westerhout, 1965). Already at that time it was apparent that the H I is in two phases, one considerably cooler than this and the other much warmer, so the assumptions behind this correction factor are clearly invalid. Kerr and Westerhout recognized these uncertainties, and so for the total H I mass of the galaxy they conservatively quoted 3 (4?) $\cdot 10^9$ M$_\odot$.

A more direct way to correct for self-absorption is to measure the optical depth and spin temperature at each velocity, and then correct the integrand above for finite optical depth, as

$$N = 1.83 \cdot 10^{18} \text{ cm}^{-2} \cdot \frac{\int T_B (v) \dfrac{\tau(v)}{1 - e^{-\tau(v)}} \, dv}{\text{K km s}^{-1}} \qquad (3)$$

This makes the assumption that all the gas contributing to each velocity channel has the same temperature, which is also generally invalid, but at least it provides for a broad range of temperatures at different velocities, which is typical of galactic spectra. Using this method Dickey and Benson (1982) and Kulkarni et al. (1984) find that the correction factor 1.4 is about right for intermediate latitudes (10° to 30°), but below about 3° latitude the correction factors get as high as 2 or more. More sophisticated assumptions about the temperature structure along the line of sight in each velocity channel get complicated because we do not know the juxtaposition of the warm, optically thin gas and the cool, absorbing gas. An illustration of this effect is given by Lockman and Dickey (1990).

There are not enough strong continuum background sources to allow measurement of optical depth everywhere, although recent surveys of low latitude directions have added to our knowledge of the abundance of optically thick clouds far from the solar neighborhood (Garwood and Dickey 1988, Colgan, Salpeter and Terzian 1988, Kuchar and Bania 1989). In addition, emission surveys specifically searching for the signature of self-absorption have shown that a significant portion of l-v space (roughly corresponding to the region with strong CO emission) has H I optical depths high enough to effect the profile shapes (Peters and Bash, 1987, and references therein). Like the CO, this H I self-absorption is mostly found in the inner galaxy, but the bulk of the Milky Way H I mass is outside the solar circle, where presumably the optical depth correction is less serious. Taken together the evidence from absorption observations to date suggests that the correction factors used 25 years ago were about right (but for the wrong reasons !). It is comforting in this regard that the gamma ray observations, which indirectly sample the total interstellar gas density, agree quite well with the predictions of the H I observations in the outer galaxy (making the old 125 K T_{spin} correction for optical depth, Bloemen *et al.* 1986, Strong *et al.* 1988).

3. 21-cm Emission from Other Galaxies

For H I observations of external galaxies the column density obtained from equation 2 is integrated over the solid angle of the galaxy to obtain the total H I mass, M_H. Thus it is the flux in the 21-cm line which determines the mass of gas, as :

$$\frac{M_H}{M_\odot} = 2.3 \cdot 10^5 \frac{d^2}{Mpc^2} \frac{\int S_v \, dv}{Jy \, km \, s^{-1}} \tag{4}$$

where d is the distance to the galaxy. If the optical depth is significant then this must be corrected as discussed above, but fortunately most spiral disks are seen at low enough inclination that this correction is small. Haynes and Giovanelli (1984) discuss this issue statistically using a large sample of galaxies, and find that it is only for Sb, Sbc, and Sc galaxies that there is a significant correlation of observed H I surface density with inclination, as shown in the figure, which is taken from their figure 13. Haynes and Giovanelli fit these data with power laws with slopes of about 0.15 to get an empirical correction factor f_H to the observed 21-cm flux :

$$f_H \equiv \frac{S_{corrected}}{S_{observed}} = r_i^{-0.15} \tag{5}$$

where r_i is the axis ratio, the cosine of the inclination. This can be used in equation 4 to empirically correct the observed hydrogen mass or surface density for self absorption.

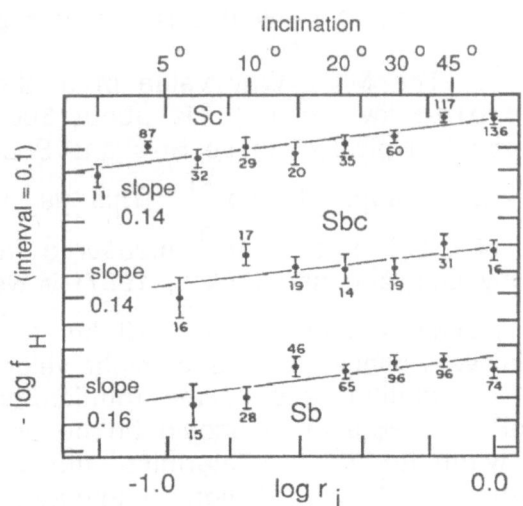

Figure : Correlation between measured H I surface density and axis ratio for spiral galaxies (taken from Haynes and Giovanelli, 1984). The ordinate scale is logarithmic with intervals of 0.1, the points are averages of samples of galaxies, with the number in the sample given next to each point. Other morphological types show little or no correlation.

These extragalactic results set an interesting limit on the typical 21-cm opacity, κ, in spiral disks. Making the same assumption of isothermal gas used in equation 3 and assuming further that galaxy disks are homogeneous with thickness h we would expect :

$$f_H = \left(\frac{\dfrac{\kappa h}{r_i}}{1 - \exp\left(-\dfrac{\kappa h}{r_i}\right)} \right) \cong \begin{cases} 1 + \dfrac{\kappa h}{2 r_i} & (\kappa h << r_i) \\ \left[\dfrac{r_i}{\kappa h}\right]^{-1} & (\kappa h >> r_i) \end{cases} \tag{6}$$

This functional form fits the data at least as well as the power laws shown in the figure, although there are not enough data at very low inclinations to determine accurately either the shape of the function or the mean value of κh for the different morphological types. For the types Sb and Sbc there is a hint of a steepening of the slope below log $r_i \cong$ -0.6 to -0.8. If this is real it suggests a value of 4 to 6 for $(\kappa h)^{-1}$. The Milky Way value of h, the effective full width of the **absorbing** layer of H I, is about 300 pc (Crovisier 1981), which is probably typical of these Sb's and Sbc's, so we may roughly estimate κ to be about 0.6 kpc^{-1}. This fits in with the solar neighborhood value of $\kappa \cong$ 5 km s^{-1} kpc^{-1} measured from absorption surveys (reviewed by Garwood and Dickey, 1987) if we assume that the line of sight velocity spread is about 10 km s^{-1} through the disks of these galaxies. Since the line of sight velocity gradient is itself a function of inclination (and of the rotation curve), a more sophisticated model than equation 6 should be be used to interpret the data, if we want to try to determine the opacity of the interstellar hydrogen from them. A more fundamental complication is that the H I is not homogeneous and isothermal in these galaxies any more than in the Milky Way, so that a model incorporating the spectrum of interstellar cloud optical depths should be used (Dickey and Garwood, 1989). With even more extragalactic observations it may be possible to study the universality of this spectrum.

4. Funny Excitation

In some extreme extragalactic environments the excitation of the 21-cm transition can deviate from the kinetic temperature so severely that equation 1 itself may break down. This has been studied for the environment of intense, varying radiation sources such as quasars by Urbaniak and Wolfe (1981) and by Bahcall and Ekers (1969). A simpler concern is that the transition may become subthermal in a low density environment where the collisition rate is less than the spontaneous lifetime of the upper state ($A^{-1} = 1.1 \cdot 10^7$ years) which occurs for densities below about $5 \cdot 10^{-3}$ cm^{-3}.

Watson and Deguchi (1984) show that redistribution of the level populations by scattering of Lyman-α photons can prevent this effect. Even the extragalactic background inferred at this wavelength by extrapolation of the soft x-ray spectrum is sufficient to keep the spin temperature near the kinetic temperature for extragalactic H I clouds such as that detected by Schneider *et al.* (1983). So in most astrophysical environments self-absorption is the only cause for correction to the simple formulae of equations 2 and 4.

5. Working from Absorption Alone

At redshifts higher than about 25,000 km s^{-1} emission in the 21-cm line becomes hard to detect, even from a gas rich galaxy. Much more distant objects have been detected in absorption at 21-cm (reviewed by Wolfe in this volume). Determining the H I column density corresponding to an observed absorption line requires some estimate of the spin temperature. There are two approaches to this, both based on solar neighborhood observations. The first is to average the emission and absorption measurements in high latitude surveys to obtain a mean value for the column density and optical depth of the Milky Way disk seen face-on. The full column density (from one side to the other) of the disk comes out to be in the range 5.4 to 6.6 · 10^{20} cm^{-2} depending on what latitude range is included in the average (the lower value corresponds to latitudes above 30°, *ie.* the area within about 300 pc of the sun, the higher value to latitudes of 10° to 30°, *ie.* distances of 300 pc to 1 kpc, Kulkarni and Heiles, 1986). The corresponding total equivalent width (integral of optical depth over velocity) determined from absorption surveys is 1.0 to 1.8 km s^{-1}. This gives an effective spin temperature (an average over warm and cool phases weighted by column density) of 200 to 300 K. M31 appears to have a similar distribution of phases of H I (Dickey and Brinks 1988). This would be an appropriate value to use to determine the mean column density for a sample of spiral galaxies studied in 21-cm absorption. On the other hand, in the case of a single **detected** absorption line this spin temperature estimate is too high, since if we restrict the averaging of the solar neighborhood lines of sight to only those directions showing absorption the ratio of column density to equivalent width is lower, the effective spin temperature is 100 to 150 K. But this depends on the threshold for optical depth detection, lines of optical depth below 10^{-2} would not be detected on most lines of sight in most absorption surveys, so the higher effective spin temperatures are more appropriate for estimating column densities corresponding to weak absorption lines.

For this problem also the correct approach would be to use the solar neighborhood spectrum of cloud optical depths to determine the mean free path between clouds corresponding to the one observed, and then multiply this by the midplane density of H I (in all phases) of 0.5 to 0.7 cm^{-3} typical of the solar circle. All this assumes that solar neighborhood conditions apply to these systems, which is unlikely. As our knowledge of the diversity of interstellar thermodynamic environments increases we will be able to improve the sophistication of our methods for estimating H I masses and column densities.

Acknowledgements

I am grateful to Martha Haynes and Ricardo Giovanelli for permission to use portions of their figure. Portions of this work were supported by the National Science Foundation under grant AST 8722990 to the University of Minnesota.

References

Bahcall, J. and Ekers, R., 1969, *Ap. J.*, **157**, 1055.
Bloemen, J.B.G.M, *et al.*, 1986, *Astron. Astrophys.*, **154**, 25.
Colgan, S., Salpeter, E.E., and Terzian, Y., 1988, *Ap. J.*, **328**, 275.
Crovisier, J., 1981, *Astron. Astrophys.*, **94**, 162.
Dickey, J.M. and Benson, J.M., 1982, *Astron. J.*, **87**, 278.

Dickey, J.M. and Brinks, E., 1988, *M.N.R.A.S*, **233**, 781.

Dickey, J.M. and Garwood, R.W., 1989, *Ap. J.*, **341**, 201.

Garwood, R.W. and Dickey, J.M., 1988, *Ap. J.*, **338**, 841.

Haynes, M.P. and Giovanelli, R. 1984, *Astron. J.*, **89**, 758.

Kerr, F.J. and Westerhout, G., 1965, in **Stars and Stellar Systems, vol. V: Galactic Structure,** ed.s : A. Blaauw and M. Schmidt , (Chicago : University of Chicago Press).

Kuchar, T.A. and Bania, T.M., 1989, *Ap. J.*, in press.

Kulkarni, S.R., *et al.*, 1984, *NASA Conference Publ. #2345* (IAU Colloq. # 81), eds. : Y. Kondo, F.C. Bruhweiler adn B.D. Savage, pp. 269 - 273.

Kulkarni, S.R. and Heiles, C., 1987, in **Interstellar Processes**, eds. D.J. Hollenbach and H.A. Thronson, Jr., (Reidel: Dordrecht).

Lockman, F.J. and Dickey, J.M, 1990, *Ann. Rev. Astr. Astrop.* in press.

Peters, W.L. and Bash, F.N., 1987, *Ap. J.*, **317**, 646.

Schneider, S.E. *et al.*, 1983, *Ap. J. (Lett.)*, **273**, L1.

Strong, A.W., *et al.*, 1988, *Astron. Astrophys.*, **207**, 1.

Urbaniak, J.J. and Wolfe, A.M., 1981, *Ap. J.*, **244**, 406.

Watson, W.D. and Deguchi, S., 1984, *Ap. J.*, **281**, L5.

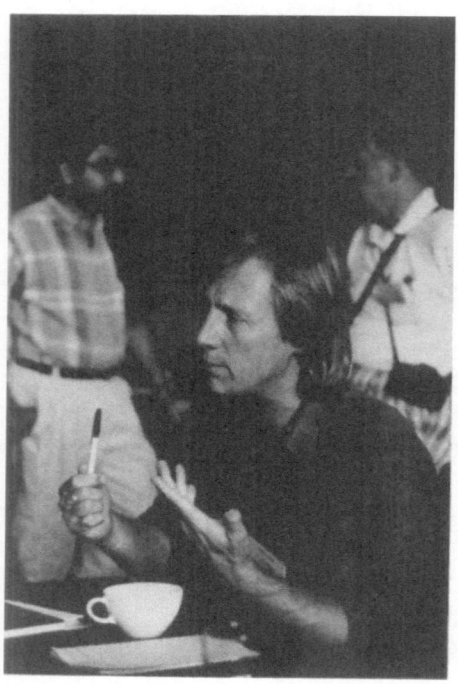

Mass Determinations from Far-Infrared Observations

Bruce T. Draine
Princeton University Observatory
Peyton Hall
Princeton NJ 08544 USA

ABSTRACT. The estimation of dust masses, and inferred gas masses, from observations of far-infrared emission is reviewed. Special attention is given to the uncertainties resulting from lack of knowledge of (a) dust emissivities; and (b) dust temperature distributions. To illustrate the uncertainties two examples are considered: far-infrared emission from Galactic diffuse clouds, and far-infrared emission from the starburst galaxy NGC6240. The importance of observations at $\lambda \gtrsim 300\,\mu$m is emphasized.

1. Introduction

Because interstellar clouds are optically-thin in the far-infrared (FIR), observations of FIR emission from dust sample the entire cloud, and offer the prospect of being able to estimate the cloud mass. In practice, the path from flux measurement to mass estimation is fraught with uncertainty, as I shall try to emphasize here.

The paper is organized as follows: The basic theory will be reviewed in §2. The state of our knowledge of grain opacities is the subject of §3. In §4 the complicating effects of temperature distributions are considered – including spatial variations within the "beam", as well as temporal variations due to temperature "fluctuations" in individual grains. The uncertainties will be illustrated in §5 using observations of Galactic diffuse clouds, and of the starburst galaxy NGC6240. The conclusions are summarized in §6.

2. Theory

2.1 General Results

Assuming the cloud to be optically-thin in the FIR, the intensity I_λ is given by

$$I_\lambda = \sum_i \int_0^\infty ds\, \rho\, \kappa_i(\lambda) B_\lambda(T_i) \quad , \tag{2.01}$$

where T_i is the temperature of grains of type i, ρ is the total mass density, and $B_\lambda \equiv 2hc^2\lambda^{-5}[\exp(hc/\lambda kT) - 1]^{-1}$. For simplicity the opacity $\kappa_i(\lambda)$ due to component i has

483

H. A. Thronson, Jr. and J. M. Shull (eds.), The Interstellar Medium in Galaxies, 483–492.
© 1990 *Kluwer Academic Publishers.*

been assumed to be independent of T_i. If the cloud is at distance D, and we integrate over the telescope beam Ω, then the flux density F_λ is

$$F_\lambda = \int_\Omega d\Omega \, I_\lambda = \frac{M}{D^2} \sum_i \kappa_i(\lambda) B_\lambda(T_i) \quad , \tag{2.02}$$

where M is the cloud mass. Thus

$$M = \frac{D^2 F_\lambda}{\sum_i \kappa_i(\lambda) B_\lambda(T_i)} \quad . \tag{2.03}$$

It is sometimes convenient to relate M to the total flux $F = \int d\lambda F_\lambda$:

$$M = \frac{\pi D^2 F}{\sum_i < \kappa_i >_{T_i} \sigma T_i^4} \quad , \tag{2.04}$$

where σ is the Stefan-Boltzmann constant and the Planck-average opacity due to component i is

$$< \kappa_i >_T = \frac{\int d\lambda \kappa_i(\lambda) B_\lambda(T)}{\int d\lambda B_\lambda(T)} \quad . \tag{2.05}$$

If component i has a temperature distribution, with probability distribution function dP_i/dT, then we replace B_λ in (2.03-2.05) with the time-and-space-average of B_λ for component i:

$$M = \frac{D^2 F_\lambda}{\sum_i \kappa_i(\lambda) < B_\lambda >_i} \quad , \tag{2.06}$$

where

$$< B_\lambda >_i \equiv \int dT (dP_i/dT) B_\lambda(T) \quad ; \tag{2.07}$$

or, for the total flux F:

$$M = \frac{\pi D^2 F}{\sum_i \int dT (dP_i/dT) < \kappa_i >_T \sigma T^4} \quad . \tag{2.08}$$

2.2 Long Wavelength Limit

Suppose we have reason to believe that most of the dust is at temperatures $T > T_{min}$. If we are able to measure F_λ at some wavelength $\lambda > hc/kT_{min} = 720 \, \mu\text{m}(20 \, \text{K}/T_{min})$, then we may approximate $< B_\lambda >_i \approx 2kc\lambda^{-4} < T_i >$, where $< T_i >$ is simply the time-average temperature of grains of type i, so that

$$M \approx \frac{D^2 \lambda^4 F_\lambda}{2kc \sum_i \kappa_i(\lambda) < T_i >} \tag{2.09}$$

The advantage of working at these long wavelengths is that we probably can estimate the mean grain temperature $< T >$ to within a factor of 2 or better, whereas at shorter wavelengths our lack of knowledge regarding the temperature distribution function may introduce order-of-magnitude uncertainties in $< B_\lambda >$ or $<< \kappa >_T T^4 >$. The disadvantage of working at long wavelengths is that uncertainties in grain opacities are greatest there.

3. Grain Opacities

Since $a/\lambda \ll 1$ for interstellar dust (radii $a \lesssim 0.25\,\mu m$) in the FIR ($\lambda \gtrsim 10\,\mu m$), the absorption cross sections $C_{abs} \propto a^3$ and the opacity κ_λ depends only on the total volume of dust per H atom (and its composition), but *not* on the details of the size distribution. [1] In Figure 1 I show the far-infrared opacities advocated in various recent papers; it is clear that there are order-of-magnitude disagreements.

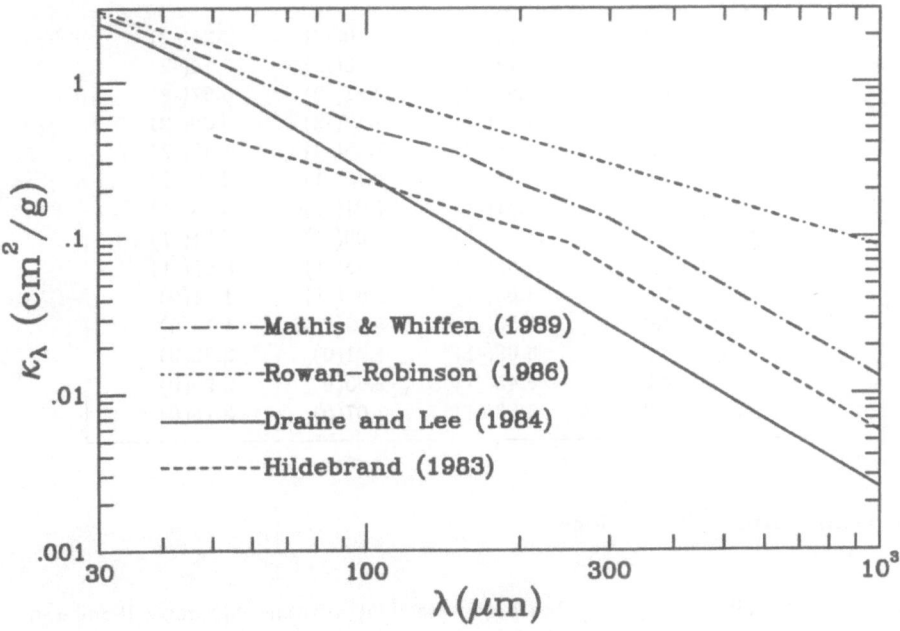

Fig. 1. – Far-infrared opacity estimates for local Galactic dust.

As our knowledge of the far-infrared spectra of clouds improves, more accurate observational determinations of far-infrared opacities for diffuse clouds should become possible. In this connection it is of interest to note that the Berkeley-Nagoya rocket experiment (Matsumoto *et al.* 1987; Lange 1989) measured far-infrared emission at $\lambda = 100$, 140, and $260\,\mu m$ in quite good agreement with what would be predicted using the Draine and Lee (1984) opacity estimates for a graphite-silicate grain mixture. *COBE* is expected to provide data which should help resolve these controversies.

---

[1] In principle the dielectric function ϵ can depend on the grain size (e.g., Seki and Yamamoto 1980; Hecht 1981; Draine and Lee 1984; Koike, Hasegawa and Hattori 1987), but the effects are probably not large compared to other uncertainties.

For the time being, however, one must make one's own educated guess. I still prefer the Draine and Lee (1984) estimates, based on a mixture with graphite and silicate volumes per H atom $V_{gr} = 2.5 \times 10^{-27}\,cm^3$, $V_{sil} = 2.8 \times 10^{-27}\,cm^3$ (Draine and Lee 1984); with densities $\rho_{gr} = 2.26\,gcm^{-3}$ and $\rho_{sil} = 3.3\,gcm^{-3}$ this gives a grain-to-gas mass ratio of .0063 (assuming He/H=0.1). Planck-averages of the Draine and Lee (1984) opacities are given in Table 1.

TABLE 1

Planck-Average Opacities $< \kappa >$ ($cm^2\,g^{-1}$)

$\log_{10} T(\,K)$	$< \kappa >_{gr}$	$< \kappa >_{sil}$	$< \kappa >$
1.0	1.11(-2)	1.10(-2)	2.21(-2)
1.1	1.78(-2)	1.76(-2)	3.54(-2)
1.2	2.86(-2)	2.81(-2)	5.67(-2)
1.3	4.57(-2)	4.52(-2)	9.08(-2)
1.4	7.26(-2)	7.28(-2)	1.45(-2)
1.5	1.14(-1)	1.18(-1)	2.32(-1)
1.6	1.77(-1)	1.91(-1)	3.68(-1)
1.7	2.61(-1)	3.09(-1)	5.71(-1)
1.8	3.62(-1)	4.98(-1)	8.60(-1)
1.9	4.65(-1)	7.85(-1)	1.25(0)
2.0	5.50(-1)	1.19(-1)	1.74(0)
2.1	6.09(-1)	1.71(0)	2.32(0)
2.2	6.45(-1)	2.35(0)	2.99(0)
2.3	6.83(-1)	3.07(0)	3.75(0)

4. Temperature Fluctuations

Because of the discrete nature of heating by starlight, very small grains will undergo a sudden rise in temperature immediately following absorption of a visible or ultraviolet photon, followed by a nearly-continuous decline in temperature as the heat is radiated away in the form of many far-infrared photons (Greenberg 1968). The grain temperature is therefore a time-dependent quantity, and we require the probability distribution function dP_i/dT, which will be a function of grain size, composition, and ambient radiation field. Grains with radii $a \gtrsim 100\,\text{Å}$ exposed to ambient starlight have temperature distribution functions which are very narrow so it is reasonable to approximate the temperature distribution function by a delta function. Based on this approximation, and estimated "average" grain temperatures $T \lesssim 20K$, it came as a surprise when $IRAS$ discovered emission from interstellar dust at 12 and 25 μm, and stronger-than-expected emission at 60 μm. The power in these bands is appreciable: $(\lambda I_\lambda)/(\lambda I_\lambda)_{100} = 0.35, 0.22$, and 0.35 for the 60, 25, and 12 μm bands, respectively (Boulanger and Perault 1988). Therefore the small grains/molecules responsible for this emission must account for a significant fraction of the total interstellar absorption of starlight. Consequently there must be very large numbers of ultrasmall grains. Similar conclusions regarding the numbers of ultrasmall grains/molecules have been reached based on attribution of observed near-infrared continuum emission in reflection nebulae to thermal emission from ultrasmall grains (Sellgren, Werner and Dinerstein 1983) and the proposal that the "unidentified infrared bands" are due to essentially thermal emission from polycyclic aromatic hydrocarbon molecules (Leger and Puget 1984).

Draine and Anderson (1985) have computed the expected emission spectrum from a graphite-silicate grain mixture, with various size distributions, exposed to interstellar starlight; more recently, Guhathakurta and Draine (1989) have developed a method for determining dP_i/dT which is both more accurate and much more efficient. Some examples of temperature distribution functions are shown in Fig. 2. It is seen, for example, that $a = 15$ Å graphite grains are heated to $T \gtrsim 200$ K following absorption of individual photons. While the distribution function $dP/d\ln T$ may seem to be small at the high-temperature tail of the distribution (e.g., $dP/d\ln T \approx 4 \times 10^{-6}$ at $T = 200$ K for $a = 15$ Å graphite grains heated by local Galactic starlight), these tails are of overwhelming importance from an energetic standpoint because when an individual grain is heated to a temperature T_{spike} by absorbing a photon, it radiates more than half of the absorbed energy while at temperatures $T > T_{spike}/2$. The function $dP/d\ln T$ is numerically small only because this energy is reradiated on a time short compared to the average interval between photon absorptions.

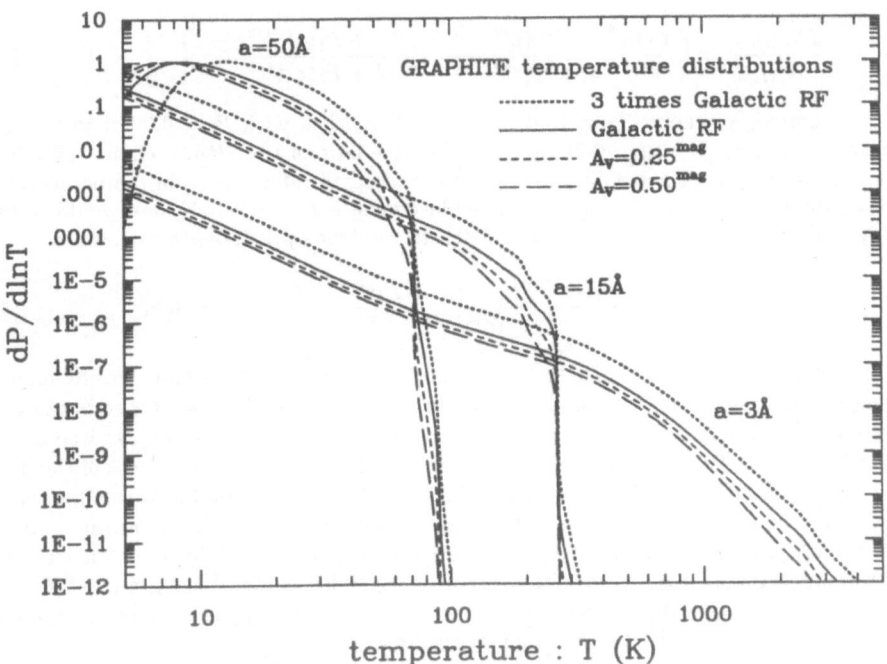

Fig. 2. – Temperature distribution functions $T dP/dT$ for graphite grains of radii $a = 50$, 15, and 5 Å, exposed to various radiation fields (from Guhathakurta and Draine 1989).

The details of the size distribution at radii $a \lesssim 100\,\text{Å}$ are not yet well-determined, nor is the composition of the small particles established (although it seems likely that polycyclic aromatic hydrocarbons must be a significant constituent), but there seems to be no way to avoid the conclusion that the ultrasmall grains must be sufficiently numerous to account for a substantial fraction of starlight absorption in diffuse clouds.

5. Two Examples: Galactic Cirrus and NGC6240

5.1 Galactic Cirrus

Serious errors can result if temperature distributions are present but are not recognized. For example, the *IRAS* $60/100\,\mu$m flux ratio is often used to estimate the "dust temperature"; the resulting "temperature" is then used, together with the $100\,\mu$m flux F_{100}, to estimate the cloud mass M using equation (2.03) and some plausible estimate of the $100\,\mu$m opacity κ_{100} (or, if M is known, κ_{100} may be inferred). The result, of course, is an (often severe) underestimate of M.

To illustrate this point, suppose the dust opacity varies as $\kappa_\lambda \propto \lambda^{-n}$, and suppose that a fraction P_2 of the dust is at T_2, and the remaining $P_1 = 1 - P_2$ of the dust is at T_1. The $60/100$ flux ratio is then

$$\frac{(\lambda F_\lambda)_{60}}{(\lambda F_\lambda)_{100}} = \left(\frac{100}{60}\right)^{4+n} \left[\frac{P_1(e^{240/T_1} - 1)^{-1} + P_2(e^{240/T_2} - 1)^{-1}}{P_1(e^{144/T_1} - 1)^{-1} + P_2(e^{144/T_2} - 1)^{-1}}\right] \quad . \tag{5.01}$$

If, for example, we choose $T_1 = 18\,\text{K}$ and $T_2 = 36\,\text{K}$, then with $P_2 = 0.004$ and $n = 2$ one obtains $(\lambda F_\lambda)_{60}/(\lambda F_\lambda)_{100} = 0.35$, a value characteristic of the *IRAS* "cirrus" (Boulanger and Perault 1988). A naive observer who attempted to infer the grain temperature from the $60/100$ flux ratio (assuming $n = 2$) would estimate $T_{est} = 23.3\,\text{K}$, and would estimate an optical depth τ_{est} which would be related to the true optical depth τ_{true} by

$$\frac{\tau_{est:60/100}}{\tau_{true}} = \frac{P_1(e^{144/T_1} - 1)^{-1} + P_2(e^{144/T_2} - 1)^{-1}}{(e^{144/T_{est}} - 1)^{-1}} = 0.20 \quad , \tag{5.02}$$

so he will *underestimate* the mass of the cloud by a factor 5! Much more severe numerical examples could obviously be found; the numbers in this example were chosen because they correspond to flux ratios actually observed for diffuse clouds in the Galaxy, as well as to the $60/100\,\mu$m flux ratios seen for other normal spirals. In the example the 0.4% of the dust at $T = 36\,\text{K}$ accounts for 76% of the $60\,\mu$m flux (and 20% of the total infrared power). The $T = 36\,\text{K}$ grains in the above example in fact approximately mimic the infrared emission of $a \approx 50\,\text{Å}$ grains heated by starlight, since starlight photons will heat such a grain up to peak temperatures $T_{spike} \approx 40 - 70\,\text{K}$ (cf. Fig. 2). The 99.6% of the grains at $T = 18\,\text{K}$ are at approximately the temperature expected for $a \gtrsim 100\,\text{Å}$ graphite or silicate grains (Draine and Lee 1984) heated by ambient starlight.

As noted in §2, using observations at longer wavelengths can decrease the sensitivity of the results to details of the temperature distributions. Suppose that instead of using the $60/100\,\mu$m flux ratio, the observer was able to use observations at $300\,\mu$m. Our two-temperature (18 and 36K) model has $(\lambda F_\lambda)_{300}/(\lambda F_\lambda)_{100} = 0.25$, from which the inferred grain temperature would be 18.7K, and the estimated optical depth $\tau_{est:300/100} = 0.92\tau_{true}$

– the error resulting from the single-temperature assumption has been reduced from a factor of five to merely 8%! This demonstrates the importance of photometry of diffuse clouds at wavelengths $\gtrsim 300\,\mu$m.

5.2 NGC6240

Consider the ultraluminous "starburst" galaxy NGC6240, at an estimated distance $D = 92$Mpc (assuming $H_0 = 80\,\mathrm{km\,s^{-1}\,Mpc^{-1}}$). Fluxes from the *IRAS* Point Source Catalog (1986) have been color-corrected (Beichman *et al.* 1988). We fit the optical and infrared emission by

$$F_\lambda = a_1 B_\lambda(6000\,\mathrm{K}) + a_2 B_\lambda(2000\,\mathrm{K}) + \sum_{i=1}^{N} b_i \lambda^{-n_i} B_\lambda(T_i) \ . \tag{5.03}$$

To determine the coefficients a_i and a_2 for the two components representing reddened starlight we simply fit by eye to the UBV photometry of Fosbury and Wall (1979) and the JHK photometry of Rieke *et al.* (1985). To determine the coefficients b_i we forced F_λ to match the *IRAS* fluxes at $\lambda = 12, 25, 60$, and $100\,\mu$m (only slightly different coefficients b_i would have been obtained if we had properly convolved with the *IRAS* bandpass functions). From the b_i we estimate the luminosity L_i associated with each component:

$$L_i = 4\pi D^2 b_i \int_0^\infty d\lambda\, \lambda^{-n_i} B_\lambda(T_i) = 8\pi D^2 b_i hc^2 \left(\frac{kT_i}{hc}\right)^{4+n_i} \Gamma(4+n_i)\zeta(4+n_i) \ , \tag{5.04}$$

where Γ and ζ are the usual gamma and Riemann zeta functions.

TABLE 2
Models for the FIR Emission from NGC6240

quantity	Model A	Model B	Model C	Model D
T_1	36	25	20	20
T_2	–	50	40	50
T_3	100	100	100	100
T_4	200	200	200	200
n_1	2	2	2	2
n_2	–	2	2	2
n_3	1	1	1	1
n_4	1	1	1	1
$L_1(\mathrm{L_\odot})$	3.1(11)	1.3(11)	6.2(10)	1.2(11)
L_2	–	3.0(11)	2.9(11)	3.4(11)
L_3	1.2(11)	5.2(10)	1.1(11)	4.0(10)
L_4	4.6(10)	5.2(10)	4.7(10)	5.3(10)
$\sum_{i=1}^{4} L_i(\mathrm{L_\odot})$	4.7(11)	5.3(11)	5.1(11)	5.6(11)
$M_1(\mathrm{M_\odot})$	5.6(9)	1.9(10)	3.7(10)	7.4(10)
M_2	–	7.2(8)	2.6(9)	8.4(8)
M_3	6.0(6)	2.6(6)	5.5(6)	2.0(6)
M_4	6.7(4)	7.5(4)	6.8(4)	7.6(4)
$\sum_{i=1}^{4} M_i(\mathrm{M_\odot})$	5.6(9)	2.0(10)	4.0(10)	7.5(10)

490

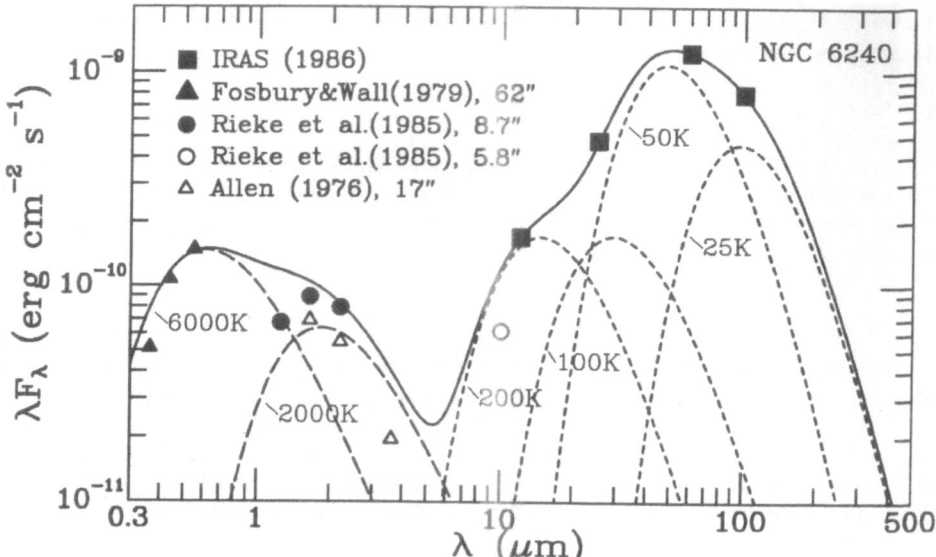

Fig. 3. – Observed spectrum of NGC6240 (symbols) together with the Model B fit (solid curve); see Table 2 for fit parameters. Separate contributions from the six fit components are shown (broken curves).

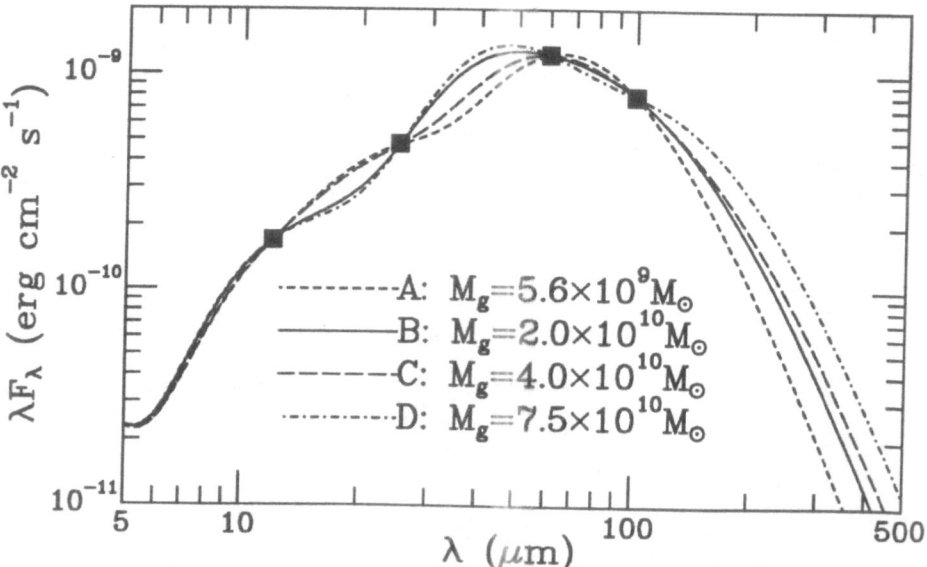

Fig. 4. – Four different model fits to the *IRAS* observations of NGC6240 (see text).

Figure 3 shows one possible fit (B), together with separate contributions from each component. Figure 4 shows four fits (A-D) differing in the choices for T_i; numerical details are given in Table 2. To estimate the gas mass associated with each component i, we use the Planck-average opacities of Table 1, which are estimated for local Galactic dust. The mass of gas associated with each component i is given by eq. (2.04): $M_i = L_i/4 < \kappa_i >_{T_i} \sigma T_i^4$. The minimum mass ($M = 5.6 \times 10^9 M_\odot$) is obtained in model A by assuming relatively warm dust, with only three dust components needed to reproduce the four fluxes if the temperatures are appropriately chosen. With the two coolest dust components at $T = 20$ and 50K in model D, the mass is found to be $M = 7.5 \times 10^{10} M_\odot$ – a factor of 13 times larger than the mass in model A!

It is disappointing that even for a galaxy with λF_λ peaking at $\lambda_p \approx 60\,\mu$m, high quality *IRAS* observations cannot distinguish between models with dust masses differing by over an order of magnitude. As seen in Fig. 4, the largest differences between the different fits occur at wavelengths $\lambda \gtrsim 150\,\mu$m where there are as yet no observational constraints for NGC6240: for example, at $300\,\mu$m Model D is brigher than Model A by a factor 3.8 . Therefore photometry at $\lambda \gtrsim 300\,\mu$m can be used to significantly narrow the range of possible models. Fortunately, this is now possible: $150 - 300\,\mu$m balloon observations (Hauser *et al.* 1984) have been used to estimate the mass of gas in the Galaxy (Broadbent, MacLaren and Wolfendale 1989). Submillimeter photometry from UKIRT at 350-$1100\,\mu$m (Eales, Wynn-Williams and Duncan 1989) and the IRTF at $360\,\mu$m (Stark *et al.* 1989) have recently been used to estimate dust masses in other galaxies.

6. Conclusions

Methods for estimating gas masses from FIR observations have been reviewed. There are two principal sources of uncertainty: *(i)* the dust opacity $\kappa(\lambda)$ at wavelengths $\lambda \gtrsim 50\,\mu$m is still controversial; *(ii)* the actual temperature distribution of the dust may be poorly known.

Increasingly accurate far-infrared observations of diffuse clouds in the Galaxy – including crucially-important observations at wavelengths $\lambda \gtrsim 300\,\mu$m – will probably be able to resolve the controversy regarding the opacity of Galactic dust, at least in diffuse regions where we know how much gas is present. Thus the uncertainties regarding dust opacities are expected to diminish in coming years.

Even with precisely-known opacities, interpretation of observations is often highly-dependent on knowledge of the grain temperature distribution. Using the galactic cirrus as an example, in §5.1 it is shown that serious errors can result if the $60/100\,\mu$m flux ratio is used to naively estimate the dust temperature, since the true dust temperature distribution is *not* a delta function. Such naive estimates can easily underestimate dust masses by a factor of ~ 5, as shown for an example. Reliable estimation of dust masses using photometry at wavelengths $\lambda \leq 100\,\mu$m is possible only if accurate estimates of the dust grain temperature distribution are available, e.g., from theoretical models. Dependence on theoretical modelling can be minimized by obtaining photometry at submillimeter wavelengths $\lambda \gtrsim 300\,\mu$m where the emission is not highly sensitive to details of the dust temperature distribution.

Progress on the determination of dust temperature distributions will require advances in both theoretical modelling of interstellar dust (in order to have better estimates of the size distribution and temperature fluctuations) and improvement in multiwavelength FIR photometry in order to observationally determine the emission spectrum.

In order to estimate the mass of dust from FIR observations of extragalactic objects, it will be necessary (for the foreseeable future) to assume the dust opacity to be the same as for local Galactic dust. With this assumption, the inferred dust masses will still have appreciable uncertainties due to lack of knowledge of the dust temperature distribution unless very long wavelength photometry ($\lambda \gtrsim 300\,\mu$m) is available, in which case precise knowledge of the dust temperature distribution is not essential.

This work was supported in part by grant AST86-12013 from the National Science Foundation.

7. References

Allen, D. A. 1976, *Ap. J.*, **207**, 367.

Beichman, C. A., Neugebauer, G., Habing, H. J., Clegg, P. E., and Chester, T. J. 1988, *IRAS Catalogs and Atlases, Vol. 1. Explanatory Supplement* (Washington: NASA).

Boulanger, F., and Perault, M. 1988, *Ap. J.*, **330**, 964.

Broadbent, A., MacLaren, I., and Wolfendale, A. W. 1989, *M.N.R.A.S.*, **237**, 1075.

Draine, B. T., and Anderson, N. 1985, *Ap. J.*, **292**, 494.

Draine, B. T., and Lee, H. M. 1984, *Ap. J.*, **285**, 89.

Eales, S. A., Wynn-Williams, C. G., and Duncan, W. D. 1989, *Ap. J.*, **339**, 859.

Fosbury, R. A. E., and Wall, J. V. 1979, *M.N.R.A.S.*, **189**, 79.

Greenberg, J. M. 1968, in *Stars and Stellar Systems*, Vol. **7**, *Nebulae and Interstellar Matter*, ed. B. M. Middlehurst and L. H. Aller (Chicago: University of Chicago Press), p. 221.

Guhathakurta, P., and Draine, B. T. 1989, *Ap. J.*, **345**, in press.

Hauser, M. G., Silverberg, R. F., Stier, M. T., Kelsall, T., Gezari, D. Y., Dwek, E., Walser, D., Mather, J. C., and Cheung, L. H. 1984, *Ap. J.*, **285**, 74.

Hecht, J. 1981, *Ap. J.*, **246**, 794.

Hildebrand, R. H. 1983, *Quart.J.R.A.S.*, **24**, 267.

Koike, C., Hasegawa, H., and Hattori, T. 1987, *Astr. Sp. Sci.*, **134**, 95.

Lange, A. E. 1989, private communication.

Leger, A., and Puget, J. L. 1984, *Astr. Ap.*, **137**, L5.

Mathis, J. S., and Whiffen, G. 1989, *Ap. J.*, **341**, 808.

Matsumoto, T. Hayakawa, S., Matsuo, H., Murakami, H., Sato, S., Lange, A. E., and Richards, P. L. 1988, *Ap. J.*, **329**, 567.

Rieke, G. H., Cutri, R. M., Black, J. H., Kailey, W. F., McAlaray, C. W., Lebofsky, M. J., and Elston, R. 1985, *Ap. J.*, **290**, 116.

Rowan-Robinson, M. 1986, *M.N.R.A.S.*, **219**, 737.

Seki, J., and Yamamoto, T. 1980, *Astr. Sp. Sci.*, **72**, 79.

Sellgren, K., Werner, M. W., and Dinerstein, H. W. 1983, *Ap. J. (Letters)*, **271**, L13.

Stark, A. A., Davidson, J. A., Harper, D. A., Pernic, R., Loewenstein, R., and Casey S. 1989, preprint.

Mass Determinations from CO Observations

Phil Maloney
Sterrewacht Leiden
Postbus 9513
2300 RA Leiden
The Netherlands

ABSTRACT. Determinations of molecular gas masses from CO observations are reviewed. The underlying basis for the method and the dependence of the ratio of $I_{CO}/N(H_2)$ on physical conditions are discussed. Estimates of the actual value for the conversion from CO integrated intensity to H_2 column density are reviewed, as well as the evidence for variations. It is concluded that the standard conversion factor is probably accurate to a factor of 2 or so for the disks of normal, metal-rich spiral galaxies; however, use of this conversion can lead to substantial errors when applied to low-metallicity systems, such as irregular galaxies, and to molecular gas in extreme environments, such as galactic nuclei and ultra-luminous galaxies.

I. Introduction

Molecular hydrogen is one of the major components, by mass, of the interstellar medium, and its abundance is directly related to the rate of star formation in a galaxy. Unfortunately, the symmetry of the H_2 molecule makes direct detection of molecular hydrogen impossible under most circumstances, since electric dipole transitions are strictly forbidden. To estimate the amount of molecular gas that is present in some region, we generally have to rely on observations of a trace molecule, which have been calibrated in some way to provide a mass estimate. In fact, this is how the molecular cloud component of the ISM was discovered, from observations of the $J = 1 - 0$ transition of the CO molecule at 115.271 GHz. A great deal of effort has gone into determining the distribution and mass of molecular gas in the Galaxy (see Blitz 1987 and references therein). Here I concentrate on the use of CO observations to determine molecular gas masses in other galaxies.

Bright emission in the pure rotation lines of CO is characteristic of molecular clouds in the Galaxy, due to the fortunate combination of three parameters: the large abundance of CO, its small rotational-energy spacing, and its small dipole moment (van Dishoeck and Black 1987). Under what circumstances can CO emission be expected to trace molecular mass? Can it reliably be used to determine gas masses in other galaxies, where physical conditions in the interstellar medium may be very different? The rest of this paper addresses these issues. The organization is as follows: §2 is a brief history of the use of CO

H. A. Thronson, Jr. and J. M. Shull (eds.), The Interstellar Medium in Galaxies, 493–523.
© *1990 Kluwer Academic Publishers.*

as an extragalactic molecular gas tracer. §3 describes the physical basis for conversions from CO emission to molecular gas mass. The evidence for variations in the value of the conversion factor and the likely uncertainties are discussed in §4. §5 discusses methods by which knowledge of the physical conditions in the ISM of other galaxies can be obtained, and the results are summarized in §6.

The use of CO in extragalactic gas mass determinations has been discussed by Rickard, Harvey and Blitz (1984), Israel (1985), Dickman, Snell and Schloerb (1986), Verter (1987), van Dishoeck and Black (1987), Maloney and Black (1988), and Israel (1988a), among others.

2. A Brief History

It is a tribute to the efforts of millimeter-wave engineers that the first detection of extra-galactic CO emission occurred only five years after the first detection of the $J = 1 - 0$ line in the Milky Way (Rickard et al. 1975). The early observational work is summarized in Morris and Rickard (1982), who also present one of the first discussions of CO mass determinations and uncertainties. Of particular importance is the concept of a 'standard' molecular cloud, which provides the physical basis for CO emission to molecular mass conversions, although this hasn't always been recognized. Since the $J = 1 - 0$ transition of ^{12}CO is ordinarily optically thick, it is not immediately obvious why the intensity of this line should be related in any way to the mass of molecular gas in the emitting region.

An important early paper is Young and Scoville (1982), who studied the distribution of CO emission in the Scd galaxies NGC 6946 and IC 342. Young and Scoville attempted to calibrate the use of ^{12}CO $J = 1 - 0$ emission as a molecular mass tracer by deriving an empirical relation between the integrated intensity in the line, I_{CO} (units of K km s^{-1}), and the column density of molecular hydrogen (as derived from extinction measurements, LTE analysis of ^{12}CO and ^{13}CO data, or virial theorem mass estimates) for a sample of Galactic molecular clouds. Young and Scoville obtained a value of $N(H_2)/I_{CO} = 4 \times 10^{20}$ cm^{-2} K km s^{-1} which they used in analyzing their observations; the same value was assumed to hold at all galactocentric radii. NB: in light of the consistent notation that has been established by the gamma-ray analysts, the conversion factor $N(H_2)/I_{CO}$ will be imaginatively denoted as X in the rest of this chapter.

The Young and Scoville value for X, or the nearly identical value derived from the UMass/Stony Brook Galactic plane survey using virial theorem analyses (Sanders, Scoville and Solomon 1985), has been used to interpret the CO emission from a large number of galaxies, and to draw broad conclusions about the interstellar medium in galaxies. The total number of galaxies that have been detected in CO is now well in excess of 100 (Verter 1985, and updates thereof), including dwarf irregular galaxies, early and late-type spirals, and extremely active galaxies such as the 'ultra-luminous galaxies' discovered by IRAS. Doubts have occasionally been raised, however, as to whether a conversion factor derived from observations of Galactic molecular clouds is appropriate for all these objects. While studies of γ-ray emission in the Galaxy have provided strong evidence for a uniform value of X (within the uncertainties) throughout most of the Galactic disk (although this is still a matter of some debate: see Bloemen 1989 and references therein), they provide equally strong evidence for a very different value of the conversion in the Galactic center. Blitz et al. (1985) concluded that use of the disk value for X overestimates the amount of molecular gas in the inner 400 pc of the Galaxy by at least a factor of 6 or 7 (unless cosmic rays are preferentially excluded from Galactic center clouds, e.g. by magnetic fields). This result raises serious doubts about the applicability of the Galactic value of X to molecular clouds in similarly extreme environments, such as galactic nuclei.

Other doubts as to the validity of interpreting all extragalactic CO values with the standard conversion factor have been raised in the study of irregular galaxies. Very few irregular galaxies have been detected in CO, and the emission is quite weak. Israel *et al.* (1986) suggested that the paucity of CO emission from the Large and Small Magellanic Clouds was due not to an absence of molecular gas, but was instead a consequence of the low metallicities of these galaxies. Similar conclusions were reached by Thronson *et al.* (1987) in a multi-wavelength study of the Magellanic irregular galaxy NGC 4449. As low metallicity is characteristic of irregular galaxies, it is possible that the use of the Galactic X value will systematically underestimate the amount of molecular gas present.

In the next section I discuss the physical basis for use of I_{CO} as a tracer of molecular mass, and the expected dependence of X on the physical conditions in the molecular gas.

3. The Physical Basis of Conversion Factors

3.1 DERIVATION

Molecular clouds in other galaxies are unresolved by existing millimeter-wavelength antennas, even interferometers, except for the largest molecular clouds in the Local Group galaxies. The observed CO emisson from a galaxy is then the convolution of the antenna power pattern with the emission from the unresolved molecular cloud ensemble. The observable CO integrated intensity can be expressed as the integral of this convolution over velocity space:

$$I_{CO} = \Omega_A^{-1} \iiint T_R(\theta, \phi) P(\theta, \phi) \, d\Omega \, dV \quad \text{K km s}^{-1} \tag{1}$$

where $\Omega_A \equiv \iint P(\theta, \phi) \, d\Omega$. (For a gaussian antenna with FWHM $\theta_A \ll (4 \ln 2)^{1/2}$, which is true for all millimeter antennas, $\Omega_A = 1.13\theta_A^2$.) For a population of discrete clouds and no cloud-cloud shielding, the integral in equation (1) can be replaced by the sum

$$I_{CO} = \Omega_A^{-1} \sum_i T_R^i A^i P^i (2\pi)^{1/2} \sigma_v^i \quad \text{K km s}^{-1} \tag{2}$$

where the superscript i denotes the properties of the ith cloud, T_R^i is the cloud area-averaged radiation temperature (note that antenna temperatures are defined as Rayleigh-Jeans temperatures; see Kutner and Ulich 1981), A^i is the area of a cloud, P^i is the value of the antenna response at the position of the ith cloud, σ_v^i is the internal velocity dispersion of a cloud, and a gaussian line profile has been assumed. For simplicity, consider an antenna whose power pattern is a uniform disk of diameter θ_A, i.e., $P(\theta)$ equals 1, $\theta \leq \theta_A/2$, and 0 otherwise. This idealization will not affect the results significantly. The beam area is then $\Omega_A = \pi \theta_A^2 / 4$.

Equation (2) is most easily treated statistically (Dickman, Snell and Schloerb 1986, hereafter DSS), by replacing the sum over individual cloud parameters by the ensemble-averaged values. Performing this replacement, and writing the cloud area as $A = \pi D^2 / 4$, where D is the mean diameter, the integrated intensity becomes

$$I_{CO} = \frac{\pi (2\pi)^{1/2}}{4\Omega_A} N \langle T_R \rangle \langle D \rangle \langle \sigma_v \rangle \quad \text{K km s}^{-1} \tag{3}$$

where $\langle \; \rangle$ denotes an ensemble-averaged quantity, and N is the total number of clouds within the beam.

Similarly, the beam-averaged column density of H_2 can be expressed as

$$N(H_2) = \frac{\pi}{6\Omega_A} N \langle \bar{n} \rangle \langle D^3 \rangle \;\; cm^{-2} \tag{4}$$

where $\langle \bar{n} \rangle$ is the mean H_2 number density in cm^{-3}. The ratio of equations (4) and (3) is then

$$X = \frac{2}{3} \frac{\langle \bar{n} \rangle \langle D^3 \rangle}{(2\pi)^{1/2} \langle T_R \rangle \langle D^2 \rangle \langle \sigma_v \rangle}$$

$$= 8.2 \times 10^{17} \frac{\langle \bar{n} \rangle \langle D \rangle}{\langle T_R \rangle \langle \sigma_v \rangle} \;\; cm^{-2} \; (K \; km \; s^{-1})^{-1} \tag{5}$$

where the numerical constant is for D in parsecs and σ_v in $km \; s^{-1}$, and it has also been assumed that $\langle D^n \rangle = \langle D \rangle^n$.

Equation (5) as it stands does not contain any physics. To go further, we have to assume that there is some relation between the velocity dispersion and the size and mass of a molecular cloud. This is usually done by supposing that clouds are in virial equilibrium. A stable cloud which is dominated by its own self-gravity and which has had time to reach virial equilibrium will have a one-dimensional velocity dispersion given by

$$\sigma_v = \left(\frac{2\varphi GM}{3D} \right)^{1/2} \tag{6}$$

where φ is a parameter of order unity which depends on the mass distribution in the cloud. For a homogeneous spherical cloud, $\varphi = 0.6$, while for a spherical cloud with an r^{-2} density profile, $\varphi = 1.0$. In addition, for an optically thick transition like the $J = 1 - 0$ line of ^{12}CO, the apparent velocity dispersion will be larger than that given by equation (6), due to saturation of the line emission. This effect can be described by including an additional factor γ in equation (6), so that $\sigma_v = (2\varphi\gamma GM/3D)^{1/2}$. For the ^{12}CO $J = 1 - 0$ line, γ probably lies in the range 1.5–2 (DSS). For the rest of this discussion it will be assumed that $\varphi\gamma = 1$.

In terms of mean density $\bar{n}$ and diameter D, equation (6) becomes

$$\sigma_v = (\pi G \bar{n} \mu D^2 / 9)^{1/2}$$

$$= 9.7 \times 10^{-3} (\bar{n} D^2)^{1/2} \;\; km \; s^{-1} \tag{7}$$

where μ is the mean mass per H_2 molecule ($\mu = 2.56 m_H$ for the cosmic He/H ratio), and the numerical coefficient is for D in pc, σ_v in $km \; s^{-1}$. With the assumption of virial equilibrium, equation (5) then becomes

$$X = 8.43 \times 10^{19} \frac{\langle \bar{n} \rangle^{1/2}}{\langle T_R \rangle} \;\; cm^{-2} \; (K \; km \; s^{-1})^{-1}. \tag{8}$$

For the canonical Galactic values of $\langle \bar{n} \rangle = 200 \; cm^{-3}$, $\langle T_R \rangle = 6.7 \; K$, $X = 1.8 \times 10^{20}$. Equation (8) can also be expressed in terms of mass surface density of molecular gas,

$$\frac{\sigma_{H_2}}{I_{CO}} = 1.73 \frac{\langle \bar{n} \rangle^{1/2}}{\langle T_R \rangle} \;\; M_\odot \; pc^{-2}; \tag{9}$$

the Galactic values for $\langle \bar{n} \rangle$ and $\langle T_R \rangle$ give $\sigma_{H_2}/I_{CO} = 3.6$. For reference, the Young and Scoville (1982) X-value of 4×10^{20} corresponds to $\sigma_{H_2}/I_{CO} = 8.2$, including the correction for helium.

The simple derivation above yields a value for X that is about 50% of the Young and Scoville (1982) value. It is in close agreement with the estimates for X obtained from the analysis of Galactic γ-ray emission, which provide what is undoubtedly the best estimate of the large-scale value of X in the Milky Way: $X \lesssim 2.5 \times 10^{20}$ (Bloemen et al. 1986; Bhat, Mayer and Wolfendale 1987, Bloemen 1989). For a variety of reasons the γ-ray X-value is an upper limit; the Durham group advocate a somewhat lower value than the COS-B group (see Bloemen 1989, and references therein). The agreement is encouraging, and suggests that it is reasonable to use the simple model of this section to estimate the sensitivity of X to variations in the parameters describing the molecular cloud ensemble.

3.2 PHYSICAL DEPENDENCES

Under the assumptions of the previous section, the dependence of X on temperature and density is given by equation (8): $X \propto \langle \bar{n} \rangle^{1/2} \langle T_R \rangle^{-1}$. The radiation temperature T_R is related to the excitation temperature T_{ex} of the transition, which describes the relative populations of the upper and lower states of the transition, by

$$ T_R \equiv \frac{h\nu}{k} \left[(e^{h\nu/kT_{ex}} - 1)^{-1} - (e^{h\nu/kT_{bg}} - 1)^{-1} \right] (1 - e^{-\tau}) \quad \text{K} \qquad (10) $$

where h is Planck's constant, ν the frequency of the transition, k is Boltzmann's constant, T_{bg} is the temperature of the background radiation field (ordinarily 2.7 K), and τ is the optical depth of the transition. (For the $J = 1 - 0$ transition of ^{12}CO, $h\nu/k = 5.5$ K.) The $J = 1 - 0$ line of ^{12}CO is generally very optically thick and thermalized, so that $T_{ex} \approx T_k$, the gas kinetic temperature, and the optical depth factor is unity. Since T_R is defined as a Rayleigh-Jeans temperature, the relation between T_R and T_{ex} will not be linear for small values of T_{ex}. For the $J = 1 - 0$ line, $T_R = 6.7$ for $T_{ex} = 10$ K. However, for $T_{ex} \gtrsim 15$ K, this non-linearity is not important, and T_R scales as T_{ex}. All other things being equal, the appropriate value of X will decrease nearly linearly with T_{ex}. This temperature dependence is likely to be quite important, as we shall see later.

It is important to note that the correct temperature scale should be used in estimating H_2 masses from observed CO emission. The coupling between the antenna response and the source brightness distribution will in general be quite different for observations of galaxies, where the extent of the emission is small compared to the size of the telescope error pattern, and observations of Galactic molecular clouds, for which the source may fill the telescope error pattern as well as the diffraction beam (see Ulich and Haas, 1976; Kutner and Ulich 1981). No correction for this is made in the standard temperature scales T_R^* and T_A^*. In general the correct temperature scale to use will be T_{mb}, the antenna temperature corrected for everything except the coupling of the main (diffraction) beam to the source brightness distribution.

Under the assumption of virial equilibrium, $X \propto \langle \bar{n} \rangle^{1/2}$. Thus X is less sensitive to variations in mean density than it is to temperature variations; such variations may nonetheless be important, for example in galactic nuclei (see §4.3).

There is an additional physical dependence of X which has been ignored in the derivation above, and which has been generally either disregarded or treated improperly.

That is the metallicity of the gas, Z. The effect of metallicity comes into the derivation of X in a fairly subtle way. In taking the ratio of equation (4), the beam-averaged H_2 column density, and equation (3), the integrated CO emission from the clouds contained in the beam, we implicitly assumed that the projected area of a molecular cloud, as given by the actual cloud boundaries, and the projected area of the $J = 1 - 0$ emission region are the same. It does not, in fact, matter whether this ratio is actually unity, *provided that it is always the same for all molecular clouds.*

For molecular clouds in the solar neighborhood, the ratio of CO-size to molecular (H_2) size is probably close to unity: CO becomes the dominant gas phase carbon species not very much deeper into a cloud than the depth at which H_2 dominates the hydrogen abundance (Maloney and Black 1988). The increase in gas-phase metallicity towards the Galactic center (e.g., Shaver *et al.* 1983) will thus have very little effect on X, as the minor increase in the CO-size that might be expected is insignificant, and the emergent intensity from a cloud surface will not increase since the $J = 1 - 0$ line of ^{12}CO is already optically thick. *It is not expected that the value of X will depend on the gas-phase [CO]/[H_2] ratio for metallicities $Z \gtrsim Z_\odot$.* One must be cautious, however, when deriving H_2 column densities from CO *column densities* determined in some fashion, e.g. analysis of ^{12}CO and ^{13}CO data, since then the calculated hydrogen column density depends directly on the assumed [CO]/[H_2] ratio, which may be higher by $Z/Z_\odot$ than the solar neighborhood value.

In low-metallicity environments, the dependence of CO luminosity on metallicity may be very different. When the gas-phase abundances of carbon and oxygen are substantially smaller than the solar neighborhood values, the CO-size of a cloud may become very much smaller than the H_2-size, which means the use of a solar neighborhood-calibrated X-value will drastically underestimate the amount of molecular gas present. This is discussed in detail in §4.3.

The derivation in this section shows that, under the assumption that molecular clouds are in virial equilibrium and have $T_{ex} \approx 10$ K, it is possible to derive a value for X that is about 50% of the widely-used value of Young and Scoville (1982), and is in good agreement with the estimate provided by analysis of γ-ray emission. It has also made clear, I hope, that use of an X-value derived from observations of molecular clouds in the Milky Way implicitly assumes that the molecular cloud ensembles in other galaxies are characterized by similar average values of cloud properties, i.e., kinetic temperatures of ≈ 10 K and mean densities of a few hundred cm^{-3}, as well as roughly solar (or higher) metallicity. Major deviations from these values will produce corresponding changes in X; such changes are the subject of the next section.

4. Variations and Uncertainties

4.1 INTRODUCTION

A single value of X has been used to interpret a large number of observations of other galaxies, ranging in luminosity from dwarf irregular galaxies to the ultraluminous infrared galaxies discovered by *IRAS*. Can we reasonably expect the constant of proportionality between H_2 column density and I_{CO} to remain the same over such a large range of physical conditions? In this section I discuss the evidence and present arguments for believing that there is considerable variation in X among galaxies of different types, and possibly within individual galaxies.

Ensembles of molecular clouds characterized by parameters similar to those found in the disk of the Milky Way will also be characterized by a similar value of X. This is probably true of the disks of the majority of normal, late-type giant spiral galaxies, and quite possibly of galaxies of earlier Hubble type, although exceptions undoubtedly exist. In the absence of a theory that explains why the molecular cloud component of the interstellar medium in the Milky Way exhibits the physical parameters we observe, and what values those parameters would have under other conditions, it is a little risky to make sweeping statements, but it is certainly reasonable to expect that galaxies that have similar metallicities, energy injection rates into the ISM, etc. to the Milky Way will also have molecular cloud ensembles with similar average properties, and hence approximately the same X-value.

However, there are galaxies in which conditions in the interstellar medium must be very different from the solar neighborhood, and a number of arguments indicate that X has a very different value than in the Galaxy; use of the Galactic value will then lead to erroneous results. There are several classes of galaxies (or regions of galaxies) for which this is probably true, and these objects are discussed in this section. They have been split into two groups, largely on the basis of the relevant physics. §2 discusses galactic nuclei, starburst galaxies, and ultraluminous galaxies, while §3 considers the effect of low metallicity, which is characteristic of irregular galaxies, on X.

4.2 MOLECULAR GAS IN GALACTIC NUCLEI, STARBURST GALAXIES, AND UL- TRALUMINOUS GALAXIES

To some extent astronomers exhibit a very parochial view of the universe: conditions in the solar neighborhood are generally assumed to be typical of nearly all galaxies, at least for redshift $z \approx 0$. However, we don't have to look very far to find an environment that is very different: the nucleus of the Milky Way, at a distance of 8.5 kpc (or perhaps less, as the Galaxy continues to shrink at the behest of the *IAU*). Detailed observations (for recent reviews, see Genzel and Townes 1987; Liszt 1988) show that the inner few hundred parsecs of the Galaxy are characterized by intense radiation fields, high gas temperatures, and complicated gas dynamics. That X may be very different in the inner few hundred parsecs from the disk value is suggested by the γ-ray analysis of Blitz *et al.* (1985), who found that the γ-ray peak, which should be present if the disk X-value applies to the nuclear CO emission, is absent; the upper limit to γ-ray emission from the inner 400 pc implies that the amount of molecular gas in this region is overestimated by $\gtrsim 6 - 7$, unless γ-rays are preferentially excluded from Galactic center clouds.

We possess a much more detailed picture of the nucleus of our own Galaxy than it is possible to obtain for other galaxies; nevertheless it is likely that physical conditions in most galactic nuclei will be very different than in the disk of the Galaxy. This is especially true of galaxies that exhibit either high rates of nuclear star formation or an AGN, or both. It undoubtedly applies to the *ultraluminous galaxies* discovered by *IRAS*, which emit most of their radiation in the far-infrared and have luminosities $L_{IR} \gtrsim 5 \times 10^{11} L_\odot$. Some of these galaxies appear to be starbursts, while others are dominated by an active nucleus. The ultraluminous galaxies, starburst galaxies, and infrared-bright galactic nuclei have infrared surface brightnesses in the range $10^5 - 10^7 L_\odot$ pc^{-2} (Lo *et al.* 1987), several orders of magnitude higher than the typical disk value ($\approx 10 L_\odot$ pc^{-2}) in the Milky Way. The very intense radiation fields, high stellar densities, and possible large-scale hydrodynamic phenomena in these galaxies may have drastic effects on the state of molecular gas, and make the blanket application of a single X-value very risky.

A number of lines of evidence suggest that the use of the Galactic X-value considerably overestimates the amount of molecular gas in the high energy-density environments of these galaxies; this evidence is discussed in detail in the following sections.

4.2.1 *Extinctions and Dust Masses.* The column density of molecular hydrogen inferred from I_{CO} by use of a conversion factor will have an associated amount of extinction due to dust. If we assume that the ratio of hydrogen column density to visual extinction is the same in other galaxies as it is in the solar neighborhood (i.e., that the dust-to-gas ratio and grain properties are similar), we can estimate the amount of extinction expected from the CO observations. Similarly, the amount of gas can be determined from submillimeter observations of thermal emission from dust if we assume values for the dust emissivity and the dust-to-gas ratio (Hildebrand 1983; see §3.2). The CO-derived H_2 column densities and visual extinctions for a sample of infrared-bright galaxies are given in Table 1; the column densities are calculated using $X = 2.5 \times 10^{20}$ (the value derived from the γ-ray analysis); this value will be used throughout the rest of this paper. The galaxies in Table 1 comprise all the galaxies for which C^+ 158μm as well as CO $J = 1 - 0$ data are available; not coincidentally, they all have high far-infrared fluxes. The C^+ data are discussed in §4.2.4. All the CO data in Table 1 were obtained with the FCRAO 14m or the NRAO 12m, so the beam FWHM is 45″ or 60″. The measured I_{CO} values have been scaled up by a factor of 1.5 as an approximate correction for beam efficiency (§3.2). The CO observations are for the nuclear position, or the entire galaxy if it is unresolved. The values of A_V are the nuclear extinctions (assumed to be 1/2 the total extinction) and are calculated from $N(H_2)$ using $N(H_2)/A_V = 1.59 \times 10^{21}$ cm^{-2} mag^{-1} (Savage, Bohlin, Drake and Budich 1977). If the Young and Scoville (1982) X-value were used, the extinctions and column densities would be larger by a factor of 1.6.

It is to be emphasized that the extinctions and column densities in Table 1 are *beam-averaged* quantities, that is, they assume that the emission is spread uniformly across the beam. Concentration of the emission will produce a corresponding increase in the column densities and extinction. For example, CO $J = 1 - 0$ observations of NGC 6946 and Maffei 2 by Weliachew, Casoli and Combes (1988) with 23″ resolution give central column densities and extinctions that are about twice as large as the CO data at 45″ resolution of the same galaxies. A similar result was found for the edge-on spiral NGC 3628 by Boisse, Casoli and Combes (1987): their peak I_{CO} at 23″ resolution is 3.7 times larger than the value of Young, Tacconi and Scoville (1983). The beam sizes projected onto the galaxy (for a 55″ beam) are given in Table 1; $\theta_B \gtrsim 1$ kpc for all the galaxies.

The extinctions in Table 1 are quite large for most of the galaxies, especially considering the large spatial scales being averaged over. While some of the A_V values appear reasonable (NGC 4565, NGC 4736, NGC 891), others are uncomfortably large. The benchmark for "uncomfortable" in this context is as follows: Estimates of the average column density of giant molecular clouds (GMCs) in the Milky Way are of the order of 10^{22} (Solomon *et al.* 1987). Thus for 12 of the 16 galaxies in Table 1, the area filling-factor of molecular clouds within the beam is ≥ 1: the entire region of the galaxy within the beam is obscured by giant molecular clouds if the derived column densities are correct. The implied values for A_V for the late-type face-on spirals in the sample are all large, $A_V \approx 10$. There

Table 1
CO and C II Observations of Infrared Bright Galaxies

Galaxy	I_{CO}[a]	$I_{[C II]}$[b]	θ_B (Kpc)	$N(H_2)$[c] $(10^{21}$ cm$^{-2})$	A_V^d (mag)	ϕ_b^e	I_{CO}^{PD}[f]	f_{PD}	Ref
NGC 891	26.7	11.[h]	1.9	6.7	4.2	0.07	13.3	0.50	1
NGC 1068	87.2	72.[g]	4.0	21.8	13.7	0.48	91.2	1.05	2
NGC 2146	60.2	39.[h]	3.1	15.1	9.5	0.26	49.4	0.82	3
M 82	252.	170.[g]	0.9	63.0	39.6	1.	190.	0.75	4
NGC 3079	72.0	11.[h]	4.0	18.0	11.3	0.07	13.3	0.18	5
NGC 3628	60.6	8.[h]	3.0	15.2	9.5	0.05	9.5	0.15	6
Arp 299	6.0	9.[h]	11.2	1.5	1.0	0.06	11.4	1.90	7
NGC 4565	9.7	3.[h]	4.3	2.2	1.4	0.02	3.8	0.44	1
NGC 4736	15.3	7.[h]	1.1	3.8	2.4	0.05	9.5	0.62	8
NGC 5128	65.	30.[g]	1.9	16.3	10.2	0.20	38.0	0.58	9
M 51	46.7	17.[g]	2.6	11.7	7.3	0.11	20.9	0.45	10
NGC 6240	13.5	3.[h]	25.6	3.4	2.1	0.02	3.8	0.28	7
NGC 6946	72.6	8.[h]	2.7	17.4	10.9	0.05	9.5	0.13	11
M 83	88.5	50.[g]	1.9	10.1	6.4	0.33	62.7	0.71	12
Maffei 2	103.	19.[h]	1.8	25.8	16.2	0.13	24.7	0.24	13
IC 342	57.2	40.[g]	1.2	14.3	9.0	0.27	51.3	0.90	14

[a] In K km s^{-1}; all intensities on the T_R^* scale multiplied by 1.5 to correct for beam efficiency. Beam FWHM $\theta_B = 45 - 60''$.
[b] In units of 10^{-5} ergs cm^{-2} s^{-1} sr^{-1}. $\theta_B = 55''$.
[c] Total $N(H_2)$, assuming $N(H_2)/I_{CO} = 2.5 \times 10^{20}$ cm^{-2} (K km s^{-1})$^{-1}$
[d] Nuclear extinction ($\equiv A_V^{tot}/2$), assuming $N(H_2)/A_V = 1.59 \times 10^{21}$ cm^{-2} mag^{-1}
[e] Calculated from the observed $I_{[C II]}$ assuming an emergent intensity in the 158μm line of 1.5×10^{-3} ergs cm^{-2} s^{-1} sr^{-1} from a typical PDR (THa; WHT)
[f] Calculated from ϕ_b assuming an emergent intensity in the $J = 1 - 0$ line of 190 K km s^{-1} from a typical PDR (WHT)
[g] Crawford et al. 1985
[h] Stacey et al. 1989

References for CO Observations: 1) Maloney, unpublished; 2) Scoville, Young and Lucy 1983; 3) Sanders et al. . 1986; 4) Young and Scoville 1984; 5) Young, Claussen and Scoville 1988; 6) Young, Tacconi and Scoville 1983; 7) Sanders and Mirabel 1985; 8) Garman and Young 1986; 9) Israel et al. 1989; 10) Scoville and Young 1983; 11) Tacconi and Young 1989; 12) Lord 1987; 13) Sargent et al. 1985; 14) Young and Scoville 1982

is no tendency for the edge-on galaxies to have larger extinctions. It is hard to reconcile such large extinctions with optical spectra of the galaxies, and with the fact that they typically have roughly equal optical and infrared luminosities.

The highest extinction is derived for the peculiar, dusty galaxy M82, the canonical starburst galaxy. Nuclear extinction of $A_V \approx 25$ has been estimated by Rieke et al. (1980) for this galaxy, but this is at a spatial scale of a few arc seconds. Observations of M82 by Jaffe, Becklin, and Hildebrand (1984) at 400μm indicate a nuclear $A_V \approx 5$ and a corresponding total H_2 column density of 8×10^{21} within their $42''$ beam. This is 8 times smaller than the value in Table 1, and the discrepancy is just as bad at smaller spatial scales: using the standard conversion factor, the interferometer data of Lo *et al.* 1987 imply a mean nuclear extinction of $A_V \sim 160$ averaged over 110 pc ($7''$) throughout the entire central 700 pc $\times$ 200 pc. Lester *et al.* (1989) argue that the nuclear extinction is only $A_V \approx 5$ at $5''$ resolution, which makes the discordancy with the CO results even worse if the Galactic X-value is used. M82 has now been mapped at 450μm with $10''$ resolution by Smith *et al.* (1989) using the JCMT. Their continuum map is centrally peaked, unlike the CO maps, which leads them to suggest that the CO emission cannot be accurately tracing the distribution of gas mass.

Additional evidence that the Galactic X-value frequently leads to overestimates of the amount of molecular gas present is provided by dust mass determinations of other galaxies. Using submillimeter observations of a sample of IR-bright galaxies, Eales, Wynn-Williams and Duncan (1989) derive gas masses which are typically a few times smaller than the CO-derived masses; the ratio of $M_{H_2}(dust)/M_{H_2}(CO)$ ranges from about 0.15 to 1. Since their beam size at 350μm ($86''$) is considerably larger than the CO beam size ($45''$), the dust-derived masses need to be corrected downwards for the galaxies which are resolved by the IR beam.

Although the extinction in Table 1 for the interacting galaxy pair Arp 299 (IC 694/NGC 3690) is small (1 magnitude), the CO emission is completely unresolved by the beam. Interferometric observations (Sargent *et al.* 1987) find that the emission is centered on two of the near-IR peaks, A and C, and that the source sizes are smaller than 500 pc (the emission is unresolved by the interferometer). Interpreted with the Galactic X-value, their data imply a lower limit to the beam-averaged column density of 7.6×10^{22}, and a corresponding minimum extinction of 48 magnitudes towards A and C. However, Gehrz, Sramek and Weedman (1983) measured the optical depth of the 10μm silicate feature in absorption towards components A, B, C, finding values of 0.9, 0.5, and 0.2, respectively. Assuming a ratio $A_V/\tau_{10} \approx 15$ (Rieke and Lebofsky 1984), the inferred A_V are 14 and 3 towards A and C, values which are 3.6 and 16 times smaller than the CO-derived extinctions. From an extensive near-IR study of Arp 299, Telesco, Decher and Gatley (1985) derived $A_V \approx 4.6$ towards A, even smaller than the Gehrz, Sramek and Weedman value.

Thus we see that while the derived extinctions and H_2 column densities are plausible for some galaxies (generally the lower-luminosity ones), they are in many cases in clear disagreement with other determinations of the extinction and gas masses. This assumes that the dust-to-gas ratio in these galaxies is the same as the solar neighborhood, so that one way out of the discrepancy would be to assume a lower dust-to-gas ratio. However, this requires nearly all of the galaxies in the sample to have lower (sometimes much lower) dust-to-gas ratios than the solar neighborhood. Furthermore, since galaxies generally show metallicities that increase with decreasing radius, for the nuclei of the spiral galaxies in the sample we expect that, if anything, the dust-to-gas ratio will be higher than the solar neighborhood value, which would make the discrepancy even worse. The data strongly suggests that use of the Galactic X-value results in a considerable overestimate of the amount of molecular gas present in a significant fraction of galaxies/galactic nuclei.

4.2.2 *Surface Densities and Dynamical Masses.* The very large gas column densities derived using the Galactic X-value for some of the very luminous galaxies imply extremely high gas mass surface densities. For example, interferometric observations of Arp 220 (Scoville *et al.*

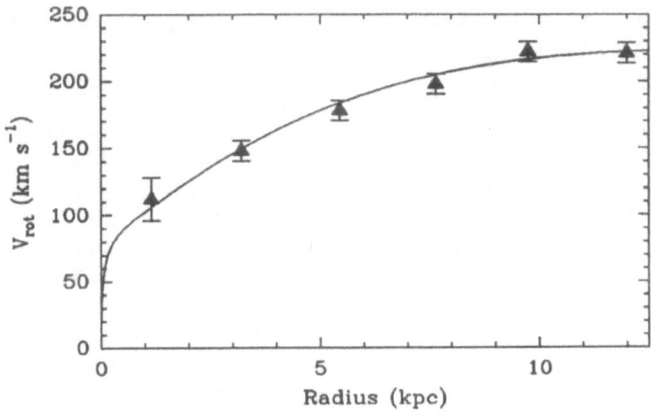

Figure 1. *H I rotation curve for NGC 6946 (data of Tacconi and Young 1986) and mass model fit. The error bars on the data are 1σ. The two halves of the rotation curve were averaged. The central surface density of the disk component is 750 $M_\odot$ pc^{-2}; see text for details.*

1986) and Arp 299 (discussed above) give lower limits to the gas surface densities of a few thousand $M_\odot$ pc^{-2}. As pointed out by Shu (1987), this is comparable to the densest regions of the Rho Ophiuchi molecular cloud, so it is difficult to see how the molecular clouds could resemble typical Galactic disk clouds. This inconsistency makes use of the Galactic conversion factor suspect. Furthermore, the gas mass would dominate the gravitational potential in the inner regions of these galaxies; the implied velocity dispersions are several hundred km s^{-1}.

In lower-luminosity, less disturbed systems, where the galaxy exhibits ordered rotation, it is possible to calculate dynamical masses from the rotation curve, and compare these values with the inferred molecular gas masses. For example, Meixner *et al.* (1989), in an interferometric study of two Seyfert galaxies, NGC 3227 and NGC 7469, found ratios of M_{H_2}/M_{dyn} of 0.26 and 0.85, respectively (including correction for helium). These numbers are again rather disturbingly large, especially for NGC 7469, at least for those of us who believe that galaxy nuclei contain stars as well as gas.

Even for normal, infrared-bright spirals, the implied gas mass fractions can get quite high, as can be seen by estimating the total disk mass surface density from the rotation curve. For example, the Scd spiral galaxy NGC 6946 has a fairly well-determined H I rotation curve (Tacconi and Young 1986). Figure 1 shows the rotation curve (the two sides of the galaxy have been averaged) along with a mass model fit. The model assumes a disk with an exponential surface density distribution and a thickness/radius ratio of 0.1, and an $r^{1/4}$ bulge (Monnet and Simien 1977); both components are assumed to have constant mass to light ratios. The disk scale length is 2.2' = 6.3 kpc for an assumed distance of 10.1 Mpc (Tacconi and Young 1986). The bulge component has an effective radius of 3.9 kpc. The scale sizes are taken from the photometric decomposition of Simien and de Vaucouleurs (1987). The disk scale length is 40% larger than that derived by Ables (1971),

Table 2
Gas Mass Surface Densities and Mass Fractions in NGC 6946

R (Kpc)	σ_{HI}[a] ($M_\odot$ pc^{-2})	σ_{H_2}[a] ($M_\odot$ pc^{-2})	σ_d ($M_\odot$ pc^{-2})	f_{H_2}	M_d^b ($10^9\ M_\odot$)	M_{H_2}/M_d^c
1	7.4	231.	640.	0.36	2.12	0.57
2	7.5	134.	546.	0.24	7.65	0.37
3	7.6	93.	466.	0.20	15.5	0.30
4	8.4	60.	397.	0.15	25.0	0.25
5	9.2	44.	339.	0.13	35.3	0.22
6	9.8	36.	289.	0.12	46.1	0.20
7	10.7	27.	247.	0.11	57.1	0.18
8	11.5	23.	211.	0.11	67.8	0.17
9	12.3	18.	180.	0.10	78.2	0.16
10	11.7	14.	153.	0.09	88.1	0.16
11	11.1	9.	131.	0.07	97.4	0.15
12	10.	9.	112.	0.08	106.	0.14

[a] Tacconi and Young (1986)
[b] Disk mass within radius R
[c] Ratio of molecular mass to disk mass within radius R

but is identical to the I-band value of Elmegreen and Elmegreen (1984). The fit shown in figure 1 has a disk central surface density of 750 $M_\odot$pc^{-2}. The bulge parameters are rather uncertain, but these uncertainties have little effect on the disk parameters; the disk surface density is constrained to about 10%. (*NB*: For a pure exponential disk with scale-length r_d, the frequently used (Keplerian) expression for calculating dynamical masses, $M_{dyn} = V^2R/G = 2.33 \times 10^5 V^2 R$ for V in km s^{-1} and R in kpc, will underestimate the disk mass for $R < r_d$, by a factor of 0.4 at $r_d/5$, and will overestimate the disk mass slightly for $R > r_d$, by about 20% at $2r_d$.)

The molecular gas surface densities have been calculated using the Galactic conversion factor from the data of Tacconi and Young (1986, 1989). Comparison with the total disk mass surface densities is made in Table 2, which gives the disk mass surface densities, gas surface densities, and gas mass fractions as a function of radius. Both the HI and H$_2$ masses have been corrected for helium. The derived molecular gas mass fractions are extremely large in the inner galaxy. The nuclear CO spike implies $10^9\ M_\odot$ of gas at $R \leq 1$ kpc if the standard value of X is used. The contribution from this gas can be seen in the values of $M_{H_2}(R)/M_d(R)$, the ratio of the molecular gas mass to the disk mass within radius R. At $R = 1$ kpc the inferred molecular mass is 57% of the disk mass. Including the bulge contribution would lower the molecular gas contribution to the total surface density by about 40% at 1 kpc, and decrease $M_{H_2}(R)/M_d(R)$ by about a factor of two. However, since the molecular gas layer is almost certainly thin, the mass in the bulge is not really relevant to this discussion.

It is difficult to see how such large gas mass fractions can be correct. The disks of spiral galaxies are generally fairly well described by exponential radial surface brightness distributions, which for a constant (M/L) disk implies an exponential mass surface density.

If the underlying stellar disk in NGC 6946 has an exponential radial surface density, then the large molecular gas mass at $R \lesssim$ a few kpc represents a large mass perturbation, which should be reflected in the rotation curve; the contribution from the gas would be of order 70 km s^{-1} at 1 kpc. Such a perturbation in the rotation curve is not seen, requiring an explanation as to why the total surface density is not very different from exponential when the inferred gas contribution varies so strongly with radius. Such large gas mass fractions would also be very destabilizing unless the velocity dispersion of the clouds is quite high. These considerations suggest that the disk gas mass fractions in normal spiral galaxies cannot be higher than $10 - 20\%$. For comparison, the Milky Way, which is a moderately infrared-luminous spiral, has a molecular gas mass to disk mass ratio of ≈ 0.05 within 8.5 kpc, and a peak molecular gas surface density (in the molecular ring) of $\lesssim 10$ percent.

Since the disk mass derived from the rotation curve and the molecular mass have different dependences on the assumed distance ($M_d \propto D$, $M_{H_2} \propto D^2$), the extremely high gas mass fractions will be reduced if the distance to NGC 6946 has been overestimated. The distance is not well-determined, with values from 4–10 kpc being suggested, so this is a possibility. However, similar conclusions are also reached for the disks of other galaxies, for example, M51, IC 342, and M83, and the nuclear region of M82. This suggests rather strongly that the mass of molecular gas within the inner couple of kiloparsecs may have been substantially overestimated by the use of the Galactic conversion factor from I_{CO} to $N(H_2)$.

4.2.3 *Temperatures and Radiation Fields.* One of the characteristics of the starburst and ultraluminous galaxies surveyed by *IRAS* is high dust temperature. The disks of spiral galaxies typically have *IRAS* $60/100\mu m$ temperatures of 25–30 K, which is about the lowest temperature dust to which *IRAS* was sensitive. The very IR-luminous galaxies, however, generally have dust color temperatures $T_d \approx 40$–50 K. Because dust grains radiate efficiently, with the emitted flux scaling like T_d^{5-6}, depending on the behavior of grain emissivity with wavelength, the dust temperature is not very sensitive to the intensity of the radiation field to which the grain is exposed. Thus very large increases in the intensity of the radiation field are required to produce modest increases in T_d. Contrary to popular belief, the dust temperature does not set an upper limit to the gas temperature: when exposed to intense ultraviolet radiation fields, the dominant gas heating mechanism is photoelectron ejection from grains (Draine 1978). Because dust grains cool much more efficiently than gas, the gas temperature may considerably exceed the dust temperature (Tielens and Hollenbach 1985), even at high gas densities. The high dust temperatures typically found for IR-luminous galaxies suggest that the energy density in the ISM is much higher in these objects than it is in normal galaxies, and raise the possibility that the gas temperatures are also high. Note that if $T_{gas} \approx T_d$ in the high-luminosity galaxies, the error in gas mass determination is not simply the ratio of $\sim 50/30$, the ratio of the dust temperature in a high-luminosity galaxy to that in a normal spiral, which is less than a factor of two. This is because it is believed that the *gas* temperature in normal spirals with *IRAS* dust temperatures $T_d \approx 30$ is only $T_{gas} \approx 10$ K, as in the Milky Way disk clouds; this value is of course implicit in the use of the Galactic X-value. In fact, analysis of the *IRAS* data of the Galactic plane by Sodroski *et al.* (1987) shows that the observed dust temperature is 24 K, with very little variation. (Interpreting the $60/100\mu m$ color temperature as a physical temperature is itself fraught with peril: see Draine's chapter.)

Direct evidence for high gas temperatures is seen in the nuclei of several galaxies.

Ammonia observations of the Galactic center (Mauersberger *et al.* 1986) detect molecular gas with temperatures in excess of 200 K. Morris *et al.* (1983) find ammonia temperatures of order 50–60 K for the inner few hundred parsecs of the Galaxy. Similarly, ammonia observations of the nucleus of IC 342 (Martin and Ho 1986) indicate gas kinetic temperatures $T_k \approx 50$ K on a size scale of several hundred parsecs.

The center of M82, which has a dust temperature $T_d \approx 45$ K, has been suggested to have optically thin CO emission. If this is so, the excitation temperature of the $J = 1 - 0$ transition must be in excess of 40 K, in order to explain the observed brightness of the emission. In fact, CO observations of M82 provide direct evidence for high molecular gas temperatures, because its proximity makes it possible to obtain very high spatial resolution using interferometers. Lo *et al.* (1987) find main beam temperatures that are so large that the gas temperatures must be considerably higher than typical Galactic values, of the order of 30 K or more. In any case, measurements of the ionizing photon flux in the inner 450 pc of M82 show that the average intensity of the UV radiation field is at least several hundred times higher than the solar neighborhood value, so high gas temperatures are not surprising. Since the infrared surface brightnesses of most IR-bright galactic nuclei and the IR-luminous galaxies are typically comparable to or larger than that of M82, similarly intense radiation fields are to be expected.

The difficulty caused by gas temperature variations in galaxies can be seen directly from equation (3): the observed CO integrated intensity from a galaxy is roughly $I_{CO} = f\bar{I}$, where f is the areal filling factor of clouds in the beam and $\bar{I}$ is the average integrated intensity from a cloud. Since it is impossible to separately determine f and $\bar{I}$, the same observed I_{CO} can be produced by cold gas or a much smaller mass of hot gas. In general, CO observations alone do not place constraints on either of these quantities that are strong enough to be useful. For example, equation (9) may be rearranged somewhat, to

$$\frac{\bar{n}^{1/2}}{\langle T_R \rangle} = 594 \frac{(M_{H_2}/10^9 \, M_\odot)}{(A_B/\text{kpc}^2) T_{mb} \Delta V} \, \text{cm}^{-3/2} \, \text{K}^{-1} \qquad (11)$$

where A_B is the beam area projected on the galaxy in kpc^2 and T_{mb} is the antenna temperature, T_R^*, corrected for main beam efficiency (typically about a factor of 1.5). The CO observations of NGC 6946 of Tacconi and Young (1989) indicate a molecular gas mass of $1.1 \times 10^9 \, M_\odot$ in the inner 1.1 kpc, using the Galactic X-value. If this gas were distributed uniformly in a disk 200 pc thick, the average density would be $\langle n_{H_2} \rangle \approx 28 \, \text{cm}^{-3}$. Since Galactic disk clouds have average densities of roughly $\bar{n} \approx 200 \, \text{cm}^{-3}$, the implied volume filling factor $f_v = \langle n_{H_2} \rangle / \bar{n}$ would be about 0.1, which is not a terribly interesting value: although larger than the volume filling factor of GMCs in the Milky Way ($\approx 1\%$), it is not so large as to require that the clouds have considerably larger densities than Galactic GMCs (assuming that the CO-derived mass is correct). From equation (11), it is clear that the ensemble-averaged cloud temperature and the molecular gas mass can be varied inversely at a given $\bar{n}$ without affecting the observed emission. As discussed earlier, the very large beam *areal*-filling factors required (roughly unity) to explain the observed CO flux assuming Galactic disk-like clouds suggest that the the molecular gas mass has been overestimated in the nuclei of many galaxies.

Even in the case of Arp 220, for which an enormous mass of molecular gas (nearly $10^{10} \, M_\odot$) is inferred to be present in the inner 1.5 kpc (Scoville *et al.* 1986), the constraints on the density barely become interesting: the average density inferred from the data is $N(H_2) \approx 100 \, \text{cm}^{-3}$ for a spherical distribution, or $N(H_2) \approx 450 \, \text{cm}^{-3}$ for a 200 pc thick layer. Obviously, the average cloud density would have to be several times these values in order to reduce f_V to a reasonable value. However, once again the inferred mass (and thus the average density) can be decreased by raising the gas temperature. *If* the gas temperature is as high as the dust temperature, $T_d \approx 50$ K, which seems a reasonable assumption, the amount of molecular gas present is overestimated by about a factor of 6 *unless* the molecular clouds have densities $\bar{n} \approx 7000$ cm^{-3}. There are no *a priori* reasons for favoring a particular value of $\bar{n}$, so this is a possibility. In fact, for Arp 220 the gas mass derived from 350μm observations using the usual assumptions about dust properties (Eales, Wynn-Williams and Duncan 1989) and with the usual uncertainties, is only about a factor of two smaller than the H$_2$ mass derived from CO, suggesting that the clouds are denser than Galactic GMCs if they are in fact hotter. In general, however, it is clear that a cloud population with very different values of $\langle \bar{n} \rangle$ and $\langle T_R \rangle$ from the Galactic disk clouds will have an identical X-value only by a very unlikely coincidence. There are additional data which suggest that much of the CO emission from many IR-bright galaxies comes from warm gas with $T_k \approx 40 - 50$ K that is exposed to intense UV radiation. These data are discussed in the following section.

4.2.4 *Far-Infrared Fine Structure Lines.* Additional evidence that the CO emission from some galaxies and galactic nuclei is not dominated by Galactic disk-like clouds is provided by observations of a number of atomic fine-structure lines in the far-infrared. These lines, such as C II 158μm, O I 63μm, and O III 52 and 88μm, are collisionally-excited fine-structure transitions. They are very important coolants in both the atomic and ionized ISM, and have the advantage of not suffering from extinction. These lines have been extensively observed in the Galaxy, where they predominantly arise in the interfaces between H II regions and molecular gas (see, e.g., Genzel, Harris and Stutzki 1989).

The pioneering study of the C II 158μm line in a sample of infrared-bright galaxies by Crawford *et al.* (1985) produced two extremely important results: 1) The 158μm line is extremely bright: on the order of 0.5% of the bolometric luminosity of the galaxies they observed is radiated in the 158μm line; 2) There is a very good correlation, both in flux and intensity, between the C II line and the ^{12}CO $J = 1 - 0$ line. These results have been substantiated by further observations of galaxies (Stacey *et al.* 1989). Crawford *et al.* concluded that the only sources capable of explaining the brightness of the 158μm emission they observed are *photodissociation regions (PDRs)*: the interfaces produced when molecular gas is exposed to very intense UV radiation fields. Detailed models of PDRs have been constructed by Tielens and Hollenbach (1985a,b) (THa,b). Grain photoelectric heating is the dominant heat source to $A_V \approx 5$; the gas temperature considerably exceeds the dust temperature throughout. PDRs produce strong CO emission in addition to radiation in the fine-structure lines like C II; the CO emission in the lower rotational levels arises in the warm ($T \approx 50 - 200$ K), partially photodissociated region (containing H$_2$, O, C, and CO) at depths of $A_V \approx 3 - 5$. The lower CO transitions are optically thick.

The very good correlation between the CO $J = 1 - 0$ and C II 158μm emission led Crawford *et al.* to suggest that a major fraction of the CO emission in their sample of galaxy nuclei arose in photodissociation regions, so that the CO emission is in fact tracing the energy density of UV radiation, rather than gas mass. This correlation, which also holds for a large variety of Galactic objects (H II regions, planetary nebulae, and reflection nebulae) has been examined in detail by Wolfire, Hollenbach, and Tielens (1989) (WHT). They find that this correlation is most easily understood by postulating a common origin for the C II and CO emission, in PDRs.

We can examine quantitatively whether PDRs can contribute significantly to the observed CO emission from galaxies. Table 1 lists the observed $158\mu m$ fluxes for sixteen galaxies for which there is also CO data; the data are taken from the compilation of Genzel, Harris and Stutzki (1989). If we assume a typical emergent intensity from a PDR in the $158\mu m$ line of 1.5×10^{-3}ergs cm^{-2} s^{-1} sr^{-1} (the high-density, high-intensity limit: THa,b; WHT), then we can calculate the beam-filling factor of PDRs for the observed galaxies in Table 1. If we then use the canonical value for the emergent intensity in the ^{12}CO $J = 1 - 0$ line of 3×10^{-7}ergs cm^{-2} s^{-1} sr^{-1} = 190 K km s^{-1} (WHT), we can estimate the contribution of PDRs to the observed CO emission. The estimates of I_{CO}^{PD} and the fraction of the observed emission, f_{PD}, that this represents are also given in Table 1. Nine out of the 16 galaxies have $f_{PD} \geq 0.5$. Note also that the value of $I_{[C\ II]}$ used to calculate the filling factors is the high-intensity limit; for lower UV radiation field intensities the ratio of $I_{[C\ II]}/I_{CO}$ is lower. This may explain why three of the galaxies with low values of f_{PD} are galaxies with active nuclei (NGC 3079, NGC 3628, and probably NGC 6240): a power-law continuum has a much lower ratio of UV flux to bolometric flux. NGC 1068, which also has an AGN, has $f_{PD} = 1$; however, about half of the far-infrared luminosity of NGC 1068 arises in an extended region of star formation, not from the AGN. Given the uncertainties, 2/3 of the galaxies have values of f_{PD} consistent with all of the observed CO in the same region arising from PDRs. Since the ^{12}CO $J = 1 - 0$ emission is produced in warm ($T \gtrsim 50$) gas and is optically thick, interpretation of the emission using the standard X-value will overestimate the amount of molecular gas present substantially.

Because the gas temperatures in PDRs increase towards the exterior of the cloud, the brightness temperature of the lower CO rotational transitions will increase with J; this is due to the increase of the Einstein A coefficients with increasing J, so that the higher J transitions become optically thick closer to the surface, where the gas is warmer. Thus it is possible to get ratios of antenna temperature (such as $J = 2 - 1/J = 1 - 0$) that are somewhat in excess of unity (although less than ≈ 2); this may be the cause of the high $J = 2 - 1/J = 1 - 0$ ratio in M82.

4.2.5. *Gas Dynamics, Shocks, and Virialization.* It is worth commenting briefly on the question of whether one of the assumptions behind the use of conversion factors, namely, that clouds are self-gravitating objects in virial equilibrium, is likely to be true in all galaxies. In the Milky Way, it is very likely that most giant molecular clouds are at least self-gravitating, which, within the usual factor of two uncertainties, is equivalent to being virialized. There are probably exceptions; for example, Scoville et al. (1987), in a sample of Milky Way molecular clouds, find no difference in masses calculated from the virial theorem (assuming that only self-gravity is important in binding the cloud, which is the usual assumption made in virial theorem mass estimates for molecular clouds) and from the CO luminosity for clouds associated with H II regions and clouds that are not, even though the H II region clouds have temperatures that are systematically higher. However, the H II region clouds have systematically higher velocity dispersions for a given cloud diameter, for clouds smaller than about 30 pc. This increase in velocity dispersion is probably a reflection of the injection of energy into the clouds by embedded young stars, so that the masses of these clouds are overestimated by use of the virial theorem. Also, the high-latitude diffuse molecular clouds have virial theorem masses which are much larger than the masses estimated from CO or extinction (Magnani, Blitz and Mundy 1985; Keto and Myers 1986); these clouds are probably either confined by the pressure of the ISM, or are transient shock relicts (Blitz 1987), and may contribute a significant fraction of the CO luminosity of the Galaxy (Polk, Knapp, Stark and Wilson 1988). Nevertheless, the molecular clouds which dominate the emission in the disks of normal spirals are probably virialized, at least to an adequate approximation (but see Maloney 1989 and Issa, MacLaren and Wolfendale 1989 for criticism of the virial theorem analyses).

It is not at all certain, however, that clouds in highly energetic environments, like galactic nuclei, or very dynamic environments, like merging galaxies, are going to be viri-alized. The measured planar velocity dispersion of molecular clouds in the disk of the Milky Way is only a few km s^{-1} (Stark 1984; Stark and Brand 1989). The CO linewidths observed for the very luminous *IRAS* galaxies are often several hundred km s^{-1} (Sanders and Mirabel 1985; Sanders *et al.* 1986). Such large linewidths suggest that non-circular motions are very likely. Further evidence of this is provided by the H I study of a sample of 92 infrared-luminous galaxies by Mirabel and Sanders (1988), who found that the fraction of galaxies showing H I in absorption increases with infrared luminosity, reaching 100% for $L_{IR} \gtrsim 10^{12} L_{\odot}$. In order to have absorption dominate over emission, the gas must be concentrated near the nuclear continuum source. Nonetheless, the absorption linewidths that Mirabel and Sanders found were similar to the CO linewidths, typically several hundred km s^{-1}. This implies that turbulent and/or non-circular motions of several hundred km s^{-1} are common in the nuclear regions of very IR-luminous galaxies. At less extreme luminosities, evidence for non-circular gas motions is often seen in galactic nuclei, including that of the Milky Way (Liszt 1988); the measured linewidths are often much larger than can be explained by averaging over the rotation curve (e.g., NGC 6946, Weliachew *et al.* 1988).

It is obvious that increasing the velocity dispersion of a population of virialized clouds will not affect the observed integrated CO emission, provided that cloud-cloud shielding is unimportant: the peak temperature simply decreases as the linewidth goes up. However, it is not at all clear that the cloud population will remain in virial equilibrium when a major fraction of the gas possesses enormous non-circular or turbulent velocities, since cloud-cloud collisions at high velocities, which will generally be very disruptive (Gilden 1984) will be common. Many of the ultraluminous *IRAS* galaxies have turned out to show very strong emission in the near-infrared rotation-vibration lines of molecular hydrogen; this emission is probably mostly produced by shocks. The most extreme example, NGC 6240, has a luminosity in the $v = 1 - 0$ S(1) line of H$_2$ that is so large that producing it via shock excitation essentially requires the face-on collision of two gas disks (Rieke *et al.* 1985; Lester, Harvey and Carr 1988). Since many of the most luminous *IRAS* galaxies are believed to be two galaxies in the process of merging, such large-scale gas dynamic processes may be quite common in these galaxies, leading to very disturbed molecular cloud ensembles. Some theoretical study of this topic has been done by Scalo and Struck-Marcell (1986, 1987) and Struck-Marcell and Scalo (1987). Similar effects may result from the effects of bursts of star formation (Chevalier and Clegg 1985; Tanaka and Ikeuchi 1988) or an AGN (Krolik and Begelmann 1986, 1988) on the ISM. In such galaxies, the ISM may be dominated by the effects of winds (powered either by an AGN or intense massive star formation), supernovae, stellar winds, and large-scale gas flows, so that the molecular gas does not exist in the form of well-defined individual clouds, but has been swept up into sheets and filaments. There is good evidence for this in the nucleus of M82 (Lo *et al.* 1987; Lugten *et al.* 1986).

Non-virialized cloud velocity dispersions can lead to very large overestimates of cloud masses: if the velocity dispersion σ_v of a cloud is larger by a factor β than the value it would have in equilibrium, its mass will be overestimated by a factor β^2 by assuming virial equilibrium. Modest increases in σ_v above equilibrium values can lead to order-of-magnitude mass overestimates.

4.2.6 *Summary*

A large amount of data suggests that the molecular gas masses in galactic nuclei and infrared-luminous galaxies has been substantially overestimated by use of the Galactic X-value. Gas column densities determined from submillimeter observations of dust emission are typically a few times smaller than inferred from CO observations using the standard conversion factor. Taking the CO-derived gas masses at face value implies that molecular gas comprises a very large fraction ($\approx 50\%$) of the disk mass in the inner regions of IR-bright, normal galaxies, a value that is difficult to reconcile with galaxy dynamics. The high dust temperatures seen in very luminous galaxies and starburst systems like M82 suggest that the ambient radiation fields are very intense, and may substantially affect the conditions in molecular clouds. This suggestion is confirmed by the correlation seen between $^{12}CO\ J = 1 - 0$ emission and C^+ 158μm emission: this correlation is most easily explained by a common origin for the emission from both species, in photodissociation regions produced when molecular gas is exposed to intense UV radiation fields. The very large linewidths that are seen in most of the very infrared-luminous galaxies, and in some galactic nuclei, make the assumption of virial equilibrium of molecular clouds—one of the basic assumptions behind the use of conversion factors—very dubious; the amount of gas present will then be overestimated by a substantial factor by assuming virial equilibrium.

4.3 MOLECULAR GAS IN LOW-METALLICITY GALAXIES

The majority (by number density) of actively star-forming galaxies are probably irregular galaxies, which are typified by lower masses and less organized structure than giant spirals like the Milky Way (Gallagher and Hunter 1984, 1986). In fact, some of the largest complexes of young stars that are known are in irregular galaxies: 30 Doradus in the Large Magellanic Cloud, the giant H II region complex in NGC 4449, and the immense complexes of OB stars which comprise the 'clumps' in clumpy irregular galaxies (Heidmann 1987). Since observations of the Galactic disk show that star formation occurs in molecular clouds, we would expect that irregular galaxies with high rates of star formation would show evidence for a corresponding abundance of molecular gas. This has not proved to be the case, however, if we assume that I_{CO} is a good tracer of molecular gas: irregular galaxies have turned out to be remarkably deficient in CO emission. The early survey of Elmegreen, Elmegreen and Morris (1980) reached conclusions that have been supported by subsequent CO studies of irregular galaxies: the intensity of CO emission is much less than would be expected from a population of Milky Way-like giant molecular clouds; most surveys of irregular dwarf galaxies have produced only upper limits (Gordon *et al.* 1982; Israel and Burton 1986).

More recent observations made with higher sensitivity have detected CO in a number of irregular galaxies (Tacconi and Young 1985, 1987; Dettmar and Heithausen 1989) including the Magellanic clouds (Israel *et al.* 1982, 1986) and the starburst irregular galaxy

NGC 1569 (Young, Gallagher and Hunter 1984). The paucity of CO emission from these galaxies has led to suggestions that star formation in these galaxies must be a much more efficient process than in spiral galaxies (Young, Gallagher and Hunter 1984; Young 1987). However, there is a considerable amount of evidence suggesting that the amount of gas in these galaxies has been substantially underestimated by the use of the Galactic conversion factor; this evidence is discussed below, followed by a discussion of how dissociation of CO in clumpy, low-metallicity molecular clouds will affect the observed emission and X-value.

4.3.1 *CO Observations of the Large And Small Magellanic Clouds.* The Magellanic Clouds are the nearest galaxies to our own (distances of 53 and 63 kpc, respectively; Humphreys 1984), so higher spatial resolution ($1' = 15$ pc and 18 pc, respectively) can be obtained than for any other galaxy, and it is possible to just barely resolve individual GMCs. Israel *et al.* (1986) searched for ^{12}CO $J = 2 - 1$ emission from the LMC and the SMC using the ESO 3.6m telescope, with a beam FWHM of 2.'0. They found peak antenna temperatures $T_R^* \approx 1$ K, about 1/4 of the expected signal from typical Galactic disk clouds (with $T_k = 10$ and mean projected surface areas of 2300 pc^2; Blitz 1978) at Magellanic Cloud distances. Israel *et al.* argued that this weakness is probably not the result of an actual deficit of molecular gas, but is instead due to a much higher rate of destruction of CO in the Magellanic Clouds, caused by the lower gas-phase abundances, dust-to-gas ratios, and higher UV radiation fields that characterize the clouds.

As was discussed earlier (§3.2), a decrease of the effective CO-emitting area of a molecular cloud may result from lower metallicity. This effect was examined in detail by Maloney and Black (1988), who constructed models of molecular clouds with parameters appropriate to the solar neighborhood, the LMC and the SMC. For the latter models the abundances of carbon and oxygen with respect to hydrogen and $A_V/N(\mathrm{H})$ were lowered by factors of 4 and 17, respectively. All the models had total H$_2$ column densities of 10^{22} cm^{-2}, and were calculated in plane-parallel geometry. The effect of the lower abundances and dust-to-gas ratios on the column densities of CO are quite dramatic: the ^{12}CO column density in the SMC model is ~ 2 orders of magnitude smaller than in the Galactic cloud model. Because of the importance of self-shielding in the survival of CO in molecular clouds, the dependence of CO column density on metallicity is very nonlinear. Thus the claim that has occasionally been made that lowering the abundances of carbon and oxygen, even by substantial factors, will not affect the observed emission is not correct, since the CO column density does *not* drop linearly with abundance. Conversely, the depth dependence of molecular hydrogen is very similar in the three models: lowering the H$_2$ formation rate by a factor of 17 (the SMC model) only reduces the peak H$_2$ number density by about 10%, and the depth at which H$_2$ becomes the dominant hydrogen species is still only a couple of percent of the cloud thickness. Thus the Magellanic cloud models, although nearly identical in molecular hydrogen to the Galactic model, would have very different observational signatures in CO, due to the much smaller emitting area. Reduction of the emitting areas of giant molecular clouds by a factor of four would explain the low peak temperatures found by Israel *et al.* (1986). The plausibility of this explanation is further reinforced by estimates of the sizes of Magellanic GMCs. The dark cloud studies of Hodge (1972, 1974) show that the mean surface area of cloud complexes are 3.3×10^3 pc^2 and 1.2×10^3 pc^2 for the LMC and SMC, respectively. These values are comparable to Galactic GMCs. In addition, the CO detections of Israel *et al.* show typical linewidths of 7 km s^{-1}, so that the sizes and linewidths (and therefore masses, assuming virial equilibrium) of

Magellanic GMCs are similar to their Galactic counterparts.

4.3.2 *Dust Masses and* I_{CO} *Masses.* The low metallicity of the Magellanic clouds is not atypical; in fact, low metallicity is characteristic of irregular galaxies (Gallagher and Hunter 1984). It thus seems very likely that the differences between low-metallicity molecular clouds and Galactic disk clouds will cause the amount of molecular gas in these galaxies to be generally underestimated. That this is the case is indicated by observations of dust emission from low-metallicity galaxies.

Mass determinations from observations of dust are reviewed by Draine elsewhere in this volume. Briefly, the mass of dust corresponding to a measured flux density f_ν is

$$M_d = \frac{f_\nu D^2}{B_\nu(T_d)} \frac{4a}{3Q_\nu} \rho_d \qquad (12)$$

(Hildebrand 1983) where D is the distance of the source, T_d the dust temperature, a is the radius of a dust grain, Q_ν is the grain emissivity at frequency ν, and ρ_d is the grain density. For $\lambda \gg a$, a/Q_ν is independent of a. The gas mass associated with this is just $R_{gd}M_d$, where R_{gd} is the gas-to-dust mass ratio. To make use of this expression we need to determine the relevant parameters; such determinations are of course restricted to the solar neighborhood. We can write the derived gas mass as

$$M_{gas} = 4000 D^2 f_\nu \lambda_{250}^3 R_{gd} \frac{[N(\mathrm{H})/\tau_\nu]}{10^{25}} \left(e^{57.6/\lambda_{250}T_d} - 1 \right) M_\odot \qquad (13)$$

where D is in Mpc, f_ν in Jansky, $\lambda_{250} = \lambda/250\mu m$, and $N(\mathrm{H})/\tau_\nu$ is the ratio of total hydrogen column density to dust optical depth at the wavelength being observed. If the widely-used dust parameters suggested by Hildebrand (1983) are inserted, equation (12) becomes

$$M_{gas} = 1.9 \times 10^4 D^2 f_\nu \lambda_{250}^5 \left(e^{57.6/\lambda_{250}T_d} - 1 \right) M_\odot . \qquad (14)$$

Most of the existing data on dust emission from irregular galaxies are from *IRAS*. At wavelengths as short as $100\mu m$, it is necessary to worry about what fraction of the dust present is hot enough to be detectable. Empirically, comparisons of *IRAS*-derived dust masses with gas masses derived from CO and H I observations find $M_{gas}/M_d \approx 500 - 700$, considerably larger than the estimated value of $R_{gd} \approx 100$ found from submillimeter observations of solar neighborhood objects (Hildebrand 1983). In a detailed analysis of the emission from the Galactic plane as observed by *IRAS*, Sodroski *et al.* (1989) find $M_{gas}/M_d \approx 330 - 400$. This difference probably results from a) the presence of dust which is too cold to emit at $100\mu m$; b) the uncertainties in deriving dust and gas masses; c) variations in dust-to-gas ratios. In regard to this last point, it is worth noting that since most irregular galaxies are low-metallicity systems, it is expected that the dust-to-gas ratio will be lower than in the solar neighborhood; this is directly observed in the Magellanic Clouds (Koorneef 1984; Schwering 1988). For a sample of irregular galaxies detected by *IRAS*, Hunter *et al.* (1989) find an average $M_{\mathrm{H\,I}}/M_d \approx 10^4$, about twenty times larger than the values found for giant spirals. Thus even if these galaxies had as much molecular as atomic gas, which seems a reasonable upper limit, the ratio of dust mass to gas mass is roughly an order of magnitude smaller than for spiral galaxies.

In a detailed analysis of the Magellanic irregular galaxy NGC 4449, Thronson *et al.* (1989) compared CO observations with $150\mu m$ data obtained using the Kuiper Airborne

Observatory, with comparable spatial resolution. The dust mass derived for the nucleus (where there is very little H I emission) is 1/30 the molecular mass inferred from the CO. Thus either the dust-to-gas mass ratio in the nucleus is an order of magnitude or more *larger* than is typical for giant spirals, or, far more likely, use of the Galactic X-value leads to an underestimate of the molecular gas mass by a substantial factor. Similar conclusions were reached by Dettmar and Heithausen (1989) in a study of the Magellanic irregular NGC 55: they found that virial mass estimates and infrared mass estimates (using $R_{gd} = 500$) for the largest star-forming complex in the galaxy agreed to a factor of two, whereas the integrated CO emission and the Galactic X-value indicate a mass 20 times smaller. As the mass of H I in this complex is only about 1/4 of the estimated gas mass, it is very likely that the CO emission is a poor tracer of molecular gas in this galaxy, as in NGC 4449.

The extreme weakness of CO emission from irregular galaxies, even those which have high current rates of star formation such as NGC 1569 (Israel 1988b) and II Zw 40 (Joy and Lester 1988), suggest that CO cannot be used to make quantitative estimates of molecular gas mass in such low-metallicity objects, due to the combination of reduced gas-phase abundances and lower extinction. The following section presents a simplified model for how the CO luminosity of realistic molecular clouds may be affected by low metallicity.

4.3.3 *Destruction of CO in Clumpy Giant Molecular Clouds.* It is a well-established fact that real molecular clouds, far from being the spherical, constant density objects so beloved by theorists, are usually messy, irregularly shaped, and very lumpy: most of the mass of a cloud is contributed by clumps of much smaller size, with much higher densities than the volume average of the cloud (e.g., Goldsmith 1987; Blitz 1988; Falgarone and Perault 1988). Observations of widespread fine-structure emission of lines such as C II 158μm in molecular clouds also require that clouds are very clumpy (Stutzki *et al.* 1988).

Consider a GMC that is made up of a large number of clumps spread throughout some volume; the space between the clumps is filled with interclump gas. The mass fraction of the GMC that is in the clumps is f; observations suggest that $f \sim 1$. We will assume that the clumps can be regarded as spherical, constant density objects, although all clumps do not necessarily have the same density: in other words, we'll commit the same crime referred to above, but on a much smaller spatial scale. The clumps will have some range of sizes and densities; equivalently, they have a spectrum of column densities, which we will assume is a power-law: $N(N_p) \propto N_p^{-\alpha}$, where N_p is the peak column density through a clump. The relation between N_p and the mass of a clump, m_c, will depend on the density of the clumps and their density structure. Under the simplifying assumption of constant-density, spherical clumps, the mass of a clump is $m_c = (2\pi/3)r_c^2\mu N_p$. The total mass in clumps is $f M_{GMC}$, and equals

$$f M_{GMC} = \int_{N_p^{min}}^{N_p^{max}} N(N_p) m_c(N_p) \, dN_p$$

$$= \frac{2\pi}{3}\mu N_0 \int_{N_p^{min}}^{N_p^{max}} r_c^2(N_p) N_p^{1-\alpha} \, dN_p \tag{14}$$

where N_0 is the normalization constant in the column density distribution. If all the clumps have the same density, (14) simplifies to

$$f M_{GMC} = \frac{\pi}{6}\frac{\mu}{n_c^2} N_0 \int_{N_p^{min}}^{N_p^{max}} N_p^{3-\alpha} \, dN_p \, . \tag{15}$$

The specification of the maximum and minimum column densities is rather uncertain. The minimum column density which is molecular in the solar neighborhood appears to be roughly $A_V \sim 1$, corresponding to a total hydrogen column density of $N(\mathrm{H}) \approx 2 \times 10^{21}$; since survival of H_2 is determined by line self-shielding, this will be independent of metallicity. This value does depend, however, on the radiation field seen by clumps within a molecular cloud. The upper column density limit is even fuzzier. One could use, e.g., the Jeans mass, but this requires assumptions about the support of clumps. For simplicity, assume that the largest clump has a mass that is y times fM_{GMC}; y is probably of the order of 10^{-2}. The mass of a clump is

$$m_c = 113 \frac{N_{p21}^{\;3}}{n_{c,2}^2} \; M_\odot \tag{15}$$

where N_{p21} is the H_2 clump column density in units of $10^{21}\,\mathrm{cm}^{-2}$, and $n_{c,2}$ is the clump density in units of $100\,\mathrm{cm}^{-3}$. If all clumps have the same density, the maximum column density is then

$$N_{p21}^{max} = \left(\frac{yfM_{GMC}n_{c,2}^2}{113} \right)^{1/3} \mathrm{cm}^{-2} \tag{16}$$

for M_{GMC} in $M_\odot$.

Assume the clumps dominate the CO emission from the cloud. Now, suppose that we have two GMCs that are identical in size and mass, and in clump parameters, but not in metallicity. As discussed previously, the dominant effect of low metallicity will be to reduce the effective size of a clump (as measured in CO). We will assume that all other clump properties stay the same: density, excitation, etc., and that we can ignore any optical depth effects aside from the decrease in size. How will the CO luminosities of the two GMCs differ?

We can approximate this effect by assuming that all CO is destroyed for column densities less than some critical value, N_{21}^{cr}, i.e., one has to have a column density larger than N_{21}^{cr} before CO is present; thus, for example, a clump with total column density $4N_{21}^{cr}$ would have a 'CO-size' which is one-half its diameter. It will also be assumed that in the solar metallicity case we can take the CO-size to be the same as the clump size; N_{21}^{cr} is the *increase* in column density required for CO to survive in a low-metallicity GMC.

The contribution to the total luminosity from a given clump will also depend on the scaling of linewidth with size. For simplicity, we will consider two cases: $\sigma_v = $ constant, and $\sigma_v = \sigma_0 r_c$. The first gives constant-pressure clumps, the second virialized clumps. The CO luminosity of a clump in the two cases is given by

$$\begin{aligned} L_{CO}^c &= (2\pi)^{1/2} \pi r_c^2 \bar{T}_A \sigma_v \\ &= 20.7 \frac{(N_{p21} - N_{21}^{cr})^2}{n_{c,2}^2} \bar{T}_A \sigma_c \; \mathrm{K\;km\;s\;pc^2} \end{aligned} \tag{16a}$$

and

$$L_{CO}^c = 33.5 \frac{(N_{p21} - N_{21}^{cr})^3}{n_{c,2}^3} \bar{T}_A \sigma_0 \; \mathrm{K\;km\;s\;pc^2} \tag{16b}$$

for the two cases, respectively, where $\bar{T}_A$ is the area-averaged antenna temperature from a cloud. Compared to the solar metallicity GMC, the low-metallicity GMC will have a

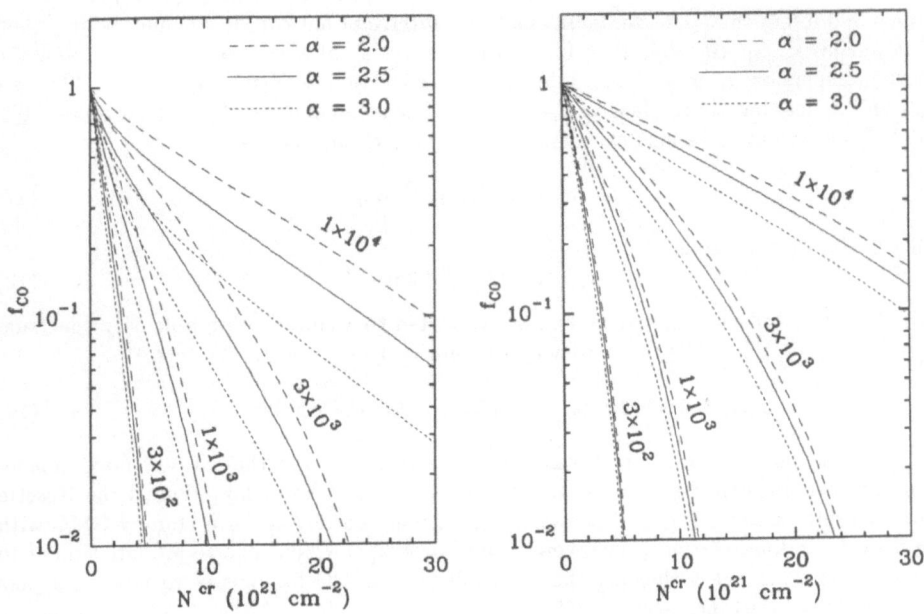

Figure 2. *Fractional CO luminosities for low-metallicity, clumpy molecular clouds. The fractional luminosity (compared to a Galactic GMC) is plotted against the increase in H_2 column density that is required for survival of CO. The left-hand plot assumes that the clump velocity dispersion is independent of clump size; the right-hand plot assumes that $\sigma_v \propto r_c$. Results are shown for three different column density spectra; the curves are labelled with the clump density n_c.*

CO luminosity which is smaller by a factor f_{CO}. This factor is given by the ratio of equation (16a) or (16b), integrated from $N_{21}^L \equiv max(N_{p21}^{min}, N_{21}^{cr})$ to N_{p21}^{max}, to the same equation integrated from N_{p21}^{min} to N_{p21}^{max} with $N_{21}^{cr} = 0$. The integrals are trivial; the results are shown in Figure 2 for several values of the column density spectral index α and clump density n_c, for a GMC with a total mass of $5 \times 10^5\ M_\odot$, $y = 10^{-2}$, and $f = 0.8$. $N_{p21}^{max}/N_{p21}^{min}$ varies from 7 to 71 as n_c increases from 3×10^2 to 10^4. The minimum column density N_{p21}^{min} has been taken to be 10^{21} molecules cm^{-2}.

As is obvious from the figure, the results can be quite dramatic, especially for modest clump densities or large values of N_{21}^{cr}. For a clump density $n_c = 10^3$ cm^{-3}, the CO luminosity drops by a factor of two for $N_{21}^{cr} = 1.4$, and by an order of magnitude for $N_{21}^{cr} = 5.4$. The results are not terribly sensitive to the index of the column density spectrum for the $\sigma_v \propto r_c$ models; the $\sigma_v = $ constant models are somewhat more sensitive to α for $n_c \gtrsim 1000$ cm^{-3}. As models of low metallicity clouds indicate that N_{21}^{cr} can be substantial, it is evident that the CO luminosities of low-metallicity GMCs can be much smaller than that of Galactic disk clouds of the same mass. Similar arguments were presented by Israel *et al.* (1986) to explain the weakness of CO emission from the Magellanic clouds.

Data on clump size and mass spectra for Galactic GMCs are still very inadequate. In a detailed study of the Rosette molecular cloud, Blitz, Stark and Long (1989) catalogued clumps and found that the size, mass and line-width relations can be adequately described by power-laws. Specifically, they found that the mass of a clump $m_c \propto r_c^{2.6}$ and the velocity dispersion $\sigma_v \propto m_c^{0.12}$, while the mass spectrum is $N(m_c) \propto m_c^{-1.5}$. The first relation implies that clump density $n_c \propto r_c^{-0.4}$. The mean density of clumps is about 1000 cm^{-3}. Their scaling relations can be used to express clump mass as

$$m_c = 0.38 N_{p21}^{4.3} \, M_\odot \tag{17}$$

and velocity dispersion as

$$\sigma_v = 0.39 N_{p21}^{0.5} \text{ km s}^{-1} \tag{18}$$

while their derived mass spectrum can be converted to a clump column density spectrum $N(N_p) \propto N_p^{-3.2}$. The CO luminosity of a clump is then given by

$$L_{CO}^c = 2.8 \times 10^{-2} N_{p21}^{1.9} \left(N_{p21} - N_{21}^{cr} \right)^2 \bar{T}_A \text{ K km s pc}^2 \, . \tag{19}$$

The clumps found by Blitz *et al.* range from 18 to 2800 $M_\odot$; the ratio of maximum to minimum column densities is only 3. Of course, we don't know how typical the Rosette is of molecular clouds. Figure 3 shows f_{CO} as a function of N_{21}^{cr} for a clumpy GMC with the Rosette scaling relations, for several values of $N_{p21}^{max}/N_{p21}^{min}$. The results are similar to those in Figure 2, again showing the sensitivity of the CO luminosity to moderate (less then a factor of ten) increases in N_{21}^{cr}.

5. Determination of Physical Conditions from Observations

The use of CO as an extragalactic molecular mass tracer is hampered by our lack of knowledge of the physical conditions in the interstellar medium of other galaxies. In galaxies where the temperatures, densities, abundances, etc. approximate those of the disk of the Milky Way, it is reasonable to expect that–at least to first order–the ensemble-averaged properties of molecular clouds will also be similar, so that the conversion between integrated CO intensity and molecular gas mass will be close to the Galactic value. However, there are many environments, such as galactic nuclei and ultraluminous galaxies, where the physical conditions in the ISM may be very different, and our lack of knowledge of $\langle T_R \rangle$ and $\langle \bar{n} \rangle$ in these galaxies makes the application of the Galactic X-value rather risky. Getting an accurate determination of the amount of molecular gas present in such objects requires additional information. In this section I briefly discuss what observations can be used to constrain the physical conditions in molecular clouds in other galaxies.

5.1 OTHER ROTATIONAL TRANSITIONS OF CO

An obvious way to try to estimate the temperature of the gas dominating the CO emission is to look at emission from higher rotational transitions of CO. If we assume that the emission in both transitions is optically thick, the ratio of antenna temperatures (assuming identical beam sizes and efficiencies at the different frequencies) of two rotational transitions is just

$$\frac{T_R(2)}{T_R(1)} = \frac{\nu_2}{\nu_1} \frac{F_\nu(T_{ex}(2)) - F_\nu(T_{bg})}{F_\nu(T_{ex}(1)) - F_\nu(T_{bg})} \tag{20}$$

Figure 3. *As Figure 2, for GMCs with clump parameters derived from observations of the Rosette molecular cloud by Blitz, Stark and Long (1989). The curves are labeled with the ratio of maximum to minimum clump column densities; for the Rosette clumps this is 3.*

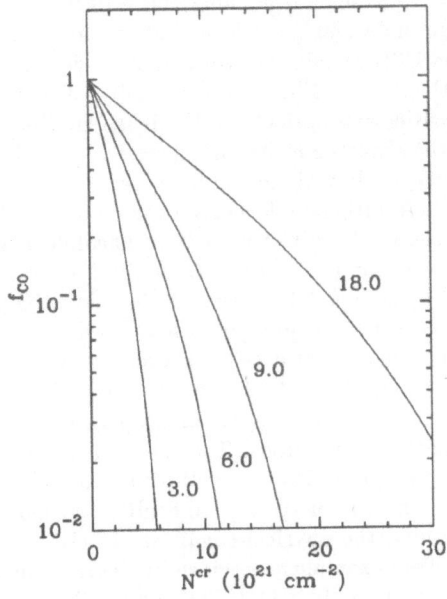

where $F_\nu(T) \equiv (e^{h\nu/kT} - 1)^{-1}$. If we *assume* that the excitation temperatures of the two transitions are the same, then it is possible to determine T_{ex} from the observed ratio of antenna temperatures, and thus determine what correction, if any, must be made to X.

In practice such determinations will be difficult. The ratio of antenna temperatures of the $J = 2 - 1$ to $J = 1 - 0$ transition changes by only about 20% from $T_{ex} = 10$ to $T_{ex} = 40$ K, from 0.8 to 0.95. Since the calibration uncertainties of millimeter telescopes are typically 20%, comparing data from different telescopes is difficult. While the expected ratios change by larger factors for higher J transitions, the assumption of identical excitation temperatures is probably poorer at higher J. In addition, great care must be taken in correcting the temperatures for telescope-dependent effects, such as beam efficiencies. For example, the conclusion of Casoli *et al.* (1989) that the molecular gas in Arp 220 must be cold is based on their measured ratio of $T_R^*(2 - 1)/T_R^*(1 - 0) \approx 0.5$. However, since the $J = 2 - 1$ and $J = 1 - 0$ data were both obtained on the IRAM 30m, the temperature scales need to be corrected for the different beam efficiencies at 115 and 230 GHz, which makes the actual ratio of temperatures slightly greater than unity.

Observations of higher CO rotational transitions provide more useful temperature information than the $2 - 1$ line. This is because a transition whose lower level is well above the ground state requires a high gas temperature for excitation. For example, the $J = 7 - 6$ line of ^{12}CO has been detected from the nucleus of M82 by Wild *et al.* (1989). Since the $J = 7$ level lies 155 K above the ground state, gas with temperature of a few tens of Kelvin or more must be present for the line to be detectable.

Additional complications arise because of subthermal excitation of higher transitions, and differences in emitting areas in the different transitions. In a survey of Galactic plane emission in the $^{12}CO J = 2 - 1$ line, Kutner, Leous and Verter (1989) find that while the ratio of peak temperatures in the $J = 2-1$ and $J = 1-0$ lines are usually the same, the integrated intensity in the $J = 2-1$ line is typically only about half of that in the $J = 1-0$ line. Presumably, this is because about half of the total $J = 1-0$ emission comes from molecular gas which is at too low a density to emit strongly in the $J = 2-1$ line. Non-LTE calculations show that at densities $n_{H_2} \lesssim 300\,\mathrm{cm}^{-3}$ the excitation temperature of the $J = 2-1$ line is substantially below that of the $J = 1-0$ line, even for large column densities and optical depths. It is quite likely that more than one component contributes to the observed emission, especially since the projected beam sizes of millimeter antennas are typically a kiloparsec or more.

Additional information can be obtained by observations of the less abundant isotopic species of CO, ^{13}CO and $C^{18}O$. The difficulty here is simply that the lines are considerably weaker than those of ^{12}CO, requiring very substantial amounts of observing time to attain adequate signal to noise. Also, the ^{13}CO line may be optically thick a large fraction of the time, complicating the analysis. Since the $C^{18}O$ line is almost certainly optically thin, subthermal excitation even at fairly high densities is very likely, making detailed modelling of the data necessary: the expected line intensites in the different transitions are calculated for a range of temperatures, densities, and column densities, and compared with the observations. The usual LTE analysis (e.g. Dickman 1978) will generally not be adequate, especially as multiple components may be necessary.

In spite of all these difficulties, such observations offer considerable promise for constraining the physical conditions in the molecular gas component of other galaxies. An excellent example is provided by Eckart et al. (1989), who obtained $J = 1-0$ and $J = 2-1$ observations of ^{12}CO, ^{13}CO, and $C^{18}O$ from the nucleus of IC 342. The observational data were compared with detailed non-LTE models to place limits on the allowed temperatures and gas densities. From their models they conclude that the gas in the inner kpc must be warm ($T_k \gtrsim 30$) with density $n \approx 10^3$. With the improving technology of submillimeter receivers and the development of new observing facilities, such as SOFIA, a substantial improvement in the quality and amount of such data in the next several years is expected.

5.2 OTHER SPECIES

A large number of molecules have been detected in molecular clouds in the Galaxy, and many of these can potentially provide very useful information on molecular gas in other galaxies. Unfortunately, all of these lines will be much weaker than CO, making observation of them prohibitively expensive in telescope time. For bright galaxies, a few exceptions exist:

CS, HCN, and HCO$^+$: All of these species require considerably higher densities than CO to excite, so that they provide information on higher-density gas ($n \gtrsim 10^4$) than is typically probed by CO. All have been detected in a few galaxies; HCN and HCO$^+$ have been mapped in detail in the nucleus of NGC 253 by Carlstrom et al. (1989). CS has been detected in a number of galaxies, following the initial detection in M82 and IC 342 by Henkel and Bally (1985). Walker et al. (1989) have mapped the CS $2-1$ emission from the nucleus of M82. Note that the mere *detection* of CS (or HCN or HCO$^+$) emission does not indicate that the *average* density of molecular clouds is high; what is relevant is the ratio of CS to CO integrated intensities. Bally et al. (1987) have mapped the Galactic plane in CS $2-1$, and find that the ratio of CS to CO integrated intensities is much higher in the inner few hundred pc of the Galaxy than in the rest of the disk, suggesting that clouds in the Galactic center are unusually dense.

NH₃: Observations of the centimeter wavelength hyperfine transitions can provide very useful estimates of gas temperature, provided the density is high (Ho and Martin 1986). Due to the weakness of the lines, this is very difficult, and only IC 342 has been well studied.

A great deal of information can also be obtained from observations of various atomic ions. The far-infrared fine-structure transitions of a number of abundant elements are readily detectable in infrared-bright galaxies (see §4.2.4). Various line ratios are sensitive to density and temperature, and provide probably the best diagnostics of conditions in gas closely associated with regions of massive star formation (Genzel, Harris and Stutzki 1989; Wolfire, Tielens and Hollenbach 1989).

6. Summary and Concluding Remarks

The use of conversion factors to derive molecular gas masses from observations of the $J = 1 - 0$ line of ^{12}CO has generated far more debate and acrimony than the topic (or any other topic in astronomy) deserves. While application of the Galactic conversion factor to other galaxies is often justified by statements such as "All determinations of the conversion factor agree to within a factor of two", this is not only not true (remember the Galactic center) but is also completely irrelevant: since all of these determinations apply to Milky Way disk clouds, they provide no information on possible variations of the conversion factor for molecular clouds in environments very different from the Galactic disk. Within the usual factor of two uncertainties, the Galactic X-value is probably appropriate for the disks of normal, giant spiral galaxies. Observations of low-metallicity galaxies, such as irregulars and generally low-luminosity galaxies, suggest that use of the Galactic conversion factor underestimates the amount of molecular gas present by a substantial amount. There are excellent theoretical reasons for expecting this. At the other extreme, the amount of energy that is pumped into the interstellar medium in a starburst galaxy, ultraluminous galaxy, or many galactic nuclei is so much larger than in the solar neighborhood that application of the local X-value seems questionable. It is hard to see how molecular clouds in galaxies with infrared surface brightnesses five or six orders of magnitude higher than the typical disk value in the Milky Way could have temperatures of 10 K. While higher densities, if typical, will tend to counteract the higher temperatures, once we allow both of these parameters to vary there is no reason to expect a conversion factor whose value is close to the Galactic one, in general. A substantial amount of observational data suggests that for some fraction of these galaxies, the amount of molecular gas present has been considerably overestimated. Observations of other transitions and species which are curently possible, or are almost within reach of current technology, will place more stringent limits on conditions in molecular gas in other galaxies.

Acknowledgements. This paper has benefitted from discussions with many people, especially Frank Israel. Leo Blitz kindly provided data in advance of publication. I would like to thank Ewine van Dishoeck, Frank Israel, and especially Fran Verter for their comments on the manuscript. I am grateful to the ASTRON foundation of the Netherlands Organization for Pure Research (NWO) for support under grant number 782-372-025, and to Harley Thronson and the good people of Wyoming for partial support of my attendance

520

at the meeting. This paper is dedicated to the memory of my father, Edward Lawrence Maloney. "Nothing is ever forgotten".

References

Ables, H.D. 1971, *Pub. US Naval Observatory, Series II*, **20**, Part 4
Bhat, C.L., Mayer, C.J., and Wolfendale, A.W. 1986, *Phil. Trans. R. Soc. London A*, **319**, 249
Blitz, L. 1978, Ph.D. Thesis, Columbia University
Blitz, L. 1987, in *Physical Processes in Interstellar Clouds*, ed. G.E. Morfill and M. Scholer, (Dordrecht: Reidel), p. 35
Blitz, L., Stark, A.A., and Long, K. 1989, in preparation
Blitz, L., Bloemen, J.B.G.M., Hermsen, W., and Bania, T. 1985, *Astr. Ap.*, **143**, 267
Bloemen, H. 1989, *Ann. Rev. Astr. Ap.*, **27**, in press.
Bloemen, J.B.G.M., *et al.* 1986, *Astr. Ap.*, **154**, 25
Boisse, P., Casoli, F., and Combes, F. 1987, *Astr. Ap.*, **173**, 229
Carlstrom, J.E., Jackson, J., Ho, P.T.P., and Turner, J.L. 1989, in *The Interstellar Medium in Galaxies: Abstracts of Contributed Papers*, ed. D. Hollenbach, (NASA SP), in press.
Casoli, F., Combes, F., Dupraz, Ch., Gerin, M., Encrenaz, P., and Salez, M. 1988, *Astr. Ap.*, **192**, L17
Chevalier, R.A., and Clegg, A.W. 1985, *Nature*, **317**, 44
Crawford, M.K., Genzel, R., Townes, C.H., and Watson, D.M. 1985, *Ap. J.*, **291**, 755
De Vries, H.W., Heithausen, A., and Thaddeus, P. 1987, *Ap. J.*, **319**, 723
Dickman, R.L. 1978, *Ap. J. Suppl.*, **37**, 407
Dickman, R.L., Snell, R.L., and Schloerb, F.P. 1986, *Ap. J.*, **309**, 326 (DSS)
Dettmar, R.J., and Heithausen, A. 1989, *Ap. J. (Letters)*, in press
Draine, B.D. 1978, *Ap. J. Suppl.*, **36**, 595
Eales, S.A., Wynn-Williams, C.G., and Duncan, W.D. 1989, *Ap. J.*, **339**, 859
Eckart, A., Downes, D., Genzel, R., Harris, A.I., Jaffe, D.T., and Wild, W. 1989, preprint.
Elmegreen, B.G., and Elmegreen, D.M. 1984, *Ap. J. Suppl.*, **54**, 127
Elmegreen, B.G., Elmegreen, D.M., and Morris, M. 1980, *Ap. J.*, **240**, 455
Falgarone, E., and Perault, M. 1987, in *Physical Processes in Interstellar Clouds*, ed. G.E. Morfill and M. Scholer, (Dordrecht: Reidel), p. 59
Gallagher, J.S., and Hunter, D.A. 1984, *Ann. Rev. Astr. Ap.*, **22**, 37
Garman, L.E., and Young, J.S. 1986, *Astr. Ap.*, **154**, 8
Gehrz, R.D., Sramek, R.A., and Weedman, D.W. 1983, *Ap. J.*, **267**, 551
Genzel, R., and Townes, C.H. 1987, *Ann. Rev. Astr. Ap.*, **25**, 377
Genzel, R., Harris, A.I., and Stuztki, J. 1989, in *Infrared Spectroscopy in Astronomy*, 22nd ESLAB Symposium, ed. B.H. Kaldeich, (ESA SP-290), p. 115
Gilden, D.L. 1984, *Ap. J.*, **279**, 335
Goldsmith, P. 1987, in *Interstellar Processes*, ed. H.A. Thronson and D.J. Hollenbach, (Dordrecht: Reidel), p. 51
Gordon, M.A., Heidmann, J., and Epstein, E.E. 1982, *Publ. Astr. Soc. Pacific*, **94**, 415
Hodge, P.W. 1972, *Publ. Astr. Soc. Pacific*, **84**, 365
Hodge, P.W. 1974, *Publ. Astr. Soc. Pacific*, **86**, 263

Humphreys, R.M. 1984, in *IAU Symposium 108, Structure and Evolution of the Magellanic Clouds*, ed. S. van den Bergh and K.S. de Boer, (Dordrecht: Reidel), p. 145

Hunter, D.A., and Gallagher, J.S. 1985, *Ap. J. Suppl.*, **58**, 533

Hunter, D.A., Gallagher, J.S., Rice, W.L., and Gillett, F.C. 1989, *Ap. J.*, **336**, 152

Israel, F.P. 1984, in *IAU Symposium 108, Structure and Evolution of the Magellanic Clouds*, ed. S. van den Bergh and K.S. de Boer, (Dordrecht: Reidel), p. 319

Israel, F.P. 1985, in *New Aspects of Galaxy Photometry*, ed. J.-L. Nieto, (Berlin: Springer-Verlag), p. 101

Israel, F.P. 1988a, in *Millimetre and Submillimetre Astronomy*, ed. R.D. Wolstencroft and W.B. Burton, (Dordrecht: Reidel), p. 281

Israel, F.P. 1988b, *Astr. Ap.*, **194**, 24

Israel, F.P., and Burton, W.B. 1986, *Astr. Ap.*, **168**, 369

Israel, F.P., de Graauw, Th., Lidholm, S., van de Stadt, H., and de Vries, C.P. 1982, *Ap. J.*, **262**, 100

Israel, F.P., de Graauw, Th., van de Stadt, H., and de Vries, C.P. 1986, *Ap. J.*, **303**, 186

Issa, M., MacLaren, I., and Wolfendale, A.W. 1989, *Ap. J.*, in press

Jaffe, D.T., Becklin, E.E., and Hildebrand, R.H. 1984, *Ap. J. (Letters)*, **285**, L31

Joy, M., and Lester, D.F. 1988, *Ap. J.*, **331**, 145

Keto, E.R., and Myers, P.C. 1986, *Ap. J.*, **304**, 466

Koornneef, J. 1984, in *IAU Symposium 108, Structure and Evolution of the Magellanic Clouds*, ed. S. van den Bergh and K.S. de Boer, (Dordrecht: Reidel), p. 333

Krolik, J.H., and Begelman, M.C. 1986, *Ap. J. (Letters)*, **308**, L55

Krolik, J.H., and Begelman, M.C. 1988, *Ap. J.*, **329**, 702

Kutner, M.L., and Ulich, B.L. 1981, *Ap. J.*, **250**, 341

Kutner, M.L., Leous, J., and Verter, F. 1989, in preparation

Lester, D.F., Harvey, P.M., and Carr, J. 1988, *Ap. J.*, **329**, 641

Lester, D., Gaffney, N., Carr, J., and Joy, M. 1989, in *The Interstellar Medium in Galaxies: Abstracts of Contributed Papers*, ed. D. Hollenbach, (NASA SP), in press.

Liszt, H.S. 1988, in *Galactic and Extragalactic Radio Astronomy*, ed. G.L. Verschuur and K.I. Kellerman, (Berlin: Springer-Verlag), p. 359

Lo, K.Y., Cheung, K.W., Masson, C.R., Phillips, T.G., Scott, S.L., and Woody, D.P. 1987, *Ap. J.*, **312**, 574

Lord, S.D. 1987, Ph.D. Thesis, University of Massachusetts

Lugten, J.B., Watson, D.M., Crawford, M.K., and Genzel, R. 1986, *Ap. J. (Letters)*, **311**, L51

Maloney, P. 1989, *Ap. J. (Letters)*, in press

Maloney, P., and Black, J.H. 1988, *Ap. J.*, **325**, 389

Magnani, L., Blitz, L., and Mundy, L. 1985, *Ap. J.*, **295**, 402

Martin, R.N., and Ho, P.T.P. 1986, *Ap. J. (Letters)*, **308**, L7

Mauersberger, R., Henkel, C., Wilson, T.L., and Walmsley, C.M. 1986, *Astr. Ap.*, **162**, 199

Meixner, M., Puchalsky, R., Blitz, L., and Wright, M. 1989, in *The Interstellar Medium in Galaxies: Abstracts of Contributed Papers*, ed. D. Hollenbach, (NASA SP), in press.

Mirabel, I.F., and Sanders, D.B. 1988, *Ap. J.*, **335**, 104

Monnet, G., and Simien, F. 1977, *Astr. Ap.*, **56**, 173

Morris, M., and Rickard, L.J 1982, *Ann. Rev. Astr. Ap.*, **20**, 517

Morris, M., Polish, N., Zuckerman, B., and Kaifu, N. 1983, *Astr. J.*, **88**, 1228

522

Polk, K.S., Knapp, G.R., Stark, A.A., and Wilson, R.W. 1988, *Ap. J.*, **332**, 432

Rickard, L.J, Harvey, P.M., and Blitz, L. 1984, in *Airborne Astronomy Symposium*, ed. H.A. Thronson and E.F. Erickson, (NASA CP 2353), p. 287

Rickard, L.J, Palmer, P., Morris, M., Zuckerman, B., and Turner, B.E. 1975, *Ap. J. (Letters)*, **199**, L75

Rieke, G.H., and Lebofsky, M.J. 1985, *Ap. J.*, **288**, 618

Rieke, G.H., Lebofsky, M.J., Thompson, R.I., Low, F.J., and Tokunaga, A.T. 1980, *Ap. J.*, **238**, 24

Rieke, G.H., Cutri, R.M., Black, J.H., Kailey, W.F., McAlary, C.W., Lebofsky, M.J., and Elston, R. 1985, *Ap. J.*, **290**, 116

Sanders, D.B., and Mirabel, I.F. 1985, *Ap. J. (Letters)*, **298**, L31

Sanders, D.B., Scoville, N.Z., and Solomon, P.M. 1985, *Ap. J.*, **289**, 373

Sanders, D.B., Scoville, N.Z., Young, J.S., Soifer, B.T., Schloerb, F.P., Rice, W.L., and Danielson, G.E. 1986, *Ap. J. (Letters)*, **305**, L45

Sargent, A.I., Sanders, D.B., Scoville, N.Z., and Soifer, B.T. 1987, *Ap. J. (Letters)*, **312**, L35

Sargent, A.I., Sutton, E.C., Masson, C.R., Lo, K.Y., and Phillips, T.G. 1985, *Ap. J.*, **289**, 150

Savage, B.D., Bohlin, R.C., Drake, J.F., and Budich, W. 1977, *Ap. J.*, **216**, 291

Scalo, J.M, and Struck-Marcell, C. 1984, *Ap. J.*, **276**, 60

Scalo, J.M, and Struck-Marcell, C. 1986, *Ap. J.*, **301**, 77

Schwering, P.B. 1988, Ph.D. Thesis, University of Leiden

Scoville, N.Z., Young, J.S., and Lucy, L.B. 1983, *Ap. J.*, **270**, 443

Scoville, N.Z., Sanders, D.B., Sargent, A.I., Soifer, B.T., Scott, S.L., and Lo, K.Y. 1986, *Ap. J. (Letters)*, **311**, L47

Scoville, N.Z., Yun, M.S., Clemens, D.P., Sanders, D.B., and Waller, W.H. 1987, *Ap. J. Suppl.*, **63**, 821

Shaver, P.A., McGee, R.X., Newton, L.M., Danks, A.C., and Pottasch, S.R. 1983, *M. N. R. A. S.*, **204**, 53

Shu, F. 1987, in *Star Formation in Galaxies*, ed. C. J. Lonsdale Persson, (NASA SP 2466), p. 743

Simien, F., and de Vaucouleurs, G. 1986, *Ap. J.*, **302**, 564

Smith, P.A., Brand, P.W.J.L., Puxley, P.J., Mountain, C.M., Gear, W.K., and Nakai, N. 1989, *M. N. R. A. S.*, submitted

Sodroski, T.J., Dwek, E., Hauser, M.G., and Kerr, F.J. 1987, *Ap. J.*, **322**, 101

Sodroski, T.J., Dwek, E., Hauser, M.G., and Kerr, F.J. 1989, *Ap. J.*, **336**, 762

Solomon, P.M., Rivolo, A.R., Barrett, J., and Yahil, A. 1987, *Ap. J.*, **319**, 730

Stacey, G., Lugten, J., Genzel, R., and Townes, C. 1989, *Ap. J.*, submitted

Stark, A.A. 1984, *Ap. J.*, **281**, 624

Stark, A.A., and Brand, J. 1989, *Ap. J.*, **339**, 763

Stutzki, J., Stacey, G.J., Genzel, R., Harris, A.I., Jaffe, D.T., and Lugten, J.B. 1988, *Ap. J.*, **332**, 379

Tacconi, L.J., and Young, J.S. 1985, *Ap. J.*, **290**, 602

Tacconi, L.J., and Young, J.S. 1986, *Ap. J.*, **308**, 600

Tacconi, L.J., and Young, J.S. 1987, *Ap. J.*, **322**, 681

Tacconi, L.J., and Young, J.S. 1989, *Ap. J.*, submitted

Thronson, H.A., Hunter, D.A., Telesco, C.M., Harper, D.A., and Decher, R. 1987, *Ap. J.*,

317, 180

Tielens, A.G.G.M., and Hollenbach, D. 1985a, *Ap. J.*, **291**, 722 (THa)

Tielens, A.G.G.M., and Hollenbach, D. 1985b, *Ap. J.*, **291**, 747 (THb)

Tomisaka, K., and Ikeuchi, S. 1988, *Ap. J.*, **330**, 695

Ulich, B.L., and Haas, R.W. 1976, *Ap. J. Suppl.*, **30**, 247

van Dishoeck, E.F., and Black, J.H. 1987, in *Physical Processes in Interstellar Clouds*, ed. G.E. Morfill and M. Scholer, (Dordrecht: Reidel), p. 241

Verter, F. 1985, *Ap. J. Suppl.*, **57**, 261

Verter, F. 1987, *Ap. J. Suppl.*, **65**, 555

Walker, C.E., Walker, C.K., Martin, R.N., and Carlstrom, J.E. 1989, in *The Interstellar Medium in Galaxies: Abstracts of Contributed Papers*, ed. D. Hollenbach, (NASA SP), in press.

Weliachew, L., Casoli, F., and Combes, F. 1988, *Astr. Ap.*, **199**, 29

Wild, W., Eckart, A., Genzel, R., Harris, A.I., Jackson, J.M., Jaffe, D.T., Lugten, J.B., and Stutzki, J. 1989, in *The Interstellar Medium in Galaxies: Abstracts of Contributed Papers*, ed. D. Hollenbach, (NASA SP), in press.

Wolfire, M.G., Hollenbach, D., and Tielens, A.G.G.M. 1989, *Ap. J.*, in press

Wolfire, M.G., Tielens, A.G.G.M., and Hollenbach, D. 1989, *Ap. J.*, submitted

Young, J.S., in *IAU Symposium 115, Star Forming Regions*, ed. M. Peimbert and J. Jugaku, (Dordrecht: Reidel), p. 557

Young, J.S. and Scoville, N.Z. 1982, *Ap. J.*, **258**, 467

Young, J.S., and Scoville, N.Z. 1984, *Ap. J.*, **287**, 153

Young, J.S., Gallagher, J.S., and Hunter, D.A. 1984, *Ap. J.*, **276**, 476

Young, J.S., Tacconi, L.J., and Scoville, N.Z. 1983, *Ap. J.*, **269**, 136

Young, J.S., Claussen, M.J., and Scoville, N.Z. 1988, *Ap. J.*, **324**, 115

Young, J.S., Schloerb, F.P., Kenney, J.D., and Lord, S.D. 1986, *Ap. J.*, **304**, 443

The M51 System: Review of the Interstellar Medium

Richard J. Rand
Department of Astronomy,
California Institute of Technology

Remo P. J. Tilanus
Institute for Astronomy,
University of Hawaii

ABSTRACT. We review observations of the radio continuum, the atomic and molecular gas, the optical emission lines, X-ray and infrared emission in the disk, spiral arms, and nucleus of NGC 5194 and in the companion, NGC 5195. The composition, distribution and kinematics of the Interstellar Medium (ISM) are discussed.

1. Introduction

The binary system M51 consists of a late-type spiral galaxy (NGC 5194) and a peculiar companion (NGC 5195). NGC 5194 has been classified as a Sc by Sandage (1961), a Sc I by van den Bergh (1960), and as a Sbc by Holmberg (1958) with photometric diameters of $14\overset{'}{.}2 \times 9\overset{'}{.}5$. The galaxy shows one of the most well-defined and pronounced spiral structures of any galaxy. The companion galaxy NGC 5195 has been classified variously as an irregular of the M82 type (Sandage 1961), an I0 (Holmberg 1958), a dwarf S0 (Burbidge and Burbidge 1964), and an SB0p (Sandage and Tammann 1987). The infrared plate of Spinrad and Harlan (1973) shows that NGC 5195 is morphologically similar to a SB(r)a. The distance to M51 is given 4.6 Mpc by de Vaucouleurs *et al.* (1976), but throughout this paper we will adopt the larger distance of 9.7 Mpc arrived at by Sandage and Tammann (1975). More recent determinations based on the infrared and optical Tully-Fisher relation yielded distances of 8.7 Mpc (Aaronson and Mould 1983) and 7.6 Mpc (Bottinelli *et al.* 1983). Some general properties of M51 have been listed in Table 1 (Tully 1974a, b, c; Goad, De Veny, and Goad 1979; Monnet, Paturel, and Simien 1981). Impressive images of M51 are presented by Burkhead (1978), showing the truly interactive nature of the system. Faint "streamers" around the two galaxies give the system a very disturbed appearance. The encounter between NGC 5194 and NGC 5195 has been simulated numerically by Toomre and Toomre (1972, 1978). New simulations of the effects of the encounter on the distribution and kinematics of both the gas and stars are being carried out by Hernquist (private communication).

Work on the ISM of M51 divides naturally into three areas: i) the disk and spiral structure, which we discuss in §2; ii) the nuclear region, discussed in §3; and iii) NGC 5195, discussed in §4.

H. A. Thronson, Jr. and J. M. Shull (eds.), The Interstellar Medium in Galaxies, 525–542.
© 1990 *Kluwer Academic Publishers.*

Table 1. M51: General properties

R.A. (1950.0)[a]	$13^h 27^m 46\overset{s}{.}327$	Ford *et al.* (1985)[a]
Decl. (1950.0)[a]	$47°27'10\overset{''}{.}25$	Ford *et al.* (1985)[a]
Distance	9.7 Mpc	Sandage and Tammann (1975)
Systemic velocity	464 km s^{-1}	Tully (1974b)
Holmberg diameter	$14\overset{'}{.}2 \times 9\overset{'}{.}5$	Holmberg (1958)
P.A. major axis	170°	Tully (1974b)
Inclination angle	20°	Tully (1974b)
M_{HI} (NGC 5194)	5×10^9 M$_\odot$	Rots (1980)
M_{H2} (NGC 5194)	9×10^9 M$_\odot$	Scoville and Young (1983)
L_B (NGC 5194)	3.8×10^{10} L$_\odot$	de Vaucouleurs *et al.* (1976)
L_{FIR} (NGC 5194)	2×10^{10} L$_\odot$	Smith (1982)
L_X (0.2–4 keV) (NGC 5194)	3×10^{40} erg s^{-1}	Palumbo *et al.* (1985)
S_{20cm} (NGC 5194)	1500 mJy	Hummel (1980)
$L_{H\alpha}$ (NGC 5194)	3.8×10^{41} erg s^{-1}	Kennicutt and Kent (1983)
$S_{60\mu m}$ (M51)	98.8 Jy	Soifer *et al.* (1989)
L_{FIR} (NGC 5195)	$\gtrsim 3 \times 10^9$ L$_\odot$	Smith (1982)
L_X (0.2–4 keV) (NGC 5195)	2.1×10^{39} erg s^{-1}	Palumbo *et al.* (1985)

[a] nuclear radio-continuum source

2. The Disk and Spiral Arms

2.1 RADIO CONTINUUM

Even the first radio synthesis observations of M51, which provided the very first detection of radio-continuum emission from spiral arms outside of the Local Group, revealed unique information which is unsurpassed by observations of any other galaxy to date. Using the Westerbork Synthesis Radio Telescope (WSRT) at 21 cm with a resolution of 24″×32″, Mathewson, van der Kruit, and Brouw (1972; MKB hereafter) observed a prominent two-armed spiral structure and an underlying "base" disk. To the limit of their resolution, the radio ridges are located along the inside edges of the bright optical arms and are generally coincident with the dust lanes. The detection of linear polarization in the emission confirmed an origin due to the synchrotron mechanism. MKB interpreted the coincidence of the radio arms with the inner edges of the optical arms as first-class observational evidence both for the existence in M51 of density waves (*e.g.* Lin and Shu 1964), and for the nonlinear response of the interstellar gas to these waves (*e.g.* Roberts 1969).

Following the pioneering paper by MKB, Segalovitz (1976, 1977a) observed M51 at 6, 21, and 49 cm. His 6 cm observations show the nucleus to be elongated roughly in the NS direction with the approximate dimensions of 6″×18″ (see §3). He also noted that the inner radio arms (within the range 1 to 2.5 kpc) are coincident with and as wide as the dust lanes. Segalovitz constructed detailed diffusion models

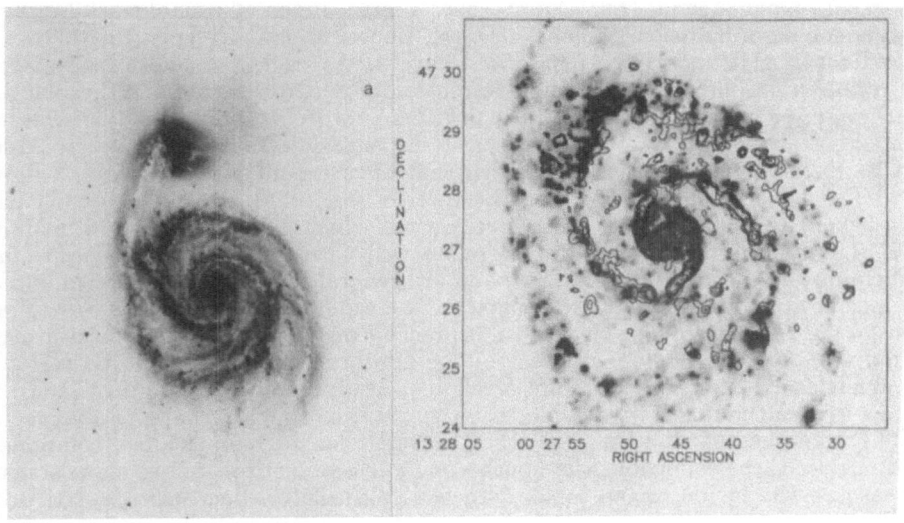

Figure 1 (a) Hale 5-m plate of M51. The scale is roughly 9″/mm.
(b) New VLA map of HI column densities in M51 (Rots *et al.* 1989).

for the cosmic rays in M51 (Segalovitz 1977b), based on the assumption that the variation of the spectral index (from the 21 and 49 cm data) across the disk of the galaxy was due only to a variation of the *nonthermal* contribution. However, based on the same radio data and optical emission line observations, van der Kruit (1977) argued that a considerable fraction of the 6 cm emission is thermal, and that thermal emission can account for the radial variations of the spectral index between 21 and 49 cm. Whereas in Segalovitz's models the cosmic rays are injected from the spiral arms or sites of recent star formation, van der Kruit proposed that its sources are widely distributed through the disk and are linked to the total stellar disk population. This issue is remarkably relevant in light of the current discussion of the radio/infrared correlation in spiral galaxies (*e.g.* Dickey, Garwood, and Helou 1987).

Using the Effelsberg 100-m telescope, Klein, Wielebinski and Beck (1984) observed M51 at wavelenghts of 1.3, 2.0, and 2.8 cm, with resolutions ranging from 40–70″. They conclude that the amount of thermal emission is lower than estimated by *e.g.* van der Kruit. In agreement with previous estimates they find a typical extinction of about 1.5 mag from a comparison of the thermal emission to the integrated Hα flux. The average value of the nonthermal spectral index they find is about −0.9, which agrees with the value found by Klein and Emerson (1981), but is somewhat steeper than the average spectral index of roughly −0.8 derived from a sample of 31 nearby galaxies (Duric, Bourneuf, and Gregory 1988).

Recently, Tilanus and Allen (1988) have used 6 and 20 cm Very Large Array (VLA) observations with a common resolution of 8″, to separate the thermal and nonthermal radio continuum emission in M51 (Figure 4a and 4d). The thermal component accounts for 5% of the total radio emission from the galaxy at 20 cm, and its distribution correlates well with that of the giant H II complexes (Figure 4d). The nonthermal emission can be further separated into an extended base disk, which contributes 60%, and spiral arms which account for the remaining 35%. The nonthermal spiral arms correlate well with the dust lanes (Figure 5a), confirming the earlier result of MKB, and are generally not coincident with and on the inside of the ridges of H II complexes. The separation they find is 5″−10″ on average (235−470 pc). The detailed morphology of the nonthermal spiral arms is asymmetric, with a shallow gradient of the nonthermal emission on the *inside* of the arms, and a rapid drop to faint levels just beyond (*outside*) the dust lanes. In the context of the density-wave model for spiral arms, such a shape is unexpected if the synchrotron gas flows along with the interstellar medium as a one-component fluid. Under these conditions the nonthermal emission is expected to show a sharp increase on the inside of the arms, followed by a shallow decrease into the post-shock region, precisely the opposite shape of what is observed. The authors suggest that the synchrotron-emitting gas does not behave as a cold fluid, and separates kinematically from the cool-cloud component on approach of the arm. The well-defined dust lanes indicate that the latter component experiences a shock as it falls into the potential well.

It should be noted that *e.g.* the near-IR image published by Schweizer (1976) also shows a shallow increase in brightness on the inside of the dust lanes. These "red arms" are interpreted as being the result of an increased surface brightness of the underlying stellar disk, due to the density wave.

2.1.1 Magnetic fields.

Based on their 1.3−2.8 cm observations, and assuming equipartition between the magnetic field and the cosmic ray energy, Klein, Wielebinski and Beck (1984) derive a magnetic field strength for M51 of $11 \pm 3 \, \mu G$, about 2−3 times stronger than in our Galaxy and M31. A similar value for the magnetic field is derived by the same authors based on 6.3 cm polarization observations on M51 (Beck, Klein, and Wielebinski 1987). They find a high degree of polarization and a magnetic field with the field lines uniformly following the spiral arms. Optical polarization measurements covering the whole galaxy from a radius of 200 pc outward have been published by Scarrott, Ward-Thompson, and Warren-Smith (1987). Outside the nuclear regions the optical polarization vectors follow the spiral arms. Both Scarrott *et al.* and Klein *et al.* find that the radio and optical vectors are perpendicular over a large region of the galaxy, showing that the same magnetic field that gives rise to the radio synchrotron emission also aligns the dust grains. However, in the south-western quadrant of M51 the vectors diverge by up to 60°.

Recent VLA D-array observations show the magnetic field of M51 to have a bisymmetric spiral (BSS) structure (Horellou, Beck, and Klein 1989) as in M81. M31 and IC342 on the other hand show an axisymmetric structure. The galactic dynamo models (Krasheninnikova *et al.* 1989; Lesch *et al.* 1988) predict for a galaxy like NGC 5194 a strong dynamo-effect and favor the BSS mode over the axisymmetric mode. Lesch *et al.* argue that the decisive difference between the two possible field configurations is the shear between the gas and the density wave. Krasheninnikova *et al.* suggest that the BSS mode may become the dominant mode in interacting galaxies due to an asymmetric gravitational disturbance by the companion.

Horellou *et al.* find the maxima of the polarized emission at 20.5 and 18.0 cm not to be located on the optical spiral arms. A similar situation is found to exist in M81 (Krause, Hummel, and Beck 1989) and M83 (Sukumar and Allen 1989): the maxima of the polarized emission lie in the the dark interarm regions or in the dark outer regions of the galaxy. Krause *et al.* suggest two alternative explanations for this phenomenon: either the interstellar magnetic field is less disturbed in the dark regions, or the dynamo-effect works more effectively. On the basis of an absence of an enhancement of the total synchrotron emission and a low star-formation activity in the dark regions, Sukumar and Allen consider the former explanation to be the most straightforward. They suggest the kinetic energy associated with star formation to be the source of the disorder. Horellou *et al.* also suggest that the enhanced star formation in the spiral arms could induce an enhanced disorder in the interstellar magnetic field, and that the density waves in M51 are unable to align the field. However, at the scales of the width of the spiral arms (300 pc) their resolution (2 kpc) may be too poor to detect a narrow ridge of high polarization (MKB) at the position of the dust lanes.

2.2 ATOMIC HYDROGEN

In contrast with the radio continuum observations, relatively few observations of the distribution and kinematics of the H I in M51 have been published. The total H I mass of M51 is $5 \times 10^9 \, M_\odot$ (Rots 1980) with an H I diameter of $\sim 11'$ at a standard isophote of 1.82×10^{20} atoms cm^{-2} (Bosma 1981). Using the 76-m radiotelescope at Jodrell Bank, Appleton, Foster, and Davies (1986) derived a total H I mass of $5.1 \times 10^9 \, M_\odot$ and an extent of the H I emission of 50' measured at a column density of 4.1×10^{18} atoms cm^{-2}.

The first high-resolution (2') synthesis image of M51 in H I was presented by Weliachew and Gottesman (1973). Although their observations do not resolve the spiral structure of the disk, it reveals the striking absence of atomic hydrogen in the central region of M51, a feature often seen in late-type spirals (*see also* Figure 2). Their velocity field shows a good deal of large-scale symmetry. However, they find evidence for strong non-circular motions along the positions of the outer spiral arms, which are in quantitative agreement with the tidal encounter model of Toomre and Toomre (1972). The H I map at a resolution of 25" × 35" presented by Shane (1975) shows a well-defined spiral structure in H I, with the outermost parts of the arms being brighter than the inner arms. The radial distribution of the H I is rather constant, in contrast with the exponential disk seen in CO, the optical, and the FIR. Small-scale "wiggles" in the velocity field are found across the outer spiral arms, similar to the features which were found in the velocity field of M81 (Rots 1975) and attributed to density wave perturbations (Visser 1980a, b).

The above H I observation also showed evidence for extended H I outside the south-western edge of the galaxy. Shane (1975) finds this gas in apparent counter-

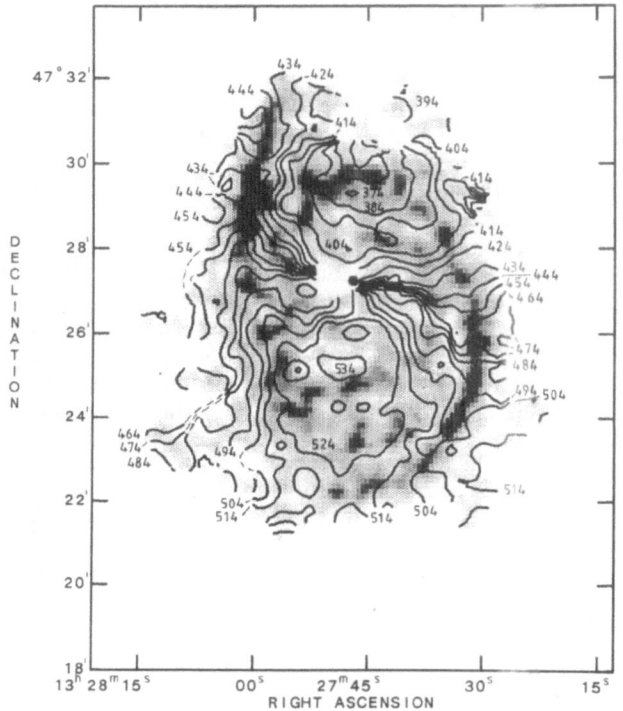

Figure 2
Intensity-weighted mean velocity field (contours) overlaying the HI distribution
(gray-scales) showing
the spiral arms.
The resolution is 20″.
(Tilanus and Allen 1989b)

rotation with respect to the rest of the galaxy. New VLA observations reveal
a broad spiral arm extending eastward and running couterclockwise around the
southern half of M51 (Rots *et al.* 1989; Figure 1b). Appleton, Foster, and Davies
(1986) conclude that the velocity reversal is caused by a warping of the plane of
the disk through the plane of the sky in the outermost regions of M51, reversing
its inclination. Extending the encounter models of Toomre and Toomre further
outward, they find that the warped, outer arm may be explained if NGC 5194
originally posessed an H I disk having a diameter of 100 kpc, which was subsequently
disrupted by the passage of NGC 5195 through the disk. A warping of the disk
of NGC 5194 is consistent with the apparent decline of the rotation curve of the
galaxy beyond a radius of ~2.5′ (Tully 1974b; Shane 1975).

Tilanus and Allen (1989a, b) have observed M51 at a resolution of 12″×18″
with the WSRT (Figure 2). They compare the density distribution of the atomic
hydrogen in the inner region of the galaxy with other tracers of spiral structure: the
dust lanes, H II regions and the nonthermal radio continuum. Over major sections
of the inner spiral arms the H I ridge is parallel to the ridge of the nonthermal
emission but not coincident with it (Figures 4a and 4c). The displacement of the
H I arms downstream from the nonthermal emission and the dust lanes is equal to
the separation seen between the nonthermal radio continuum and the H II complexes

(see §2.1). This phenomenon, but more pronounced, has also been observed in a 7 kpc section of a spiral arm of M83 (Allen, Atherton, and Tilanus 1986). The authors propose a picture which describes the observations in the framework of density-wave streaming of the interstellar gas. They conclude that, in molecular-rich galaxies, rather than H I clouds collapsing to produce molecular clouds, the H I is the product of dissociation of the predominantly molecular interstellar medium by the star formation process. This conclusion has been greatly strengthened by high-resolution CO observations of the spiral arms of M51, which find the molecular gas to be coincident with the dust lanes and upstream of the H I (Figure 4; see §2.3).

H I observations with better resolution are needed to study the relation between the atomic and molecular gas in detail. If the atomic gas is confirmed to be the product of dissociation it will substantially alter our interpretations of H I observations of gas- and dust-rich spirals in general.

2.3 MOLECULAR GAS AND STAR FORMATION

M51 is the best-studied external galaxy in terms of its molecular gas distribution and kinematics. It has a total gas mass of about 10^{10} M$_\odot$ – about 5 times that in our own Galaxy (Scoville and Sanders 1988) – of which 75% of the gas within 10 kpc of the center is in molecular form (Scoville and Young 1983, hereafter SY). This fraction increases with decreasing galactocentric radius. The CO emission, and presumably the H$_2$ surface density, falls off exponentially with radius – similar to the falloffs in optical light (red), and FIR, Hα and non-thermal radio emission (SY). A comparison of CO 2−1 and 1−0 spectra by SY showed that the CO emission is optically thick. New insight into the arm/interarm excitation conditions and optical depth variations should be provided by an ongoing, high-resolution, single-dish study of the 2−1 and 1−0 lines of ^{13}CO and ^{12}CO by the IRAM group. An early result of note from this study is a radial decrease in the $(2-1)/(1-0)$ ratio of ^{12}CO (Guélin et al. 1988).

Interferometric observations of M51 have been carried out by Lo et al. (1987), Vogel, Scoville and Kulkarni (1988, hereafter VKS) and Rand and Kulkarni (1989a; 1989b, hereafter RK). VKS mapped the inner and outer arms in the NW quadrant using the OVRO Millimeter Interferometer. They detected narrow, coherent spiral structure in the inner arm and patchy spiral structure in the outer arm. The CO emission shows excellent coincidence with the dust lanes, but is offset upstream from the peak ridge of Hα emission. This offset is not due to extinction since a similar offset is seen between the CO and thermal radio emision of Tilanus and Allen (1988; Figure 4). The offset implies a delay of 3×10^7 years between the peak compression of the molecular gas and the peak of massive star formation.

Taking into account the single-dish flux not seen by the interferometer, VKS estimate an arm-interarm contrast in molecular gas mass of 2.4−3.0. The contrast in the number of H II regions is larger (about 10), implying that the star formation efficiency (SFE) is higher on the arms than off. Such a contrast has also been recently seen in NGC 6946 by Tacconi and Xie (1989). VKS conclude that the density wave has triggered an excess of star formation on the arms for a given amount of molecular gas. This conclusion would seem to disagree with that of Elmegreen and Elmegreen (1986), who found from the global data on spirals that there is no correlation between strength of the density wave and the SFE. However, the *global* enhancement in the SFE due to the density wave in M51 is estimated by

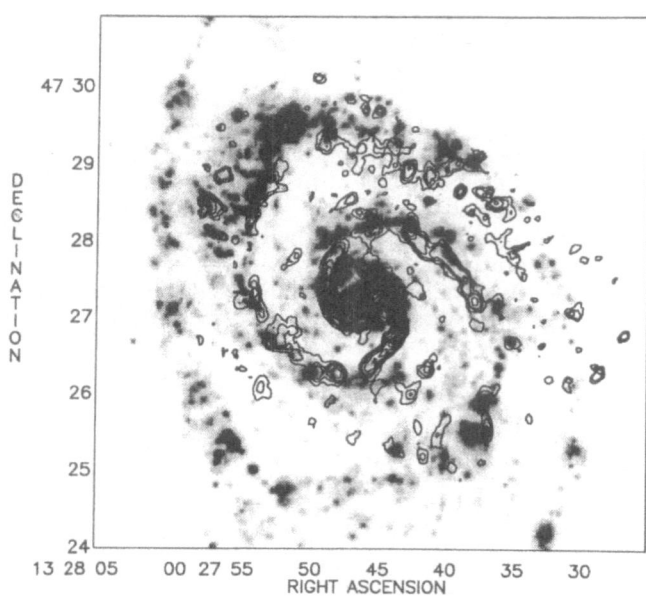

Figure 3
The CO map of RK overlaid on an Hα CCD image. The contour levels are 4.5, 9.0, 13.5, 22.5, 31.5 and 40 Jy km s^{-1}.

VKS to be much lower than the scatter in the Elmegreens' data.

Lord and Young (1989, hereafter LY) provided an important confirmation of density wave triggering of star formation with their 50$''$-resolution FCRAO data and Hα images. Their finding of an arm-interarm SFE contrast using single-dish data shows that VKS's conclusion cannot be explained by the missing flux in their interferometric observations. LY concluded that the SFE contrast is consistent with the idea of massive star formation occurring due to collisions of molecular clouds (*e.g.* Scoville, Sanders, and Clemens 1986; Shu, Adams, and Lizano 1987). The constancy of the azimuthally-averaged radial profile of massive SFE does not have an explanation in terms of simple star formation models.

RK presented an interferometric mosaic map of CO emission made at OVRO covering most of the optical disk of M51 (Figures 3 and 4b). A two-arm spiral structure is clearly seen. VKS had noted that the arm emission tends to break up into discrete features, termed "Giant Molecular Associations" (GMAs) with masses $\gtrsim 10^7 M_\odot$, equivalent to $\gtrsim 10$ typical Galactic GMCs. The GMAs must be transient features because of their concentration to the arms. RK identified 26 GMAs in their mosaic, including 6 with somewhat lower masses *between* the arms. Most significantly, they found that most of the on-arm GMAs are gravitationally bound units, while the interarm GMAs are unbound. RK showed that interarm GMAs do not owe their existence to random superpositions of a normal population of GMCs. The GMAs may be agglomerations of smaller clouds built up through collisions (*e.g.* Scoville, Sanders, and Clemens 1986) or may form due to enhanced gravitational instability (*e.g.* Elmegreen 1987; Balbus 1988). In the latter case, collisions may still provide the energy dissipation necessary for the growth of instabilities (Elmegreen 1988a). The one interarm GMA near the minor axis shows an inward radial streaming motion as predicted by density wave theory (see also §2.4), and RK suggest, as had Rydbeck, Hjalmarson and Rydbeck (1985), that there may be a weak secondary compression of the density wave between the main arms. In support of this explanation are the faint, narrow spiral-shaped strings of H II regions seen between the arms. RK suggest that the GMAs may be destroyed by galactic tidal forces as they leave the arms (Elmegreen 1988b) or may be disrupted by star formation.

Thus, a picture of star formation triggered by a density wave has emerged in which the density wave compresses the molecular gas (giving rise to CO emission in the map of RK and the optical dust lanes), causing an increase in the efficiency of the star-formation triggering mechanism, and, after a delay, leading to the formation of massive stars (resulting in H II regions, free-free emission and the dissociation of H_2 into H I; Figure 4).

2.4 STREAMING MOTIONS

Streaming motions in M51 indicative of a strong density wave, in the form of tangential and radial velocity shifts (*e.g.* Roberts 1969), were first seen in Hα by Tully (1974c) and in the star-forming molecular gas by Rydbeck, Hjalmarson and Rydbeck (1985, hereafter RHR) with the Onsala 20-m telescope (see also Rydbeck *et al.* 1987). The magnitudes of the shifts ($\sim 70 \, \mathrm{km \, s^{-1}}$ assuming a 20° inclination) are a significant fraction of the rotational velocity, indicating a very strong density wave. The sense of the shifts are in agreement with density wave theory. VKS found a resolved velocity shift in CO all along the inner NW arm, with the same large amplitude seen by RHR. RHR also found some weak evidence for velocity shifts *between* the arms, speculating that these may represent a higher order mode of the density wave, as seen in M81 by Visser (1980a, b). RK have found further evidence for such a mode (see §2.3). Using new Fabry-Perôt Hα observations of the central 3' of M51, Tilanus and Allen (1989b) detected tangential and radial velocity gradients across the inner spiral arms in the ionized gas, similar to the velocity gradients seen in CO. The streaming motions in M51 are about 2−3 times as large as the ones found in H I by Rots (1975) in M81, which were successfully modelled by Visser (1980a, b) with a self-consistent density wave model. The large streaming motions may indicate that the density-wave in NGC 5194 is not purely self-excited, but has been modified by the gravitational interaction with the companion. Tilanus and Allen have not been able to conclusively detect streaming motions in the H I emission from the arms, perhaps due to the relatively poor angular resolution ($\sim 15''$) of their H I observations.

2.5 H II REGIONS

A detailed study of the properties and extinctions of the H II regions in M51 has been carried out by van der Hulst, Kennicutt, Crane, and Rots (1988, hereafter HKCR), using 6 cm and 20 cm radio continuum images at 8'' resolution, Hα surface photometry, and 9''-aperture optical spectrophotometry. Special care was given to ensuring that the radio and Balmer fluxes were measured through the same

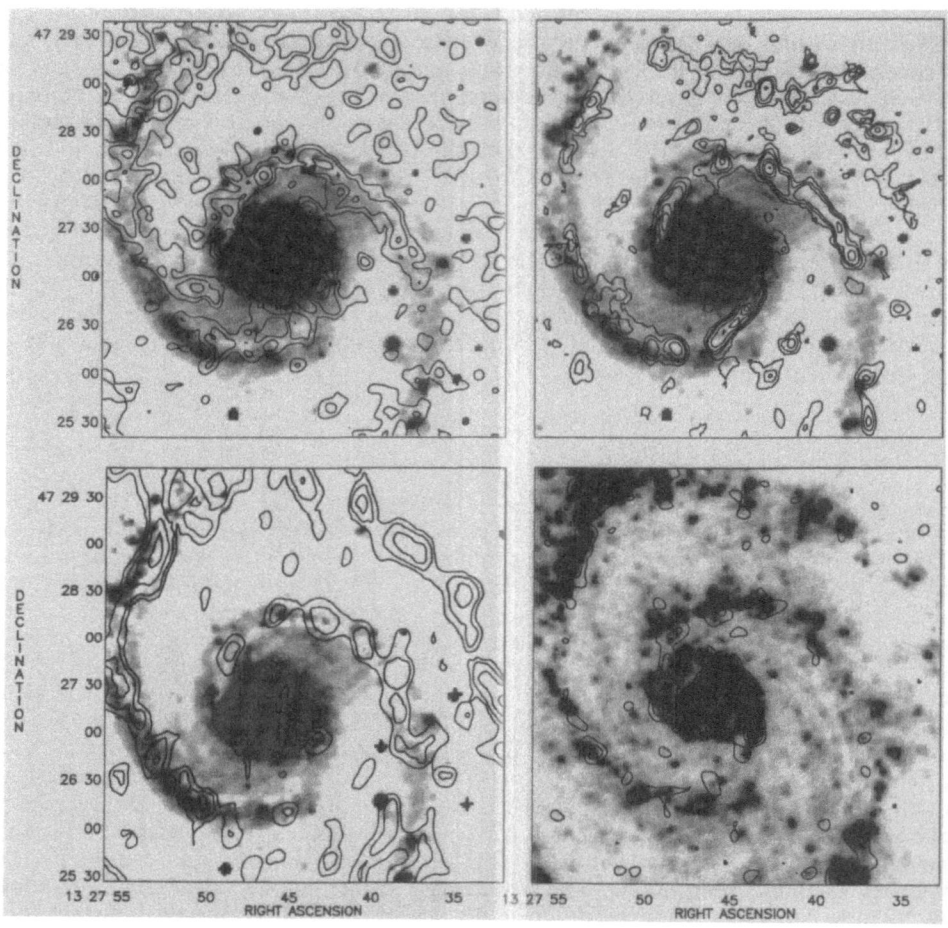

Figure 4 Overlaying the red continuum (gray-scale) are: (a) the nonthermal radio-continuum emission at 8″. (b) the CO emission at 8″. (c) the HI distribution at ∼15″. (d) the thermal radio continuum (contours) overlaying the Hα emission (gray-scale).

apertures. Their results indicate that extinctions derived from Balmer-decrements are much better indicators of the foreground extinction than had been generally thought. The extinctions derived from the ratio of thermal radio flux to Hα flux exceed the Balmer-decrement extinctions by 0.5 mag at the most, much less than the previously found discrepancies of more than a magnitude. Similar results have been obtained for the LMC (Caplan and Deharveng 1986) and M33 (Viallefond and Goss 1986). The remaining discrepancy can be easily accounted for by clumpiness of the dust in and around the H II regions. HKCR also conclude that in the H II regions of M51 homogeneously distributed internal dust does not appear to make a large contribution to the total extinction.

HKCR find that the extinction towards the H II regions does not correlate well with the H I column densities, in contrast to the good correlation found in our own Galaxy (Bohlin et $al.$ 1978). The N_{HI}/A_v ratio increases with galactocentric radius, although the dispersion is quite substantial. HKCR give as a likely explanation a variation of the gas-to-dust ratio with radius, which can account partly for the observed variations: the oxygen abundance in M51 may drop by a factor of 4 from the inner to the outer regions (Pagel and Edmunds 1981). An additional effect could be the increasing fraction of H_2 in the total gas content at smaller galactocentric radii.

A rather unexpected result is the presence of small-scale nonthermal emission contaminating the thermal emission of several H II regions. This nonthermal emission could be caused by small-scale structure in the underlying disk and spiral-arm emission, like local shock enhancements, or in some cases could result from embedded supernovae or supernova remnants.

2.6 X-RAY EMISSION

Palumbo et al (1985, hereafter PFFT) studied the Einstein HRI data on NGC 5194 and found that it has an luminosity in 0.2−4.0 keV X-rays of $L_x = 3 \times 10^{40}$ erg s^{-1}. This L_x, the X-ray-to-optical and X-ray-to-radio flux ratios are typical of normal spirals. Three point sources are found in the disk, all in the spiral arms.

The X-ray radial profile follows that of the blue light, radio continuum, CO, FIR and Hα, implying an association with young, massive stars. PFFT conclude, however, that most of the X-rays originate in the "older, smooth disk Population I component", because the radial profile does not follow that of the optical emission from the arms, which is roughly constant with radius (Schweizer 1976). However, we see this as weak evidence for an older population as the source of the bulk of the X-rays. Even though massive star formation is concentrated on the arms, such azimuthally-averaged radial profiles smear out the known spiral pattern in all star formation tracers. Thus such a comparison does not provide direct evidence against an association of X-rays with ongoing star formation.

The best evidence for an older population origin (PFFT favor low mass X-ray binaries) of the X-rays is the lack of a spiral pattern in their 24″-resolution contour plot.

3. THE NUCLEAR REGION

Just as high spatial resolution has made M51 a unique laboratory for studying density waves and star formation in the disk of a spiral, so has it allowed us to gain a deeper understanding of processes which are likely to be occurring in the nuclei of active galaxies.

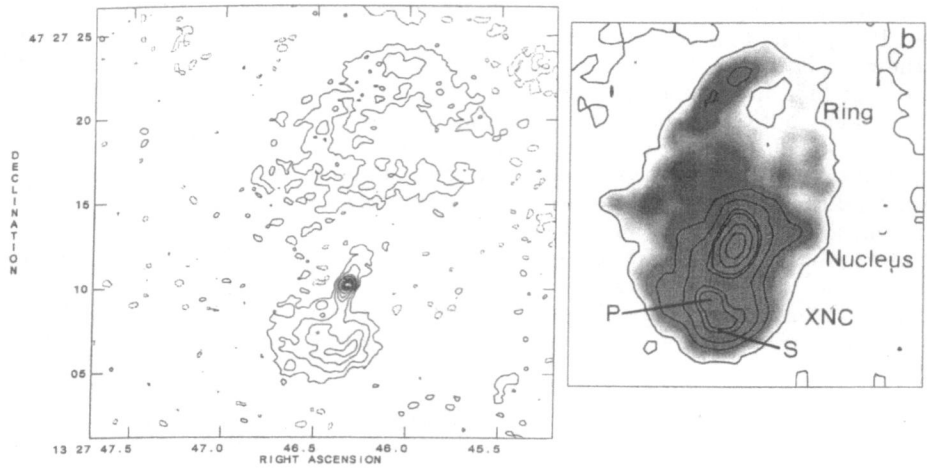

Figure 5 (a) New VLA 6 cm map of the nuclear region with a resolution of ∼0.5″ (Crane *et al.* 1988). (b) The V_{sys}+40 km s^{-1} channel of Cecil's (1988) [NII] λ6583 Fabry-Perôt data cube. The N—S extent is 21″.

3.1 SEYFERT-LIKE ACTIVITY

Large non-circular motions in the nuclear region of M51 have been known for 25 years (Burbidge and Burbidge 1964). Since then many studies of the nuclear region have revealed Seyfert-like activity (Searle 1976; Goad, deVeny and Goad 1979; Rose and Searle 1982, herafter RS; Rose and Cecil 1983, hereafter RC; Cecil and Rose 1984; Goad and Gallagher 1985, herafter GG; Ford *et al.* 1985, hereafter FCJLV; Cecil 1988).

High [NII]/Hα ratios, which are not characteristic of ionization by thermal sources, were first seen by Burbidge and Burbidge (1965). RS found a radial gradient of this ratio in the nuclear region, going from 5–7 in the inner 5″ to 0.5 at 12″ radius. They find that the lines are broad (250 − 450 km s^{-1} FWHM). To explain the [NII]/Hα, [OIII]/[OI] and [OIII]/Hα line ratios and their gradients, RS preferred a non-thermal nuclear source of ionizing photons. RC gave support to the photoionization explanation by measuring a temperature from the [OIII] λ4363 line in the central 5″of T ≲ 15000 K. Such a low temperature would rule out collisional ionization, *e.g.* shocks, as the ionization mechanism in the central 5″.

A great advance in our understanding of the nuclear activity in M51 was provided by the high-resolution (≲1″) observations of FCJLV. VLA maps at 20 cm and 6 cm (Figure 5a) and images in Hα +[NII] (see also Figure 5b) reveal a filled extranuclear cloud ("XNC") 3″ SSE of the nucleus and a ring-shaped cloud (the "ring") centered 9″ NNW of the nucleus. Thus the nucleus, XNC and ring form a linear structure in the VLA maps and narrow-band images, with the nucleus being elongated in the directions of the XNC and ring.

The edge of the XNC furthest from the nucleus is limb-brightened and bow-shaped, and shows the flattest spectral index (-0.7) in the cloud, all of which was taken by FCJLV as evidence for some type of working surface. Spectroscopy at $2''$ resolution revealed that line ratios of O, N and S relative to Hα in the XNC and ring are better fit by a shock model than a photoionization model, seemingly in contrast to the conclusions of RS and RC. In support of the shock explanation are the observed sudden changes in the line ratios and increases in the line widths at the edges of the XNC and ring away from the nucleus, as would be expected if shocks were occurring there. RS and RC did not have the spatial resolution to find these discontinuities or even the "bipolar" morphology of the nuclear region, and thus preferred the photoionization model. FCJLV pointed out that an important test to distinguish between shock-ionization and photoionization is the measurement of the temperature in the XNC from the lines of [OIII]. RC's low value of $\lesssim 15000\,\mathrm{K}$ was for the entire central $5''$. However, it may be significantly higher in the postulated shock regions. FCJLV agreed with RS and RC that the nucleus itself is likely to be photoionized.

As for the origin of these two clouds, FCJLV suggest two possibilities. The first is that they are due to a nuclear two-sided jet interacting with the disk of M51. The second is that they are discrete, expanding "plasmoids" of different ages which were ejected from the nucleus and are now moving through the ISM. Limb-brightened far edges are expected in either case.

All in all the nuclear region of M51, with its optical luminosity of $\sim 10^{42}$ erg s^{-1}, shows many characteristics of LINERS (Low Ionization Nuclear Emission-line Regions) and Seyferts, including broad optical line widths characteristic of "Narrow Line Regions" (NLRs), line ratios characteristic of a non-thermal ionization source, line profile asymmetries suggestive of outflows, and a quasi-linear distribution of active regions seen in optical line and radio emission (GG; in particular, see the study of NGC 1068 by Wilson and Ulvestad 1983). M51 is also an example of the statistical correlations of interactions with nuclear activity (*e.g.* Stocke 1978).

Cecil's (1988) Fabry-Perôt study of the [NII] λ6583 emission and spectroscopic work provided yet more insight into the nature of the nuclear activity, and in particular, the XNC. The Fabry-Perôt data reveal that the broad wings in the [NII] line profiles at the flux peak of the XNC (feature "P" in Figure 5b) are actually due to discrete kinematic components with dispersions $\sim 140\,\mathrm{km\,s^{-1}}$ and centroids that differ by $\sim 630\,\mathrm{km\,s^{-1}}$. These components join smoothly onto the low velocity component both spatially and spectrally, such that the entire structure strongly suggests an expanding shell. This explanation may apply to NLRs in general.

Cecil finds that the observed detailed morphology of the XNC, including the brightened back edge (feature "S" in Figure 5b) and "P" on the front edge, is well reproduced by a shock model in which a nuclear outflow obiquely strikes a dense concentration of gas ("P") before forming the main working surface at "S", which is the boundary of a previously inflated cocoon (Williams and Gull 1985). This situation could occur if the jet interacts with rotating, clumpy disk gas, such that different dense concentrations are continuously rotated into the jet path.

In the XNC, the [NII]/Hα ratio is found to have a velocity-dependent component and is highest at "S", providing strong evidence against photoionization by a central source. Cecil concludes that shocks must be present, but shock ionization can't completely explain the high [NII]/Hα ratio (5.0) at "S" – heating by relativistic particles is also necessary.

The electron density in the XNC, as estimated from the SII doublet, is $\sim 230\,\mathrm{cm}^{-3}$, and the total mass of ionized gas is $14{,}000\,\mathrm{M}_\odot$ in the XNC, and $48{,}000\,\mathrm{M}_\odot$ in the entire active region.

3.2 STELLAR OVALS AND STAR FORMATION IN RINGS AND OTHERWISE

Both Zaritsky and Lo (1986) and Pierce (1986) presented evidence for the existence of an oval distribution of old stars in the central $40''$ of M51. Both papers pointed out the simulations of gaseous response to a non-axisymmetric potential by, for example, Sanders and Huntley (1976), which predict radial inflow of gas inside the corotation radius, possibly resulting in a build-up of gas at the Inner Lindblad Resonance (ILR; Schwarz 1981). Streaming motions in the inner regions, indicative of elliptical orbits, have been seen in the $H\alpha$ Fabry-Perôt study of Tully (1974c). The modelled density perturbation responsible for the motions agrees well with the observed stellar oval. Non-circular motions are also seen in the CO data of RHR.

Pierce suggested that the pile-up of gas at the ILR may result in a ring of star formation. Such a ring, with radius $15''-20''$, was found in blue light by Worden (1974), and later hinted at in $H\alpha$ by GG, FIR by Lester, Harvey and Joy (1986), and CO by RHR. Both Pierce and Zaritsky and Lo pointed out that if inflow continues into the very central regions, it could be a fuel source for the nuclear activity as suggested by Simkin, Su and Schwarz (1980).

The $10\,\mu m$ emission cut through the nuclear region by Telesco, Decher and Gatley (1986) is bumpy, showing peaks at the inner spiral arms and the possible ring. Through $1-4\,\mu m$ photometry, these authors are able to isolate a component of the $3.5\,\mu m$ emission which is due to dust reradiation, and thus derive a high $10\,\mu m/3.5\,\mu m$ color temperature of $\sim 400\,\mathrm{K}$ (with large spatial variations), implying that most of the $10\,\mu m$ emission is due to dust heated by young stars. A simple starburst model implied a $\mathrm{SFR} \approx 3\,\mathrm{M}_\odot\,\mathrm{yr}^{-1}$ in the central $1'$.

Lester, Harvey and Joy (1986), through the use of "superresolution" techniques, were able to resolve a double peaked profile in their $100\,\mu m$ scan through the nuclear region. This profile may represent a cut through a ring as mentioned above. The authors concluded that the inner $20''$ is therefore relatively devoid of gas, dust and star formation, and has perhaps been swept clean by the nuclear outflow. Certainly there is a deficit of CO emission within $10''$ of the nucleus, as can be seen in the mosaic map of RK.

From the $1-2.5\,\mu m$ emission in the central $50''$, Thronson and Greenhouse (1988) estimated the total stellar mass. Taken together, the derived total mass of old stars ($1.2 \times 10^{10}\,\mathrm{M}_\odot$), the SFR as derived from the L_{FIR} of $6 \times 10^9\,\mathrm{L}_\odot$ (Smith 1982), and the mass of molecular gas of $8 \times 10^8\,\mathrm{M}_\odot$ (SY), imply that the current SFR in the central $50''$ is enhanced ($4\,\mathrm{M}_\odot\,\mathrm{yr}^{-1}$, equal to the entire Milky Way), but the true current SFE in this burst is not extreme (1%). The authors concluded that bursts like the current one last $\sim 10^8\,\mathrm{yr}$ and occur only a small fraction of the time. The burst ends when the gas has been locked up in stars and/or driven away by SNe. During the long quiescent periods, gas is accumulated in the central regions in preparation for another burst. Thus, star formation is not particularly efficient, but relies on effective accumulation of gas in the central region.

PFFT found that the X-ray luminosity of the nucleus, $L_x = 8.4 \times 10^{39}\,\mathrm{erg\,s}^{-1}$, and the $L_x/L_{H\alpha}$ ratio are intermediate between typical values for starburst and Seyfert nuclei. The nuclear emission is resolved, so only a small fraction can originate in the non-thermal nucleus. The data allows an unresolved source at the nucleus with at most 20% of the above X-ray emission. Most of the X-rays must therefore be due to starburst activity and/or outflow.

4. NGC 5195, The Companion

Just as the encounter has almost certainly altered the morphology, star formation properties and nuclear activity of M51, so has it had an effect on the companion, NGC 5195. No H I has been observed to be associated with NGC 5195. However, recently Rots *et al.* (1989) detected H I north and south-east of this galaxy at column densities of $1 - 3 \times 10^{20}$ atoms cm^{-2}, which they suggest could originally have been part of the companion.

Schweizer (1977) estimates the total mass of NGC 5195 to be one-third to one-half that of M51. Its X-ray luminosity is $L_x = 2.1 \times 10^{39}$ erg s^{-1}, which is in the range of typical irregulars (PFFT). Heckman (1980) found it to be a member of the LINER class. Nuclear activity is indicated in this galaxy by widths and ratios of lines of O, N, and S. Roche and Aitken (1985) carried out $8-13\,\mu$m spectrophotometry of the companion, and found spectral structure typical of galaxies with enhanced nuclear star formation.

There is mounting evidence that, for its class, NGC 5195 is somewhat extreme in its star forming properties. Very recently, Sage and Wrobel (1989) completed a study of CO emission from high-L_{FIR} SO galaxies using the NRAO 12-m antenna. Smith (1982) found an L_{FIR} of $\gtrsim 3 \times 10^9$ L$_\odot$, and an unusually warm dust temperature of 65 K. NGC 5195 was found to have the highest CO luminosity of all the galaxies studied by Sage and Wrobel, and to be the only galaxy with extended CO emission. Sage (1989) further discussed in detail the extended CO emission from NGC 5195. The spectra generally show two components: a wide component at $v_{LSR} \approx 660$ km s^{-1}, and a narrow component at $v_{LSR} \approx 430$ km s^{-1}. From their morphology and velocity, Sage identifies the two components as being the companion and the overlapping spiral arm of M51, respectively.

CO is only found on the southern side of the bar in NGC 5195. Such an asymmetry is also seen in the radio continuum emission (Segalovitz 1976, 1977a), but is unique among barred galaxies in which CO has been mapped. Sage suggests the interaction may have altered the distribution of molecular gas.

From Sage and Wrobel, one can calculate a rough molecular mass of M$_{H_2} \sim 10^9$ M$_\odot$, but Sage cautions against interpretation of L_{CO} in terms of M$_{H_2}$, because the interaction may have changed the gas temperature or density, making the standard conversion factor questionable.

In summary, the nuclear activity seen in both M51 and NGC 5195 may well implicate galactic encounters as a source of such activity, as noted by GG. This conclusion agrees well with the statistical evidence for enhanced activity in interacting spiral galaxies (*e.g.* Stocke 1978).

Acknowledgements. The authors express special thanks to N. Scoville for stimulating the writing of this review. We also thank G. Cecil for rushing an electronic version of his image to us on short notice, and R. Lupton for transforming it into a readable format. RPJT thanks A. Broeils and K. Begeman for digging up the author's data in the Groningen database. RPJT also thanks S. Kulkarni for travel support to visit Caltech, and both authors appreciated his spirited encouragement during the final stages of writing. The authors thank G. Wynn-Williams and R. Allen for proofreading the manuscript. RPJT's research at the University of Hawaii is supported by NSF grant AST-8615684.

5. References

Aaronson, M, and Mould, J. 1983, *Ap. J.*, **265**, 1.
Allen, R. J., Atherton, P. D., and Tilanus, R. P. J. 1986, *Nature*, **319**, 296.
Appleton, P. N., Foster, P. A., and Davies, R. D. 1986, *M.N.R.A.S.*, **221**, 393.
Balbus, S. A. 1988, *Ap. J.*, **324**, 60.
Beck, R., Klein, U., and Wielebinski, R. 1987, *Astr. Ap.*, **186**, 95.
Bohlin, R. C., Savage, B. D., and Drake, J. F. 1978, *Ap. J.*, **224**, 132.
Bosma, A. 1981, *A. J.*, **86**, 1825.
Bottinelli, L., Gouguenheim, L., Paturel, G., and de Vaucouleurs, G. 1983, *Astr. Ap.*, **118**, 4.
Burbidge, E. M., and Burbidge, G. R. 1964, *Ap. J.*, **140**, 1445.
Burbidge, E. M., and Burbidge, G. R. 1965, *Ap. J.*, **142**, 634.
Burkhead, M. S. 1978, *Ap. J. Suppl.*, **38**, 147.
Caplan, J., and Deharveng, L. 1986, *Astr. Ap.*, **155**, 297.
Cecil, G. 1988, *Ap. J.*, **329**, 38.
Cecil, G., and Rose, J. A. 1984, *Ap. J.*, **287**, 131.
Crane, P. C., van der Hulst, J. M., Ford, H. C., Lawrie, D. G., and Jacoby, G. H. (1988), preprint.
de Vaucouleurs, G., de Vaucouleurs, A., and Corwin, H. G. 1976 in *Second Reference Cataloque of Bright Galaxies*, (Austin: University of Texas Press).
Dickey, J. M., Garwood, R. W., and Helou, G. 1987, in Star Formation in Galaxies, ed. C. J. Lonsdale Persson, (Washington: NASA), p. 575.
Duric, N., Bourneuf, E., and Gregory, P. C. 1988, *A. J.*, **96**, 81.
Elmegreen, B. G. 1987, in *I.A.U. Symposium No. 115, Star Forming Regions*, ed. M. Peimbert and J. Jugaku, (Dordrecht: Reidel), p. 457.
Elmegreen, B. G. 1988a, in *Comets to Cosmology*, ed. A. Lawrence (Berlin: Springer Verlag), p. 186.
Elmegreen, B. G. 1988b, in *Galactic and Extragalactic Star Formation* ed. R. E. Pudritz and M. Fich (Kluwer), p. 215.
Elmegreen, B. G., and Elmegreen, D. M. 1986, *Ap. J.*, **311**, 554.
Ford, H. C., Crane, P. C., Jacoby, G. H., Lawrie, D. G., and van der Hulst, J. M. 1985, *Ap. J.*, **293**, 132.
Goad, J. W., De Veny, J. B., and Goad, L. E. 1979, *Ap. J. Suppl.*, **39**, 439.
Goad, J. W., and Gallagher J. S. 1985, *Ap. J.*, **297**, 98.
Guélin, M. Garcia-Burillo, S. Blundell, R., Cernicharo, J., Despois, D., and Steppe, H. 1988, preprint.
Heckman, T. M. 1980, *Astr. Ap.*, **87**, 152.
Holmberg, E. 1958, *Medd. Lund Astron. Obs. Ser. II*, No. **136**.
Horellou, C., Beck, R., and Klein, U. 1989, in *Galactic and Extragalactic Magnetic Fields*, ed. R. Beck, P. Kronberg, and R. Wielebinski, in press.
Hummel, E. 1980, *Astr. Ap. Suppl.*, **41**, 151.
Kennicutt, R. C., and Kent, S. M. 1983, *A. J.*, **88**, 1094.
Klein, U., and Emerson, D. T. 1981, *Astr. Ap.*, **94**, 29.

541

Klein, U., Wielebinski, R., and Beck, R. 1984, *Astr. Ap.*, **135**, 213.
Krasheninnikova, Y., Ruzmaikin, A., Sokoloff, D., and Shukurov, A. 1989, *Astr. Ap.*, **213**, 19.
Krause, M., Hummel, E., and Beck, U. 1989, *Astr. Ap.*, in press.
Lesch, H., Sawa, T., Krause, M., Beck, R., Fujimoto, M., and Biermann, P. L. 1988, *Ap. J.*, **192**, 19.
Lester, D. F., Harvey, P. M., and Joy, M. 1986, *Ap. J.*, **302**, 280.
Lin, C. C., and Shu, F. H. 1964, *Ap. J.*, **140**, 646.
Lo, K. Y., Ball, R., Masson, C. R, Phillips, T. G, Scott, S., and Woody, D. P. 1987, *Ap. J.*, **317**, L63.
Lord, S. D., and Young, J. S. 1989, preprint.
Mathewson, D. S., van der Kruit, P. C., and Brouw, W. N. 1972, *Astr. Ap.*, **17**, 468. **MKB**
Monnet, G. , Paturel, G., and Simien, F. 1981, *Astr. Ap.*, **102**, 119.
Pagel, B. E. J., and Edmunds, M. G. 1981, *Ann. Rev. Astr. Ap.*, **19**, 77.
Palumbo, G. G. C., Fabbiano, G., Fransson, C., and Trinchieri, G. 1985, *Ap. J.*, **298**, 259.
Pierce, M. J. 1986, *A. J.*, **92**, 285.
Rand, R. J. and Kulkarni, S. R 1989a, in *Millimeter and Sub-millimeter Astronomy*, ed. A. Webster (Kluwer), in press.
Rand, R. J., and Kulkarni, S. R. 1989b, *Ap. J. Lett.*, in press.
Roberts, W. W. 1969, *Ap. J.*, **158**, 123.
Roche, P. F., and Aitken, D. K. 1985, *M.N.R.A.S.*, **213**, 789.
Rose, J. A., and Cecil, G. 1983, *Ap. J.*, **266**, 531.
Rose, J. A., and Searle, L. 1982, *Ap. J.*, **253**, 556.
Rots, A. H., Bosma, A., van der Hulst, J. M., Athanassoula, E., and Crane, P. C. 1989, in 1988 *Ann. Report Neth. Foun. for Radio Astr.*, p. 91; also in *Dynamics and Interactions of Galaxies*, ed. R. Wielen, in press.
Rots, A. H., 1975, *Astr. Ap.*, **45**, 43.
Rots, A. H. 1980, *Ap. J. Suppl.*, **41**, 189.
Rydbeck, G., Hjalmarson, A., and Rydbeck, O. E. H. 1985, *Astr. Ap.*, **144**, 282.
Rydbeck, G., Hjalmarson, A., Wiklind, T., and Rydbeck, O. E. H. 1988, preprint.
Sage, L. J. 1989, *Ap. J.*, **344**, 200.
Sage, L. J., and Wrobel, J. M. 1989, *Ap. J.*, **344**, 204.
Sandage, A., and Tammann, G. A. 1975, *Ap. J.*, **196**, 313.
Sandage, A., and Tammann, G. A. 1987, *Revised Shapley-Ames Catalog of Bright Galaxies*, (2d ed.; Washington: Carnegie Institute).
Sandage A. R., 1961, *Hubble Atlas of Galaxies*, (Washington: Carnegie Institute), p. 26.
Sanders, R. H., and Huntley, J. M. 1976, *Ap. J.*, **209**, 53.
Scarrott, S. M., Ward-Thompson, D., and Warren-Smith, R. F. 1987, *M.N.R.A.S.*, **224**, 299.
Schwarz, M. P. 1981, *Ap. J.*, **247**, 77.
Schweizer, F. 1976, *Ap. J. Suppl.*, **31**, 313.
Schweizer, F. 1977, *Ap. J.*, **211**, 324.
Scoville, N. Z., and Sanders, D. B. 1988, in *Interstellar Processes*, ed. D. Hollenbach and H. A. Thronson (Dordrecht: Reidel), p. 21.
Scoville, N. Z., Sanders, D. B., and Clemens, D. P. 1986, *Ap. J. (Letters)*, **310**, L77.
Scoville, N. Z., and Young, J. S. 1983, *Ap. J.*, **265**, 148.

542

Searle, L. 1976, in *The Galaxy and the Local Group*, ed. R. J. Dickens and J. E. Perry (*RGO Bull.*, No. 182).

Segalovitz, A. 1976, *Astr. Ap.*, **52**, 167.

Segalovitz, A. 1977a, *Astr. Ap.*, **54**, 703.

Segalovitz, A. 1977b, *Astr. Ap.*, **61**, 59.

Shane, W. W. 1975, in *La Dynamique des Galaxies Spirale*, ed. L. Weliachew, (Paris: Centre National de la Recherche Scientifique), p. 217.

Shu, F. H., Adams, F. C., and Lizano, S. 1987, *Ann. Rev. Astr. Ap.*, **25**, 23.

Simkin, S. M., Su, H. J., and Schwarz, M. P. 1980, *Ap. J.*, **237**, 404.

Smith, J. 1982, *Ap. J.*, **261**, 463.

Soifer, B. T, Boehmer, L., Neugebauer, G., and Sanders, D. B., *A. J.*, **98**, 766.

Spinrad, H., and Harlan, E. 1973, *Pub. A.S.P.*, **85**, 815.

Stocke, J. T. 1978, *A. J.*, **83**, 348.

Sukumar, S., and Allen, R. J. 1989, *Nature*, **340**, 537.

Tacconi, L. J., and Xie, S. 1989, preprint.

Telesco, C. M., Decher, R., and Gatley, I 1986, *Ap. J.*, **302**, 632.

Thronson, H. A., and Greenhouse, M. A. 1988, *Ap. J.*, **327**, 671.

Tilanus, R. P. J., and Allen, R. J. 1988, *Ap. J.*, **330**, 667.

Tilanus, R. P. J., and Allen, R. J. 1989a, *Ap. J.*, **339**, L57.

Tilanus, R. P. J., and Allen, R. J. 1989b, *Astr. Ap.*, submitted.

Toomre, A., and Toomre, J. 1972, *Ap. J.*, **178**, 623.

Toomre, A. 1978, in *IAU Symposium No. 79, The Large-Scale Structure of the Universe*, ed. M. S. Longair and J. Einasto, (Dordrecht: Reidel), p. 109.

Tully, R. B. 1974a, *Ap. J. Suppl.*, **27**, 415.

Tully, R. B. 1974b, *Ap. J. Suppl.*, **27**, 437.

Tully, R. B. 1974c, *Ap. J. Suppl.*, **27**, 449.

van den Bergh, S. 1960, *Publ. David Dunlop Obs.* **II**, 6.

van der Hulst, J. M., Kennicutt, R. C., Crane, P. C., and Rots, A. H. 1988, *Astr. Ap.*, **195**, 38.

van der Kruit, P. C. 1977, *Astr. Ap.*, **59**, 359.

Viallefond, F., and Goss, W. M. 1986, *Astr. Ap.*, **154**, 357.

Visser, H. C. D. 1980a, *Astr. Ap.*, **88**, 149.

Visser, H. C. D. 1980b, *Astr. Ap.*, **88**, 159.

Vogel, S. N., Kulkarni, S. R., and Scoville, N. Z. 1988, *Nature*, **334**, 402.

Weliachew, L., and Gottesman, S. T., *Astr. Ap.*, **24**, 59.

Williams, A. G., and Gull, S. F. 1985, *Nature*, **313**, 34.

Wilson, A. S., and Ulvestad, J. S. 1982, *Ap. J.*, **263**, 576.

Worden, S. P. 1974, *Pub. A.S.P.*, **86**, 92.

Zaritsky, D., and Lo, K. Y. 1986, *Ap. J.*, **303**, 66.

Interstellar Medium in External Galaxies

Jeremiah P. Ostriker
Princeton University Observatory

ABSTRACT. Certain questions are isolated as particularly vital to our understanding of the interstellar medium. What is distribution of magnetic field values? What is the coherence length? What fraction of the volume in the plane of typical spiral galaxies is in the hot gaseous phase? What forces maintain large amounts of warm, neutral medium far from the galactic plane? In molecular clouds, what precisely is measured by the CO luminosity? What physical processes account for the random energy of the molecular clouds? Is star formation a uni- or bimodal process? Are interstellar clouds really filaments or sheets and what is the overall topological structure? Does a typical spiral galaxy have a net inflow or outflow of gas?

1. Introduction

It is now hard to believe that only 40 years ago Lyman Spitzer wrote a paper (1948) asserting that star formation must be an ongoing process in the Galaxy. He argued that, since there are many stars with ages young compared to the estimated Galactic age, star formation must be a contemporary, continuing process if we are not to live at a special epoch. He further noted that since the young stars are seen in regions rich in gas and dust, it was reasonable to suppose that by some mysterious process stars were made out of interstellar gas clouds. This may all seem fairly obvious now, but at that time the conclusions were novel and important. Up until that point, stars were pretty much taken for granted. They existed; therefore they had to have been made on some occasion in the past, but the process was not considered to be of particular relevance, since it was presumed to be far back into an unobservable realm.

There was an similar situation with respect to galaxies when I went to graduate school in the mid 60's. Galaxies existed and presumably were made at some time, some how, but that time was long ago and the epoch and process was not of very pressing importance. It was only when it was realized (through the work of Tinsley (1972) and others) that they could not be treated as standard candles and meter sticks, that we astronomers, as a group, faced the fact that we *must* care about the formation and evolution of galaxies. Facing the further distressing fact that galaxies were formed by processes analogous to star formation processes has taken even longer. For a long time "models" of galaxy formation treated the process as if the galaxies were made of non-interacting particles - massive neutrinos, for example. As Simon White's talk at this conference showed, this is no longer the case. Simulations now allow for gas, dynamical physics such as heating and cooling and also, necessarily, for the energy input from stars. As

543

H. A. Thronson, Jr. and J. M. Shull (eds.), The Interstellar Medium in Galaxies, 543–549.

Dopita and others emphasized, the evolution of our Galaxy and others is intimately connected with the evolution of the ISM - young star system. It is that system which evolves. If only stellar dynamics were involved, the evolution would be infinitesimal on the 10^{10} year scale.

As our intellectual horizons have broadened, however reluctantly, our techniques have also improved so that we can now study nearby galaxies in great detail, often better than our own. And, from work performed in the last decade, we even have the first preliminary information concerning the Universe at large redshift, the types of galaxies, the changing of their state between then and now, and even the gaseous environment from which galaxies were formed.

I will try in this brief review to summarize the state of our knowledge as presented by this conference, laying primary stress on the theoretical papers because I understood them better than I do those which have an exclusively observational orientation.

My method will seem strange to some. I will not spend much time on the papers which might be considered best, those which showed excellent analysis of a well understood subject, where the questions are well defined and accurate observational results are known. My mentality is the extreme of one type of scientist. The most important question, I believe, is "what are the important questions?" The pathbreaking papers set out those questions. Then understanding occurs when we have good preliminary answers. That is exciting - to understand the first outlines of a phenomena. If we understand something *well*, well enough to use it as a tool, then the subject approaches the state of engineering, and, to me, while it remains important, interest declines. So I will stress the areas where the questions are most widely open and where even our qualitative understanding is poor or disputed. A topical, rather than chronological, organization will be followed.

The following table presents one arbitrary cut through the parameter space by which one could organize this material.

Component Label -- Physical Composition		n(atom/cc)	u(ev/cc)	ε(ev/cc/My)
Relativistic -- Magnetic Fields		NR	1	$\ll 1$?
Fluid -- Cosmic Ray Gas		$\ll 1$	1	1
Gaseous Phases -- Hot Ionized Medium				
in Statistical -- Warm Ionized Medium		1	1	$(+1-1)=0$
Hydrodynamic -- Warm Neutral Medium				
Equilibrium -- Cold Neutral Medium				
Self-Gravitating -- Molecular Cloud				
Gaseous Phase -- Complexes		1	(0.1) NR	(0.1) ?
Transitional -- Young Stars and				
Population I -- Associated ISM:		1	NR	1
-- Bipolar flows				
-- Winds				
-- SN remnants				

2. Open Questions

2.1 RELATIVISTIC FLUID

2.1.1 *Magnetic Fields*. Wielebinski's talk left me hungrier at the end than at the beginning. He quite persuaded me of the importance of magnetic fields and stimulated my imagination with fascinating images, pictures showing field strength and polarization, but he left my appetite for *numbers* unslaked. What are the characteristic values of $(\vec{B}, B_{rms})$ averaged over small volumes, i. e., should we consider $\langle|B|\rangle$ or $\langle B^2\rangle^{1/2}$ to be 3μG, 5μG, or 10μG? As Don Cox showed, it makes a huge difference for the dynamic equilibrium of the gaseous phases if the values are at the low end or at the high end of this range. How are the actual values distributed in the sense of f(r, B)dB giving the fractional volume of galaxy, at radius r with B_{rms} in the range of B→B+dB? What is the correlation length for the field in different parts of the galaxy? And what is its overall geometric structure (e.g., parallel or perpendicular to the galactic plane at high latitudes, a question that was addressed by Salpeter)? Does the Galaxy (or other galaxies) have a significant net magnetic flux? I understand that at the recent Heidelberg conference, there was a *vote* on whether the galactic field was, in the main part, primordial or due to dynamo processes. Presumably, there are better methods for answering this question. While dynamos are attractive, Kulsrud (1989) has pointed out that they amplify the chaotic part of the field more rapidly than the mean part, and that for fields larger than 10^{-10} gauss, dynamo amplification results only in chaotic fields at variance with observations. Are there systematic variations in the typical field strengths from galaxy to galaxy dependent on environment or Hubble type? I understand that recent work (Kulkarni and Frick, 1985) is beginning to provide quantitative answers to some of these questions for our own Galaxy, but for this all important subject, one must believe that future knowledge will far surpass what we now know.

2.1.2 *Galactic Cosmic Rays*. Here our knowledge (and that presented at this conference) is still more incomplete. We know that the energy density is large enough (~ 1ev/cc) and lifetime short enough ($10^{6.5}$ yrs), so that the energy input is very substantial, an important driver for all other phenomena, but our detailed knowledge is poor. The energy and composition distributions in the vicinity of the Sun are known best. Information from gamma-ray satellite experiments (Bloemen 1987) is vastly increasing our knowledge of the galactic CR distribution. But our knowledge of the sources and sinks of cosmic rays is still quite poor, so that even a rough breakdown of acceleration into stellar origin vs. SN (and SNR) origin, vs. distributed shock acceleration cannot be made with any confidence.

2.2 GASEOUS PHASES

2.2.1 *HIM*. D. Cox argued at this conference that the filling factor of this component was negligible. C. McKee countered that the filling factor was significant. Which view is correct? My own views are well known, but arguments among theoreticians are inconclusive. How can matter be resolved? Good X-ray pictures of other face-on galaxies would definitely help. Can coronal lines be used?

2.2.2 *WIM*. J. Kenny and others stressed the great extent of HI discs and that the density drops off where they are ionized. This leads me to question how much more WIM is there surrounding galaxies. In the galactic disc, pulsar dispersion measurements give us a good picture. It will be interesting when we can make the same measurements for other galaxies; perhaps there is much more gas in the planes of these systems at larger radii. Are all galaxies like A. Wolfe discs?

2.2.3 *WNM*. Is it a distinct phase? This question was not addressed in M. Begelman's elegant talk. What keeps it warm? Photoelectric heating? MHD heating was not discussed at this conference, but the Chernoff poster was indirectly related to it. This issue is vital. Cloud motions dissipate on a collision timescale, but presumably the energy goes in comparable parts into radiation (lost) and MHD waves (recyclable). The latter is, in turn, available for subsequent heating of the gas. What holds up all this gas at large distances above the galactic plane? The fact that it is there at large scale height and the importance, theoretically, of that fact was discussed by many (see, for example, papers by Cox, McKee) at this meeting. How this gas is levitated is not clear. My guess is that turbulent pressures or C.R. streaming is responsible. This perplexing issue is also addressed at some length in Spitzer's forthcoming review "Theories of the Hot Interstellar Gas" (1990).

2.2.4 *CNM*. This gas is still our best kinematic tracer (c.f., paper by Brinks, at this conference). If we are to understand the dynamics of warps, motions in bars, and infall, we must use this material. My reading is that we still do not understand these important phenomena. Higher resolution is perhaps not so important (now) as is higher sensitivity. And of course, more attention to solving the underlying dynamical problems is vital.

2.3 MOLECULAR GAS IN CLOUD COMPLEXES

The beautiful observations summarized by J. Young (this conference) indicate that important systematic correlations have been found. But what do they signify? I learned a great deal, but I also learned that I did not know if L_{co} measures, in the first instance, $M(H_2)$, or energy input, or energy output, or something else, or some combination of these phenomena. It is *very* important to straighten this matter out. When that is achieved, the wealth of data and the order apparent in that data will provide a Rosetta stone for comprehending star formation.

But a still more fundamental question was addressed by McKee, Dopita, and indirectly by Balbus. Why are there molecular clouds at all? What physical process sustains them? Are they permanent, transient or optical illusions due to crowding in velocity space?

What accounts for their velocity dispersion? Jog and Ostriker (1988) note that they are far from equipartition in that their masses are several orders of magnitude larger than the CNM and WNM clouds but that their characteristic velocity dispersion is only smaller by about a factor of two. (We are discussing here the velocity with respect to the local circular galactic rotation frame, not the internal velocity dispersion). It seems to be quite impossible for supernova singly or in groups to accelerate these massive objects, and external perturbations, which may be important for long lived components like the disc stars, are too infrequent to accelerate these clouds. Jog and Ostriker (1988) suggest that viscosity (i.e., cloud-cloud interactions) acts in the differentially rotating cloud fluid to maintain the momentum in the same way that Goldreich and Tremaine (1982) find that it operates in planetary rings. Energy removed from the overall rotation leads to a flow of mass inwards and angular momentum outwards in the differentially rotating disc.

2.4   STAR FORMATION REGIONS

One important question is whether the initial mass function is bimodal or unimodal (c.f. papers by Kennicutt, Dopita, at this conference). My own work several years ago indicated bimodal. This matter should be resolvable on an empirical basis by looking at clusters and associations of different types and ages.

How can we put together the information we have obtained on local star formation with cosmological information? We now know how dust obscuration, Ly continuum emission, IR emission, etc., are related to star formation rates in local galaxies due to the work of Kennicutt and others by harmonizing the different methods of evaluating star formation rates. It would be very valuable if this information were synthesized into a form useful to S. White, T. Tyson and others who are addressing galaxy formation at earlier epochs and other related cosmological problems.

There was a fascinating relation between star formation rate and Toomre's Q parameter reported by Kennicutt. It seemed to me a very plausible and very important relation. It is so physically interesting that it might be worth following S. White's suggestion to see if it could be used as a distance indicator.

Since stars do the stirring, it is vital to improve our knowledge of the mechanical work they do. Bipolar flows, winds, SN, novae, stellar flares, etc., provide the mechanical energy input to the ISM. The latter two were not mentioned at this meeting, but my impression is that the output from these sources is significant.

3.   Global Questions

The schematic approach I have adopted to address the various components in turn must be supplemented by examining some issues that transcend this narrow categorization.

3.1   GEOMETRY-TOPOLOGY

In addition to the questions of filling factor, there are other geometrical questions of importance. The hot component: is it tunnel-like or pervasive, as Cox, McKee and I originally proposed, or is it best thought of as isolated islands around SN as in the older picture? Or perhaps Gott's (1986) sponge-like geometry is more appropriate than either of these pictures with both phases interconnected. Interstellar "clouds": are they really clouds, or would filaments or sheets describe them better? It does make a difference for many purposes. My own guess is that, if we could see this gas in sufficient detail, we would see the clouds as a tangled skein of filaments. Gott and associates (1989) have emphasized the importance of topology in cosmology, but it is just as important here. For example, the evaporation rate of filaments or sheets is quite different (and less) than the evaporation rate of "clouds." Help from the observers is needed and an imaginative analysis of observations is just as necessary.

3.2   MASS TRANSFER

The question posed by Jill Knapp (at this conference) was perfect. Follow an H atom. How long does it spend in each state, going from component to component including grain mantles? The demography of molecular clouds is a subject I have not seen addressed. The cloud densities in the arm and inter-arm regions must, physically, be connected by an appropriate continuity equation

Topics tend to be addressed in isolation. Molecular clouds will be destroyed by shocks, embedded SN, etc. But, what *forms* clouds at an equal rate? This was the major dangling piece left out of the MO (McKee, Ostriker 1977) synthesis. I thought that we did not answer it satisfactorily. Somewhere we should see molecular clouds forming.

3.3   OVERALL INPUT AND OUTPUT

3.3.1   *Cooling Flows*. These are probably seen in some giant ellipticals. But, does the gas in normal spirals flow in or out? Sarazin stressed the evolution of our views on this, but many problems remain.

3.3.2   *Normal Galaxies* . Do both infall and outflow occur at the same time, or alternately? Is there infall, a fountain or a wind? Which picture is appropriate? Cox, Salpeter, and White offered different perspectives. I am convinced from these that, for ellipticals, the flow, after formation, was primarily outwards and remains so except in unusual circumstances. For spirals, it has been primarily inward but it is at a low rate in the current era or else X-ray emission would be far higher than is observed. To put the question simply, is the net flow from our Galaxy now in or out, or are both phenomena occurring?

3.4   ENVIRONMENTAL INFLUENCES

This brings us to the general question of nature vs. nurture in the evolution of galaxies. This issue, which has produced such idiotic debates among psychologists and sociologists, we may hope to avoid. The answer, of course, is *yes*. Would anyone think of asking this question of a gardener? Is it nature or nurture that makes your garden grow? The gardener knows that, no matter how he tends a rose bush, it will not become an oak tree., i. e., nature is important. But he also knows that water, fertilizer, etc., will affect and alter growth. So one must be quantitative. It seems that there is a "natural" evolution of galaxies (c.f., presentations by Dinerstein, Dopita, etc., at this conference), but that cluster environment (c.f., poster papers) merger (Noguchi, at this conference), etc., can affect this evolution. The question is, are environmental influences a primary or secondary matter? Simon White would say they are primary. I think that the observed regularities of galaxies are too strong to cede primacy to any stochastic process. For a few percent of galaxies, environmental influences are of primary importance, but for most galaxies, they are important at only the few percent level. Time will tell.

4.   Conclusions

Our inventory of questions seems robust in that no totally new phenomena are still being added to the list of processes to be studied. Also, it seems that our tools to study the interstellar medium have grown impressively, and here I include as a "tool" the powerful new computational programs underway, which can begin to simulate the complex, fully three-dimensional ISM phenomena. Growth of knowledge seems slowest with respect to the all important relativistic fluid, the magnetic field and cosmic ray components.

But finally, alas, I must say that we remain ignorant concerning many of the most fundamental issues, such as the typical value of B, the fractional distribution of the various gaseous components, the physical processes that maintain molecular clouds, the mechanism of star formation (or even an accurate description of it) and the overall issue of transfer of mass and energy between galaxies and their surroundings. These areas of fundamental ignorance of course make the subject exciting; discovering the truth provides the fun in science! Let us also hope that our humility in this relatively old subject may provide an example to our colleagues in allied fields. If it has been painfully learned how complex are star formation and the interstellar medium, then is it likely that galaxy formation and the intergalactic medium will succumb to modelling based on much more simplified assumptions?

5. References

Bloemen, J. B.1987, *Ap. J.*, **322**, 694.
Brinks, E., this conference.
Begelman, M., this conference.
Balbus, S., this conference.
Chernoff, D.F., this conference
Cox. D., this conference.
Dinerstein, H., this conference.
Dopita, M., this conference
Goldreich, P., and Tremaine, S.C. 1982, *Ann. Rev. Astr. Ap.*, **20**, 249.
Gott, J.R., Melott, A.L. and Dickinson, M. 1986, *Ap.J.*, **306**, 341.
Gott, J.R. *et al.* 1989, *Ap.J.*, **340**, 625.
Jog, C., and Ostriker, J. P. 1988, *Ap. J.*, **328**, 404.
Kennicutt, R., this conference.
Kenny, J., this conference.
Knapp, G.R., this conference.
Kulkarni, S.R., and Fick, M. 1985, *Ap. J.*, **289**, 792.
Kulsrud, R. M.1989, "The Origin of Galactic Magnetic Fields" in the proceedings of the NATO workshop *Physics of Hot Cosmic Plasmas*, Vulcano Island, Italy (May 1989).
McKee, C.F., this conference.
McKee, C.F., and Ostriker, J.P. 1977, *Ap.J.*, **218**, 148.
Noguchi, M., this conference.
Tinsley, B.M. 1972, *Ap.J. (Letters)*, **178**, L39.
Salpeter, E.E., this conference.
Spitzer, L. 1948, *Physics Today*, **1**, 6.
_____. 1990, *Annual Reviews of Astron. & Astro.*, **29**.
White, S., this conference.
Wielebinski, R., this conference.
Wolfe, A., this conference.
Young, J., this conference.

OBJECT INDEX